Advances in Delivery Science and Technology

Series Editor

Michael Rathbone

For further volumes:
http://www.springer.com/series/8875

Hugh D.C. Smyth · Anthony J. Hickey
Editors

Controlled Pulmonary Drug Delivery

Springer

Editors
Hugh D.C. Smyth, Ph.D.
College of Pharmacy
University of Texas
Austin, TX
USA
hsmyth@mail.utexas.edu

Anthony J. Hickey, Ph.D.
Eshelman School of Pharmacy
University of North Carolina at Chapel Hill
Chapel Hill, NC
USA
ahickey@unc.edu

ISBN 978-1-4419-9744-9 e-ISBN 978-1-4419-9745-6
DOI 10.1007/978-1-4419-9745-6
Springer New York Dordrecht Heidelberg London

Library of Congress Control Number: 2011929875

Printed on acid-free paper

Springer is part of Springer Science+Business Media (www.springer.com)

Preface

The pace of new research and level of innovation repeatedly introduced into the field of drug delivery to the lung are surprising given its state of maturity since the introduction of the pressurized metered dose inhaler over a half a century ago. It is clear that our understanding of pulmonary drug delivery has now evolved to the point that inhalation aerosols can be controlled both spatially and temporally to optimize their biological effects. These abilities include controlling lung deposition, by adopting formulation strategies or device technologies, and controlling drug uptake and release through sophisticated particle technologies. The large number of contributions to the scientific literature and variety of excellent texts published in recent years are evidence for the continued interest in pulmonary drug delivery research. This reference text endeavors to bring together the fundamental theory and practice of controlled drug delivery to the airways that is unavailable elsewhere. Collating and synthesizing the material in this rapidly evolving field presented a challenge and ultimately a sense of achievement that is hopefully reflected in the content of the volume.

The spatial and temporal control of drug delivery to the airways as a general theme runs through the entire volume from discussions of micro and macro structure of the lung, particle engineering and polymer science, device design, to regulatory perspectives and science. The initial chapter topics were selected to provide a fundamental background to the problems and opportunities for controlled pulmonary drug delivery. In addition to providing an anatomical, physiological, and metabolic overview of the airways, the book provides unique guidance on specific microenvironments that exist in both health and disease within the airways – opening possible avenues to allow for targeted, triggered, or modulated delivery systems based on the physicochemical differences between target and bystander tissues and cells. The latter sections of the book explore technologies and tools available to facilitate controlled drug delivery to the airways, specifically covering topics such as, aerosol delivery technologies, materials and excipients, particle science, gene delivery, in vitro and in vivo tools including imaging. Finally, regulatory approval perspectives and the development of performance specifications complete the "tool box" that is provided by the text as a whole.

The authors who kindly agreed to contribute to *Controlled Pulmonary Drug Delivery* are acknowledged leaders in their respective fields, and many have initiated research programs in new and emerging research areas of relevance to the title of the volume. As a result, we hope that this text will provide a framework for interested researchers to find solutions to their drug delivery questions. The contents of the book should provide bridges between the multiple disciplines needed to successfully achieve controlled pulmonary drug delivery.

Austin, TX Hugh D.C. Smyth
Chapel Hill, NC Anthony J. Hickey

Acknowledgments

The editors would like to express thanks to all the authors who devoted significant time and effort and without which, this book would not have been possible. We also would like to thank Renata Hutter and Carolyn Honour of Springer Science+ Business Media, LLC for their assistance and patience during the preparation of the manuscript. Lastly, we would like to thank Mike Rathbone and the Controlled Release Society for providing the opportunity for this book to come to fruition.

Contents

Contributors

Mark M. Bailey Department of Chemical and Petroleum Engineering, Department of Pharmaceutical Chemistry, University of Kansas, Lawrence, KS, USA

Cory Berkland Department of Chemical and Petroleum Engineering, Department of Pharmaceutical Chemistry, University of Kansas, Lawrence, KS, USA

Balaji Bharatwaj Chemical Engineering and Materials Science, Wayne State University, Detroit, MI, USA

Eva Bondesson AstraZeneca R&D Lund, Lund, Sweden; Presently at Semcon Drug Development Consulting, Lund, Sweden

Gerrit Borchard School of Pharmaceutical Sciences, University of Geneva, University of Lausanne, Geneva, Switzerland

Lars Borgström AstraZeneca R&D Lund, Lund, Sweden

Timothy Brenza Department of Pharmaceutical Sciences and Experimental Therapeutics, University of Iowa, Iowa City, IA, USA

Stephen T. Buckley School of Pharmacy and Pharmaceutical Sciences, Trinity College Dublin, Dublin, Ireland

Timothy M. Crowder GlaxoSmithKline, Durham, NC, USA

Sandro R.P. da Rocha Chemical Engineering and Materials Science, Wayne State University, Detroit, MI, USA

Martin J. Donovan University of Texas at Austin, TX, USA

Staffan Edsbäcker AstraZeneca R&D Lund, Lund, Sweden

Carsten Ehrhardt School of Pharmacy and Pharmaceutical Sciences, Trinity College Dublin, Panoz Institute, Dublin, Ireland

Stefan Eirefelt AstraZeneca R&D Lund, Lund, Sweden;
Presently at LEO Pharma A/S, Ballerup, Denmark

Katarina Ekelund AstraZeneca R&D Lund, Lund, Sweden

Nashwa El-Gendy Department of Chemical and Petroleum Engineering, Department of Pharmaceutical Chemistry, University of Kansas, Lawrence, KS, USA

Ibrahim M. El-Sherbiny College of Pharmacy, University of Texas, Austin, TX, USA; Mansoura University, Mansoura, Egypt

Jennifer Fiegel Department of Pharmaceutical Sciences and Experimental Therapeutics; Department of Chemical and Biochemical Engineering, The University of Iowa, Iowa City, IA, USA

Lucila Garcia-Contreras Department of Pharmaceutical Sciences, College of Pharmacy, The University of Oklahoma, Oklahoma City, OK, USA

Aileen Gibbons College of Pharmacy, The University of Texas at Austin, Austin, TX, USA

Mark Gumbleton Welsh School of Pharmacy, Cardiff University, Cardiff, Wales, United Kingdom

Lena Gustavsson AstraZeneca R&D Lund, Lund, Sweden

Rania Hamed Department of Chemical and Biochemical Engineering, The University of Iowa, Iowa City, IA, USA

Tove Hegelund-Myrbäck AstraZeneca R&D Lund, Lund, Sweden

Dea Herrera Mansoura University, Mansoura, Egypt

Simon Heuking Vaccine Formulation Laboratory, WHO Collaborating Centre, University of Lausanne, Department of Biochemistry, Chemin des Boveresses, Switzerland

Anthony J. Hickey Eshelman School of Pharmacy, University of North Carolina, Chapel Hill, NC, USA

Stephen T. Horhota Boehringer Ingelheim Pharmaceuticals, Inc., Ridgefield, CT, USA

Kwang-Jin Kim Keck School of Medicine, University of Southern California, Los Angeles, CA, USA

Philip J. Kuehl Lovelace Respiratory Research Institute, Albuquerque, NM, USA

Stefan Leiner Boehringer Ingelheim Pharma GmbH & Co KG, Ingelheim am Rhein, Germany

Jason T. McConville College of Pharmacy, University of Texas at Austin, Austin, TX, USA

Jolyon P. Mitchell Trudell Medical International, London, ON, Canada

Paul B. Myrdal University of Arizona, College of Pharmacy, Tucson, AZ, USA

Kevin P. O'Donnell College of Pharmacy, University of Texas at Austin, Austin, TX, USA

Bo Olsson AstraZeneca R&D Lund, Lund, Sweden

Guirag Poochikian Poochikian Pharma Consulting, Rockville, MD, USA

Sowmya Saiprasad Chemical Engineering and Materials Science, Wayne State University, Detroit, MI, USA

Masahiro Sakagami Department of Pharmaceutics, School of Pharmacy, Virginia Commonwealth University, Richmond, VA, USA

Mark Sanders Clement Clarke International Ltd, Dunstable, Bedfordshire, United Kingdom

Poonam Sheth University of Arizona, College of Pharmacy, Tucson, AZ, USA

Gur Jai Pal Singh Axar Pharmaceuticals, Irvine, CA, USA

Hugh D.C. Smyth College of Pharmacy, University of Texas at Austin, Austin, TX, USA

Yoen-Ju Son College of Pharmacy, University of Texas at Austin, Austin, TX, USA

Janani Swaminathan School of Pharmacy and Pharmaceutical Sciences, Trinity College Dublin, Dublin, Ireland

Diana G. Villanueva UAEM, Morelos, Mexico

Alan B. Watts MicroDose Therapeutx, Monmouth Junction, NJ, USA

Robert O. Williams III Division of Pharmaceutics, College of Pharmacy, University of Texas at Austin, Austin, TX, USA

Zhen Xu Eshelman School of Pharmacy, University of North Carolina, Chapel Hill, NC, USA

Chapter 1
Macro- and Microstructure of the Airways for Drug Delivery

Kevin P. O'Donnell and Hugh D.C. Smyth

Abstract Both anatomy and physiology of the airways are critical for understanding and predicting the dynamics of drug delivery systems that are inhaled. This theme that intimately links the biology of the airways to the response of pulmonary drug delivery systems is present throughout other following chapters in this book. Therefore, it is ideal to introduce these concepts in this chapter by first addressing the lung architecture on the macroscale and how it influences drug delivery. Then, we discuss the microscale interactions between the airway environment and drug delivery system. By discussing the anatomy and physiology at these scales in the direct context of pulmonary drug delivery, we believe this chapter is unique and, hopefully, useful for those seeking the controlled release in the respiratory tract.

Keywords Anatomy • Cell biology • Deposition • Physiology

1.1 Macrostructure and Function

When considering drug delivery to the lung, one must first understand the general composite structure of the lung. The airway can be broken into two distinct zones: the conducting airway and the respiratory airway.

H.D.C. Smyth (✉)
College of Pharmacy, University of Texas at Austin, Austin, TX 78746, USA
e-mail: hsmyth@mail.utexas.edu

H.D.C. Smyth and A.J. Hickey (eds.), *Controlled Pulmonary Drug Delivery*, Advances in Delivery Science and Technology, DOI 10.1007/978-1-4419-9745-6_1,

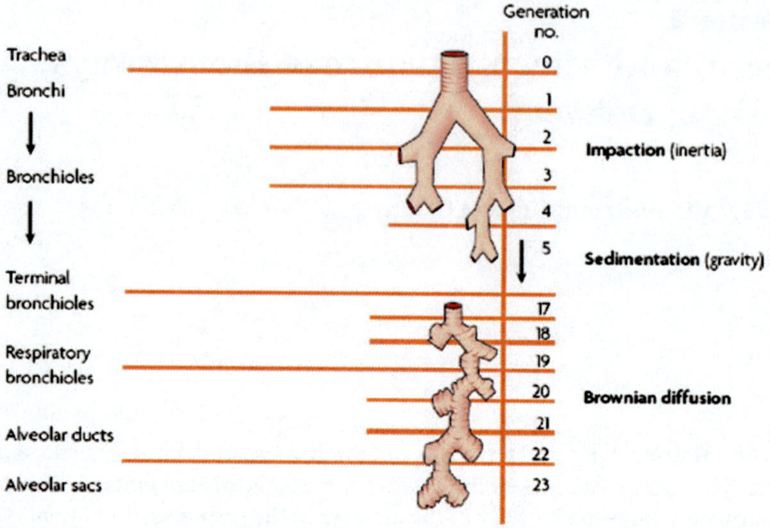

Fig. 1.1 Factors influencing lung deposition. Reproduced from Patton and Byron [53] with permissions

1.1.1 Conducting Airways

The conducting zone constitutes the upper portion of the airways. It begins at the mouth/nose, and comprises the trachea, bronchi, bronchioles, and terminal bronchioles. This portion of the airway bifurcates approximately 17 times prior to reaching the respiratory zone of the deep lungs, which branches further and is discussed in the following sections. This branching allows for a progressive increase in surface area and a corresponding decrease in air velocity (Fig. 1.1) [43]. The presence of smooth muscle gradually increases from the trachea to the terminal bronchioles, connecting the incomplete cartilage rings of the trachea and eventually becoming a complete layer. Within the conducting region, no gas exchange takes place; its primary purpose, rather, is to transport the gas to the respiratory zone [1]. A secondary function of the conducting airways is to ensure that inspired gasses are humidified and heated so as to provide the alveoli with air identical to the preexisting environment. The dangers associated with the inspiration of dry and improperly heated gasses have been demonstrated [65]. The obvious problems associated with cold air inspiration are the loss of body heat (i.e., drop in core body temperature) due to heat transfer between the body and respired air, as well as water loss due to humid air expiration. Accurate humidification of respiratory gasses is crucial to ensure proper function of the airways. Improper humidity content of respired air can lead to extensive dehydration and loss of body weight [46]. Additionally, functional impairments may be rapidly observed including extensive impairment of the mucociliary escalator. The following list describes the potential damages caused by dry gas inspiration [65]:

(a) Destruction of cilia and damage to mucous glands
(b) Disorganization and flattening of pseudostratified columnar epithelium and cuboidal epithelium
(c) Disorganization of basement membrane
(d) Cytoplasmic and nuclear degeneration
(e) Desquamation of cells
(f) Mucosal ulceration
(g) Reactive hyperemia following damage

Furthermore, overhumidified air poses dangers as well [71]. In fact, overhumid air may lead to water intoxication with the final effects being analogous to those listed for dehumidified air, in the opposing direction. Ultimately, improperly functioning conducting airways may lead to impaired respiration through increased surface tension and bronchoconstriction. This in turn will lead to inefficient drug delivery to the lung or, potentially, impaired pulmonary absorption ability. While damage to the ciliated epithelium may limit clearance of particles impacted in the upper airway, the body's natural defense system will also be harmed, resulting in a higher propensity for infection.

1.1.2 Respiratory Airways

Distal to the terminal bronchioles of the conducting zone lies the respiratory zone which consists of the respiratory bronchioles, alveolar ducts, and alveolar sacs. The alveolar ducts are typically 1 mm in length formed via connected groups of alveoli, polyhedral chambers with an average diameter of 250 μm characterized by a 0.1–0.4-μm epithelium and 70 nm liquid lining layer. Itoh et al. [38] provide excellent microscopy and 3D modeling images of the alveolar locality [38]. The primary function of this region is gas exchange, which may take place throughout the listed bifurcations. This region is ideally suited for gas exchange due to its inherent physical characteristics. The surface area of the distal airway is approximately 102 m^2, while the conducting airway is a mere 2–3 m^2, allowing for much greater contact with the inspired gas or therapeutic aerosol [52]. Second, the thickness of the cell layer, which makes up the respiratory region, is progressively reduced from approximately 60 μm in the upper airway to the aforementioned submicron thickness in the alveoli [53]. Similarly, the fluid layer at the cell surface decreases from 8 μm to approximately 70 nm in direct correlation with the decrease in cell thickness. Figure 1.2 depicts this change in breadth. The cell types allowing for this reduction in thickness are discussed in detail in later sections. Lastly, the partial pressure of oxygen within the alveoli is far less than that within the CO_2-rich blood present in pulmonary circulation. The pressure gradient coupled with a markedly thin diffusion pathway

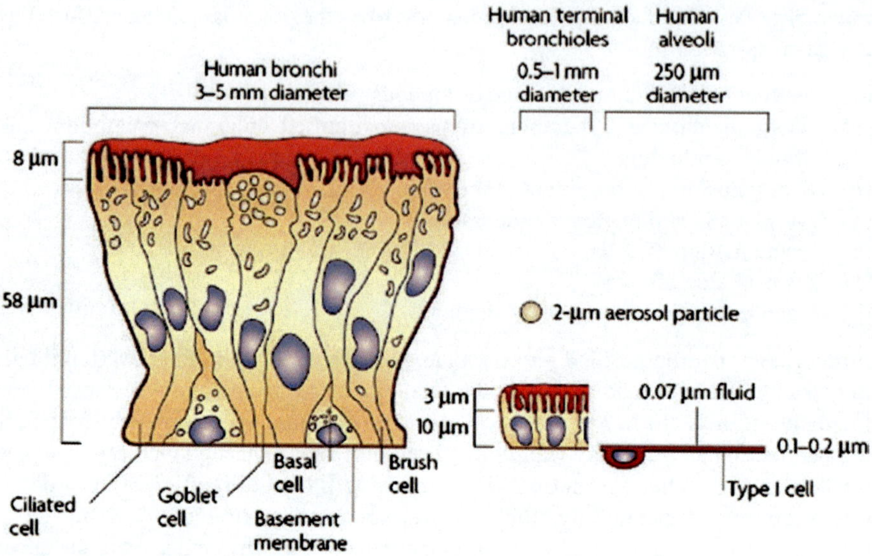

Fig. 1.2 Cell type and regional thickness across the regions of the airway. Reproduced from Patton and Byron [53] with permission

allows oxygen to diffuse from the alveoli into the blood, while CO_2 diffuses in the opposite direction. For a detailed explanation on the mechanisms of gas transport and development of the alveolar gas exchange, the reader is referred to Massoro (1996) and Hickey and Thompson (2003) [31, 47].

1.2 Targeted Anatomical Sites for Aerosol Delivery

As can be derived from the above description of the airway structure, the premier target for drug delivery will be the alveolar region of the deep lung, which presents the largest surface area and thinnest diffusion pathway for dissolved material. Studies have demonstrated that the trachea presents >90% resistance to transport, limiting its potential in drug delivery [78]. While the nonrespiratory bronchioles are involved in many disease states, they too exhibit poor drug absorption; though due to the thinner cell layer, absorption may be marginally increased. This is most likely due to a reduction in intercellular and intracellular transport path lengths. Many disease states, such as asthma and chronic bronchitis, elicit their effects in the bronchioles or trachea; however, because of their poor drug reception ability, they are typically not a primary target. Drugs delivered via inhalation seldom achieve a dose of greater than 20% to the alveoli when administering via a pressurized metered dose inhaler (pMDI) [27]. While other devices, such as spacers, and patient parameters (synchronization of breath/actuation) can increase the dose well above 20%, a potentially large amount of drug may deposit within the conducting airways and will subsequently be transported up the airway via the mucociliary escalator, ultimately resulting in the swallowing of a large portion of the dose by the patient. For a detailed description of the mucociliary system, the reader is referred to [54] in International Review of Physiology; Respiratory Physiology III. Following chapters within this publication discuss in great detail the clearance of particles from the lungs, as well as methods of targeting to various sites within the airway. Many targeting techniques often utilize receptor–ligand conjugation. Targeting of specific areas rich in the desired receptor can lead to greater therapeutic efficiency. A discussion of such receptors can be found in later sections.

1.2.1 Aerosol Particle Deposition Mechanisms

In order for a physiological effect to occur, inhaled particles must first deposit within the respiratory system. A large number of factors contribute to particle deposition within the airway including particle size, shape and density, airflow velocity and volume, interpatient physiological variations, and pause time between inspiration and expiration. While all of these factors and more contribute to total deposition, there are three generally accepted mechanisms by which particle deposition within the airway occurs: impaction, sedimentation, and diffusion. Two more mechanisms, interception and electrostatic precipitation, may also result in particle deposition; however, they are not discussed in detail here [43]. Impaction of particles upon airway surfaces is influenced by particle size, density, and velocity. As a result of inertial forces, it is most likely to occur in the upper conducting airways characterized by high particle velocities and drastic changes in airflow direction. This mechanism is most influential upon particles of sizes greater than 2 μm [64]. Sedimentation via gravitational force of particles within the airway is dependent upon particle mass. Particles will be subject to sedimentation if the product of their settling velocity and residence time is greater than the distance required for contact of the surface airway. Sedimentation typically influences particles between 0.5 and 2 μm. Particles less than 0.5 μm are subject to deposition via diffusion based on Brownian motion. A reduction in particle size and increase in residence time increase the probability of a particle to deposit through diffusion, and thus, breath holding may increase deposition via this mechanism [62].

1.3 Physiological Factors Influencing Deposition

1.3.1 Mode of Inhalation

While a large number of formulation factors will influence the mechanism and extent of particle deposition, patient dependent dynamics must also be considered. Regardless of the method of aerosol generation, the method of inhalation by the patient will strongly influence the degree and location of particle deposition.

The airflow pattern instigated by the patient will determine the extent of throat deposition, ultimately swallowed and considered a nontherapeutic dose, which will in turn determine the dose available to reach the lung [8]. Deposition within the throat and mouth can lead to a high degree of interpatient variability, and thus, adequate breath patterns can increase efficacy. In fact, it has been demonstrated that slow inhalation by the patient reduces throat deposition, increases lung deposition, and ultimately reduces patient variability [70]. While this slow inhalation may be applied during passive inhalation methods (i.e., nebulizers), it is not reasonable for devices such as pMDIs, which produce a high velocity aerosol with which the patient must time their breathing. Indeed, proper breath timing and inspiration pressure can increase the lung deposition of pMDI aerosols from 7 to 20% [51]. Similarly, slow air velocities may not be appropriate for dry-powder inhalers where sufficient airflow must be applied to generate the aerosol [83]. Therefore, an appropriate method of inhalation must be applied that is device specific to minimize conducting zone deposition and increase the therapeutic dose to the deep lung. The inspiration–breath hold–expiration cycle creates enormous variations in airflow patterns. These variations may be compounded by contractions or restrictions created by the aforementioned smooth muscle. These patterns will navigate inspired particles in conjunction with the physical properties of the particle. A detailed description of the influence and characteristics of airflow across the complex bifurcating human lung is given by [37], and the reader is referred there for further information.

1.3.2 Oropharyngeal Deposition

Aside from the method of inhalation, the individual's morphology will determine the success of drug delivery to the lung. Variations in diameter and length of individual generations create inherent volumetric and structural differences between patients [22, 34]. Figure 1.1 depicts this progressive change across the generations, while Table 1.1 demonstrates intersubject variation in oropharyngeal geometries.

Table 1.1 Oropharyngeal morphology in multiple subjects

Oropharyngeal designation[a]	Oropharyngeal volume/cm^3	Oropharyngeal opening[b]	Oropharyngeal centerline length/cm	Gender	Oral cavity inlet cross-section shape
1C	37.6	C	17.1	F	Rectangular
2C	53.4	C	18.7	F	Circular
3A	55.9	A	19.9	M	Rectangular
4A	61.8	A	17.8	F	Circular
5A	68.4	A	19.9	M	Circular
6B	75.1	B	21.6	M	Circular
7B	80.8	B	22.3	M	Rectangular

Reproduced from Ehtezazi et al. [22] with permission
[a] The designation sequence is rank of oropharyngeal volume/oropharyngeal configuration category
[b] *A* wide open space; *B* a moderate narrowing; *C* a marked constriction; *F* female; *M* male

Furthermore, these anatomical differences may be altered based on the disease state of the patient, a matter discussed in the following section. This physical difference in structure coupled with variations in respiration rate creates tidal volumes ranging from 460 to 900 ml in adults, capable of producing a twofold difference in particle deposition [33]. While larger particles are deposited by inertia at the 90° bend transition between the mouth and throat, small (i.e., submicron) particles may utilize the transition period between inspiration and expiration and sediment within the oropharyngeal space. It should be noted that residence times within any given portion of the respiratory tract may be subsecond, and the final dynamics influencing particle deposition within the selected region is a result of local aerodynamics, determined by local morphology and airflow changes throughout breathing [67].

1.3.3 Morphological and Deposition Changes Due to Disease

1.3.3.1 Obstructive Diseases

Obstruction of the airway due to disease may drastically reduce the ability to deliver drug to the lungs. Chronic pulmonary obstructive diseases (COPD) such as chronic bronchitis (CB), obstructive sleep apnea (OSA), and emphysema restrict airflow in an irreversible manner. CB is characterized by thickening and inflammation of the bronchial walls and increased mucus production, creating a restricted pathway for airflow and aerosols and an increased diffusion path for therapeutic entities. Emphysema, on the contrary, is the permanent enlargement of the gas exchange zone of the lungs, resulting in destruction of the alveolar walls [16, 24]. OSA is characterized by a reoccurring temporary stoppage of breathing during REM sleep and is prevalent in patients with a narrow oropharyngeal airway and low sleeping lung volume [79]. COPDs characteristically cause productive coughing in the patient, which may increase the clearance of delivered particles or further hinder inhalation ability. For a detailed description of the structural changes associated with COPD, the reader is referred to [17]. Asthma is a complex pulmonary disorder similar to CB in that bronchial inflammation and hyperactivity are observed [42]. Since particle deposition is highly dependent upon airflow patterns, any obstructive disease will alter the deliverability of drugs. For example, an obstructed throat causes recirculation in both directions, greatly altering flow characteristics [44]. It has also been demonstrated that patients suffering an asthmatic attack show a narrowing of the pharyngeal airway, which results in increased oropharyngeal deposition of an aerosolized dose and consequently, due to clearance and impaction, a reduction in total dose delivered to the lungs [69] While obstructions may lead to greater bronchial and central airway deposition due to impaction caused by altered flow patterns, it is possible to achieve a high degree of deep lung penetration in COPD patients. Maeyer et al. [45] demonstrated that low airflow rates could provide above 50% dose delivery to the deep lung for particles between 2 and 4 μm (1–5 μm is considered the ideal particle size range). Furthermore, breath holding following reception of a dose may significantly increase deposition against COPD [45, 63]. Thus, while obstructions may greatly hinder the ability of the patient to breathe, it may still be possible to treat the condition via deep lung penetration. Owing to the chronic and reoccurring nature of these disease states, the ability to prolong release in a local manner would be of extreme benefit to the patient population by potentially reducing not only the number of doses required but also the occurrences of impaired breathing.

1.3.3.2 Infectious Disease

Owing to direct tissue contact with the external environment via inspired air, the lung is susceptible to bacterial infections. While many pulmonary defense systems exist, infection is still possible, especially within immunocompromised patients. Owing to a mutated membrane embedded chloride channel (CFTR), patients suffering from cystic fibrosis exhibit a dehydrated airway liquid surface, resulting in an increase in mucus viscosity and poor mucociliary clearance. As a result of this altered physiological environment, this patient population has an increased incidence of pulmonary infection. Such infections include, but are not limited to, aspergillosis, zygomycosis, cryptococcosis, histoplasmosis, pneumocystis pneumonia, *Pseudomonas aeruginosa* infection, and tuberculosis (TB) [60, 66, 74]. In any case, pulmonary function may be impaired with common symptoms including acute dyspnea, pleuritic chest pain, chronic inflammation, and tachypnea. Tidal volume may be compromised, leading to insufficient inhalation ability. *Pseudomonas aeruginosa* rhamnolipid has been reported to cause cell membrane damage and inhibit epithelial ion transport, resulting in deteriorated mucociliary clearance mechanisms [2, 56]. TB can cause serious tissue damage, ultimately liquefying infected regions as a result of the host immune response creating an ideal growth medium. This liquid caseous material does not promote the growth or survival of alveolar macrophages, further reducing the innate immune ability [50]. It has been reported that targeted delivery directly to the site of infections results in high localized concentrations while maintaining relatively low systemic concentrations of therapeutic agents [75, 82]. This is desirable when delivering antibiotic or antifungal agents due to a high prevalence of adverse side effects associated with high systemic plasma concentrations. Furthermore, systemic delivery via the lung is still possible for infections that have disseminated to multiple locations within the host [68]. However, due to the aforementioned pulmonary impairments, sufficient delivery of inhaled therapeutics may be difficult. Reduced tidal volumes, altered mucosal physiology, altered epithelial function, and surface blockage due to colonization may all inhibit proper deposition. In cases of insufficient mucus clearance, the additional liquid yields a thicker diffusion pathway, presenting increased difficulty for diffusion as well as increased adhesive properties in undesired locations.

1.4 Airway Cells

1.4.1 Relevant Cells of the Conducting Airways

The columnar epithelium of the conducting airways is a gradually thinning barrier comprising a variety of cells. Below the luminal epithelium, and in no contact with inspired gasses, lie the basal cells. Basal cells are pyramidal progenitor cells that differentiate into the cell types found within the tracheobronchial epithelium [3]. Also, below the surface lie the neuroendocrine cells, such as K-cells, which contain and secrete peptide hormones. While their secretions may reach the surface, the cell walls themselves rarely do. Ciliated columnar cells make up a large portion of the conducting airway epithelium. These cells are a major component of the mucociliary escalator responsible for the clearing of foreign material from the airways, including deposited therapeutics. Interdispersed between the ciliated cells, and comprising approximately 25% of the tracheobronchial epithelium, are nonciliated goblet cells. Goblet cells are nonglandular cells involved in mucus secretion creating the viscoelastic layer lining the bronchial region. The most prevalent cell type in the epithelium is Clara cells. These nonciliated cells have a number of key roles including production of surfactant components, production of protease inhibitors, and metabolic detoxification [20, 21]. The cells of the epithelium are connected via tight junctions, preventing intercellular penetration of inhaled matter into the body. These junctions, however, may be broken or damaged by aerosolized pharmaceutical constituents such as chitosan. While such a component may serve to enhance the delivery platform, it may also be exposing the body to potential antigens while also inhibiting the body's natural clearance mechanisms [39, 73]. Other cells within the tracheobronchial epithelium include smooth muscle cells and mast cells, whose primary functions are contraction/relaxation of the airway and antigen recognition and response, respectively, and secretory gland cells (mucus and serous), which are discussed in a later section. Mast cell response must be considered during formulation due to the fact that upon recognition of an antigen or allergen, mast cells release inflammatory mediators creating a variety of biological responses including, but not limited to, airway constriction, swelling of tissue, blood vessel dilation, and permeation [40]. Particle size is of concern as well in avoidance of uptake by patrolling macrophages.

1.4.2 Relevant Cells of the Respiratory Airways

Entry into the respiratory zone is marked by a gradual transition as the bronchial wall is partially replaced by alveoli, with eventual complete replacement. The drastically thinner epithelium within this region is responsible for gas exchange and will be the primary target of most pharmaceutical scientists in drug delivery. As such, the two primary cell types of interest within this region are those responsible for gas and material exchange: alveolar type 1 (AT1) and alveolar type 2 cells (AT2).

Capillary endothelial cells and alveolar macrophages are also present and may be found close to the alveolar epithelium. AT1 cells cover over 95% of the alveolar surface area with AT2 cells accounting for the remainder, yet AT2 cells are far more numerous than AT1 cells, present at a ratio of approximately 2:1 [15, 20, 21, 58]. AT1 cells are approximately 50–100 μm in diameter while remaining extremely thin, 2 μm at the nucleus and 0.2 μm in cytoplasmic regions. These cells have shown the highest water permeability of any cell type. Furthermore, transport of macromolecules may be possible due to the presence of vesicles and caveolae invaginations, further reducing the membrane thickness. By contrast, AT2 cells are 10 μm in diameter, exhibiting a cuboidal shape. The primary role of AT2 cells is the production, secretion, and recycling of lung surfactant material. A secondary function of AT2 cells is the generation of AT1 cells. Upon division, AT2 cells may either proliferate into additional AT2 cells or may differentiate into AT1 cells to aid in repair and surface maintenance [14, 30, 47]. The alveolar cells are connected by both tight and gap junctions, creating a protective boundary between the environment and body while maintaining potential as a therapeutic target.

1.5 Airway Receptors

Therapeutic entities may elicit their cellular effect both pre and post internalization via cell receptor binding. As a combinatory example, many tumors overexpress receptors for hyaluronic acid, which upon coupling with a ligand results in rapid internalization of the bound molecule, which may further conjugate with an internal target. Thus, an understanding of cellular receptors within the pulmonary system creates potential for more effective inhalation therapy. Presented herein are only some of the most pertinent airway receptors in drug delivery. One of the most studied pulmonary receptor classes is β adrenergic receptors (βAR). β_2 adrenergic receptors may be found throughout the airway on the epithelium and smooth muscle; however, the greatest concentration may be found in the alveoli (on both AT1 and AT2 cells). More specifically, β_2 adrenergic receptors have been observed on the vascular endothelium, ciliated epithelium, mast cells, circulating inflammatory cells, Clara cells, and others. β_1 receptors lie primarily on alveolar walls [4, 9, 28]. Complexation with a β adrenergic agonist can lead to a number of physiological responses (Table 1.2), including bronchodilation via smooth muscle relaxation, which has lead to their targeting in the treatment of asthma and COPD. βAR activation also upregulates Na+ transport and accelerates clearance of fluid from the alveolar airspace, allowing for treatment of edema. Hanania and Moore [28] describe the influence on particular cell lineages and the reader is referred there if such information is desired. Desensitization of βAR can occur over prolonged or continuous exposure to agonists, for example via an uncoupling from the corresponding G protein responsible for stimulation [29, 49]. Activation of muscarinic receptors has proven to be more effective than βAR agonists in the treatment of COPD, but not asthma. Acetylcholine-lad bronchoconstriction is

Table 1.2 Documented physiological effects of β_2-adrenergic receptor stimulation in human lung

Airway smooth muscle relaxation
Prejunctional inhibition of acetylcholine release from parasympathetic neurons in airway smooth muscle
Stimulation of mucous and serous cell secretion
Stimulation of chloride ion secretion across the apical membrane of airway epithelial cells
Increase in ciliary beat frequency
Stimulation of surfactant secretion from alveolar type II cells
Inhibition of mediator release from lung mast cells and neutrophils

Reproduced from Bai [4] with permissions

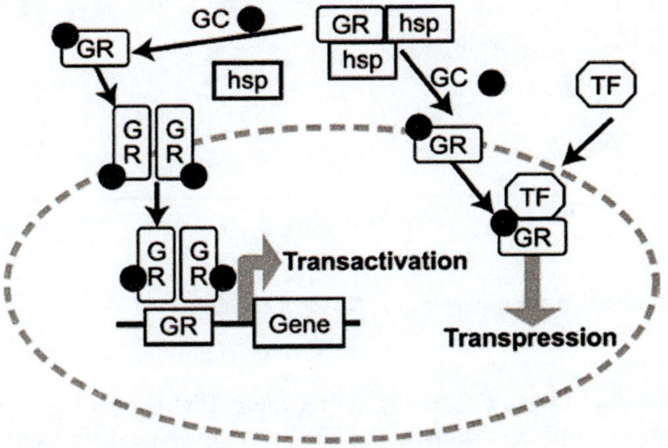

Fig. 1.3 Transactivation and transrepression pathways. *GC* glucocorticoid; *GR* glucocorticoid receptor; *hsp* heat shock protein; *TF* transcription factor. Reproduced from Hochhaus [32] with permissions

mediated by the M_3 receptor. Anticholinergics block M_1 and M_3 receptors, resulting in bronchodilation; however, selectivity away from the M_2 receptor (which mediates acetylcholine release) has proven to be challenging [6]. Endothelin receptors A and B (ET_A and ET_B) may be targeted with endothelin-1, a vasoconstrictor, for the treatment of pulmonary arterial hypertension [57]. Glucocorticoid receptors (GR) reside in the cytoplasm of pulmonary cells in an inactive complexed form. Upon binding to a corticosteroid ligand, its cytoplasmic complex disaggregates and is transported to the nucleus where dimerization occurs followed by interaction with response elements. Alternatively, bound monomers may interact with protein transcription factors potentially reducing proinflammatory cytokine production (Fig. 1.3) [32]. Many interstitial pulmonary diseases can result in an increase in lung GR, typically due to the increase in parenchyma cellular density associated with the disease state, leading to necessary alterations in treatment [61]. Figure 1.4 [23] presents binding affinities of various corticosteroids

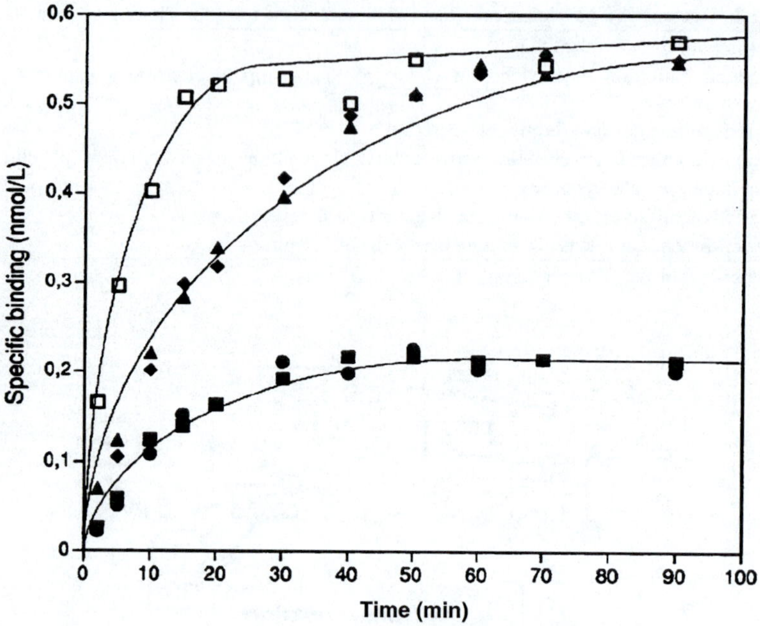

Fig. 1.4 Reproduced from Esmailpour et al. [23] with permission

used in inhalation therapy. Prostacyclin receptor (PR) agonists provide an anti-inflammatory response in asthma and COPD patients. Agonistic binding of bronchial epithelial PRs enhances the anti-inflammatory cellular response generated by glucocorticoids. Thus, combination therapies employing PR agonists may prove to be superior to steroid treatment alone [81]. A vast array of pulmonary receptors mediates clearance patterns, immunological responses, and particle uptake. Targeting such receptors or employing them synergistically can be highly beneficial in inhalation therapies.

1.6 Blood Flow

1.6.1 Bronchial and Pulmonary Circulation Systems

Airway circulation serves a number of roles in addition to providing nutrients to the region. These functions include heat–water exchange, regulation of airway caliber in the peripheral lung, clearance of biological substances, and recruitment of inflammatory mediators. Blood flow may be influenced by a number of factors including hyperventilation, airway pressure, inspired air temperature, and alterations in airway fluid content [13]. Blood is supplied to the pulmonary system via the right ventricle of the

Table 1.3 Blood flow and physiological parameters of various species

	Rats	Dogs	Humans
Body weight (kg)	0.25	10.0	70.0
Respiratory rate (per min)	97	21	16
Tidal volume (ml)	1.55	114	400
Dimensions of upper and lower airways, and of alveolar region			
Upper and lower airways			
Alveolar region		see Table 1.2	
Luminal volume (ml)	3.55	140	2,670
Surface area (cm^2)	45.7×10^4	754×10^4	$3{,}310 \times 10^4$
Blood flow (ml/min/whole tissue)			
Upper airway	1.05	11.0	44.2
Lower airway (1% of cardic output)	0.61	12.1	58.4
Alveolar region (95% of cardiac output)	60.8	1,150	5,550
Diffusion distance (cm)			
Upper airway (cm)	0.02	0.04	0.07
Lower airway (cm)	5% of lumminal diameter		
Alveolar region (μm)	0.4	0.48	0.62

Reproduced from Tsujino et al. [76] with permission

heart to the pulmonary arteries. In humans, the bronchial arterial system stems from the thoracic aorta with one bronchial artery supplying the right lung and two supplying the left lung. Intrapulmonary bronchial arteries provide at least two arteries for each bronchus. Branches enter the bronchial muscular layer to provide a submucosal vascular system. At the terminal bronchiole, arterioles branch to form a pulmonary capillary network to perfuse the alveoli. The diameter of the pulmonary capillaries is between 4.3 and 8.6 μm. Here, the capillary walls are in close proximity to the alveolar membrane and in many cases are fused directly to it to facilitate rapid diffusion. In such cases, the cumulative diffusion pathway from luminal air to the blood is less than 400 nm. For this reason, alveolar deposition has become a primary target for systemic drug delivery via inhalation [82]. Two pulmonary veins stem from each lung to transport oxygenated blood to the left heart via the left atrium and subsequently the left ventricle for systemic circulation. Table 1.3 defines the blood flow with corresponding cardiac output of the different airway regions in ml/min/whole tissue. For a more in-depth look at pulmonary circulation, the reader is referred to Chediak and Wanner [10, 11, 24, 55, 76].

1.6.2 Blood Flow Influence on Drug Delivery

The disease state of the patient may greatly influence perfusion of the airway. Asthma and COPD have shown to increase airway blood flow as a result of an increase in the number of blood vessels in the mucosa as well as the dilation of resistance arteries. Furthermore, increased blood vessel wall thickness, increased microvascular permeability, and edema formation may all be observed. Lung hyperinflation potentially compresses vessels, resulting in increased arterial flow resistance, ultimately resulting in vascular remodeling. An increase in blood flow and permeability is beneficial for drugs meant for systemic treatment in that they will be transported throughout the body more rapidly. Indeed, transport processes are enhanced in the presence of flow. However, for drugs intended to act locally, this presents the problem of increased clearance and potential metabolism, making the treatment difficult. Furthermore, unintentional/undesired systemic exposure may occur for intended local treatments, especially in cases with elevated blood flow, creating potential adverse events. This is most prevalent for small-molecule drugs capable of quickly navigating the diffusion pathway into the blood stream. The extent of perfusion may inhibit long-acting drugs from eliciting a long-term effect if they are cleared rapidly, or conversely, may prolong the effect of intended short-action drugs [10, 17, 35, 36, 48].

1.7 Airway Secretions

As mentioned, while discussing respiratory cellular composition, two primary cell types are responsible for the majority of secretions found in the airways: serous cells and mucous cells. In the cartilaginous tracheobronchial region, serous and mucous cells form glands beneath the epithelial surface, connecting to the lumen via ducts [25]. Glands comprised primarily of serous cells have been shown to secrete both antimicrobial and immunological agents, including lysozyme, IgA, and antimicrobial peptide LL-37, making them crucial in certain disease states. Glands formed of mucous cells as described by their nomenclature are the primary producers of mucus and mucoproteins [5]. Produced mucins may be either secreted by these glands into the mucus layer or may be membrane-associated, in which case antigen recognition and cellular signaling may be of primary purpose. Secreted mucus, along with the perciliary fluid produced by the epithelial goblet cells, creates an adhesive, viscoelastic fluid layer covering the airway surface. The viscosity of the mucus layer decreases with increasing strain in a time-dependent manner, allowing for proper control of clearance rates. Mucus is composed of 95% water and approximately 5% solids, which contribute to its gel-like nature. These values vary regionally within the lung. The solids of the mucus include the mucin glycoproteins, lipids, DNA, actin, and minerals. The mucin glycoproteins, which vary greatly in length, are approximately 75% glycosylated and provide binding sites for large carbohydrate formations [7, 77, 80]. The variation in protein length creates changes in the swelling capability of the mucus network, and thus, directed cleaving of the S:S bond of apomucin polymers may drastically accelerate mucus distribution and clearance [19]. Primary functions of the mucus layer include binding (via adhesion) and clearance of particles (mucociliary clearance), hydration of the cell surface, humidification of respired air, and lubrication of the airways [18, 41]. A list of the primary glandular secretions may be found in

Table 1.4 Major secretory products of airway gland cells

Serous cells	Mucous cells
Mucin (MUC 2, MUC 7)	Mucin (MUC 2, MUC 5B)
Proteoglycans	Antimicrobial peptide LL-37/hCAP-18
Lysozyme	
Lactoferrin	
Secretory IgA	
Antimicrobial peptide LL-37/hCAP-18	
Antileukoprotease	
β-defensin 1	
β-defensin 1	
Proline-rich proteins	
Albumin	

Reproduced from Finkbeiner [25] with permission

Table 1.4, and for a more detailed discussion on individual components of the mucus, the reader is referred to [7]. The extent of mucus production is dependent upon individual physiology of the patient, including disease state, and is primarily activated via cholinergic stimulation mediated by the aforementioned $M_{1\text{-}3}$ receptors, though innervations via other means may also be influential. An in-depth discussion of glandular and goblet cell motor control is provided by [59]. Cholinergic innervations of the glands may also influence the epithelium by instigating active ion transport with associated passive water flux, as well as secretion of large molecules such as albumin. For example, CB is characterized by hypersecretion of mucous with concurrent enlargement of the secretory glands due to hyperplasia. This may increase not only the duration but also the frequency of pulmonary infection and can result in airway plugging [16]. Cystic fibrosis exhibits improper secretory gland function as well owing to the fact that the glands are unresponsive to the synergistic effect of vasoactive intestinal peptide (VIP) acting with acetylcholine. This improper mucus production, in conjunction with the epithelial damage discussed previously, creates an environment with increased susceptibility to infection [12].

1.7.1 Importance in Drug Delivery

Particles delivered to the lung will, unless expired, inevitably contact the lung lining fluid. As such, its production, properties, and interaction with the inhaled material will all impact therapeutic delivery. If soluble within the mucosal matrix, dissolution kinetics will determine the extent of solubilization and absorption of the drug. However, poorly soluble materials with low dissolution velocities, i.e., hydrophobic formulations, as well as material with a high affinity for mucus protein binding will be subjected to clearance, preventing absorption [53, 82]. The clearance mechanisms vary based on pulmonary location and are beyond the scope of this chapter but are discussed in Chaps. 2 and 6. Drugs that do enter solution can cross the epithelial membrane via passive diffusion with the rate controlled by physical properties of the therapeutic entity (i.e., hydrophobicity, molecular weight). Material that is not dissolved, nor cleared from the system, may be physically translocated to the epithelial surface allowing for particle–cell interactions and potential internalization. In either case, the mucus presents a physical barrier. Variations in surface tension created by various disease states as well as particle morphology will determine the extent of submersion into the liquid lining layer [26]. Thus, surface and formulation modifications that increase dissolution (i.e., cyclodextrin complexation) or translocation (i.e., particle size control, surface wetting agent) may be extremely beneficial in enhancing therapeutic efficiencies of inhaled particles [72].

References

1. Altiere RJ, Tompson DC (2007) Physiology and pharmacology of the airways. In: Hickey AJ (ed) Inhalation aerosols. Informa Healthcare, New York, pp 83–126
2. Arnold MM, Gorman EM, Schieber LJ, Munson EJ, Berkland C (2007) NanoCipro encapsulation in monodisperse large porous PLGA microparticles. J Control Release 121:100–109
3. Ayers MM, Jeffery PK (1988) Proliferation adn differentiation in mammalian airway epithelium. Eur Respir J 1(1):58–80
4. Bai TR (1992) Beta2 adrenergic receptors in asthma: a current perspective. Lung 170:125–141
5. Ballard ST, Inglis SK (2004) Liquid secretion properties of airway submucosal glands. J Physiol 556(1):1–10
6. Barnes PJ (1998) Muscarinic receptor subtypes in airways. Respir Immunol 149:201–202
7. Boat TF, Cheng PW, Leigh MW (1994) Biochemistry of mucus. In: Takishima T, Shimura S (eds) Airway secretion – physiological bases for the control of mucous hypersecretion. Marcel Dekker, New York, pp 217–282
8. Borgström L, Olsoon B, Thorsson L (2006) Degree of throat deposition can explain the variability in lung deposition of inhaled drugs. J Aerosol Med 19(4):473–483
9. Carstairs JR, Nimmo AJ, Barnes PJ (1985) Autoradiographic visualization of beta-adrenoceptor subtypes in human lung. Am Rev Respir Dis 132:541–547
10. Charan NB, Thompson WH, Carvalho P (2007) Functional anatomy of bronchial veins. Pulm Pharmacol Ther 20:100–103
11. Chediak AD, Wanner A (1990) The circulation of the airways: anatomy, physiology and potential role in drug delivery to the respiratory tract. Adv Drug Deliv Rev 5:11–18
12. Choi JY, Joo NS, Krouse ME, Wu JV, Robbins RC, Ianowski JP, Hanrahan JW, Wine JJ (2007) Synergistic airway gland mucus secretion in response to vasoactive intestinal peptide and carbachol is lost in cystic fibrosis. J Clin Invest 117(10):3118–3127
13. Coleridge HM, Coleridge JCG (1994) Neural regulation of bronchial blood flow. Respir Physiol 98:1–13
14. Crandall ED, Matthay MA (2001) Alveolar epithelial transport: basic science to clinical medicine. Am J Respir Crit Care Med 163:1021–1029
15. Crapo JD, Young SL, Fram EK, Pinkerton KE, Barry BE, Crapo RO (1983) Morphometric characteristics of cells in the alveolar region of mammalian lungs. Am Rev Respir Dis 128(2):S42–S46

16. Danahay HJ, Jackson AD (2005) Epithelial mucus-hypersecretion and respiratory disease. Curr Drug Targets Inflamm Allergy 4:651–664
17. Davidson W, Bai TR (2005) Lung structural changes in chronic obstructive pulmonary diseases. Curr Drug Targets Inflamm Allergy 4(6):643–649
18. Davis CW, Dickey BF (2008) Regulated airway goblet cell mucin secretion. Annu Rev Physiol 70:487–512
19. Edwards SF (1986) The theory of macromolecular networks. Biorheology 23:589–603
20. Ehrhardt C, Forbes B, Kim KJ (2008) In vitro models of the tracheo-bronchial epithelium. In: Ehrhardt C, Kim KJ (eds) Drug absorption studies. Springer, New York, p 235
21. Ehrhardt C, Laue M, Kim KJ (2008) In vitro models of the alveolar epithelial barrier. In: Ehrhardt C, Kim KJ (eds) Drug absorption studies. Springer, New York, pp 258–282
22. Ehtezazi T, Saleem I, Shrubb I, Allanson DR, Jenkinson ID, O'Callaghan C (2010) The interaction between teh oropharyngeal geometry and aerosols via pressurised metered dose inhalers. Pharm Res 27:175–186
23. Esmailpour N, Hogger P, Rohdewald P (1998) Binding kinetics of budesonide to teh human glucocorticoid receptor. Eur J Pharm Sci 6:219–223
24. Farquhar SL, Fantasia L (2005) Pulmonary anatomy and physiology and the effects of COPD. Home Healthc Nurse 23(3):167–174
25. Finkbeiner WE (1999) Physiology and pathology of tracheobronchial glands. Respir Physiol 118:77–83
26. Geiser M, Schurch S, Gehr P (2003) Influence of surface chemistry and topography of particles on their immersion into the lung's surface-lining layer. J Appl Physiol 94:1793–1801
27. Gonda I (2003) Targeting by deposition. In: Hickey AJ (ed) Pharmaceutical inhalation aerosol technology. Informa Healthcare, Hoboken, pp 65–87
28. Hanania NA, Moore RH (2004) Anti-inflammatory activities of B2-Agonists. Curr Drug Targets Inflamm Allergy 3:271–277
29. Hausdorff WP, Caron MC, Lefkowitz RJ (1990) Turning off the signal: desensitization of β-adrenergic receptor function. FASEB J 4(11):2881–2889
30. Helms MN, Jaln L, Self JL, Eaton DC (2008) Redox regulation of epithelial sodium channels examined in alveolar type 1 and 2 cells patch-clamped in lung slice tissue. J Biol Chem 283(33):22875–22883
31. Hickey AJ, Thompson DC (2003) Physiology of the airways. In: Hickey AJ (ed) Pharmaceutical inhalation aerosol technology. Informa Healthcare, Hoboken, pp 1–29
32. Hochhaus G (2007) Pharmacokinetic and pharmacodynamic properties important for inhaled corticosteroids. Ann Allergy Asthma Immunol 98(Suppl 2):S7–S15
33. Hofmann W, Morawska L, Bergmann R (2001) Environmental tobacco smoke deposition in the human respiratory tract: Differences between experimental and theoretical approaches. J Aerosol Med 14:317–326
34. Hofmann W, Asgharian B, Winkler-Heil R (2002) Modeling intersubject variability of particle deposition in human lungs. Aerosol Sci 33:219–235
35. Horvath G, Wanner A (2006) Inhaled corticosteroids: effects on the airway vasculature in bronchial asthma. Eur Respir J 27(1):172–187
36. Horvath G, Vasas S, Wanner A (2007) Inhaled corticosteroids reduce asthma-associated airway hyperfusion through genomic and nongenomic mechanisms. Pulm Pharmacol Ther 20:157–162
37. Isaacs K, Rosati JA, Martonen TB (2004) Mechanisms of particle deposition. In: Ruzer SL, Harley NH (eds) Aerosols handbook: measurement, dosimetry, and health effects. CRC Press, Boca Raton, pp 85–115
38. Itoh H, Nishino M, Hatabu H (2004) Architecture of the lung: morphology and function. J Thorac Imaging 19(4):221–227
39. Junginger HE, Verhoef JC (1998) Macromolecules as safe penetration enhancers for hydrophilic drugs – a fiction? Pharm Sci Technol Today 1(9):370–376
40. Kalin TA, Meliton L, Meliton AY, Zhu X, Whitsett JA, Kalinichenko VV (2008) Pulmonary mastocytosis and enhanced lung inflammation in mice heterozygous null for the Foxf1 gene. Am J Respir Cell Mol Biol 39:390–399

41. King M, Rubin BK (1994) Rheology of airway mucus. In: Takishima T, Shimura S (eds) Airway secretion – physiological bases for the control of mucous hypersecretion. Marcel Dekker, New York, pp 283–314
42. Lilly CM (2005) Diversity of asthma: evolving concepts of pathophysiology and lessons from genetics. J Allergy Immunol 115(4):S526–S531
43. Lippmann M, Yeates DB, Albert RE (1980) Deposition, retention, and clearance of inhaled particles. Br J Ind Med 37:337–362
44. Luo HY, Liu Y, Yang XL (2007) Particle deposition in obstructed airways. J Biomech 40:3096–3104
45. Maeyer T, Mullinger B, Sommerer K, Scheuch G, Brand P, Beckmann H, Haussing K, Weber N, Siekmeier R (2003) Pulmonary deposition of monodisperse aerosols in patients with chronic obstructive pulmonary disease. Exp Lung Res 29:475–484
46. Marfatia S, Donahoe PK, Hendren WH (1975) Effect of dry and humidified gases on the respiratory epithelium in rabbits. J Paediatr Surg 10:583
47. Massaro GD, Massaro D (1996) Formation of pulmonary alveoli and gas-exchange surface area: Quantitation and Regulation. Annu Rev Physiol 58:73–92
48. Merrikh AA, Lage JL (2005) The role of red cell movement on alveolar gas diffusion. Materialwiss Werkstofftech 36(10):497–504
49. Mutlu GM, Adir Y, Jameel M, Akhmedov AT, Welch L, Dumasius V, Meng FJ, Zabner J, Koenig C, Lewis ER, Balagani R, Traver G, Sznajder JI, Factor P (2005) Interdependency of β-adrenergic receptors and CFTR in regulation of alveolar active Na^+ transport. Circ Res 96:999–1005
50. Muttil P, Wang C, Hickey AJ (2009) Inhaled drug delivery for tuberculosis therapy. Pharm Res 26(11):2401–2416
51. Newman S, Weisz A, Talaee N, Clarke SW (1991) Improvement of drug delivery with a breath actuated pressurized aerosol for patients with poor inhaler technique. Thorax 46(10):712–716
52. Patton JS (1996) Mechanisms of macromolecule absorption by the lungs. Adv Drug Deliv Rev 19:3–36
53. Patton JS, Byron PR (2007) Inhaling medicines: delivering drugs to the body through the lungs. Nat Rev 6:67–74
54. Phipps RJ (1981) The airway mucociliary system. Respir Physiol III 23:215–249
55. Pump KK (1972) Distribution of bronchial arteries in the human lung. Chest 62:447–451
56. Read RC, Roberts P, Munro N, Rutman A, Hastie A, Shryock T, Hall R, McDonlad-Gibson W, Lund V, Taylor G, Cole PJ, Wilson R (1992) Effect of Pseudomonas aeruginosa rhamnolipids on mucociliary transport and ciliary beating. J Appl Physiol 72(6):2271–2277
57. Rhodes CJ, Davidson A, Gibbs SR, Wharton J, Wilkins MR (2009) Therapeutic targets in pulmonary arterial hypertension. Pharmacol Ther 121:69–88
58. Ridge KM, Olivera WG, Saldias F, Azzam Z, Horowitz S, Rutschman DH, Dumasius V, Factor P, Sznajder JI (2003) Alveolar type 1 cells express the a2 Na, K-ATPase, which contributes to lung liquid clearance. Circ Res 92(4):453–460
59. Rogers DF (2000) Motor control of airway goblet cells and glands. Respir Physiol 125:129–144
60. Saldías SM, Valvano MA (2009) Interactions of Burkholderia cenocepacia and other Burkholderia cepacia complex bacteria with epithelial and phagocytic cells. Microbiology 155:2809–2817
61. Saldiva PH, Brentani MM, de Carvalho CR, Auler Junior JO, Calheiros DF, Pacheco MM (1985) Changes in the pulmonary glucocorticoid receptor content in the course of interstitial disease. Chest 88:417–419
62. Scheuch G, Kohlhaeufl MJ, Brand P, Siekmeier R (2006) Clinical perspectives on pulmonary systemic and macromolecular delivery. Adv Drug Deliv Rev 58:996–1008
63. Scheuch G, Kohlhaeufl M, Moller W, Brand P, Meyer T, Haussinger K, Sommerer K, Heyder J (2008) Particle clearance from the airways of subjects with bronchial hyperresponsiveness and with chronic obstructive pulmonary disease. Exp Lung Res 34:531–549
64. Schulz H (1998) Mechanisms and factors affecting intrapulmonary particle deposition: implications for efficient inhalation therapies. Pharm Sci Technol Today 1(8):336–344

65. Shelly MP, Lloyd GM, Park GR (1988) A review of the mechanisms and methods of humidification of inspired gases. Intensive Care Med 14:1–9
66. Silveira F, Paterson DL (2005) Pulmonary fungal infections. Curr Opin Pulm Med 11:242–246
67. Sosnowski TR, Arkadiusz M, Gradon L (2007) Mechanisms of aerosol particle deposition in the oro-pharynx under non-steady airflow. Ann Occup Hygene 51(1):19–25
68. Sung JC, Garcia-Contreras L, VerBerkmoes JL, Peloquin CA, Elbert KJ, Hickey AJ, Edwards DA (2009) Dry powder nitroimidazopyran antibiotic PA-824 aerosol for inhalation. Antimicrob Agents Chemother 53(4):1338–1343
69. Svartengren K, Anderson M, Svartengren M, Philipson K, Camner P (1996) Oropharyngeal deposition of 3.5 um particles inhaled through an elongated mouthpiece. Eur Respir J 9:1556–1559
70. Svartengren M, Svartengren K, Aghaie F, Philipson K, Camner P (1999) Lung deposition and extremely slow inhalations of particles. Limited effect of induced airway obstruction. Exp Lung Res 25:353–366
71. Tamer MA, Modell JH, Rieffel CN (1970) Hyponatremia secondary to ultrasonic aerosol therapy in the newborn infant. J Pediatr 77(6):1051–1054
72. Tewes F, Brillault J, Couet W, Oliver JC (2008) Formulation of rifampicin-cyclodextrin complexes for lung nebulization. J Control Release 129:93–99
73. Thanou MM, Verhoef JC, Romeijn SG, Nagelkerke JF, Merkus FWHM, Junginger HE (1999) Effects of N-trimethylchitosan chloride, a novel absorption enhancer, on Caco-2 intestinal epithelia adn teh ciliary beat frequency of chicken embryo trachea. Int J Pharm 185(1):73–82
74. Thomas CF, Limper AH (2004) Pneumoncystis pneumonia. N Engl J Med 350:2487–2498
75. Tolman JA, Nelson NA, Son YJ, Bosselmann S, Wiederhold NP, Peters JI, McConville JT, Williams RO (2009) Characterization and pharmacokinetic analysis of aerosolized aqueous voriconazole solution. Eur J Pharm Biopharm 72:199–205
76. Tsujino I, Kawakami Y, Kaneko A (2005) Comparative simulation of gas transport in airway models of rat, dog, and human. Inhal Toxicol 17:475–485
77. Voynow JA, Rubin BK (2009) Mucins, mucus, and sputum. Chest 135(2):505
78. Wangensteen OD, Schneider LA, Fahrenkrug SC, Brottman GM, Maynard RC (1993) Tracheal epithelial permeability to nonelectrolytes: species differences. J Appl Physiol 75:1009–1018
79. White DP (2006) The pathogenesis of obstructive sleep apnea: advances in the past 100 years. Am J Respir Cell Mol Biol 34:1–6
80. Widdicombe JG, Webber SE (1990) Airway mucus secretion. News Physiol Sci 5:2–5
81. Wilson SM, Shen P, Rider CF, Traves SL, Proud D, Newton R, Giembycz MA (2009) Selective prostacyclin receptor agonism augments glucocorticoid-induced gene expression in human bronchial epithelial cells. J Immunol 183:6788–6799
82. Yang W, Peters JI, Williams RO (2008) Inhaled nanoparticles – a current review. Int J Pharm 356:239–247
83. Young PM, Traini D, Stephen E (2008) Advances in pulmonary therapy. In: Williams RO, Taft DR, McConville JT (eds) Advanced drug formulation design to optimize therapeutic outcomes. Informa Healthcare, New York, pp 1–52

Chapter 2
Pulmonary Drug Metabolism, Clearance, and Absorption

Bo Olsson, Eva Bondesson, Lars Borgström, Staffan Edsbäcker, Stefan Eirefelt, Katarina Ekelund, Lena Gustavsson, and Tove Hegelund-Myrbäck

Abstract Delivering therapeutic agents to the lungs requires a deep understanding of the kinetics and dynamics of drugs in this biologically and physiologically complex system. In this chapter these concepts are discussed and include drug dissolution rates in the airways, physical clearance mechanisms of the mucociliary escalator and cough, alveolar macrophage clearance, pulmonary metabolism, and pulmonary absorption. Finally, these aspects are considered together with drug and formulation aspects as determinants of duration of effects of inhaled products. The mechanisms of elimination of drug activity in the lungs by the various clearance processes described here are important factors to consider both in the development of new drugs and in understanding the relative merits of existing therapies.

Keywords Airway selectivity • Lung absorption • Lung metabolism • Lung retention • Mucociliary clearance • Prodrugs • Pulmonary drug dissolution • Pulmonary drug transporters • Soft drugs

2.1 Introduction

There are essentially three major benefits that may be attained by delivering medication to the lungs via the inhaled route: rapid onset of action, high local concentration by delivery directly to the airways (and hence high therapeutic ratio and increased selectivity), and needle-free systemic delivery of drugs with poor oral bioavailability. If any of these benefits can be achieved and are of therapeutic importance, inhalation will be a reasonable delivery route. These benefits are dependent

B. Olsson (✉)
AstraZeneca R&D Lund, SE-221 87 Lund, Sweden
e-mail: bo.l.olsson@astrazeneca.com

H.D.C. Smyth and A.J. Hickey (eds.), *Controlled Pulmonary Drug Delivery*, Advances in Delivery Science and Technology, DOI 10.1007/978-1-4419-9745-6_2,

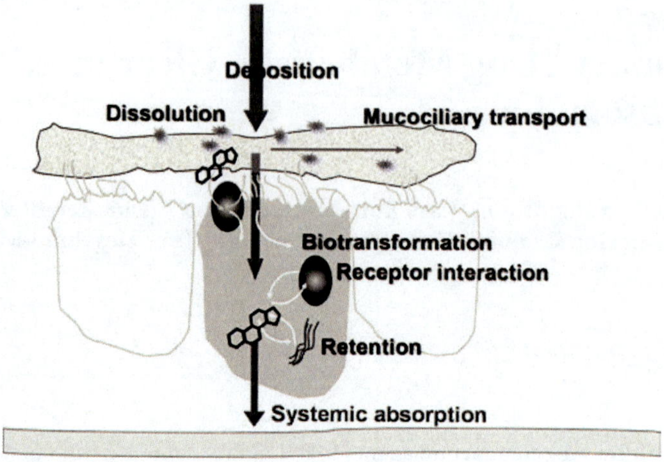

Fig. 2.1 Possible routes of elimination of drug from the lungs. Pulmonary drug metabolism, clearance, and absorption depend on an interplay between factors occurring prior to target interaction (deposition, dissolution, and mucociliary transport), and events competing with target interaction (biotransformation, receptor interaction, and nonspecific retention), and following target interaction (systemic absorption) (modified from [34], with permission)

on the mode and rate of elimination of the delivered drug from the lungs by the various clearance mechanisms, which are the subject of this chapter.

An inhaled drug substance may be eliminated from the lung by mucociliary or cough clearance to the gastrointestinal tract, by passive or active absorption into the capillary blood network, or by metabolism in the mucus or lung tissue (Fig. 2.1). These mechanisms may act in parallel and are responsible for the disposition and dissipation of the initially high local drug concentration in the lungs over time. Hence, these mechanisms have an important role, e.g., in determining the drug's duration of action in the lungs and the airway selectivity of the inhaled drug. Since elimination by absorption and metabolism, and pharmacodynamic activity require the drug to be in solution, dissolution kinetics may be an important factor in this context.

2.2 Dissolution

Once the drug aerosol has been deposited onto the lung surface, the immediate fate of the drug depends on its physical state. A free, solubilized drug will rapidly diffuse into the epithelial lining fluid and become available for absorption, while a drug deposited as particulate material has to be dissolved prior to absorption and may be subject to clearance by other mechanisms such as mucociliary or cough clearance (see below).

The physicochemical properties of inhaled drugs vary considerably, from very hydrophilic to very hydrophobic (log P from −2 for the β_2-agonist salbutamol [albuterol] sulfate and 5 for the corticosteroid fluticasone propionate, where P is the octanol:water partition coefficient) with low aqueous solubility from submicrogram per milliliter (0.1 μg/mL fluticasone propionate) to that of hundreds of milligram per milliliter (250 mg/mL for salbutamol sulfate). For compounds with high aqueous solubility, dissolution is not considered to impact the lung clearance rate, and no or only small differences in pharmacokinetics are expected for different types of formulations unless the regional deposition is substantially different and/or absorption is altered by excipients. The very poorly soluble compounds show fairly rapid onset of absorption followed by sustained absorption over time, which is thought to be dissolution limited. For micronized lipophilic drugs, time of peak concentration has been suggested to correlate with intrinsic solubility [151].

Assuming diffusion-controlled dissolution, the rate of dissolution is proportional to the drug's *solubility*, the *concentration* of the drug in the surrounding liquid film and the *area* of the solid–liquid interface. The *solubility* of a drug depends on the compound, the formulation and physical form of the drug, as well as on the composition of the dissolving media in the lung the epithelial lining fluid. The composition of this fluid is mainly water (96%), salts, phospholipids, proteins, and mucins with a pH about 6.6 in healthy individuals [7, 44], while the surface-lining layer in the alveoli is composed of a thin layer of alveolar surfactant (phospholipids and proteins). The lipids and proteins in the lining fluid will increase the wetting, the solubility, and hence the dissolution rate of poorly soluble drugs [115, 166]. Generally, the solubility, and hence dissolution rate, is higher for a less thermodynamically stable material (crystalline polymorph or amorphous form) than that of a molecular high-order crystalline state. The concentration in the liquid film surrounding an amorphous particle may thus become supersaturated, which can promote crystallization into a more stable material unless the rate of disappearance of the solute from the surrounding liquid is sufficiently rapid. The particle surface properties can, therefore, change with time, potentially affecting both the solubility and the surface area. An example was presented by Freiwald et al. [47], where amorphous beclomethasone dipropionate (BDP) particles delivered by hydrofluoroalkane (HFA)-propelled aerosols re-crystallized in contact with bronchial fluid in vitro as shown by scanning electron microscopy images.

The total liquid volume available for dissolution in the human lung is approximately 10–30 mL. Considering that a clinically relevant dose of fluticasone propionate, as an example of a poorly soluble drug, would require a volume in excess of 1 L for complete dissolution in a stationary system, the liquid volume in the lung is small. The thickness of the lining fluid varies from about 5–10 μm in the conducting airways and gradually decreases distally to about 0.01–0.08 μm in the alveoli (although in pooled areas it may be several microns thick) [38, 107]. A drug particle deposited in the conducting airways can thus be immersed in the lining fluid while the lining fluid film may be much thinner than the diameter of a deposited drug particle in the alveoli. Consequently, the *area* of the solid liquid interface between the particles and the fluid is proportional to particle surface area in the conducting airways but limited by the thickness of the fluid in the alveoli. This suggests that particles deposited in the upper airways could dissolve more rapidly than particles deposited in the alveoli. However, other factors such as greater solubility, larger total interfacial surface area, and/or more rapid absorption in the periphery could arguably lead to the opposite.

Assessing dissolution in the lungs is very complex as each of the governing parameters will be different in the different regions of the lungs, leading to several different dissolution processes occurring in parallel. By contrast, dissolution in the gastrointestinal tract can be described as a continuous process over a sequence of tanks with different properties [170].

Dissolution in the lungs has not been as systematically explored as dissolution in the gastrointestinal tract; rather the knowledge is derived from a number of diverse studies investigating in vitro dissolution models in the environmental and drug research area. Hence, there are presently no established in vivo predictive in vitro dissolution models for pulmonary formulations, in contrast to the standardized dissolution test methods available for oral solid dosage forms [51]. The lack of standardized methods may also be linked to the fact that there are no controlled release products for inhalation on the market, although the field has been intensely researched for decades. Still, in attempts to identify dissolution methods to characterize inhaled drugs, several dissolution models have been investigated, for a review see Salama et al. [127]. Most models use large liquid volumes in which the powder is immersed [30, 138], but models mimicking the air–liquid surface have also been published [24]. In these models, samples of well-defined aerodynamic particle size distributions can be prepared by deposition of the aerosol in a cascade impactor, but the mass of drug per surface area presently needs to be far higher than in the in vivo situation for analytical reasons. As a consequence, an important aspect of in vivo relevance is lost. Even though the dissolution models are fairly simple compared to the physiological complexity of the lungs, there is evidence suggesting correlations between in vitro dissolution and in vivo exposure and efficacy, especially in the field of modified-release formulations. An example is the correlation between therapeutic effect duration in an in vivo bronchoconstriction animal model and in vitro dissolution profiles presented for salbutamol sulfate controlled-release formulations [123].

In conclusion, there is no reason to doubt that the dissolution rate of formulations or compounds may affect the rate and mode of clearance from the lungs, and therefore influence the pharmacodynamic properties of a drug. As a result of the difficulty in accounting for all the confounding factors, direct evidence is scarce, which limits our ability to predict how variations in formulation that affect solubility and/or dissolution will affect the pharmacokinetic and pharmacodynamic properties of a product.

2.3 Mucociliary and Cough Clearance

Mucociliary clearance is of fundamental importance for the removal of secretions and foreign particles that have been deposited in the airways. The overall principle of the system is simple: the ciliated cells in the epithelium transport the mucus together with any deposited particles in a proximal direction and eventually the mucus is expectorated or swallowed. Inhaled aerosolized drug deposited in the lung will either penetrate the mucus, dissolve, and become absorbed or follow the mucus and eventually become swallowed (Fig. 2.2). After deposition in the airways, the vast majority of insoluble particles larger than about 6 μm in geometric diameter is eliminated from the airways by mucociliary clearance [140]. Smaller particles are able to penetrate the mucus and enter the bronchial epithelium. The smaller the particles, the faster they reach the epithelium and thus escape mucociliary clearance. Small particles will be preferably deposited in the alveolar part of the lungs and will be dissolved or retained for substantially longer periods of time in the lung than larger insoluble particles that are more proximally deposited.

The mucus is produced by secretory (goblet) cells within the epithelium and by submucosal glands. Traditionally, the fluid is considered to consist of two phases [90]: the watery periciliary sol phase in which the cilia can beat without too high resistance, and the gel phase overlying the sol phase that contains mucins and other glycoproteins. The rapid strokes of the cilia bring them in contact with the gel, thereby propelling it in the proximal direction (Fig. 2.2) [145, 146]. In the lungs of healthy individuals, the production of mucus reaches about 10–20 mL/day [152]; in patients with chronic bronchitis 10 times these volumes may be produced.

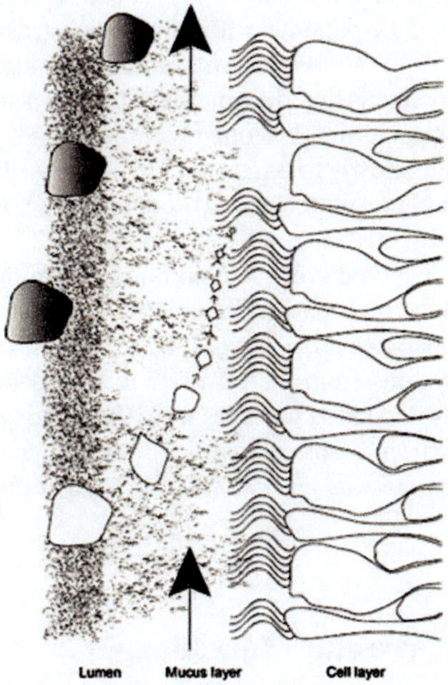

Fig. 2.2 Mucociliary clearance competing with particle dissolution and absorption. Particles deposited on the mucus layer will gradually dissolve and diffuse toward the cell layer where the drug substance eventually may get absorbed. Particles of slowly dissolving compounds will be partly cleared by ciliary action thus reducing the amount absorbed (reprinted with permission from [34])

Mucociliary clearance of insoluble particles has been studied extensively in vivo. The mucociliary transport of such particles can be examined by direct observation through a bronchoscope, by radiography or by external monitoring of radiolabeled particles. It is fair to assume that total lung deposition does not differ to a clinically meaningful degree between healthy volunteers and patients for any of the common inhalation products on the market. However, the regional distribution of the total lung dose is likely to differ substantially between a healthy individual and a patient suffering from asthma or chronic obstructive pulmonary disease (COPD). In asthma, and also in COPD, the cross-sectional area of the airways will be smaller and thus an inhaled aerosol is more prone to impact in the more proximal parts of the lung [147]. The more central deposition of inhaled drug in constrained lungs means that any mechanism that clears the central lungs of drug into a site from which it is not systemically available (e.g., mucociliary clearance) could explain a reduced systemic exposure in patients – on the condition that the rate of absorption is sufficiently slow compared with the rate of clearance.

Mucus velocity increases from the peripheral toward the central airways [46], as is expected from the anatomical arrangement. In healthy individuals, the mucus moves upward at a rate of about 1 mm/min in the small peripheral airways but can be as quick as 20 mm/min in the trachea [164]. The rate of mucociliary clearance decreases with age in healthy individuals and can also be affected by airway disease. Patients with acute asthma have markedly reduced clearance [94]. Mucociliary clearance is also reduced in patients with chronic obstructive lung disease [19, 163] and is reduced by smoking [49, 83]. β-Agonists are known to improve pathologically reduced mucociliary function in patients with asthma [111] or chronic bronchitis [124]. The long-acting β-agonist formoterol is a powerful ciliary stimulant [86] that has been shown to increase mucociliary clearance by 46% after 6 days of treatment in patients with chronic bronchitis [93]. Inhaled corticosteroids do not appear to affect directly mucociliary clearance [32, 125], and the improved mucociliary clearance observed with inhaled corticosteroids in asthma patients is thus most likely a result of their anti-inflammatory properties [1, 94].

When mucociliary clearance is decreased, cough becomes increasingly important for the removal of secretions from the airways. A considerably larger proportion, around 60%, of centrally deposited particles were shown to be eliminated by coughing in patients with chronic obstructive lung disease compared with healthy individuals, where the value was around 8% [117]. Total clearance was similar in the two groups.

2.4 Clearance by Alveolar Macrophages

Slowly dissolving drug deposited in the alveolar region will tend to be phagocytosed by alveolar macrophages, which slowly dispose of particles either by transporting them along the alveolar surface to the mucociliary escalator, or by translocation to tracheobronchial lymph or by internal enzymatic degradation

[81, 88, 103]. Phagocytosis appears optimal for particles of 1.5–3 μm in size [102]. This size discriminating property of the macrophages has been a basis for controlled-release formulations of inhaled drugs [35]. Large porous particles will reach also the alveoli because of their relatively small aerodynamic size but clearance by alveolar macrophages is reduced as a result of their large geometric diameter.

2.5 Pulmonary Drug Metabolism

The body's primary detoxification enzymes, the cytochrome P450 (CYP) families, show the highest expression levels in hepatocytes and enterocytes but are also expressed in the lungs and other organs, providing a line of defense against ingested or inhaled xenobiotics. There are 57 active human CYP genes divided into families and subfamilies mainly based on sequence (Cytochrome P450 Homepage, http://drnelson.utmem.edu/cytochromeP450.html). The human drug metabolizing enzymes belong to the families CYP1, CYP2, and CYP3. These enzymes catalyze an oxidation reaction in which a functional group is introduced to the molecule, serving as a site of conjugation, increasing hydrophilicity, and thereby facilitating elimination [106]. After this initial phase I conversion, the compound is excreted or conjugated to an endogenous group by a phase II enzyme. This serves to further increase hydrophilicity, e.g., through acetylation, methylation, sulfation, or conjugation with glucuronic acid, glutathione, or amino acids [106]. Even though detoxification and facilitation of excretion is the primary scope of CYP biotransformations, sometimes an activation of a compound can occur. Reactive or toxic metabolites, which could potentially harm the tissue, may be formed.

In the lungs, several CYP isoforms are expressed as well as other biotransformation enzymes such as sulfotransferases (SULT), UDP glucuronosyl transferases (UGT), glutathione S-transferases (GST), esterases, peptidases, cyclo-oxygenases, flavine mono-oxygenases (FMO) [69, 109, 169]. The wide range of biotransformation enzymes enables metabolism of a broad spectrum of chemically different substrates. Local metabolism of several inhaled compounds, pharmaceutical drugs as well as tobacco-smoke components, pollutants, and toxicants have been demonstrated in lung tissue [73, 101]. Table 2.1 gives an overview of the expression of metabolic enzymes detected in the lungs.

Metabolism in lungs differs substantially from the intestinal–hepatic metabolism. The expression levels of enzymes are generally lower and the expression patterns of drug-metabolizing enzymes differ. In the lungs, CYP1B1, CYP2B6, CYP2E1, CYP2J2, CYP3A5, and CYP1A1 (the latter being highly induced in smokers) appear to be the most common CYP enzymes (Table 2.1), whereas the major CYP enzymes in human liver are CYP1A2, CYP2C9, CYP2C19, CYP2D6, and CYP3A4 [31, 69, 119, 134, 171]. The most abundant liver CYP enzyme, CYP3A4, is expressed to a lower degree in pulmonary tissue and evidence points toward it only being expressed in 20% of the individuals tested, whereas the isoform CYP3A5 is considered more important in lung tissue [3].

There is less information available regarding the phase II enzymes compared with the CYPs. However, the expression and metabolic activity of SULT enzymes appear to be about the same in the lungs and in the liver, whereas epoxide hydrolase and esterases show an intermediate metabolic activity. The UGT enzyme activities appear to be low in human lungs. Somers et al. [137] estimated the metabolic activity of different enzymes in the lungs compared with those in the liver and showed that CYP enzymes displayed between 1 and 10% of the activity in the liver, epoxide hydrolase and esterases around 20% and SULT enzymes a similar activity in lungs, and for some isoforms, substantially higher than that in the liver [137]. Peptidase activity is high in the lung, as found elsewhere in the body [109].

Pulmonary tissue consists of several cell types with different expression patterns of metabolizing enzymes. In combination with low expression levels of certain enzymes, this results in technical and methodological challenges when studying lung metabolism, which may be one reason for the disparate results on enzyme expression levels that are reported in the literature (Table 2.1). Low expression levels of mRNA can be detected using polymerase chain reaction (PCR) techniques, giving an indication that a functional enzyme could be active in some cell types in the tissue, even though metabolic activity has not been demonstrated. In addition to these challenges, there are substantial differences between human and the most common laboratory animals in the expression pattern of biotransformation enzymes, which further complicates investigations and data interpretation.

The drug-metabolizing capacity of the lungs is in general substantially lower than that of the liver, and there is little evidence of a major contribution to systemic clearance. Many small molecules have near complete bioavailability via lung absorption as a consequence of low metabolic activity and relatively rapid absorption [12, 107, 128, 154, 155]. Still, several drugs, such as budesonide, ciclesonide, salmeterol, fluticasone propionate, and theophylline are substrates to enzymes present in the lungs [21, 57, 99, 112, 156]. Since formation of local metabolites cannot be excluded, and are sometimes anticipated, studies are conducted during drug development to screen for lung metabolism and assess the risks for metabolic interactions and toxicity, e.g., by incubating candidate drugs with lung subcellular fractions. However, local metabolism in the lungs can be taken advantage of, as several of the drugs listed above demonstrate. One example is the use of prodrugs where the administered form of the drug is activated in situ by metabolic enzymes. Ciclesonide, e.g., is metabolized into its active form by esterases in the lungs and further reversibly conjugated to fatty acids [99]. The prodrug beclomethasone dipropionate (BDP) is metabolized to the more potent 17-beclamethasone monopropionate (BMP) by esterases in the lungs [167]. For budesonide, the conjugation to fatty acids in the lungs results in the formation of a compound that has a substantially lower elimination rate. This reversible biotransformation contributes to the prolonged lung retention and duration of effect observed with this compound [156, 158]. The concept of "soft drugs" is another approach where differences in metabolic stability between the lungs (stable, to retain efficacy) and in circulating blood (labile, to minimize side effects) are the target properties.

Table 2.1 Overview of the expression of metabolic enzymes detected in lung tissue (compiled from [3, 68, 69, 138, 149, 171, 173])

	Isoform	mRNA expression	Protein expression	Metabolic activity	Comment
Phase I enzymes	CYP1A1	Yes	Yes	Yes	Only in smokers, decrease to normal levels within 2 months after cessation of smoking
	CYP1A2	Disparate data	Disparate data	Not reported	
	CYP1B1	Yes	Yes	Yes	Induced by smoking
	CYP2A6	Yes	Disparate data	Not reported	
	CYP2A13	Yes		Not reported	No clear report on activity
	CYP2B6	Yes	Yes	Yes	Splicing variant, previous 2B7
	CYP2C	Disparate data		Yes	
	CYP2D6	Yes	Yes	Disparate data	
	CYP2E1	Yes	Yes	Yes	
	CYP2F1	Yes		Not reported	
	CYP2J2	Yes	Yes	Yes	Strongly expressed, possible endogenous role
	CYP2S1	Yes	Yes	Unclear	
	CYP3A4	(Yes)	(Yes)		Protein detected in 20% of investigated samples
	CYP3A5	Yes	Yes	Yes	
	CYP4	Yes		Unclear	
	FMO	Yes			Varied expression among isoforms
Phase II enzymes	UGT	Yes	Yes	(Yes)	Varied expression among isoforms, generally low metabolic capacity
	GST	Yes		Yes	
	Esterases	Yes		Yes	High metabolic capacity
	Epoxide hydrolase	Yes		Yes	Low metabolic activity
	Peptidases			Yes	High metabolic capacity
	SULT	Yes		Yes	High metabolic capacity

Most drug-metabolizing enzymes (and several transporters, discussed below) are inducible as a response to increased exposure to their substrates. An organism adapts to the new situation with an increased metabolic capacity in order to handle the "threat" [58, 64, 110]. By contrast, suppression of enzyme levels can also occur as a response to xenobiotics or disease [96, 172]. Information around the regulation of metabolizing enzymes in pulmonary tissue is scarce. An important example of enzyme induction that changes the metabolizing capacity of the lungs is the effects of tobacco smoke. Tobacco smoke contains several 1,000 different chemicals and several of these affect metabolizing enzymes in the lung. The induction of drug-metabolizing enzymes may result in increased metabolism of drugs, which could result in impaired therapeutic effect. Clinically relevant induction of CYP1A1 activity has been observed in patients who smoke during treatment with theophylline compared with nonsmokers [80].

Several compounds have the ability to inhibit drug-metabolizing enzymes and thereby cause clinically adverse interactions [54]. Drug interactions resulting from inhibition are, however, not likely to have a major significance in the lungs because of the limited contribution of lung metabolism to systemic clearance, although the local metabolism pattern could be affected by such interactions.

2.6 Pulmonary Drug Absorption

The optimal absorption characteristics of a pulmonary drug depend on the site of drug action. For locally acting drugs, the drug absorption process may determine the removal and consequently the termination of action of the drug in the lungs, as well as the onset of any systemically mediated adverse effects. For systemically acting drugs, absorption from the lungs determines the therapeutic effect profile (onset, intensity, and duration of action) of the drug. Therefore, when designing drugs for pulmonary delivery, it is important to consider both lung–tissue retention and permeability, irrespective of site of action.

The air-to-blood transfer always begins with an interaction between the drug and the surfactant: following deposition onto the mucosa of the tracheobronchial airways or alveolar region, the drug solute or particle encounters at least a monolayer of surface-active agents in which the fatty acid tails of lipids project into the air. For a drug compound of macromolecular size (e.g., a peptide or a protein), this lung surfactant may induce aggregation and, thus, potentially compromise dissolution or enhance macrophage engulfment and digestion [107]. By contrast, lung surfactant can enhance solubility of small, lipophilic drug molecules, as demonstrated with glucocorticosteroids and a number of cationic compounds [85, 166], which may potentially increase the rate and extent of absorption. Immediately below the molecular layer(s) of lung surfactant lies the 0.01–10 μm thick lining fluid through which drug must diffuse to get to the epithelium. The routes of drug absorption across the epithelium include passive and active transport mechanisms involving paracellular and transcellular transport, pore formation, vesicular transport, and drainage into the lymphatics – depending on the drug and site of absorption. The drug solute will pass through a cellular barrier that varies from a monolayer of thick (about 60 μm) columnar cells in the bronchi to a monolayer of thin (0.2 μm) broad cells in the alveoli. The epithelial cells are attached to a basement membrane, a thin matrix of fibers, and below that, there is the interstitium of the lungs containing a variety of cells, collagen, elastic fibers, interstitial fluid, and lymphatic vessels. Since plasma proteins and most solutes are thought to diffuse relatively unhindered through both the epithelial basement membrane and the interstitium [38], it is reasonable to believe that neither of these tissues is a significant restrictor of transport of drug solutes. Drug absorbed from the air spaces into the blood must traverse a final barrier after the surfactant layer, the lining fluid, the epithelium, its basement membrane, and the interstitium: the cell monolayer that makes up the walls of the microvessels, the endothelium. Also, the endothelial cells are attached to a basement membrane, but where the endothelium comes into contact with the epithelium, which is frequent throughout the alveoli, their basement membranes fuse to form one common basement membrane. The alveolar–capillary endothelium is extremely thin (0.03–0.2 μm) and has a relatively large number of endocytic vesicles. Although generally the alveolar epithelial cells and not the underlying endothelial cells are considered the major barrier to transport, the contribution of the endothelium as a barrier for drug absorption is uncertain and needs to be further investigated.

2.6.1 *Passive Diffusion*

The tendency of a solute to pass from a point of higher concentration to a point of lower concentration is called passive diffusion. An inhaled drug solute diffuses through the epithelial cells or via paracellular pores or junctions into the submucosa along a concentration gradient. The slope of the gradient is dependent upon the physicochemical properties of the drug, the thickness of the air–blood barrier and the rate of blood perfusion in the submucosa, i.e., the diffusion rate is both compound specific and region selective. Absorption of lipophilic compounds is generally considered to occur through transcellular diffusion [36]. Hydrophilic compounds appear to be absorbed via paracellular diffusion through intercellular junction pores [132]. Based on data from extensive in vivo preclinical research performed by Schanker and colleagues (e.g., [12, 128]), within the molecular weight (MW) range of 100–1,000 Da pulmonary drug absorption rate is size independent but dependent on aqueous solubility at physiological pH such that lipophilic drugs get absorbed rapidly (with absorption half-lives in the range of minutes) and hydrophilic drugs more slowly (with absorption half-lives in the range of hours) [109]. The exact mechanisms and pathways underlying pulmonary absorption of macromolecules remain largely unknown. It is uncertain whether peptides, such as insulin, and proteins are absorbed primarily paracellularly through tight junctions or transcellularly via receptor- or cavelae-mediated transport [15, 70, 75, 76, 108]. Likewise, the mechanisms and the relative importance on drug absorption in health and disease of the interrelations between paracellular and transcellular pathways in lung epithelia and endothelia [162] remain to be fully elucidated.

2.6.2 *Drug Transporters*

During the last decade, it has become evident that drug transporters play an important role in drug disposition [61, 135]. This has recently been highlighted in a review of clinically relevant drug transporters and recommendations of their assessment in drug development [71]. Transporters of the solute carrier (SLC) family facilitate transport across the cell membrane and most commonly enhance uptake of compounds into cells. Depending on subtype of SLC, the translocation of compound across the membrane is driven by different mechanisms, e.g., counter-transport of another ion or the membrane potential. Examples of drug-transporting SLCs are organic anion transporters (OAT and OATP) and organic cation transporters (OCT). By contrast, drug transporters of the ATP-binding cassette family (ABC transporters) efflux compounds out of cells via an ATP-dependent mechanism. Examples of ABC transporters are multidrug-resistance proteins (MDR), multidrug-resistance-associated proteins (MRP), and breast cancer resistance protein (BCRP). In different barriers of the body, e.g., intestine and blood–brain barrier, ABC transporters hinder foreign molecules from entering the systemic circulation or the brain, respectively. The highest impact of SLC transporters is observed for compounds with low passive permeability as the transporter-mediated uptake may be the rate limiting step in the transepithelial transport of such compounds [135]. Uptake and efflux transporters work in concert with passive diffusion to influence the absorption of drugs through an epithelial cell layer [142]. In addition, drug transporters may be important in the regulation of the intracellular concentration of a drug and consequently in influencing efficacy as well as toxicity [105] (Fig. 2.3). Although relatively little is known about the functional impact of drug transporters in the lungs, it would be expected that they would play a central role in protecting the lungs from any entering xenobiotic.

Like other barriers in the body, cells in the lung express a variety of drug transporters of the SLC/SLCO and ABC families although with a different organ-specific profile (Table 2.2, for review see [9]). Organic cation transporter expression

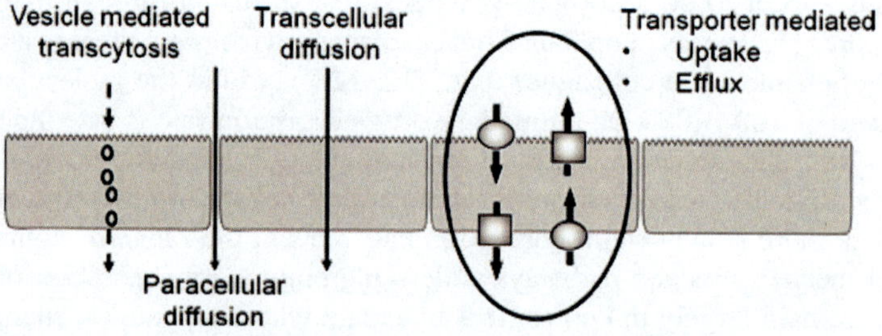

Fig. 2.3 Schematic of drug transporters in lung cells. SLC transporters in lung cell membranes may mediate drug uptake into the intracellular space with the highest impact on low passive permeability drugs. ABC transporters in the lungs may mediate drug efflux out of cells. Together uptake and efflux transporters may influence the intracellular concentration of drugs

Table 2.2 Overview of major drug transporters expressed in the lungs

Transporter	Gene	Localization with comments	References
OCT1	SLC22A1	Apical on ciliated epithelial cells in bronchi	[88]
OCT2	SLC22A2	"	
OCT3	SLC22A3	Smooth muscle cells	[66]
OCTN1	SLC22A4	Apical on epithelial cells in trachea, OCTN2 also in alveolar epithelia. High expression	[67]
OCTN2	SLC22A5	"	
OATP2B1	SLCO2B1	Only mRNA data published. No cellular localization reports to date.	[6, 23]
OATP3A1	SLCO3A1	"	
OATP4A1	SLCO4A1	"	
OATP4C1	SLCO4C1	"	
PEPT2	SLC15A2	Apical, bronchial epithelium, and alveolar type II pneumocytes	[53, 54]
MDR1/Pgp	ABCB1	Apical, bronchial epithelium, alveolar type I, endothelium	[21, 85, 130, 160]
MRP1	ABCC1	Basolateral/lateral, bronchial epithelium, goblet cells, peripheral epithelium. High expression	[11, 130, 160]
Other MRP	ABCC	Several MRPs expressed in lung	[160]
BCRP	ABCG2	Basolateral, bronchial epithelium, endothelium	[130]

Includes transporters for which mRNA, protein, and immunohistochemistry have been demonstrated

has been demonstrated in human lungs. OCT1 [87], OCTN1, and OCTN2 [66] expression have been demonstrated both at the mRNA and protein level. All three transporters are localized in the apical membrane of lung epithelial cells. Solute carriers may play a key role in transporting compounds with low passive permeability through the epithelial cell barrier. A number of inhaled drugs, e.g., β_2-adrenergic agonists and anticholinergic bronchodilators, are polar and contain a positive charge at physiological pH, and OCT may therefore play a role in their absorption and distribution. Data from Ehrhardt and coworkers indicated that the polar basic β_2-adrenergic agonist salbutamol was actively transported in lung epithelial cell lines [37]. Based on transport studies in human airway primary epithelial cells, Horvath and coworkers proposed that OCTN1 or OCTN2 is involved in the translocation of inhaled organic cation bronchodilators across the pulmonary epithelial cell layer, therefore being an important part in their delivery to the site of action [66]. Similarly, Nakamura et al. [100] demonstrated that ipratropium, a muscarinic antagonist, was taken up into human bronchial epithelial cells by OCTNs.

Although a clear functional role of organic anion transporters has been demonstrated in organs such as the liver [135], there are very limited data available in the literature on expression of this subgroup of transporters in lung tissue. Bleasby et al. [6] detected mRNA of several OATPs in human lung, but no protein or cellular localization data have been published.

The peptide transporter PEPT2 is another example of an SLC expressed in lung tissue. PEPT2 is transporting di- and tripeptides, as well as some peptidomimetic drugs, and is expressed on the apical membrane of epithelial cells in bronchi and alveoli [53]. PEPT2 substrates of relevance to lung disease are antibiotics such as beta-lactams and cefadroxil, and PEPT2 has been suggested as a way of targeting drugs to their site of action in the airways.

In terms of ABC transporters, MDR1 (Pgp), BCRP, and several MRP are expressed in the lungs (Table 2.2, for review see [9, 159]). Among the ABC transporters, MRP1 is the most highly expressed in lung tissue. MRP1 is localized to the basolateral side of pulmonary epithelial cells and has been detected both in the bronchial as well as the alveolar epithelium. MRP1 transports a broad range of anionic drugs as well as metabolites following glutathione, glucuronide, and sulfate conjugation. Interestingly, it has been reported that MRP1 expression is decreased in COPD [160]. Cigarette smoke extract (CSE) has been demonstrated to increase the expression of MRP1 in pulmonary epithelial cells and at the same time inhibit MRP1 leading to an increased toxicity of CSE, indicating a protective role of MRP1 [161]. MDR1 and BCRP expression have been demonstrated on both pulmonary epithelial as well as vascular endothelial cells in the lungs (Table 2.2). Because of the heterogeneity of lung tissue, there is still a need to further investigate the drug transporter expression in different cell types in the lungs. In terms of functional activity, most studies so far are based on cell models and very limited data are available on the functional impact in vivo. Thus, the relevance of drug transporters for pulmonary pharmacokinetics is currently unclear [9].

2.6.3 *Vesicle-Mediated Transport*

Membrane vesicles within the alveolar epithelial type I and type II cells have been suggested to be involved in macromolecule transport across the alveolar epithelium, and caveolae have been shown to be present in the alveolar type I cells [26, 55]. The main route of alveolar epithelial protein transport is through transcytosis involving caveolae and clathrin-coated pits, although the relative contributions of these internalization steps to overall protein handling remain to be determined [56, 75]. For particles (e.g., consisting of liposomes for drug delivery) the relative contribution of caveolae- and clathrin-mediated endocytosis has been shown to be size-dependent with an upper size limit of approximately 200 nm for the clathrin-mediated pathway [120].

2.6.4 *Nonspecific Particle Trapping*

There is evidence that nanoparticles, defined as particles less than 100 nm in at least one dimension, are taken up by alveolar epithelial type I cells via well-known pathways of endocytosis and also via other, less well-understood, mechanisms [74, 97]. The quantitative relationships between mechanisms remain to be established. For the translocation of particles from the lungs into the systemic circulation lymph drainage is likely to contribute [78, 144]. Opportunities of aerosolized nanoparticles of drug for pulmonary administration are being reviewed [16], although currently available technology does not appear to allow for large-scale production of such pharmaceutical formulations.

2.6.5 Importance of Lung Physiology and Pathophysiology

It is well known that different conditions alter pulmonary drug absorption. Stretching of the lungs by, e.g., exercise-induced deep ventilation, high-altitude pulmonary edema, or mechanical ventilation at high tidal volumes causing more permeable transcellular pores and possible expansion of the surface of the alveolar epithelial type I cells and caveolae of the capillary endothelium has been reported to increase the absorption of hydrophilic solutes [92, 114, 132, 153, 165, 169]. Likewise, active or passive cigarette smoking, exposure to surface-active agents used as absorption enhancers in experimental drug aerosols, or alcohol abuse, all appear to interfere with the functional integrity of the air–blood barrier resulting in increased uptake of hydrophilic solutes [8, 18, 59, 60, 130]. By contrast, physiological changes in the lung parenchyma that occur during aging, e.g., smaller total alveolar surface area, impaired ventilation–perfusion matching, larger small airway closing volumes, and smaller effective lung volumes, would more likely result in decreased absorption of inhaled drugs.

The presence of pathological lung conditions is expected to have an effect on pulmonary drug absorption. Chronic bronchitis and cystic fibrosis are pulmonary diseases characterized by the presence of a viscoelastic mucus layer in the upper airways and bronchi. As the mucus layer may retard or even block air-to-blood transfer of drug solutes and solids, this barrier must be overcome for pulmonary drug delivery to be effective. By contrast, the presence of inflammation in the lungs would be expected to facilitate paracellular transport of hydrophilic solutes. In models of allergic airway inflammation, various inflammatory mediators have been shown to dysregulate intercellular junction pores, thereby contributing to epithelial barrier dysfunction and injury [104]. Likewise, permeability edema – a major complication of acute lung injury, severe pneumonia, and the acute respiratory distress syndrome (ARDS) – is associated with an epithelial and endothelial hyperpermeability and a disruption of the epithelial and endothelial barriers, which may increase the paracellular transport of hydrophilic drugs. Emphysema, on the other hand, is characterized by structural modifications involving loss of inter-alveolar septa and thus would be likely to decrease pulmonary uptake of drug because of the reduction in total alveolar surface area.

Caveolae-mediated transcytosis can be upregulated in response to pathological stimuli [50, 116, 150]. Indeed, as caveolae and caveolins seem to play important roles for the airway epithelium, airway smooth muscle, airway fibroblasts, airway inflammatory cells, and the pulmonary vasculature, it has become clear that aberrant regulation of their expression and function may trigger pulmonary defects, including pulmonary fibrosis, pulmonary hypertension, and lung cancer. However, further examination of caveolae and caveolins in obstructive airways diseases, including asthma and COPD, is warranted.

Airway hyperperfusion associated with asthma is suppressed temporarily by inhaled corticosteroids via their acute vasoconstrictor action. The decrease in airway blood flow is likely to retard the uptake of inhaled bronchodilators and thereby enhance their action.

2.6.6 Assessing Pulmonary Drug Absorption

Further elucidation of the mechanisms involved in drug uptake after aerosol delivery is needed. In vitro models can be used to clarify the interactions between drug formulations and the epithelial barriers within the trachea, the bronchial airways, and the alveoli. A variety of airway and alveolar epithelial cell culture systems have been established as in vitro absorption models [14, 45, 139, 141]. Recently, drug transporter expression in different pulmonary epithelial cell models was reported [39]. Models of tracheobronchial epithelium include not only primary cell cultures and cell lines of healthy human and animal phenotypes but also airway cells with characteristics of lung disease such as cystic fibrosis (e.g., CFBE41o- and CuFi cell lines). Primary cell cultures more closely resemble the native epithelia, but are less reproducible, convenient, and economical compared with the cell lines, which makes them less suitable for permeability screening purposes. In contrast to gastrointestinal in vitro testing, where Caco-2 cells have emerged as the gold standard, there is no such consensus to date on the preferred cell line(s) for modeling the tracheobronchial epithelium in vitro. Since alveolar epithelium cell lines available to date do not form functional tight junctions, most in vitro studies of alveolar epithelial function have been performed using primary cultures of alveolar type II cells, which under appropriate conditions differentiate into type-I-like cells and form tight epithelial barriers that are morphologically similar to the in vivo alveolar epithelium. The isolation of alveolar epithelial type II cells, predominantly from rat and rabbit lung tissue, and their culture over time leading to a primary culture of type I-like cells is now an established technique. The use of human alveolar epithelial type II cells obtained from patients undergoing lung resection or derived from embryonic stem cells is, however, limited by lack of material, and a complex, time-consuming and cost-intensive isolation procedure. Methods for delivering aerosols directly to the surface of air-interface epithelial cell layers have been developed [17, 25, 43]. Recent research is addressing more complex cell culture systems in order to generate models that are closer to the in vivo situation. These include co-cultures of different cell types to investigate the cellular interplay after particle deposition [48]. Although suitable cell culture models of the cellular part of the human air–blood barrier are established and well characterized, the physical barriers on top of the cellular barriers – the surfactant and lining fluid – are less well understood and their influence on the safety and efficacy of aerosol medicines may be underestimated [14].

More advanced models than cell cultures are presently required to maintain the structural and cellular integrity of the lung tissue, the interaction between different cell types, and the biochemical activity. Recently, a human breathing lung-on-a-chip microdevice has been developed to study various physiological and pathological lung functions at the alveolar-capillary interface that could provide a new capability of cell culture models [67]. The isolated perfused lung model (IPL) is the most commonly used ex vivo method [126]. Tronde and coworkers investigated a wide range of drugs administered as aerosolized aqueous solutions in the IPL rat model and found pulmonary absorption to correlate with physicochemical descriptors, in vitro assessed permeability, and pulmonary absorption determined in vivo in rats [154, 155]. The IPL rat model has also been successfully used in the study of dry powder aerosols of drug [42]. The IPL model is a valuable complement to in vivo whole animal studies, despite limitations such as short viable periods, the level of technical expertise required to set up the IPL, low-throughput and absence of tracheobronchial circulation because this model allows drug absorption to be studied in absence of the confounding factors obviously present in vivo. A variant of the IPL is the human lung reperfusion model [47].

When using in vivo animal models, the anatomical complexities and interspecies differences in the lungs have to be considered [27, 114, 126]. For instance, mice and rats lack respiratory bronchioles and appear to have a relatively fast early alveolar clearance of insoluble particles, whereas guinea pigs, dogs, monkeys, and humans have much slower alveolar clearance.

Although various approaches can be used to study factors influencing the pharmacokinetics of an orally inhaled drug aerosol in humans, no method will measure directly the transepithelial transfer of drug. Compartmental modeling of blood level data obtained for inhaled drug has been used to estimate the rate of absorption into the systemic circulation [79]. Another approach is to quantify the initial pulmonary drug distribution and subsequent disappearance from the lungs using radioisotopically labeled drug and an imaging technique [5, 21, 136]. The pulmonary drug disappearance rate assessed from a time-series of images would then make a composite measure of both transepithelial transfer and nonabsorptive clearance mechanisms.

2.7 Duration of Effect and Airway Selectivity

The aim of local respiratory treatment is to attain maximal exposure at specific lung targets (receptors, transporters, or enzymes), while concentrations elsewhere are kept at a minimum. If this is achieved, maximum therapeutic effect can be attained by low inhaled doses, whereby risk of systemic side effects is minimized. *Airway selectivity* was introduced as a concept by Hochhaus in 1997 to describe these features in a pharmacokinetic context. Hochhaus showed that lung-related properties such as pulmonary deposition, pulmonary residence time, and pulmonary drug release (dissolution) are at least as important with regard to therapeutic index as intrinsic potency and traditionally recognized pharmacokinetic parameters including systemic clearance and oral bioavailability.

This chapter, so far, has discussed the various mechanisms that contribute to the elimination of inhaled drug activity in the lung: dissolution, mucociliary clearance, metabolism, passive and active transport, and trapping. In this concluding section, the perspective is changed to that of *effect duration*, and the crucial importance that dissolution rate, tissue affinity, and biotransformation may have on airway selectivity.

2.7.1 Dissolution Rate

Slow dissolution can be used as a strategy to increase the retention of drug in the lungs and thereby prolong the duration of effect. However, if the rate of dissolution is too slow, the drug is more susceptible to mucociliary clearance and/or phagocytic elimination, which in turn will reduce retention [95]. Hence, the solid state property of the drug is an important feature in optimizing airway selectivity. The inhaled corticosteroid (ICS) fluticasone propionate (FP), which is highly lipophilic and slowly dissolving in sputum, is retained in the lungs up to 20 h after inhalation [41]. The mean absorption time (MAT) after inhalation is prolonged compared with the more soluble ICS budesonide (MAT after inhalation is 5–7 h vs. approximately 1 h for budesonide) [148]. Increased mucociliary clearance of undissolved FP resulting from a more central deposition in asthmatic patients than in healthy volunteers has been proposed to explain reduced bioavailability in these patients [13]. Interestingly, in a recent study in COPD patients, significantly more FP was eliminated via mucociliary and cough clearance than budesonide [28]. Thus, it appears that slow dissolution of FP causes its long lung retention, which at the same time is limited in the central lung due to mucociliary and cough clearance. This may serve as one plausible explanation of why FP shows shorter effect duration than that required for once-daily dosing [118].

2.7.2 Formulation Approaches

In part, shortcomings in terms of solid-state properties can theoretically be overcome by using appropriate formulation strategies such as reducing particle size in nano-suspensions, entrapment of active drug in liposomes, or formulating large porous particles [35, 133, 168]. One such application is budesonide embedded in stealth liposomes, which significantly prolonged effect duration in experimental asthma [77]. This and other formulation approaches to improve inhaled drug properties are discussed in Chaps. 6 and 13. Assuming that Chaps 6 and 13 still refer to overcoming lung clearance mechanisms for controlled release drug delivery, and particle engineering technologies for pulmonary drug delivery, respectively.

2.7.3 Tissue Affinity

Tissue affinity is a loosely defined term used to describe binding to different substructures in the tissue (such as structures in the cell membrane, target receptors, cell organelles) by multiple mechanisms, thereby slowing down lung clearance and possibly increasing local effect duration. It has been demonstrated that cationic lipophilic molecules (generally basic amines with pKa > 8) can accumulate in lung tissue. The mechanism behind this has been suggested to be a combination of lysosomal trapping and accumulation in membranes [91].

The inhaled β_2-agonists salmeterol, formoterol, and indacaterol have prolonged effect duration [82, 89, 157]. Formoterol, which has intermediary lipophilicity, is believed to be retained in the membrane and from this position be able to interact with the β_2-receptor in a rapid and prolonged manner. The more lipophilic salmeterol is retained by the same mechanism but is less available for fast onset of action [2]. The inhaled muscarinic receptor antagonist tiotropium has demonstrated prolonged effect duration; the mechanism for this has been proposed to be prolonged binding to the M3-receptor [4].

2.7.4 Biotransformation

The intracellular esterification of budesonide and ciclesonide active metabolite (and to some extent BDP) increases the retention time of these drugs in the airways and thereby prolongs their durations of action [10, 33, 100]. Free parent compound becomes available when these esterified forms are slowly hydrolyzed back to their active form and this mechanism contributes to the prolonged effect duration seen with these compounds.

Both inhaled steroids and bronchodilators are treatments with very favorable airway selectivity, which has improved the quality of life of millions of asthmatic and COPD patients. As exemplified above, drug developers have been successful in maintaining drug in the lung by slow dissolution, tissue retention, and/or reversible biotransformation [2, 33]. Most of the inhaled drugs on the market are high clearance compounds, characterized by a low systemic availability of the swallowed "waste" fraction and an extensive liver clearance. Many are highly bound to plasma proteins, thus reducing systemic circulation of free, active drug. Hence, it is a challenge for drug developers to improve on these already favorable therapeutic properties. Yet, some pharmacokinetic "tools" remain to be fully explored, such as the development of prodrugs and soft drugs. Others have to be better understood, such as the impact of protein binding on airway selectivity.

2.7.5 *Prodrugs and Soft Drugs*

Attempts have been made to improve airway selectivity by esterification to either an inactive prodrug that is locally activated before or at the target in the lung, or a pharmacologically active "soft drug," which is readily inactivated by hydrolysis in the lungs or blood. Both these structural alterations may theoretically improve airway selectivity. By inhaling an inactive prodrug, the risk of local side effects in the oropharyngeal tract can be reduced. Also, depending on the physicochemical properties of the active metabolite vs. the parent compound, the uptake and lung retention characteristics can be improved [143].

Currently approved steroid prodrugs for airway delivery include BDP and ciclesonide (CIC). Although oropharyngeal side effects have been demonstrated to diminish for CIC [40, 121], it is still unclear to what extent hydrolysis of the inactive prodrugs CIC and BDP to their respective active metabolites, CIC-AM and BMP, improves airway selectivity. Since one third of lung deposited CIC is systemically absorbed as intact prodrug, and since both absorption and airway hydrolysis are fast, the partial activation in the systemic circulation rather than in the lung will reduce rather than improve airway selectivity. Hence, the favorable airway selectivity that has been demonstrated clinically for CIC is not likely to be primarily as a result of its prodrug features [72].

Several soft-drug steroids (butixocort propionate, fluocortin butyl-ester, itrocinonide and γ-butyrolactone steroids) have been in clinical development as inhalation products, with the purpose of maintaining a high local intrinsic activity, but where the parent compound then is supposed to be readily inactivated in the target organ or blood to avoid systemic spillover and side effects. However, none of the soft-drug inhaled steroids has yet been found to retain sufficient clinical potency and improved airway selectivity to be taken into late stage clinical development.

2.7.6 *Protein Binding*

A basic presumption in pharmacokinetic science is that only unbound molecules are available for pharmacological effect. For inhaled drugs, high protein binding has been regarded as beneficial, since it will theoretically reduce unwanted systemic side effects and hence improve the therapeutic ratio of the drug [122]. Lung effects have, however, generally been assumed not to be affected by protein binding [29]. Evidence for the opposite was recently provided by Hochhaus [63]: in this study total and free drug levels as well as receptor occupancies were assessed after infusion of CIC-AM (the active metabolite of ciclesonide) and budesonide in rats. The two ICSs have similar potencies and basic PK properties but CIC-AM is about tenfold more bound to plasma proteins than budesonide [33]. While the total levels of the two drugs were similar both in the lung and in all other organs assessed in the study, CIC-AM showed significantly lower free levels as well as receptor occupancies (a measure related to effect) than budesonide also in the lungs. The authors concluded that high protein binding will reduce free drug concentrations and consequently the pharmacologic effect, but that airway selectivity is not necessarily improved by high protein binding.

2.8 Concluding Remarks

Drug activity in the lung is eliminated and modified by several mechanisms, most importantly mucociliary clearance, metabolism in the lung tissue itself, and absorption into the systemic circulation. As has been discussed, these mechanisms play an important role in the overall pharmacokinetic and pharmacodynamic characteristics of any inhaled drug, influencing onset of action, duration of effect, and therapeutic index. These mechanisms act in parallel so that the rate and extent of lung clearance by any particular mechanism is dependent on the rate and extent of clearance by the other processes. It is, therefore, very difficult to study a particular mechanism in humans since what is observed is the sum total, with confounding factors influencing the results to an often unknown degree. What happens between the "particle has landed" until the drug is detected in the systemic circulation is a major "black box" in the science of inhaled medication. Various levels of abstraction have been used to shine a light into the box, e.g., animal in vivo studies, ex vivo systems, cell cultures, and pure in vitro set-ups; but rational drug and formulation design will be hampered by our lack of detailed understanding for some time yet.

Since metabolism and absorption act only on solutes, dissolution kinetics of solid particles is important for lung clearance but probably only for relatively poorly soluble formulations. Lung retention may be increased by decreasing solubility, but this strategy for increasing duration of effect may be limited by the clearance mechanisms that act on undissolved drug; mucociliary and cough clearance in the central lungs and phagocytosis by alveolar macrophages in the lung periphery.

Although metabolic activity in the lungs is generally much lower than in the gut or liver, making the contribution from lung metabolism to the overall elimination of activity small, biotransformation in the lung has been shown to be important for the duration of effect of some corticosteroids that can undergo reversible esterification. Prodrugs are dependent on enzymatic activation in the lung, and airway selectivity may be improved by optimizing the balance between the kinetics of activation and the kinetics of absorption of both the prodrug and the active form. Soft drugs may also theoretically improve airway selectivity, but to date it appears that inactivation in the lungs of these labile molecules is too extensive for efficacy to be sufficiently retained.

Desired properties of pulmonary drug absorption depend on whether the target is local or systemic. For a local target, prolonged lung retention is usually desired to increase the duration of effect. Therefore, absorption into the systemic circulation should not be too fast. Both for polar and especially for lipophilic compounds, unhindered absorption seems generally to be faster than desired. For prolonged duration, other properties, such as low solubility, reversible biotransformation, or high tissue affinity, need to be present to achieve a suitably slow elimination from the lungs. By contrast, for systemic targets of peptides and proteins, the challenge seems to be to achieve absorption at a rate that is competitive relative to the rate of other clearance mechanisms in order to secure sufficient bioavailability. The very rapid absorption of many small, lipophilic molecules offers great potential for fast onset of action for systemic targets.

In conclusion, the mechanisms of elimination of drug activity in the lungs by the various clearance processes described here are important factors to consider both in the development of new drugs and in understanding the relative merits of existing therapies. What constitute optimal characteristics in each particular case, and how these may be predicted from abstracted properties, remains far from clear cut with our present understanding.

2.9 Contributions

Although a collaborative work, major contributions were as follows: dissolution, Katarina; mucociliary clearance, Lars; metabolism, Tove; absorption, Eva and Lena; effect duration and airway selectivity, Staffan and Stefan. Bo was the coordinator and local editor.

References

1. Agnew JF, Bateman JRM, Pavia D, Clarke SW (1984) Peripheral airways mucous clearance in stable asthma is improved by oral glucocorticoid therapy. Bull Eur Physiopathol Respir 20:295–301
2. Anderson GP (1993) Formoterol: pharmacology, molecular basis of agonism, and mechanism of long duration of a highly potent and selective beta2-adrenoceptor agonist bronchodilator. Life Sci 52:2145–2160
3. Anttila S, Hukkanen J, Hakkola J, Stjernvall T, Beaune P, Edwards RJ, Boobis AR, Pelkonen O, Raunio H (1997) Expression and localization of CYP3A4 and CYP3A5 in human lung. Am J Respir Cell Mol Biol 16:242–249
4. Barnes PJ (2000) The pharmacological properties of tiotropium. Chest 117:635–665
5. Berridge MS, Heald DL (1999) In vivo characterization of inhaled pharmaceuticals using quantitative positron emission tomography. J Clin Pharmacol 39:25S–29S
6. Bleasby K, Castle JC, Roberts CJ, Cheng C, Bailey WJ, Sina JF, Kulkarni AV, Hafey MJ, Evers R, Johnson JM, Ulrich RG, Slatter JG (2006) Expression profiles of 50 xenobiotic transporter genes in humans and pre-clinical species: a resource for investigations into drug disposition. Xenobiotica 36:963–988
7. Boat TF, Cheng PW (1980) Biochemistry of airway mucus secretions. Fed Proc 39:3067–3074
8. Bondesson E, Bengtsson T, Nilsson LE, Wollmer P (2006) Effects of sodium taurocholate on the absorption of inhaled 99mTc-DTPA. Pharm Res 23:2122–2128

9. Bosquillon C (2010) Drug transporters in the lung – do they play a role in the biopharmaceutics of inhaled drugs? J Pharm Sci 99:2240–2255
10. Brattsand R, Miller-Larsson A (2003) The role of intracellular esterification in budesonide once-daily dosing and airway selectivity. Clin Ther 25(suppl 3):C28–C41
11. Brechot JM, Hurbain I, Fajac A, Daty N, Bernaudin JF (1998) Different pattern of MRP localization in ciliated and basal cells from human bronchial epithelium. J Histochem Cytochem 46:513–517
12. Brown RA, Schanker LS (1983) Absorption of aerosolized drugs from the rat lung. Drug Metab Dispos 11:355–360
13. Brutsche MH, Brutsche IC, Munawar M, Langley SJ, Masterson CM, Daley-Yates PT, Brown R, Custovic A, Woodcock A (2000) Comparison of pharmacokinetics and systemic effects of inhaled fluticasone propionate in patients with asthma and healthy volunteers: a randomised crossover study. Lancet 356(9229):556–561
14. Bur M, Lehr C-M (2008) Pulmonary cell culture models to study the safety and efficacy of innovative aerosol medicines. Expert Opin Drug Deliv 5:641–652
15. Bur M, Huwer H, Lehr C-M, Hagen N, Guldbrandt M, Kim K-J, Ehrhardt C (2006) Assessment of transport rates of proteins and peptides across primary human alveolar epithelial cell monolayers. Eur J Pharm Sci 28:196–203
16. Bur M, Henning A, Hein S, Schneider M, Lehr C-M (2009) Inhalative nanomedicine-opportunities and challenges. Inhalation Tox 21(S1):137–143
17. Bur M, Rothen-Rutishauser B, Huwer H, Lehr C-M (2009) A novel cell compatible impingement system to study in vitro drug absorption from dry powder aerosol formulations. Eur J Pharm Biopharm 72:350–357
18. Burnham EL, Halkar R, Burks M, Moss M (2009) The effects of alcohol abuse on pulmonary alveolar-capillary barrier function in humans. Alcohol Alcohol 44:8–12
19. Camner P, Mossberg B, Philipson K (1973) Tracheobronchial clearance and chronic obstructive lung disease. Scand J Respir Dis 54:272–281
20. Campbell L, Abulrob A-G, Kandalaft LE, Plummer S, Hollins AJ, Gibbs A et al (2003) Constitutive expression of P-glycoprotein in normal lung alveolar epithelium and functionality in primary alveolar epithelial cultures. J Pharm Exp Ther 304:441–452
21. Cazzola M, Testi R, Matera MG (2002) Clinical pharmacokinetics of salmeterol. Clin Pharmacokinet 41:19–30
22. Cheng X, Maher J, Chen C, Klaassen CD (2005) Tissue distribution and ontogeny of mouse organic anion transporting polypeptides (oatps). Drug Metab Disp 33:1062–1073
23. Constantinou M, Waters SL, Steel CJ, Poole KG, Marino PS, Ind PW, Rhodes CG, Aigbirhio FI, Moore A, Pike VW, Luthra SK (2004) Flixotide™-pressurized metered-dose inhalers leaded with [18F]fluticasone propionate particles for drug deposition studies in humans with PET-formulation and analysis. J Label Compd Radiopharm 47:55–70
24. Cook RO, Pannu RK, Kellaway IW (2005) Novel sustained release microspheres for pulmonary drug delivery. J Control Release 104:79–90
25. Cooney D, Kazantseva M, Hickey AJ (2004) Development of a size-dependent aerosol deposition model utilising human airway epithelial cells for evaluating aerosol drug delivery. Alter Lab Anim 32:581–590
26. Crandall ED, Matthay MA (2001) Alveolar epithelial transport. Basic science to clinical medicine. Am J Respir Crit Care Med 163:1021–1029
27. Cryan S-A, Sivadas N, Garcia-Contreras L (2007) In vivo animal models for drug delivery across the lung mucosal barrier. Adv Drug Del Rev 59:1133–1151
28. Dalby C, Polanowski T, Larsson T, Borgström L, Edsbäcker S, Harrison TW (2009) The bioavailability and airway clearance of the steroid component of budesonide/formoterol and salmeterol/fluticasone after inhaled administration in patients with COPD and healthy subjects: a randomized controlled trial. Respir Res 10:104
29. Daley-Yates PT, Harker AJ, Taylor S, Daniel MJ (2005) Plasma protein binding (PPB) of corticosteroids (CS): reappraisal of its significance in systemic pharmacological activity. J Allergy Clin Immunol 115:S4

30. Davies NM, Feddah MR (2003) A novel method for assessing dissolution of aerosol inhaler products. Int J Pharm 255:175–187
31. Ding X, Kaminsky LS (2003) Human extrahepatic cytochromes P450: function in xenobiotic metabolism and tissue-selective chemical toxicity in the respiratory and gastrointestinal tracts. Annu Rev Pharmacol Toxicol 43:149–173
32. Duchateau GS, Zuidema J, Merkus FW (1986) The in vitro and in vivo effect of a new non-halogenated corticosteroid – budesonide – aerosol on human ciliary epithelial function. Allergy 41:260–265
33. Edsbäcker S, Johansson C-J (2006) Airway selectivity: an update of pharmacokinetic factors affecting local and systemic disposition of inhaled steroids. Basic Clin Pharmacol Toxicol 98:523–536
34. Edsbäcker S, Wollmer P, Selroos O, Borgström L, Olsson B, Ingelf J (2008) Do airway clearance mechanisms influence the local and systemic effects of inhaled corticosteroids? Pulm Pharmacol Ther 21:247–258
35. Edwards DA, Hanes J, Caponetti G, Hrkach J, Ben-Jebria A, Lou Eskew M, Mintzes J, Deaver D, Lotan N, Langer R (1997) Large porous particles for pulmonary drug delivery. Science 276:1868–1872
36. Effros RM, Mason GR (1983) Measurements of pulmonary epithelial permeability in vivo. Am Rev Respir Dis 127(suppl 5):S59–S65
37. Ehrhardt C, Kneuer C, Bies C, Lehr C, Kim K, Bakowsky U (2005) Salbutamol is actively absorbed across human bronchial epithelial cell layers. Pulm Pharm Ther 18:165–170
38. Eljamal M, Nagarajan S, Patton JS (1996) In situ and in vivo methods for pulmonary delivery. Pharm Biotechnol 8:361–374
39. Endter S, Francombe D, Ehrhardt C, Gumbleton M (2009) RT-PCR analysis of ABC, SLC and SLCO drug transporters in human lung epithelial cell models. J Pharm Pharmacol 61:583–591
40. Engelstätter R, Banerji D, Steinijans VW, Wurst W (2004) Low incidence of oropharyngeal adverse events in asthma patients treated with ciclesonide: results from a pooled analysis. Am J Resp Crit Care Med 169(7 Suppl):A92
41. Esmailpour N, Högger P, Rabe KF, Heitmann U, Nakashima M, Rohdewald P (1997) Distribution of inhaled fluticasone propionate between human lung tissue and serum in vivo. Eur Respir J 10:1496–1499
42. Ewing P, Eirefelt SJ, Andersson P, Blomgren A, Ryrfeldt A, Gerde P (2008) Short inhalation exposures of the isolated and perfused rat lung to respirable dry particle aerosols; the detailed pharmacokinetics of budesonide, formoterol, and terbutaline. J Aerosol Med Pulm Drug Del 21:169–180
43. Fiegel J, Ehrhardt C, Schaefer UF, Lehr C, Hanes J (2003) Large porous particle impingement on lung epithelial cell monolayers: toward improved particle characterization in the lung. Pharm Res 20:788–796
44. Fischer H, Widdicombe JH (2006) Mechanisms of acid and base secretion by the airway epithelium. J Membr Biol 211:139–150
45. Forbes B, Ehrhardt C (2005) Human respiratory epithelial cell culture for drug delivery applications. Eur J Pharm Biopharm 60:193–205
46. Foster WM, Langenback EG, Bergofsky EH (1980) Measurement of tracheal and bronchial mucus velocities in man: relation to lung clearance. J Appl Physiol 48:965–971
47. Freiwald M, Valotis A, Kirschbaum A, McClellan M, Mürdter T, Fritz P, Friedel G, Thomas M, Högger P (2005) Monitoring the initial pulmonary absorption of two different beclomethasone dipropionate aerosols employing a human lung reperfusion model. Respir Res 6:21–33
48. Gehr P, Blank F, Rothen-Rutishauser BM (2006) Fate of inhaled particles after interaction with the lung surface. Paediatr Respir Rev 7(Suppl1):S73–S75
49. Goodman RM, Yergin BM, Landa JF, Golinvaux MH, Sackner MA (1978) Relationship of smoking history and pulmonary function tests to tracheal mucous velocity in nonsmokers, young smokers, ex-smokers and patients with chronic bronchitis. Am Rev Respir Dis 117:205–214

50. Gosens R, Mutawe M, Martin S, Basu S, Bos ST, Tran T, Halayko AJ (2008) Caveolae and caveolins in the respiratory system. Current Mol Med 8:741–753
51. Grey VA, Hickey AJ, Ballmer P, Davies NM, Dunbar C, Foster TS, Olsson BL, Sakagami M, Shah VP, Smurthwaite MJ, Veranth JM, Zaidi K (2008) The inhalation Ad Hoc advisory panel for the usp performance tests of inhalation dosage forms. Pharm Forum 34: 1068–1074
52. Groneberg DA, Nickolaus M, Springer J, Doring F, Daniel H, Fischer A (2001) Localization of the peptide transporter PEPT2 in the lung: implications for pulmonary oligopeptide uptake. Am J Pathol 158:707–714
53. Groneberg DA, Fischer A, Chung KF, Daniel H (2004) Molecular mechanisms of pulmonary peptidomimetic drug and peptide transport. Am J Respir Cell Mol Biol 30:251–260
54. Guengerich FP (1997) Role of cytochrome P450 enzymes in drug-drug interactions. Adv Pharmacol 43:7–35
55. Gumbleton M (2001) Caveolae as potential macromolecule trafficking compartments within alveolar epithelium. Adv Drug Deliv Rev 49:281–300
56. Gumbleton M, Hollins AJ, Omidi Y, Campbell L, Taylor G (2003) Targeting caveolae for vesicular drug transport. J Control Release 87:139–151
57. Ha HR, Chen J, Freiburghaus AU, Follath F (1995) Metabolism of theophylline by cDNA-expressed human cytochromes P-450. Br J Clin Pharmacol 39:321–326
58. Hahn ME (2002) Aryl hydrocarbon receptors: diversity and evolution. Chem Biol Interact 141:131–160
59. Heinemann L, Klappoth W, Rave K, Hompesch B, Linkeschowa R, Heise T (2000) Intra-individual variability of the metabolic effect of inhaled insulin together with an absorption enhancer. Diabetes Care 23:1343–1347
60. Himmelmann A, Jendle J, Mellén A, Petersen AH, Dahl UL, Wollmer P (2003) The impact of smoking on inhaled insulin. Diabetes Care 26:677–682
61. Ho RH, Kim RB (2005) Transporters and drug therapy: implications for drug disposition and disease. Clin Pharm Ther 78:260–277
62. Hochhaus G, Möllmann H, Derendorf H, Gonzalez-Rothi RJ (1997) Pharmacokinetic/pharmacodynamic aspects of aerosol therapy using glucocorticoids as a model. J Clin Pharmacol 37:881–892
63. Hochhaus G, Hall S, Wu K, Blomgren A, Ekholm K, Weber B, Edsbäcker S (2009) Budesonide and Ciclesonide: effect of tissue binding on pulmonary receptor binding. Drug Metab Dispos 37:1421–1426
64. Honkakoski P, Negishi M (2000) Regulation of cytochrome P450 (CYP) genes by nuclear receptors. Biochem J 347:321–337
65. Horvath G, Sutto Z, Torbati A, Conner GE, Salathe M, Wanner A (2003) Norepinephrine transport by the extraneuronal monoamine transporter in human bronchial arterial smooth muscle cells. Am J Physiol Lung Cell Mol Physiol 285:829–837
66. Horvath G, Schmid N, Fragoso MA, Schmid A, Conner GE, Salathe M et al (2007) Epithelial organic cation transporters ensure pH-dependent drug absorption in the airway. Am J Respir Cell Mol Biol 36:53–60
67. Huh D, Matthews BD, Mammoto A, Montoya-Zavala M, Hsin HY, Ingber DE (2010) Reconstituting organ-level lung functions on a chip. Science 328:1662–1668
68. Hukkanen J, Hakkola J, Anttila S, Piipari R, Karjalainen A, Pelkonen O, Raunio H (1997) Detection of mRNA encoding xenobiotic-metabolizing cytochrome P450s in human bronchoalveolar macrophages and peripheral blood lymphocytes. Mol Carcinog 20:224–230
69. Hukkanen J, Pelkonen O, Hakkola J, Raunio H (2002) Expression and regulation of xenobiotic-metabolizing cytochrome P450 (CYP) enzymes in human lung. Crit Rev Toxicol 32:391–411
70. Hussain A, Ahsan F (2006) Indication of transcytotic movement of insulin across human bronchial epithelial cells. J Drug Target 14:181–190
71. International Transporter Consortium (2010) Membrane transporters in drug development. Nat Rev Drug Discov 9:215–236

72. Johansson C-J, Sternebring O, Edsbacker S (2004) Pharmacokinetic factors affecting airway selectivity of inhaled steroids – role of prodrug conversion and absorption. Am J Respir Crit Care Med 169:A37
73. Kelly JD, Eaton DL, Guengerich FP, Coulombe RA Jr (1997) Aflatoxin B1 activation in human lung. Toxicol Appl Pharmacol 144:88–95
74. Kemp SJ, Thorley AJ, Gorelik J, Seckl MJ, O'Hare MJ, Arcaro A, Korchev Y, Goldstraw P, Tetley TD (2008) Immortalization of human alveolar epithelial cells to investigate nanoparticle uptake. Am J Respir Cell Mol Bio 39:591–597
75. Kim KJ, Malik AB (2003) Protein transport across the lung epithelial barrier. Am J Physiol Lung Cell Mol Physiol 284:L247–L259
76. Kim K-J, Fandy TE, Lee VHL, Ann DK, Borok Z, Crandall ED (2004) Net absorption of IgG via FcRn-mediated transcytosis across rat alveolar epithelial cell monolayers. Am J Physiol Lung Cell Mol Physiol 287:L616–L622
77. Konduri KM, Nandedkar S, Düzgünes N, Suzara V, Artwohl J, Bunte R, Gangadharam PRJ (2003) Efficacy of liposomal budesonide in experimental asthma. J Allergy Clin Immunol 111:321–327
78. Kreyling WG, Semmler M, Erbe F, Mayer P, Takenaka S, Schulz H, Oberdorster G, Ziesenis A (2002) Translocation of ultrafine insoluble iridium particles from lung epithelium to extrapulmonary organs is size dependent but very low. J Toxicol Environ Health A 65:1513–1530
79. Krishnaswami S, Hockhaus G, Mollmann H, Barth J, Derendorf H (2005) Interpretation of absorption rate data for inhaled fluticasone propionate obtained in compartmental pharmacokinetic modeling. Int J Clin Pharmacol Ther 43:117–122
80. Kroon LA (2007) Drug interactions with smoking. Am J Health-System Pharm 64:1917–1921
81. Labiris NR, Dolovich MB (2003) Pulmonary drug delivery part I: physiological factors affecting therapeutic effectiveness of aerosolized medications. Br J Clin Pharmacol 56:588–599
82. LaForce C, Korenblat P, Osborne P, Dong F, Higgins M (2009) 24-hour bronchodilator efficacy of single doses of indacaterol in patients with persistent asthma: comparison with placebo and formoterol. Curr Med Res Opin 25:2353–2359
83. Laurenco RV, Klimek MF, Borowski CJ (1971) Deposition and clearance of 2 mm particles in the tracheobronchial tree of normal subjects – smokers and nonsmokers. J Clin Invest 50:1411–1420
84. Lechapt-Zalcman E, Hurbain I, Lacave R, Commo F, Urban T, Antoine M et al (1997) MDR1-pgp 170 expression in human bronchus. Eur Respir J 10:1837–1843
85. Liao X, Wiedmann TS (2003) Solubilization of cationic drugs in lung surfactant. Pharm Res 20:1858–1863
86. Lindberg S, Khan R, Runer T (1995) The effects of formoterol, a long-acting beta2-adrenoceptor agonist, on mucociliary activity. Eur J Pharmacol 285:275–280
87. Lips KS, Volk C, Schmitt BM, Pfeil U, Arndt P, Miska D et al (2005) Polyspecific cation transporters mediate luminal release of acetylcholine from bronchial epithelium. Am J Respir Cell Mol Biol 33:79–88
88. Lombry C, Edwards DA, Preat V, Vanbever R (2004) Alveolar macrophages are a primary barrier to pulmonary absorption of macromolecules. Am J Physiol Lung Cell Mol Physiol 286:L1002–L1008
89. Lötvall J, Ankerst J (2008) Long duration of airway but not systemic effects of inhaled formoterol in asthmatic patients. Respir Med 102:449–456
90. Lucas AM, Douglas LC (1934) Principles underlying ciliary activity in the respiratory tract. II. A comparison of nasal clearance in man, monkey and other mammals. Arch Otolaryngol 20:518–541
91. Macintyre AC, Cutler DJ (1988) The potential role of lysosomes in tissue distribution of weak bases. Biopharm Drug Dispos 9:513–526
92. Mason GR, Peters AM, Bagdades E, Myers MJ, Snook D, Hughes JMB (2001) Evaluation of pulmonary alveolar epithelial integrity by the detection of restriction to diffusion of hydrophilic solutes of different molecular sizes. Clin Sci 100:231–236
93. Melloni B, Germouty J (1992) The influence on mucociliary function of formoterol: a new beta-agonist. Rev Mal Respir 9:503–507

94. Messina MS, O'Riordan TG, Smaldone GC (1993) Changes in mucociliary clearance during acute exacerbations of asthma. Am Rev Respir Dis 143:993–997
95. Mobley C, Hochhaus G (2001) Methods used to assess pulmonary deposition and absorption of drugs. Drug Discov Today 6:367–375
96. Morgan ET (2001) Regulation of cytochrome P450 by inflammatory mediators: why and how? Drug Metab Dispos 29:207–212
97. Mühlfield C, Gehr P, Rothen-Rutishauser B (2008) Translocation and cellular entering mechanisms of nanoparticles in the respiratory tract. Swiss Med Weekly 138:387–391
98. Nakamura T, Nakanishi T, Haruta T, Shirasaka Y, Keogh JP, Tamai I (2010) Transport of ipratropium, an anti-chronic obstructive pulmonary disease drug, is mediated by organic cation/carnitine transporters in human bronchial epithelial cells: implications for carrier-mediated pulmonary absorption. Mol Pharm 7:187–195
99. Nave R, Fisher R, Zech K (2006) In vitro metabolism of ciclesonide in human lung and liver precision-cut tissue slices. Biopharm Drug Dispos 27:197–207
100. Nave R, Fisher R, McCracken N (2007) In vitro metabolism of beclomethasone dipropionate, budesonide, ciclesonide anf fluticasone propionate in human lung precision-cut tissue slices. Respir Res 8:65
101. Nebert DW, Dalton TP, Okey AB, Gonzalez FJ (2004) Role of aryl hydrocarbon receptor-mediated induction of the CYP1 enzymes in environmental toxicity and cancer. J Biol Chem 279:23847–23850
102. Oberdörster G (1998) Lung clearance of inhaled insoluble and soluble particles. J Aerosol Med 1:289–330
103. Oberdorster G, Cox C, Gelein R (1997) Intratracheal instillation versus intratracheal-inhalation of tracer particles for measuring lung clearance function. Exp Lung Res 231:17–34
104. Olson N, Greul A-K, Hristova M, Bove PF, Kasahara DI, van der Vliet A (2009) Nitric oxide and airway epithelial barrier function: regulation of tight junction proteins and epithelial permeability. Arch Biochem Biophys 484:205–213
105. Oswald S, Grube M, Siegmund W, Kroemer HK (2007) Transporter-mediated uptake into cellular compartments. Xenobiotica, 37(modelling and simulation approaches to in vitro-in vivo, extrapolation of ADME properties in drug development, editor: Gordon Gibson, guest editor: Amin Rostami-Hodjegan):1171–1195
106. Parkinson A (1996) Biotransformation of xenobiotics. In: Curtis Klaassen D (ed) Casarett & doull's toxicology – the basic science of poisons. McGraw-Hill, ON, Canada
107. Patton JS (1996) Mechanisms of macromolecule absorption by the lungs. Adv Drug Deliv Rev 19:3–36
108. Patton JS, Byron PR (2007) Inhaling medicines: delivering drugs to the body through the lungs. Nat Rev Drug Discov 6:67–74
109. Patton JS, Fishburn CS, Weers JG (2004) The lungs as a portal of entry for systemic drug delivery. Proc Am Thorac Soc 1:338–344
110. Pavek P, Dvorak Z (2008) Xenobiotic-induced transcriptional regulation of xenobiotic metabolizing enzymes of the cytochrome P450 superfamily in human extrahepatic tissues. Curr Drug Metab 9:129–143
111. Pavia D, Agnew JE, Sutton PP, Lopez-Vidriero MT, Clay MM, Killip M, Clarke SW (1987) Effect of terbutaline administered from metered dose inhaler (2 mg) and subcutaneously (0.25 mg) on tracheobronchial clearance in mild asthma. Br J Dis Chest 81:361–370
112. Pearce RE, Leeder JS, Kearns GL (2006) Biotransformation of fluticasone: in vitro characterization. Drug Metab Dispos 34:1035–1040
113. Petersen AH, Köhler GH, Korsatko S, Wutte A, Wonisch M, Mautner A, Ronn BB, Clauson P, Laursen T, Wollmer P, Rieber TR (2007) The effect of exercise on the absorption of inhaled human insulin in healthy volunteers. Br J Clin Pharmacol 65:165–171
114. Phalen RF, Oldham MJ, Wolff RK (2008) The relevance of animal models for aerosol studies. J Aerosol Med 21:113–124
115. Pham S, Wiedmann TS (2001) Dissolution of aerosol particles of budesonide in Survanta, a model lung surfactant. J Pharm Sci 90:98–104

116. Predescu SA, Predescu DN, Malik AB (2007) Molecular determinants of endothelial transcytosis and their role in endothelial permeability. Am J Physiol Lung Cell Mol Physiol 293:L823–L842
117. Puchelle E, Zahm JM, Girard F, Bertrand A, Polu JM, Aug F, Sadoul P (1980) Mucociliary transport in vivo and in vitro. Relations to sputum properties in chronic bronchitis. Eur J Respir Dis 61:254–264
118. Purucker ME, Rosebraugh CJ, Zhou F, Meyer RJ (2003) Inhaled fluticasone propionate by Diskus in the treatment of asthma: a comparison of the efficacy and safety of the same nominal dose given either once or twice a day. Chest 124:1584–1593
119. Raunio H, Hakkola J, Hukkanen J, Lassila A, Paivarinta K, Pelkonen O, Anttila S, Piipari R, Boobis A, Edwards RJ (1999) Expression of xenobiotic-metabolizing CYPs in human pulmonary tissue. Exp Toxicol Pathol 51:412–417
120. Rejman J, Oberle V, Zuhorn IS, Hoekstra D (2004) Size-dependent internalization of particles via the pathways of clathrin- and caveolae-mediated endocytosis. Biochem J 377: 159–169
121. Richter K, Kanniess F, Biberger C, Nave R, Magnussen H (2005) Comparison of the oropharyngeal deposition of inhaled ciclesonide and fluticasone propionate in patients with asthma. J Clin Pharmacol 45:146–152
122. Rohatagi S, Luo Y, Shen L, Guo Z, Schemm C, Huang Y, Chen K, David M, Nave R, King SP (2005) Protein binding and its potential for eliciting minimal systemic side effects with a novel inhaled corticosteroid, ciclesonide. Am J Ther 12:201–209
123. Rytting E, Bur M, Cartier R, Bouyssou T, Wang X, Krüger M, Lehr C-M, Kissel T (2010) In vitro and in vivo performance of biocompatible negatively-charged salbutamol-loaded nanoparticles. J Contr Rel 141:101–107
124. Sackner MA (1978) Effect of respiratory drugs on mucociliary clearance. Chest 73: 958–966
125. Sackner MA, Reinhart M, Arkin B (1977) Effects of beclomethasone dipropionate on tracheal mucous velocity. Am Rev Respir Dis 115:1069–1070
126. Sakagami M (2006) In vivo, in vitro and ex vivo models to assess pulmonary absorption and disposition of inhaled therapeutics for systemic delivery. Adv Drug Del Rev 58: 1030–1060
127. Salama R, Traini D, Chan H-K, Young PM (2009) Recent advances in controlled release pulmonary therapy. Curr Drug Deliv 6:404–414
128. Schanker LS, Mitchell EW, Brown RA (1986) Species comparison of drug absorption from the lung after aerosol inhalation or intratracheal injection. Drug Metab Dispos 14:79–88
129. Scheffer GL, Pijnenborg ACLM, Smit EF, Muller M, Postma DS, Timens W et al (2002) Multidrug resistance related molecules in human and murine lung. J Clin Pathol 55: 332–339
130. Schmekel B, Borgström L, Wollmer P (1991) Difference in pulmonary absorption of inhaled terbutaline in healthy smokers and non-smokers. Thorax 46:225–228
131. Schmekel B, Borgström L, Wollmer P (1992) Exercise increases the rate of pulmonary absorption of inhaled terbutaline. Chest 101:742–745
132. Schneeberger EE (1991) Airway and alveolar epithelial cell junctions. In: Crystal RG, West JB et al (eds) The lung. Raven Press, New York, pp 205–214
133. Schreier H (1994) Pulmonary delivery of liposomal drugs. J Lipos Res 1:229–238
134. Shimada T, Yamazaki H, Mimura M, Inui Y, Guengerich FP (1994) Interindividual variations in human liver cytochrome P-450 enzymes involved in the oxidation of drugs, carcinogens and toxic chemicals: studies with liver microsomes of 30 japanese and 30 caucasians. J Pharmacol Exp Ther 270:414–423
135. Shitara Y, Horie T, Sugiyama Y (2006) Transporters as a determinant of drug clearance and tissue distribution. Eur J Pharm Sci 27:425–446
136. Short MD, Singh CA, Few JD, Studdy PR, Heaf PJD, Spiro SG (1981) The labelling and monitoring of lung deposition of an inhaled synthetic anticholinergic bronchodilating agent. Chest 80(6 suppl):918–921

137. Somers GI, Lindsay N, Lowdon BM, Jones AE, Freathy C, Ho S, Woodrooffe AJM, Bayliss MK, Manchee GR (2007) A comparison of the expression and metabolizing activities of phase I and II enzymes in freshly isolated human lung parenchymal cells and cryopreserved human hepatocytes. Drug Metab Dispos 35:1797–1805
138. Son Y-J, McConville JT (2009) Development of a standardized dissolution test method for inhaled pharmaceutical formulations. Int J Pharm 382:15–22
139. Sporty JL, Horálková L, Ehrhardt C (2008) In vitro cell culture models for the assessment of pulmonary drug disposition. Expert Opin Drug Metab Toxicol 4:333–345
140. Stahlhofen W, Köbrich R, Rudolf G, Scheuch G (1990) Short-term and long-term clearance of particles from the upper human respiratory tract as function of particle size. J Aerosol Sci 21:S407–S41
141. Steimer A, Haltner E, Lehr CM (2005) Cell culture models of the respiratory tract relevant to pulmonary drug delivery. J Aerosol Med 18:137–182
142. Sugano K, Kansy M, Artursson P, Avdeef A, Bendels S, Di L, Ecker G-F, Faller B, Fisher H, Gerebtzoff G, Lennernaes H, Senner F (2010) Coexistence of passive and carrier-mediated processes in drug transport. Nat Rev Drug Disc 9:597–614
143. Svensson L-Å (1991) Mechanism of action of bambuterol: a beta-agonist prodrug with sustained lung affinity. Agents Actions Suppl 34:71–78
144. Takenaka S, Karg E, Roth C, Schulz H, Ziesenis A, Heinzmann U, Schramel P, Heyder J (2001) Pulmonary and systemic distribution of inhaled ultrafine silver particles in rats. Environ Health Persp 109(suppl 4):547–551
145. Tarran R, Button B, Boucher RC (2006) Regulation of normal and cystic fibrosis airway surface liquid volume by phasic shear stress. Annu Rev Physiol 68:543–561
146. Tarran R, Trout L, Donaldson SH, Boucher RC (2006) Soluble mediators, not cilia, determine airway surface liquid volume in normal and cystic fibrosis superficial airway epithelia. J Gen Physiol 127:591–604
147. Thorsson L, Kenyon C, Newman SP, Borgström L (1998) Lung deposition of budesonide in asthmatics: a comparison of different formulations. Int J Pharm 168:119–127
148. Thorsson L, Edsbäcker S, Källén A, Löfdahl CG (2001) Pharmacokinetics and systemic activity of fluticasone via Diskus and pMDI, and of budesonide via Turbuhaler. Br J Clin Pharmacol 52:529–538
149. Thum T, Erpenbeck VJ, Moeller J, Hohlfeld JM, Krug N, Borlak J (2006) Expression of xenobiotic metabolizing enzymes in different lung compartments of smokers and nonsmokers. Environ Health Persp 114:1655–1661
150. Tiruppathi C, Shimizu J, Miyawaki-Shimizu K, Vogel SM, Bair AM, Minshall RD, Predescu D, Malik AB (2008) Role of NF-κB-dependent caveolin-1 expression in the mechanism of increased endothelial permeability induced by lipopolysaccharide. J Biol Chem 283:4210–4218
151. Tolman JA, Williams RO (2010) Advances in the pulmonary delivery of poorly water-soluble drugs: influence of solubilization on pharmacokinetic properties. Drug Dev Ind Pharm 36:1–30
152. Toremalm NH (1960) The daily amount of tracheobronchial secretion in man: a method for continuous tracheal aspiration in laryngectomized and tracheostomized patients. Acta Otolaryngol Suppl 158:43–53
153. Trepat X, Farré R (2008) Alveolar permeability and stretch: too far, too fast. Eur Respir J 32:826–828
154. Tronde A, Nordén B, Jeppsson AB, Brunmark P, Nilsson E, Lennernäs H, Bengtsson UH (2003) Drug absorption from the isolated perfused rat lung – correlations with drug physicochemical properties and epithelial permeability. J Drug Target 11:61–74
155. Tronde A, Nordén B, Marchner H, Wendel A-K, Lennernäs H, Bengtsson UH (2003) Pulmonary absorption rate and bioavailability of drugs in vivo in rats: structure-absorption relationships and physicochemical profiling of inhaled drugs. J Pharm Sci 92:1216–1233
156. Tunek A, Sjödin K, Hallström G (1997) Reversible formation of fatty acid esters of budesonide, an antiasthma glucocorticoid, in human lung and liver microsomes. Drug Metab Dispos 25:1311–1317

157. Ullman A, Svedmyr N (1988) Salmeterol, a new long acting inhaled beta 2 adrenoceptor agonist: comparison with salbutamol in adult asthmatic patients. Thorax 43:674–678
158. Van Den Brink KIM, Boorsma M, Staal-Van Den Brekel AJ, Edsbäcker S, Wouters EF, Thorsson L (2008) Evidence of the in vivo esterification of budesonide in human airways. Br J Clin Pharmacol 66:27–35
159. van der Deen M, de Vries EGE, Timens W, Scheper RJ, Timmer-Bosscha H, Postma DS (2005) ATP-binding cassette (ABC) transporters in normal and pathological lung. Respir Res 6:59
160. van der Deen M, Marks H, Willemse BWM, Postma DS, Muller M, Smit EF et al (2006) Diminished expression of multidrug resistance-associated protein 1 (MRP1) in bronchial epithelium of COPD patients. Virchows Archiv 449:682–688
161. van der Deen M, de Vries EGE, Visserman H, Zandbergen W, Postma DS, Timens W et al (2007) Cigarette smoke extract affects functional activity of MRP1 in bronchial epithelial cells. J Biochem Mol Toxicol 21:243–251
162. van Drissche W, Kreindler JL, Malik EA, Margulies S, Lewis SA, Kim K-J (2007) Interrelations/cross talk between transcellular transport function and paracellular tight junctional properties in lung epithelial and endothelial barriers. Am J Physiol Lung Cell Mol Physiol 293:L520–L524
163. Vastag E, Matthys H, Zsamboki G, Kohler D (1986) Mucociliary clearance in smokers. Eur J Respir Dis 68:107–113
164. West JB (1992) Pulmonary patophysiology, The essentials, 4th edn. Williams & Wilkins, Baltimore, p 137
165. West JB (2000) Cellular responses to mechanical stress: invited review: pulmonary capillary stress failure. J Appl Physiol 89:2483–2489
166. Wiedmann TS, Bhatia R, Wattenberg LW (2000) Drug solubilization in lung surfactant. J Contr Release 65:43–47
167. Wurthwein G, Rohdewald P (1990) Activation of beclomethasone dipropionate by hydrolysis to beclomethasone-17-monopropionate. Biopharm Drug Dispos 11:381–394
168. Yang JZ, Young AL, Chiang P-O, Thurston A, Pretzer DK (2008) Fluticasone and budesonide nanosuspensions for pulmonary delivery: preparation, characterization and pharmacokinetic studies. J Pharm Sci 97:4869–4878
169. Yu J, Chien YW (2002) Significance of respiratory dynamics of the lung tissue in pulmonary drug permeation. Pharm Dev Technol 7:215–225
170. Yu LX, Lipka E, Crison JR, Amidon GL (1996) Transport approaches to the biopharmaceutical design of oral drug delivery systems: prediction of intestinal absorption. Adv Drug Deliv Rev 19:359–376
171. Zhang JY, Wang YF, Prakash C (2006) Xenobiotic-metabolizing enzymes in human lung. Curr Drug Metab 7:939–948
172. Zhang K, Kuroha M, Shibata Y, Kokue E, Shimoda M (2006) Effect of oral administration of clinically relevant doses of dexamethasone on regulation of cytochrome P450 subfamilies in hepatic microsomes from dogs and rats. Am J Vet Res 67:329–334
173. Zheng Z, Fang J-L, Lazarus P (2002) Glucuronidation: an important mechanism for detoxification of benzo[a]pyrene metabolites in aerodigestive tract tissues. Drug Metab Dispos 30:397–403

Chapter 3
Pulmonary Drug Delivery: An Historical Overview

Mark Sanders

Abstract The history of pulmonary drug delivery is nearly as old as recorded history itself. The journey from these ancient times to today's therapeutics and delivery systems is fascinating and provides an understanding of what has driven innovation. It is also critical to understand the evolving science and technology of pulmonary drug delivery and its trajectory from the current state of the art.

Keywords Atomizer • Dry powder inhaler • History • Inhalation • Inhaler • Nebulizer • pMDI

3.1 The Earliest History of Pulmonary Drug Delivery

The earliest recordings of inhalation therapy are from ancient Egypt, in approximately 1554 BC. Asthma had been first described in China much earlier than this, in 2600 BC, but these accounts do not describe inhalation. The author, Huang-Ti, does however identify the use of Ma Huang and this was later shown to contain ephedrine [11].

The Ebers papyrus from ancient Egypt (1554 BC) describes how the breathless may be treated by the inhalation of the vapor of black henbane. There was not a specific inhaler device, instead the leaves would be thrown onto hot bricks, causing the alkaloids to vaporize and the patient was to inhale the free vapor. We do not know how much henbane was used or with what frequency [7].

Inhalation treatments emerged independently in different cultures. From old Assyria (c650 BC), we learn from cuneiform texts that inhaled cromones were used in asthma [26]. In ancient Greece Hippocrates (460–377 BC), who gave us the

M. Sanders (✉)
Clement Clarke International Ltd, 15 Friars Walk, Dunstable, Bedfordshire LU6 3JA, United Kingdom
e-mail: ms@bethere.co.uk

H.D.C. Smyth and A.J. Hickey (eds.), *Controlled Pulmonary Drug Delivery*, Advances in Delivery Science and Technology, DOI 10.1007/978-1-4419-9745-6_3,

Fig. 3.1 Native smoking pipes

word "asthma," tells us that he employed an apparatus which "consisted of a pot, the lid of which had an opening for the reception of a reed, through which the vapor escaped and was inhaled through the open mouth: the latter being protected from scalding by moist sponges." In the pot he boiled herbs and resins in vinegar and oil [4, 15, 28]. In India the Ayurvedic practice of making inhalation of stramonium and hemp by pipe was developing but may even have had its origins earlier than the ancient Egyptian use of inhalers [16]. In China the burning of opium was developing. In Rome, in the second century, Galen recommended that his breathless patients should go to the foothills of Mount Vesuvius to take the sulfurous air from the volcano. In South and Central America, natives developed the practice of smoking and fashioned ornate pipes from bone in order to smoke tobacco and similar plants. These were perhaps the earliest of the surviving inhaler technologies, dating from about 2,000 years ago (Fig. 3.1).

Rhazes (c850 AD–c923 AD) a physician in Baghdad advocated the inhalation of the vapor of arsenic, working in Baghdad, he wrote, "patients die from consumption because the lungs cannot be treated like external parts." We know from his writing that like the ancient Egyptians he also used leaves thrown onto hot bricks as a treatment [4, 15].

The first book to be written about asthma was by Maimonides (1135–1204 AD), the Spanish-born physician to the Arab King Saladin. Saladin's son was asthmatic and Maimonides was required to treat him. His book, A treatise on Asthma (1190 AD), recommended numerous measures and treatments, including abstention from sexual activity and imbibing chicken soup. Among his treatments, Maimonides also described the inhalation of vapors from throwing herbs onto a fire (Maimonides, edited by Muntuner [25]).

The advent of publishing revolutionized medical practice and with it inhalation. But it was not asthma that led us to the first published inhaler device, instead it was tuberculosis, variously called consumption or phthisis.

3.2 Pulmonary Drug Delivery in the Seventeenth and Eighteenth Centuries

In 1654 Christopher Bennet (1617–1655), an English physician, suffering from tuberculosis wrote Theatri Tabidorum and gave us four wood cut drawings of an inhalation device, with measurements enabling the reader to have their own inhaler made [2]. The treatment was balsam. There is no evidence that Bennet's inhaler was ever manufactured. Bennet succumbed to tuberculosis the following year (Fig. 3.2).

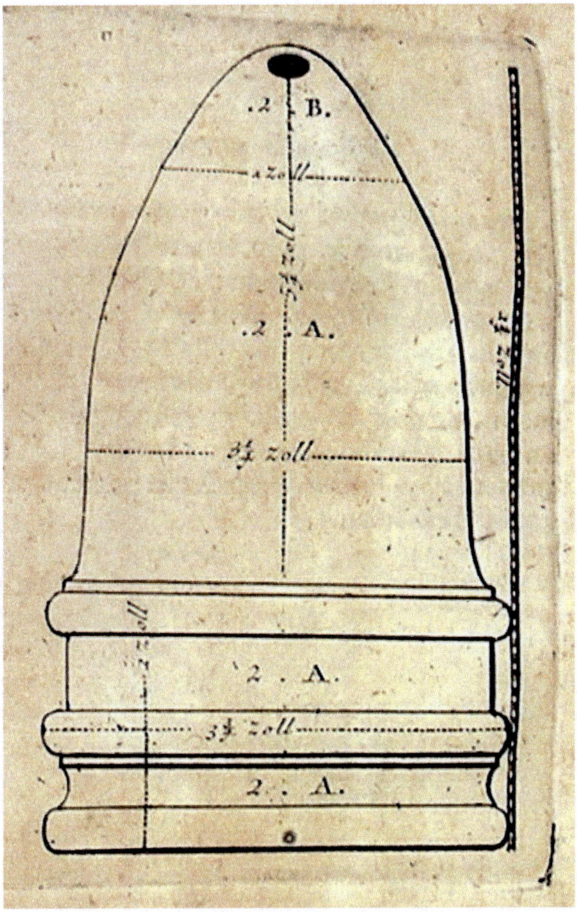

Fig. 3.2 Woodcut of Bennet inhaler

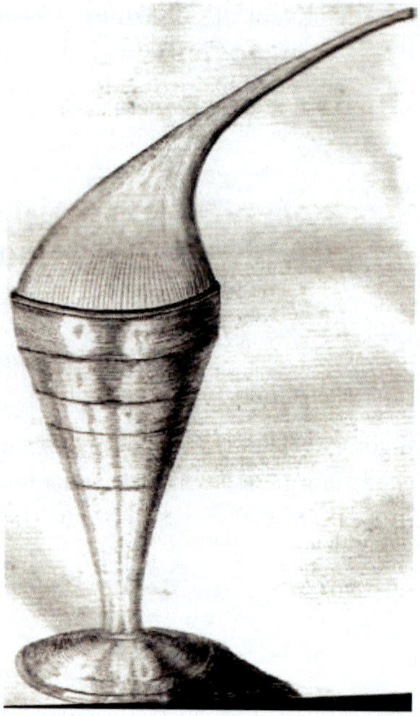

Fig. 3.3 Woodcut of Stern inhaler

Another English physician, Philip Stern, published a small booklet in 1767 which caused him to be alienated by his fellow doctors. His pamphlet, *Medical Advice to the Consumptive and Asthmatic People of England*, was clearly aimed at the patient and not, as was customary, to his peers. He described and illustrated an inhalation device for use with his own proprietary formulation of balsam. His writings rationalize that "the only possible way of applying medicines directly to the lung is through the windpipe." He did not, however, use the word "inhaler," instead he referred to his device as a vessel or urn [44] (Fig. 3.3).

The first use of the word "inhaler" is attributed to John Mudge, brother of the famous horologist Thomas Mudge. John Mudge was a physician and astronomer living in Plymouth. As an astronomer he won the Copley Medal in 1777 for his work on the reflecting telescope, but as a physician he gave us an inhaler which was to be successfully manufactured and sold over a period of 160 years!

Mudge's [29] book, A *Radical* and *Expeditious* Cure for a Recent Catarrhous Cough, describes and illustrates the Mudge Inhaler in precise detail. It was first manufactured by William Barnes, a pewterer, of Fleet Street, London and because it was made from robust pewter there are still examples which survive. The inhaler was an adaptation to a standard tankard, holes were made in the handle and where it joined the tankard to permit air to enter and be drawn through hot water impregnated with the medication. The patient was instructed to inhale and exhale through the device, a small valve in the lid permitted the exhalate to escape. Mudge's cure was elixir paregoricum, a mixture of opium, benzoic acid and camphor. He discouraged the use of his inhaler with other medications as their possible inefficacy might

Fig. 3.4 Mudge inhaler manufactured by Wm Barnes

diminish the perception of his device, but clearly he recognized the broad utility of a portable inhaler device. The Mudge inhaler proved popular and was carried in pharmacies all the way through the Victorian era and was still listed in a pharmacy catalogue in 1938 – over that period of 160 years, and despite Mudge's insistence, his inhaler was used with many different medications [29] (Fig. 3.4).

3.3 Pulmonary Drug Delivery in the Nineteenth Century

In 1802 General Gent, armed with the research of Dr. James Anderson in Madras, returned to England with knowledge and samples of the ancient Indian practice of inhaling the leaves of Datura Stramonium. Gent gave samples to Dr. Sims who successfully trialed it and then described his findings in Monthly Magazine (1812 and later in the Edinburgh Medical and Surgical Magazine) [39]. Gent later died of over-using stramonium. Obtaining precise doses from different Datura Stramonium species was by no means easy. Gradually, this practice became popular during the 1800s and eventually led to a wide range of proprietary powders containing stramonium, belladonna, lobelia and sometimes opium. This practice usually entailed putting a spoonful of powder onto a plate and igniting it. The patient sat leaning over the plate and inhaling the fumes. Later examples suggest the use of a funnel to direct the vapor towards the patient's airways.

Up to this point the medications being delivered were volatile substances, the technology did not exist to force nonvolatile substance to be delivered to the airway. Additionally, many of the descriptions of treatments were not precise or sufficiently detailed to permit others to adopt the treatment.

Fig. 3.5 Humphrey's inhaler

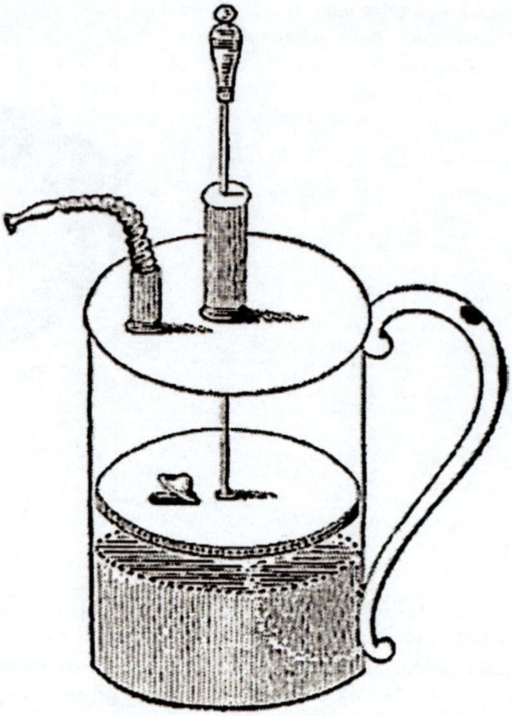

In 1817 Humphrey described a steam-driven inhaler, but there is doubt that it was ever produced. If it was then it was the first of the pressurized inhalers [20] (Fig. 3.5).

In 1834 Sir Charles Scudamore described the inhalation of the vapor of iodine and hemlock for the treatment of consumption and advises that the treatment be given 3 times a day, for 15–20 min using a vapor inhaler with water at 120 F [37].

In 1844 Maddock provided a useful review of the state of knowledge on inhalations and provided an illustration of a glass vapor inhaler in use at the time [24]. At about the same time, inhalation received a huge leap in credibility as the use of inhaled anesthesia became known. With its huge benefits to surgery, inhaled anesthesia sparked a great interest in inhaled delivery across the world.

Work in London had begun on dry powder inhalation in the late 1840s. Dr. T.K Chambers was working on the application of silver nitrate and copper sulfate to the lungs. His solution was ingenious and represents possibly the first use of particle engineering in inhaled delivery [9]. Working without any of the sophisticated equipment we take for granted these days, he used lycopodium spores as a form of carrier, soaking the spores in saturated solutions of copper sulfate and silver nitrate, drying and reducing the loaded spores, presumably by pestle and mortar, to an impalpable powder. He found that 2.5 grains of pollen could be made to contain 1 grain of silver nitrate.

Chambers' work was extended by William Cornell in Boston, who reported in the Boston Medical & Surgical Journal in 1850 that he had found the powder serviceable in bronchitis, laryngitis and other conditions [12]. Chambers' and Cornell's technique involved a glass funnel introduced to the patient's mouth while an attendant dusted the powder into the funnel during inhalation.

The first useful suggestion for the atomization of nonvolatile substances in solution was from M. Auphan at the Spa of Euzet-les-Bains in 1849. With the knowledge that water existed as a spray when it was violently crashed against hard surfaces, e.g., at waterfalls or the sea-shore, he developed the idea that at the mineral spa a jet of water could be directed at a hard surface to create a spray. His method was ado-pted by many spa houses but was unsuitable for private medical practice [1].

Auphan's ideas were developed by another French physician, M. Sales-Girons, who developed a portable inhalation apparatus that applied the same principles to achieve atomization. His work, presented at the Paris Academy of Medicine in 1858, was favorably received. His device comprised: a reservoir for the medication solution, an air pump, a small jet nozzle and an impaction plate. The pump forced the solution through the nozzle where it was atomized against the plate. The Academy considered whether the fine spray was suitable to reach the bronchial tubes and after deliberation they confirmed that it would, and Trousseau stated that "Sales-Girons has rendered a great service to the world at large by his invention of the treatment by pulverisation." The device went through several iterations but was manufactured and marketed [35, 41] (Fig. 3.6).

Still in France, the following year M. Matthieu introduced the Néphogène (anglicized to nebulizer). Instead of directing a jet at a hard surface to cause atomization, Mathieu introduced the idea of using a blast of compressed air into which was dripped the medication and then projected outwards through a fine nozzle. This created a fine but cold spray [27].

Meanwhile in London, in 1860, Henry Hyde Salter wrote a detailed review of asthma, *On Asthma its Pathology and Treatment*. He systematically reviewed the treatment options of the period, including inhaled therapies, and distinguished depressants (ipecacuanha, tobacco, tartar-emetic), stimulants (coffee, alcohol), sedatives (tobacco, chloroform, opium, stramonium, lobelia, Indian hemp). The inhalation of the fumes of burning nitre papers was also discussed by Salter [36].

In Germany in 1862, Dr. Bergson introduced his idea for making Matthieu's approach workable. The idea had originally been suggested to him by Dr. Nathanson based on the principle embodied in Gifford's steam injector. But it was Bergson's efforts that brought this principle to bear in the atomization of medicines for inhalation. Known as the Bergson tubes, his approach was embodied in a device he called the "Hydrokonium." It consisted of two tubes that interfaced at right angles, with one tube taking a feed from a reservoir of liquid medication while the other has a jet of air passed through it. The air jet was generated by a double squeeze-ball arrangement made of an early form of rubber [1] (Fig. 3.7). The publication Deutsche Klinik through the 1860s carried frequent ideas for new inhalation technologies from Drs. Lewin, Man, Nathanson and others.

Fig. 3.6 Sales-Girons' portable atomiser

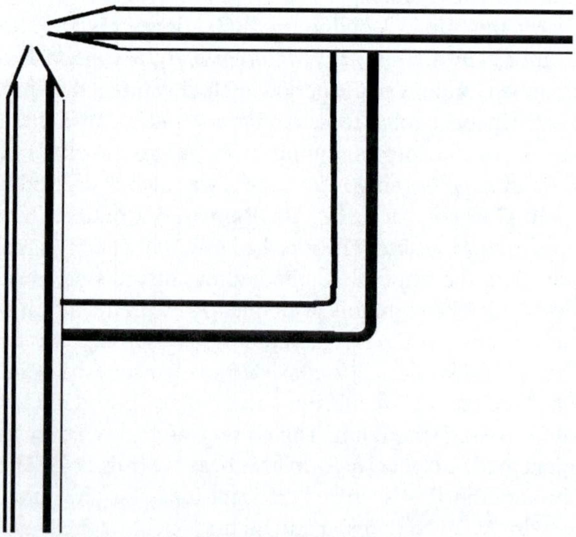

Fig. 3.7 Bergson tubes

Fig. 3.8 Siegle's inhaler

Bergson's approach was further improved by Dr. Emil Siegle of Stuttgart who applied for a patent in April 1864 for a Steam Spray Producer. Instead of relying on a squeeze ball, the jet was created using steam from a small boiler [38]. Dr. H Waldenberg, also in Germany, had also been working with steam and published his ideas in 1862 based on a different approach. The original Siegle inhaler was far from perfect and an improved boiler arrangement was proposed by Dr. Adams of Glasgow in 1868. Siegle-based inhalers were popular in many countries for many years (Figs. 3.8 and 3.9).

The use of the Bergson tubes with a double rubber squeeze-ball arrangement was improved and popularized in Britain by Dr. Andrew Clark, being manufactured by Krohne and Sesemann in 1865. This manufacturer worked with several doctors at the time to produce devices, including Dr. Richardson, whose device was applied to anesthesia [22].

Inhalation was a topic of much interest at this time, but perhaps none were so enthusiastic as Dr. Broadbent in his 1862 book "A Medical Treatise on the Causes and Curability of Consumption, Laryngitis, Chronic Catarrh and Diseases of the Air Passages – Combining the Treatment by Inhalation by Medicated Vapours." His lists of inhalants included "iodine, conium, oxygen, hydrogen, nitrogen, nitrous gas, a great variety of gums and balsamic resins, vapor of boiling tar, hydrocyanic acid, camphor, ammonia, balsam tolu, naphtha, chlorine, hyoscyamus, lactuca, belladona, digitalis, colchicum, a great variety of balsamic herbs, galbanum, vapor of vinegar, nitre, stramonium, lobelia, inflata, ipecacuanha, alcohol, hydriodate of potassa, storax, marshmallows, rose water, a great variety of emollient and narcotic herbs, etc." [6].

Fig. 3.9 Adam's inhaler

In 1864 another step forward with dry powder inhalation took place. Alfred Newton applied for a patent in London after correspondence with Dr. James in New York. The dry powder device resembled a mantle clock with an orifice at one side, inside was a mesh and a crank shaft with feather beaters. Powder was placed into the bottom of the box and the patient would put their mouth to the orifice and inhale while the handle was cranked causing the feathers to beat the powder creating a dust, which would pass through the mesh towards the mouth. It was not especially portable, but Newton recognized that the powder needed to be pulverized into a fine dust and that it should be kept dry, two essentials of modern dry powder inhalers. The device was designed for the delivery of potassium chlorate which today is recognized as a lung irritant [31] (Fig. 3.10).

Dry powder inhalation was very much in vogue at this time, though frequently for laryngitis rather than deeper lung deposition. Its exponents included Trousseau in Paris using a glass tube, Burow in Germany using an adapted steel pen, the same method employed by Pserhofer in Vienna and Ebart in Berlin, a glass retort was adopted by Lewin in Berlin, a gum squeeze-ball was used by Rauchfuss [41]. In Baltimore, Prof. Thomas used a technique developed by his brother to make fine

Fig. 3.10 Illustration from Newton's patent for dry powder inhaler

cinchona powder by rubbing the material against a rotating grind wheel as the patient sitting nearby inhaled [45].

The Lancet in 1865 carried news of a new device, known as the improved Dr. Nelson inhaler. It was an earthenware vapor inhaler with an air inlet and a mouthpiece. Originally, the medication was to be impregnated on a sponge in the mouthpiece but later it was used without the sponge, the medication being put into the hot water. These ceramic inhalers were produced in numerous different and attractive forms. Different pharmacy wholesalers commissioned their own models and sold them to pharmacies. Today, they are regarded as highly collectable. The Nelson inhaler is still available through pharmacy suppliers, over 140 years later [23] (Fig. 3.11).

An important landmark in 1867 was the inclusion of inhalants in the British Pharmacopeia for the first time. Five inhalants were listed: vapor acidi hydrocyanici, vapor chlori, vapor coniae, vapor creasoti and vapor iodi. This reflected the rising importance and popularity of inhaled medicine [5]. Overshadowing this landmark was the inclusion of inhalations in 1864 in the American "Medical Formulary" [14].

Fig. 3.11 Nelson inhaler

Also in 1867, in France, Trousseau wrote about stramonium, by which time as well as being smoked in pipes it was also being made into asthma cigarettes. Trousseau provides us with his formula which includes belladonna and opium in addition to stramonium [47]. The practice of smoking arsenic cigarettes for asthma had also been reported in the 1860s [21] (Figs. 3.12 and 3.13).

In Philadelphia in the same year, Da Costa wrote, *On Inhalations*, reviewing the treatment and the available technology. In commenting on doses of medicines for inhalation he observed, "… the dose varies with the apparatus; or rather, though the dose be the same, to obtain that dose in the mouth we may have to use more of the medicated fluid or a stronger solution with one atomiser than with the other" [13]. This issue of matching dose is one that continues to confound those developing generic inhaled products.

Also in Philadelphia, in the same year, Solis Cohen produced his *Treatise on inhalation of Gases, Vapor, Nebulized Fluids and Powders*, giving a thorough review of the state of the art at that time, including his own apparatus which incorporated a gravity-fed adaptation to the Bergson tubes [41].

In 1869 the German doctor Louis Waldenburg, writing about tuberculosis, claims to have completely cured asthma though the continued use of inhalations. He

Fig. 3.12 Asthma powders

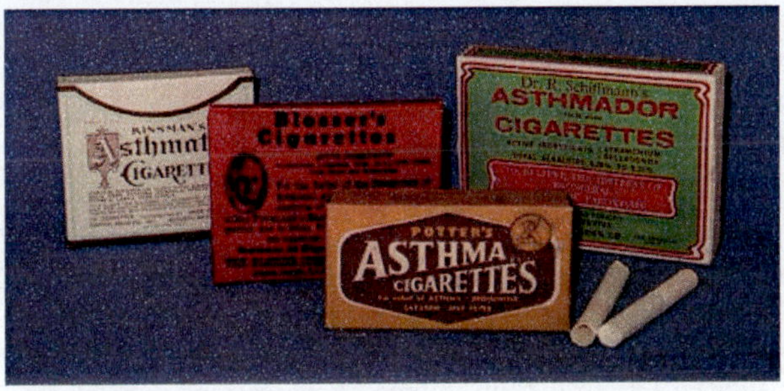

Fig. 3.13 Asthma cigarettes

employed the use of sodium chloride inhalations to soften the viscid mucus and oil of turpentine to stimulate the mucus glands [3].

In 1872 Joseph Lister, famous for his work on antiseptic surgery, adopted an adaptation to the Siegle steam spray as his preferred method of delivery. A review by George Beatson in 1880 describes the spray producers used in antisepsis and provides a good history of spray producers [1].

Berkart, writing in 1878, "On asthma: its pathology and treatment" identified the use of turpentine and assafoetida as a mucolytic, and commented how they were best administered as inhalations. He also describes a "spray-producing apparatus" and expresses some disappointment, which he rationalizes as the spray not always helping the deeper structures of the lung [3].

By 1884 Dr. G Evans, having also recognized the problem of large droplets sometimes being produced from Bergson tubes, set about inventing a device to overcome this. His work published in the New York Medical Journal in 1886 showed a more sophisticated arrangement that prevented larger droplets being emitted [15].

This had been a time when many new inhalation devices were being developed, the availability of new materials, such as hard rubber, created many possibilities. Asthma powders (stramonium and belladonna) in attractive tin packaging were widely available. Society showcased its inventions in grand exhibitions with prizes awarded for inventiveness. Advertising of these inhalers and inhalants had started to appear, and unrestricted claims were made as to the curing properties of various medications. The advertisements frequently incorporated testimonials from various dignitaries, but it came to a crunch point when one company offered a reward if their inhaled treatment did not work. The company was The Carbolic Smoke Ball Company (Fig. 3.14).

In 1889 Frederick Roe patented a dry powder inhaler known as the Carbolic Smoke Ball. It was a squeeze ball made of India rubber filled with fine carbolic acid powder. The Carbolic Smoke Ball Company promoted the inhaler with advertisements claiming it to be a preventative for influenza and offering a reward to anyone who despite using the product correctly caught influenza. They had no intention of paying any such reward, it was simply a statement of confidence. Mrs. Calill used the Carbolic Smoke Ball and sure enough caught influenza and claimed her reward. The company denied her claim and she took the company to court. The company's defence that the claim was just sales "puff" was thrown out and Mrs. Calill won in a landmark case that established the basis of contract law. Ironically, Mrs. Calill went on to live a long life and eventually died in the 1940s with influenza as a contributory cause [40] (Fig. 3.15).

The days of reckless advertising were numbered but in the early years of the 1900s one of the most curiously named inhalers was patented in the USA, manufactured and advertised. Its inventor Dr. Worst gave his name to the inhaler. The Dr. Worst Inhaler was a refillable metal canister for the inhalation of menthol, incorporating an innovative idea, a tethered mouthpiece. A couple of years later the Dr. Worst Inhaler was being marketed with a new name – perhaps recognizing the difficulty in selling a product with an unfortunate name (Fig. 3.16).

The world was also changing for the physician, as increasingly the scientist and technologist were moving towards the forefront of inhaled delivery and the larger pharmaceutical companies were beginning to emerge. Scientific techniques were making big strides too.

In the late 1890s Prof. John Abel, at John Hopkins University, isolated a hormone from the suprarenal glands of sheep. The substance he called "epinephrine" was inactive, but later found to open airways, raise blood pressure and heart rate. He published his results in 1898 [19]. Working independently in Baltimore, Jokichi Takamine isolated the same hormone in an active form which he called "adrenalin" [19].

In short order Parke Davis & Co marketed adrenalin for a variety of ailments. At that stage it was given by injection.

Suprarenal material includes not only adrenalin but also cortisone, and in 1898 Sir William Osler administered a crude preparation of adrenal cells to someone with

Fig. 3.14 Advertisements for inhalers

Addison's disease [32]. Many years later, the use of refined cortisone in Addison's and many other disease become commonplace.

In 1899 and 1900 Solis Cohen also published papers on suprarenal substance in hay-fever and asthma, but it took a further 50 years before the benefits of purified steroids could be shown in asthma [42, 43].

Fig. 3.15 Carbolic Smoke Ball advertisement

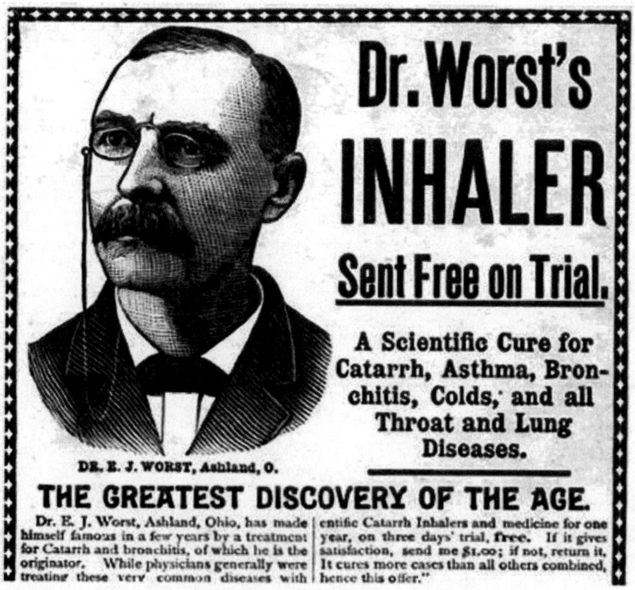

Fig. 3.16 Dr. Worst's inhaler advertisement

3.4 Pulmonary Drug Delivery in the Twentieth Century

1903/1904 Kaplan and Bullowa working in New York identified the value of hypodermic administration of adrenalin in asthma – it went on to become the foundation of what we can consider modern pharmaceutical inhalation therapy [8]. Pick, in 1911, reported successful nebulization of adrenalin in two patients and it rapidly became standard therapy, administered by hand-held nebulizer [34].

Atropine was also being explored in asthma at this time, von Terray claimed atropine hypodermically given to be a specific treatment for asthma, Weiss found good results from spraying it through the mouth.

Controls on advertising and promotion imposed by the FDA in 1906 sounded a death knell for the entrepreneurial quack cures and the scientist took over.

Sir Hiram Maxim, an American, living in London, was an inveterate inventor. Most famous for the Maxim machine gun, he also invented a steam-driven flying machine! In August 1909, plagued by the effects of the London atmosphere, he developed and patented his own inhaler. It resembled a glass retort and delivered a combination of menthol and wintergreen essence (dirigo), and was known as the Pipe of Peace. It was an early example of combination therapy (Fig. 3.17).

In 1914 Osler claimed success with injected pilocarpine (anticholinergic) in asthma [33]. Atropine had already been synthesized in 1901 and was a treatment option. In the 1920s the first mention of the inhalation of atropine from an atomizer

Fig. 3.17 Maxim's pipe of peace

was made by Hurst. In 1924 Schmidt and Chen investigating Ma Huang at Peking Union Medical College identified ephedrine – a substance similar to adrenaline but effective by mouth and longer-acting [10]. In 1926 Thomas, at the University of Pennsylvania, having received samples of ephedrine from Chen and Schmidt, described the successful use of ephedrine in cases of asthma [46].

Not only was inhalation being seen as a treatment option for respiratory conditions, but also as a whole new delivery option for other systemic medications. In 1924 first use of inhaled insulin was reported by Heubner, Jongh and Laqueur. Shortly after, in 1925, Gansslen demonstrated that inhaled insulin lowered blood glucose in five diabetic subjects [17]. Later, in 1936, insulin was just one of many possible inhalants on offer at the London Inhalatorium, for delivery by a Spiess Drager, pressurized gas-driven, nebulizer (originated in 1902, by Prof. Spiess of Frankfurt). The use of adrenalin nebulized in an oxygen stream was described by Camps in 1929.

The early 1930s saw the advent of the electric compressor nebulizer. Manufactured by Weil in Frankfurt, the 110–220 volt machine was supplied in UK by Riddell to nebulize bronchovydrin (papaverine and eumydrine) (Fig. 3.18).

In 1940 isoprenaline was first described by Heribert Konzett at the University of Vienna, but was only introduced commercially in 1951 [30]. An innovative dry powder inhaler was developed by Mack Fields in Chicago. Working for Abbott he developed the Aerohalor. It used small "sifter cartridges" of a lactose-based formulation. When the patient inhaled through the device, a ball bearing would bash the sifter cartridge and shake a little of the fine powder out, through a fine mesh, into the airstream. In 1948 Abbott launched the Aerohalor with a preparation of Penicillin for inhalation and a further presentation with the bronchodilator norethisterone was also made available (Fig. 3.19).

Fig. 3.18 Pneumostat electric compressor nebulizer

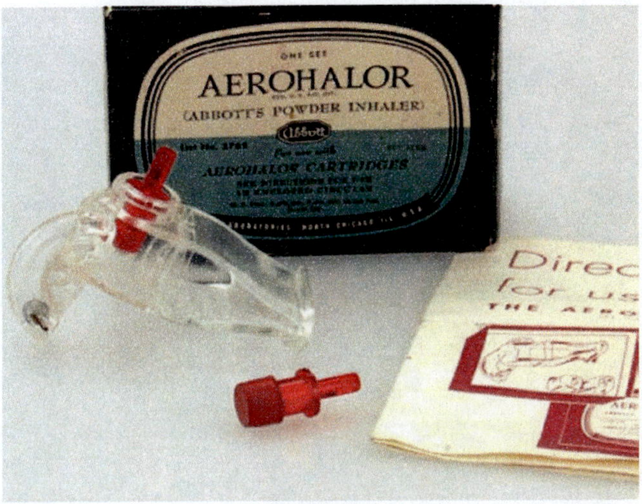

Fig. 3.19 Aerohalor

In the late 1950s development of the ultrasonic nebulizer began. Despite initially being large units, they have been successfully miniaturized such that modern ultrasonic nebulizers are readily portable. In 1950 Reeder and Mackay treated pneumonia with inhaled corticosteroid, leading to much investigation of inhaled anti-inflammatories. Gelfrand, in 1951, reported five patients with nebulized cortisone [18]. In 1955 Foulds used hydrocortisone in powder form. This work was to ultimately lead years later towards the important discovery of beclomethasone.

Undoubtedly, the most significant event of the 1950s was the development in 1955 of the pressurized metered dose inhaler (pMDI). Charles Thiel working at Riker (3 M) was one of the key workers in the team that developed cold-fill pMDIs using special valves developed by Philip Maschberg. The products were shown to be effective in studies conducted by Dr. Carr at the Veterans Administration Hospital (Long Beach, California) in June 1955, and New Drug Applications filed in January 1956 were approved in March the same year. Medihaler-iso (isoprenaline) and Medihaler-epi (adrenalin) were launched in late March 1956. Pressurized MDIs have become a very important inhalation technology with annual sales currently in excess of 400 million units. The drugs, of course, have been updated and the propellant technology improved through the use of less environmentally damaging hydrofluoroalkanes. The use of pMDIs has, however, been limited by several drawbacks; the coordination of breathe and activation has proved difficult, the inhalers have lacked dose-counters and delivery of large doses is limited, but those limitations have fuelled other innovations, such as the use of spacers and breath-actuated technologies (Fig. 3.20).

Bengers Laboratories (subsequently acquired by Fison's Pharmaceuticals) were engaged in testing the anti-allergic properties of synthetic cromone derivatives of khellin (an Egyptian plant) during the early 1960s. Dr. Roger Altounyan, the real-life Roger of Swallows and Amazons fame, himself an asthmatic, personally tested

Fig. 3.20 MediHaler advertisement

the compounds. A former spitfire pilot and very familiar with aerodynamics, he invented the Spinhaler (patented 1963), a capsule inhaler that used a small propeller to create turbulence. It was necessary to find alternatives to the standard pMDI technology because the dose to be delivered exceeded the capability of pMDI metering valves. Intal (sodium cromoglycate) was launched in 1967 with the Spinhaler. Some years later a pMDI version was created, in fact using a lower dose. The following year, 1968, the remarkable bronchodilator actions of Salbutamol were described by Lunts, working at Allen and Hanbury's. Having been first synthesized in 1966, it was marketed in 1969 as Ventolin, and rapidly became the most prescribed bronchodilator globally. Salbutamol was the first of the selective bronchodilators, relatively free from secondary effects on other systems. Salbutamol was formulated in a pMDI, as a DPI in Rotahaler and then Diskhaler, as well as in nebulizer, tablet and syrup presentations. The Rotahaler, a dry powder capsule inhaler, simply required twisting for a loaded capsule to be pulled apart and the powder inhaled through a mesh. The Diskhaler, represented an important step forward, because did not require reloading for every dose, it was the first multiunit dose dry powder inhaler. This inhaler used a cartridge with foiled sealed premeasured doses that were perforated at the time of use to permit the drug to be inhaled.

In 1972 Allen & Hanbury's marketed the first inhaled steroid, beclomethasone, in the same inhalers as salbutamol. Numerous steroids have subsequently been developed, and some have led to new inhaler developments. The Allen & Hanbury's technology was refined and substantially improved in the Diskus (Accuhaler), which has proved outstandingly popular and was presented with both bronchodilator, anti-inflammatory drugs and combinations of the two.

In 1980 Astra Zeneca launched a novel steroid, budesonide, in a new multidose dry powder inhaler that did not depend on each dose being isolated, but rather the drug substance kept in a reservoir from which individual doses were taken at administration. This device, the Turbuhaler, the work of Kjetil Wetterlin, also proved very popular and similarly has been presented with bronchodilators, a steroid and combinations thereof. Subsequently, new asthma dry powder developments have followed these three themes described above: the capsule inhalers, the multiple unit-dose inhalers and the reservoir devices.

In 1986 ipratropium bromide was introduced, revitalizing treatment with anticholinergics that had first begun with stramonium inhalation. In dry powder, pMDI and nebulized forms it created further therapeutic options.

Throughout the 1990s, the tantalizing scope for inhalation as a means of delivering proteins and peptides was being explored. Formulation technologies that permitted stabilization of the material and aerosolization of what were often quite large payloads were being developed. As a delivery route that avoided gastric enzymes and first pass metabolism, it was an appealing option for sensitive and often highly expensive biological drugs. Additionally with the prospect of rapid access to the brain, it was seen as a highly prospective route for pain relief. The introduction of a modern inhaled insulin product was seen as somewhat of a test case for inhaled delivery outside of respiratory medicine. Unfortunately, the commercial results of the Pfizer Exubera product were disappointing and this setback has had a limiting impact on those seeking to develop inhaled forms of systemic drugs, but there remains a number of pipeline endeavors that may yet restore the enthusiasm for inhaled systemic delivery.

As we have progressed through the history of inhalation we see several themes emerging:

(a) The inventor moves from the independent (often afflicted) physician to the scientist/technologist working as part of a large team.
(b) Many early inhalers are based on adaptations rather than on design, largely because the independent physician lacked design skills and know-how. Today, good design is at the core of inhaler development.
(c) Materials have extended possibilities – the advent of India rubber and plastics have opened new opportunities that have, in turn, necessitated the formation of specialist functions within inhaler development.

References

1. Beatson G (1880) On spray producers as used in lister's antiseptic system. Edinburgh: MacLachlan & Stewart, pp 1–43
2. Bennet C (1654) Theatri tabidorum vestibulum: seu exercitationes dianoeticae cum historiis et experimentis demonstratives. London: Newcomb, pp 1–126
3. Berkart J (1878) On asthma: its pathology and treatment. London: Churchill, pp 1–264
4. Brewis R (1991) Classic papers in asthma, vol 2. Science Press, London, pp 1–214
5. British Pharmacopoeia (1867) London: Spottiswoode & Co. pp 1–434
6. Broadbent C (1862) A medical treatise on the causes and curability of consumption, laryngitis, chronic catarrh, and diseases of the air passages. Combining the treatment by inhalation of medicated vapors. Damrell Welch, Boston
7. Bryan C (1930) The papyrus ebers. London: Geoffrey Bles
8. Bullowa JJM, Kaplan DM (1903) On the hypodermic use of adrenalin chloride in the treatment of asthamatic attacks. Med News (NY) 83:787
9. Chambers TK (1848) The Lancet. London.
10. Chen K, Schmidt C (1924) The action of ephedrine, the active principle of the chinese drug Ma Huang. Pharm Exp Ther 24:339–357
11. Chen K, Schmidt C (1937) The action of ephidrine: the active principle of chinese drug Ma Huang. JAMA 109:1712
12. Cornell W (1850) Boston Med Surg J
13. Da Costa J (1867) Inhalations in the treatment of disease of the respiratory passages particularly as effected by the use of atomized fluids. Lippincott, Philadelphia
14. Ellis B (1864) The medical formulary. Philadelphia, pp 109–112
15. Evans G (1891) Notes on respiratory therapeutics. Brooklyn Med J 12:787–789
16. Gandevia B (1975) Historical review of the use of parasympathetic agents in the treatment of respiratory disorders. Postgrad Med 51(suppl 7):13–20
17. Gansslen M (1925) Uber inhalation von insulin. Klin Wocchenschr 4:71
18. Gelfand M (1951) Administration of cortisone by the aerosol method in treatment of bronchial asthma. N Engl J Med 245:293
19. Goldstein DS (2006) Adrenalin and the inner world. John Hopkins University Press, Baltimore
20. Humphrey J (1819) Letter to Dr T. C. James. Eclectic Rep Anal Rev 9(2):259–260
21. Julius F (1862) On arsenic smoking in asthma, the retrospect of practical medicine and surgery. Part XLIV January
22. Krohne and Sesemann (1866) Letter in: British Med Feb 17:191

23. Lancet (1863) Dr Nelson's improved earthenware inhaler. The Lancet Feb 11 1863:152
24. Maddock AB (1844) Practical observations on the efficacy of medicated inhalations in the treatment of pulmonary consumption, asthma, bronchitis, chronic cough and other diseases of the respiratory organs. Simpkin and Marshall, London
25. Muntner S (1963) Treatise on asthma by Maimonides. Philadelphia: Lippincott, pp 1–115
26. Mann RD (1984) Modern drug use. An inquiry on historical principles. MTP, Lancaster
27. Matthieu M (1859) Gazette des Hôpitaux. Paris p 297
28. McFadden ER Jr (1986) Inhaled aerosol bronchodilators. Williams & Williams, Baltimore
29. Mudge J (1778) A radical and expeditious cure for a recent catarrhous cough. London: Allen, pp 1–252
30. Nerbrink O (2003) A history of the development of therapy by jet nebulisation. In: Gradoń L, Marijnissen J (eds) Optimization of aerosol drug delivery, p 18
31. Newton A (1864) Inhaling Apparatus UK patent 1161
32. Osler W (1898) The principles and practice of medicine. D. Appleton & Co., New York
33. Osler W, McCrae T (1914) Bronchial asthma. In: The principles and practice of medicine, 8th ed. D. Appleton & Co., New York, pp 627–631
34. Pick L (1911) The action of adrenalin preparations. Bronchial asthma. Medizinische Klinik; No. 5
35. Sales-Girons (1858) Thérapeutique Respiratoire: Traité Théorique et Pratique des Salles de Respiration Nouvelles à l'eau Minérale Pulvérisée, pour le traitement des Maladies de Poitrine. Paris: Adrien Delahaye, pp 1–194
36. Salter H (1860) On asthma: its pathology and treatment. John Churchill, London
37. Scudamore C (1834) Cases illustrating and confirming the remedial power of the inhalation of iodine and conium in tubercular phthisis and various disordered states of the lungs and air passages. London: Longman, pp 1–227
38. Siegle EA (1864) Steam spray producer for inhalation. Patent April 22nd. Kroner, Stuttgart
39. Sims J (1812) Datura stramonium or thorn apple as a cure or relief of asthma. Edinburgh Med Surg J 8:364–367
40. Snell N (2001) The carbolic smoke ball. Int J Pharm Med 15:195–196
41. Solis Cohen J (1867) Inhalation: its therapeutics and practice. Philadelphia: Lindsay & Blakiston
42. Solis Cohen S (1899) A preliminary note on the treatment of hay-fever with suprarenal substance: with a report of personal experience. Philadelphia Med J 2:341–343
43. Solis Cohen S (1900) The use of adrenal substance in the treatment of asthma. JAMA 34:1164–1166
44. Stern P (1767) Medical advice to the consumptive and asthmatic peoples of England. London: Almon
45. Thomas RH (1855) Transactions of the American Medical Association, vol. Viii, p 591
46. Thomas WS (1926) Ephedrin in asthma: a clinical report. Am J Med Sci 171:179
47. Trousseau A (1867) Lectures on clinical medicine delivered at hotel-Dieu, Paris. Vol. 1 London: Hardwicke, pp 1–712

Chapter 4
The Physics of Aerosol Droplet and Particle Generation from Inhalers

Zhen Xu and Anthony J. Hickey

Abstract The three major aerosolization systems comprised of nebulizers, propellant based systems, and dry powder inhalers, each has unique physical principles. A good understanding of these physical principles is required for the successful development. Therefore, in this chapter, several important empirical and scientific principles that include the particle generation, the formulation implication, and drug aerosolization mechanisms of each aerosolization systems are highlighted. Examples of particle engineering to achieve aerodynamically favorable dry powder aerosols are briefly discussed. Next generation inhaler products will be expedited if these physical mechanisms are understood.

Keywords Aerosol droplets • Aerosol particles • Dry powder inhalers • Nebulizers • Propellant driven metered dose inhalers • Spray drying and mixing

4.1 Introduction

4.1.1 Background to Inhaler Technology Overview

Drug delivery by the pulmonary route offers many advantages with respect to oral or parenteral routes [1]. Briefly, these include local targeting, eliminate of first pass metabolism, and rapid absorption. An aerosol is defined as a collection of solid or liquid particles suspended in a gas. Therapeutic aerosols are directly delivered into the respiratory tract for local or systemic action. Although there is a very long history of anecdotal use of inhaled drug therapy, the commercialization of inhalation

A.J. Hickey (✉)
Eshelman School of Pharmacy, University of North Carolina,
Chapel Hill, NC 27599-7360, USA
e-mail: ahickey@unc.edu

H.D.C. Smyth and A.J. Hickey (eds.), *Controlled Pulmonary Drug Delivery*,
Advances in Delivery Science and Technology, DOI 10.1007/978-1-4419-9745-6_4,

products with manufacturing standards and stringent test procedure occurred in the last 50–60 years [2].

During the development of pharmaceutical inhalation technologies, three distinct categories of delivery system were evolved gradually including nebulizers, propellant driven metered dose inhalers (pMDIs), and dry powder inhalers (DPIs) [3].

Among them, nebulizer has the longest history of use for delivery of therapeutic agents [4]. Nebulizers can be defined in terms of an aqueous solution or suspension formulation, an aerosol generating device, and an energy source (Fig. 4.1a). Unlike pMDIs and DPIs, the major components (solution/suspension formulation and device) of nebulizers are developed and marketed independently. The solution formulation conforms to requirements for parenteral products with appropriate pH, tonicity, and other physical properties. Liquid systems provide ease of filling and flexible dosing. Polydispersed droplet particles are generated from a drug solution by either compressed air or ultrasonic-assisted atomization mechanisms [5]. A baffle plate is typically installed to remove large droplet and enhance the production of smaller droplets. The aerosol mist is typically generated and inhaled for 10–20 min based on the rate of nebulization and the dose regimen. The extended period of time allows for a higher dose than pMDIs and DPIs. As nebulizers depend less on patient compliance than pMDIs and DPIs, they are often being used for hospital emergency care and/or pediatric patients.

Modern pMDIs (Fig. 4.1b) were first developed in the mid-1950s for the anti-asthma therapy [2]. pMDI formulation consists of a drug solution or suspension of a high vapor pressure propellant (typically chloro- or hydrofluorocarbon), which is liquefied in a canister. A small amount of hydrophobic surfactant including oleic acid, lecithin, or sorbitan trioleate has been added to stabilize the formulation. The device is composed of a metering valve in the canister, and an actuator mouthpiece. Upon actuation, a metered dose is emitted through an actuator orifice where a plume of heterogeneous droplets is released. The vapor pressure of the propellant equilibrating with atmospheric pressure provides the energy for shear thinning and evaporation, two mechanisms of droplet formation [6]. pMDIs are small, portable, and contain multiple doses for repeated medication (typically >100 actuation). As the phase-out of ozone-depleting propellant (implementing of the Montreal Protocol), they are being reformulated with propellant such as HFA-134 and HFA-227 [7]. The same reason also spurs the development of DPIs.

Modern DPIs first appeared in the 1960s and 1970s [2]. The first commercial DPI was the Spinhaler® for the delivery of disodium cromoglycate to treat asthma [8]. As the advances of powder technology and device design occurred, more sophisticated passive and active DPIs appeared in the market. DPIs offer an alternative to pMDIs for aerosol therapy. It has been predicted that they may supersede pMDIs as the most popular and numerous devices [9]. All DPIs contain three basic features including a dry powder formulation, a drug aerosolization mechanism, and a metering system (Fig. 4.1c) [10]. The formulation contains either micronized drug alone or a physical mixture with a carrier powder, usually α-lactose monohydrate. The carrier functions primarily to increase powder flow, facilitate drug aerosolization, and act as a bulking agent to increase volume for uniform capsule filling. The drug will be more stable in powder form. Unlike pMDIs, the drug aerosolization

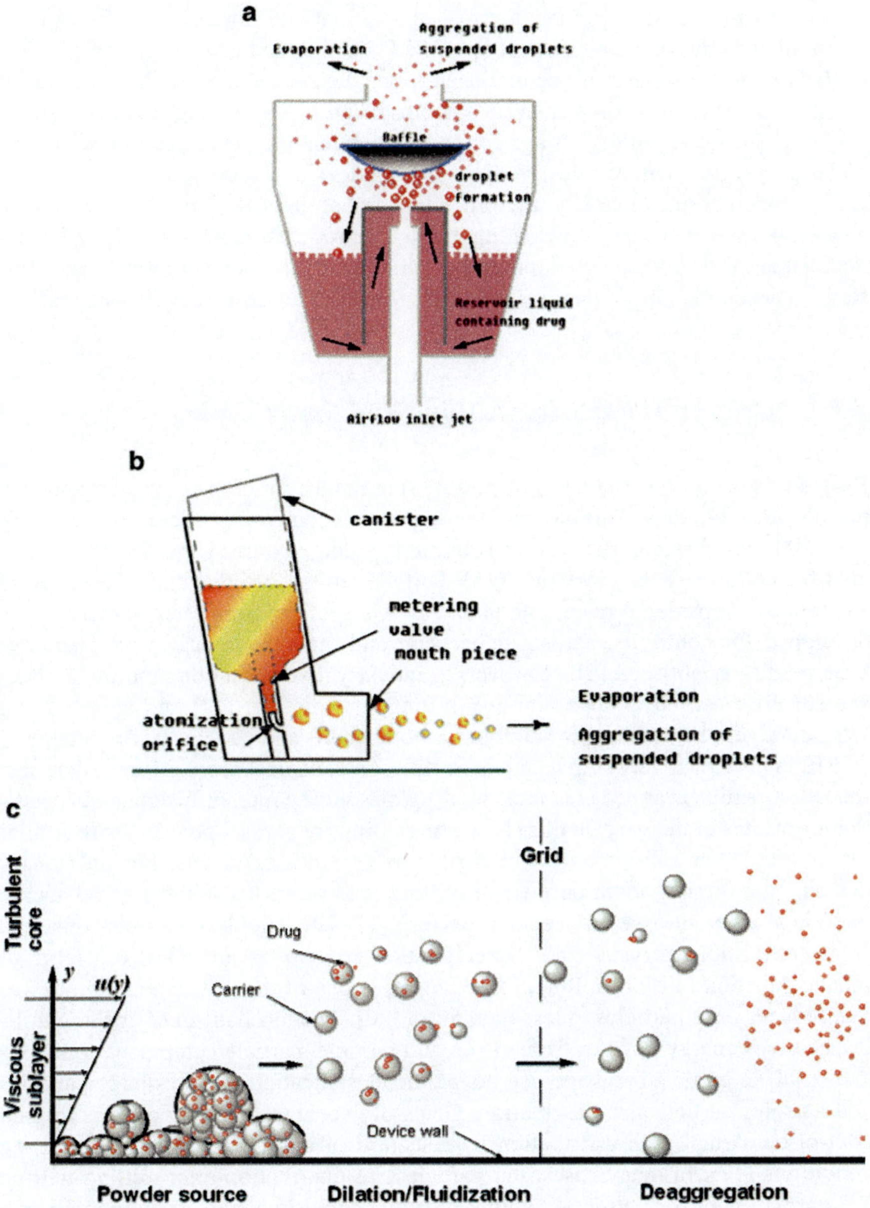

Fig. 4.1 Schematic diagrams showing key features of: (**a**) nebulizer, (**b**) pMDI, and (**c**) DPI

from DPIs does not rely on a liquefied propellant, so it is environmentally friendly and requires minimal coordination. For all currently marketed DPIs, the energy required to disperse the powder mixture is provided by the patient's inspiration. Drug aerosolization mechanisms include aerodynamic forces (drag and lift); inertial forces (impact, vibration, and rotation); and shear and friction forces [11].

As the innovation continues, a range of efficient and portable nebulizers [12], reformulated environmental friendly pMDIs [13], and next generation DPIs [9, 14] are being developed and will be marketed in the next decade. Nevertheless, optimization of the inhalation drug delivery systems is a time consuming task because the overall performance of the dosage forms is dependent on a combination of the formulation development, the design of the dosing device, and the proper aerosol generation mechanism. In reality, the procedure for product design relies heavily on empirical strategies based on existing marketed products. However, years of accumulated aerosol product development has produced valuable theoretical bases that serve to accelerate the development of advanced aerosol drug delivery systems.

4.1.2 Origins of Particles: Milling and Spray Drying

Bulk drugs must be prepared in respirable size ranges for aerosol powder and suspension formulations. Particle engineering with geometric particle sizes ranging from 100 nm (ultrafine particles) to 100 μm (e.g., large porous particles) are considered for pulmonary drug delivery [15]. Particles in the aerodynamic size range of 1–5 μm are considered respirable [16]. A variety of research methods have been developed for controlled aerosol generation and their theoretical principles have been readily established [5]. However, in industry, two most common methods of aerosol drug particle preparation are milling (destructive) [17] and spray drying (constructive) [18], which are suitable for both small- and large-scale processing.

Milling is a relatively simple and economical mechanical process for particle size reduction, which is aimed at enhancing drug dissolution rate, diffusion, and absorption from sites of delivery. In this sense, the milling for aerosol powder share similar intent with other solid dosage forms such as capsules or tablets. For pulmonary delivery, a more important purpose of milling is to achieve a suitable aerodynamic particle size for site-specific aerosol delivery [17, 19]. The drug particles obtained from conventional crystallization/precipitation techniques are often too large to inhale. Attrition (air jet) milling, for example, is one of the fundamental processes that reduces drug particles to less than 5 μm [20]. The mechanism of attrition milling involves energy transfer from air jet and particle–particle/equipment collisions that result in particle breakage. Aerodynamic classification removes large particles and narrows particle size distribution. Generally speaking, milling confers limited control of particle size distribution by selection of appropriate mills and milling conditions [17]. In many cases, the particle size distribution after milling will be log-normal, but distribution of varying breadth or multimodal distribution will also be possible [19]. Significant energies are expended in the milling process, most of which are dissipated as heat [17]. Milling is accompanied by surface area and molecular disorder increases, and potentially surface charge buildup, which may cause particle agglomeration. Some specialized milling process can be applied to overcome these problems. For example, milling at elevated humidity or wet milling may facilitate heat dissipation and in situ recrystallization [21]. As processing technologies advance, milling may achieve more controlled and predictable results.

Spray drying is another important method of generating fine particles [18, 22–24]. The spray drying techniques for producing micron-sized powders for pulmonary drug delivery has been well recognized and enormous effort has been made since 1990s [21]. A pressurized drug solution or suspension is first atomized to a spray form and then in contact with a hot drying airstream, resulting in rapid evaporation and drying of the droplet to form solid particles. Coarse particles remain in the main chamber, while most of the fine particles are collected by cyclone separator. Compared to milling, spray drying offers more control over the physicochemical properties of powders because of the diversity of manipulatible operation variables in the atomization (e.g., nozzle type and position) [4, 25], drying dynamics (e.g., airflow rate, droplet evaporation rate) [18, 23], and liquid properties (e.g., viscosity, surface tension) [18, 23]. These operation variables can be generated through interrelated parameters such as the inlet air temperature, the drying air flow rate, the liquid feeding rate, and the pressure of atomizing air [23]. For example, the diversity of particle morphology can be controlled by varying the drying dynamics [18]. Inhalation aerosols with good aerodynamic properties (e.g., flowability, dispersibility) can often be generated without excipients, while co-spray drying with a wide range of excipients adds more flexibility for particle engineering of this technique [22]. Spray drying also finds its use in heat labile materials such as protein or peptides, because of the cooling effect during rapid evaporation and short residence time [26]. The range of applications of spray drying is broad. Novel spray dried formulations such as from nanosuspensions and microemulsions are being developed for industrial processing [24]. A related technology, spray freeze drying, offers a drying alternative for compounds with low glass transition temperature [27–29].

Other methods of manufacture of inhalation aerosol particles include, but are not limited to controlled crystallization/precipitation [30, 31], specialized milling [19], supercritical fluid technologies [32, 33], etc. Each particle generation process has its advantages for producing desirable pulmonary drugs and drawbacks in controlling powder properties. In the next step, the pulmonary drugs are formulated into different dosage forms for further processing. The fundamental aspects of sprays and powders before and after aerosolization will be discussed below (Sect. 4.2). Some particles with small aerodynamic sizes such as low density or elongated particles will be discussed in Sects. 4.3.4 and 4.3.5 of this chapter, respectively.

4.2 Sprays (pMDI, Nebulizer)

For the solution/suspension formulations used for nebulizers and pMDIs, the first requirement is the correct dose of each ingredient for the desired function and their mutual physicochemical stability and compatibility. The second requirement is the efficient and reproducible aerosol droplet formation. The formulation parameters controlling droplet production include the viscosity, surface tension, and vapor pressure of the liquid.

4.2.1 Solution/Suspension Formulation

4.2.1.1 Nebulizer

The nebulizer formulations are similar to those for sterile products used in parenteral or ophthalmic administration. They are most commonly aqueous drug solutions. Suspensions in aqueous media have also been prepared such as corticosteroid suspension [34], macromolecules (protein and other biotherapeutics [35]), liposomes [36–38], and noisomes [39]. Different aerosolization performances were observed using aqueous solution vs. suspension formulation [34]. The formulation design is usually based on empirical methods that consider both the pharmacological/toxicological effect and the physicochemical properties (viscosity, surface tension, drug concentration) for efficient aerosolization. Ethanol is often used as a cosolvent to enhance drug solubility. Other additives such as glycerin, propylene glycol, surfactants, and preservatives (e.g., EDTA) have been incorporated to enhance solubility and stability. Isotonicity and pH (3–8.5) suitable for pulmonary delivery are usually employed to prevent respiratory irritation and bronchospasm [40]. However, caution should be exerted when incorporating these additives in the nebulizer formulation because limited compounds are approved for lung delivery. Some, generally-regarded-as-safe compounds for parenteral delivery are not suitable for lung delivery [39].

Compared to pMDIs or DPIs, nebulizers often give wider dose ranges [41]. Jet nebulizers are more popular in hospitals than ultrasonic nebulizers because they are less expensive and often potentially more efficient when nebulizing suspended spheres [41, 42]. Liposomal disruption, protein denaturation, or aggregation was previously reported when ultrasonic nebulizers were employed [42]. Some more versatile and efficient nebulizers that apply porous mesh or plates (see Sect. 4.2.2.1) can overcome the limitations associated with jet or ultrasonic nebulizers.

4.2.1.2 pMDI

In contrast to the nebulizers, the formulations for pMDIs are prepared as drug solutions or suspensions in an inert and nonaqueous propellant [39]. The original pMDI employed chlorofluorocarbon (CFC) propellants 11, 114, and 12 [43]. The use of CFC propellants was limited in the late 1980s and eventually banned by 2009 in the United States for ozone-layer protection [44, 45]. They have been reformulated with hydrofluoroalkane (HFA) propellants 134a or 227 [46]. The physicochemical properties of propellant CFC and HFA are quite different from vapor pressure, density, hydrophobicity, solubility, and interfacial behavior (e.g., CFC 11 and HFA 134a have vapor pressure of 89 and 572 kPa at 20°C, respectively [47]). In this case, the optimal doses, excipient selection, and delivery parameter all need to be reassessed and new drug applications to be filed.

The toxicological profiles of many additives which are considered safe for oral or parenteral administration may render them inappropriate for lung delivery.

A limited number of excipients are approved for enhancing drug dissolution or stabilizing suspension [40, 48]. Three surfactants including oleic acid, sorbitan trioleate (i.e., Span 85), and lecithin were previously approved to disperse respirable particles in CFC suspension, whereas oleic acid is the only surfactant currently employed in the HFA systems [46]. Surfactants in suspension formulations can aid wetting, reduce electrostatic interfacial interactions, and provide steric barriers against flocculation [48]. They also function as valve lubricants. Other excipients such as oligolactic acids [49], menthol [50], have been investigated for their use in the HFA suspension. Ethanol is now being used in HFA systems as cosolvent to increase drug solubility and modulate propellant vapor pressure [40]. Water is generally viewed as an impurity in the nonaqueous pMDI systems. The moisture ingress can significantly change the interfacial interactions causing flocculation and crystal growth, which are detrimental to the drug aerosolization.

4.2.2 *Droplet Formation*

The physical forces governing the droplet production of nebulizers and pMDIs are surface tension and viscosity vs. aerodynamic forces. The formulation implication and the drug aerosolization processes of these two aerosol delivery systems will be discussed.

4.2.2.1 Nebulizer

Three types of nebulizers exist in the market based on their aerosol droplet formation mechanisms: the compressed air-assisted [4, 51], ultrasonic energy [42, 52], and vibrating-mesh nebulizers [53, 54].

The air-jet nebulizer works when a compressed airflow passes over a short section of liquid surface supplied by a capillary tube. Venturi (Bernoulli) effect results in a pressure gradient that draws the solution/suspension from the reservoir to the tube exit [17]. Droplet formation (nebulization) occurs when a certain levels of aerodynamic forces (turbulence and shear force) reached to overcome the internal forces (surface tension and viscosity) acting on the droplets. The air-jet nebulization is influenced by the fluid feeding rate, the baffle design, the angle of entrainment of the solution/suspension into the gas, and the physicochemical properties of the solution/suspension [4]. Two types of nozzles are applied. The internal mixing type has the airflow mixing with the liquid prior to leaving the nozzle exit port, whereas the external mixing type mix after leaving the nozzle [4].

The key component of ultrasonic nebulizer is a piezoelectric crystal transducer. It generates oscillatory motion at a high frequency (1–3 MHz [42]) when alternating electric current is applied. The oscillatory energy is transmitted into the liquid and results in low velocity aerosol droplets [42]. Two mechanisms were proposed for the processes of liquid disintegration: the capillary wave and the cavitation mechanisms.

In the former mechanism, the vertical capillaries are formed when surface waves constructively interfere. When the amplitude of the applied energy reaches a threshold, the capillaries break up into droplets [55, 56]. The threshold amplitude, A, is related with the kinematic viscosity v, the acoustic frequency f, and the capillary wavelength λ [57]:

$$A = 4\frac{v}{f\lambda}, \tag{4.1}$$

where

$$\lambda = (8\pi\sigma/\rho f^2)^{1/3}, \tag{4.2}$$

where σ is the surface tension. The count median diameter, CMD, of the formed droplets was found experimentally to be proportional to capillary wavelength [55]:

$$CMD = 0.34\lambda \tag{4.3}$$

The cavitation mechanism proposed that cavitation bubbles were formed when oscillatory motion was transmitted into liquid and transcended to the air–liquid interface, causing their implosion and droplet formation [42]. It is likely that the droplet formation results from both proposed mechanisms.

The aerosol generating mechanism of vibrating-mesh nebulizer is similar to the vibrating orifice aerosol generator described by Berglund and Liu [58]. In addition, it requires a porous mesh with a large number of tapered orifices, several micrometers in diameter. Alternating pressure in the fluid near the mesh causes the liquid to be ejected through the nozzles. Two types of vibrating-mesh nebulizers exist in the market. The passively vibrating-mesh nebulizer (e.g., Omron MicroAir nebulizer) induces vibrations in a porous mesh, which "passively" break up liquid into droplets [59]. The actively vibrating-mesh nebulizer (e.g., Aeroneb Pro nebulizer) employs a vibrating element that results in upward and downward movement of the mesh by a few micrometers, extruding the fluid and generating the aerosols [53]. The vibrating-mesh nebulizers have been widely used in hospital because of their advantages including controlled droplet size (see below), narrower size distribution, low residual losses, and high output rate [54].

Droplet size is an important parameter that affects lung deposition. The primary droplet size is a function of several factors including the dimensions and geometry of nozzles, physiochemical properties of the liquid (surface tension, viscosity, and density), volumetric air/liquid feeding ratio, and the inclusion of suspended particles [54, 59].

For air-jet nebulizers, the primary droplet size, d, can be estimated using linear stability analysis of air flowing across water, given by [51]:

$$d \approx \frac{1.26\sigma}{(\rho_g V^2)}, \tag{4.4}$$

where V is the relative velocity for the air–liquid interface, σ is the surface tension of the liquid, and ρ_g is the density of the air [51]. The primary droplet size was estimated to be 3–32 μm [51]. There are several other empirical equations for primary droplet size estimation, but the estimated size range differs greatly [4].

One of the advantages of vibrating-mesh nebulizers is their versatility to modify droplet size. The primary droplet diameter, d, can be calculated directly from the ratio of the liquid feed rate, Q, to the frequency of vibration, f, assuming the same droplet density, ρ_p, as that of the bulk material [5]:

$$d = \left(\frac{cQ}{\pi f \rho_p} \right)^{1/3}, \tag{4.5}$$

where c is the mass concentration of the solute per unit volume of the solution.

The primary droplets are often unstable and tend to break up into smaller droplets. Two processes cause these primary droplets to break up into smaller droplets in jet nebulizers: the aerodynamic forces resulting from relative velocity between the droplet and its surrounding air; and impaction on baffle [51].

The critical diameter, d, of a droplet to breakup can be estimated from two nondimensional parameter, Weber number (We) of the droplet, by equating the Newton's resistance equation for aerodynamic drag and the surface tension; and Reynolds number (Re) of the droplet, as shown in [4, 60]:

$$We_{crit} = \left(\frac{\rho_g V^2 d}{\gamma_L} \right) = \frac{8}{C_D}, \tag{4.6}$$

$$Re = \frac{\rho_L dV}{\eta_L}, \tag{4.7}$$

where ρ_g and ρ_L are the density of air and droplet, respectively; γ_L is the surface tension of the liquid; η_L is the viscosity of the droplet; V is the relative velocity between the droplet and the airstream; C_D is the coefficient of drag. If the viscous force of the liquid droplet is not accounted for (which is valid for most nebulizer formulations), the estimated critical Weber number for the droplet to breakup is about 18 [4]. Finlay described several scenarios of droplet deformation such as "vibrational bagging," "sheet stripping," "crest stripping" when the droplets are subjected to abrupt aerodynamic loading. These scenarios fall into different regimes of Weber number [51, 61].

Baffles are used in the nebulizers for several purposes: to recycle the solution; to facilitate secondary droplet formation by impact; to control the size distribution by aerodynamic selection; to slow the droplet velocity. The baffle collection efficiency can be expressed using the inertial impaction theory for curvilinear motion [62]:

$$d_{50} = \left[\frac{9\eta D_j (Stk_{50})}{\rho_L V C_c} \right]^{1/2} \approx \left[\frac{9\eta D_j}{4\rho_L V C_c} \right]^{1/2}, \tag{4.8}$$

where d_{50} is the droplet diameter having 50% collection efficiency; Stk_{50} is the Stokes number having 50% collection efficiency, which is 0.24 for circular nozzle [62]; η the air viscosity, ρ_L the density of droplet, V the velocity of droplet, D_j the nozzle diameter, and C_c Cunningham slip correction. The critical diameter that a primary droplet can splash into smaller droplets when impacting on the baffle is determined by the droplet velocity, density, surface tension, and viscosity [51].

The mass output rate and the nebulizer efficiency are two more important parameters with nebulization. The total mass output rate (solution) is intrinsically determined by the nebulizer design including the nozzle system, baffle, and reservoir [4]. The nebulizer efficiency (i.e., the fraction of the solute drug in the nebulizer that is delivered to the patient) increases with increased airflow rate, increased solution concentration, or decreased liquid viscosity [4, 51]. The simplest way to determination of solution output is by weighing nebulizer during nebulization (gravimetric measurement). But the seemingly straightforward method is not simple because of evaporation. The gravimetric measurement is often compared with a "standing cloud" measurement which is an assay that determines the quantity of drug captured on the filter when nebulizing [63].

4.2.2.2 pMDI

pMDI works by quick evaporation and shear thinning of the propellant into liquid droplets that vaporize and leave the residual nonvolatile drug in the form of aerosol powder particles of respirable size range. The fluid dynamics associated with pMDI actuation is a complex, turbulent, multiphase nonequilibrium to equilibrium process. Typical Reynolds numbers are several hundred thousand at pMDI stem and nozzle [51]. The localized pressure fluctuations can result in eddy formation and cavitation [51]. As the propellant vaporizes, an appreciable heat transfer may lead to temperature drop in the metering chamber as well [51]. The actual droplet formation process is not entirely understood. A quantitative characterization of this aerosolization process from a fundamental viewpoint remains elusive.

However, there are some simplified theories derived from the engineering terms and empirical estimates. The emission forces of the droplets from the pMDI are derived from the pressure gradient between the propellant and atmosphere. High velocity emission is generated when the propellant vaporizes upon dynamic equilibrium with atmospheric pressure, leaving cavities filled with propellant vapor. This vaporization is usually referred to as "cavitation" [51]. The volumetric flow rate (Q) and the linear velocity of the fluid (V) can be determined by [64]:

$$Q = C_D a_0 \left[\frac{2\Delta P}{\rho(1 - a_0 / a_1)} \right]^{0.5}, \tag{4.9}$$

$$V = Q / a_0, \tag{4.10}$$

where C_D is the coefficient of discharge, a_0 is the orifice area, a_1 is the inlet area, ρ is the density of the propellant, and ΔP is the pressure drop across the nozzle. An empirical estimation of the critical bubble radius, R_c, for cavitation can be estimated by [51]:

$$R_c = \frac{2\gamma_{LV} \sin\theta}{p_s - p}, \tag{4.11}$$

where γ_{LV} is interfacial tension of the propellant liquid/vapor, θ is the contact angle, p_s is the saturation vapor pressure of propellant, and p is the atmospheric pressure. Among them, γ_{LV} and p_s are functions of temperature. This equation indicates that cavitation is a function of surface tension and temperature. As the primary propellant droplets are formed, an empirical relationship between mass median droplet diameter, D_i, and atomizer conditions can be demonstrated by [16, 51, 65]:

$$D_i = \frac{C_s}{Q_e^m[(P_e - P_A)/P_A]^n}, \tag{4.12}$$

where C_s, m, and n are three constants; Q_e the mass fraction of vapor phase in the expansion chamber, P_e and P_A are the pressure downstream of discharge orifice and atmospheric pressure, respectively. Estimates of Q_e and P_e can be made using adiabatic, isentropic one-dimensional analysis of the flow [25, 65]. The mass median diameter can also be derived empirically as a function of droplet Reynolds number, Re, and orifice diameter, d_0 [64]:

$$D_i = 6d_0(\text{Re})^{-0.15}. \tag{4.13}$$

Similar to the secondary droplet formation in nebulization, the pMDI secondary droplets form when rapid moving droplets traverse a stagnant gas phase. (Note the distinction that airflow travels faster than droplet in nebulization). The behavior of the secondary droplets can be predicted using empirical correlation described by Shraiber, assuming gradual aerodynamic loading [51, 61]. The diameter of droplets is dependent on the types of deformation which are associated with different regime of Weber number [61]. The velocity of the droplet as a function of time, $V(t)$, can be estimated using the empirical correlation [51]:

$$V(t) = V_0 / [1 + 3C_d t(\rho_g / \rho_L)^{1/2} / (4\tau_0)], \tag{4.14}$$

where

$$\tau_0 \approx \frac{d(\rho_L / \rho_g)^{1/2}}{V_0}, \tag{4.15}$$

where V_0 is the initial droplet velocity, C_d is the coefficient of drag, ρ_g and ρ_L are density of air and droplet, respectively, d is the primary droplet diameter before breakup, and τ_0 is the time scale for growth of the Rayleigh-Taylor instability. Droplet breakup time and the distance traveled before breakup can then be estimated correspondingly, depending on the types of deformation that can occur [51]. These parameters can influence the lung deposition efficiency. For example, if the droplet traveling distance is too long, the droplet will end up deposited in the oropharynx before breakup. High exit velocities and breakup distance smaller than the pMDI orifice-oropharynx are considered beneficial for improving lung deposition [64]. The lung deposition efficiency and reproducibility of pMDIs are typically evaluated by inertial sampling methods, similar to DPIs [66]. The term emitted dose, fine particle mass (or fraction), and mass median aerodynamic diameter are frequently used. Many pMDIs have spacers to increase pulmonary deposition and reducing oropharyngeal drug deposition [67]. This is because of the increased distance from the nozzle to the back of the oropharynx, reduced impaction in the mouth-throat by lowering the Stokes number, and prolonged time for secondary droplet formation and evaporation.

4.2.3 Evaporation

The liquid droplets generated by nebulizers and pMDIs evaporate to come into equilibrium with the surrounding air. The evaporation rate of the droplets, by diffusion, is expressed by [62]:

$$\frac{d(d_p)}{dt} = \frac{4DM}{R\rho_p d_p}\left[\frac{p_\infty}{T_\infty} - \frac{p_d}{T_d}\right], \tag{4.16}$$

where D is the diffusion coefficient of water vapor in air, M is the molecular weight, d_p is droplet diameter, ρ_p is the density of the droplet, p_∞ and p_d are the partial pressures at the ambient and droplet surface, T_∞ and T_d are the ambient and droplet surface temperatures, and R is the gas constant. Evaporation occurs when the partial pressure of the ambient is less than that at the droplet surface. The droplet is cooled by the heat of evaporation. In nebulizer, because most of the droplets are recycled, the nebulizer solutions in the reservoir are cooled and concentrated as the evaporation continues. In the pMDI, the rapid expansion of propellant vapor in the metering chamber and the continued post-nozzle evaporation cool the droplets. Finlay showed that in order to reasonably predict the behavior of the droplets generated by pMDI, a correction to account for Stefan flow is needed because of evaporation [51].

4.2.4 Aggregation of Suspension Aerosols

The nebulized or pMDI droplets containing the suspension differ from those of the solution because the suspension droplets contain relatively few particles.

The number of particles per droplet will depend on the relative size of the droplets and particles, and the concentration of the suspension [68]. Due to the evaporation of dispersant (i.e., water or propellant), the drug particles form aggregates, which will alter the particle size distribution from that of the primary drug particles. By assuming the uniform distribution, monodisperse and spherical primary drug particles in droplets, the average number of particles in the aggregate and the variance of this number can be calculated from the Poisson probability distribution [69, 70]. In systems that the primary drug particle diameters are much smaller than the droplet and relatively high average number of particles per droplet, the aggregate size distributions are essentially log-normal with the same geometric standard deviations as the droplets [69, 70]. A significant increase in the drug mass median aerodynamic diameter as a result of aggregation has been predicted [70, 71]. However, when the ratio of the droplet diameters is small and the suspension concentration is high (i.e., when the droplets are too small to accommodate the number of primary particles), the aggregates will deviate from log-normal distribution [70].

4.3 Powders

Powder formulation is one of the key components in DPIs. Unlike the situation with liquid droplets, the solid particles have limited molecular mobility. They are rigid and resistant to stress. The surface molecules of the solid cannot immediately rearrange to equilibrium with the bulk. Consequently, the surfaces of the particles are highly conditioned by their past history such as the particle formation (Sect. 4.1.2), processing (milling and blending), and storage conditions (humidity). Particle engineering (e.g., the manipulation of particle morphology), if applied correctly, can produce powders with good aerodynamic properties for drug aerosolization and deposition. A brief discussion of the fundamental particulate interactions, powder mixing, the fluid dynamic, and a few examples of particle engineering including low density porous particles (Sect. 4.3.4) and elongated particles (Sect. 4.3.5) to enhance drug delivery will be enlightening.

4.3.1 Particulate Interactions

The fundamental adhesive and cohesive physical forces involved in macrobodies such as colloids, microparticles and surfaces, namely the interparticulate interactions, are similar to those at the molecular level [72]. These attractive physical forces in DPI formulations are mainly classified as van der Waals, electrostatic, and capillary forces [10]. For rigid and surface homogeneous spheres, interparticulate van der Waals forces can be described as [10]:

$$F_{vdw} = \frac{AR}{6h^2}\frac{R_1R_2}{R_1+R_2} = \frac{AR}{12h^2} \quad (R_1 = R_2 = R, \textit{for two identical spheres}), \tag{4.17}$$

$$F_{vdw} = \frac{AR}{6h^2} \quad (R_2 \gg R_1,\ \textit{for sphere and surface}), \tag{4.18}$$

where A is the Hamaker constant (typically ~10^{-19} J), R_1 and R_2 are the particle radius, and h is the separation distance. Van der Waals forces are effective only within a separation distance less than 100 nm [73]. They decay rapidly as separation distance increases and the corresponding decrease of Hamaker constant, known as retardation effect [72]. Therefore, van der Waals forces are quite sensitive to particle shape and surface roughness change. The change of surface roughness can change van der Waals forces by several orders of magnitude.

Electrostatic forces can be attractive or repulsive. They occur by means of contact charging, coulombic interaction, and induced charging. Contact charging occurs during contact between uncharged and unlike particles by electron transfer until equilibrium state of the Fermi level is achieved. The magnitude of the contact potential force (F_c) may be expressed as [10]:

$$F_c = \frac{2\pi q^2}{A}, \tag{4.19}$$

where q is the particle charge on detachment from a substrate and A is the contact area between the particle and the substrate.

Coulombic forces occur when two charged particles are brought together. Either attraction or repulsion is possible depending on the electrical charge signs. The magnitude of Coulombic forces may be expressed as [74]:

$$F_q = \frac{q_1 q_2}{4\pi\varepsilon h^2}, \tag{4.20}$$

where q_1 and q_2 are the electrical charges on the two particles, respectively, ε is the permittivity, and h is the separation distance.

The induced charging occurs when a charged particle, q, approaches an uncharged particle of radius R. The magnitude of induced charging forces may be calculated as [74]:

$$F_e = q^2\left[1 - \frac{h}{(R^2 + h^2)^{0.5}}\right] \times \frac{1}{16\pi\varepsilon_0 h^2}, \tag{4.21}$$

where h is the separation distance and ε_0 is the permittivity.

Capillary forces rise between particles following capillary condensation of water vapor at high relative humidity (RH). The Laplace pressure develops due to the water meniscus formed in the contact zone that pulls particles together. The capillary forces between two spherical particles may be described as [72]:

$$F_c = 4\pi R^* \gamma_L \cos\theta + 4\pi R^* \gamma_{SL}, \tag{4.22}$$

where F_c is the capillary force, R^* is the harmonic mean of the particle radii, γ_L is the surface tension of water (72.8 mN/m at 20°C), γ_{SL} is the solid–liquid interfacial energy, and θ is the measured contact angle that water is supposed to form with the particle surface.

The magnitude of adhesive forces between two rigid and smooth spheres were compared [75]. At a distance of typical pharmaceutical powder (e.g., separation distance ~5 nm), van der Waals forces are about 1–2 orders of magnitude larger than electrostatic forces. Capillary forces are significantly higher than both van der Waals and electrostatic forces, but only prevail at high RH (e.g., RH>65%) [75, 76]. Electrostatic forces are "long" range forces compared to van der Waals forces, but decay quickly at high RH [77]. Although the quantification of particulate adhesion between ideal spheres appears to be straightforward, the measured forces are not directly translatable to aerosol performance. This may be attributed to the complexity of real particles, which are inherently heterogeneous. In reality, the particulate adhesion is complicated with a variety of factors including surface asperities, contact deformation, surface energy heterogeneity, and effect of elasticity and plasticity.

Alternatively, the description of particulate adhesion can be achieved by thermodynamic approach relating to surface energetics. The advantage of introducing surface energy is that elastic deformation of the particle is readily included [78, 79]. For noncomforming elastic particle surfaces, two models including the Johnson-Kendall-Roberts (JKR) [78] and the Derjaguin-Muller-Toporov (DMT) [79] have often been used to determine the pull-off forces.

$$F_{ad} = n\pi R^* W_{ad} \quad \text{when } n = 1.5 \text{ for JKR; n} = 2 \text{ for DMT} \tag{4.23}$$

where F_{ad} is the adhesion force (N), R^* is the harmonic mean of the particle radii (m), and W_{ad} is the work of adhesion (N/m). The DMT model is appropriate for hard material with low surface energies and high elastic moduli (Young's modulus >10^9 N/m^2), and considers attractive force beyond the contact region. The JKR model is for softer material and neglects the attractive force outside the contact circle.

4.3.2 *Mixing (with Lactose or Other Excipients)*

Many DPI formulations function by using interactive physical mixtures of micronized drug particles (typically 1–5 μm) and carrier particles (typically 75–200 μm α-lactose monohydrate) [40]. Mixing is a basic pharmaceutical process. However, the fundamental physics of the mixing is not yet fully understood. The performance of mixing is still based on empirical principles. Unlike molecular diffusion in gas or liquid that would occur spontaneously and completely at given time, the mixing (or demixing) of powders do not occur without an energy input [17, 80]. Mixing is a considered a statistical process that involves relative displacement of particles of one component with another component until a state of homogeneity is achieved [17]. The degree of homogeneity can be estimated by evaluating the sample variance (σ_s^2) [17, 81]:

$$\sigma_s^2 = \sum_{i=1}^{N} \frac{(x_i - \overline{x}_s)^2}{N-1}, \tag{4.24}$$

where x_i ($i=1, 2,\ldots, N$) is the composition in the ith sample taken randomly (e.g., employing sample thieves [82]) and $\overline{x}_s$ is the sample mean. More detailed discussion of sampling methods and homogeneity analysis was reviewed previously [83–86]. A concept that distinguishes randomized mixing with ordered mixing is worth mentioning [87]. In randomized mixing, two powders of identical size, shape, and surface texture are mixed. No interactions (adhesion or cohesion) are involved. Therefore, the state of uniformity that obeys the probability law is expected. However, DPI formulation mixture (e.g., a drug-carrier binary system) contains particles having two different size distributions. The heterogeneity of surface morphology and energy distribution determines the local sites of adhesion. In such mixture, the randomized distribution is disturbed [74]. This type of mixture is often called "ordered mixture" or more precisely "interactive physical mixture" [87, 88].

The mechanisms of mixing include convective (large-scale random motion), diffusive (small-scale random motion), and shear (momentum exchange due to velocity distribution) mixing depending on the mixer in use and mixing time [17, 81]. Mixing is a time-dependent process, its rate for any mechanism can be expressed by a first-order kinetics as [17]:

$$\frac{dM}{dt} = k(1-M), \tag{4.25}$$

$$M = 1 - e^{-kt}, \tag{4.26}$$

where M, the mixing index, is defined as:

$$M = \frac{s_r}{s}, \tag{4.27}$$

where s_r represents the standard deviation of a series of samples drawn from a random mixture, s is the standard deviation of samples drawn from the mixture under examination, and k is the rate constant (s^{-1}) dependent on the material, mixer geometry, and operation. The power (P) requirement for mixing is determined by the following [81]:

$$P = 2\pi\omega T, \tag{4.28}$$

$$E = \int_{t_1}^{t_2} P \, dt, \tag{4.29}$$

where T is the axial torque for rotating the mixer or impeller, ω (rps) is the rotational speed, and E (J) is the energy input between time t_1 and t_2.

Segregation, known as the competing mechanism and opposite of mixing, has been studied extensively [89, 90]. The mechanisms of segregation can be classified as: (1) percolation; (2) trajectory effect; (3) rolling effect; and (4) push-away effect [89]. Segregation may result in poor blend homogeneity, especially when the drug concentration is high and carrier surfaces are saturated [91]. A narrower size distribution of carrier particles can greatly increase the mixing efficiency and avoid segregation [91]. Increasing the interparticulate adhesion forces such as opposite charges between drug and carrier particles could also minimize segregation and stabilize the blends [92, 93].

4.3.3 Fluid Dynamics of Dry Powder Aerosols (Particle Motion in Airflow)

When a dry powder formulation turns into aerosols by inhalation, two different processes occur in the airways: drug aerosolization and deposition. The former process occurs when a static powder bed is pneumatically actuated in the DPI device and the latter, when the aerosolized particles enter and attach to airway walls. These two processes will be discussed later separately.

4.3.3.1 Drug Aerosolization

The aerosolization of carrier-based DPI formulation occurs in four concurrent phases from static powder bed, dilation, fluidization, to drug resuspension. The mechanisms of fluidization and deaggregation vary greatly, depending on the device construction and inspiratory flow maneuver. They may include capillary fluidization, laminar/turbulent shear fluidization and deaggregation, momentum transfer by collision and rotation, and turbulent deaggregation [75]. The drag (F_D) and lift forces (F_L) to overcome particle–particle or particle–wall adhesion can be expressed as below [75]:

$$F_D + F_L = \frac{\pi}{8} d_p^2 \rho C_D |\vec{U} - \vec{U}_p| (\vec{U} - \vec{U}_p) - \frac{\pi}{6} d_p^3 \frac{\partial P}{\partial \vec{x}} + \frac{\pi d_p^3}{6} \rho |\vec{U} - \vec{U}_p| |\varpi - \varpi_p| + 1.6(\mu\rho)^{0.5} d_p^2 |\vec{U} - \vec{U}_p| \left| \frac{\partial \vec{U}}{\partial \vec{x}} \right|^{0.5} + F_c - F_g, \quad (4.30)$$

where the terms on the right-hand side of the equation, in order from left to right, represent the drag force, the force due to static pressure gradients in the flow, the Magnus force, Saffman force, centrifugal force, and gravity force [75]. The predominating force acting on a particle is the drag force. The Saffman lift force is caused by velocity gradient of the fluid, the Magnus force is caused by particle contact and rebound from a wall, the centrifugal force can be expressed as $F_c = 0.5\, m_p \omega^2 D_c$, where m_p is the mass of drug particle, and D_c is the diameter of the carrier particle.

One way to incorporate different forces of deaggregation is the application of energy function. The aerodynamic aerosolization is directly related to the fluid energy dissipation rate, ε, adequate measurement of which is the magnitude of viscous turbulent shear stress [94]. The shear stress, τ_s, can be calculated by applying the Kolmogorov theory in a nozzle flow [94]:

$$\tau_s = \eta \left(\frac{\varepsilon}{\nu} \right)^{1/2}, \tag{4.31}$$

where

$$\varepsilon \cong \frac{2\langle u \rangle^3}{D} \cong \frac{2(0.2u)^3}{D}, \tag{4.32}$$

where $\langle u \rangle$ is the root-mean square velocity fluctuation of the nozzle velocity u, D is the nozzle diameter, η and ν are the dynamic and kinematic viscosity of air, respectively. Several other aerodynamic parameters such as Reynolds number, pressure drop, and power are often useful for characterizing the aerodynamic properties in the DPIs.

A more general equation that considers the whole shear field in airflow can be expressed as Newton's resistance equation and a corresponding initial kinetic energy of moving objects [75]:

$$F_{drag} = C_d \frac{\pi}{8} \rho_a d^2 V^2, \tag{4.33}$$

$$E_k = \frac{\pi}{12} \rho_p d^3 V^2, \tag{4.34}$$

where C_d is the drag coefficient, ρ_a and ρ_p are the density of air and aerosol particle, respectively, d is the particle diameter, and V is the airflow velocity. The impaction and momentum transfer play a significant role in deaggregation because they're proportional to the d^3 [95, 96]. Friction forces may also play a role at initiation of powder dispersion and to some extent during aerosol delivery [97].

4.3.3.2 Drug Deposition

Airflow in the respiratory system is complex. The understanding of the local aerodynamics derives mainly from experimental studies using physical models of lung airway. In Weibel's model, 23 generations of the airway from trachea to alveolar sacs were designated [98]. An estimate of an average adult's airflow indicates that flows in the upper airway are turbulent to transition region, while the 16 generation and below are laminar flow, as is indicated by Reynolds number (Re) [98]. The dynamics of particles in the lung airway fall in the Stokes's regime for particle diameters larger than 1.3 μm [10]. Particle movement in this size regime follows Stokes's law. Equating the aerodynamic drag force with the Stokes's law result in the terminal settling velocity, V_{TS}, in a laminar condition and neglecting slip correction [62]:

$$V_{TS} = \frac{\rho_p d_e^2 g}{18\eta\chi} \quad (for\ \mathrm{Re} < 1), \tag{4.35}$$

where ρ_p is the particle density, d_e is the equivalent volume diameter (diameter of a sphere having the same volume as that of the irregular particle), η is the viscosity of air, and χ is the dynamic shape factor. Reynolds number of particle is defined as:

$$\mathrm{Re} = \frac{\rho_p d_p V}{\eta}, \tag{4.36}$$

where d_p is the particle diameter, V is the relative velocity between particle and airflow.

Particles dispersed in airflow follow curvilinear motion. The trajectory of a particle departs slightly due to a variety of physical forces such as inertia, gravity, and electrostatic force. Mechanisms of particle deposition in the respiratory tract include: (1) Inertial impaction; (2) Gravitational settling; (3) Brownian motion; (4) Interception; (5) Electrostatic forces. Turbulent deposition may happen in the upper respiratory tract, but will not be discussed here.

Deposition by inertial impaction is the main mechanism for larger particles (>5 μm) with high velocity. This mechanism occurs at the curvature when the airflow is deflected. While smaller and slower particles can move along the streamline, the larger and faster particles are unable to follow the streamline due to inertia. A parameter that determines the deposition efficiency is the Stokes number [62, 99]:

$$Stk = \frac{\tau V}{D/2} = \frac{\rho_p d_p^2 V C_c}{9\eta D}, \tag{4.37}$$

where τ is the relaxation time, d_p and V are the particle diameter and velocity, respectively and C_c is the Cunningham correction factor.

Gravitational settling (sedimentation) is governed by Stokes's law. Particle diameters between 0.5 and 3 μm are deposited in the bronchial and alveolar regions by this mechanism. Both inertial impaction and gravitational settling are dependent on the relaxation time (τ) of the particle, which is defined as [62]:

$$\tau = \frac{d_p^2 \rho_p}{18\eta}. \tag{4.38}$$

Increased gravitational settling occurs by either breathholding or slow tidal breathing [99].

Brownian motion (diffusion) dominates the deposition of particles smaller than 0.1 μm. The rate of diffusion, according to the Stokes–Einstein equation, is proportional to the temperature and inversely proportional to the particle size and air viscosity.

Interception is mainly referred to elongated particles which are discussed in Sect. 4.3.5. It should be noted that the deposition occurs when several mechanisms act simultaneously. A thorough discussion of all the deposition mechanisms has been included in several books [62, 99, 100].

4.3.4 *Low Density Porous Particles*

As seen above, the aerodynamic particle size determines the regional deposition in the lung. There are three ways to affect the aerodynamic diameter of a particle: change of particle size, density, or dynamic shape factor. This can be clearly seen from the definition of aerodynamic diameter, d_a, given by [62]:

$$d_a = d_e\sqrt{\frac{\rho}{\rho_0 \chi}} = d_s\sqrt{\frac{\rho}{\rho_0}}, \tag{4.39}$$

where d_e and d_s are the equivalent volume diameter and stokes's diameter, respectively and ρ and ρ_0 are particle density (void included), unit density (1 g/cm^3), respectively. χ is the dynamic shape factor that applied to Stokes's law (see Sect. 4.3.5 below).

Low density particles achieve small aerodynamic diameter by reducing particle density (mass) while maintaining the equivalent volume diameter. The dynamic shape factor may also increase due to the increase in aerodynamic drag. It is hypothesized that low density particles significantly reduce particle–particle interactions because the asperities prevent close contact of particles within van der Waals region [75]. Enhanced powder fluidization and drug aerosolization may lead to increases in drug delivery efficiency and reduction in the inspiratory flow rate dependency [101, 102]. Some porous drug particles without carriers were also developed [103]. Porous particles have been evaluated for their ability to stabilize suspensions in pMDI [104, 105]. Because their projected areas are greater than ordinary particles of the same mass, they experience greater aerodynamic drag force to reduce particle velocity. Therefore, deposition to the oropharyngeal region by inertial impaction is greatly reduced [18].

Porous particles can be produced by creating void inside the particles, forming porous matrix or/and hollow spheres [106]. The particle density and geometric size can be manipulated by varying the volume fraction of the pore-forming agent and processing conditions [15]. A variety of low density particle preparation conditions have been reviewed [15].

4.3.5 Elongated Particles

Elongated particles such as fibers have high aspect ratio (length/diameter). These particles have interesting airborne and physiological characteristics. The dynamic shape factor given above is defined as:

$$\chi = \frac{\textit{The actual resistance of nonspherical particle}}{\textit{Resistance of a sphere having having the same volume \& velocity}} = \frac{F_D}{3\pi\eta V d_e} \tag{4.40}$$

where F_D is the actual resistance force of fiber, η is viscosity, and V is particle velocity.

While flowing in the airway, elongated particles are predominantly aligned with streamlines with their long axis. Therefore, they have aerodynamic diameters (d_a) that are mainly decided by their short axis (diameter). It is well known that elongated particles such as asbestos and mineral fibers are able to deposit in deep lung (alveolar region) because of their small aerodynamic diameters and the mechanism of interception. Once deposited, they have less chance of being cleared by macrophage due to large length [62]. Because of these characteristics, the use of elongated particles for drug delivery has attracted much attention. There are quite a few reports that elongated drug [107–110] or carrier [111, 112] particles enhance aerosolization performance exemplified as increase in fine particle fraction. However, the enhanced performance relies on the loose agglomeration of the fiber to facilitate fluidization and deaggregation during actuation. The performance could be poor when the fibrous particles are in contact along their length [106]. Powder flow is adversely influenced by increasing elongation ratio, which may also cause processing problems such as reduced content uniformity. Moreover, reduced surface energy due to polymorphism may also contribute to enhance the performance [113].

A key point for predicting the aerodynamic behavior of elongated particles is the estimation of their aerodynamic diameter. The theoretical expressions of aerodynamic diameters of fibers are based on the viscous drag of prolate spheroids or cylinder [69]. Relating the definition for terminal settling velocity (V_{ST}), the aerodynamic diameter (d_a) can be expressed as [69]:

$$d_a = \sqrt{\frac{6\rho V}{\pi d_D \rho_0}}, \tag{4.41}$$

where ρ and ρ_0 are particle true density and unit density (1 g/cm^3), respectively, V is particle volume, and d_D is diameter of a sphere with the same viscous drag as the fibrous particle. Assuming the aerodynamic behavior of a fiber (e.g., prolate spheroid) in the human respiratory tract follows randomly oriented flight, Eq. 4.41 is converted to [69]:

$$d_a = \sqrt{\frac{3}{2}}d\sqrt{\frac{\rho/\rho_0}{\dfrac{0.385}{\ln(2L/d)-0.5}+\dfrac{1.230}{\ln(2L/d)+0.5}}} \tag{4.42}$$

where L and d are the fiber's length and diameter, respectively.

4.4 Conclusion

Modern respiratory drug delivery has evolved three aerosolization systems: nebulizer, pMDI, and DPI. Each system has its unique field of technologies and physical principles. As a steady increase of new marketed aerosol products in recent decades, a good understanding of the aerosol formulation and device design is required for the successful development of these systems. In this chapter, several important empirical and scientific principles that include the particle generation, the formulation implication, and drug aerosolization mechanisms of each aerosolization systems have been highlighted. Examples of particle engineering to achieve aerodynamically favorable dry powder aerosols are briefly discussed. We believe that the design and development of next generation inhaler products will be expedited if these mechanistic physical understandings are correctly implemented.

References

1. Hickey AJ (1996) Inhalation aerosols: physical and biological basis for therapy. Marcel Dekker, New York
2. O'Callaghan C, Nerbrink O, Vidgren M (2001) in: Drug delivery to the lung, vol 162, 1st edn. Marcel Dekker, New York, pp 1–20
3. Wright P (2004) in: Pharmaceutical preformulation and formulation: a practical guide from candidate drug selection to commercial dosage form, 1st edn. Taylor & Francis Group/CRC Press, Boca Raton, pp 355–378
4. Niven RW, Hickey AJ (2007) in: Inhalation aerosols: physical and biological basis for therapy, vol 221, 2nd edn. Informa Healthcare USA, New York, pp 253–283
5. Leong KH (2004) in: Pharmaceutical inhalation aerosol technology, vol 134, 2nd edn. Marcel Dekker, New York, pp 253–278
6. Hickey AJ, Evans RM (1996) in: Inhalation aerosols: physical and biological basis for therapy, vol 94, 1st edn. Marcel Dekker, New York, pp 417–439
7. Partridge MR, Woodcock AA (2001) in: Drug delivery to the lung, vol 162, 1st edn. Marcel Dekker, New York, pp 371–388
8. Bell JH, Hartley PS, Cox JS (1971) Dry powder aerosols. I. a new powder inhalation device. J Pharm Sci 60:1559–1564
9. Hickey AJ, Crowder TM (2007) in: Inhalation aerosols: physical and biological basis for therapy, vol 221, 2nd edn. Informa Healthcare USA, New York, pp 445–460
10. Crowder TM, Hickey AJ, Louey MD, Orr N (2003) A guide to pharmaceutical particulate science, Interpharm Press/CRC, Boca Raton
11. Louey MD, VanOort M, Hickey AJ (2006) Standardized entrainment tubes for the evaluation of pharmaceutical dry powder dispersion. J Aerosol Sci 37:1520–1533
12. Nennis JH, Nerbrink O (2001) in: Drug delivery to the lung, vol 162, 1st edn. Marcel Dekker, New York, pp 303–336
13. Wu L, Peguin RPS, Selvam P, Chokshi U, da Rocha SRP (2007) in: Inhalation aerosols: physical and biological basis for therapy, vol 221, 2nd edn. Informa Healthcare USA, New York, pp 373–398
14. Chan HK (2003) Inhalation drug delivery devices and emerging technologies. Expert Opin on Ther Patents 13:1333–1343
15. Weers JG, Tarara TE, Clark AR (2007) Design of fine particles for pulmonary drug delivery. Expert Opin Drug Deliv 4:297–313

16. Hickey AJ, Mansour HM (2009) in: Modern pharmaceutics, vol 2, 5th edn. Informa Healthcare USA, New York, pp 191–219
17. Hickey AJ, Ganderton D (2001) Pharmaceutical process engineering. Marcel Dekker, New York
18. Van Oort M, Sacchetti M (2007) in: Inhalation aerosols: physical and biological basis for therapy, vol 221, 2nd edn. Informa Healthcare, New York, pp 307–346
19. Fisher ES (2007) in: Encyclopedia of pharmaceutical technology, vol 4, 3rd edn. Informa Healthcare, USA, New York, pp 2339–2351
20. Rasenack N, Muller BW (2004) Micron-size drug particles: common and novel micronization techniques. Pharm Dev Technol 9:1–13
21. Chan HK, Chew NY (2003) Novel alternative methods for the delivery of drugs for the treatment of asthma. Adv Drug Deliv Rev 55:793–805
22. Seville PC, Li HY, Learoyd TP (2007) Spray-dried powders for pulmonary drug delivery. Crit Rev Ther Drug Carrier Syst 24:307–360
23. Cal K, Sollohub K (2010) Spray drying technique. I: hardware and process parameters. J Pharm Sci 99:575–586
24. Sollohub K, Cal K (2010) Spray drying technique: II. current applications in pharmaceutical technology. J Pharm Sci 99:587–597
25. Dunbar CA, Concessio NM, Hickey AJ (1998) Evaluation of atomizer performance in production of respirable spray-dried particles. Pharm Dev Technol 3:433–441
26. Chan HK, Clark AR, Feeley JC, Kuo MC, Lehrman SR, Pikal-Cleland K, Miller DP, Vehring R, Lechuga-Ballesteros D (2004) Physical stability of salmon calcitonin spray-dried powders for inhalation. J Pharm Sci 93:792–804
27. van Drooge DJ, Hinrichs WL, Dickhoff BH, Elli MN, Visser MR, Zijlstra GS, Frijlink HW (2005) Spray freeze drying to produce a stable Delta(9)-tetrahydrocannabinol containing inulin-based solid dispersion powder suitable for inhalation. Eur J Pharm Sci 26:231–240
28. Maa YF, Ameri M, Shu C, Payne LG, Chen D (2004) Influenza vaccine powder formulation development: spray-freeze-drying and stability evaluation. J Pharm Sci 93:1912–1923
29. Maa YF, Nguyen PA, Sweeney T, Shire SJ, Hsu CC (1999) Protein inhalation powders: spray drying vs. spray freeze drying. Pharm Res 16:249–254
30. Steckel H, Rasenack N, Muller BW (2003) In-situ-micronization of disodium cromoglycate for pulmonary delivery. Eur J Pharm Biopharm 55:173–180
31. Westmeier R, Steckel H (2008) Combination particles containing salmeterol xinafoate and fluticasone propionate: formulation and aerodynamic assessment. J Pharm Sci 97:2299–2310
32. York P (1999) Strategies for particle design using supercritical fluid technologies. Pharm Sci Technol Today 2:430–440
33. Rehman M, Shekunov BY, York P, Lechuga-Ballesteros D, Miller DP, Tan T, Colthorpe P (2004) Optimisation of powders for pulmonary delivery using supercritical fluid technology. Eur J Pharm Sci 22:1–17
34. Cameron D, Clay M, Silverman M (1990) Evaluation of nebulizers for use in neonatal ventilator circuits. Crit Care Med 18:866–870
35. Niven RW (1995) Delivery of biotherapeutics by inhalation aerosol. Crit Rev Ther Drug Carrier Syst 12:151–231
36. Niven RW, Carvajal TM, Schreier H (1992) Nebulization of liposomes. III. The effects of operating conditions and local environment. Pharm Res 9:515–520
37. Niven RW, Schreier H (1990) Nebulization of liposomes. I. Effects of lipid composition. Pharm Res 7:1127–1133
38. Niven RW, Speer M, Schreier H (1991) Nebulization of liposomes. II. The effects of size and modeling of solute release profiles. Pharm Res 8:217–221
39. Smyth HDC, Garcia-Contreras L, Cooney DJ, Garmise RJ, Jones LD, Hickey AJ (2005) in: Aerosols handbook: measurement, dosimetry, and health effects, 1st edn. CRC Press, Boca Raton, pp 313–341

40. Smyth H (2006) in: Excipient development for pharmaceutical, biotechnology, and drug delivery systems, 1st edn. Informa Healthcare USA, New York, pp 225–249
41. Bernstein JA, Amin H, Smith SJ (2007) in: Inhalation aerosols: physical and biological basis for therapy, vol 221, 2nd edn. Informa Healthcare USA, New York, pp 219–252
42. Taylor KMG, McCallion O (2007) in: Encyclopedia of pharmaceutical technology, vol 6, 3rd edn. Informa Healthcare USA, New York, pp 3854–3861
43. Purewal TS, Grant DJ (1997) in: Metered dose inhaler technology, 1st edn. Informa Healthcare USA, New York
44. The Montreal Protocol on substances that deplete the ozone layer (1987) UNEP, Nairobi, Kenya, Federal Register 1994, 59FR5627656298
45. U.S. Food and Drug Administration (2008) Use of ozone-depleting substances; removal of essential use designation. http://federalregister.gov/a/E8-27436
46. Hickey AJ (2005) in: Drug delivery: principles and applications, 1st edn. Wiley, New Jersey, pp 341–361
47. Hindle M (2008) in: Pharmaceutical manufacturing handbook: production and processes, 1st edn. Wiley, New Jersey, pp 683–727
48. Johnson KA (2007) in: Inhalation Aerosols: Physical and biological basis for therapy, vol 221, 2nd edn. Informa Healthcare USA, Inc., NY, pp 347–371
49. Stefely JD, Duan DC, Myrdal PB, Ross DL, Schultz DW, Leach CL (2000) Respiratory drug delivery VII, vol 1. Serentec Press, Raleigh, pp 83–90
50. Tzou TZ (1999) Aerodynamic particle size of metered-dose inhalers determined by the quartz crystal microbalance and the Andersen cascade impactor. Int J Pharm 186:71–79
51. Finlay WH (2001) The mechanics of inhaled pharmaceutical aerosols: an introduction. Academic Press, London, UK
52. Greenspan BJ (2007) in: Inhalation aerosols: physical and biological basis for therapy, vol 221, 2nd edn. Informa Healthcare USA, NY, pp 285–306
53. Dhand R (2002) Nebulizers that use a vibrating mesh or plate with multiple apertures to generate aerosol. Respir Care 47:1406–16; discussion 1416–1418
54. Lass JS, Sant A, Knoch M (2006) New advances in aerosolised drug delivery: vibrating membrane nebuliser technology. Expert Opin Drug Deliv 3:693–702
55. Lang RJ (1962) Ultrasonic atomization of liquids. J Acoust Soc Am 36:6–8
56. Peskin RL, Raco RJ (1963) Ultrasonic atomization of liquids. J Acoust Soc Am 35:1378–1381
57. Mercer TT (1981) Production of therapeutic aerosols; principles and techniques. Chest 80:813–818
58. Berglund RN, Liu BY (1973) Generation of monodisperse aerosol standards. Environ Sci Technol 7:147–153
59. Ghazanfari T, Elhissi AM, Ding Z, Taylor KM (2007) The influence of fluid physicochemical properties on vibrating-mesh nebulization. Int J Pharm 339:103–111
60. Placke ME, Ding J, Zimlich WC (2007) in: Encyclopedia of pharmaceutical technology, vol 3, 3rd edn. Informa Healthcare USA, New York, pp 2092–2118
61. Shraiber AA, Podvysotsky AM, Dubrovsky VV (1996) Deformation and breakup of drops by aerodynamic forces. Atomization Sprays 6:667–692
62. Hinds WC (1999) Aerosol technology: properties, behavior, and measurement of airborne particles, 2nd edn. Wiley, New York
63. Smaldone GC, Lesouef PN (2001) in: Drug delivery to the lung, vol 162, 1st edn. Marcel Dekker, New York, pp 269–302
64. Smyth HDC, Evans RM, Hickey AJ (2007) in: Inhalation aerosols: physical and biological basis for therapy, vol 221, 2nd edn. Informa Healthcare USA, New York, pp 399–416
65. Dunbar CA, Watkins AP, Miller JF (1997) Theoretical investigation of the spray from a pressureized metered-dose inhaler. Atomization Sprays 7:417–436
66. Munro SJM, Cripps AL (2007) in: Encyclopedia of pharmaceutical technology, vol 4, 3rd edn. Informa Healthcare USA, New York, pp 2269–2285
67. Hickey AJ (2004) in: Pharmaceutical inhalation aerosol technology, vol 134, 2nd edn. Marcel Dekker, New York, pp 385–422

68. Callingham MA (1980) A statistical model for predicting the particle distribution of an atomized powder in liquid mixture. Int J Cosmet Sci 2:107–126
69. Gonda I, Khalik AFAE (1985) On the calculation of aerodynamic diameters of fibers. Aerosol Sci Tech 4:233–238
70. Chan HK, Gonda I (1988) Development of a systematic theory of suspension inhalation aerosols. II. aggregates of monodisperse particles nebulized in polydisperse droplets. Int J Pharm 41:147–157
71. Gonda I, Chan HK (1985. Suppl.) Model of aerodynamic behaviour of aggregates formed on nebulization of suspensions. J Pharm Pharmacol 37:56P
72. Israelachvili JN (1992) Intermolecular and surface forces, 2nd edn. Elsevier, London, UK
73. Erbil HY (2006) Surface chemistry: solid and liquid interfaces. Blackwell, Malden
74. Zeng XM, Martin GP, Marriott C (2001) Particulate interactions in dry powder formulations for inhalation. Taylor & Francis, New York
75. Dunbar CA, Hickey AJ and Holzner P (1998) Dispersion and characterization of pharmaceutical dry powder aerosols. Kona 16:7–45
76. Weiler C, Egen M, Trunk M and Langguth P (2010) Force control and powder dispersibility of spray dried particles for inhalation. J Pharm Sci 99:303–316
77. Elajnaf A, Carter P and Rowley G (2007) The effect of relative humidity on electrostatic charge decay of drugs and excipient used in dry powder inhaler formulation. Drug Dev Ind Pharm 33:967–974
78. Johnson KL, Kendall K, Roberts AD (1971) Surface energy and the contact of elastic solids. Proc R Soc London A 324:301–313
79. Derjaguin BV, Muller BW, Toporov YP (1975) Effect of contact deformations on the adhesion of particles. J Colloid Interface Sci 53:314–325
80. Hickey AJ, Ganderton D (2007) in: Encyclopedia of pharmaceutical technology, vol 6, 3rd edn. Informa Healthcare USA, New York, pp 3862–3907
81. Miyanami K (2006) in: Powder technology handbook, 3rd edn. Taylor & Francis, Boca Raton, pp 577–589
82. Muzzio FJ, Robinson P, Wightman C, Brone D (1997) Sampling practices in powder blending. Int J Pharm 155:153–178
83. Berman J (2001) The compliance and science of blend uniformity analysis. PDA J Pharm Sci Technol 55:209–222
84. Berman J, Elinski DE, Gonzales CR, Hofer JD, Jimenez PJ, Planchard JA, Tlachac RJ, Vogel PF (1997) Blend uniformity analysis: validation and in-process testing. PDA J Pharm Sci Technol 51(Suppl 3):i–iii; S1–S99
85. Boehm G, Clark J, Dietrick J, Foust L, Garcia T, Gavini M, Gelber L, Geoffroy JM, Hoblitzell J, Jimenez P, Mergen G, Muzzio F, Planchard J, Prescott J, Timmermans J, Takiar N (2003) The use of stratefied sampling of blend and dosage units to demonstrate adequacy of mix for powder blends. PDA J Pharm Sci Technol 57:64–74
86. Howard-Sparks M, Gawlikowski A (2004) Evaluation of blend uniformity and content uniformity based on 2003 stratified sampling guidance and 1999 blend uniformity analysis guidance: product A. PDA J Pharm Sci Technol 58:222–230
87. Hersey JA (1975) Ordered mixing: a new concept in powder mixing practice. Powder Technol 11:41–44
88. Verraes J, Kinget R (1979) Ordered mixing. J Pharm Belg 34:297–307
89. Shinohara K (2006) in: Powder technology handbook, 3st edn. Taylor & Francis, Boca Raton, pp 371–382
90. Venables HJ, Wells JI (2001) Powder mixing. Drug Dev Ind Pharm 27:599–612
91. Cartillier LG, Moes AJ (1989) Effect od drug agglomerates upon the kinetcs of mixing of low dosage cohesive powders. Drug Dev Ind Pharm 15:1911–1931
92. Staniforth JN, Rees JE (1981) Powder mixing by triboelectrification. Powder Technol 30: 255–256
93. Pu Y, Mazumder M, Cooney C (2009) Effects of electrostatic charging on pharmaceutical powder blending homogeneity. J Pharm Sci 98:2412–2421

94. Shekunov BY, Feeley JC, Chow AH, Tong HH, York P (2003) Aerosolization behavior of micronized and supercritically-processed powders. J Aerosol Sci 34:553–568
95. de Boer AH, Hagedoorn P, Gjaltema D, Goede J, Frijlink HW (2003) Air classifier technology (ACT) in dry powder inhalation. Part 1. Introduction of a novel force distribution concept (FDC) explaining the performance of a basic air classifier on adhesive mixtures. Int J Pharm 260:187–200
96. Nichols SC and Wynn E (2008) New approaches to optimizing dispersion in dry powder inhalers-dispersion force mapping and adhesion measurements. Respiratory drug delivery vol 1, pp 175–184
97. Podczeck F (1998) The relationship between physical properties of lactose monohydrate and the aerodynamic behaviour of adhered drug particles. Int J Pharm 160:119–130
98. Wang C (2005) in: Inhaled particles, vol 5, 1st edn. Elsevier, London, pp 31–54
99. Gonda I (2004) in: Pharmaceutical inhalation aerosol technology, vol 134, 2nd Edn. Marcel Dekkar, New York, pp 65–88
100. Wang C (2005) in: Inhaled particles, vol 5, 1st edn. Elsevier, London, pp 55–78
101. Duddu SP, Sisk SA, Walter YH, Tarara TE, Trimble KR, Clark AR, Eldon MA, Elton RC, Pickford M, Hirst PH, Newman SP, Weers JG (2002) Improved lung delivery from a passive dry powder inhaler using an engineered PulmoSphere powder. Pharm Res 19:689–695
102. Edwards DA, Hanes J, Caponetti G, Hrkach J, Ben-Jebria A, Eskew ML, Mintzes J, Deaver D, Lotan N, Langer R (1997) Large porous particles for pulmonary drug delivery. Science 276:1868–71
103. Newhouse MT, Hirst PH, Duddu SP, Walter YH, Tarara TE, Clark AR, Weers JG (2003) Inhalation of a dry powder tobramycin PulmoSphere formulation in healthy volunteers. Chest 124:360–366
104. Dellamary LA, Tarara TE, Smith DJ, Woelk CH, Adractas A, Costello ML, Gill H, Weers JG (2000) Hollow porous particles in metered dose inhalers. Pharm Res 17:168–174
105. Tarara TE, Hartman MS, Gill H, Kennedy AA, Weers JG (2004) Characterization of suspension-based metered dose inhaler formulations composed of spray-dried budesonide microcrystals dispersed in HFA-134a. Pharm Res 21:1607–1614
106. Crowder TM, Rosati JA, Schroeter JD, Hickey AJ, Martonen TB (2002) Fundamental effects of particle morphology on lung delivery: predictions of Stokes' law and the particular relevance to dry powder inhaler formulation and development. Pharm Res 19:239–245
107. Ikegami K, Kawashima Y, Takeuchi H, Yamamoto H, Isshiki N, Momose D, Ouchi K (2002) Improved inhalation behavior of steroid KSR-592 in vitro with Jethaler by polymorphic transformation to needle-like crystals (beta-form). Pharm Res 19:1439–1445
108. Chan HK, Gonda I (1989) Respirable form of crystals of cromoglycic acid. J Pharm Sci 78:176–180
109. Chan HK, Gonda I (1995) Physicochemical characterization of a new respirable form of nedocromil. J Pharm Sci 84:692–696
110. Ikegami K, Kawashima Y, Takeuchi H, Yamamoto H, Mimura K, Momose D, Ouchi K (2003) A new agglomerated KSR-592 beta-form crystal system for dry powder inhalation formulation to improve inhalation performance in vitro and in vivo. J Control Release 88:23–33
111. Zeng XM, Martin AP, Marriott C, Pritchard J (2000) The influence of carrier morphology on drug delivery by dry powder inhalers. Int J Pharm 200:93–106
112. Larhrib H, Martin GP, Marriott C, Prime D (2003) The influence of carrier and drug morphology on drug delivery from dry powder formulations, Int J Pharm 257:283–296
113. Chan HK (2008) What is the role of particle morphology in pharmaceutical powder aerosols?. Expert Opin Drug Deliv 5:909–914

Chapter 5
Overcoming Lung Clearance Mechanisms for Controlled Release Drug Delivery

Ibrahim M. El-Sherbiny, Diana G. Villanueva, Dea Herrera, and Hugh D.C. Smyth

Abstract The lung anatomy and physiology is well adapted to handle exogenous aerosols and rapidly process them for clearance. Only in a few circumstances in nature are these defenses breached, with dire consequences, as in the case of asbestos. For controlled drug delivery to the lung, however, the goal is to slow or evade these clearance mechanisms using biocompatible systems that elicit no inflammatory response but allow modulated drug release. In this chapter, we introduce briefly the principles of the lung clearance mechanisms including mucociliary, absorptive, phagocytic, and metabolic clearance. Then, we review the literature and present the current and emerging approaches to effectively control release in the pulmonary system. These approaches include manipulating drug deposition site, modifying drug absorption rates, eluding macrophage uptake, and controlling degradation of the therapeutic agent.

Keywords Avoiding • Controlled delivery • Clearance • Lung • Phagocytosis • PEGylation • Pulmonary • Sustained release

5.1 Introduction

Development of administration methods that allow patients to safely treat themselves and maintain good compliance with a therapeutic regimen is important for cost-effective healthcare development. This is also particularly in poor countries where doctors clean syringes, sterile needles, and sophisticated treatments are rare [81]. With conventional immediate release formulations, the concentration of the drug

I.M. El-Sherbiny (✉)
College of Pharmacy, University of Texas, Austin, TX, USA
and
Mansoura University, Mansoura, Egypt
e-mail: imelsherbiny@gmail.com

H.D.C. Smyth and A.J. Hickey (eds.), *Controlled Pulmonary Drug Delivery*, Advances in Delivery Science and Technology, DOI 10.1007/978-1-4419-9745-6_5,

has an initial high peak and then declines rapidly, often below the minimum therapeutic level [57, 81]. Consequently, for drugs whose actions correlate with their tissue and blood concentrations these fluctuations in the drug levels may cause unacceptable side effects at the peaks, followed by insufficient therapy in the troughs. By contrast, regulating the rates of drug release offers numerous advantages over conventional dosage forms [56]. Some of these advantages are described below:

1. Controlled release of drugs results in a reduction in the frequency of the drug administration and should encourage patients to comply with the dosing instructions.
2. The drug could be released in a targeted region, which could maximize drug efficacy.
3. Systemic exposure of drug could be reduced upon its targeting to the desired environment. This results in decreasing the systemic side effects of the drugs (especially for toxic drugs).
4. The controlled release formulations enable the drug to be protected from the physiological environment for longer durations. Consequently, the effective residence time of the drug should be extended.

Controlled drug release to the lung is an emerging research field that may offer new opportunities for product development, product line extensions, and enhanced clinical responses for both local and systemic treatments. However, due to the efficiency of lung clearance mechanisms, our ability to control release of drug within the lung is considered one of the major challenges in pulmonary drug delivery [65, 98]. The kinetics of drug within the airways, following deposition as an aerosol, is controlled by both the physicochemical properties of the drug and location of deposition within the lung and has been discussed in Chaps. 2 and 6. As a consequence of these clearance mechanisms, researchers are very limited in their capacity to: (i) change lung residence times, (ii) overcome clearance mechanisms, and (iii) modify drug–lung microenvironment interactions. However, utilizing the current understanding of lung–particle interactions, there are some strategies evolved to overcome these limitations. Using formulation approaches, for instance, pulmonary controlled release can be achieved using liposomes, polymeric microparticles, swellable microgels, density modified particles, complexation, and drug conjugates. These drug delivery systems will be discussed with respect to their ability to control the release of drug in the lung and their specific applications. Also, some potential issues in development and manufacture of these systems will be reviewed.

5.2 Clearance Mechanisms Within the Lung

The rate of drug release in the lung is determined by various factors. These factors include the physicochemical characteristics of the drug (liquid or solid, solubility, charge, partition coefficient, and molecular weight), the carrier particle characteristics (liposomes or polymers, size, shape, surface, and bioadhesive characteristics).

In concert with these drug and formulation factors, physiological and anatomical factors also significantly determine the retention time of a drug, its delivery rate, and the speed of clearance of drug from the lungs. Specifically, the mechanism of clearance of inhaled particulates depends on their site of deposition within the lungs. If particulates are deposited in the tracheobronchial tree, then they will be rapidly cleared by the mucociliary escalator [103]. Cough clearance may also play a significant role. Particulates deposited in the lower alveolar regions of lungs are more likely to be engulfed by alveolar macrophages (phagocytosis) and/or transported across the epithelium layer to be absorbed in the blood stream. The airway metabolic environment, due to presence of enzymes, may also play significant roles in clearance [22]. The following is a brief description of each of these clearance mechanisms:

5.2.1 Mucociliary Clearance

The conducting airways are lined by epithelial cells. This cellular layer consists mainly of two different types of cells: ciliated cells and goblet (mucus-producing) cells, which collectively form the mucociliary escalator (Fig. 5.1). As introduced in earlier chapters, this mucus-cilia system serves to entrap insoluble inhaled particulates that deposit in the airways. The mucus entrapment works in a synchronized manner with the beating cilia to efficiently sweep particulates out of the lung via the trachea where they are swallowed. This clearance mechanism, based on this moving layer of mucus, operates over time periods of up to 24 h for the particulates deposited in the deepest ciliated airways [42]. The mucociliary escalator is a skillfully

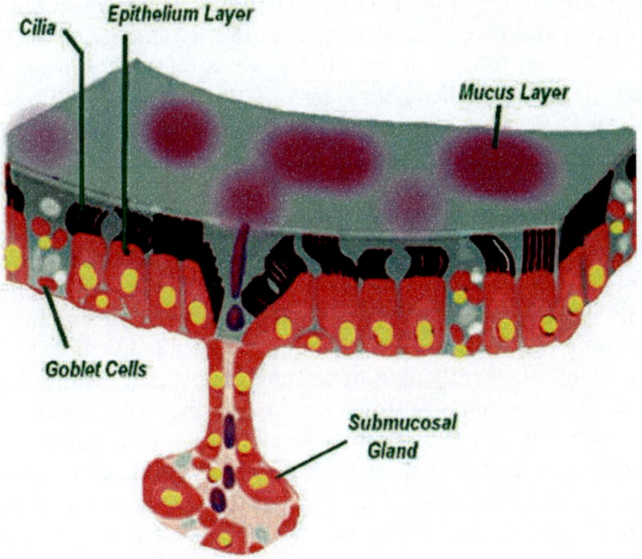

Fig. 5.1 A schematic illustration of mucociliary escalator

balanced system depending on preservation of appropriate matches of ciliated and mucus-producing cells and the normal functioning of those cells to achieve effective clearance [42, 83].

5.2.2 *Alveolar Macrophage Clearance (Phagocytosis)*

Particles penetrating down to the respiratory airways are rapidly cleared by alveolar macrophages. Alveolar macrophages are scavenger cells present in large numbers (half a billion) in human lungs and derived from monocytes in the circulation [83]. The air-side surface of each alveolus is regularly cleaned by 12–14 alveolar macrophages, which phagocytose any insoluble particles that deposit in the alveolar region [100]. Insoluble, nondigestible particulates that deposit in the alveolar region may reside in the lungs for months or even years and usually sequestered within the macrophages [63, 83].

5.2.3 *Absorption Through Epithelial Cells (Ciliated and Alveolar)*

Absorption of drugs from the lungs is extensively discussed in Chaps. 2 and 6 of this book. Briefly, the drug absorption can be rapid from the lungs given large surface area of the lungs in adult human (few meters square in the conducting airways and up to 100 m^2 of alveolar surface) [82, 121], good epithelial permeability, small aqueous volume at the absorptive surface [82] in addition to the highly dispersed nature of therapeutic aerosols (hundreds of millions of particulates per dose). Depending on the physico-chemical properties of the molecule, many inhaled therapeutic drugs that deposited in the lungs are rapidly absorbed into the systemic circulation and also cleared in order of seconds to minutes [83]. Therefore, different approaches may be required for slowing the absorption of some of the inhaled drugs so that their local and even their systemic effects can continue over a prolonged time. Several reported studies have shown that the absorption process within the lungs is primarily dependent on both nature and molecular weight of the inhaled therapeutics [2, 18, 82, 83]. For instance, small hydrophobic molecules are thought to be rapidly absorbed (within few seconds) throughout the lungs by passive diffusion through the plasma membrane. Small hydrophilic molecules may be absorbed with aid of specific transporters or through the tight junctions. In case of peptide molecules (such as insulin), it is still unsure whether they are absorbed mainly by transcytosis through caveoli or paracellularly through the tight junctions [82]. However, most researchers assume the dominance of the paracellular route. Also, it was suggested that small peptides and insulin need to be deposited in the deep lungs to accomplish optimum absorption rather than in the upper airways [2, 18]. The number of tight junctions in the airways was assumed to be roughly five times that in the alveoli [82]. The electrical resistance across the epithelium, which is a measure of the tightness of cell junctions, appears to descend from a maximum value in the trachea to a minimum in the distal airways before going up again to a high value in the alveoli. Thus, it was reported that the real site for absorption of small peptide therapeutics (such as insulin) may be the distal airways right before the alveoli [11]. Detailed discussion of diffusion and other absorptive transport mechanisms will be discussed in the next chapter.

5.2.4 Clearance Through Metabolic Degradation

Inhaled drugs and particles may be enzymatically degraded intracellularly (within alveolar macrophages) and/or extracellularly with the aid of membrane-associated enzymes (epithelial and endothelial). Mechanisms of metabolic degradation within the pulmonary system were extensively reviewed in Chap. 2.

Figure 5.2 shows a summary of the broad aspects that contribute to drug deposition and clearance. From the figure, the pulmonary clearance mechanisms and rates depend on a multitude of factors. Particles deposition site is controlled by the size and other physicochemical characteristics of particles, in addition to patient-related

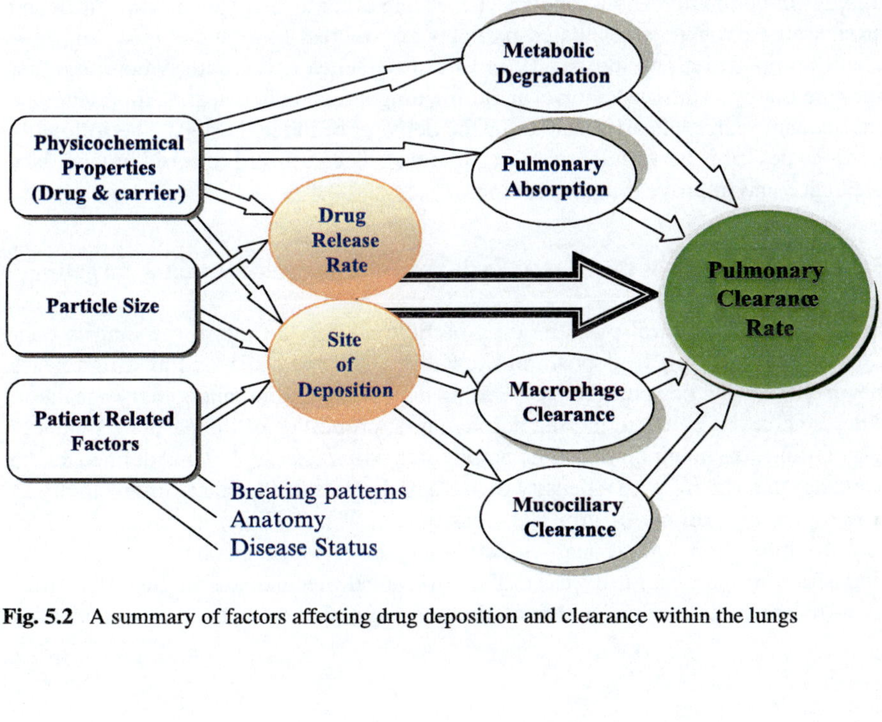

Fig. 5.2 A summary of factors affecting drug deposition and clearance within the lungs

factors such as breathing patterns, anatomy, and disease states. Moreover, both the size and physicochemical properties of the drug/carrier also control method and/or rate of drug release.

5.3 Overcoming Lung Clearance Mechanisms for Controlled Drug Release in the Lung

The efficient clearance mechanisms of the airways represent a major restriction to the controlled drug release in the lung. These mechanisms are now more deeply understood due to the significant contributions of scientists in various fields including respiratory toxicology and biomaterials sciences. In the following sections, and based on our earlier brief description of the various clearance mechanisms (more widely explored in Chaps. 2 and 6), we discuss the different reported approaches that can evade these formidable and dynamic biological responses to exogenous aerosol particulates in the airways toward achieving controlled drug release in the lung.

5.3.1 Avoiding Mucociliary Clearance

Mucociliary clearance is one the most important host-defense mechanisms [43] in both upper and lower airways. It provides protection from a variety of pollutants, allergens, and pathogens that are usually deposited in the lung and are able to trigger airways inflammation and infections [10, 90]. In order to keep the airways clean and sterile, all pathogens and inhaled particles are trapped in a mucus layer and efficiently propelled along the airway toward the trachea and mouth, where together they are finally swallowed. However, during lung infection or diseases, this clearance mechanism represents an obstacle for the delivery of inhaled drugs. The following sections describe the various strategies that have been utilized to avoid mucociliary clearance and improve drug deposition.

5.3.1.1 Avoidance of the Mucociliary Escalator via Aerodynamic Targeting

Some degree of control over the drug/particle residence time in the lung is conferred by modulating the deposition site within the airways. Extending drug release beyond a day can be achieved by avoiding deposition on the mucociliary escalator. This is achieved by changing the aerodynamic properties of the particle, typically via particle size manipulation. The aerodynamic diameter, d_a [17], is defined as the diameter of a sphere with a density of 1 g/cm^3 that has the same settling velocity as a nonspherical particle of arbitrary density [26, 60]. Both aerodynamic diameter and the inhalation flow dynamics determine the physical mechanism of particle deposition in the respiratory tract. These physical mechanisms include the diffusion by Brownian motion, sedimentation by gravitational forces, the impaction by

Table 5.1 Particle behavior in the airways based on their aerodynamic characteristics

Mechanism	Particle size (μm)	Particle density	Flow rate
Diffusional	< 0.5	Density independent	Flow rate-independent
Gravitational	> 0.5	Increase with the density	Flow rate-independent
Inertial	> 5	Increase with the density	Increase with the flow rate

inertial forces, electrostatic deposition, and interception [26, 60, 90]. In general, submicron particles with a $d_a < 0.5$ μm are usually exhaled because of the inefficient time for their diffusion to surfaces within the lung lumen although breath hold maneuvers can significantly increase deposition of "extra-fine" aerosols [17, 111]. Impaction is the predominant deposition mechanism for particles with d_a above 10 μm. Particles of this size are generally deposited in the proximal airways including the mouth, throat, and larynx. On the other hand, particles with d_a of 1–5 μm can be deposited in the alveolar airspaces [73, 113] with a limited proportion entrapped in mouth and throat (Table 5.1). Particles of this size range may escape from the mucociliary clearance if deposition occurs as deep as the alveoli. In general, particles of this size are considered optimal design for pharmaceutical inhalation aerosol for most diseases and therapeutic approaches for pulmonary drug delivery. For instance, inhalation of 1.8 and 2.8 μm aerosol of terbutaline and salbutamol was found very efficient for bronchodilation in asthmatic patients [66, 124]. Recent work by Usami et al. [118] using monodisperse drug aerosols has also demonstrated that deep lung deposition is preferable for certain drug classes. Also, systemic delivery of leuprolide acetate aerosols with d_a values of 1–5 μm in dogs has been optimal after inhalation [1].

5.3.1.2 Interactions with Extracellular Barriers

To extend drug release and drug/particle residence time in the airways, avoidance of the mucociliary escalator can also be achieved by rapidly diffusing through the mucus blanket lining the conducting airways, and accessing the periciliary fluid layer. The mucociliary clearance, a primary defense mechanism of the mammalian airways of pathogens and inhaled particles [67], is facilitated by an airway surface liquid (ASL) [122] that covers the ciliated epithelial cells. The ASL is composed of two different layers: a well-defined water layer known as periciliary liquid layer (PCL) and a mucus layer located at top of the PCL layer [99, 122]. The periciliary layer lubricates and provides a desirable liquid environment for the displacement of cilia toward an efficient mucus clearance from the lungs to the mouth [64].

Avoidance of the mucus barrier and penetration into the underlying PCL has not received much attention in the scientific literature. Mucus itself is a complex biopolymer that appears to be well adapted to entrapping a wide range of materials across different chemistries and particle sizes even to the small nanoscale. Specific surface chemistries that allow nanoparticulate systems to penetrate mucus much more rapidly than controls have been developed [58]. Technologies developed by Transave, Inc

Table 5.2 Mucoadhesion mechanisms

Mechanism	Description
Adsorption	Adhesion occurs as a result of van der Waals forces, hydrophobic interactions, electrostatic attractions, and hydrogen bonds
Diffusion	Polymer chains in the formulation diffuse into the mucus network
Electronic	Differences between the polymer and mucus electronic structure lead to the formation of a double layer at the interface where adhesion occur because of attractive forces across the double layer
Fracture	Used for calculation of adhesive bond
Wetting	Interfacial tension is used to predict adhesion on liquid preparations

Table 5.3 Mucoadhesivity in drug delivery to the lung

Drug	Mucoadhesive agent	Bioavailability (%)
Gentamicin	Chitosan	31.4
Gentamicin	Hyaluronan/chitosan	42.9
Leupolide	HPC/MCC	34.9
Ciprofloxacin	HPMC	40.2
Ciprofloxacin	HEC	19.5
Ciprofloxacin	HEC/Tween 80	25.4

(now Insmed, Inc) also describe liposome formulations that achieve higher mucus penetration [70].

Due to the specific surface characteristics of respiratory mucus such as adhesivity and wettability [43], it is able to bind to solid surfaces readily and almost non-specifically. This mucoadhesion ability thus allows an interaction between the drug and mucus layer. The contact time between a formulation and mucus may have a direct impact on the residence time and bioavailability of a delivered drug. An increase in the mucoadhesiveness of mucosal delivered drug is likely increasing the absorption of drug as a result of the prolonged contact at the site of absorption. This is a common approach in oral drug delivery where mucoadhesives can promote drug residence time. Several methods have been applied to develop pharmaceutical formulations with a mucoadhesion through the creation of an intimate contact between the dosage form and mucus (Table 5.2) [39]. This has been achieved by using polymers with mucins and hydrophobic domains. Several characteristics of the polymeric dosage forms can affect this contact and interactions including molecular weight, concentration, cross-linking density, chemical structure, and particle size [25].

Several antibiotics (Table 5.3), such as gentamicin and ciprofloxacin, have been delivered with mucoadherent properties employing common polymers. Also, lectin was linked to therapeutic molecules for epithelial adhesion and internalization. Mucolytic agents such as *N*-acetylcysteine and *S*-carboxymethylcysteine have led to a reduction in the viscosity of the mucus due to breaking of the disulphide bonds of glycoproteins presents in the mucus [25].

5.3.2 *Avoiding Rapid Drug Absorption*

In some cases, inhaled therapeutic molecules that deposited in the lung are very rapidly absorbed and cleared from the blood so quickly, in order of seconds or minutes, that their local concentrations in lung tissue can be very short. This may then require a frequent drug administration that leads to product and therapeutic failure due to patient compliance issues. The following section will discuss the various approaches that can be applied for prolonging the absorption of some inhaled therapeutic agents so that their systemic effect can continue over enough time.

5.3.2.1 Drug Structural Modification

Extending release of therapeutic entities can sometimes be achieved by modification of the chemical structure, if these modifications do not impact on the activity of the molecule. Binding to tissues may be one way of promoting drug retention. Some preclinical investigations have shown that a very small amount of some rapidly absorbed inhaled therapeutics was slowed in their pulmonary absorption due to binding with tissues in the airways [30, 71]. Moreover, it has been reported by various researchers that the prolonged retention of trace amounts of many hydrophobic drugs occurs primarily in the airway tissues [45, 71]. For instance, it was found in case of some retained steroids that the molecules have been esterified in the tissue to long-chain fatty acids, thereby rendering them very hydrophobic and insoluble. This tends to explain the extended duration of topical pharmacological effects of these hydrophobic drugs in the lungs [45, 71]. In addition, it was reported that presence of positive charges onto some moderately lipophilic drugs under physiological conditions (as in case of pentamidine and verapamil) allows these drugs to bind preferentially to lung tissues [12]. This can be one of the possible reasons for the successful development of tobramycin as an inhaled antibiotic with half-life of several hours, owing to its physiological existence as a polycation [72]. Therefore, one of the strategies that can be applied to extend drug retention in the lungs is relied on drug structural modifications that increase hydrophobicity and lung tissue binding [51].

5.3.2.2 Drug Formulations (Microencapsulation)

A prolonged absorption of inhaled drugs can be achieved via formulating them with some types of excipients/carriers that impart slow dissolution in the lungs and yet offer an extended drug release. Also, selection of appropriate carrier can help in the delivery to a specific area of the lungs. Then, once delivered, a carrier can further influence the drug distribution and rate of clearance from the site of action. In addition to prolonging the residence time of the drug at its site of action via reducing clearance, these carrier-based formulations offer many other advantages. These include improving the drug stability in vivo and also the product shelf life time, reducing irritation caused by the drug, and also decreasing toxicity due to high initial doses of the drug [89].

In general, the choice of a carrier depends on several factors, including the physicochemical characteristics of the drug to be delivered, the used inhaler device for delivery, the disease status, the site of action, and the nature and safety of the carrier. The most common carrier-based formulations, which will be covered throughout this section, are liposomes and polymeric microparticles.

Liposomes

A significant body of research has focused on the use of liposomes in drug delivery [14, 24, 33, 68, 69, 101, 102, 125, 126]. Chapter 14 specifically focuses on liposomal delivery systems for pulmonary drug delivery. Liposomal-based aerosol formulations have numerous advantages, including sustained drug release, reduced toxicity, avoidance of local irritation, enhanced stability in the large aqueous core, and the possibility to control release and targeting by manipulating the composition and also through changing the preparation method [19]. Drug loading efficiency, release rate, and the deposition site of liposomes in the lungs depend primarily on the lipid composition, liposomes size, drug to lipid ratio, net charge in addition to the delivery method [33, 102, 125].

Liposomes are mainly composed of neutral or anionic lipids (natural or synthetic). The most commonly used lipids are lecithins, phosphatidyl ethanolamines, sphingomyelins, phosphatidyl glycerols, and phatidylserines [59]. Liposomal aerosols may be used for inhalation either in liquid [35] or as dry powder form [91]. The dry powder liposomal aerosols have been prepared either by spray-drying [50, 92, 97] or via lyophilization followed by milling [46, 50, 91, 92, 97]. In the case of liquid forms of liposomal aerosols, drug can release during nebulization; however, this release can be minimized upon changing the operating conditions [75], the lipid composition [21, 76], and liposomes size [77].

One of the most sophisticated classes of liposomal aerosol formulations is the targeted (reactive) liposomes. This class contains certain targeting moieties (ligands) attached to liposomal surface, which allows liposomes to interact with certain receptors and/or cell types within the lungs [89]. These targeting moieties include lectins and monoclonal antibodies. Within the class of targeted liposomes, there is a certain type that acts as stimuli-responsive liposomes. This type tends to alter its phase and structure upon changing of some of the environmental stimuli such as pH and ionic strength [59].

Encapsulation of inhaled therapeutic agents in liposomes normally increases residence time and/or decreases toxic side effects of these therapeutics delivered to the lungs [89]. For instance, upon encapsulation of cyclosporine in liposomes and delivering to the lungs, the drug was retained for 120 min in a dog model [61]. Also, pulmonary liposomal delivery allowed an aqueous-based formulation to be prepared and, when tested in mice by Arppe et al. [4], it increased significantly the lung retention times from 17 min for free drug to 4.8 h in the case of formulated drug. In a comparative study on the pulmonary delivery of tobramycin encapsulated into PLA microspheres and liposomes following endotracheal and intravenous delivery, Poyner et al. [85] have found that the renal drug levels of the intravenously delivered microencapsulated tobramycin were significantly higher as compared to the liposomal-based formulations at 6 and 24 h. However, liposomes showed pulmonary levels three times higher than those of the free drug both at 6 and 24 h. After 24 h, renal drug levels following the endotracheal delivery were lower for both encapsulated formulations than for free drug. On the other hand, pulmonary drug levels were higher following administration of the encapsulated drug as compared to the levels following the delivery of free drug delivery at 24 h. These observations revealed that, with the aid of microencapsulated or liposomal-based formulations, tobramycin can be retained in the lung. Also, recombinant secretory leukocyte protease inhibitor (rSLPI) was found to reveal promising results in treatments of inflammatory lung conditions. However, it faces a rapid clearance from the lung after inhalation due to fast absorption and degradation. In a recent investigation, Gibbons et al. [34] encapsulated rSLPI in 1,2-Dioleoyl-sn-Glycero-3-[Phospho-L-Serine]:Cholesterol (DOPS:Chol) liposomes for inhalation. This encapsulation allows retention of 92.6% of the rSLPI activity. Moreover, the developed liposomal-based formulations of rSLPI were biocompatible, stable, and potentially reduced the required dosing frequency for therapeutic effect after inhalation.

Microparticles

Microparticles are an important class of carriers that may be used to encapsulate drugs in order to extend their retention in the lungs in addition to offering a sustained drug release. Microparticles are physically and chemically more stable than liposomes and allow for higher drug loading capacity [89]. Therefore, microparticles were considered good carrier candidates for pulmonary delivery of several drugs [28, 44, 115, 123]. Microparticles-based aerosols are produced from polymeric materials (natural or synthetic). However, some lipid-based microparticles were also developed [7, 9]. The most commonly used natural polymers are chitosan, gelatin, dextran, and albumin while, poly(lactic-co-glycolic) acid (PLGA), polylactic acid (PLA), and polyethylene glycol (PEG) represent the most used synthetic polymers. Microparticles can be prepared using a wide range of techniques including spray drying [7–9, 84], emulsion-solvent evaporation [23, 31, 32, 49], and supercritical fluid technology [16].

Drug release from microparticles depends primarily on drug characteristics such as size, concentration, and solubility and polymer characteristics such as structure, molecular weight, morphology, porosity, and mechanical properties. In addition to enhancing aerosol particle characteristics, an appropriate polymeric carrier can reduce the rapid drug absorption and consequently allow prolonged retention time of drug in the lungs. For instance, PLGA microparticles (average size, 0.4 μm) have extended the duration of action of an inhaled dose of insulin significantly from 6 h (for a nebulized aqueous solution of insulin) up to 48 h in guinea pigs [49]. Also, modification of polymeric microparticles via coating can be used to alter their characteristics

Table 5.4 Some reported examples of microparticles for prolonged local pulmonary drug delivery

Drug	Polymeric carrier	Size (μm)	Duration of effect	References
Ceftazidime and ciprofloxacin	Dipalmitoylphosphatidylcholine, albumin, and lactose	About 5		[115]
Lysozyme and doxorubicin·HCl	Poly(lactic-co-glycolic) acid (PLGA)	4.5–4.6 MMAD	Up to 4 days	[123]
Beclomethasone dipropionate (20% w/w; 0.25 mg/kg HPC/BDP vs. 1.37 mg/kg BDP alone in guinea pigs)	Hydroxypropylcellulose (HPC, 200–450 kDa)	2.5 MMAD	24 vs. 1–6 h inhibition of eosinophil infiltration	[88]
Isoprenaline (7%w/w; 0.1 mg/kg dose in rats)	Poly(lactide-co-glycolide) (PLG, 34 kDa)	4.5	≥12 vs. ≤0.5 h bronchoconstriction	[58]
Salbutamol sulfate (4% w/w; 0.29 mg/kg in guinea pigs)	Dipalmitoyl phosphatidylcholin/human serum albumin/lactose (DPPC/HSA/lactose)	1.6 MMAD	≥16 h (porous) vs. ≤5 h (nonporous) bronchoconstriction	[5]
Elastase inhibitor	Human serum albumin (HSA)	2–4		[37]
Rifampicin (30% w/w; 12 mg/kg in guinea pigs)	PLGA, 85.2 kDa	2.8	Effective for 28 days	[104]
Rifampicin/Isoniazid	Poly(lactic acid) (PLA)	6.2–6.8	Higher drug levels in macrophages compared to oral delivery	[105]
Budesonide (99% w/w)	PLGA, 5–6 kDa	< 5	12 vs. 6 h binding to receptors ex vivo (rat)	[110]

Table 5.5 Influence of PEGylation of some therapeutic agents on their duration of effect

PEGylated drug	Duration of effect	References
Interferon-2α	Increased half-life (from 3–8 to 65 h)	[127]
Tumor necrosis factor	Increased half-life (from 3–45 to 136 min)	[116]
Interleukin-6	Increased half-life (from 2.1 to 206 min)	[117]

in vivo [89]. For instance, coating of some therapeutic peptide-loaded microparticles with mucoadhesive polymers such as hydroxypropyl cellulose and chitosan has extended the residence time of these drugs in the lung [109]. Table 5.4 shows some reported examples of microparticles for prolonged local pulmonary drug delivery.

PEGylation

PEGylation is a process that involves covalent conjugation of PEG to certain types of drugs such as therapeutic proteins. PEGylation is routinely achieved by incubation of a reactive derivative of PEG with the target therapeutic macromolecule. It works as a good approach for avoiding the rapid drug absorption and consequently acts to prolong the residence time (half-life) of drugs in the lungs via reducing their degradation. At the same time, it helps in reducing the side effects through limiting the extent of systemic absorption [74]. PEGylation can also provide water solubility to hydrophobic drugs and proteins [6, 38].

Some studies that carried out onto some inhaled PEGylated drugs showed their safety [52]. It was also found by Niven et al. [78] that PEGylation process has prolonged the effect of rhG-CSF upon delivery to the airways of male rats. Moreover, Nektar Pharmaceuticals is in the very early stages of developing a long-acting inhaled insulin using PEGylation [89]. The influence of PEGylation of some therapeutic agents on their absorption rate (duration of effect) upon delivery is shown in Table 5.5.

Prodrugs

Prodrugs are pharmacological agents that are administered as inactive forms. Then, once administered, they undergo chemical conversion through metabolism into active forms. The reason behind the use of prodrugs is generally for improving their bioavailability. Also, using therapeutics agents in the form of prodrugs enhances the selectivity of these agents for its intended target which in consequence will increase the bioavailability, prolong the absorption time in addition to reducing the side effects associated with these therapeutic agents. For instance, Olsson and Svensson [80] have developed and evaluated two lipophilic terbutaline ester prodrugs. Once inhaled, these prodrugs achieved enhanced absorption and a reasonably hydrolytic stability during first-pass which led to a prolonged duration of the therapeutic effect.

5.3.3 *Avoiding Macrophage Clearance*

5.3.3.1 Particle Size

Rapid clearance of drugs from the lungs by alveolar macrophages (phagocytosis) is considered as one of the main reasons of current inability to control the pharmacokinetics of inhaled drugs beyond few hours [36]. Size of inhaled particles is a key factor that controls the delivery of therapeutics through the pulmonary route and particularly affects the rate of engulfment by alveolar macrophages. Specifically, the inhaled carrier particles, targeted to the deep lung, should be small enough to pass through the mouth, throat, and conducting airways and deposit in the deep lung, but not so small that they fail to deposit and are exhaled again [36]. These particles should have diameters in the range 0.5–5 μm to be respirable. However, microparticles of this size have rapid clearance from lung by macrophages. Various approaches have been reported in literature to overcome this major challenge in pulmonary drug delivery via developing of appropriate carrier systems with adequate aerodynamic characteristics that will allow particles to be respirable and at the same time are able to evade macrophage uptake and confer prolonged drug release once deposited in the lung.

One of the main emerged approaches is the development of large porous particles (LPPs). This type of particles has geometric diameters more than 5 μm, but their aerodynamic diameters are less than 5 μm due to their low density (around 0.1 g/l or less) [27]. These particles showed promising flowability and aerosolization characteristics relative to conventional inhaled therapeutic aerosols. In addition, LPPs possess a good potential for avoidance macrophage uptake because of their large aerodynamic size [26]. It was reported also that, LPPs, once inhaled, can deposit homogeneously on the cell surface and revealed a relative nontoxicity to the airway cells [32].

A second approach for avoiding macrophage clearance, relying on controlling particle size, is through using inhaled particles with geometric diameters less than a few hundred nanometers [79]. It has been reported that these nanoparticles (NPs), once deposited, tend to remain in the lung lining fluid and evading both phagocytosis and mucociliary clearance [48, 55, 87, 119]. Therefore, using NPs as carriers for drugs delivered to the lung tends to offer the potential for prolonged drug effect and release throughout the lumen of the lung, not only in the alveolar region, where phagocytosis occurs [48, 55, 87, 119]. However, NPs for pulmonary delivery (as detailed in Chap. 15) have delivery limitations. Specifically, their small size leads them to be predominantly exhaled from the lung after inhalation as primary particles [40, 41]. Moreover, nanoparticles generally require the large amounts of energy for their formation during inhalation. Therefore, most researchers have taken the approach whereby nanoparticles are delivered in microparticles (such as in a dispersion for nebulization). In a more recent approach, a certain class of particles that combines the benefits of both LPPs and NPs while avoiding their drawbacks has developed for pulmonary drug delivery [114]. These particles are called "Trojan" microparticles due to their ability to escape both phagocytosis and mucociliary

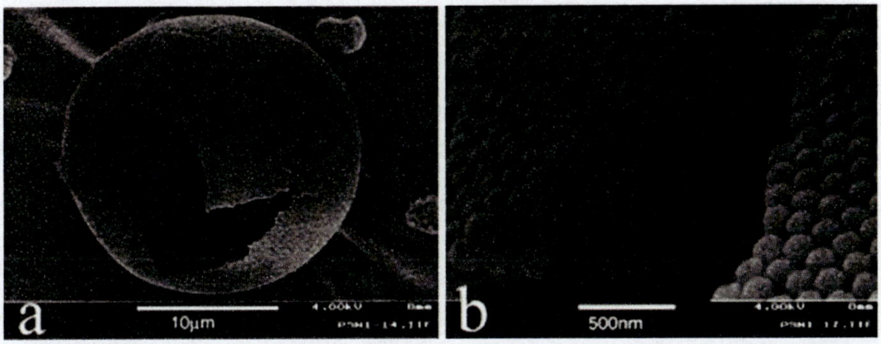

Fig. 5.3 Scanning electron micrographs of Trojan particles (reproduced, with permission, from Tsapis et al. [114])

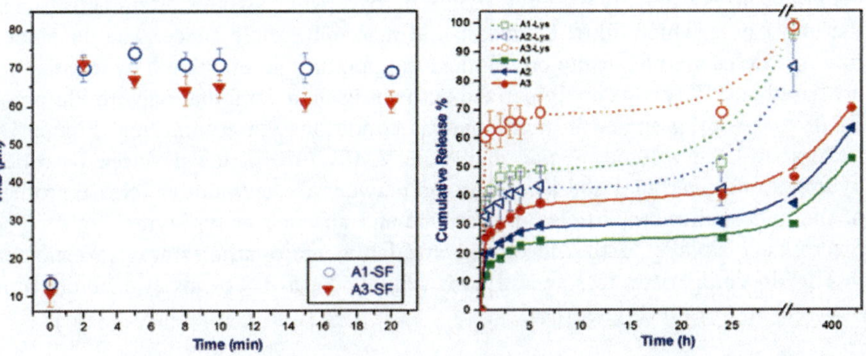

Fig. 5.4 Volume median diameter (VMD, µm) of developed swelled hydrogel microparticles (*left*) and a comparison between the cumulative release profiles of the microparticles in PBS, pH 7.4 at 37°C in presence and absence of lysozyme (*right*) [29]

clearance within the airways. Trojan particles (Fig. 5.3) are produced through spray drying of polymeric NPs (such as polystyrene, PS-NPs) where, during the spray drying process, the PS-NPs assemble into hollow micron-scale particles with low density (less than 0.1 g/l) [114]. The assembling of NPs occurs through either physical forces such as Van der Waals forces or within a matrix of some additional materials such as phospholipids or biopolymers. Trojan microparticles offer a potential for producing a dry powder inhaler system with good flowability and dispersability characteristics, which, once delivered to the peripheral airways, will release NPs that will be able to avoid pulmonary clearance and offer sustained drug release.

It has been reported by several groups [3, 65] that increasing the microparticle size reduces their clearance by macrophages. However, increasing microparticle size for pulmonary drug delivery is impractical as this would minimize the respirable fraction that can be delivered. Therefore, a very recent approach has been applied

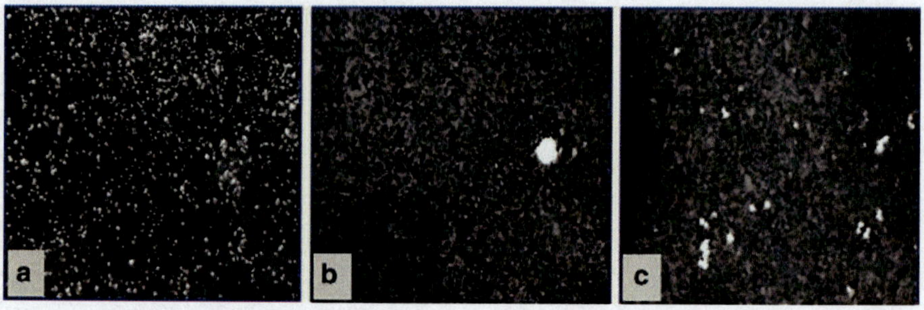

Fig. 5.5 Particle uptake by macrophages in vitro (**a**) nonswellable 1 μm PS beads at 60 min incubation time point, (**b**, **c**) Swellable microparticles at 12 h [29]

via development of swellable microparticles [29]. These microparticles have respirable sizes when dry but large geometric sizes when swollen after deposition in the moist lung which allow them to avoid macrophage clearance. The developed microparticles are physically crosslinked (amphiphilic interactions) hydrogels that are based on PEG graft copolymerized onto chitosan in combination with Pluronic® F-108 and were prepared in mild aqueous conditions via cryomilling. Figure 5.4 (left) shows the volume median diameter (VMD, μm) of the developed swelled hydrogel microparticles and a comparison between the cumulative release profiles of the hydrogel microparticles in presence and absence of lysozyme (right). The macrophage uptake of the developed swellable microparticles was investigated in vitro in comparison to standard nonswellable 1 μm PS beads at different time intervals up to 24 h (Fig. 5.5).

5.3.3.2 Particle Shape

Alveolar macrophages in the human lung encounter target particulates with a high diversity not only in their sizes but also in their shapes. Therefore, it is expected that phagocytic clearance from the airways will remain efficient despite moderate changes in particle shape. Observations made with needle-shaped particulates, such as asbestos, have hinted that aspect ratio may be important. Several studies have been carried out particularly to investigate how particle geometry affects phagocytosis [13, 48, 54, 87, 96, 108]. These studies mainly focused on particle size effects due to the difficulties of fabricating nonspherical particulates of controlled dimensions. Up until recently, the roles of particle size and shape in phagocytosis, despite their high relevance, remain largely unidentified [15]. Champion and Mitragotri [15] have reported, with aid of alveolar macrophages (as model phagocytes) and polystyrene (PS) particulates of various sizes and shapes (as model targets), that particle shape at the point of first contact by macrophages, not particle size, decisively determines whether macrophage cells will continue with phagocytosis or just spread over the particles. Particle size, on the other hand, mainly impacts the completion of phagocytosis when the particle volume is larger than the macrophage volume [15].

In this study, an array of nonopsonized and *IgG*-opsonized geometrically anisotropic PS particles representing six different geometric shapes was fabricated. These shapes include, for instance, spheres, rectangular disks, elliptical disks, and oblate ellipsoids. The test particles were incubated with alveolar macrophages and observed under a light microscope with time-lapse video microscopy. It was found that internalization of both opsonized and nonopsonized particles displayed a strong dependence on particle shape from the perspective of the macrophage cell. It was also found that points of initial contact of the macrophage on the same particle (except for spheres) influence rates of engulfment. For instance, it was found that macrophages that attached to the elliptical disks along the major axis internalized the particles very quickly, in less than 6 min. By contrast, the cells that attached to the same elliptical disks along their minor axis or through their flat surface did not engulf them, even after 2 h. They did, however, spread over the particle surface.

Similar results were observed for all other shapes of particles, where engulfment does not occur when cells attach to the concave region but engulfment occurred upon attachment to the dome or ring regions. This orientation bias was

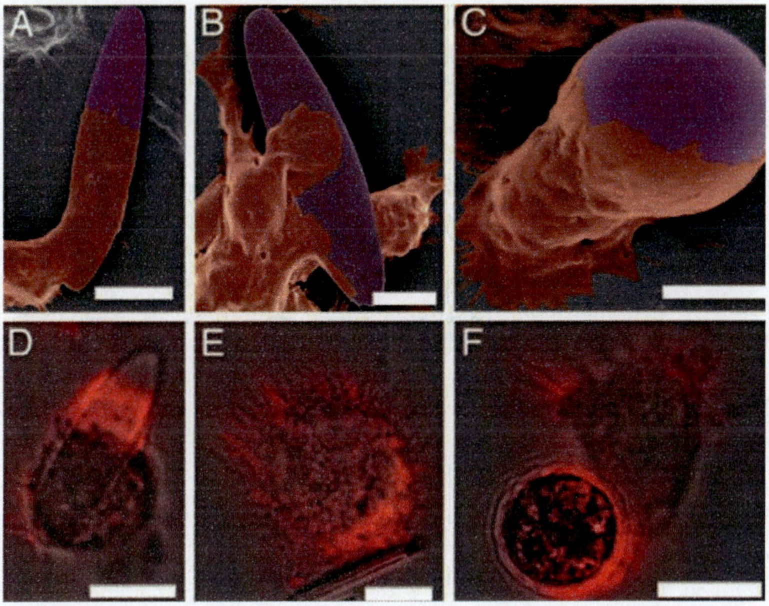

Fig. 5.6 Scanning electron micrographs **(a–c)** and actin staining **(d–f)** illustrate the effect of particle shape, at the point of first contact with macrophages, on determination whether the macrophage cells will engulf or just spread over particles (reproduced with permission from Champion and Mitragotri [15])

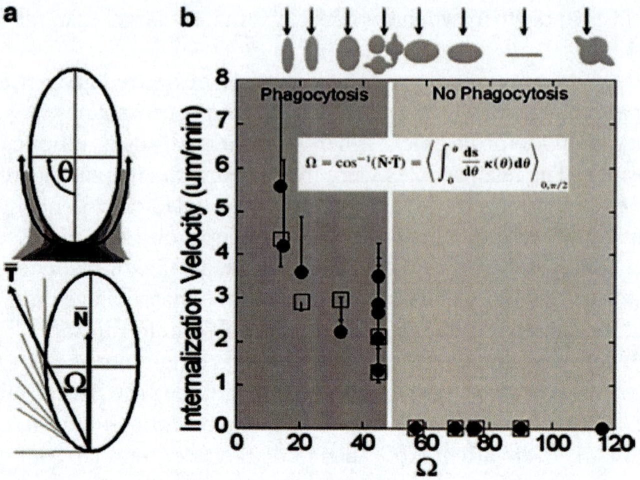

Fig. 5.7 A schematic diagram shows the dependence of phagocytosis occurrence on particle shape and its orientation upon attaching to macrophage cell. *T* represents the average of tangential angles from $\theta = 0$ to $\theta = \pi/2$. Ω is the angle between *T* and cell membrane at the site of attachment, *N* and the internalization velocity, μm/min is the distance traveled by the cell membrane divided by the time required to internalize (the *filled circles* represent the nonopsonized particles while *open squares* correspond to *IgG*-opsonized particles) (reproduced with permission from Champion and Mitragotri [15])

confirmed using scanning electron microscopy (Fig. 5.6) [15]. As shown in the figure, the cell membrane progresses on elliptical disks when the particle was approached along the major axis (Fig. 5.6a). By contrast, cells that attached to the flat side of the elliptical disks showed a kind of spreading but no internalization of particles, even after 2 h (Fig. 5.6b). Figure 5.7c shows an opsonized spherical particle that attached to the top of a macrophage cell, and the cell membrane has progressed over approximately half the particle. The Fig. 5.6d–f represents overlays of bright-field and fluorescent images after fixing the cells and staining for polymerized actin with rhodamine phalloidin [15]. A schematic representation for the dependence of phagocytosis occurrence and engulfment rate on both particle shape and particle orientation upon attaching to macrophage cells is shown in Fig. 5.7 [15].

5.3.3.3 Stealth Characteristics

Development of inhaled therapeutics with biological stealth characteristics represents one of the promising strategies that may be applied to avoid clearance by alveolar macrophages (phagocytosis). These stealth characteristics can be achieved either through attachment of the inhaled therapeutic ingredient to a material that offer this stealthiness or by coating of the therapeutics with this material. For instance, covalent attachment of PEG to inhaled drug or therapeutic protein can mask this therapeutic agent from the pulmonary immune system (reduced immunogenicity and antigenicity) and consequently prolongs its circulatory time via reducing its clearance [38]. Also, surface coating of the microparticles delivered to the lung can be used to alter their stealth characteristics in vivo. For instance, Evora et al. [31] have found that coating of protein peroxidase-loaded PLGA microspheres with the lipid dipalmitoyl phosphatidylcholine (DPPC) decreased uptake of the cargo protein peroxidase into macrophages. In some other studies, hyaluronic acid (HA), a naturally occurring mucoadhesive polysaccharide present in lung, was formulated with inhaled therapeutics in order to evade macrophage clearance. For example, it was found that phagocytosis of Alterlatex was inhibited in a dose and molecular weight dependent way by the HA [107], and this led to prolonged action of HA-based formulations close to the main absorption site in the deep lung. Also, in a separate study, Surendrakumar et al. [106] investigated the potential of HA-based dry powder formulations for pulmonary controlled release of insulin. In this study, recombinant human insulin was cospray dried with HA to form inhaled microparticles delivered to the lungs of male Beagle dogs. It was found that HA-based formulations containing 10% insulin have prolonged mean residence time (MRT) and terminal half-life as compared to spray dried pure insulin.

5.3.3.4 Drug Release within/from Macrophages

As discussed earlier, the inhaled particles reaching the deep lung are phagocytosed rapidly by alveolar macrophages. Although this phagocytosis process of inhaled powders may represent a problem for drug delivery to other cells comprising lung tissue, it is beneficial for chemotherapy of some diseases. In case of tuberculosis (TB), for instance, phagocytosed drug-loaded microparticles can potentially deliver larger amounts of the drug to the cytosol where a population of the bacteria are located. In addition, macrophage delivery may serve as a drug reservoir, aiding in reducing dosage frequency and magnitude, which is favorable for maintaining drug levels and improving patient compliance [93]. As example, Sharma et al. [93] have developed inhalable PLA microparticles that delivered a combination of two anti-TB drugs, rifampicin and isoniazid, to the macrophages in a similar proportion when phagocytosed.

Moreover, an efficient chemotherapy can be achieved in case of some pulmonary diseases such as TB via drug targeting to the alveolar macrophages using ligand-anchored microparticles or liposomes. For instance, a trial has been made by Vyas et al. [120] to develop and assess the performance of rifampicin-loaded ligand-appended inhaled liposomes for the selective targeting of infected alveolar macrophages for treatment of TB. As another example, various studies [20, 62, 86] have reported the targeting of peptides and protein therapeutics to alveolar macrophages depending on the reported facts that, mammalian macrophages can engulf glycoproteins with exposed mannose residues [86, 112] and the presence of the receptors for lactoferrin, transferrin, interleukin-2, immunoglobulins, and the granulocyte-macrophage colony-stimulating factors on the cell membranes of alveolar macrophages [47, 94].

5.3.4 Avoiding Biological and Macromolecule Degradation

Over the last decade, the advances in biotherapeutics have led to the production of various agents to be used in clinical treatment such as growth factors, hormones, monoclonal antibodies, and cytokines [95]. However, the delivery of these proteins by the pulmonary route has become a challenge because they are prone to enzymatic degradation, which threatens their physical and chemical stabilities. These enzymes include peptidases and proteases.

In order to stabilize proteins and peptides against degradation while promoting their lung adsorption, the addition of protease inhibitors or packing of the macromolecules into particles to prevent proteolysis and the inactivation of these biomolecules have been applied [53]. Nafamostat mesilate, aprotinin, and (*p*-amidinophenyl)-methanesulfonylfluoride-HCl are some examples of protease inhibitors [95].

Protein formulations have been protected from degradation using DPPC, an endogenous surfactant present in the lungs. Also, surfactants such as polysorbates, polyvinyl alcohol have been used. Polysaccharides such as trehalose, mannitol, and sucrose with a high T_g have been also used to maintain the stability of proteins and peptides [39]. Proteins and enzymes such as glutathione, superoxide dismutase, and catalase have been encapsulated by liposomes as vehicles for pulmonary drug delivery to improve their stability. Since liposomes have chemical and structural similarities to the biological cell membrane, they can merge into the cell membranes and facilitate drug delivery [95]. Another way to decrease protein degradation and improve their bioavailability is by using cyclodextrins, which interact with aromatic

Table 5.6 Strategies for stabilization and absorption enhancement of protein drugs

Stabilizer	Compound
Protease inhibitiors	Antipain, aprotinin, bestatin, nafamostat, pepstatin, phenanthroline, elastinal foroxymithin, amastatin
Liposomes	Chitosan/tripolyphospathe and phospholipids, dipalmitoyl phosphatidylcholine, human serum albumin and lactose (DPPC/HAS/lactose), poly (lactic-co-glycolic) acid (PLGA). Fatty acids and nonionic surfactants: glycerol trioleate, lauryl ether, linoleic acid, oleyl alcohol, sorbitan monooleate
Cyclodextrins	α-cyclodextrin, β-cyclodextrin, γ-cyclodextrin, dimethyl-β-cyclodextrin
Surface active agents	Bile acids and their salts: sodium cholate, sodium deoxycholate, sodium glycoholate, sodium taurocholate

amino acids present in proteins and reduce their degradation [53]. Table 5.6 [95] also shows some of the reported methods to enhance the stability of drugs in the lung.

5.4 Conclusion

Airway clearance mechanisms are formidable, but research in this area has shown them to be surmountable. The complexity of the particle–lung interactions, excipient safety and selection, control over deposition, and many other aspects have delayed the widespread use and clinical translation of controlled release delivery systems in the lung. However, given the pressing clinical need and the growing understanding of the physiological dynamics of the pulmonary system, future products will, without doubt, be deployed such that sustained, modified, and controlled release can occur.

References

1. Adjei A, Hui J, Finely R et al (1994) Pulmonary bioavailability of leuprolide acetate following multiple dosing to beagle dogs: some pharmacokinetic and preclinical issues. J Pharm Sci 107:57–66
2. Adjei AL, Carrigan PJ (1992) Pulmonary bioavailability of LH-RH analogs: some pharmaceutical guidelines. J Biopharm Sci 3:247–254
3. Ahsan F, Rivas IP, Khan MA, Súarez-Torres AI (2002) Targeting to macrophages: role of physicochemical properties of particulate carriers–liposomes and microspheres on the phagocytosis by macrophages. J Control Release 79:29–40
4. Arppe J, Vidgren M, Waldrep JC (1998) Pulmonary pharmacokinetics of cyclosporin A liposomes. Int J Pharm 161:205–214
5. Ben-Jebria A, Chen D, Eskew ML et al (1999) Large porous particles for sustained protection from carbachol-induced bronchoconstriction in guinea pigs. Pharm Res 16:555–561
6. Beyerle A, Merkel O, Stoeger T, Kissel T (2010) PEGylation affects cytotoxicity and cell-compatibility of poly(ethylene imine) for lung application: structure–function relationships. Toxicol Appl Pharmacol 242:146–154
7. Bhat M (2002) Development of a novel spray-drying technique to produce particles for aerosol delivery. In: Dalby R, Byron PR, Peart J, Farr SJ (eds) Respiratory drug delivery XIII. Davis Horwood International Publishing, Raleigh, NC, pp 427–429
8. Bittner B, Kissel T (1999) Ultrasonic atomization for spray drying: a versatile technique for the preparation of protein loaded biodegradable microspheres. J Microencapsul 16:325–341
9. Bot AI, Tarara TE, Smith DJ et al (2000) Novel lipid-based hollow-porous microparticles as a platform for immunoglobulin delivery to the respiratory tract. Pharm Res 17:275–283
10. Boucher RC (2003) Regulation of airway surface liquid volume by human airway epithelia. Eur J Physiol 445:495–498
11. Boucher RC, Stutts MJ, Gatzy JT (1981) Regional differences in bioelectric properties and ion flow in excised canine airways. J Appl Physiol 51:706–714
12. Byron PR (1993) Physicochemical effects on lung disposition of pharmaceutical aerosols. Aerosol Sci Tech 18:223–229
13. Cannon GJ, Swanson JA (1992) The macrophage capacity for phagocytosis. J Cell Sci 101: 907–913

14. Canonico AE, Plitman JD, Conary JT et al (1994) No lung toxicity after repeated aerosol or intravenous delivery of plasmid cationic liposome complexes. J Appl Physiol 77:415–419
15. Champion JA, Mitragotri S (2006) Role of target geometry in phagocytosis. PNAS 103:4930–4934
16. Cheng YS, Yazzie D, Gao J et al (2003) Particle characteristics and lung deposition patterns in a human airway replica of a dry powder formulation of polylactic acid produced using supercritical fluid technology. J Aerosol Med 16:65–73
17. Chow AH, Tong HH, Chattopadhyay P et al (2007) Particle engineering for pulmonary drug delivery. Pharma Res 24:411–432
18. Codrons V, Vanderbist F, Ucakar B et al (2004) Impact of formulation and methods of pulmonary delivery on absorption of parathyroid hormone (1–34) from rat lungs. J Pharm Sci 93:1241–1252
19. Crommelin D, Schreier H (1994) Liposomes. In: Kreuter J (ed) Colloidal drug delivery systems. Marcel Dekker, New York, pp 73–190
20. Derrien D, Midoux P, Petit C et al (1989) Muramyl dipeptide bound to poly-L-lysine substituted with mannose and gluconoyl residues as macrophage activators. Glycoconj J 6:241–255
21. Desai T, Hancock REW, Finaly WH (2002) A facile method of delivery of liposomes by nebulization. J Control Release 84:69–78
22. Devereux TR, Domin BA, Philpot RM (1989) Xenobiotic metabolism by isolated pulmonary cells. Pharmacol Ther 41:243–256
23. Dhiman N, Khuller GK (1998) Protective efficacy of mycobacterial 71-kDa cell wall associated protein using poly (DL-lactide-co-glycolide) microparticles as carrier vehicles. FEMS Immunol Med Microbiol 21:19–28
24. Eastman S, Tousignant JD, Lukason MJ et al (1997) Optimisation of formulations and conditions for the aerosol delivery of functional cationic lipids: DNA complex. Hum Gene Ther 8:313–322
25. Edsman K, Hägerström H (2005) Pharmaceutical applications of mucoadhesion for the non-oral routes. J Pharm Pharmacol 57:3–22
26. Edwards DA, Ben-Jebria A, Langer R (1998) Recent advances in pulmonary drug delivery using large, porous inhaled particles. J Appl Physiol 85:379–385
27. Edwards DA, Hanes J, Caponetti G et al (1997) Large porous particles for pulmonary drug delivery. Science 276:1868–1871
28. Ehrhardt C, Fiegel J, Fuchs S et al (2002) Drug absorption by the respiratory mucosa: cell culture models and particulate drug carriers. J Aerosol Med 15:131–139
29. El-Sherbiny IM, McGill S, Smyth HDC (2009) Swellable microparticles as carriers for sustained pulmonary drug delivery. J Pharm Sci 99:2343–2356
30. Esmailpour NH, Hogger P, Rabe KF et al (1997) Distribution of inhaled fluticasone propionate between human lung tissue and serum in vivo. Eur Respir J 10:1496–1499
31. Evora C, Soriano I, Rogers RA et al (1998) Relating the phagocytosis of microparticles by alveolar macrophages to surface chemistry: the effect of 1,2-dipalmitoylphosphatidylcholine. J Control Release 51:143–152
32. Fiegel J, Ehrhardt C, Schaefer UF, Lehr CM, Hanes J (2003) Large porous particle impingement on lung epithelial cell monolayers-toward improved particle characterization in the lung. Pharm Res 20:788–796
33. Fielding R, Ahra RM (1992) Factors affecting the release rate of terbutaline from liposome formulations after intratracheal instillation in the guinea pig. Pharm Res 9:220–223
34. Gibbons AM, McElvaney NJ, Taggart CC, Cryan SA (2009) Delivery of rSLPI in a liposomal carrier for inhalation provides protection against cathepsin L degradation. J Microencapsul 26:513–522
35. Gregoriadis G (1984) Liposome technology. CRC, Boca Raton
36. Grenha A, Grainger CI, Dailey LA et al (2007) Chitosan nanoparticles are compatible with respiratory epithelial cells in vitro. Eur J Pharm Sci 31:73–84
37. Gudapaty SR, Liener IE, Hoidal JR et al (1985) The prevention of elastase-induced emphysema in hamsters by intratracheal ad-ministration of a synthetic elastase inhibitor bound to albumin microspheres. Am Rev Respir Dis 132:159–163

38. Harris JM, Chess RB (2003) Effect of PEGylation on pharmaceuticals. Nat Rev Drug Discov 2:214–221
39. Harvey RJ, Schlosser RJ (2009) Local drug delivery. Otolaryngol Clin North Am 42:829–845
40. Heyder J, Gebhart J, Rudolf G et al (1896) Deposition of particles in the human respiratory tract in the size range 0.005-15 μm. J Aerosol Sci 17:811–825
41. Heydet J, Rudolf G (1984) Mathematical models of particle deposition in the human respiratory tract. J Aerosol Sci 17:697–707
42. Hogg JC (1985) Response of the lung to inhaled particles. Med J Aust 142:675–678
43. Houtmeyers E, Gosselink R, Gayan-Ramirez G et al (1999) Regulation of mucociliary clearance in health and disease. Eur Respir J 13:1177–1188
44. Hutchinson FG, Furr BJ (1989) Biodegradable polymers for controlled release of peptides and proteins. Horiz Biochem Biophys 9:111–129
45. Johnson M (1996) Pharmacodynamics and pharmacokinetics of inhaled glucocorticoids. J Allergy Clin Immunol 97:169–176
46. Joshi M, Misra A (2001) Dry powder inhalation of liposomal Ketotifen fumarate formulation and characterisation. Int J Pharm 223:15–27
47. Kaplan J, Ward DM (1990) Movement of receptors and ligands through the endocytic apparatus in alveolar macrophages. Am J Physiol 258:L263–L270
48. Kawaguchi H, Koiwai N, Ohtsuka Y et al (1986) Phagocytosis of latex particles by leucocytes. Dependence of phagocytosis on the size and surface potential of particles. Biomaterials 7:61–66
49. Kawashima Y, Yamamoto H, Takeuchi H, Fujioka S, Hino T (1999) Pulmonary delivery of insulin with nebulized DL-lactide/glycolide copolymer (PLGA) nanospheres to prolong hypoglycemic effect. J Control Release 62:279–287
50. Kim JC, Kim JD (2001) Preparation by spray-drying of amphotericin B phospholipid composite particles and their anticellular activity. Drug Deliv 8:143–147
51. Kips JC, Pauwels RA (2001) Long acting inhaled β2-agonist therapy in asthma. Am J Respir Crit Care Med 164:923–932
52. Klonne DR, Dodd DE, Losco PE et al (1989) Two-week aerosol inhalation study on polyethylene glycol (PEG) 3350 in F-344 rats. Drug Chem Toxicol 12:39–48
53. Koushik K, Dhanda DS, Cheruvu NPS, Kompella UB (2004) Pulmonary Delivery of Deslorelin: Large-porous PLGA particles and HPBCD complexes. Pharm Res 21:1119–1126
54. Koval M, Preiter K, Adled C et al (1998) Size of IgG-opsonized particles determines macrophage response during internalization. Exp Cell Res 242:265–273
55. Krenis LJ, Strauss B (1961) Effect of size and concentration of latex particles on respiration of human blood leucocytes. Proc Soc Exp Med 107:748–750
56. Kumar N, Chaubal M, Domb A et al (2003) Controlled release technology. In: Encyclopedia of polymer science and technology, vol. 5. Wiley, Hoboken, p 697.
57. Kumar RMNV (2000) A review of chitin and chitosan applications. React Funct Polym 46:1–27
58. Lai Y, Mehta RC, Thacker AA et al (1993) Sustained bronchodilation with isoproterenol poly(glycolide-co-lactide) microspheres. Pharm Res 10:119–125
59. Lasic D (1997) Liposomes in gene delivery. CRC, Boca Raton
60. Lee SL, Adams WP, Li BV et al (2009) In vitro considerations to support bioequivalence of locally acting drug in dry powder inhalers for lung diseases. AAPS J 11:414–423
61. Letsou GV, Safi HJ, Reardon MJ et al (1999) Pharmacokinetics of liposomal aerosolized cyclosporine A for pulmonary immunosuppression. Ann Thorac Surg 68:2044–2048
62. Liang WW, Shi X, Deshpande D et al (1996) Oligonucleotide targeting to alveolar macrophages by mannose receptormediated endocytosis. Biochim Biophys Acta 1279:227–234
63. Lombry C, Edwards DA, Preat V et al (2004) Alveolar macrophages are a primary barrier to pulmonary absorption of macromolecules. Am J Physiol Lung Cell Mol Physiol 286:L1002–L1008
64. Mall MA (2008) Role of the cilia, mucus, and airway surface liquid in mucociliary dysfunction: lessons from mouse models. J Aerosol Med Pulm Drug Deliv 21:12–24

65. Martonen T, Smyth HDC, Isacs K et al (2005) Issues in drug delivery: dry powder inhaler performance and lung deposition. Respir Care 50:1228–1252
66. Mathias NR, Hussainn MA (2010) Non-invasive systemic drug delivery: developability considerations for alternate routes of administration. J Pharm Sci 99:1–20
67. Matsui H, Randell SH, Peretti SW et al (1998) Coordinated cleareance of periciliary liquid and mucus from airway surfaces. J Clin Invest 102:1125–1132
68. McCluskie MJ, Chu Y, Xia JL et al (1998) Direct gene transfer to the respiratory tract of mice with pure plasmid and lipid-formulated DNA. Antisense Nucleic Acid Drug Dev 8:401–414
69. McLachlan G, Davidson DJ, Stevenson BJ et al (1995) Evaluation in vitro and in vivo of cationic liposome-expression construct complexes for cystic fibrosis gene therapy. Gene Ther 2:614–622
70. Meers P, Neville M, Malinin V et al (2008) Biofilm preparation, triggered release and in vivo activity of inhaled liposomal amikacin in chronic Pseudomonas aerudinosa ling infections. J Antimicrob Chemother 61:859–868
71. Miller-Larsson A (2004) Concluding clinical synergism from preclinical study data. Respir Drug Deliv 1:87–98
72. Montgomery AB, Pitlick WH, Nardella P et al (2000) Sputum concentrations and systemic pharmacokinetics of aerosolized tobramycin (tobi) in diseased lungs. Respir Drug Deliv 1:19–24
73. Musante CJ, Schroeter JD, Rosati JA et al (2002) Factors affecting the deposition of inhaled porous drug particles. J Pharm Sci 91:1590–1600
74. Niven R (2003) Modulated drug therapy with inhalation aerosols. In: Hickey AJ (ed) Pharmaceutical inhalation aerosols technology. Marcel Dekker, New York, pp 551–570
75. Niven R, Carvajal TM, Schreier H (1992) Nebulization of liposomes. III. The effects of operating conditions and local environment. Pharm Res 9:515–520
76. Niven R, Schreier H (1990) Nebulization of liposomes. I. Effects of lipid composition. Pharm Res 7:1127–1133
77. Niven R, Speer M, Schreier H (1991) Nebulization of liposomes. II. The effects of size and modeling of solute release profiles. Pharm Res 8:217–221
78. Niven RW, Whitcomb KL, Shaner L et al (1995) The pulmonary absorption of aerosolized and intratracheally instilled rhG-CSF and monoPEGylated rhG-CSF. Pharm Res 12:1343–1349
79. Oberdorster G (2001) Pulmonary effects of inhaled ultrafine particles. Int Arch Occup Environ Health 74:1–8
80. Olsson T, Svensson LA (1984) New lipophilic terbutaline ester prodrugs with long effect duration. Pharm Res 3:19–23
81. Park K (1997) Controlled drug delivery: challenges and strategies. American Chemical Society, Washington, 629
82. Patton JS (1996) Mechanisms of macromolecule absorption by the lungs. Adv Drug Del Rev 19:3–36
83. Patton JS, Byron PR (2007) Inhaling medicines: delivering drugs to the body through the lungs. Nat Rev Drug Discov 6:67–74
84. Plowman S, Langner E, Blair J (2002) Elucidation of insulin release mechanism from OED microparticles using ATR-FTIR. In: Dalby R, Byron PR, Peart J, Farr SJ (eds) Respiratory drug delivery VIII. Davis Horwood International Publishing, Raleigh, pp 423–426
85. Poyner EA, Alpar HO, Almeida AJ et al (1995) A comparative study on the pulmonary delivery of tobramycin encapsulated into liposomes and PLA microspheres following intravenous and endotracheal delivery. J Control Release 35:41–48
86. Robbins JC, Lam MH, Tripp CS et al (1981) Synthetic glycopeptide substrates for receptor-mediated endocytosis by macrophages. Proc Natl Acad Sci USA 78:7294–7298
87. Rudt S, Muller RH (1992) In vitro phagocytosis assay of nano- and microparticles by chemiluminescence. III. Uptake of differently sized surface-modified particles and its correlation to particle properties and in vivo distribution. Eur J Pharm Sci 1:31–39
88. Sakagami M, Kinoshita W, Sakon K et al (2002) Mucoadhesive beclomethasone microspheres for powder inhalation: their pharmacokinetics and pharmacodynamics evaluation. J Control Release 80:207–218

89. Sally-Ann C (2005) Carrier-based strategies for targeting protein and peptide drugs to the lungs. AAPS J 7:E20–E41
90. Scheuch G, Siekmeier R (2007) Novel approacches to enhance pulmonary delivery of proteins and peptides. J Physiol Pharmacol 58:615–625
91. Schreier H, Mobley WC, Concessio N et al (1994) Formulation and in vitro performance of liposome powder aerosols. STP Pharma Sci 4:38–44
92. Seville PC, Kellaway IW, Birchall JC (2002) Preparation of dry powder dispersions for non-viral gene delivery by freeze-drying and spray-drying. J Gene Med 4:428–437
93. Sharma R, Saxena D, Dwivedi AK, Misra A (2001) Inhalable microparticles containing drug combinations to target alveolar macrophages for treatment of pulmonary tuberculosis. Pharm Res 18:1405–1410
94. Sibille Y, Reynolds HY (1990) Macrophages and polymorphonuclear neutrophils in lung defense and injury. Am Rev Respir Dis 141:471–501
95. Siekmeier R, Scheuch G (2008) Systemic treatment by inhalation of macromolecules-principles, problems, and examples. J Physiol Pharmacol 59:53–79
96. Simon SI, Schmid-Chongeing GW (1988) Biophysical aspects of microsphere engulfment by human neutrophils. Biophys J 53:163–173
97. Skalko-Basnet N, Pavelic Z, Becirevic-Lacan M (2000) Liposomes containing drug and cyclodextrin prepared by the one-step spray-drying method. Drug Dev Ind Pharm 26:1279–1284
98. Smyth HDC, Hickey AJ (2005) Carriers in drug powder delivery: implications for inhalation system design. Am J Drug Deliv 3:117–132
99. Standard W, O'callaghan C (2006) Ciliary function and the role of cilia in clearance. J Aerosol Med 19:110–115
100. Stone KC, Mercer RR, Gehr P et al (1992) Allometric relationships of cell numbers and size in the mammalian lung. Am J Respir Cell Mol Biol 6:235–243
101. Stribling R, Brunette E, Liggitt D et al (1992) Aerosol gene delivery in vivo. Proc Natl Acad Sci USA 89:11277–11281
102. Suarez S, Gonzalez-Rothi RJ, Schreier H, Hochhaus G (1998) Effect of dose and release rate on pulmonary targeting of liposomal triamcinolone acetonide phosphate. Pharm Res 15:461–465
103. Suarez S, Hickey AJ (2000) Drug properties affecting aerosol behavior. Respir Care 45:652–666
104. Suarez S, O'Hara P, Kazantseva M et al (2001) Airways delivery of rifampicin microparticles for the treatment of tuberculosis. J Antimicrob Chemother 48:431–434
105. Suarez S, O'Hara P, Kazantseva M et al (2001) Respirable PLGA microspheres containing rifampicin for the treatment of tuber-culosis: screening in an infectious disease model. Pharm Res 18:1315–1319
106. Surendrakumar K, Martyn GP, Hodgers ECM et al (2003) Sustained release of insulin from sodium hyaluronate based dry powder formulations after pulmonary delivery to beagle dogs. J Control Release 91:385–394
107. Suzuki Y, Yamaguchi T (1993) Effects of hyaluronic acid on macrophage phagocytosis and active oxygen release. Agents Actions 38:32–37
108. Tabata Y, Ikada Y (1988) Effect of the size and surface charge of polymer microspheres on their phagocytosis by macrophage. Biomaterials 9:356–362
109. Takeuchi H, Yamamoto H, Kawashima Y (2001) Mucoadhesive nanoparticulate systems for peptide drug delivery. Adv Drug Deliv Rev 47:39–54
110. Talton J, Fitz-Gerald J, Singh R et al (2000) Nano-thin coating for improved lung targeting of glucocorticoid dry powders: in-vitro and in-vivo characteristics. In: Dalby RN, Byron PR, Farr SJ et al (eds) Respiratory drug delivery VII. Serentec Press, Raleigh, pp 67–74
111. Telko MJ, Hickey AJ (2005) Dry powder inhaler formulation. Respir Care 50:1209–1227
112. Tietze C, Schlesinger P, Stahl P (1982) Mannose-specific endocytosis receptor of alveolar macrophages: demonstration of two functionally distinct intracellular pools of receptor and their roles in receptor recycling. J Cell Biol 92:417–424
113. Traini D, Young PM (2009) Delivery of antibiotics to the respiratory tract: an update. Expert Opin Drug Deliv 6:897–905

114. Tsapis N, Bennett D, Jackson B et al (2002) Trojan particles: large porous carriers of nanoparticles for drug delivery. PNAS 99:12001–12005
115. Tsifansky MD, Yeo Y, Evgenov OV et al (2008) Microparticles for inhalational delivery of antipseudomonal antibiotics. AAPS J 10:254–260
116. Tsutsumi Y, Kihira T, Tsunoda S et al (1995) Molecular design of hybrid tumour necrosis factor alfa with polyethylene glycol increases its anti-tumor potency. Br J Cancer 71:963–968
117. Tsutsumi Y, Tsunoda S, Kamada H et al (1997) PEGylation of interleukin-6 effectively increases its thrombopoietic potency. Thromb Haemost 77:168–173
118. Usami OS, Biddiscombe MF, Underwood RS et al (2004) Characterization of the generation of radiolabeled monodisperse albuterol particles using the spinning-top aerosol generator. J Nucl Med 45:69–73
119. Venbever R, Ben-Jebria A, Mintzes JD et al (1999) sustained-release of insulin from insoluble inhaled particles. Drug Dev Res 48:178–185
120. Vyas SP, Kannan ME, Jain S, Mishra V et al (2004) Design of liposomal aerosols for improved delivery of rifampicin to alveolar macrophages. Int J Pharm 269:37–49
121. Weibel ER (1963) Morphometry of the human lung. Academic, New York
122. Widdicombe JH (2002) Regulation of the depth and composition of airway surface liquid. J Anat 201:313–318
123. Yanga Y, Bajajc N, Xua P, Ohnd K et al (2009) Development of highly porous large PLGA microparticles for pulmonary drug delivery. Biomaterials 30:1947–1953
124. Zanen P, Go LT, Lammers JWJ (1994) The optimal particle size for β-Adrenergic aerosols in mild asthmatics. Int J Pharm 107:211–217
125. Zeng X, Martin CG, Marriott C (1995) The controlled delivery of drugs to the lung. Int J Pharm 124:149–164
126. Zou Y, Zong G, Ling YH, Perez-Soler R (2000) Development of cationic liposome formulations for intratracheal gene therapy of early lung cancer. Cancer Gene Ther 7:683–696
127. Algranati NE, Sy S, Modi M (1999) A branched methoxy 40 KDA polyethylene glycol (PEG) moiety optimizes the pharmacokinetics (PK) of peginterferon alpha-2a (peginterferon) and may explain its enhanced efficacy in chronic hepatitis C (CHC). Hepatology 30: 190A

Chapter 6
Targeted Drug Delivery Through the Respiratory System: Molecular Control on Lung Absorption and Disposition

Masahiro Sakagami and Mark Gumbleton

Abstract Drug delivery through the respiratory system is important for both locally and systemically acting drugs delivered as inhalation aerosols. This chapter reviews our current knowledge of diffusion and transporter controls on lung epithelial absorption of inhaled drug molecules. Moreover, competing nonabsorptive clearance via local metabolism, phagocytosis, and mucociliary escalator, and its impact and control on lung absorption will be also discussed.

Keywords Diffusion • Epithelium • Inhaled therapeutics • Lung absorption • Metabolism • Mucociliary clearance • Phagocytosis • Transporters

6.1 Introduction

It becomes clear that the lung is naturally permeable to most small molecular weight drugs as well as many therapeutic peptides and proteins [49]. This is primarily by virtue of high diffusive permeability of its epithelium, compared to other organs, e.g., intestinal and nasal epithelia. Even so, a protracted question remains in lung physiology, as to the mechanisms by which certain ions, nutrients, and proteins appear to be transported the airway-to-blood barrier at rates differing from those expected for their size in diffusion [28]. In this context, evidence has emerged that a variety of membrane transporter proteins are expressed in lung epithelium in efflux, uptake, and/or endo-/trans-cytotic fashions [5, 7, 28]. Although their current pursuits have not yet ensured all of the functionality in the intact lung and rather often provided perplexity, it would be of great interest to exploit such specialized

M. Sakagami (✉)
Department of Pharmaceutics, School of Pharmacy, Virginia Commonwealth University, 410 N 12th Street, P.O. Box 980533, Richmond, VA 23298-0533, USA
e-mail: msakagam@vcu.edu

H.D.C. Smyth and A.J. Hickey (eds.), *Controlled Pulmonary Drug Delivery*, Advances in Delivery Science and Technology, DOI 10.1007/978-1-4419-9745-6_6,

mechanisms, if any, to develop new or improved drug molecules for inhalation. Hence, this chapter will review our current knowledge of diffusion and transporter controls on lung (epithelial) absorption of inhaled drug molecules, with a focus on the kinetics in the intact lung. Moreover, because the lung is equipped with other than absorption for removal of inhaled drug molecules following deposition [28, 56], such a parallel competing nonabsorptive clearance via local metabolism, phagocytosis and mucociliary escalator, and its impact and control on lung absorption will be also discussed.

6.2 Lung Disposition: Absorption vs. Nonabsorptive Clearance

Aerosol drug deposition and lung-regional distribution are predominantly controlled by aerodynamic diameter and its distribution [29, 44]. Their differences, even within the "respirable" size range of 1–5 μm, can still impact lung-regional distribution, as greater deposition within the lung periphery (i.e., alveolar region) is achieved by decreasing aerosol size to a range of 1–2 μm [29, 44]. Therefore, in the context of conventional wisdom that the lung's alveolar region is the most favorable for drug absorption via diffusion by virtue of its thin (~1 μm) and large (~140 m^2) epithelial surface to the pulmonary circulation [47], efforts to increase systemic absorption via the lung should be first focused on generating drug aerosols in this size range. Clearly, this logic would not always work for drug molecules that undergo specialized absorption mechanisms via membrane transporter proteins in efflux, uptake, and endo-/trans-cytotic fashions. The kinetics of such active transport mechanisms should depend upon lung-regional expression and localization of responsible transporters or receptors. For instance, the neonatal constant fragment (Fc) region receptor (FcRn) has been shown to localize more abundantly in the bronchial airways than the alveoli [5, 64]. Hence, its substrates, immunoglobulin G (IgG) and antibodies, and Fc-fusion molecules are to be delivered to such rather upper airways for more efficient and greater transcytotic absorption to the systemic circulation, despite their unfavorable barrier feature for diffusion, i.e., thicker epithelium with a smaller surface area [5, 64]. Meanwhile, for drugs intended for local lung actions, their target receptor localization to maximize the therapeutic benefits should be of primary interest for delivery. In recent studies by Usmani et al. [66], a short-acting bronchodilator, albuterol, was shown to benefit in its therapeutic effects from the larger 3 and 6 μm aerosols than the smaller 1.5 μm aerosols, by virtue of better targeting to the airway muscle β_2 receptors. By contrast, in the case of an inhaled corticosteroid, beclomethasone dipropionate, the smaller 1.1 μm aerosols enabled 62% reduction of the therapeutically effective dose, compared to the 3.5 μm aerosols, attributed to an increased distribution in the lung periphery where inflammation was presumably most significant [36].

In addition to absorption, the lung is naturally equipped with a variety of efficient clearance mechanisms for inhaled drug molecules following deposition [28, 56].

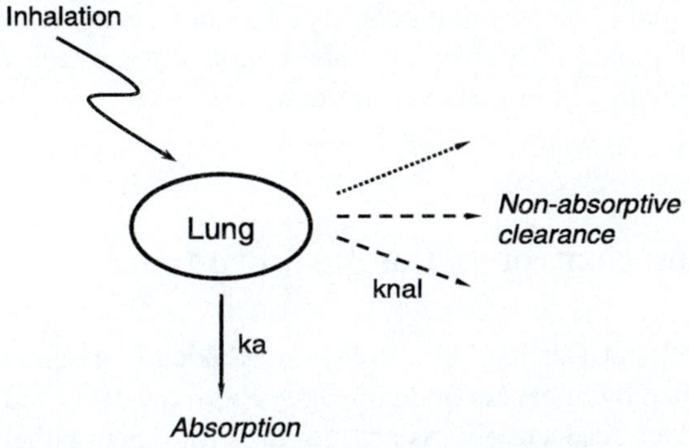

Fig. 6.1 Lung absorption and disposition kinetic model for inhaled drug molecules, incorporating parallel and competitive absorption and nonabsorptive clearance with their respective first-order rate constants, k_a and k_{nal}. The lung's nonabsorptive clearance is composed of local enzymatic metabolism, alveolar macrophagic phagocytosis, and/or sweeping mucociliary clearance

These include local enzymatic metabolism, alveolar macrophagic phagocytosis, and sweeping clearance by the mucociliary escalator [10, 28, 56]. From a drug delivery perspective, however, these can be simplified kinetically to a single mechanism competing with absorption, and collectively referred to as nonabsorptive clearance [56]. Hence, Fig. 6.1 illustrates a practical working concept of biopharmaceutical lung where the first-order kinetic rate constants for lung absorption and nonabsorptive clearance are assigned in a parallel competing fashion as k_a and k_{nal}, respectively. By so doing, the rate and extent of lung absorption and residence following deposition can be derived and discussed with their respective half-lives ($T_{1/2,ab}$ and $T_{1/2,re}$), and the dose fraction of lung absorption, i.e., absolute bioavailability (F), can be determined from:

$$T_{1/2,ab} = 0.693/k_a$$

$$T_{1/2,re} = 0.693/(k_a + k_{nal})$$

$$F = k_a/(k_a + k_{nal})$$

It is clear therefore that, for a given inhaled drug molecule, decreasing the k_{nal} value should be as important as increasing the k_a value, such that its lung absorption would become faster and greater. It should be also noted that each of these lung disposition processes was shown to depend on lung-regional distribution, which therefore must be first controlled and matched, in order to fairly assess how lung absorption for inhaled drug molecules can be improved. In the literature however, it was only recently that comparable lung-regional distribution was ensured in animals and humans, prior to the assessment and comparison of lung absorption. Otherwise, faster and greater lung absorption for given test molecules or formulations could be caused by increased k_a value and/or decreased k_{nal} value, simply resulting from more favorable lung-regional distribution for absorption, e.g., greater alveolar distribution would enable greater diffusive absorption.

6.3 Diffusion Control on Lung Absorption

Patton et al. [49] and Hastings et al. [28] independently assembled many animal data sets published by Schanker and colleagues between 1973 and 1986 and Effors and Mason in 1983, respectively. As are the cases for other epithelium, the kinetics of diffusive absorption from the lung should follow the Fick's diffusion theory across biological lipid cell membranes [10]. Hence, given comparable lung-regional distribution, the greater the lipophilicity and the smaller the molecular weight, the faster and greater the lung absorption. Nevertheless, as shown in Fig. 6.2a, for small molecular weight (≤781 Da) drugs, the half-lives derived from their lung disappearance

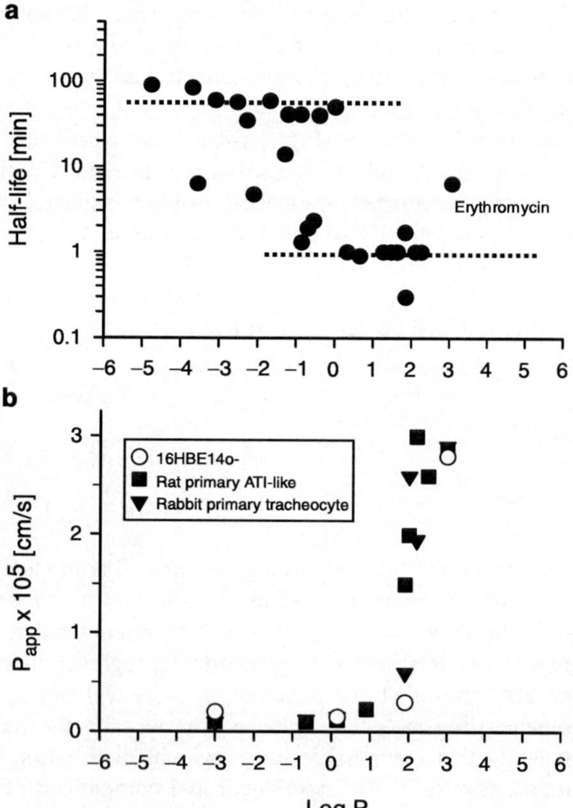

Fig. 6.2 (**a**) In vivo half-life dependence on the drug's lipophilicity in rats, assembled by Patton et al. [49]. The half-life was derived from the drug disappearance profiles from the rat lung taken by Schanker and colleagues between 1973 and 1986, whereas the drug's lipophilicity is represented with logarithmic partition coefficient (log *P*) between octanol and water; (**b**) in vitro apparent permeability (P_{app}) dependence on the drug's lipophilicity (log *P*) in various lung epithelial cell monolayer systems, reported by Forbes et al. [23]

profiles in rats seem to be not so dependent upon the drug's lipophilicity represented with octanol-water partition coefficients ($\log P$). As a result, Patton et al. [49] advocated a natural break at $\log P=0$, such that lipophilic ($\log P>0$) and hydrophilic ($\log P<0$) drug molecules cluster around a half-life of ~1 min and ~1 h, respectively. Accordingly, the half-lives would change substantially, only when the $\log P$ values change from negative values to positive values or vice versa. Notably however, this natural break at $\log P=0$ in rats was inconsistent with that seen in the in vitro lung epithelial cell monolayers [23]. As shown in Fig. 6.2b, their natural break causing substantial increase in the apparent permeability (P_{app}; representing the rate of diffusion) was $\log P=2$ [23]. Hence, this would raise a question how accurately the break at $\log P=0$ in rats has been derived and how representative these in vitro cell monolayers can be, as to the effect of lipophilicity on the rate of lung epithelial diffusion. Meanwhile, there appeared to be deviations from these clustered half-lives in Fig. 6.2a, when drug molecules were suggested to interact with the lung lipid membranes significantly, e.g., erythromycin, or to undergo nonlinear "active" transport, e.g., cycloleucine and disodium cromoglycate [38, 49, 62]; however their responsible mechanisms and impact on systemic appearance have yet to be fully clarified and resolved kinetically.

Figure 6.3 shows the half-life dependence on the molecular weight from two independent data assembled by Patton et al. [49] and Hastings et al. [28]. In both cases, substantial variability in the half-lives observed for small molecular weight (<1,000 Da) drugs masked to derive the clear dependence of the half-life on the molecular weight; this was more evident in Fig. 6.3a due to its inclusion of rather lipophilic molecules [49], despite the cross-species analysis in Fig. 6.3b. By contrast, it became obvious in both studies that the half-life was increased with increasing the molecular weight at 1,000 Da or greater. Provided that such large molecules tested in these studies were highly hydrophilic peptides, proteins, or polysaccharides, the mechanism of their lung epithelial absorption was most likely paracellular diffusion through the aqueous pores of the intercellular tight junctions. In fact, it was only growth hormone (GH; 22,000 Da) that showed nonlinear absorption kinetics in the animal studies [22, 28, 49]. On the contrary, several microscopic studies reported that alveolar epithelial cells contained endogenous peptides and proteins, e.g., insulin, GH, albumin, and IgG in their noncoated or coated vesicles [4, 30–32]. While this implied their active intracellular uptake or transport via receptors and/or channels, the kinetic contributions in overall transport in the intact lung have not yet fully established, as discussed in the below section.

It should be critical to recognize that the half-lives obtained from the lung disappearance or clearance profiles above were the half-lives of lung residence ($T_{1/2,re}$) by definition and thus do not necessarily represent the half-lives of lung absorption ($T_{1/2,ab}$). All of the deductions made from Figs. 6.2 and 6.3a must have assumed negligible or at least consistent involvement of lung's nonabsorptive clearance (i.e., k_{nal}) along with comparable lung-regional distribution within and across studies. Moreover, as many peptides and proteins like insulin, GH, and IgG have been shown to form molecular aggregates in solution [37, 46, 48], their diffusive rates could become slower than those expected from the molecular weights

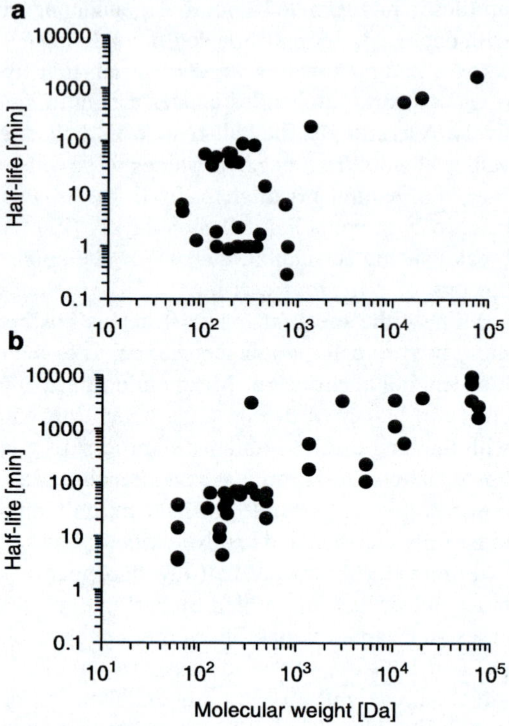

Fig. 6.3 In vivo lung disappearance half-life dependence on the drug's molecular weight in animals, assembled by (**a**) Patton et al. [49] and (**b**) Hastings et al. [28]

of the monomeric species, which may have caused deviation and variability, especially in Fig. 6.3a. Indeed, a recent study by Pang et al. [46] has demonstrated that the rate constant for lung absorption (k_a) of insulin in the isolated perfused rat lung (IPRL) was decreased with increasing % hexameric formation, due to the increased apparent molecular size for diffusion. Likewise, Patton et al. [48] implied in a rat study that higher molecular weight aggregates of GH accounted, at least in part, for ~30% of the dose remaining in the lung at 24 h following intratracheal instillation.

6.4 Transporter Control on Lung Absorption

The relevance of transporters in drug absorption, distribution, and elimination as well as in drug–drug interactions has been mostly demonstrated in the intestine, liver, kidney, and brain as well as their interfaces with the systemic circulation [3, 54]. It has been accepted by now that intestinal absorption and brain uptake (i.e., distribution) of many lipophilic drugs are limited by P-glycoprotein (PgP) and other ATP-dependent efflux transporters [3, 12]. In addition, successful development of valacyclovir to increase intestinal absorption of acyclovir via the peptide transporter

Table 6.1 Immunohistological drug transporter expression in human lungs

Protein	Expression	Cellular distribution	Cellular localization	References
ABC transporters				
PgP	Moderate	Alveoli	Apical	[11]
		Bronchi	Apical	
	Undetectable			[14]
MRP1	High	Bronchi	Basolateral	[8, 21, 63]
MRP2–9	Undetectable			[34]
BCRP	High	Alveoli	Apical	[20]
	Low-undetectable	Bronchi	Apical	
SLC transporters				
OCT1–3	Detectable	Trachea	Apical	[39]
	Detectable	Bronchi	Apical	
OCTN1&2	Detectable	Alveoli	Apical	[39]
PEPT1	Undetectable	Alveoli		[7]
PEPT2	High	Alveoli	Cytoplasmic	[26]
OAT1–4	Undetectable	Alveoli		[7]
SLCO transporters				
OATPs	Undetectable			[7]
Endo-/trans-cytotic transporters				
gp60	Detectable	Alveoli	Apical	[28]
FcRn	Undetectable	Alveoli	Apical	[64]
	High	Bronchi	Apical	[64]
pIgR	High	Bronchi		[43]

1 (PEPT1) was a premier example on this venture on the solute-linked carrier (SLC or SLCO) transporters [41]. In physiology, it was in 2006 that the two most abundant proteins in the body, albumin and IgG, were shown to bind the same transporter, FcRn on the vascular endothelium, such that their endocytosis, followed by recycling back to the circulation seemed to enable, at least in part, their unusually long half-lives of 3 weeks [13]. By contrast, studies on such transporters on lung disposition of inhaled drug molecules have just recently emerged, yet primarily focused on their expression and localization within the lung [5, 7, 19]. The lungs were shown to express a variety of efflux, uptake, and endo-/trans-cytotic drug transporters at the protein and/or mRNA level, as summarized in Table 6.1. However, only little is presently known as to their functionalities in, and kinetic contributions to, lung disposition (e.g., absorption) of inhaled drug molecules in health and disease.

6.4.1 *Efflux Transporter Control: ATP-Binding Cassette (ABC) Transporters*

As have been the cases in other organs, the ATP-binding cassette (ABC) transporters are believed to serve for efflux in the lung [7, 19], as one of the defensive functions against certain exogenous substances. Among them, the lung resistance-related protein (LRP) and PgP have been fairly studied in the lung, as related to the multidrug resistance (MDR) of many chemotherapeutic agents in cancer cells [7, 16]. In recent years however, an interest has been extended to the impact on disposition of inhaled drug molecules within normal or noncancer disease lungs, especially after Campbell et al. [11] first confirmed PgP protein expression in the intact lung peripheries.

It appears clear that PgP functions in pulmonary vascular endothelium, reducing cellular uptake of its substrate drugs from the circulation. Several studies in the isolated perfused animal lung models showed increased uptake of the substrate drugs, such as rhodamine 6G and idarubicin, from the perfusate to the lung in the presence of PgP inhibitors [33, 51]. By contrast however, its functionality in lung epithelial absorption has been controversial due to conflicting results in the IPRL model and in vivo animals. Lung absorption of a well-known PgP substrate, rhodamine 123, in the IPRL was shown to increase by the PgP inhibitor, GF120918, which suggested its PgP-limited lung absorption [24]. However, the same inhibitor, GF120918, failed to increase lung absorption of another known PgP substrate, digoxin in the IPRL, countering the former finding of PgP-limited lung absorption [42]. Likewise, evidence for PgP expression in the alveolar regions has been also controversial. Campbell et al. [11] reported moderate immunostaining of PgP protein along the alveolar type I epithelial cell membranes in both rat and human lung tissues. However, Cordon-Cardo et al. [14] noted undetectable immunostaining of PgP within normal human lung alveolar tissues. Clearly, these results should be interpreted with caution and presently remain inconclusive, especially as the substrate specificity to other transporters, the use of appropriate dose or concentration of the substrates and inhibitors, and the antibody sensitivity for PgP protein detection have been raised as misleading concerns in this debate [7]. In reality however, lung absorption of PgP substrate drugs has been generally high (e.g., 60–90%), which would call into question how significantly this epithelial efflux compromises lung absorption of inhaled substrate drug molecules.

Meanwhile, almost nothing is known as to the functionality of multidrug resistance proteins (MRPs) and breast cancer resistance protein (BCRP) in the lung. Strong immunostaining of certain MRP protein (e.g., MRP1) was reported in the bronchial sections of normal human lung, but not in the pneumocytes of alveolar epithelium [8, 21, 63]. This bronchial expression of MRP1 has been later confirmed with the mRNA data, while it seems likely that the expression is primarily on the basolateral membranes [34, 63]. By contrast, other MRPs, i.e., MRP2 through MRP9, were reported to be essentially undetectable at the protein level in the intact lung, even though some of them were shown positively in RT-PCR and gene microarray assays [6, 34]. Meanwhile, Fetsch et al. [20] reported BCRP staining in the alveolar pneumocytes of normal human lung archival sections and little to no staining in the bronchial epithelial membranes. This has been supported by recent gene microarrays [6], while RT-PCR has resulted in conflicting low-to-negative expression at the mRNA level [17]. Most importantly, however, there have been no studies demonstrating MRPs- or BCRP-limited drug absorption from the intact lung.

6.4.2 Uptake Transporter Control: Solute-Linked Carrier (SLC/SLCO) Transporters

Like MRPs and BCRP, evidence for the SLC/SLCO transporter's functionality within the intact lung is lacking, while their protein, gene, and/or mRNA expression has been often positive, as summarized in Table 6.1. At the protein level, all five main types of the organic cation transporters (OCTs, i.e., OCT1, OCT2, OCT3, OCTN1, and OCTN2) were shown to express in healthy human lungs [7, 39]. In addition, PEPT2 protein expression has been shown in the intact lung [26], while other SLC/SLCO transporters like PEPT1 and organic anion transporters (OATs and OATPs) remain negative, which has speculated no or minor roles in drug absorption from the lung. Likewise, at the mRNA level, OCTs and PEPT2 were reported to be positive [7]. However, recent RT-PCR and gene microarray studies have found relatively high levels of several OATs (e.g., OAT2 and certain OATPs) in human lung [6, 65], leaving possibly conflicting evidence for these transporters in the intact lung. Overall, it is still premature to even speculate the kinetic contribution of these SLC/SLCO transporters on drug absorption from the lung, especially as earlier findings of dose-dependent saturable lung disappearance kinetics for certain molecules, such as phenol red, cycloleucine, and disodium cromoglycate [38, 62], have not yet been linked and/or resolved with any of these transporters.

6.4.3 Endocytotic Transporter Control: Albumin, IgG, and IgA

It has been long recognized that endogenous serum proteins, such as albumin, IgG, and IgA, appear at relatively high concentrations into the lung luminal lining fluid, presumably from the systemic circulation, despite their large molecular sizes [4, 28]. In addition, these proteins were also found within the lung epithelial cells, specifically in the noncoated vesicles (e.g., caveolae), which has raised a question in physiology how their transport and cellular trafficking occur. As a result, by 2006, it has been discovered that these proteins could be transcytosed via their specific receptors expressed on the lung epithelium, gp60 [28, 30], FcRn [32, 61], and pIgR [52], respectively. Both gp60 and FcRn were shown to drive albumin and IgG, respectively, from the apical to the basolateral (i.e., absorptive) direction, while pIgR was believed to deliver IgA dimers and IgM pentamers in an opposite (i.e., secretive) direction until Sacaan et al. [53] showed that the receptor also functioned in the absorptive direction. Moreover, it was also in 2006 that FcRn was shown to transport albumin via the mechanism apparently differing from that for IgG, in order to protect this most abundant serum protein from intracellular catabolic degradation [13]. It should be noted, however, that these specialized mechanisms appear not to be always predominant in these protein clearance kinetics in the intact lung, as inhibition of such endo-/trans-cytosis has often failed to alter their rates [27].

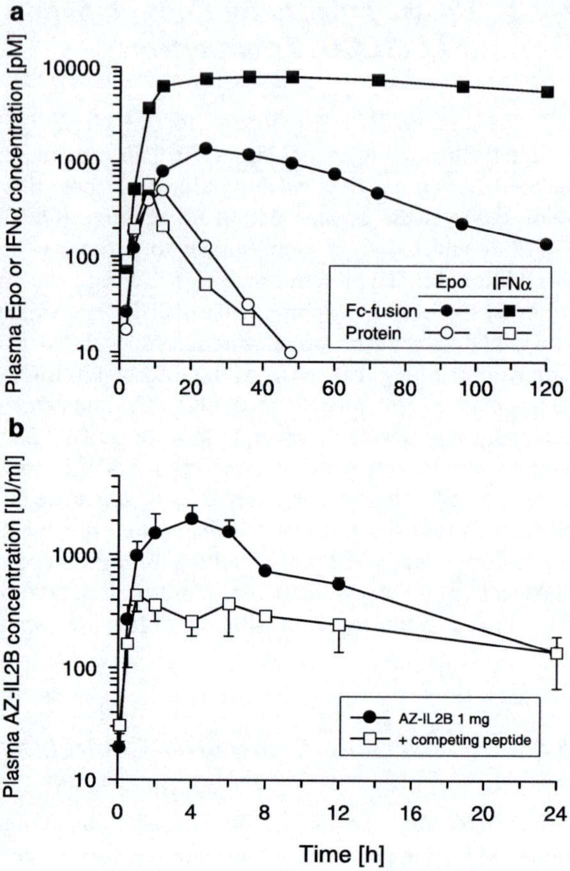

Fig. 6.4 (**a**) Plasma erythropoietin (Epo) or interferon α (IFNα) concentration vs. time profiles in nonhuman primates following aerosol administration of their Fc-fusion proteins or intact proteins by alone at the identical molar doses to the lung. The plasma concentrations were taken from Bitonti and Dumont [5] and replotted as mean data from $n = 2$–4; (**b**) plasma AZ-IL2B concentration vs. time profiles in rats following intratracheal instillation of AZ-IL2B in the absence or presence of a competing peptide blocking the AZ-IL2B's binding site to pIgR. The plasma concentrations were taken from Sacaan et al. [53]

While gp60-mediated transcytosis has not yet been exploited for inhaled drug delivery through the lung, FcRn- and pIgR-mediated transcytotic lung absorption has been recently explored, specifically to deliver certain fusion protein molecules through the lung by inhalation [5, 53]. As shown in Fig. 6.4a, Fc-fusion proteins of erythropoietin (EPO) and interferon α (IFNα) were absorbed through the lung in nonhuman primates following aerosol inhalation much faster and greater than the intact proteins themselves, attributed to the FcRn-mediated transcytotic lung absorption [5]. This increased lung absorption has been also confirmed in humans with the Fc-fusion EPO, exhibiting successfully increased reticulocyte counts following single inhalation at a dose of only 30 μg/kg [18]. Likewise, as shown in Fig. 6.4b, lung absorption of a dimeric fusion protein of interleukin-2 with single chain Fv (scFv), AZ-IL2B, was shown to decrease in rats, when its chimeric protein was blocked with a small peptide inhibiting its binding to pIgR [53]. This was logically by virtue of the transcytotic lung absorption of AZ-IL2B via pIgR, thereby successfully resulting in significantly fewer tumor nodules following inhalation in a rat model of lung metastases [53]. Even so, it appears that these FcRn- and pIgR-mediated transcytotic lung absorptions are kinetically high affinity and low capacity systems and thus may not be universally applicable to all classes of the drug to aid for increased lung absorption [57]. Indeed, their kinetic analysis in the intact lung has suggested that these transcytotic systems are limited to only highly potent fusion proteins that would require ≤1 nM of the systemic level, but not to most of the systemically acting monoclonal antibody therapeutics, as they require ≥10 nM, unless much higher dose inhalation systems, e.g., ≥10 mg to the lung, become readily available [57]. Even so, exploiting these naturally functioning absorptive transcytosis in the lung is still of great interest for pulmonary delivery, especially for large protein therapeutics, as they generally suffer from unrealistically low (≤10%) bioavailability in absorption, resulting from slow diffusive absorption and substantial nonabsorptive clearance in the lung, i.e., low k_a and high k_{nal} values in the model shown in Fig. 6.1.

6.5 Nonabsorptive Clearance Control on Lung Absorption

Given the working concept of biopharmaceutical lung shown in Fig. 6.1, it is clear that decreasing the lung's nonabsorptive clearance (k_{nal}) is as important as increasing the absorption (k_a) for faster and greater lung absorption. Because the lung's nonabsorptive clearance is generally caused by local enzymatic metabolism, alveolar macrophagic phagocytosis, and/or sweeping clearance by the mucociliary escalator [10, 28, 56], it should be worth attempting their control and improvement for inhaled drug molecules and/or formulations. Among various catalytic (e.g., hydrolytic, proteolytic, and oxidoreductive) enzymes shown to function in the lung, peptidases are of importance for protein and peptide drug molecules [50, 56]. For instance, vasoactive intestinal peptide (VIP) exhibited only <1% of the bioavailable fraction due to degradation, and Leu-enkephalin was degraded in 10 min in the lung [1, 25]. Inhaled insulin appears to undergo substantial metabolic loss in the lung, leading to only <10% of bioavailability in animals and humans [45, 55]. Hence, the use of metabolic inhibitors enabled identification of such enzymatic degradation, potentially leading to the development of new analog molecules resistant or less susceptible to degradation. Currently, such activities are underway, yet primarily for oral drug delivery [2], but not for pulmonary drug delivery. Meanwhile, alveolar macrophagic phagocytosis seems to be critical only for limited classes of molecules, such as IgG and chorionic gonadotropin [40]. Hence, despite the efficient transcytotic FcRn-mediated absorption, the Fc-fusion protein molecules shown in Fig. 6.4a were also likely subjected to simultaneous macrophagic phagocytosis primarily via the Fcγ receptors (FcγR), which may have reduced their lung absorption [57]. It is logically beneficial therefore to design IgG or Fc-fusion molecules to selectively bind FcRn but not FcγR and in fact, such an activity has been recently reported, yet in aiming for prolonged systemic circulation [15], again, not for pulmonary drug delivery. Finally, mucociliary clearance functions to physically remove lung-deposited drugs out of the lung with relatively fast kinetics with 0.5–4.2 h of the half-life [9, 35, 59]. While this clearance appears to become slower and less significant kinetically by increasing deposition into the lung peripheries [9, 45], its retardation by the use of mucoadhesive microspheres was shown to modulate drug absorption in animals by virtue of longer retention within the lung [58, 60]. Likewise, liposome delivery has been shown to prolong the drug retention in lung, which can be logically more beneficial for certain classes of drug molecules, such as fluoroquinolones for lung infection [67]. It should be noted however that these controls on the lung's nonabsorptive clearance have been so far demonstrated to be feasible in animals and cells, but not yet in humans.

6.6 Conclusions

Diffusion continues to control lung absorption kinetics of most inhaled drug molecules, while there are presently quite limited evidences for transporter control on lung absorption. Specifically, as both efflux and uptake transporters typically recognize relatively small molecular weight (e.g., ≤1 kDa) drugs, it appears likely that diffusion kinetically overpowers such specialized transport mechanisms by virtue of highly permeable and thin epithelia with a huge surface area. Nevertheless, certain transporters, especially those serving for endo-/trans-cytosis, are of great interest in macromolecular drug delivery, provided that such large molecules by alone generally suffer from unrealistically low (≤10%) bioavailability due to slow diffusive absorption and substantial nonabsorptive clearance in the lung, i.e., low k_a and high k_{nal} values. Even so, their applications should be limited to only highly potent molecules like Fc- or scFv-fusion protein therapeutics, primarily due to their high affinity and low capacity kinetics for lung absorption. Meanwhile, reducing the kinetics of lung's nonabsorptive clearance via metabolism, phagocytosis and/or mucociliary escalator should also be an important control to increase and/or sustain lung absorption for inhaled drug molecules. In this sense, with an increasing development of novel and potent drug molecules potentially for inhalation, it becomes quite pivotal to first accurately assess their kinetic controllers for lung absorption and disposition, such that medicinal chemistry and pharmaceutics can together efficiently strive for improved lung absorption and pharmacological actions for novel inhaled drug molecules.

References

1. Altiere RJ (1991) Peptide metabolism in the airways: possible therapeutic implications. In: Byron PR (ed) Respiratory drug delivery II. University of Kentucky College of Pharmacy, Lexington, pp 588–601
2. Antosova Z, Mackova M, Kral V, Macek T (2009) Therapeutic application of peptides and proteins: parenteral forever? Trends Biotechnol 27:628–635
3. Ayrton A, Morgan P (2008) Role of transport proteins in drug discovery and development: a pharmaceutical perspective. Xenobiotica 38:676–708

4. Bignon J, Jaurand MC, Pinchon MC, Sapin C, Warnet JM (1976) Immunoelectron microscopic and immunochemical demonstrations of serum proteins in the alveolar lining material of the rat lung. Am Rev Respir Dis 113:109–120
5. Bitonti AJ, Dumont JA (2006) Pulmonary administration of therapeutic proteins using an immunoglobulin transport pathway. Adv Drug Deliv Rev 58:1106–1118
6. Bleasby K, Castle JC, Roberts CJ, Cheng C, Bailey WJ, Sina JF, Kulkarni AV, Hafey MJ, Evers R, Johnson JM, Ulrich RG, Slatter JG (2006) Expression profiles of 50 xenobiotic transporter genes in humans and pre-clinical species: a resource for investigations into drug disposition. Xenobiotica 36:963–988
7. Bosquillon C (2010) Drug transporters in the lung-do they play a role in the biopharmaceutics of inhaled drugs? J Pharm Sci 99:2240–2255
8. Brechot JM, Hurbain I, Fajac A, Daty N, Bernaudin JF (1998) Different pattern of MRP localization in ciliated and basal cells from human bronchial epithelium. J Histochem Cytochem 46:513–517
9. Byron PR (1986) Prediction of drug residence times in regions of the human respiratory tract following aerosol inhalation. J Pharm Sci 75:433–438
10. Byron PR, Patton JS (1994) Drug delivery via the respiratory tract. J Aerosol Med 7:49–75
11. Campbell L, Abulrob A-NG, Kandalaft LE, Plummer S, Hollins AJ, Gibbs A, Gumbleton M (2003) Constitutive expression of p-glycoprotein in normal lung alveolar epithelium and functionality in primary alveolar epithelial cultures. J Pharmacol Exp Ther 304:441–452
12. Chan LMS, Lowes S, Hirst BH (2004) The ABCs of drug transport in intestine and liver: efflux proteins limiting drug absorption and bioavailability. Eur J Pharm Sci 21:25–51
13. Chaudhury C, Brooks CL, Carter DC, Robinson JM, Anderson CL (2006) Albumin binding to FcRn: distinct from the FcRn-IgG interaction. Biochemistry 45:4983–4990
14. Cordon-Cardo C, O'Brien JP, Boccia J, Casals D, Bertino JR, Melamed MR (1990) Expression of the multidrug resistance gene product (P-glycoprotein) in human normal and tumor tissues. J Histochem Cytochem 38:1277–1287
15. Dall'Acqua WF, Kiener PA, Wu H (2006) Properties of human IgG1s engineered for enhanced binding to the neonatal Fc receptor (FcRn). J Biol Chem 281:23514–23524
16. Dalton WS, Scheper RJ (1999) Lung resistance-related protein: determining its role in multidrug resistance. J Natl Cancer Inst 91:1604–1605
17. Doyle LA, Yang W, Abruzzo LV, Krogmann T, Gao Y, Rishi AK, Ross DD (1998) A multidrug resistance transporter from human MCF-7 breast cancer cells. Proc Natl Acad Sci USA 95:15665–15670
18. Dumont JA, Bitonti AJ, Clark D, Evans S, Pickford M, Newman SP (2005) Delivery of an erythropoietin-Fc fusion protein by inhalation in humans through an immunoglobulin transport pathway. J Aerosol Med 18:294–303
19. Endter S, Francombe D, Ehrhardt C, Gumbleton M (2009) RT-PCR analysis of ABC, SLC and SLCO drug transporters in human lung epithelial cell models. J Pharm Pharmacol 61:583–591
20. Fetsch P, Abati A, Litman T, Morisaki K, Honjo Y, Mittal K, Bates S (2006) Localization of the ABCG2 mitoxantrone resistance-associated protein in normal tissues. Cancer Lett 235: 84–92
21. Flens MJ, Zaman GJ, van der Valk P, Izquierdo MA, Schroeijers AB, Scheffer GL, van der Groep P, de Haas M, Meijer CJ, Scheper RJ (1996) Tissue distribution of the multidrug resistance protein. Am J Pathol 148:1237–1247
22. Folkesson HG, Hedin L, Weström BR (1992) Lung to blood passage of human growth hormone after intratracheal instillation: stimulation of growth in hypophysectomized rats. J Endocrinol 134:197–203
23. Forbes B, Shah A, Martin GP, Lansley AB (2003) The human bronchial epithelial cell line 16HBE14o – as a model system of the airways for studying drug transport. Int J Pharm 257:161–167
24. Francombe D, Taylor G, Taylor S, Somers G, Edwards CD, Gumbleton M (2008) Functional role of P-gp efflux in limiting pulmonary drug absorption within an intact lung: application of an isolated perfused rat lung model. In: Dalby RN, Byron PR, Peart J, Suman JD, Farr SJ,

Young PM (eds) Respiratory drug delivery 2008 volume II. Davis Healthcare International Publishing, River Grove, pp 461–464
25. Gillespie MN, Krechniak JW, Crooks PA, Altiere RJ, Olson JW (1985) Pulmonary metabolism of exogenous enkephalins in isolated perfused rat lungs. J Pharmacol Exp Ther 232:675–681
26. Groneberg DA, Eynott PR, Döring F, Dinh QT, Oates T, Barnes PJ, Chung KF, Daniel H, Fischer A (2002) Distribution and function of the peptide transporter PEPT2 in normal and cystic fibrosis human lung. Thorax 57:55–60
27. Hastings RH, Wright JR, Albertine KH, Ciriales R, Matthay MA (1994) Effect of endocytosis inhibitors on alveolar clearance of albumin, immunoglobulin G and SP-A in rabbits. Am J Physiol Lung Cell Mol Physiol 266:L544–L552
28. Hastings RH, Folkesson HG, Matthay MA (2004) Mechanisms of alveolar protein clearance in the intact lung. Am J Physiol Lung Cell Mol Physiol 286:L679–L689
29. Heyder J, Gebhart J, Rudolf G, Schiller CF, Stahlhofen W (1986) Deposition of particles in the human respiratory tract in the size range 0.005 to 15 μm. J Aerosol Sci 17:811–825
30. John TA, Vogel SM, Minshall RD, Ridge K, Tiruppathi C, Malik AB (2001) Evidence for the role of alveolar epithelial gp60 in active transalveolar albumin transport in the rat lung. J Physiol 533:547–559
31. Kim K-J, Malik AB (2003) Protein transport across the lung epithelial barrier. Am J Physiol Lung Cell Mol Physiol 284:L247–L259
32. Kim K-J, Fandy TE, Lee VHL, Ann DK, Borok Z, Crandall ED (2004) Net absorption of IgG via FcRn-mediated transcytosis across rat alveolar epithelial cell monolayers. Am J Physiol Lung Cell Mol Physiol 287:L616–L622
33. Kuhlmann O, Hofmann HS, Muller SP, Weiss M (2003) Pharmacokinetics of idarubicin in the isolated perfused rat lung: effect of cinchonine and rutin. Anticancer Drugs 14:411–416
34. Langmann T, Mauerer R, Zahn A, Moehle C, Probst M, Stremmel W, Schmitz G (2003) Real-time reverse transcription-PCR expression profiling of the complete human ATP-binding cassette transporter superfamily in various tissues. Clin Chem 49:230–238
35. Lanseley AB (1993) Mucociliary clearance and drug delivery via the respiratory tract. Adv Drug Deliv Rev 11:299–327
36. Leach CL, Davidson PJ, Boudreau RJ (1998) Improved airway targeting with the CFC-free HFA-beclomethasone metered-dose inhaler compared with CFC-beclomethasone. Eur Respir J 12:1346–1353
37. Lencer WI, Blumberg RS (2005) A passionate kiss, then run: exocytosis and recycling of IgG by FcRn. Trends Cell Biol 15:5–9
38. Lin YJ, Schanker LS (1981) Pulmonary absorption of amino acids in the rat: evidence of carrier transport. Am J Physiol 240:C215–C221
39. Lips KS, Volk C, Schmitt BM, Pfeil U, Arndt P, Miska D, Ermert L, Kummer W, Koepsell H (2005) Polyspecific cation transporters mediate luminal release of acetylcholine from bronchial epithelium. Am J Respir Cell Mol Biol 33:79–88
40. Lombry C, Edwards DA, Prëat V, Vanbever R (2004) Alveolar macrophages are a primary barrier to pulmonary absorption of macromolecules. Am J Physiol 286:L1002–L1008
41. MacDougall C, Guglielmo BJ (2004) Pharmacokinetics of valaciclovir. J Antimicrob Chemother 53:899–901
42. Madlova M, Bosquillon C, Asker D, Dolezal P, Forbes B (2009) In-vitro respiratory drug absorption models possess nominal functional P-glycoprotein activity. J Pharm Pharmacol 61:293–301
43. Mostov KE, Kaetzel CS (1999) Immunoglobulin transport and the polymeric immunoglobulin receptor. In: Mestecky J, Bienenstock J, Lamm ME, Strober W, McGhee JR, Mayer L (eds) Mucosal immunology. Academic, New York, pp 181–211
44. Newman S (2009) Aerosol properties and deposition principles. In: Newman S (ed) Respiratory drug delivery: essential theory & practice. Davis Healthcare International Publishing, River Grove, pp 29–58
45. Pang Y, Sakagami M, Byron PR (2005) The pharmacokinetics of pulmonary insulin in the isolated perfused rat lung: implications of metabolism and regional deposition. Eur J Pharm Sci 25:369–378

46. Pang Y, Sakagami M, Byron PR (2007) Insulin self-association: effects of lung disposition kinetics in the airways of the isolated perfused rat lung. Pharm Res 24:1636–1644
47. Patton JS (1996) Mechanisms of macromolecule absorption by the lung. Adv Drug Deliv Rev 19:3–36
48. Patton JS, McCabe JG, Hansen SE, Daugherty AL (1990) Absorption of human growth hormone from the rat lung. Biotechnol Ther 1:213–228
49. Patton JS, Fishburn CS, Weers JG (2004) The lung as a portal of entry for systemic drug delivery. Proc Am Thorac Soc 1:338–344
50. Qiu Y, Gupta PK, Adjei AL (1997) Absorption and bioavailability of inhaled peptides and proteins. In: Adjei AL, Gupta PK (eds) Inhalation delivery of therapeutic peptides and proteins. Marcel Dekker, New York, pp 89–132
51. Roerig DL, Audi SH, Ahlf SB (2004) Kinetic characterization of P-glycoprotein-mediated efflux of rhodamine 6G in the intact rabbit lung. Drug Metab Dispos 32:953–958
52. Rojas R, Apodaca G (2002) Immunoglobulin transport across polarised epithelial cells. Nat Rev Mol Cell Biol 3:944–956
53. Sacaan AI, Edwards WR, Domingo D, Glynn JM, Hawley S, Pierce L, Shabram P, Palmer DT (2004) Transport of aerosolized IL-2 chimeric protein using polymeric immunoglobulin receptor in the lung. In: Dalby RN, Byron PR, Peart J, Suman JD, Farr SJ (eds) Respiratory drug delivery IX volume II. Davis Healthcare International Publishing, River Grove, pp 357–360
54. Sai Y, Tsuji A (2004) Transporter-mediated drug delivery: recent progress and experimental approaches. Drug Discov Today 9:712–720
55. Sakagami M (2004) Insulin disposition in the lung following oral inhalation in humans: a meta-analysis of its pharmacokinetics. Clin Pharmacokinet 43:539–552
56. Sakagami M (2006) In vivo, in vitro and ex vivo models to assess pulmonary absorption and disposition of inhaled therapeutics for systemic delivery. Adv Drug Deliv Rev 58:1030–1060
57. Sakagami M, Gumbleton M (2006) Endogenous and exogenous IgG transfer in the airways: a prospective evaluation. In: Dalby RN, Byron PR, Peart J, Suman JD, Farr SJ (eds) Respiratory drug delivery, vol I. Davis Healthcare International Publishing, River Grove, pp 57–64
58. Sakagami M, Sakon K, Kinoshita W, Makino Y (2001) Enhanced pulmonary absorption following aerosol administration of mucoadhesive powder microspheres. J Control Release 77: 117–129
59. Sakagami M, Byron PR, Venitz J, Rypacek F (2002) Solute disposition in the rat lung in vivo and in vitro: determining regional absorption kinetics in the presence of mucociliary clearance. J Pharm Sci 91:594–694
60. Sakagami M, Kinoshita W, Sakon K, Makino Y (2002) Mucoadhesive beclomethasone microspheres for powder inhalation: their pharmacokinetics and pharmacodynamics evaluation. J Control Release 80:207–218
61. Sakagami M, Omidi Y, Campbell L, Kandalaft LE, Morris CJ, Barar J, Gumbleton M (2006) Expression and transport functionality of FcRn within rat alveolar epithelium: a study in primary cell culture and in the isolated perfused lung. Pharm Res 23:270–279
62. Schanker LS (1978) Drug absorption from the lung. Biochem Pharmacol 27:381–385
63. Scheffer GL, Pijnenborg AC, Smit EF, Müller M, Postma DS, Timens W, van der Valk P, de Vries EG, Scheper RJ (2002) Multidrug resistance related molecules in human and murine lung. J Clin Pathol 55:332–339
64. Spiekermann GM, Finn PW, Ward ES, Dumont J, Dickinson BL, Blumberg RS, Lencer WI (2002) Receptor-mediated immunoglobulin G transport across mucosal barriers in adult life: functional expression of FcRn in the mammalian lung. J Exp Med 196:303–310
65. Tamai I, Nezu J, Uchino H, Sai Y, Oku A, Shimane M, Tsuji A (2000) Molecular identification and characterization of novel members of the human organic anion transporter (OATP) family. Biochem Biophys Res Commun 273:251–260
66. Usmani OS, Biddiscombe MF, Barnes PJ (2005) Regional lung deposition and bronchodilator response as a function of β_2-agonist particle size. Am J Respir Crit Care Med 172:1497–1504
67. Wong JP, Yang H, Blasetti KL, Schnell G, Conley J, Schofield LN (2003) Liposome delivery of ciprofloxacin against intracellular Francisella tularensis infection. J Control Release 92:265–273

Chapter 7
Controlled Transport for Pulmonary Drug Delivery

Jennifer Fiegel, Timothy Brenza, and Rania Hamed

Abstract Interactions between inhaled particles and the respiratory tract fluids are important for inhaled drug delivery systems. In particular, controlling the transport of aerosol particles after deposition in the respiratory tract may improve drug retention time in the lungs, allow targeting, and facilitate optimal transport through innate defense mechanisms of the lung. In this chapter the mechanisms by which particles can transport in the lungs and the aerosol design criteria for improving particle residence times or promoting uniform distribution across or within the lung space are discussed.

Keywords Diffusion • Mucus • Particle • Submersion • Surface tension • Surfactant • Transport

7.1 Introduction

The therapeutic delivery of drugs directly to the respiratory tract offers several advantages over traditional drug delivery. These include the direct application of therapeutics to the site of respiratory disease and the ability to avoid first pass metabolism of sensitive drugs. It has been reported, though, that using marketed inhaled drug delivery devices typically leads to less than 10% of inhaled drugs reaching the target region in the lungs [1]. In cases of airway obstruction, lung inflammation, and infection, the amount of drug deposited in the lungs may be even less than 10% [2]. Improvements in inhaler design and particle engineering are

J. Fiegel (✉)
Department of Pharmaceutical Sciences and Experimental Therapeutics
and
Department of Chemical and Biochemical Engineering, The University of Iowa,
115 S. Grand Avenue, S215 PHAR, Iowa City, IA 52242, USA
e-mail: jennifer-fiegel@uiowa.edu

H.D.C. Smyth and A.J. Hickey (eds.), *Controlled Pulmonary Drug Delivery*,
Advances in Delivery Science and Technology, DOI 10.1007/978-1-4419-9745-6_7,

improving this situation, with as much as 60% deposition being reported for newer technologies. Still, these technological advances have typically ignored the innate defense mechanisms that operate in the respiratory tract and limit therapeutic efficacy. For example, deposited foreign material is efficiently trapped by mucosal fluids and cleared via the mucociliary escalator on relatively short time scales, thereby limiting the residence time of the aerosols in the respiratory tract. In addition, small molecules and many biomolecules are often absorbed too quickly for effective therapy, and drug aerosols are not uniformly deposited and distributed in the lung, leading to areas of high and low drug concentrations within the respiratory tract [3]. Therefore, achieving appropriate drug retention and distribution after deposition remains a challenge.

Controlling the transport of aerosol particles after deposition in the respiratory tract is one potential method for improving drug retention time in the lungs and to target drugs to the desired site of action. While the respiratory tract fluids act as a significant barrier to aerosol particle and drug transport, a variety of strategies have been developed to overcome these limitations. Therefore, this chapter will focus on the interactions of aerosol particles with respiratory fluids and methods that have been developed to improve the spatial and temporal distribution of particles within the airways. Particle transport in the lungs involves highly complex processes which depend on the physical and chemical properties of the particles, as well as the properties of the fluids themselves. Hence we will review our current knowledge on the mechanisms by which particles can transport in the lungs and discuss aerosol design criteria for improving particle residence times and for promoting uniform distribution across or within the lung spaces. The chapter is written to reflect the order in which various modes of particle transport can occur; with surface spreading and submersion first, as they typically occur over relatively short time scales, followed by mucociliary clearance and diffusion, which occur over longer time scales.

7.2 Surfactant-Assisted Spreading for Enhanced Spatial Distribution of Aerosols

Upon deposition, aerosol particles first interact with lung fluids at the air-fluid interface. This interface is characterized by relatively low surface tension, indicating the presence of surface-active agents. The surface tension of lung fluids decreases with each successive generation, with that in the trachea measured at ~32 mN/m and the deep lungs (i.e., alveoli) reaching values less than 10 mN/m [4, 5]. For comparison, the surface tension of water is ~72 mN/m. These low surface tensions help stabilize the lungs from collapse during exhalation, decrease the amount of work required to inflate the lungs, and facilitate the spreading of mucus after its secretion from goblet cells and the submucosal glands [6–12].

The natural surface tension gradient in the respiratory tract can aid the spreading of aerosol particles and drug compounds across the lung surface [13, 14]. In theory, convective flow is generated as surfactant moves from areas of high surfactant concentration (low surface tension, such as in the deep lungs) to areas of low surfactant concentration (high surface tension, such as in the tracheobronchial region). Surfactant can also be added to an aerosolized therapy to enhance surface-tension driven spreading within the respiratory tract [1, 3]. Surfactant release from the deposited aerosol causes a local decrease in the fluid surface tension, which drives convective flow (Marangoni flow) away from the aerosol and to the surrounding fluid [15]. This method has been used to improve the uniformity of drug distribution into unreachable "deeper" areas of diseased lungs which are plugged by mucus [3]. Surfactant replacement therapies such as Calfactant (ONY Inc.) and Survanta (Abbott Labs Inc.) have been shown to disperse drugs deposited in the tracheobronchial region to the peripheral regions of the lungs [1, 16]. These exogenous surfactants have additionally been used to deliver antibiotics, such as tobramycin, and immune-suppressive agents, such as cyclosporine and rapamycin, via tracheal instillation to the alveolar compartment in rodents [16]. A recent study has shown that delivering various antitubercular drugs loaded into dipalmitoylphosphatidylcholine (DPPC) liposomes via nebulization leads to a uniform drug distribution in vitro [17]. Corcoran et al. studied the distribution of fluorescently labeled aerosols (1–4 μm) on model airway surfaces using four surfactants: sodium dodecyl sulfate (SDS), cetyl trimethyl ammonium bromide (CTAB), tyloxapol, and Calfactant (a natural lung surfactant extract) [3]. They observed a significant increase in aerosol dispersion after deposition compared to saline (control), which was attributed to convective transport (i.e., surface-tension driven flow). Kharasch et al. observed a more uniform and wider distribution pattern when a surfactant vehicle was used to deliver technetium sulfur colloid and pentamidine intratracheally compared to saline [13]. Other nonantimicrobial compounds such as recombinant adenoviral vectors and corticosteroids have also revealed a better spatial distribution profile when delivered intratracheally using surfactant carriers [2]. A mathematical derivation of the effects of endogenous surfactant on particle dispersion is discussed in detail in the work of Grotberg et al. [15].

Zhang et al. developed a model to investigate the final distribution of a chemical delivered with surfactant as a bolus driven by Marangoni flow [18]. Mass transport equations were applied to the geometry of a symmetric, dichotomous branching respiratory model. In this work the authors investigated the sensitivity of the pulmonary chemical absorption due to the following parameters: dose of surfactant, air flow rate, viscosity, and chemical absorption rate. An increase in dose of surfactant was found to both improve the uniformity of the chemical deposition within the airways and moved the peak position of the chemical distribution more distally in their model. Increased air flow rate also led to improved deposition uniformity and deeper chemical penetration. Their model also suggested that increased formulation viscosity will lead to a more distal chemical deposition and improved uniformity at the expense of chemical absorption time. The chemical absorption rate in their model strongly influenced the final deposition of the chemical, with highly absorptive chemicals failing to progress beyond the upper airways. This work suggested that the choice of a surfactant-chemical system can greatly aid in the uniform distribution of the chemical in the respiratory tract.

7.3 Aerosol Submersion Within Respiratory Fluids

In addition to lateral flow across the air-fluid interface, deposited aerosols can be wetted and displaced into respiratory fluids. In vivo animal studies have demonstrated that many deposited particles submerse, partially or fully, within this interface (Fig. 7.1) [5, 19–23]. The submersion process can, theoretically, be controlled by altering the interfacial tension of the fluid's or particle's physicochemical properties such as size, shape, and surface chemistry. Thus, submersion could be used to push particles past the mucus barrier and to force particles to interact with the underlying epithelium.

The process of particle submersion is primarily controlled by the need to maintain mechanical equilibrium at the interface and can be described under static conditions by the Young equation:

$$\gamma_{\mathrm{lv}} \cos\theta = \gamma_{\mathrm{sv}} - \gamma_{\mathrm{sl}}, \tag{7.1}$$

where γ is the surface tension or surface energy at the liquid-vapor (lv), solid-vapor (sv), or solid-liquid (sl) interface and θ is the contact angle of the particle. A schematic of this process is depicted in Fig. 7.2. Since the surface tension of the lung fluids (γ_{lv} in (7.1)) varies based on location, the ability of a particle to submerse

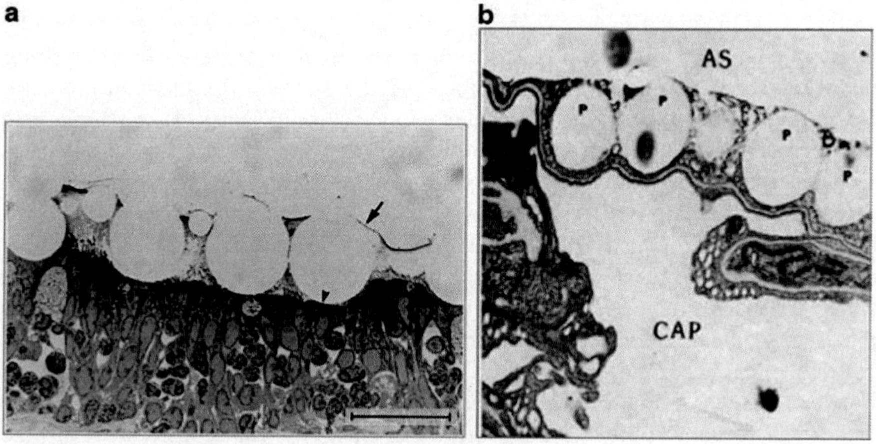

Fig. 7.1 (**a**) Light micrograph of a guinea pig trachea fixed by immersion in osmium tetroxide/perfluorocarbon 5 min after exposure to a polydispersed aerosol of polymethylmethacrylate (PMMA; diameter 10–100 μm). The microspheres are partially or totally (smaller particles) submersed beneath an osmiophilic film (*arrow*) and indent the underlying epithelium (*arrowhead*). Bar = 30 μm. Reprinted from Schürch et al. [24], Copyright (1999), with permission from Elsevier. (**b**) Transmission electron microscopy (TEM) image of alveolar wall showing 1 μm diameter polystyrene beads in an alveolar space (AS). The particles have been displaced toward the epithelium and are completely covered by an osmiophilic layer (surfactant). Surface tension forces have caused deformation of the underlying capillary (CAP) by the particles. Reprinted from Schürch et al. [5], Copyright (1990), with permission from Elsevier

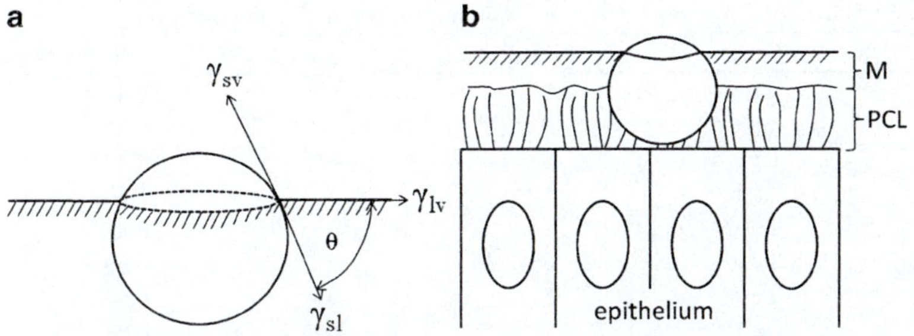

Fig. 7.2 Schematic of particle wetting and submersion of (**a**) a solid spherical particle deposited onto the upper respiratory tract surface where the particle diameter is of similar size to the depth of the mucus layer (*M*), and (**b**) a larger particle which is displaced into the periciliary fluid layer (PCL). In each case, the particle is wetted and displaced into the fluid until its contact angle θ is satisfied

within the interface changes within the respiratory tract. The Young equation predicts that as surface tension is decreased, particles will become further submersed within the fluids. As the surface tension approaches 0, particles become completely submersed. Therefore, in the alveoli, where surface tensions reach very low values, particles are expected to be completely submersed. Conversely, in the tracheobronchial region, where surface tensions as high as 32 mN/m have been reported, particles are expected to be partially or fully wetted.

The three remaining parameters in the Young equation (θ, γ_{sv}, and γ_{sl}) depend on the properties of the solid particle. Particle contact angle is a material attribute of the particle and, therefore, can be easily changed by surface modifications. Increasing the hydrophobicity of a particle increases its contact angle, generally causing a particle to be less submersed in a fluid. For hydrophilic particles, the high humidity in the respiratory tract (close to 100%) can hydrate the surface during inhalation, causing the particles to become substantially more wetted and displaced into the interface than hydrophobic particles [24]. Particle shape can also be altered to change a particle's contact angle. The influence of sharp edges and highly curved surfaces of different particles on the spreading of a surfactant film was investigated by Gerber et al. [25]. The results showed that the resistance of a surfactant film to spread on particle surfaces increased with particle curvature.

The size of a particle compared to the depth of the fluid upon which it deposits can alter the particle's ability to become submersed. Typical therapeutic aerosol particles range in size from about 1–20 μm. The depth of respiratory fluids decreases with each successive generation into the respiratory tract (Table 7.1). It has been reported that in nonpathological conditions the depth varies between 10 and 30 μm in the trachea and between 2 and 5 μm in the bronchi [26, 27]. While actual measurements of the in vivo depth of the airway fluids is difficult to obtain, the upper respiratory tract fluid in excised human airways has been measured between 7 and 55 μm, illustrating this variation between subjects [28, 29]. Reports of mucus depth

Table 7.1 Physical properties of human respiratory tract (conducting airways)

Generation	Length (cm)[a]	Mucus (μm)[b, c]	Sol (μm)[b, c]	Velocity (mm/min)[d]
0	9.1	27	8	6.46
1	3.8	5	6	6.59
2	1.5	5	6	2.47
3	0.83	5	6	1.20
4	0.90	5	6	0.89
5	0.81	5	6	0.69
6	0.66	5	6	0.54
7	0.60	5	6	0.43
8	0.53	5	6	0.36
9	0.43	4	2	0.29
10	0.36	4	2	0.21
11	0.30	4	2	0.15
12	0.25	4	2	0.11
13	0.21	4	2	0.08
14	0.17	4	2	0.05
15	0.14	4	2	0.04

[a] Dimensional model of tracheobronchiolar tree in adult male [97]
[b] Estimates of dimensions of trachea, bronchus, and bronchiole [98]
[c] Morphometric model [99]
[d] Yu et al. [44]

as large as 260 μm have been reported [10]. This discrepancy in observed mucus layer thickness may be due to the release of mucus granules during the fixation process or due to subject-to-subject variation. To small particles deposited in the upper respiratory tract, the lung fluids appear to be infinitely deep and the Young equation can be used to estimate particle submersion. Larger aerosol particles, when completely submerged, will penetrate through the mucus gel layer and into the periciliary fluid. This is likely to limit clearance by the mucociliary escalator, thereby increasing particle residence time in the lungs (see section on the mucociliary clearance below for further detail). As the size of a deposited particle gets close to the depth of the fluid, interactions with the underlying epithelium become important. At the more distal generations, the aqueous fluid is quite thin. The alveoli fluids have an average depth ranging from 0.18 μm above flat alveolar walls to 0.89 μm in the alveolar corners [30]. Therefore, particles deposited in the alveolar region, as well as very large particles in the upper respiratory tract, can make direct contact with the epithelium upon submersion, as observed in Fig. 7.1a. Mathematical modeling of particle submersion in thin films laying on top of an inflexible membrane has shown that the surface tension force in much of the respiratory tract can be large enough to cause the fluids to climb up the particle surface to enable complete submersion [31]. In vivo experiments in animals have shown not only that respiratory fluids can coat particles that are much larger than the fluid depth, but also that the lung epithelium is a flexible membrane which can be indented significantly upon particle submersion, as observed in Fig. 7.1b [5, 20, 32]. This phenomenon can be utilized to enhance particle–cell interactions and limit particle clearance.

Limitations to accurately predict particle submersion are primarily due to the dynamics of the lung from expansion and contraction during breathing, as well as mucociliary clearance. Mucociliary clearance is likely to promote particle rolling, which could promote surfactant adsorption onto particle surfaces and lead to enhanced particle submersion. Breathing causes the depth in alveoli to change significantly, therefore particle wetting could be promoted during exhalation when fluid depth is at a maximum and reduced during inhalation. An example of the effects of lung dynamics has been observed with puffball spores. These spores are decorated externally with tiny warts which increase the local surface curvature and are expected to hinder submersion [32]. However, in vivo experiments showed that the spores experienced the same submersion as microspheres with a relatively smooth surface [19]. These results suggest that these dynamic events may substantially alter particle submersion, which could impact the therapeutic effectiveness of aerosols.

7.4 Controlling Particle Transport via the Mucociliary Escalator

Submersion within the conducting airway fluids may be an innate defense mechanism designed to efficiently trap foreign material for clearance from the airways via the mucociliary escalator. Airway fluids, and any trapped material, are continuously propelled toward the larynx by the coordinated movement of the cilia on the lung epithelium [33–35]. Fig. 7.3 shows a histological image of an airway tissue culture in which the tips of the cilia are touching a viscous mucus layer, which sits above a dilute, low viscosity, "watery" periciliary fluid. The beating of the cilia in the periciliary layer transmits energy to the mucus layer, which is then used for bulk transport. Effective mucociliary clearance requires fine control over ciliary function, periciliary fluid depth, and mucus rheological properties [36–38]. Therefore, any disturbance in these properties can significantly alter mucociliary clearance. For example, if the mucus lacks elasticity, beating cilia cannot transmit energy to the mucus and mucociliary clearance is impaired [36]. If the periciliary layer is too thin or too thick, the cilia cannot beat properly or cannot reach the mucus layer, respectively, thereby limiting clearance [35]. Chronic pulmonary diseases, such as CF, COPD, and asthma, result in mucus hypersecretion and fluid dehydration [39–42]. These alterations reduce the ability of the ciliated cells to efficiently move the mucosal secretions, leading to reduced mucociliary clearance and retention of mucus in the lungs [43].

The rate at which mucociliary clearance occurs determines the residence time of fluids and aerosol particles in the conducting airways. Transport of respiratory mucus up and out of healthy lungs occurs at rates of ~5–100 μm/s (Table 7.1) [35, 44]. Therefore, particles are typically cleared from the most distal ciliated airways in healthy lungs within 1 day, although values between individuals can vary from 0.1 to 1 day [37, 45]. In diseased lungs, transient time has been shown to be significantly

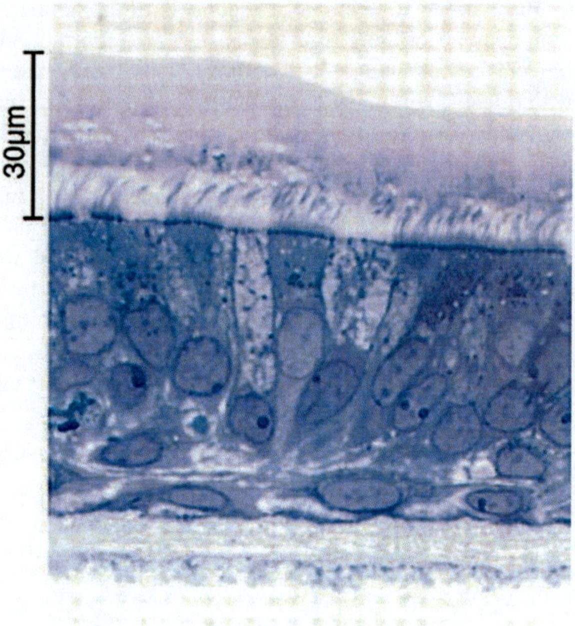

Fig. 7.3 Histology of an airway epithelial culture fixed by PFC/OsO_4. Note that the mucus network is in contact with the cilia tips above the periciliary fluid. The total airway surface fluid depth is indicated by the *scale bar*. Reprinted from Matsui et al. [39], Copyright (1998), with permission from Elsevier

reduced. For example, a reduction from 53.5 ± 2.8% for healthy patients to 40.7 ± 2.5 for CF patients in the percentage of tracer particles cleared from the lungs in 24 h has been observed [46]. Clearance by the mucociliary escalator can be altered to either enhance or further limit transport through the addition of various mucoactive compounds or through modifications to particle properties.

7.4.1 Altering Particle Clearance via Mucoactive Compounds

Mucolytic agents are used to break down or depolymerize the mucus gel network. By reducing the viscosity, adherence, and tenacity of the mucosal fluids, mucociliary clearance can be improved [47, 48]. It is likely that improving mucociliary clearance would lead to enhanced aerosol particle clearance. Classic mucolytic agents include DNase, an enzyme which cleaves DNA, and thiol agents such as *N*-acetylcysteine (NAC), which breaks disulfide bonds between the glycoproteins [6, 47, 49]. Surfactants are currently under development as mucolytic agents as they have been seen to reduce sputum adhesivity and consequently enhance mucus clearability [47]. Adrenergic agonists such as isoproterenol [50, 51], epinephrine [50], salbutamol [52], terbutaline [53, 54], and fenoterol [55] have been shown to augment mucociliary clearance significantly by increasing the ciliary beat frequency [50, 55]. Olivier et al. showed uridine 5′-triphosphate (UTP), which has been shown to increase ciliary beat frequency [33, 56–58], caused a 2.5-fold increase in mucociliary clearance in normal human adults [59]. The cholinergic agonist bethanechol hydrochloride was also shown to result in a statistically significant enhancement of mucociliary clearance from the proximal airways [60]. Hypertonic saline and mannitol are used to induce an influx of water into the lungs, thereby rehydrating the mucosal fluids and enabling ciliary action [6, 35, 47]. Hypertonic saline has a secondary function of disrupting the ionic bonds within the mucin gel, leading to a reduction in mucus viscosity and elasticity [36]. In these cases the residence time of inhalable therapeutics bound within the mucus layer would be reduced.

On the other hand, an array of compounds has been shown to directly disrupt mucociliary clearance and could potentially be used as a means of increasing the residence time of aerosol particles in the conducting airways. Tertiary ammonium compounds (anticholinergics), aspirin, anesthetic agents, and benzodiazepines have been shown to be capable of depressing the mucociliary transport system by decreasing cilia beat frequency [48]. Compounds produced by infectious bacteria have also been shown to disturb the mucociliary system by slowing and disorganizing the beating of cilia [61, 62]. These include pyocyanin, 1-hydroxyphenazine, and rhamnolipid produced by *Pseudomonas aeruginosa* [62, 63], low molecular weight glycopeptides produced by *Haemophilus influenza* [64], and pneumolysin produced by *Staphylococcus pneumonia* [65].

7.4.2 *Effect of Particle Physicochemical Properties on Clearance*

The probability of particles becoming trapped in mucus is dependent on physical and chemical interactions between the particles and mucus (Fig. 7.4). Moller et al. showed a fast and slow phase of mucociliary clearance of particles in the conducting airways, suggesting that particles are capable of temporarily avoiding mucociliary clearance [66]. Using bolus inhalation techniques, they limited deposition of 4.2 μm aerodynamic diameter ferromagnetic iron oxide particles to the conducting airways. The first 49% of the deposited particles cleared with a mean half-time of 3 h. The slow phase showed a significantly longer mean half-time of 109 days. It is likely that the slow phase reflects particles that have submersed or diffused below the mucus layer and thus are less likely to be transported out of the respiratory tract via mucociliary clearance. The size of microparticles has also been shown to play a role in the rate of particle clearance. Stahlhofen et al. used the same bolus inhalation technique to deposit ferromagnetic iron oxide particles with differing aerodynamic diameters. They found that the smaller aerodynamic diameter particles (1.8 μm) were about 60% retained in the respiratory tract after 24 h, compared to 25% retention for larger (5.9 μm) particles [67]. This data suggests that the smaller particles are able to diffuse into the periciliary layer to avoid the mucociliary escalator.

Particle adhesion to the mucus (mucoadhesion) has also been exploited as a method to overcome the limitations imposed by mucociliary clearance. Long polymer chains attached to the surface of particles may enable the particles to become intertwined and entangled within the mucus network. Mucoadhesive particles typically exhibit slower

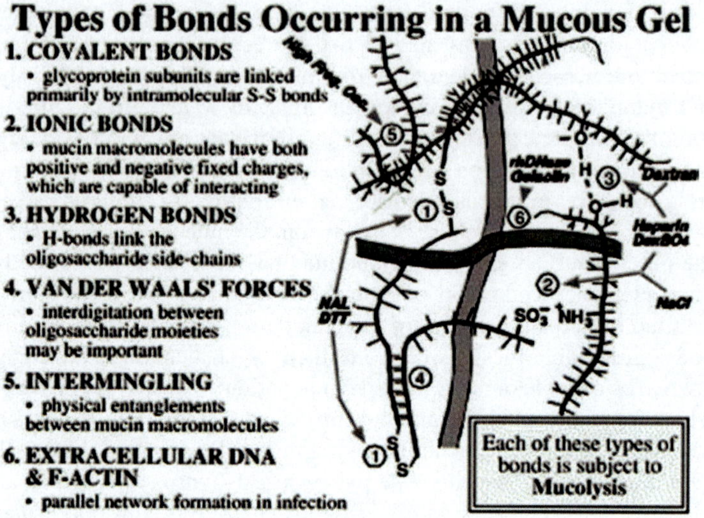

Fig. 7.4 Schematic of interactions leading to the formation of the mucus gel network [75]

particle transit time from the mucosal site, likely due to an increase in the viscoelastic properties of the mucosal fluids which inhibits the mucociliary escalator. In one study, it was shown that grafting the copolymer polyethylene glycol (PEG) monomethacrylate to methacrylic acid particles enhanced their adhesion to gelled bovine submaxillary mucins [68]. This strategy has been used extensively to enhance the mucoadhesion of particles to the gastrointestinal tract and nasal passages. It is clear from these studies with other mucosal surfaces that there is a molecular weight dependence on adhesion, where polymer chains that are too short or too long do not effectively entangle in the mucus layer [69]. The ability of mucoadhesive microspheres to limit clearance in vivo has been indirectly measured in a few cases. One group showed prolonged drug retention in asthmatic guinea pigs after delivery from hydroxypropylcellulose microspheres [70]. In addition, increased bioavailability of ofloxacin and sustained release of insulin from hyaluronic acid-based aerosols has been demonstrated in an animal model [71, 72]. Limitations to this strategy include the time scale for mucus renewal and concern over potential negative health consequences of inhibiting mucociliary clearance.

7.5 Diffusional Transport Within Respiratory Fluids

One of the most promising methods to improve particle transport within respiratory fluids is to enhance diffusion. In the upper respiratory tract, diffusion can enable particles to penetrate into the periciliary layer and reach the underlying epithelium, thereby increasing particle residence times in the lungs. In the lower airways, the diffusion of particles plays a role in the spreading and distribution on the airway surface. Simple random (Brownian) diffusion of a dilute suspension of spherical particles in a homogeneous fluid can be described by the Stokes–Einstein equation:

$$D_{AB} = \frac{\kappa T}{6\pi\mu_B R_A}, \tag{7.2}$$

where D_{AB} is the diffusion constant (cm^2/s), κ is the Boltzmann constant ($1.38 \times 10^{-23}\ m^2 kg/s^2\ K$), T is temperature (K), μ_B is the viscosity of the bulk fluid (kg/s m), and R_A is the radius of the particle (cm). From the Stokes–Einstein equation, we see that the ability of a particle to diffuse is inversely proportional to particle size and fluid viscosity. Based on purely Brownian diffusion, therefore, we would expect that smaller particles can diffuse more quickly through the mucosal fluids, but will be significantly impeded by higher viscosity fluids particularly in the tracheobronchial region. This equation, however, ignores the complex structure and interactions between the diffusing particles and components of the mucus network. This mucus network has been a major barrier to particle and drug transport for therapeutic drug delivery [35]. Due to the recent progress in controlling diffusional transport of aerosol particles in lung fluids, the remainder of this chapter will focus on describing key characteristics of the mucus gel which inhibit particle transport, techniques used to quantify transport and methods developed to enhance particle movement through these fluids.

7.5.1 Respiratory Tract Mucus as a Barrier to Particle Diffusion

The tracheobronchial region fluids are composed of approximately 95% water and 5% solids, which include 2–3% proteins and glycoproteins, 1% lipids and 1% minerals [73]. Mucin glycoproteins (mucins) are the major macromolecular components of the respiratory mucus. They are high molecular weight ($2–20 \times 10^5$ Da), highly glycosylated (50–90% by weight) proteins which form a cross-linked viscoelastic mucus gel [6, 36, 74]. Intramolecular disulfide bonds, in addition to noncovalent linkages (hydrogen bonds, electrostatic and hydrophobic interactions, and Van der Waals forces), give mucus an entangled and extended polymeric gel network [49, 75]. Particle diffusion within this network is limited both physically due to the highly cross-linked nature of the mucus as well as chemically due to interactions between the particles and mucus components.

The mucus network generates nanoscopic pores through which particles can diffuse. The transport of a variety of compounds and virus particles through respiratory mucus has been used as an indication of the mesh spacing between mucin fibers (Fig. 7.5). Many small proteins (1.2–150 kDa), including immunoglobulin (IgG), diffuse in mucus nearly as fast as they diffuse in water [76–78]. Human papilloma virus (55 nm, ~20,000 kDa), Norwalk virus (38 nm, ~10,000 kDa), and many globular proteins (~2–20 nm, 15–650 kDa) have also been shown to diffuse as rapidly in mucus as in saline [79]. Viruses known to infect epithelial surfaces are ~30–200 nm in diameter [35]. Therefore, particle sizes less than about 200 nm are typically considered appropriate for potential diffusion through nondiseased respiratory tract mucus.

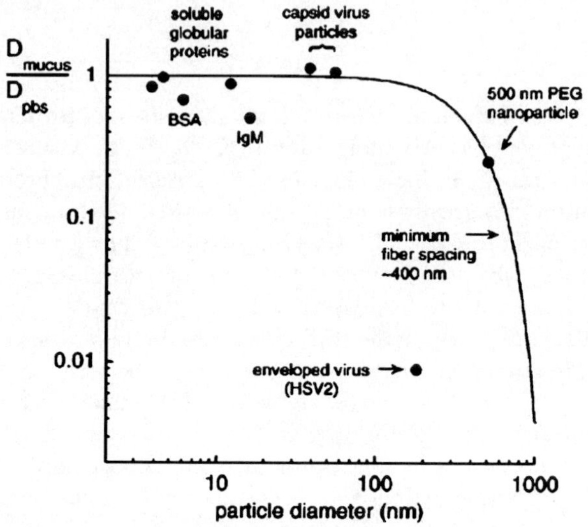

Fig. 7.5 Diffusion constants in mucus compared to phosphate buffered saline as a function of hydrodynamic diameter of the diffusing particles (modified from [79, 100]). The *curve* shows the predicted relationship using Amsden and Turner's model if the average mesh spacing is 400 nm, a spacing much larger than predicted if mucin monomers did not aggregate into large, multi-mucin "cables." Reprinted from Cone [35], Copyright (2009), with permission from Elsevier

Disease state can significantly alter the mucus mesh size. Direct measurement of pore size in diseased mucus samples has provided an indication of the maximum size of particles that can penetrate the network. Atomic force microscopy images of unprocessed CF sputa illustrate a dense porous structure with an average pore diameter of 517±76.5 nm (three adult patients, one sample each, ten measurements) [80]. Images of CF sputum obtained via scanning electron microscopy have led to estimated pore sizes in the range of 100–400 nm [81]. The larger pore size observed in CF patients has been attributed to mucin fiber bundling, which opens the pores for diffusion of larger compounds and particles [82]. Indeed, efficient diffusion of particles as large as 500 nm has been observed in CF sputum [77]. The viscosity of mucus in disease states also influences the ability of particles to diffuse. In diffusion chamber studies, the introduction of DNA to a 4% w/v mucin solution was shown to significantly decrease the diffusion rate of dissolved ticarcillin, a β-lactam antibiotic, in a dose dependent manner [83]. The addition of DNA (calf thymus at 5 mg/mL) to porcine gastric mucin (30 mg/mL) reduced the amount of 200 nm carboxylate-modified FITC-labeled nanospheres transported across a diffusion chamber from 72±5.9% of PBS for mucin to 44±1.5% of PBS for the combination [80]. Recent studies, though, have suggested that the "local" viscosity within the pores of CF sputum may be very low (close to that of water) [77]. Therefore, particles which exhibit no binding to mucus components should readily diffuse through the pores.

Mucins and other components can limit particle diffusion by forming multiple low energy bonds, which trap the particles in the mucus layer (Fig. 7.4). The negative charge and hydrophobic regions of mucins lead to significant electrostatic and hydrophobic interactions with particles. The transport of negatively charged polystyrene nanospheres and cationic lipoplexes has been investigated through sputa from CF patients [81, 84, 85]. In these studies, the number of charged particles transported after 150 min through CF sputum layer was very low (<0.5% of total number of particles present at the donor side) and strongly dependent on the size of the particles (with smaller particles transporting more readily through sputum). Studies comparing the transport of 200 nm negatively charged polystyrene nanospheres in CF sputum and buffer have shown that transport in CF sputum was 10–20-fold slower, suggesting van der Waals interactions of the sputum components with the hydrophobic polystyrene core [80, 86]. Positively charged nanoparticles, such as those coated with chitosan, have been observed to have limited transport in mucus gels [87]. One proposed mechanism for this hindered diffusion is particle aggregation caused by interaction with negatively charged CF mucus components (albumin, linear DNA, and mucin) [88].

7.5.2 *Methods to Measure Particle Diffusion Through Mucus*

Diffusion of particles in mucus has been investigated through the use of two distinct methods: diffusion chambers and single particle tracking. With diffusion chamber studies, the bulk quantity of particles which transfer through a layer of mucus is measured. These measurements rely on the mass flow of particles across a uniform thickness of fluid. In the steady state case, the mass flow rate per unit area (or molecular mass flux) is proportional to the mass fraction difference divided by the plate thickness and can be approximated by:

$$j_{\mathrm{A}} = \rho D_{\mathrm{AB}} \frac{\omega_{\mathrm{Ao}}}{L}, \tag{7.3}$$

where j_{A} is the mass flux of the particle, ρ is the fluid density, D_{AB} is the diffusion coefficient for the particle in the fluid, ω_{Ao} is the initial mass concentration of the particle, L is the thickness of the fluid layer, and the assumption is made that the mass concentration of particles in the receiver is zero. This approach has been used by Bolister et al. to show the decrease in ticarcillin diffusion through layers with increasing mucin concentration [83]. They showed that the diffusion rate of ticarcillin was 3 times greater with a mucin concentration of 1.475% (w/w) than at a concentration closer to 5%. Broughton-Head et al. showed significant reduction in transport of 200 nm fluorescent spheres at porcine gastric mucin concentrations of 25 and 50 mg/mL, but not at 10 mg/mL [80].

In single particle tracking, a subpopulation of particles is followed within the local environment and their displacement over a period of time is determined. The population of particles is then combined in order to give a population average movement of particles on that time scale. This process is depicted in Fig. 7.6. Here the diffusion coefficient is estimated as the mean-squared displacement divided by the

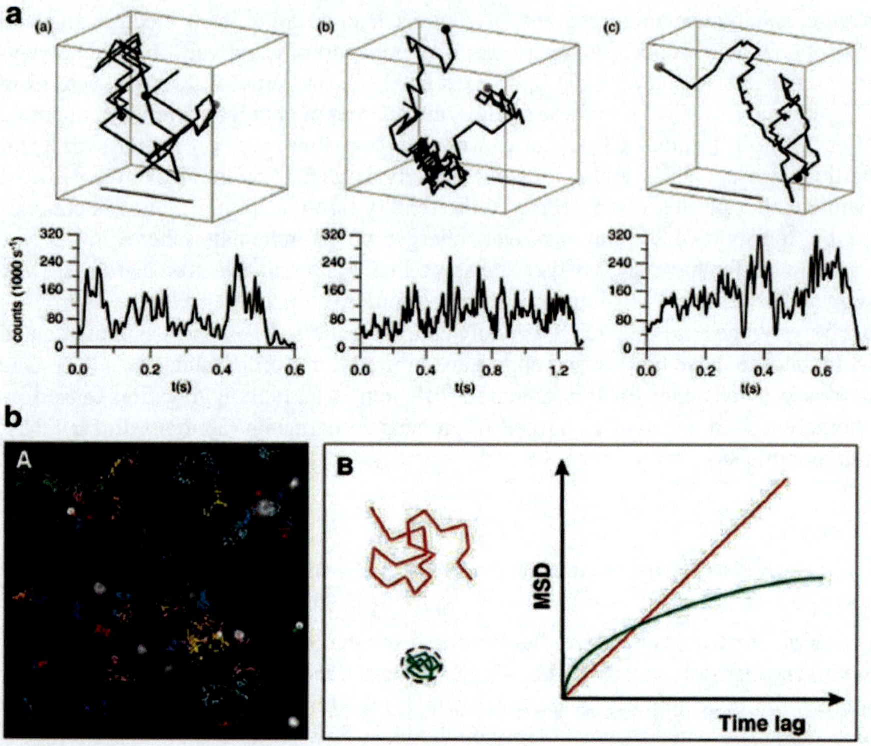

Fig. 7.6 Principle of single-particle tracking. (**a**) The *top frames* show projections of the trajectories of three individual particles, scale bar is 1 μm. The *gray dot* is the initial position of the particle at the start of the data collection period and the *black dot* is the final position. The distance between the *gray* and *black dots* were (a) 1.64, (b) 1.35, and (c) 2.61 μm. The *bottom panels* show the total number of detected photons for each 5 ms time-step during the data acquisition of the trajectory above. Reprinted with permission from Lessard et al. [101]. Copyright 2007, American Institute of Physics. (**b**) Trajectories are shown for each individual particle by colored lines in (**a**). In (**b**) the trajectories of free-diffusing (*red*) and confined (*green*) particles are shown. The graph on the *right* depicts the corresponding MSD of a particle as a function of lag time. Reprinted from Sanders et al. [30], Copyright (2009) with permission from Elsevier

time scale (7.4). One advantage to this method is that inferences can be made about local viscosity and barriers due to subpopulations of particles which may exhibit different transport characteristics than the ensemble of particles.

$$MSD = 4D_o\tau^{\alpha}, \tag{7.4}$$

where D_o is the time scale (τ) dependent diffusivity and α is the diffusion exponent reflecting the extent of particle impedance in the fluid ($\alpha = 1$ for pure Brownian diffusion; α decreases in fluids where obstructions are present). Single particle tracking experiments have shown that the population averaged movement of particles in mucus is increased when the diameter of the particles is decreased. However, the mean diffusion coefficient becomes dominated by relatively few but fast-moving particles as particle size is reduced from 500 to 100 nm [89]. This indicates that there still exists heterogeneity in the mucus even on this length scale.

7.5.3 *Methods to Enhance Particle Diffusion Through Respiratory Tract Fluids*

Altering particle surface chemistry is one strategy to minimize mucoadhesion and enhance diffusion in mucosal fluids. As mentioned above, it has been observed that the movement of many soluble proteins and viruses is not hindered by mucus. The fact that both of their surfaces are densely coated with equal quantities of positive and negative charges suggests that neutrally charged particles should exhibit free movement in mucus [90]. This finding has led to the development of a few key strategies for tailoring particle surfaces for enhanced diffusional transport. Polystyrene particles that have been densely coated with a layer of PEG have been observed to more rapidly move in CF sputum than charged particles of the same size (Fig. 7.7) [89, 91]. The covalent coupling of PEG to the surface active groups on the polystyrene particles neutralizes the charge on particles, thereby reducing electrostatic interactions, and also to increase the hydrophilicity of particles, thus limiting hydrophobic interactions with mucin fibers. While the covalent attachment of PEG to drug compounds and particles has been used extensively to limit interactions with blood serum proteins, in mucosal fluids it can potentially act as a mucoadhesive (see section above on mucociliary clearance). Optimization of PEG chain length is required to limit mucus

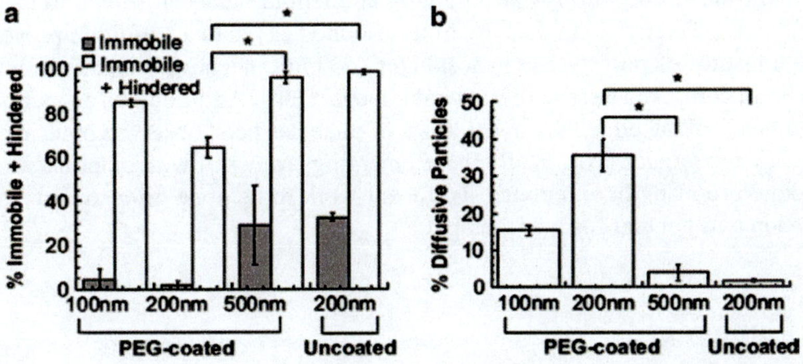

Fig. 7.7 Transport mechanism distribution of various sized particles in CF sputum, with or without PEG coating ($n=3$ experiments): (**a**) immobile and immobile + hindered and (**b**) diffusive particles. Data represents mean ± SEM of three experiments with $n>200$ particles for each experiment. Immobile particles have an MSD below the microscope detection limit (10 nm). Differences in percentages of immobile, immobile + hindered and diffusive particles are statistically significant (*asterisks*) for 200 nm PEG-coated particles as compared to either 200 nm uncoated particles or 500 nm PEG-coated particles ($p<0.01$). Reprinted from Suk et al. [91], Copyright (2009), with permission from Elsevier

interactions. Based on studies conducted to date, low molecular weight PEG within the range of 2–5 kDa is optimal for limiting mucoadhesion [92]. Coating particles with soluble proteins has also been investigated as a strategy to minimize the adhesion of particles to the respiratory mucosa. Carrabino et al. investigated the effect of the addition of human serum albumin (HSA) to polyethylenimine (PEI)-DNA complexes for gene therapy [93]. The in vitro transfection rate of PEI-DNA complexes was observed to increase by 5–10 times with absorption of HSA to the surface of the complex. The presence of CF sputum inhibited in vitro gene expression of all complexes, but HSA-PEI-DNA complexes exhibited 18–84 times higher expression than complexes without HSA on the surface.

Mucolytic compounds have also been studied for their ability to reduce mucus viscosity and therefore improve particle diffusion. DNase has been shown to enhance gene transfer, in vitro, from liposomes in the presence of CF sputum, suggesting improved particle mobility due to the digestion of DNA in the CF sputum [94]. However, the presence of actin in a CF mucus mimetic eliminated the improvement in negatively charged polystyrene nanosphere diffusion observed by DNase [80]. Interestingly, the traditional mucolytic Nacystelyn showed no improvement in gene transfer or nanosphere diffusion [80, 94]. It has been theorized that breaking up the intermolecular and intramolecular bounds of the mucin chains causes an increase in the viscosity of the mucus pores, which reduces particle diffusion through those pores. Therefore, it is important to consider the effect a particular mucolytic has on the microenvironment before implementing them as a strategy to control particle transport.

Enhancement of particle transport by external stimuli is a new strategy currently being explored. These methods attempt to overcome the limitations of diffusion by controlling the spatial location of particles in the lungs. The use of an external magnetic field to direct the transport of superparamagnetic iron oxide vectors coated with PEI (400–1,000 nm) toward the lung epithelium has been shown to enhance transfection efficiency [95]. The use of ultrasound has led to a tenfold improvement in the transport of polystyrene nanospheres (500 nm, negatively charged) through CF sputum compared to pure diffusional transport [30]. Additionally, an increase in gene transfer of naked pDNA in the lungs of mice has been observed by using high frequency ultrasound [96]. While these emerging strategies with external stimulation show promise, there remains significant work to be done prior to their implementation as a general form of therapy.

7.6 Summary

While the respiratory tract fluids act as a significant barrier to aerosol and drug transport, a variety of strategies have been developed to overcome these limitations. Controlling the physical and chemical properties of the particle provides one method to alter particle transport in the lungs. The size, shape, and hydrophilicity of the particle surface are all properties which have been observed to alter the submersion of particles, retention time, and diffusivity of the particle in the mucus layer.

Mechanisms which alter the properties of the respiratory fluids have also shown some promise in improving particle transport. These include the redistribution of particles through the altering of surface tension with surfactants and mucoactive compounds to alter the residence time of particles in the respiratory tract or diffusion of particles within the mucus layer. Finally, external stimuli have been shown to provide control of particle distribution within the lungs. While these methods have shed light on the process of controlling particle transport, there is still much to be learned about the interaction of particles with the components which make up the respiratory fluid and how these interactions change the local and global respiratory environment. Improvements in understanding and predicting these complex interactions will lead to significant therapeutic advances.

References

1. Ganguly S, Moolchandani V, Roche JA, Shapiro PS, Somaraju S, Eddington ND et al (2008) Phospholipid-induced in-vivo particle migration to enhance pulmonary deposition. J Aerosol Med Pulm Drug Deliv 4:343–350
2. Haitsma JJ, Lachmann U, Lachmann B (2001) Exogenous surfactant as a drug delivery agent. Adv Drug Deliv Rev 47(2–3):197–207
3. Marcinkowski AL, Garoff S, Tilton RD, Pilewski JM, Corcoran TE (2008) Postdeposition dispersion of aerosol medications using surfactant carriers. J Aerosol Med Pulm Drug Deliv 4:361–369
4. Im Hof V, Gehr P, Gerber V, Lee MM, Schürch S (1997) In vivo determination of surface tension in the horse trachea and in vitro model studies. Respir Physiol 109(1):81–93
5. Schürch S, Gehr P, Im Hof V, Geiser M, Green F (1990) Surfactant displaces particles toward the epithelium in airways and alveoli. Respir Physiol 80(1):17–32
6. Rubin BK (2009) Mucus, phlegm, and sputum in cystic fibrosis. Respir Care 54(6):726–732
7. King M, Rubin BK (1999) Mucus-controlling agents: past and present. Respir Care Clin N Am 5(4):575–594
8. Hohlfeld JM (2002) The role of surfactant in asthma. Respir Res 3:4
9. Banerjee R, Puniyani RR (2000) Exogenous surfactant therapy and mucus rheology in chronic obstructive airway diseases. J Biomater Appl 14(3):243–272
10. Widdicombe JG (1997) Airway liquid: a barrier to drug diffusion? Eur Respir J 10(10): 2194–2197
11. Widdicombe JH, Bastacky SJ, Wu DX, Lee CY (1997) Regulation of depth and composition of airway surface liquid. Eur Respir J 10(12):2892–2897
12. Morgenroth K (1985) Morphology of the bronchial lining layer and its alteration in IRDS, ARDS and COLD. J Respir Dis Suppl 142:7–18
13. Kharasch VS, Sweeney TD, Fredberg J, Lehr J, Damokosh AI, Avery ME et al (1991) Pulmonary surfactant as a vehicle for intratracheal delivery of technetium sulfur colloid and pentamidine in hamster lungs. Am Rev Respir Dis 144(4):909–913
14. van't Veen A, Mouton JW, Gommers D, Lachmann B (1996) Pulmonary surfactant as vehicle for intratracheally instilled tobramycin in mice infected with *Klebsiella pneumoniae*. Br J Pharmacol 119(6):1145–1148
15. Grotberg JB, Halpern J, Jensen OE (1995) Interaction of exogenous and endogenous surfactant: spreading-rate effects. J Appl Physiol 78(2):750–756
16. Gommers D, Haitsma JJ, Lachmann B (2006) Surfactant as a carrier: influence of immunosuppressive agents on surfactant activity. Clin Physiol Funct Imaging 26(6):357–361

17. Chimote G, Banerjee R (2009) Evaluation of antitubercular drug-loaded surfactants as inhalable drug-delivery systems for pulmonary tuberculosis. J Biomed Mater Res A 89(2):281–292
18. Zhang YL, Matar OK, Craster RV (2003) A theoretical study of chemical delivery within the lung using exogenous surfactant. Med Eng Phys 25:115–132
19. Geiser M, Leupin N, Maye I (2000) Im Hof V, Gehr P. Interaction of fungal spores with the lungs: distribution and retention of inhaled puffball (calvatia excipuliformis) spores. J Allergy Clin Immunol 106:92–100
20. Gehr P, Schürch S, Berthiaume Y, Im Hof V, Geiser M (1990) Particle retention in airways by surfactant. J Aerosol Med 3:27–43
21. Geiser M, Cruz-Orive LM, Im Hof V, Gehr P (1990) Assessment of particle retention and clearance in the intrapulmonary conducting airways of hamster lungs with the fractionator. J Microsc 160:75–88
22. Geiser M, Cruz-Orive LM, Waber U, Im Hof V, Gehr P (1995) Particle retention in airways estimated by unbiased stereology and photometry. J Aerosol Med 8:149–165
23. Geiser M, Gerber P, Maye I, Im Hof V, Gehr P (2000) Retention of teflon particles in hamster lungs: a stereological study. J Aerosol Med 13:43–55
24. Schürch S, Geiser M, Lee MM, Gehr P (1999) Particles at the airway interfaces of the lung. Colloids Surf B Biointerfaces 15(3–4):339–353
25. Gerber P, Lehmann C, Gehr P, Schürch S (2006) Wetting and spreading of a surfactant film on solid particles: influence of sharp edges and surface irregularities. Langmuir 22(12): 5273–5281
26. Wine JJ (1999) The genesis of cystic fibrosis lung disease. J Clin Invest 103:309–312
27. Patton JS (1996) Mechanisms of macromolecule absorption by the lungs. Adv Drug Del Rev 19:3–36
28. Song Y, Namkung W, Nielson DW, Lee J, Finkbeiner WE, Verkman AS (2009) Surface liquid depth measured in ex vivo fragments of pig and human trachea: dependence on Na+ and Cl– channel function. Am J Physiol Lung Cell Mol Physiol 297:L1131–L1140
29. Jayaraman S, Song Y, Vetrivel L, Shankar L, Verkman AS (2001) Noninvasive in vivo fluorescence measurement of airway-surface liquid depth, salt concentration, and pH. J Clin Invest 107:317–324
30. Sanders N, Rudolph C, Braeckmans K, De Smedt SC, Demeester J (2009) Extracellular barriers in respiratory gene therapy. Adv Drug Deliv Rev 61:115–127
31. Fiegel J, Jin F, Hanes J, Stebe K (2005) Wetting of a particle in a thin film. J Coll Int Sci 291:507–514
32. Geiser M, Schurch S, Gehr P (2003) Influence of surface chemistry and topography of particles on their immersion into the lung's surface-lining layer. J Appl Physiol 94:1793–1801
33. Lansley AB, Sanderson MJ, Dirksen ER (1992) Control of the beat cycle of respiratory tract cilia by Ca2+ and cAMP. Am J Physiol Lung Cell Mol Physiol 263(2):L323–L324
34. Tarran R, Trout L, Donaldson SH, Boucher RC (2006) Soluble mediators, not cilia, determine airway surface liquid volume in normal and cystic fibrosis superficial airway epithelia. J Gen Physiol 127(5):591–604
35. Cone RA (2009) Barrier properties of mucus. Adv Drug Del Rev 61:75–85
36. King M, Rubin BK (2002) Pharmacological approaches to discovery and development of new mucolytic agents. Adv Drug Del Rev 54:1475–1490
37. Wolff RK (1986) Effects of airborne pollutants on mucociliary clearance. Environ Health Perspect 66:223–237
38. Hermens WAJJ, Merkus FWHM (1987) The influence of drugs on nasal ciliary movement. Pharm Res 4(6):445–449
39. Matsui H, Grubb BR, Tarran R, Randell SH, Gatzy JT, Davis CW et al (1998) Evidence for periciliary liquid layer depletion, not abnormal ion composition, in the pathogenesis of cystic fibrosis airways disease. Cell 95:1005–1015
40. Kuperman DA, Huang X, Koth LL, Chang GH, Doganov GM, Zhu Z et al (2002) Direct effects of interleukin-13 on epithelial cells cause airway hyperreactivity and mucus overproduction in asthma. Nat Med 8:885–889

41. Vestbo J, Prescott E, Lange P (1996) Association of chronic mucus hypersecretion with FEV1 decline and chronic obstructive pulmonary disease morbidity. copenhagen city heart study group. Am J Respir Crit Care Med 153(5):1530–1535
42. Aikawa T, Shimura S, Sasaki H, Ebina M, Takishima T (1992) Marked goblet cell hyperplasia with mucus accumulation in the airways of patients who died of severe acute asthma attack. Chest 101(4):916–921
43. Robinson M, Eberl S, Tomlinson C, Daviskas E, Regnis JA, Bailey DL et al (2000) Regional mucociliary clearance in patients with cystic fibrosis. J Aerosol Med 13(2):73–86
44. Yu CP, Hu JP, Yen BM, Spektor DM, Lippmann M (1986) Models for mucociliary particle clearance in lung airways. In: Lee SD, Schneider T, Grant LD, Verkerk PJ (eds) Aerosols. Lewis Publishers, Michigan
45. Lippmann M (1986) Respiratory tract deposition and clearance of aerosols. In: Lee SD, Schneider T, Grant LD, Verkerk PJ (eds) Aerosols. Lewis Publishers, Michigan
46. Donaldson SH, Bennett WD, Zeman KL, Knowles MR, Tarran R, Boucher RC (2006) Mucus clearance and lung function in cystic fibrosis with hypertonic saline. N Engl J Med 354(3):241–250
47. Rubin BK (2007) Mucolytics, expectorants, and mucokinetic medications. Respir Care 52(7):859–865
48. Houtmeyers E, Gosselink R, Gayan-Ramirez G, Decramer M (1999) Effects of drugs on mucus clearance. Eur Respir J 14:452–467
49. Lai SK, Wang Y, Wirtz D, Hanes J (2009) Micro- and macrorheology of mucus. Am J Respir Cell Mol Biol 61:86–100
50. Foster WM, Bergofsky EH, Bohning DE, Lippmann M, Albert RE (1976) Effect of adrenergic agents and their mode of action on mucociliary clearance in man. J Appl Physiol 41(2):146–152
51. Foster WM, Langenback E, Bergofsky EH (1980) Measurement of tracheal and bronchial mucus velocities in man: relation to lung clearance. J Appl Physiol 48(6):965–971
52. Lafortuna CL, Fazio F (1984) Acute effect of inhaled salbutamol on mucociliary clearance in health and chronic bronchitis. Respiration 45:111–123
53. Mortensen J, Lange P, Nyboe J, Groth S (1994) Lung mucociliary clearance. Eur J Nucl Med 21(9):953–961
54. Groth S, Mortensen J, Lange P, Munch EP, Sorensen PG, Rossing N (1988) Imaging of the airways by bronchoscintigraphy for the study of mucociliary clearance. Thorax 43(5): 360–365
55. Yeates DB, Aspin N, Levison H, Jones MT, Bryan AC (1975) Mucociliary tracheal transport rates in man. J Appl Physiol 39(3):487–495
56. Knowles MR, Clarke LL, Boucher RC (1991) Activation by extracellular nucleotides of chloride secretion in the airway epithelia of patients with cystic fibrosis. N Engl J Med 325(8): 533–538
57. Benali R, Pierrot D, Zahm JM, de Bentzmann S, Puchelle E (1994) Effects of extracellular ATP and UTP on fluid transport by human nasal epithelial cells in culture. Am J Respir Cell Mol Biol 10:363–368
58. Merten MD, Breittmayer JP, Figarella C, Frelin C (1993) ATP and UTP increase secretion of bronchial inhibitor by human tracheal gland cells in culture. Am J Physiol Lung Cell Mol Physiol 265(5):L479–L484
59. Olivier KN, Bennett WD, Hohneker KW, Boucher RC, Knowles MR (1996) Acute safety and effects on mucociliary clearance of aerosolized uridine 5′-triphosphateamiloride in normal human adults. Am J Respir Crit Care Med 154:217–223
60. Cammer P, Strandberg K, Philipson K (1974) Increased mucociliary transport by cholinergic stimulation. Arch Environ Health 29:220–224
61. Wilson R (1988) Secondary ciliary dysfunction. Clin Sci 75:113–120
62. Wilson R, Pitt T, Taylor G, Watson D, MacDermot J, Sykes D et al (1987) Pyocyanin and 1-hydroxyphenazine produced by pseudomonas aeruginosa inhibit the beating of human respiratory cilia in vitro. J Clin Invest 79(1):221–229

63. Read RC, Roberts P, Munro N, Rutman A, Hastie A, Shryock T et al (1992) Effect of pseudomonas aeruginosa rhamnolipids on mucociliary transport and ciliary beating. J Appl Physiol 72(6):2271–2277
64. Wilson R, Moxon BR (1988) Molecular basis of *Haemophilus influenzae* pathogenicity in the respiratory tract. In: Griffiths E, Donachie W, Stephen J (eds) Bacterial infections of respiratory and gastrointestinal mucosae. IRL Press, Oxford, pp 29–40
65. Steinfort C, Wilson R, Mitchell T, Feldman C, Rutman A, Todd H et al (1989) Effect of *Streptococcus pneumoniae* on human respiratory epithelium in vitro. Infect Immun 57(7): 2006–2013
66. Moller W, Haussinger K, Winkler-Heil R, Stahlhofen W, Meyer T, Hofmann W et al (2004) Mucociliary and long-term particle clearance in the airways of healthy nonsmoker subjects. J Appl Physiol 97(6):2200–2206
67. Stahlhofen W, Koebrich R, Rudolf G, Scheuch G (1990) Short-term and long-term clearance of particles from the upper human respiratory tract as function of particle size. J Aerosol Sci 21(Suppl 1):S407–S410
68. Huang Y, Leobandung W, Foss A, Peppas NA (2000) Molecular aspects of muco- and bioadhesion: tethered structures and site-specific surfaces. J Control Release 65(1–2):63–71
69. De Ascentiis A, deGrazia JL, Bowman CN, Colombo P, Peppas NA (1995) Mucoadhesion of poly(2-hydroxyethyl methacrylate) is improved when linear poly(ethylene oxide) chains are added to the polymer network. J Control Release 33(1):197–201
70. Sakagami M, Kinoshita W, Sakon K, Sato J, Makino Y (2002) Mucoadhesive beclomethasone microspheres for powder inhalation: their pharmacokinetics and pharmacodynamics evaluation. J Control Release 80(1–3):207–218
71. Surendrakumar K, Martyn GP, Hodgers ECM, Jansen M, Blair JA (2003) Sustained release of insulin from sodium hyaluronate based dry powder formulations after pulmonary delivery to beagle dogs. J Control Release 91(3):385–394
72. Hwang SM, Kim DD, Chung SJ, Shim CK (2008) Delivery of ofloxacin to the lung and alveolar macrophages via hyaluronan microspheres for the treatment of tuberculosis. J Control Release 129(2):100–106
73. Boat TF, Cheng PW (1980) Biochemistry of airway mucus secretions. Fed Proc 39(13): 3067–3074
74. Thornton DJ, Sheehan JK (2004) From mucins to mucus: toward a more coherent understanding of this essential barrier. Proc Am Thorac Soc 1(1):54–61
75. Zayas G, Dimitry J, Zayas A, O'Brien D, King M (2005) A new paradigm in respiratory hygiene: Increasing the cohesivity of airway secretions to improve cough interaction and reduce aerosol dispersion. BMC Pulm Med 5:11
76. Radomsky ML, Whaley KJ, Cone RA, Saltzman WM (1990) Macromolecules released from polymers: diffusion into unstirred fluids. Biomaterials 11:619–624
77. Henry BT, Hibberd AS, Cheema MS, Davis SS, Rogers TG (1992) Epi-fluorescence microscopy and image analysis used to measure diffusion coefficients in gel systems. J Pharm Pharmacol 44:543–549
78. Saltzman WM, Radomsky ML, Whaley KJ, Cone RA (1994) Antibody diffusion in human cervical mucus. Biophys J 66(2, Part 1):508–815
79. Olmsted SS, Padgett JL, Yudin AI, Whaley KJ, Moench TR, Cone RA (2001) Diffusion of macromolecules and virus-like particles in human cervical mucus. Biophys J 81(4): 1930–1937
80. Broughton-Head VJ, Smith JR, Shur J, Shute JK (2007) Actin limits enhancement of nanoparticle diffusion through cystic fibrosis sputum by mucolytics. Pulm Pharmacol Ther 20(6): 708–717
81. Sanders NN, DeSmedt SC, Van Rompaev E, Simoens P, De Baets F, Demeester J (2000) Cystic fibrosis sputum – a barrier to the transport of nanospheres. Am J Respir Crit Care Med 162:1905–1911
82. Voynow JA, Rubin BK (2009) Mucins, mucus, and sputum. Chest 135(2):505–512

83. Bolister N, Basker M, Hodges NA, Marriott C (1991) The diffusion of B-lactam antibiotics through mixed gels of cystic fibrosis-derived mucin and *pseudomonas aeruginosa* alginate. J Antimicrob Chemother 27:285–293
84. Sanders NN, Van Rompaev E, DeSmedt SC, Demeester J (2002) On the transport of lipoplexes through cystic fibrosis sputum. Pharm Res 19(4):451–456
85. Sanders NN, DeSmedt SC, Demeester J (2003) Mobility and stability of gene complexes in biogels. J Control Release 87:117–129
86. Broughton-Head VJ, Shur J, Carroll MP, Smith JR, Shute JK (2007) Unfractionated heparin reduces the elasticity of sputum from patients with cystic fibrosis. Am J Physiol Lung Cell Mol Physiol 293:L1240–L1249
87. Kas HS (1997) Chitosan: properties, preparations and application to microparticulate systems. J Microencapsul 14(6):689–711
88. Sanders NN, Van Rompaev E, DeSmedt SC, Demeester J (2001) Structural alterations of gene complexes by cystic fibrosis sputum. Am J Respir Crit Care Med 164:486–493
89. Dawson M, Wirtz D, Hanes J (2003) Enhanced viscoelasticity of human cystic fibrotic sputum correlates with increasing microheterogeneity in particle transport. J Biol Chem 278: 50393–50401
90. Wada A, Nakamura H (1981) Nature of the charge distribution in proteins. Nature 5835: 757–758
91. Suk JS, Lai SK, Wang Y, Ensign LM, Zeitlin PL, Boyle MP et al (2009) The penetration of fresh undiluted sputum expectorated by cystic fibrosis patients by non-adhesive polymer nanoparticles. Biomaterials 30(13):2591–2597
92. Wang Y, Lai SK, Suk JS, Pace A, Cone R, Hanes J (2008) Addressing the PEG mucoadhesivity paradox to engineer nanoparticles that "slip" through the human mucus barrier. Angew Chem Int Ed 47:9726–9729
93. Carrabino S, Di Gioia S, Copreni E, Conese M (2005) Serum albumin enhanced polyethyleneimine-mediated gene delivery to human respiratory epithelial cells. J Gene Med 7: 1555–1564
94. Stern M, Caplen NJ, Browning JE, Griesenbach U, Sorgi F, Huang L et al (1998) The effect of mucolytic agents on gene transfer across a CF sputum barrier in vitro. Gene Therapy 5:91–98
95. Scherer F, Anton M, Schillinger U, Henke J, Kruger A, Gansbacher B (2002) Magnetofection: enhancing and targeting gene delivery by magnetic force in vitro and in vivo. Gene Therapy 9:102–109
96. Xenariou S, Griesenbach U, Liang H, Zhu J, Farley R, Somerton L et al (2007) Use of ultrasound to enhance nonviral lung gene transfer in vivo. Gene Therapy 14:768–774
97. James AC (1988) Lung dosimetry. In: Nazaroff WW, Nero AV Jr (eds) Radon and its decay products in indoor air. Wiley, New York
98. Gehr P (1994) Anatomy and morphology of the respiratory tract. In: ICRP (eds) Human respiratory tract model for radiological protection. ICRP publication 66. Ann ICRP 24(1–3): 121–166
99. ICRP (1999) Human respiratory tract model for radiological protection. ICRP publication 66. Ann ICRP 24(1–3):1–482
100. Lai SK, O'Hanlon DE, Harrold S, Man ST, Wang Y, Cone R et al (2007) Rapid transport of large polymeric nanoparticles in fresh undiluted human mucus. Proc Natl Acad Sci 104(5):1482–1487
101. Lessard GA, Goodwin PM, Werner JH (2007) Three-dimensional tracking of individual quantum dots. Appl Phys Lett 91(22):224106

Chapter 8
Science and Technology of Pressurized Metered-Dose Inhalers

Sandro R.P. da Rocha, Balaji Bharatwaj, and Sowmya Saiprasad

Abstract The pressurized, propellant driven metered-dose inhaler is an essential technology for pulmonary drug delivery. In this chapter, both the basic scientific principles and technological advances of these systems will be discussed. A particular focus will be the physical chemistry of formulation and how this understanding allows predictable product development of these products. Lastly, based on this science, future technological advances will be discussed.

Keywords Formulation development • Hydrofluoroalkane propellants • Nanomedicine • Pressurized metered-dose inhalers • Pulmonary drug delivery

8.1 Introduction

Before delving into the scientific aspects of this chapter, we will reflect on the relevance and speculate on the future of pressurized metered-dose inhalers (pMDIs) in the respiratory drug delivery market. Such discourse will unavoidably have a certain bias as our group works in this area. However, we strove to include the most appropriate references to support our claims. The introduction to the chapter will be divided in three parts: (1) comparing and contrasting pMDIs with dry powder inhalers (DPIs); (2) the environmental impact of propellants used in pMDIs; and (3) current challenges in the formulation of pMDIs.

S.R.P. da Rocha (✉)
Chemical Engineering and Materials Science, Wayne State University, Detroit, MI 48202, USA
e-mail: sdr@eng.wayne.edu

H.D.C. Smyth and A.J. Hickey (eds.), *Controlled Pulmonary Drug Delivery*, Advances in Delivery Science and Technology, DOI 10.1007/978-1-4419-9745-6_8,

8.1.1 pMDIs vs. DPIs

The two most commonly used devices in the treatment of pulmonary diseases are the pMDIs and the DPIs, accounting for approximately 65% of the total sales – tablets/granules and nebulizers occupy most of the remaining market [62]. pMDIs are the most widely used portable oral inhalation (OI) devices, with sales that reached an estimated 4.6 billion dollars in 2007. DPIs are the second most common portable inhalers, with total sales of more than nine billion dollars in the same year. The higher total sales of DPIs reflect the much higher cost per actuation – treatment costs are at least double for DPIs, and the types of drugs being commercialized in DPI platforms – more branded drugs are sold in DPIs, which may cost 10 times more per actuation than pMDIs [4, 62, 63]. It is also interesting to note that DPIs have steadily gained market share over pMDIs during the past decade [118]. While there are known advantages and disadvantages in using pMDIs, the same is true for DPIs [9]. For example, while coordination of actuation and inspiration may be a problem when using pMDIs, their performance is not as dependent on the inspirational flow rate of the patient as is the case for DPIs. Therefore pMDIs may be better suited for use by children and elderly [94]. There are several other advantages and disadvantages scenarios like the one just described that could be discussed in terms of pMDIs vs. DPIs [9]. However, what is perhaps most important to understand is that there is no single delivery system that is expected to be universally accepted by patients/doctors, and that not all drugs can be expected to be efficiently formulated in both devices, or simply available in all markets for both devices [21, 62]. From that perspective alone, therefore, pMDIs and DPIs are expected to coexist for quite some time, and the balance between them in the OI market will depend on new scientific and technological advances that are being discussed today (as in this book), and those innovations yet to come.

8.1.2 Environmental Impact of the Propellants Used in pMDIs

Another important consideration regarding the future relevance of pMDIs is the fact that propellants are required in such devices. The propellant is the solvent medium for the drug (solution or dispersion), and is what generates the aerosol (the propellant fluid). DPIs, on the other hand, do not require a propellant, as they are usually breath actuated devices [43, 59]. This is an important issue because the hydrofluoroalkane (HFA) propellants while non-ozone depleting, have a large global warming potential (GWP) [52]. GWP is a measure of the future radiative effect of an emission of a substance relative to that of the same amount of CO_2. GWP for HFA134a (CF_3CH_2F), one of the two propellants used in pMDIs is 1,300, and that for HFA227 (CF_3CHFCF_3), the other propellant acceptable for use in pMDIs, is 2,900 [21]. Careful analysis of the global market for propellants (including HFAs) reveal, however, that only 1% of the total HFA market is used in medicinal aerosols [22]. Of course, this argument alone is not enough to predict the future of HFA propellants and thus of pMDIs, as one gets in the realm of policy making [4]. Our society, on one hand, is invested in supporting sustainable processes and products. On the other hand, we are also very sensitive to the availability (or lack) of drugs that we may require to treat important pulmonary disorders, which may be further exacerbated as the environmental conditions deteriorate. Apart from that discussion, it has been concluded that the most reduction of GWP in this area (pMDIs) will come with the complete phase out of chlorofluorocarbons (CFCs) from pMDIs across the globe, with little environmental benefits to be gained even if all HFA from pMDI formulations were to be eliminated in the future [21], as for example by switching (if at all possible) all such formulations to DPIs.

8.1.3 Formulation Development

One final remark with regards to the current and future relevance of pMDIs, which may have implications not only in the pulmonary drug delivery market, but also in the overall drug delivery market when considering that pMDIs may also serve in the future as vehicles for the systemic delivery of drugs, is regarding existing challenges in the formulation stages of the product development. The HFA propellants used in pMDIs were selected based on several criteria, including their inertness and biological compatibility [51, 93]. While their vapor pressure and density, two important physicochemical characteristics that greatly impact the aerosol characteristics and overall formulation strategy [112], are not the same as those in CFCs, they represented a good compromise, and the old pMDI hardware has thus required relatively little modification to accommodate the new propellants [8, 49]. However, the solvent properties of HFAs are dramatically different from CFCs [84]. HFAs are significantly more polar, but at the same time do not have a high solubilization capacity to polar compounds [54] – they are somewhat hydrophobic; i.e., have poor solubilization capacity for most hydrophilic compounds including salts and biomacromolecules, and at the same time oleophobic; i.e., have poor solubilization capacity for hydrophobic compounds including those with long alkyl groups, such as the tails of typical surfactant molecules. As a consequence, most drugs of interest have very low solubility in HFAs [107, 117]. Moreover, many of the excipients used in the old CFC formulations are also poorly soluble in HFAs [11, 97, 120]. This has caused significant challenges in the reformulation of pMDIs. Drugs have to either be formulated as dispersions or their solubility enhanced with cosolvents (ethanol) [94, 120]. Cosolvents are also needed in dispersion-based pMDIs to enhance the solubility of surfactants, which themselves are needed for particle stabilization and valve lubrication [98]. However, there are many issues associated with using cosolvents, including the chemical stability of the drug formulation [17, 97], and changes in the aerosol characteristics upon the addition of nonvolatiles [98]. Moreover, while solute solubility of surfactants may be increased with cosolvents, solvation of the moieties required to stabilize the particles dispersions is still not optimal even in the presence of ethanol [122]. The lack of understanding of solvation forces in these semi-fluorinated propellant systems has significantly hindered the development of novel pMDI formulations, and has in turn incentivized the development of novel technologies for DPIs.

In this chapter we will focus on recent advances in the *formulation* aspects of pMDIs, which is one of the main challenges to be overcome if pMDIs are to remain competitive in the pulmonary drug delivery market, and perhaps to gain market share in the overall drug delivery market with formulations that target the systemic circulation. It is clear, however, that device design (hardware) is also expected to greatly impact the relevance of pMDIs as we go forward. The quality of the final product is an optimization problem that should address formulation development and hardware design simultaneously [8, 114], as no quantitative correlation exists between the characteristics of the formulation and the aerodynamic characteristics of the aerosol alone,[1] without consideration of the hardware [112]. It is sensible to assume, however, that if the size of the particles and chemical stability of the drug inside the container cannot be controlled during the proposed lifetime of the formulation, the characteristics of the aerosol and product performance will be affected with time, and thus certain aspects of the formulation development need be addressed to pave the way for a new generation of pMDI formulations.

8.2 Commercial pMDIs: Essential Components

Essential components of the pMDIs include the hardware, the solvent/propellant, and the active and (often times) nonactive ingredients.

8.2.1 The Hardware

The hardware of a typical pMDI, shown as a schematic diagram in Fig. 8.1, includes: (1) the canister and a metering valve, which offer a pressure proof, sealed environment where the propellant, drug, and excipients are housed; (2) the metering chamber, located in the valve assembly, and designed to release a fixed volume of the drug and propellant upon actuation; and (3) the actuator that serves to actuate the metering valve; it also provides the orifice through which the metering valve discharges the spray emitted from the canister, and directs the spray into the patient's mouth [8, 53]. Spacers may also be used to improve the efficiency of the delivery by reducing oropharyngeal deposition, as portion of the drug is retained in the spacer instead of reaching mouth and throat, thereby reducing undesired systemic drug levels and

[1] Correlations for solution-based pMDIs have indeed been developed (with cosolvents and other nonvolatiles), but for specific hardware sets (e.g., a certain valve and actuator) (Stein 2004).

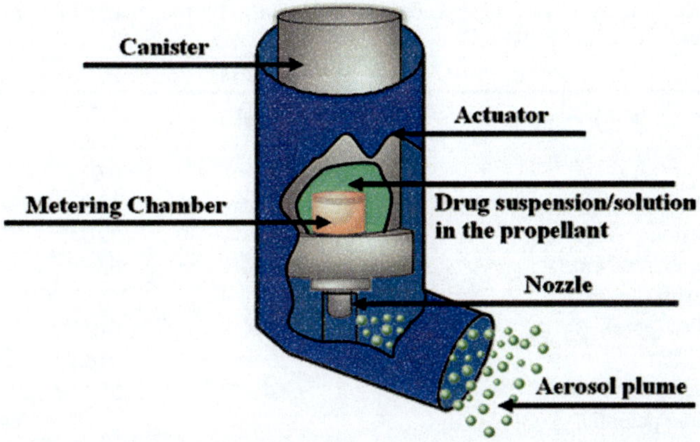

Fig. 8.1 A schematic diagram of a pMDI and its essential components. Reprinted from Wu and da Rocha [120], with permission from Hosokawa Powder Technology Foundation (KONA)

local side effects. Spacers also increase deposition in the lungs as there is more time for the evaporation of the propellant released in the plume, and thus the formation of smaller (more favorable) droplet sizes [19, 48]. While the main characteristics of the hardware has not changed significantly during the transition from CFC-based to HFA-based pMDIs, making this process seemingly simple at first, it is now recognized that there are many technological and scientific issues that need to be addressed regarding the hardware for HFA-based pMDIs that were not considered at first, including the selection and design of appropriate materials whose properties are not affected upon contacting the more polar HFAs [2, 17, 56, 73, 90, 115, 120].

8.2.2 The Formulation

The canister and metering valve hold the formulation in a sealed environment [94], or at least partially sealed, as moisture may gain access to the inside of the container – the ingress of water may affect the characteristics of the formulation and thus need to be accounted for during formulation development [116]. The formulation consists in a compressed solvent (propellant), the active pharmaceutical ingredient(s) (APIs), and excipient(s). Excipients may have different functions, including particle stabilization and valve lubrication [38, 54]. The compressed HFA gas, which makes for the largest fraction (typically >90%) of a formulation, is a liquid in equilibrium with its vapor at saturation pressure, and has dual function: it is the solvent and the propellant [120]. The system must be at saturation pressure so that the solvent quality and the pressure of the system do not change throughout the lifetime of the formulation, which could otherwise affect the physical stability and the aerosol characteristics of the same [93]. The compressed liquid also works as a solvent in which the API is dispersed or solubilized along with other excipients. The propellants used in

Table 8.1 Selected physicochemical properties of the hydrofluoroalkane (HFA) propellants used in pMDI formulations, and of HPFP

Property	HFA134a	HFA227	HPFP
Molecular weight (g mol^{-1})	102.03	170.03	252.05
Density (g cm^{-3})	1.21	1.40	1.58
Dipole moment (*D*)	2.06	0.93	1.90
Dielectric constant (ε)	9.46	4.07	15.05
Interfacial tension (γ, mN m^{-1})	31.8	32.7	33.7
Surface tension (σ, mN m^{-1})	8.69	6.96	13.59
Solubility in water (ppm)	193	58	140
Water solubility (ppm)	2,200	610	390
Boiling point (°C)	−26.3	−16.5	53.6
Vapor pressure (psig at 20°C)	68.4	56	–
Ozone depletion potential	0	0	–
Global warming potential	1,300	2,900	–
Solubility parameter (δ, $(MPa)^{0.5}$)	13.5	13.5	17.85[a]
Polarizability (α, 10^{-24} cm^3/molecule, vapor)	5.4	5.8	–

Adapted from Stein [98], Vervaet [112], Selvam [84], van Krevelen [111], Blondino [11], Traini [107], Peguin [68], Wu [122], de Jager [21], Chokshi et al. [13], and the IPCC/TEAP Special Report; Summary of Policy makers (safeguarding the ozone layer and the global climate system)
Note: The GWP is reported under the UNFCCC assessed value, and it is relative to that of CO_2. Density and surface tension at 293 K, saturation pressure; water solubility at 298 K; interfacial tension at 298 K, saturation pressure; dielectric constant at saturation pressure
[a] Estimated from group contribution method (Van Krevelen 1990)

commercial pMDI formulations and some of their relevant physicochemical properties, along with those for 2H,3H-perfluoropentane (HPFP), a liquid at ambient pressure and a mimic of HFA propellants, are shown in Table 8.1.

The propellant generates a pressure of approximately 50–80 psig inside the sealed unit [94]. Upon pressing the actuator, the propellant mixture and active drug are exposed to atmospheric pressure [105, 120]. The propellant flash evaporates to form an aerosol cloud containing the active and nonactive excipients, which are inhaled by the patient. The propellant is biocompatible and inert. Commercial pMDI formulations contain low molecular weight (MW) APIs either in solution (drugs dissolved in the propellant, usually with the help of a cosolvent) or in suspension (micronized drug crystals with or without the help of other excipients) [8, 55]. A list of HFA-based pMDIs currently commercialized in the US, with their brand name, APIs, nonactive compounds (excipients), and solvent/propellant used, is listed in Table 8.2.

The role of cosolvents and excipients in pMDIs will be discussed in the following.

8.3 Formulations in HFA-Based pMDIs: Current Efforts

While little has been promoted in terms of replacement propellants to the existing HFAs, perhaps aiming at achieving enhanced properties, and/or lower GWP, a significant amount of effort has been poured into the development of new particle

Table 8.2 Commercial HFA-based pMDI formulations available in the US market

Product name	API (concentration)	Propellant	Formulation	Excipient
ADVAIR HFA®	Fluticasone propionate (45/115/230 μg) + salmeterol xinafoate (21 μg)	HFA-134a	Suspension	–
AEROSPAN HFA®	Flunisolide (80 μg)	HFA-134a	Solution	Ethanol (10%)
ATROVENT HFA®	Ipratropium bromide (21 μg)	HFA-134a	Solution	Purified water; dehydrated alcohol; anhydrous citric acid
FLOVENT HFA®	Fluticasone propionate (44/110/220 μg)	HFA-134a	Suspension	–
NASACORT HFA®	Triamcinolone acetonide (15 μg)	HFA-134a	Suspension	Dehydrated alcohol (0.7%)
PROAIR HFA®	Albuterol sulfate (108 μg)	HFA-134a	Suspension	Ethanol
PROVENTIL-HFA®	Albuterol sulfate (120 μg)	HFA-134a	Suspension	Ethanol; oleic acid
VENTOLIN HFA®	Albuterol sulfate (120 μg)	HFA-134a	Suspension	–
XOPENEX HFA®	Levalbuterol sulfate (59 μg)	HFA-134a	Suspension	Dehydrated alcohol; oleic acid
SYMBICORT®	Budesonide (80/160 μg); formoterol fumarate dehydrate (4.5 μg)	HFA 227	Suspension	Povidone K25 USP; polyethylene glycol 1000 NF
QVAR®	Beclomethasone dipropionate (40/80 μg)	HFA-134a	Solution	Ethanol
ALVESCO®	Ciclesonide (80/160 μg)	HFA-134a	Solution	Ethanol

Source: FDA (www.fda.gov; then go to drugs on the menu) and Rx (www.rxlist.com)

Note: *USP* United States Pharmacopeia; *NF* National Formulary; *Povidone* poly(vinylpyrrolidone) or PVP; *PEG* poly(ethylene glycol)

technologies and the design of new excipients that may allow us to augment the relatively small portfolio of drugs currently being delivered with pMDIs [76, 120]. In this section we will discuss some new advances in this area, focusing to a large extent on the fundamentals in excipient design, and some recent particle engineering technologies applied to pMDI formulations.

8.3.1 Excipients in pMDIs

Excipients are used in pMDI formulations as: (1) cosolvents for enhancing the solubility of drugs in solution formulations, and for improving the solubility of other excipients as well, such as surfactants – ethanol is a classical example as shown in Table 8.2; (2) stabilizers/surfactants/lubricants for imparting stability to colloidal drug particles when dispersed in the low dielectric HFA propellants, for modulating particle aerosol size [12], and for valve lubrication – poly(vinylpyrrolidone) (PVP) (claimed a suspension agent in Symbicort®), PEG (claimed a lubricant in Symbicort®), and oleic acid are examples, as shown in Table 8.2 [108]; (3) flavoring agents, for example menthol and saccharin; and (4) antioxidants – for example ascorbic acid [94, 132]. In order to achieve their function, however, the excipients must be soluble in and/or well solvated by the semi-fluorinated propellants used in pMDIs, at a particular density of the solvent/propellant, which is that at around ambient temperature, and corresponding saturation pressure. This requirement is a major problem in the formulation of HFA-based pMDIs as the solubility of excipients commonly used in the old CFC-based pMDIs as well as most active components is typically very low in the semi-fluorinated HFA propellants. Selected studies on the solubility and solvation of excipients related to pMDIs will be discussed further [11, 15, 52, 97].

8.3.2 Cosolvents: Solubility Enhancers in HFAs

Ethanol, which is completely miscible with HFA-134a and HFA227 [54], has been widely used to either enhance the solubility of drugs in solution formulations, or to aid in the solubilization of other excipients required in solution and dispersion formulations, such as surfactants [8, 15]. The presence of ethanol, however, may have undesirable effects [8, 54, 97, 99]. The chemical and physical stability of the formulation may be impacted [76, 97]. In dispersion formulations, the solubility of the drugs in the propellant-ethanol solution may enhance the growth of larger particles in detriment to smaller particles (Ostwald ripening) [54, 97], which is clearly an undesirable effect as the intent is to enhance the solubility of another excipient in this case, and not modify the size of the drug particles. Ethanol may thus impact the physical stability of the formulation [97]. In solution formulations, while increasing the solubility of the drug in the propellant solution (desired effect in this case), the chemical stability of the system may be reduced in the presence of ethanol [99]. It has been also shown that while the solubility of drugs may be increased in solution formulations upon the addition of ethanol, the presence of such low volatility compounds (ethanol compared to HFAs) reduces the respirable fraction of formulations [32, 85], with a potentially overall decrease in the delivery dose, depending on the concentration of ethanol and the type of drug.

Other low MW compounds, as for example cineole (contains a cyclic ether) and citral (contains a carboxylic group), which have enhanced solubility in HFAs [57], may also be utilized as cosolvents. Visual observations also demonstrate that such compounds can help enhance the stability of particle dispersions in HFAs [57]. Oligomers of lactic acid (LA), functional poly(ethylene glycols), and acyl amid-acids have been also suggested as solubility enhancers for HFA-based pMDIs [97]. Different from ethanol, which acts by modifying the solvent environment, the suggested mechanism for these cosolvents is based on their ion-pairing ability with the drug of interest. Significant solubilization enhancement was shown possible (for particular drug-cosolvents), even in the absence of ethanol. Similar advantages and disadvantages could be envisioned for these other classes of cosolvents as discussed in the case of ethanol, depending on the concentration range, as the presence of nonvolatiles affects the aerosol characteristics of the formulation [112]. The ability to eliminate this extra component (cosolvent) in the formulation if possible, perhaps via the design of highly soluble excipients would, therefore, be of great interest.

8.3.3 Other Excipients: Solubility and Solvation in pMDIs

One major aspect in the design of novel HFA-based pMDI formulations is the need for chemistries that can be well solvated by HFAs, so called HFA-philes. Such chemistries will clearly have important consequences not only in terms of dispersion formulations, but also for solution-based formulations [76, 94, 97, 120, 122]. New developments in this area are also expected to provide new market opportunities for pMDI-based OI formulations, as for example in the development of systemic drug delivery technologies [6, 8, 15, 76]. While solubility of surface-active species is not strictly required for the stabilization of dispersions [112], a certain degree of solubility facilitates the formulation development process, and the solvation of the moieties responsible for preventing the particles from reaching a primary attractive minimum (irreversible flocculation) in the low dielectric HFAs is certainly a requirement for steric stabilization, the mechanism believed to be the dominant in the stabilization of dispersions in HFA – at least for the systems investigated to date [40, 68].[2] The prediction and determination of the solubility of APIs is also of great relevance in the design of pMDI formulations, and will also be discussed.

[2] We note several publications that have discussed the possibility of electrostatic stability in non-aqueous solvents (Vakarelski 2010; Patel 2010), and this may be an important area of research for HFA-based pMDIs as we go forward. However, at this time, there is not enough fundamental studies in this area to rationalize experimental results observed to date that have argued in favor of electrostatic stabilization of colloidal domains in HFAs (Peguin 2008).

8.3.3.1 Experimental Determination and Modeling of Solute Solubility in HFAs

The solubility of a wide variety of APIs and various excipients has been reported in HFA propellants. Selected solubility results are shown in Table 8.3.

The development of new experimental techniques has provided facile routes for the determination of in situ solubility of compounds in HFAs propellants [24, 29, 75, 107]. Such approaches are of great relevance as they contribute to the development of important insight in this field of research, as for example the ability to test thermodynamic models for phase equilibria involving HFA propellants, and also to develop useful correlations between relevant molecular parameters such as hydrogen(H)-donating/accepting capability, hydrogen bonding number (HBN), and log P, and the solubility of solutes in HFA [35, 68, 75, 76, 107]. One can envision the creation of large libraries that may help generate more quantitative structure-property (e.g., solubility) relationships with the help of sophisticated mathematical algorithms (e.g., neural networks) for systems that include this important class of solvents/propellants [104].

There has not been a concerted effort to develop accurate thermodynamic models to predict the solubility of solutes in HFA propellants so far. The modeling of a range of (low MW) solid-HFA propellant equilibria results shows that, as expected, correction factors need to be added to the ideal solid–liquid equilibrium equations in order to correctly estimate equilibrium concentrations [98]. However, even when (relatively simple) activity coefficient models are used, such as regular solution theory (solubility parameter), such models still lack enough (quantitative) predictive capability [34, 39, 98]. Thermodynamic modeling of hydrofluorocarbon (HFC) mixtures, including HFAs, mostly in the context of lubricants for refrigerants, as for example polyol esters, has been approached using equation of state (EOS) models, including cubic EOS, simplified statistical associating fluid theory (SSAFT), Helmholtz energy models, and Sanchez-LaCombe EOS, and some of those thermodynamic models may be extended to systems of interest to pMDIs [1, 27, 36, 42]. The modeling of the phase equilibria of solutes in HFAs for systems relevant to pMDIs is clearly in its early stages of development, and significant opportunities lie in this field of research. Computer simulations of phase equilibria and ab initio calculations also have the potential to be used as quantitative predictive tools for systems involving HFA propellants.

Surfactants and stabilizers are typically used in both solution and dispersion formulations [11], even though their use is not strictly necessary, as seen in Table 8.2. In suspension formulations, surfactants are used to prevent irreversible aggregation of otherwise unstable drug particles due to strong particle–particle (van der Waals attractive) forces, and to minimize the adhesion of drug particles onto the container walls and valve components, among others, as discussed earlier [109, 120, 128]. Poly(vinyl alcohol) (PVA) and PVP, which have some solubility in HFA134a (>0.1% w/w), have been investigated as means of preserving the activity of biomacromolecules during particle formation relevant to pMDI formulations, and to enhance the stability of such particles as dispersions in HFAs [40]. PEGs are hydrophilic nonionic polymers that are liquids up to MWs of around 600, at room temperature. The solubility of PEGs in HFAs is significant, and was shown to decrease with an increase in their MW [75, 108, 112].

Table 8.3 Solubility of selected excipients and APIs in HFA134a and HFA227

Solute	Chemistry	Solubility in HFA (%w/w)		Condition
		134a	227a	
BRIJ 30	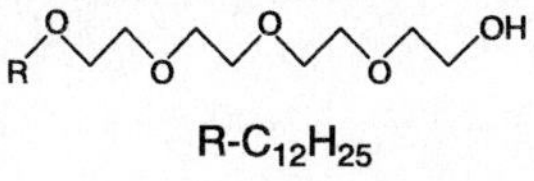	~1.8	0.8–1.2	19.5–25°C, P^{sat}
Aerosol OT	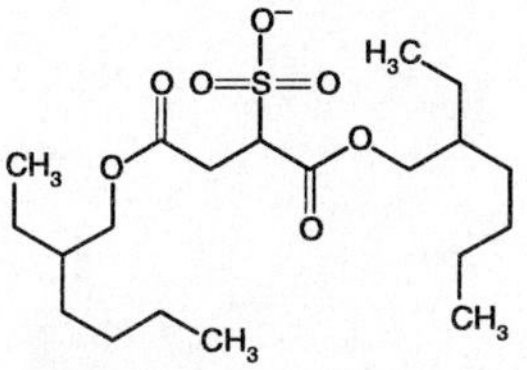	<0.01	<0.02	19.5–25°C, P^{sat}
Lecithin	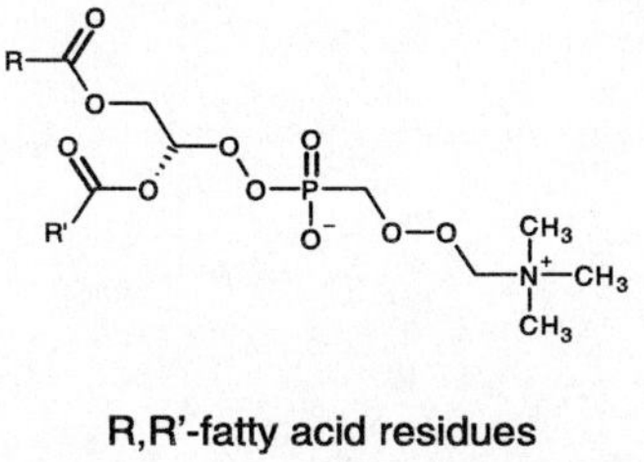	<0.01	<0.01	19.5–25°C, P^{sat}
Oleic acid	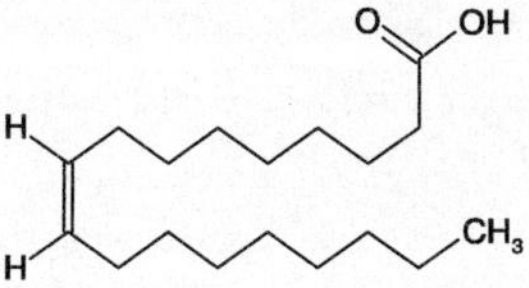	<0.02	<0.02	19.5–25°C, P^{sat}
Oligolactic acid	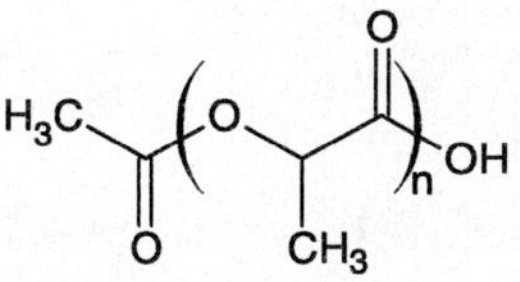	~2.7	NA	T_{room}, P^{sat}
PEG 300	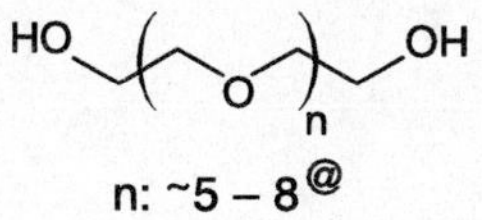	~4.0	1.5–4.3	19.5–25°C, P^{sat}
PEG 600	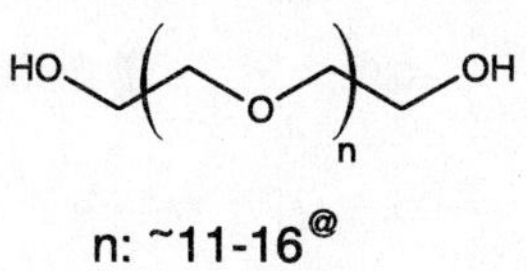	~4	>50[a]	23–25°C, P^{sat}
PEG 1000	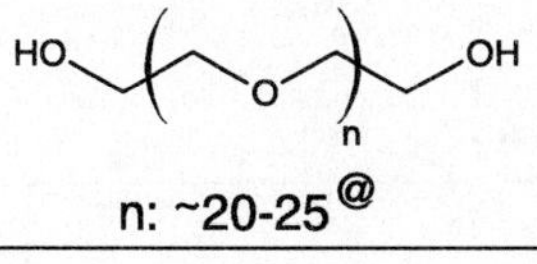	~2	~2	23–25°C, P^{sat}

(continued)

Table 8.3 (continued)

Solute	Chemistry	Solubility in HFA (%w/w) 134a	227a	Condition
PEG 2000	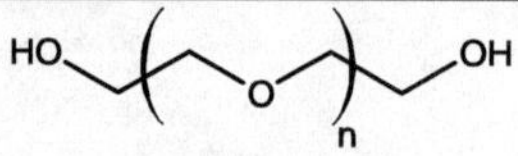 n: ~40-50@	<0.01	<0.01	23–25°C, P^{sat}
PPO 2000	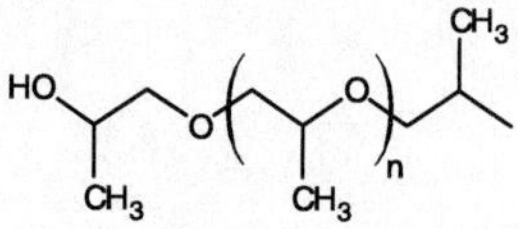	~2	>50[a]	23–25°C, P^{sat}
PVP	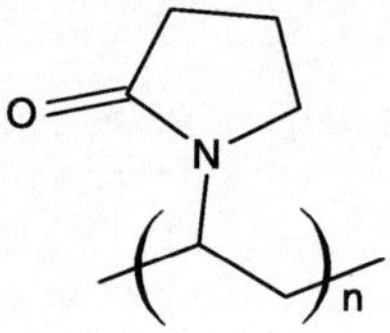	>0.1	NA	19.5–25°C, P^{sat}
PVA	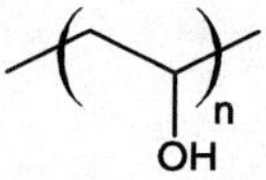	>0.1	NA	19.5–25°C, P^{sat}
Span 85®	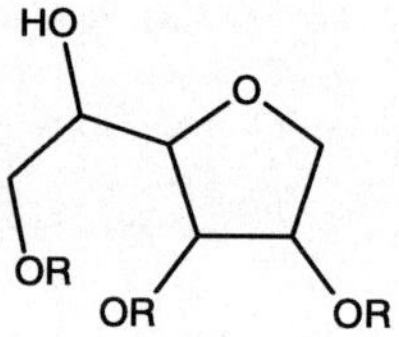 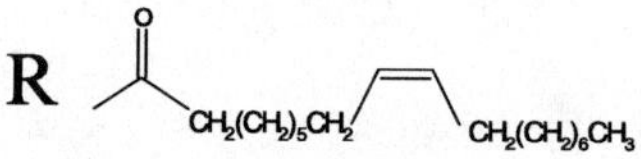	<0.02	<0.01	19.5–25°C, P^{sat}
Tween 20®	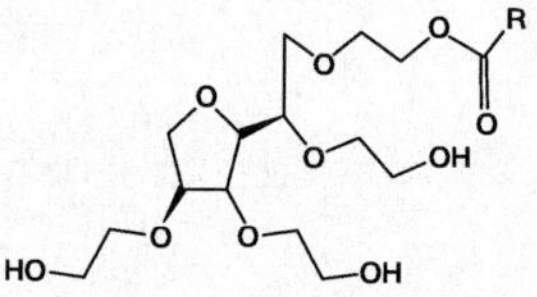 R– CH2(CH2)$_9$CH$_3$	~0.1	1.4–3.5	19.5–25°C, P^{sat}
Tween 80®	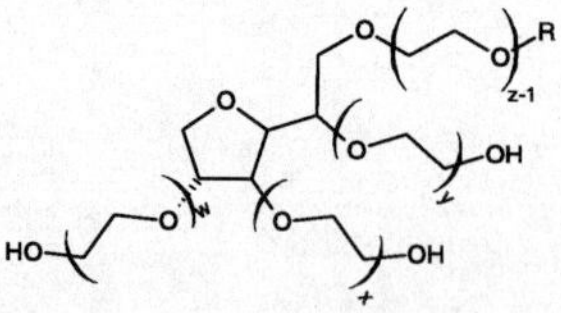 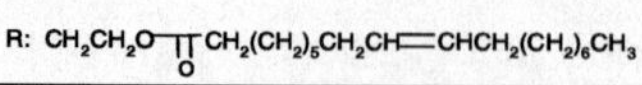	<0.03	0–10.0	19.5–25°C, P^{sat}

(continued)

Table 8.3 (continued)

Solute	Chemistry	Solubility in HFA (%w/w) 134a	227a	Condition
Cyclodextrin[b] (acylated)	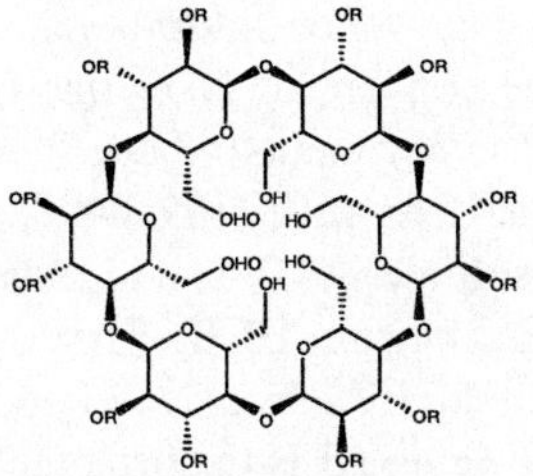 $R - COCH_3$	1	0.5	T_{room}, P^{sat}
Salbutamol sulfate	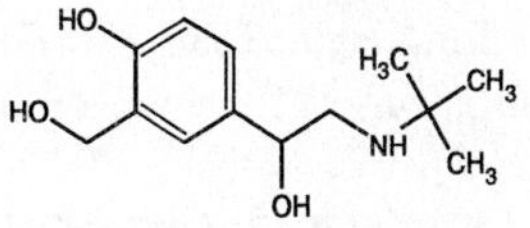	<0.000	NA	25°C, P^{sat}
Budesonide	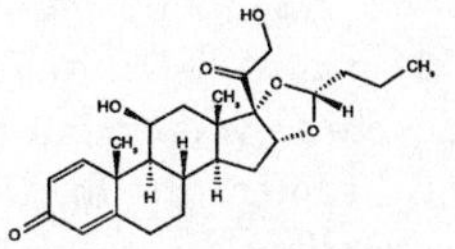	23.13 ± 2.9[c]	NA	25°C, P^{sat}
Formoterol fumarate	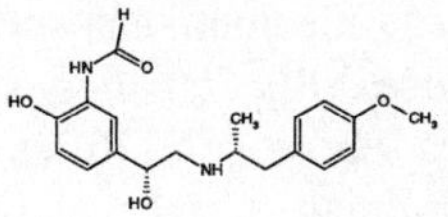	0.776 ± 1.023[c]	NA	25°C, P^{sat}
Beclomethasone dipropionate	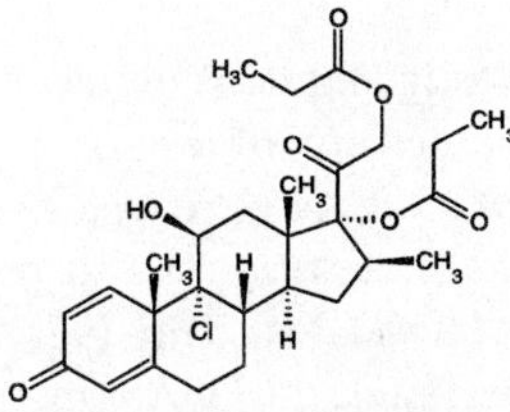	80[c]	80[c]	25°C, P^{sat}
Triamcinolone acetonide	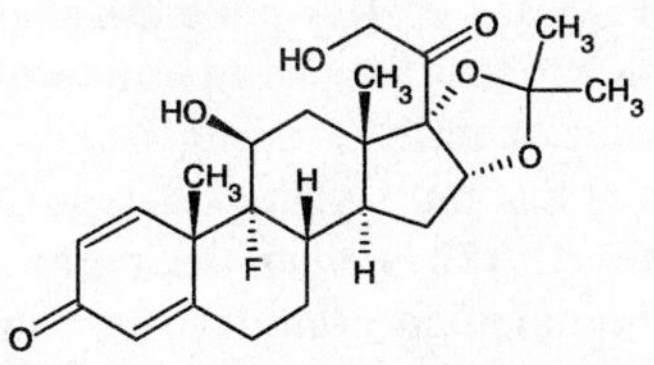	4[c]	3.5[c]	25°C, P^{sat}

Adapted from Blondino [11], Traini [107], Vervaet [112], and Williams and Hu [117]

[a] Existed as one clear phase in that concentration range, perhaps HFA solubilized in PPO instead of actual solubility; at approximated using a monomer mass ~44 g mol^{-1}

[b] Data shown for the α-form

[c] μg/mass of Propellant

PEG is insoluble at MWs>1,000 [75]. Both PEG and PVP are part of an FDA-approved pMDI formulation (Symbicort®) as shown in Table 8.2.

The solubility of a series of surfactants in HFA134a and HFA227 has been related to the surfactant hydrophilic-lipophilic balance (HLB) [11]. It was observed, in general, that surfactants with higher HLB had better solubility in HFA227 compared to HFA134a [75, 112]. However, the concept of HLB has to be used with care for HFA-based systems, as shown in experiments where the tension of the surfactant-modified HFA-water interface was determined [69, 84, 85]. For example, PEG is highly water soluble, but it is also soluble in HFA, and thus has interfacial activity by itself at that interface (implying a certain HLB). However, PEG is not truly a surfactant molecule. On the other hand, HFAs interact so poorly with alkyl-based groups (oleophobic) [84] that attributing an HLB to surfactants such as oleic acid for interfaces with HFAs is not a very plausible surfactant classification strategy. Tween 20, 80, and PEG 300 have higher solubility in HFAs when compared to surfactants oleic acid and Span 85 [112]. Along with Span 85 and lecithin, oleic acid was primarily used in CFC-based formulations as lubricants during depression and release cycle associated with actuation [93].

In addition to the solubility of selected excipients, the extent of dissolution of some relevant APIs, including salbutamol sulfate (SS), budesonide, and formoterol fumarate at ambient conditions in HFA134a is also listed in Table 8.3 [107]. Among the three APIs mentioned above, budesonide was found to be the most soluble in the propellant, followed by formoterol fumarate – the solubility of SS was undetectably small. Such results may have implications regarding the formulation of HFA suspensions as particle growth mechanisms such as Ostwald Ripening and it must also be considered for those APIs that show some solubility in HFAs [107]. Other selected drug candidates whose solubilities in HFAs have been reported include beclomethasone dipropionate (BDP) and triamcinolone acetonide (TAA) [117].

Ester-based polymers have also been investigated as stabilizers in HFAs [97]. Oligolactic acid (OLA), containing a high density of esters on the oligomer backbone, have appreciable solubility in HFAs. OLAs were tested as a suspension aid in an HFA134a formulation of pirbuterol acetate [97]. The suspensions formed using OLA as an excipient showed excellent physical stability [97]. The formation of stable dispersions of salbutamol and budesonide in HFAs, in the presence of copolymers containing short blocks of LA as the HFA-phile has also been reported [119, 124]. Recent studies have also shown that the incorporation of cyclodextrins in HFA can greatly improve the physical stability of drug suspensions [77, 96]. In a recently filed patent, it was reported that the acylated analogs of cyclodextrins were more soluble in HFA134a and HFA227a than the parent cyclodextrin, exhibiting solubilities >0.1% w/w. The modified cyclodextrins were able to help form stable suspensions of several drug particles candidates used in propellant driven formulations including budesonide, triamcinolone, and fluticasone [77]. Additionally, the complexation of the aforementioned drugs to the acylated analogs of cyclodextrin was discussed as another approach to achieve excellent suspension stability in HFA propellants [77].

Surfactants may also be used to form water-in-HFA microemulsions that could potentially find applications in the delivery of a range of therapeutics using pMDIs, including high-potency drugs, polar APIs, and biomacromolecules [65, 66, 85, 100]. The ability of copolymers of PEO and poly(propylene oxide) (PPO) to aggregate in HFAs has been demonstrated, as well as their capacity to form stable microemulsions that can uptake biomacromolecules [85]. A series of other amphiphiles, including fluorinated [65, 66, 100] and alkylated tail groups have been also tested [13, 84]. High-pressure tensiometry of the modified HFA-water interface has been shown to be a particularly relevant tool in the design of such amphiphiles [84, 85]. Computer simulations have also been used to obtain a microscopic picture of the HFA-water interface, which is relevant in the design of surfactants for that interface, as the properties of the bare interface dictate, to a large extent, the surfactant behavior [68–70, 86]. PEO-PPO surfactants have been also used to form topical foams formulated in HFA propellants [129–131], some of which may incorporate controlled-release capabilities [130].

While the solubility of a variety of excipients and APIs has been reported in propellant HFAs, it is difficult to rationalize those parsed solubility trends, and to use such information to guide the design and selection of compounds with enhanced HFA-philicity. In the next section, we discuss how microscopic information from ab initio calculations and chemical force microscopy (CFM) can be used to understand solvation forces in these semi-fluorinated propellants, and how such information can in turn be used in the rational design of HFA-philes (excipients in general).

8.3.3.2 Solvation in HFA Propellants: HFA-Philicity

At the microscopic level, the energetics of solute-HFA interactions can be quantitatively determined from ab initio calculations [69]. When combined with experimental microscopic techniques such as CFM, a very detailed and quantitative picture of the solvent environment in HFA-based systems can be obtained [122]. There are several advantages in performing ab initio calculations vs. a series of solubility measurements and property correlation based on solubility measurements. For example, one is not limited by any synthetic boundaries/time commitment involved in synthetic procedures, and a detailed quantitative description of the energetic of the system and bonding geometry naturally emerges from ab initio calculations. Moreover, such information can be used in the development of other correlations [68, 69]. However, there is an obvious computational price and limitation in such calculations in terms of the system size. Even at low levels of theory, calculations are limited to relatively small solutes and a reduced number of solvent molecules, perhaps equivalent (in number) to that which exists as the solute's first solvation shell [68].

The use of atomistic computer simulations such as molecular dynamics (MD) and Monte Carlo (MC) [68, 69, 71] may provide yet another route to understand such solvent environment, and may serve as the link between first principle calculations and bulk measurements in the quest for developing a fundamental understanding of the solvation properties of HFAs. It has been recently demonstrated that accurate force fields may be developed to describe not only the pure component properties of HFAs, but also of HFA systems, such as that of HFA and water [71]. In what follows, we show how ab initio calculations can be used to develop an HFA-philicity scale, and how such results may provide fundamental insight to further guide the design of HFA-philes. We will also discuss complementary CFM experimental measurements.

The interaction (binding) energy (E_b) between HFA134a and HFA227 with various moieties of interest has been determined via ab initio calculations. The fragments are shown in Table 8.4 [68]. They are analog fragments containing the same number of heavy atoms along their backbone, but of varying polarity. E_b is defined as the in interaction energy difference between that of the HFA solvent (s) and the moiety/tail (t), E^{st}, relative to that of the HFA (E^{s}) and moiety (E^{t}) alone, i.e., $E_b^{st} = E^{st} - E^{s} - E^{t}$. These results provide not only a quantitative account of the degree of interaction between HFA and the chosen moiety, i.e., an HFA-philicity scale, but also an opportunity to understand the details of the bonding environment, thus providing a molecular level tool for the design of HFA-philic chemistries.

Effect of the Polarity of the Fragment

The interaction between HFAs with fragments containing CH_2-based moieties (CH_2), which is representative of the hydrophobic portion of surface-active species such as oleic acid known to have poor solubility in/poorly solvated by HFAs, is significantly lower (less negative) than with fragments containing more polar moieties, such as ether (COC), representative of ethylene oxide (EO) and propylene oxide (PO), and ester (CO(O)C), representative of LA based groups, as expected. This observation agrees not only with experimental information from solubility measurements (surfactants with alkyl-based tail groups to have of poor solubility in HFAs) [11, 24, 75, 112], and the ability of the more polar tails groups (based on LA and EO) to stabilize drug particles in HFAs [119], but also with a simple dipole moment argument (see Table 8.4): the more polar moieties interact better with the (more) polar HFAs (more polar compared to CFCs).

However, dipole moment alone cannot be used to screen HFA-philes among the polar moieties. For example, even though the polarity for LA is significantly larger than that of PO, the binding energy of those fragments with HFA134a is very similar in magnitude. This apparent discrepancy can be somewhat reconciled by evaluating the interaction energies of clusters of HFAs around a central fragment. For example the energy of the HFA_4–LA cluster (the subscript stands for the number of HFA molecules) is greater than that between HFA_4 and PO (E_b^{st} HFA_4–LA = –62 kJ/mol > E_b^{st} HFA_4–PO = –42.3 kJ/mol) [68]. One important piece of information that can be obtained from such calculations is that steric constraints in the fragment can play a major role in the systems' energy (besides its dipole moment). While both oxygens (carbonyl and ester) are accessible to individual HFA molecules in the LA

Table 8.4 Binding energy of the HFA–tail fragment pairs (E_b^{st}) at MP2/aug-cc-pVDZ//MP2/6-31g+(d,p), in kJ/mol

	Tail fragments			
	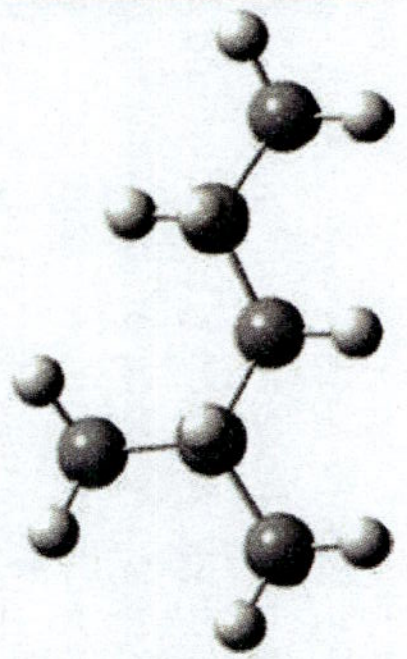		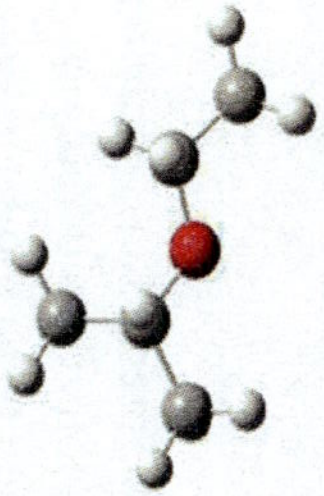	
HFA	***CH2* (0.15)**	***EO* (1.30) *(COC)***	***PO* (1.23) *(COC)***	***LA* (1.69) *(CO(O)C)***
HFA134a (2.12)	−11.3	−25.1	−26.4	−26.8
HFA 227 (1.69)	−18.4	−33.1	−35.6	−33.5

Calculated dipole moments of the tail fragments and HFAs are given in parenthesis, and are in Debye
Adapted from Peguin [68]

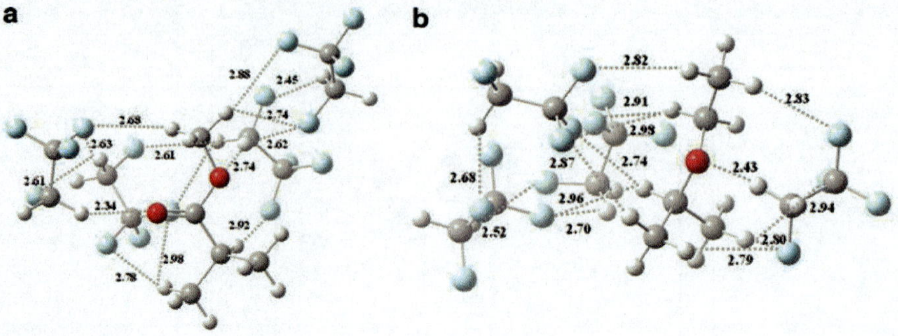

Fig. 8.2 Optimized geometry at HF/6-31g(d) for: (**a**) (HFA134a)$_4$–LA and (**b**) (HFA134a)$_4$–PO. The interatomic distances shown here are in Angstroms. Reprinted from Peguin and da Rocha [68], with permission from ACS

fragment – those are the strongest interactions – only one HFA can interact with the ether group in the PO fragment, thus considerably reducing the ability of HFAs to solvate the ether group in that moiety, as shown in Fig. 8.2 [68].

Another interesting fact that may be noted from the results shown in Table 8.4 is that the interaction energy of fragments with HFA227 is always larger than those with HFA134a, in spite of the fact that the dipole moment of HFA134a is greater than that of HFA227. This can be attributed to a large extent to the fact that the H atom in HFA227, which is the one participating in the strongest bond with the (oxygen atoms of the) fragment, has significant charge concentration, being twice as positive as the equivalent H atom (that H interacting with the oxygen atoms in the fragments) in HFA134a.

Effect of the Shape of the Fragment

E_b results from HFA-EO and HFA-PO moieties provide some insight into the effect of molecular shape of the fragment, as those two chemistries differ by a single $-CH_3$ pendant group. The interaction energy between HFA(134a and 227) and PO is larger than that with EO. This is due to the presence of the weaker, but relevant C–H…F interactions. The HFA227 molecule loses in average three such interactions by removing the $-CH_3$ group (PO to EO), while that of HFA134a looses only one. This is reflected in the greater decrease in binding energy for HFA227 from PO to EO than that observed in HFA134a. Such observations are in agreement and support literature results which show that the polypropylene oxide-based (PPO) homopolymers to be more soluble than their poly(ethylene oxide) (PEO) counterparts [75]. Solute–solute interactions in the PPO/PEO solubility in HFAs are also expected to play a role in the solute solubility, as reduced fragment–fragment interaction is expected for more the branched chemistry (PPO) – thus enhanced solute solubility in HFA.

Electrostatic or Dispersive

Which force is dominant in solute-HFA interactions? This question turns out not to have a trivial answer or universality in HFA-based systems, as the magnitude of electrostatic/dispersive contribution cannot be predicted a priori [37, 68, 80, 106]. Such energy decomposition can be achieved by careful selection of the level of theory used in the ab initio calculations; i.e., the correlation or dispersive energy can be obtained as the difference between the binding energy at the MP2 level of theory minus that at the HF level of theory (same basis set level), as HF does not include the correlation energy component, and MP2 fully takes that energy into account [68, 123]. The difference is the electrostatic contribution to the stabilization of the pair. These results are summarized in Table 8.5.

One can observe from Table 8.5 that even for polar fragments as PO, the interaction with HFAs is still largely dominated by dispersive forces. It is also surprising to note that, for that system, this holds true even for the system with the smaller and more (di)polar HFA, the HFA134a – here again we see the more acidic H in HFA227 has a significant impact in the overall energy of the system. As the fragment polarity increases even further (from PO to LA), we then observe a dominant electrostatic contribution to the stabilization of the HFA-fragment pair (61% electrostatic), but only for HFA227. Dispersion forces are still larger than electrostatic forces for the HFA134a–LA pair.

These results provide a guide to the design of HFA-philes that can be used to develop new excipients with enhanced solubility and ability to impart stability to dispersions (both liquid [68] and solid [124]) in HFA propellants, and in the development of new particles technologies, besides helping to explain some experimental results reported in the literature [122, 124].

Atomic Force Microscopy and Solvation Forces in HFAs

CFM has also been used to gain insight of the solvation forces in HFAs [120, 122]. CFM studies are relevant not only in the context of the traditional solution- and

Table 8.5 Interaction energies obtained at HF and MP2 levels of theory and 6-31g+(d,p) basis set, in kJ/mol

	CP-corrected E_b^{st} (kJ/mol)		Dispersive energy
Complex	HF/6-31g+(d,p)	MP2/6-31g+(d,p)	E_{corr}^{st}
HFA134a–LA	−7.2	−16.2	−9.0 (55%)
HFA227–LA	−13.3	−21.9	−8.6 (39%)
HFA134a–PO	−4.6	−17.1	−12.4 (72%)
HFA227–PO	−8.9	−24.0	−15.1 (63%)

CP-corrected E_b^{st} = total binding energy with BSSE; dispersive energy (E_{corr}^{st}) = correlation energy = CP-corrected E_b^{st} at MP2 – CP-corrected E_b^{st} at HF
Adapted from Peguin [68]

dispersion-based pMDI formulations, which generally require surfactants as excipients, but also for the development of novel HFA formulations (pMDIs and other medical sprays) that may find applications not only for the regional delivery of therapeutics to the lungs, but also for systemic drug delivery and topical applications, as for example in reverse aqueous microemulsions [25, 85] and emulsions/foams in HFAs [129], and in particle-based approaches that require the presence of well-solvated stabilizing moieties [108, 109, 119, 124, 128]. CFM results have been used to evaluate the forces of interaction between a chemically modified AFM tip and substrate in HPFP [70, 120, 122]. The tip and substrate were modified by silane chemistry, using both solution and vapor phase approaches [69, 122]. Several different types of chemistries were investigated, including an alkyl-based (CH_2) moiety (octyltrichlorosilane, $CH_3(CH_2)_7SiCl_3$), and more polar ether- (COC, EO/PO *like*) (3-methoxypropyltrimethoxysilane, $CH_3O(CH_2)_3Si(OCH_3)_3$), and ester- (C(O)OC, LA-*like*) (2-acetoxyethyltrichlorosilane, $CH_3COO(CH_2)_2SiCl_3$) terminated silanes [122]. The CH_2 moiety represents the tails of surfactants such as oleic acid, which has low solubility in HFAs [11, 97]. It thus represents the baseline moiety. COC and C(O)OC contain polar sites that may provide for an opportunity for enhanced interactions with the HFA propellants, and have the same chemical functionalities as those discussed in the ab initio calculations shown above. The rationale behind such CFM measurements is that, upon retraction of the tip from contact with the substrate, the adhesive (or in this case more accurately described as cohesive force as both tip and substrate have the same chemistry) force (F_{ad}) detected in the measurement indicates the enthalpic penalty that arises due to the formation of the two interfaces – that between the tail of the silanes (chemistry of interest) in the tip and that in the substrate, and the intervening liquid (HPFP, a mimicking HFA). The lower the enthalpic penalty (the lower the F_{ad}), the better the match between the solvent and the chemistry of interest. Such results should serve, therefore, as a quantitative microscopic experimental tool to screen moieties of HFAs, in the same way as the ab initio calculations discussed earlier did computationally – CFM perhaps being one of the best experimental analogs to those calculations.

The adhesive forces from CFM shown in Fig. 8.3b have been normalized by the radius of curvature of the AFM tip; i.e., F_{ad}/R. For an alkyl-modified tip and substrate in HPFP, F_{ad}/R was found to be 68.2 mN m^{-1} [122]. The F_{ad}/R for ether and ester-based moieties are 24.8 and 4.8 mN m^{-1}, respectively [70]. A lower F_{ad} indicates a smaller enthalpic penalty for creating the two interfaces between the intervening liquid (HPFP) and the chemical groups of interest. The results show, therefore, that the ester-based moiety is the most HFA-philic, followed by the ether group. The moiety that is least solvated, as expected, is the CH_2. Such favorable enthalpic interactions between HFA and the ester and (to a less extent) ether groups is important for the development of novel pMDI formulations as those moieties are biodegradable and/or biocompatible, and have been used as excipients in several formulations, including (PEG) pMDIs – see Table 8.2 [3, 76, 94, 97]. The results indicate, however, that the solvation of the ester-based moiety is still far away from ideal; i.e., there is still room for improvement in terms of an optimal candidate chemistry. The ideal solvation is represented by $F_{ad}/R \sim 0$ mN m^{-1}, which is the

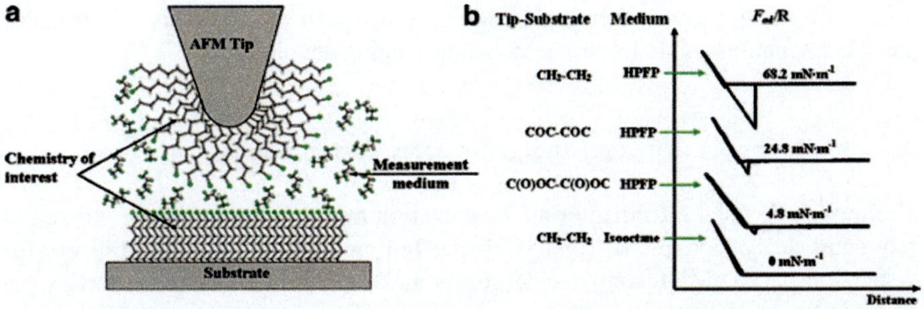

Fig. 8.3 (**a**) Schematic representation of a chemically modified CFM probe and substrate with a monolayer containing terminal groups presenting the chemistry of interest. (**b**) Normalized F_{ad} with respect to AFM tip radius (R) for candidate chemistries (CH_2, COC and C(O)OC). The measurements were done in HPFP and isooctane at 298 K. Reprinted from Wu and da Rocha [120], with permission from Hosokawa Powder Technology Foundation (KONA)

adhesive force measured between an alkyl-modified tip and substrate in an alkane (isooctane) solvent [120].

The F_{ad} of fluorinated (CF_2) (1H,1H,2H,2H-perfluorooctyltrichlorosilane, $CF_3(CF2)_5CH_2CH_2SiCl_3$) tip and substrate has also been determined in HPFP. The F_{ad}/R of 6.4 mN m^{-1} was comparable to that observed for the ester group [122], indicating that the fluorinated moiety can be well solvated by HPFP. The potential for application of fluorinated moieties in pMDI formulations is limited, however, owing to its potential toxicity [16]. Nonetheless, this is a good model system as there is detailed information on the nature and characteristics of self-assembled monolayers of such fluorinated molecules. Similar information is available for alkyl-based silanes as well, thus allowing a direct comparison between those two systems. Using the JKR theory, the F_{ad} was further normalized for the CF_2 and CH_2 systems in HPFP down to the single molecule (pair) force (F_s) [70]. F_s for the fluorinated group in HPFP was found to be 86 pN, which is significantly less than that observed between a pair of equivalent alkyl-based tails in HPFP, of 156 pN. These results correlate very well with ab initio calculations, that show that the binding energy between HFA and CH_2 is approximately half as strong as that between HFA and CF_2 [68].

The discussion above illustrates how F_{ad} results can provide an absolute scale for HFA-philicity, and are thus of great relevance for the rational design of HFA-philes. Such results can, therefore, be used in the development of novel HFA-philes for pMDI formulations. An HFA-philicity scale from such measurements has been successfully utilized to identify and develop novel amphiphiles capable of forming stable reverse aggregates of water in HFA [13, 85], and to aid in the stabilization of solid-based dispersion in the form of traditional dispersion formulations [124], and novel particle engineering technologies [121, 124]. It is also worth pointing out that the CFM results have been corroborated and complemented by the ab initio calculations as discussed earlier [68, 70]. These were the first CFM studies with relevance to the design of HFA-philes for HFA-pMDIs. There are many opportunities in this area of research, as new chemistries that may better fit new delivery platforms or particle formation technologies are constantly being sought.

8.3.3.3 Rational Surfactant Design for HFA Formulations

The molecular level information on the solvation forces in HFAs can be utilized to rationally design excipients for pMDIs, including surfactants and stabilizers for dispersion-based pMDI. Here, we purposely make a distinction between surfactants and stabilizers. Surfactants contain a group well solvated by HFA, or an HFA-phile, bonded to another group, called the anchor moiety, which is designed to interact well with the particle surface. Stabilizers may be surfactants, but they may also be simply an HFA-philic moiety which could be anchored or somewhat physically trapped onto the particle surface, thus providing the required steric hindrance for particle stabilization in the propellant HFA. When designing surface-active species for HFAs, not only does an HFA-philic moiety needs to be identified, but also an appropriate balance must be attained between the HFA-phile and the anchor moiety, which we assume to be well defined, but typically would require some previous insight or ingenuity to be identified for each drug formulation.

One approach that can be used to identify the optimum surfactant balance for (solid) dispersions, as for example drug particles including micronized crystals, is the colloidal probe microscopy (CPM) [41]. The technique consists in measuring the force of interaction (AFM) between two drug particles or a drug particle and a drug surface in liquid HFA (usually HPFP), in the presence of the surfactant/polymer chemistry of interest. The images shown in Fig. 8.4 illustrate the technique.

This approach has been used to compare surfactants with different HFA-philic moieties – same as those discussed in the ab initio and CFM studies described earlier

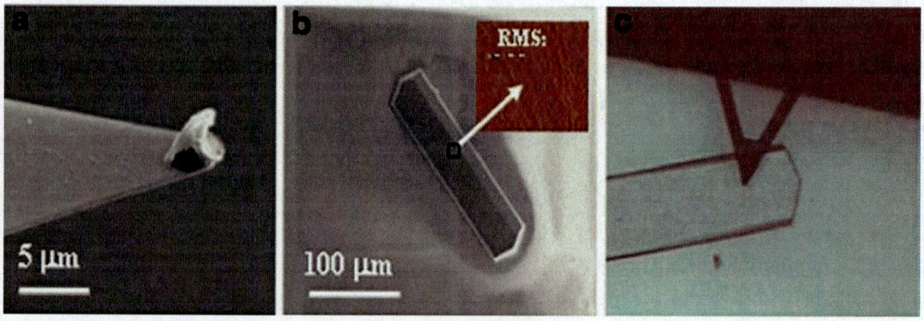

Fig. 8.4 **(a)** Scanning electron micrograph (SEM) showing the presence of the drug particle on the AFM tip – the probe. **(b)** Representative SEM image of a drug particle recrystallized from solution; *inset*: Tapping mode AFM image of a smooth single salbutamol crystal over of 9×9 μm area. **(c)** Optical microscope image of a modified AFM tip over a smooth drug crystal captured during an adhesion force measurement. Reprinted from Wu and da Rocha [119], with permission from ACS

Table 8.6 Adhesive force (F_{ad}) between an AFM tip modified with a salbutamol particle and a smooth salbutamol crystal, as determined by colloidal probe microscopy (CPM) in HPFP at 298 K, and in presence of surfactants and/or cosolvents

				F_{ad} (nN)	
Surfactant	Cosolvent	EO (wt.%)	Surf. Conc. (mM)	Probe 1	Probe 2
None	None			12.13±0.50	8.43±0.30
None	Ethanol 7% (v/v)			6.31±0.47	5.01±0.26
Oleic acid	Ethanol 7% (v/v)		1.0	1.54±0.35	1.64±0.21
$EO_6PO_{34}EO_6$	None	20	1.0	3.80±0.41	2.91±0.27
$LA_{18}EO_{14}LA_{18}$	None	20	1.0	0	0

Adapted from Wu [122]

and also to identify the optimum HFA-phile/anchor balance in a surfactant series. In Table 8.6, the results for surfactants containing alkyl- (CH_2), ether- (PO), and an ester-based (LA) tail group are summarized. The results show, as expected from the ab initio and CFM results, that the LA moiety is the most efficient in reducing the cohesive forces (F_{ad}) between particles in this mimicking HFA. We also notice that even in the presence of ethanol, oleic acid is not capable of completely screening of the F_{ad} between particles. One very important aspect of this research is that F_{ad} results have been shown to correlate well with physical stability observations in propellant HFAs [119] and those in turn usually translate to an enhancement in the aerosol characteristics of the corresponding formulation [119, 124].

The surfactant balance can also be addressed by CPM. A series of $LA_nEO_{14}LA_n$ surfactants, with varying degrees of polymerization of the LA group has been investigated [119]. At 1 mM surfactant concentration, the F_{ad} measured in HPFP decreased from ~8.42 nN, to 5.9 nN, to ~0 nN as the surfactant tail group increased from $n=1$, to 4, to 18, respectively. The CPM results can also indicate the optimum surfactant balance i.e., what size LA tail group is needed for a specific EO head-group size so that the surfactant partitions to and provides enough surface coverage for the drug surface, and the optimum tail size for a particular anchor group (in that case EO_{14}) [119]. The minimum required surfactant concentration can also be determined with such technique.

As the drug of interest is varied from a hydrophilic to a nonpolar API (SS to budesonide), a change in the anchor group from PEO to poly(caprolactone), while keeping the same stabilizing moiety (LA), was enough to provide appropriate steric stability to the drug dispersion [119, 120], reinforcing the concept that the LA group is indeed capable of provide steric stability to dispersions in HFAs.

The force of interaction between SS crystals and three pMDI canister materials (borosilicate glass, Teflon and aluminum) were determined using CPM. It was concluded that the Teflon was the most suitable material to be employed for canister lining purpose [128]. This technique was also applied in the quantification of the cohesive forces of SS in HPFP in presence of increasing PEG molecular weights (MW 200, 400, and 600) and concentrations. It was determined that the increase in the concentration of PEG (up to 0.5% v/w) caused a significant reduction in the force of interaction between the drug particles, and this was heavily dependent on both the concentration and the MW of PEG. The study also evaluated the effect of the addition of PVP to the system and found that it did not have any noteworthy impact on the force of cohesion between drug particles [108].

CPM has been also used to understand the ability of a particle shell in screening interactions between particles containing small polar APIs and biomolecules as the core [124]. It serves to demonstrate the versatility of the technique in the screening not only surfactant systems, but all forms of stabilizers. In summary, CPM is a very powerful technique in the design of amphiphiles and particle engineering systems for HFA-based formulations, and may be easily implemented in research and industrial laboratories, especially as AFM technology becomes more easily accessible both in terms of cost and also ease of use. We expect, therefore, AFM-based techniques to become of increasing relevance to technologies involving HFA-based pMDIs.

8.3.4 Particle Technologies

In recent years, several alternatives to the traditional solution and dispersion (micronized crystals) pMDI formulations have been reported in the literature [76, 121, 124]. Particle technologies can be broken up into two major groups, which encompass the majority of the work on particle formation for pMDIs: (1) those where the particle morphology is modified [23], and (2) those where the particle surface chemistry is altered [40, 121, 124]. These approaches have been employed to improve the colloidal stability of the particles (containing the APIs) in HFAs, and/or (especially in the case of chemical modification) to enhance the stability of the active ingredient itself, as for example in the case of proteins, which can be denatured during particle formation [40, 58, 83, 91]. Particle surface morphology and/or chemistry can be altered to suit the requirements of the formulation using a variety of techniques. Milling, spray drying, spray freeze-drying, supercritical fluid (SCF) technology, emulsification diffusion, and microcrystallization are some of the approaches that have been employed to formulate particles for OI purposes [40, 91, 92].

One of the earliest contributions in this area was the development of large porous particles [23, 26]. Particles with the unique porous morphology allowed for the propellant to penetrate into the pores, decreasing the density gradient between the particles and the propellant, thus reducing the sedimentation velocity and the cohesive forces between the particles. As a consequence, the dispersion stability and aerosol characteristics of the formulation greatly improved [23]. Large porous particles (large geometric size but appropriate aerodynamic size) also have the opportunity to evade clearance mechanisms of the lungs. Albuterol sulfate, Formoterol fumarate, and Cromolyn sodium were prepared as large porous particles [23]. Particles with a high degree of roughness may also be thought of as a means for reducing the adhesive contact in propellant HFA.

A new technique called thin film freezing (TFF) has been recently employed to form novel particles of APIs [28, 61, 102]. The specific shape and size of the particles help in the stabilization in HFAs, and in the case of nonpolar APIs, the dissolution rate is enhanced by reducing the size (down to nano) of one of the particle dimensions. This technique has been shown to be applicable to both small MW nonpolar APIs and biomacromolecules. The spherical API particles thus produced morphed into anisotropic particles having dimensions both in nanoscale and micron scale upon contact with the propellant. For instance, spherical particles of itraconazole produced using TFF had an initial size of 30 nm. Upon adding these particles to the propellant, it was found that these particles transformed into thin plates, with average dimensions of $600 \times 300 \times 60$ nm, as determined by SEM [102].

Altering the crystal state of the API is another interesting approach that has been recently reported. In one study, poorly water soluble budesonide and indomethacin were complexed with hydroxypropyl-β-cyclodextrin (HPBC) using a single step solvent-free SCF process [7]. Following SCF processing, the particle morphology changed from needle and plate like crystals to fused spherical structures. The drug content in the final product was greater than 85%. X-ray diffraction (XRD) studies of the supercritical carbon dioxide-processed drug revealed a loss in the crystalline peaks of the drug and the complexation of the drugs to HPBC as determined using Fourier transform-infrared spectroscopy (FT-IR). These complexes exhibited enhanced dissolution rates compared to the pure drug as determined by solubility experiments [7].

Several approaches have been devised regarding the changes in the surface chemistry of particles [40, 50, 76, 121, 124]. For example, the physical trapping of HFA-philes on the surface of drug particles formed via emulsification diffusion may be used to avoid the expensive (time and synthetic efforts) design of balanced surfactants that would otherwise be required for particle stabilization [121]. This approach has been shown to be successful in the formation of particles of polar APIs, and may be potentially extended to proteins and other larger macromolecules. It has several advantages, including the fact that there is a small fraction of excipient remaining on the particle – and in that case the excipient was PEG, and no excess surfactant is present/necessary for particle stabilization in HFA, thus decreasing potential toxic effects [121]. Deoxyribonuclease I (DNase I), a glycosylated protein commonly used in the treatment of cystic fibrosis, was also formulated as a stable dispersion in pMDIs [40]. DNase I was spray dried with biocompatible vinyl polymers PVA and PVP either in presence or absence trehalose dihydrate to form microparticles. Microparticles manufactured without the presence of either trehalose, PVA, or PVP lost over 40% activity after spray drying, whereas those spray dried in combination retained up to 100% activity. The addition of these spray dried particles to HFA134a did not induce any further reduction in the DNase activity. The formulations exhibited an aerosol performance (fine particle fraction) of around 53% after 24 weeks of storage [40].

There have been several other studies that have incorporated small APIs and biomolecules in pMDIs upon modifying the surface chemistry of the drug particle [38, 57, 58]. In one such investigation, nanoparticles (NPs) of a model protein (lysozyme) with lactose as an excipient were produced using emulsification and nanoprecipitation techniques in conjunction with freeze drying. The lysozyme particles thus formed retained 99% of the activity and formed good dispersions in HFA134a (in presence of ethanol and DPPC) at a concentration of 80% lysozyme and 20% lactose NPs (both w/w) in the formulation. Most of the biomolecular activity was retained even after contact with HFA134a [58]. Another interesting study utilized SS and salmon calcitonin as model APIs [37]. Core-shell particles comprised of an HFA-philic shell encapsulating the API as the core is another interesting strategy that can be used to form stable dispersions of APIs in HFAs [120, 124]. The shell is designed so as to include an HFA-philic moiety to enhance the dispersability of the particles in HFAs. The core-shell particles thus formed showed excellent physical stability in HFAs, and enhanced aerosol performance [124]. This technique is very versatile in that it can be applied to water soluble and water dispersible APIs, including small polar solutes and biomacromolecules. More recently, it has been demonstrated that this approach can be extended to water dispersible NCs as well [10], and can potentially work for water dispersible NPs of nonpolar APIs.

8.4 Future Directions of pMDIs

This transition period for pMDIs, where product reformulation has been necessary to accommodate a new class of propellants, has offered plenty of challenges to the industry – some still being recognized and/or worked out. However, this transition has also offered a unique opportunity to rethink the product and its potential, and has ultimately become a time of exciting new scientific and technological advances in this field, where the infusion of new ideas combined with recent advances in nanomedicine makes this a very exciting area of research to be in. We have discussed some of this work in the previous sections. We reserved this last part of the chapter to discuss some recent exciting technologies that we believe have transformative potential as they may provide opportunities for pMDIs to serve as a truly viable, efficient, and desirable platform for the regional and systemic delivery of drugs to and through the lungs. The areas highlighted next are related to nanotechnology and biomacromolecular therapeutics, and are broken down into three subsections, which are discussed next.

8.4.1 Nanosuspensions of Poorly Water Soluble APIs in pMDIs

The effectiveness of nonpolar APIs delivered to the lungs (and other routes of administration for that matter) may be severely hindered due to their limited aqueous solubility [60, 61, 102]. The poor aqueous solubility of nonpolar APIs may also lead to unwanted toxic effects. The formulation of nonpolar APIs in the form of NPs dispersed in HFA-based pMDIs (nanosuspensions) may help address such issues, and at the same time improve the dispersibility of the particles in suspension and dose uniformity, and also increase the total delivered dose [102].

NPs of nonpolar APIs have improved bioavailability due to a large extent to the higher dissolution rates (due to larger surface areas) of such particles compared to their micron-sized counterparts [127]. The potential evasion of such small particles to clearance mechanisms of the respiratory system, as for example by evading alveolar macrophages if the target is the deep regions of the lungs, may also be realized by designing NP-based APIs as they may quickly gain access to the intracellular milieu [45, 95, 101]. Nanosuspensions have also been shown to be statistically more homogeneous compared to their micron-sized counterparts, and hence can deliver dosages with better uniformity [89]. Recent studies have shown that excellent dispersion stability may be achieved in HFA-based pMDIs if APIs are formulated as nanosuspensions, with a resulting increase in the aerosol characteristic as indicated by higher fractions being delivered to the deep lungs [28, 102].

NPs of itraconazole (Itr), a poorly water soluble drug, have been prepared by a TFF process to yield drug NPs [102]. Upon addition of these NPs to the propellant, they morphed to thin nanoplates of 60 nm in diameter and 600 nm in length, which has excellent physical stability in HFA227 for over 2 years, in formulations with drug concentrations up to 10 mg mL^{-1}. The excellent stability of these formulations was attributed to the inefficient packing of the particles that form low density flocs that stack upon each other, thus preventing settling [102]. Furthermore, it was reported that the suspensions prepared using these Itr NPs in HFA 227 had a FPF of 56%, which is well above those seen the commercial formulations of pMDIs. Studies conducted to determine the dissolution characteristics of the NPs in vitro indicated that over 55% of the formulation dissolved in just over 2 h, suggesting that the dissolution of such drugs can be greatly improved by formulating them as nanosuspensions [102]. TFF process was also used to prepare nanoflocs of a model protein. BSA was also shown to form interconnected nanorods of ~50–100 nm in diameter and 2.4 μm in length. These protein nanoflocs also had excellent stability in propellant HFAs and an FPF of 47% [28].

Another methodology that can be utilized in the preparation of NPs of nonpolar APIs is the SCF extraction of emulsions [6, 14]. Model nonpolar compounds like griseofulvin, cholesterol acetate, and megestrol acetate have been converted to NPs within the size range of 100–1,000 nm. These NPs exhibited a 5–10-fold increase in their aqueous solubility compared to the micronized form [88]. Such techniques capable of preparing NPs of nonpolar APIs hold a significant promise in the development of novel pMDI formulations, provided their aerodynamic size is appropriate – in the TFF, one of the dimensions was in the nano-range and the other in the micron range. There are many unanswered questions in this area besides how to make the particles with a desired morphology and stable formulations, including the effect of size, shape, and chemistry of these nonpolar nano APIs on their interaction with the lung tissue.

8.4.2 Polymeric Nanocarriers as a Platform for the Controlled and Targeted Drug Delivery to and Through the Lungs Using pMDIs

Polymeric nanocarriers (NCs) have several distinct advantages as drug delivery vehicles [20, 31, 101]. They are versatile, i.e., they can be tailored to encapsulate a wide variety of therapeutics including biomolecules; they serve as protective depots for fragile molecules; they may enhance the bioavailability of encapsulated therapeutics; they have the ability to adapt to different administration routes; they can modulate the release of the encapsulant; and they may improve permeation and retention at inflammatory sites [20, 64, 72]. NCs in the form of particles can be prepared by a number of different techniques. Some extensively studied processes include spray drying, emulsification-solvent evaporation, and homogenization [64, 81]. All those are suitable for the formation of polymeric NCs for OI delivery, will be discussed later, need some ingenuity to be formulated in pMDIs.

One characteristic of polymeric NCs of particular importance in the field of pulmonary delivery is the fact that they have appropriate sizes to avoid macrophage clearance in the alveolar region, and may in principle be designed to control the delivery of the therapeutic of interest by either promoting transepithelial transport [79, 81, 126], or be transported itself (for systemic delivery) [33, 113, 125], via translocation or transcytosis [78], or to improve the retention of the API in the lungs in case the target is the lungs (regional delivery).

Several biocompatible and biodegradable polymers, including polyesters (polylactides [PLA] and poly(D,L-lactide-co-glycolide) [PLGA]), polyanhydrides, chitosan, polybutylcyanoacrylates, and polyacrylates have been investigated as NCs for pulmonary administration [5, 6, 81, 101]. Additionally, research groups have reported the chemical modification of existing biodegradable polymers in an attempt to control their residence time, modulate the transport across biological barriers that exist in the lungs, and to avoid particle clearance [81]. For example, a series of amine-modified, PVA-based branched polyesters with PLGA side chains were shown to have good protein compatibility and better temperature stability than PLA or PLGA. Furthermore, the biodegradation rates of these copolymers could be easily controlled by varying the polymer chemistry. For instance, the polymer with the highest degree of amine substitution had the fastest degradation rate. These copolymers were formulated as nanoparticulate suspensions in nebulizers. The suspensions thus formed did not show aggregation during the process of nebulization (both jet and ultrasonic), when compared to formulations prepared from PLGA alone. This study also shows that the NPs from these suspensions were uptaken successfully into alveolar epithelial cells, although the extent of uptake was very low [18]. PEGylation of NPs has been proposed in order to facilitate their efficient transport across mucosal barriers [46, 47]. The effective diffusion coefficients of polystyrene NPs of varying diameters coated with PEG (M_n ~ 2 KDa) were studied in mucus [46, 47, 103]. Recent in vivo studies of the

NC-lung tissue interactions are also very encouraging, suggesting favorable biological responses [5, 44, 86, 101].

While significant progress and understanding has recently emerged in the use of polymeric NCs for delivering drugs to and through the lungs, most studies have focused on the use of either DPIs or nebulizers as the delivery vehicles [30, 87, 101, 110]. Our group has recently proposed the use of pMDIs for delivering NCs to the lungs [10]. One additional challenge in using the pMDIs as delivery vehicles arises due to the fact that the chemistry of the particles to be dispersed in the propellants need to be well solvated by HFAs in order to prepare stable formulations. We have observed for example that PLGA NCs quickly flocculate in HFA propellants. An alternative approach for delivering such particles must therefore be developed. Another complicating factor in delivering NCs to the lungs is that the size of typical polymeric NCs falls outside the desired range ~0.4–5.0 μm optimal for OI delivery [67, 78]. While the delivery of polymeric NCs in aggregated form has been proposed in the case of DPIs as a means of achieving an appropriate aerodynamic diameter [87, 110], the restriction with respect to appropriateness of the chemistry for dispersing the NCs in the propellant still holds for pMDIs even for aggregated particles (e.g., PLGA NCs are not dispersible in HFAs). Moreover, aggregates of NCs may be problematic as they may arrive at the lung tissue and present themselves to the cells as such (micron size aggregates), and this may have implications as to whether the lung tissue "sees" these aggregates as NPs or as they are (large, micron-sized particles), which would have severe implications in terms of the efficiency of such carriers.

Our approach in the development of a general platform for the delivery of polymeric NCs to and through the lungs, considering such limitations, consists the development of a core-shell formulation where the NCs are trapped inside a shell that provides for the appropriate aerodynamic diameter, and at the same time helps enhance the dispersibility of the particles in the HFA propellant, and thus the characteristics of the corresponding pMDI formulation – see Fig. 8.5.

Core-shell formulations in HFA227 containing 2 mg of particles/mL of propellant, with a loading of NCs of 10% in weight with respect to the total particle mass resulted in a formulation with excellent physical stability in the propellant (as can be seen in Fig. 8.5c) and aerosol characteristics, with fine particle fractions as high as 55%. In vitro experiments revealed that the NCs released from such core-shell formulations were successfully internalized into model airway epithelial (Calu-3) cells, and that they can carry their load not only to healthy cells, but also to the infected (bacterial infected) airway epithelium. We have also shown that a similar approach may be successfully used for the formulation of polyplexes for the delivery of genes to the lungs, and that such constructs are capable of transfecting lung alveolar cells (A549) in vitro.

The development of such general platforms for the delivery of drugs using polymeric NCs to the lungs offer a wide-range of opportunities for pMDIs for both regional and systemic drug delivery. There is a vast area of research to be explored for OI using pMDIs, especially considering recent advances in the design and preparation of polymeric NCs [45, 95]. Some of the areas of relevance intersect with those of interest for systems that involve other forms of OI delivery vehicles, as for example

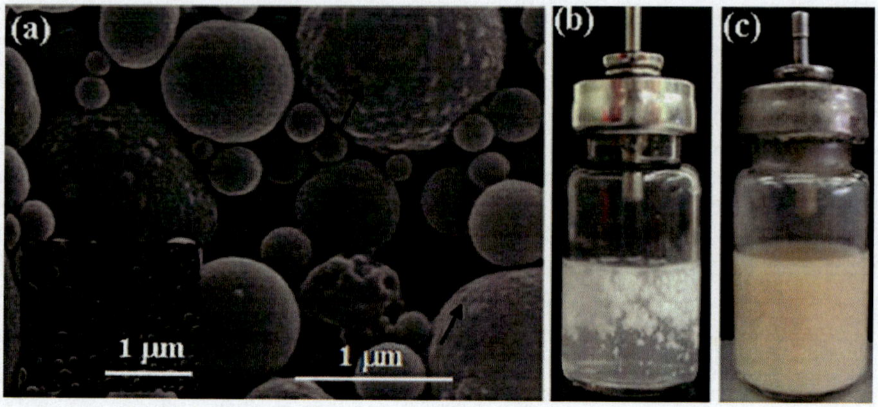

Fig. 8.5 **(a)** SEM micrograph of core-shell particles containing polymeric NCs trapped within the core. The *arrows* indicated in the image depict the presence of the NCs. (*Inset*) An SEM snapshot of PLGA NCs used in formulating the core-shell particles. **(b)** PLGA NC dispersion in HFA227. The NC concentration in the formulation was 2 mg mL^{-1} of the propellant. The image was captured 5 min after mechanical energy input. **(c)** Dispersion of the core-shell particles containing the NCs in HFA227. The particle concentration here was 2 mg mL^{-1} as well. The image was captured 15 min after mechanical energy input. Both formulations were prepared at ambient temperature and saturation pressure of the propellant

the study of the interaction of polymeric NCs with lung tissue and their biodistribution; some others are unique to pMDIs, as they require specialized particle formation technologies and biomaterials for appropriate formulation in HFA propellants.

8.4.3 *Oral Inhalation Delivery of Biomacromolecules with pMDIs*

The delivery of biomacromolecules to and through the lungs offers many opportunities and advantages compared to alternative routes of administration due to the physiology of the lung tissue [67]. OI has been suggested as a viable route for the administration of a range of therapeutic biomacromolecules including peptides, oligonucleotides, and proteins that can be used in the treatment of medically relevant diseases including diabetes, cystic fibrosis, and cancer [67, 78]. However, there are many challenges that need to be overcome in the process of formulating such therapeutics in pMDIs. Biomacromolecules need to be formulated as dispersions, as they are not soluble in HFA propellants. In the case of proteins, their exposure to harsh environments (e.g., high temperature, organic solvents) has to be minimized during the particle formation step, so as to prevent losses in activity that are typically observed during processing [91, 92]. Even in those cases when biomacromolecules may be formulated as particles (with minimum loss in activity), their physical stability in HFA propellants and aerosol characteristics may be difficult [40, 58]. While the issues discussed above represent real challenges, recent studies suggest potential strategies capable of addressing such shortcomings, thus pointing to pMDIs as emerging and viable devices for the delivery of biomacromolecules to and through the lungs.

One important question in this area is the potential impact of the compressed propellant in the structure of the biomacromolecules. It has been recently shown that the structure of a model protein remains stable when in contact with HFA propellants (both HFA123a and HFA227), with little impact on the bioactivity [74]. While more detailed investigations (e.g., under different temperature conditions, and water concentration in the propellant) need to be preformed, and a wider range of proteins need to be studied; this is an important observation, which points to the feasibility of pMDIs as vehicles for the delivery of biomacromolecules. One can expect that the structure of the more robust peptides and oligonucleotides would be even less susceptible to change in the presence of HFAs [57, 58, 74].

The stability of biomacromolecules during the particle engineering process applied to pMDIs has been approached using different particle formation methodologies, including spray drying [40, 57], emulsification diffusion [124], and nanoprecipitation [25]. For example, a study conducted on insulin shows that the conformation of the peptide may be preserved during particle formation when using an emulsification process followed by freeze drying [57]. Spherical, biostable insulin particles were produced with the help of cryoprotectants. Stabilization of the particles in HFA was possible with the help of stabilizing agents [58]. NPs of a model protein (lysozyme) were prepared by nanoprecipitation. The activity was preserved due to the presence of sugars and surfactants during particle formation, and the resulting particles could be dispersed in HFA with the help of oleic acid and ethanol [58]. Other relevant protein particles such as DNase I have also been shown amenable to formulation in pMDIs. DNase I particles have been formed in the presence of cryoprotectants and PVA/PVP using spray drying [40], and of chitosan-PLA co-oligomers (core-shell morphology) through emulsification diffusion [82]. These particles were shown to preserve the activity of DNase (even though at low loadings in the case of PVA/PVP + cryoprotectant), and to impart good stability in HFA and enhanced aerosol characteristics, with FPFs >50% [40, 82].

It is clear, therefore, that pMDIs have the potential to serve as devices for the OI delivery of biomacromolecules, and can do so in an efficient manner. This may be very relevant in the field where designer drugs (peptides, DNA, si-RNA, and others) are expected to play a significant role not only in the treatment of lung diseases, but also for systemic delivery. This area of research has seen a recent increase in activity, and new opportunities are expected to be realized as novel particle technologies continue to be developed, and a broader range of biomacromolecules are shown amenable to formulating in pMDIs.

Acknowledgments The authors would like to acknowledge the Institute of Manufacturing Research and the Graduate School at Wayne State University (WSU) for research assistantships, National Science Foundation (NSF, CBET 0933144) and Michigan University Commercialization Initiative (MUCI) for financial support.

References

1. Akasaka R (2008) An assessment of thermodynamic models for HFC refrigerant mixtures through the critical-point calculation. Int J Thermophys 29:1328–1341
2. Ashayer R, Luckham PF, Manimaaran S, Rogueda P (2004) Investigation of the molecular interactions in a pMDI formulation by atomic force microscopy. Eur J Pharm Sci 21(4):533–543
3. AstraZeneca (2009) Symbicort product information booklet
4. Atkins PJ, Woodcock A, Blinova O, Khan J, Stechert R, Wright P, Yizhong Y (2006) Medical aerosols, in IPCC/TEAP special report: safeguarding the ozone layer and the global climate system, Fakes D, Seki M, Editors. pp 351–360
5. Azarmi S, Roa WH, Lobenberg R (2008) Targeted delivery of nanoparticles for the treatment of lung diseases. Adv Drug Deliv Rev 60(8):863–875
6. Bailey MM, Berkland CJ (2009) Nanoparticle formulations in pulmonary drug delivery. Med Res Rev 29(1):196–212
7. Bandi N, Wei W, Roberts CB, Kotra LP, Kompella UB (2004) Preparation of budesonide- and indomethacin-hydroxypropyl-beta-cyclodextrin (HPBCD) complexes using a single-step, organic-solvent-free supercritical fluid process. Eur J Pharm Sci 23(2):159–168
8. Bell J, Newman S (2007) The rejuvenated pressurised metered dose inhaler. Expert Opin Drug Deliv 4(3):215–234
9. Berger W (2009) Aerosol devices and asthma therapy. Curr Drug Deliv 6(1):38–49
10. Bharatwaj B, Wu L, Whittum-Hudson J, da Rocha SRP (2010) The potential for the non-invasive delivery of polymeric nanocarriers using propellant-based inhalers in the treatment of Chlamydial respiratory infections. Biomaterials 31:7376–7385
11. Blondino FE, Byron PR (1998) Surfactant dissolution and water solubilization in chlorine-free liquified gas propellants. Drug Dev Ind Pharm 24(10):935–945
12. Brambilla G, Ganderton D, Garzia R, Lewis D, Meakin B, Ventura P (1999) Modulation of aerosol clouds produced by pressurised inhalation aerosols. Int J Pharm 186(1):53–61
13. Chokshi U, Selvam P, Porcar L, da Rocha SRP (2009) Reverse aqueous emulsions and micro-emulsions in HFA227 propellant stabilized by non-ionic ethoxylated amphiphiles. Int J Pharm 369:176–184
14. Chow AH, Tong HH, Chattopadhyay P, Shekunov BY (2007) Particle engineering for pulmonary drug delivery. Pharm Res 24(3):411–437
15. Courrier HM, Butz N, Vandamme TF (2002) Pulmonary drug delivery systems: recent developments and prospects. Crit Rev Ther Drug Carrier Syst 19(4–5):425–498
16. Courrier HM, Pons F, Lessinger JM, Frossard N, Krafft MP, Vandamme TF (2004) In vivo evaluation of a reverse water-in-fluorocarbon emulsion stabilized with a semifluorinated amphiphile as a drug delivery system through the pulmonary route. Int J Pharm 282(1–2):131–140
17. Cummings RH (1999) Pressurized metered dose inhalers: chlorofluorocarbon to hydrofluoroalkane transition-valve performance. J Allergy Clin Immunol 104(6):S230–S236
18. Dailey LA, Wittmar M, Kissel T (2005) The role of branched polyesters and their modifications in the development of modern drug delivery vehicles. J Control Release 101(1–3):137–149
19. Dalby RN, Tiano SL, Hickey AJ (2007) Medical devices for the delivery of therapeutic aerosols to the lungs. In: Hickey AJ (ed) Inhalation aerosols – physical and biological basis for therapy, 2nd edn. Informa Healthcare, New York
20. Davis ME, Chen ZG, Shin DM (2008) Nanoparticle therapeutics: an emerging treatment modality for cancer. Nat Rev Drug Discov 7(9):771–782
21. de Jager D, Manning M, Kuijpers L (2006) Safeguarding the ozone layer and the global climate system: issues related to hydrofluorocarbons and perfluorocarbons, in IPCC/TEAP special report, Davidson O, McFarland M, Midgley P, Editors. pp 19–81
22. de Jager D, Manning M, Kuijpers L (2006) Safeguarding the ozone layer and the global climate system: issues related to hydrofluorocarbons and perfluorocarbons – summary for policymakers, in IPCC/TEAP special Report, Davidson O, McFarland M, Midgley P, Editors. pp 1–13

23. Dellamary LA, Tarara TE, Smith DJ, Woelk CH (2000) Hollow porous particles in metered dose inhalers. Pharm Res 17(2):168–174
24. Dickinson PA, Seville PC, McHale H, Perkins NC, Taylor G (2000) An investigation of the solubility of various compounds in the hydrofluoroalkane propellants and possible model liquid propellants. J Aerosol Med 13(3):179–186
25. Dickinson PA, Howells SW, Kellaway IW (2001) Novel nanoparticles for pulmonary drug administration. J Drug Target 9(4):295–302
26. Edwards DA, Hanes J, Caponetti G (1997) Large porous particles for pulmonary drug delivery. Science 276(5320):1868–1871
27. Elvassore N, Bertucco A, Wahlstrom A (1999) A cubic equation of state with group contributions for the calculation of vapor-liquid equilibria of mixtures of hydrofluorocarbons and lubricant oils. Ind Eng Chem Res 38:2110–2118
28. Engstrom J, Tam J, Miller M, Williams RO, Johnston K (2009) Templated open flocs of nanorods for enhanced pulmonary delivery with pressurized metered dose inhalers. Pharm Res 26(1):101–117
29. Gupta A, Myrdal PB (2004) Novel method for the determination of solubility in aerosol propellants. J Pharm Sci 93(10):2411–2419
30. Hadinoto K, Phanapavudikul P, Kewu Z, Tan BHT (2007) Drug release study of large hollow nanoparticulate aggregates carrier particles for pulmonary delivery. Int J Pharm 341:195–206
31. Hans ML, Lowman AM (2002) Biodegradable nanoparticles for drug delivery and targeting. Curr Opin Solid State Mater Sci 6:319–327
32. Harris JA, Stein SW, Myrdal PB (2006) Evaluation of the TSI aerosol impactor 3306/3321 system using a redesigned impactor stage with solution and suspension metered-dose inhalers. AAPS PharmSciTech 7(1):E20
33. Hitzman CJ, Wattenberg LW, Wiedmann TS (2006) Pharmacokinetics of 5-fluorouracil in the hamster following inhalation delivery of lipid-coated nanoparticles. J Pharm Sci 95(6): 1196–1211
34. Hoye JA, Myrdal PB (2008) Measurement and correlation of solute solubility in HFA-134a/ ethanol systems. Int J Pharm 362(1–2):184–188
35. Hoye JA, Gupta A, Myrdal PB (2008) Solubility of solid solutes in HFA-134a with a correlation to physico-chemical properties. J Pharm Sci 97(1):198–208
36. Huber A, Lemmon EW, Friend DG (2002) Modeling bubble points of mixtures of hydrofluorocarbon refrigerants and polyol ester lubricants. Fluid Phase Equilibria 194:511–519
37. James J, Crean B, Davies M, Toon R, Jinks P, Roberts CJ (2008) The surface characterisation and comparison of two potential sub-micron, sugar bulking excipients for use in low-dose, suspension formulations in metered dose inhalers. Int J Pharm 361(1–2):209–221
38. James J, Davies M, Toon R, Jinks P, Roberts CJ (2009) Particulate drug interactions with polymeric and elastomeric valve components in suspension formulations for metered dose inhalers. Int J Pharm 366(1–2):124–132
39. Johnson KA (2007) Interfacial phenomena and phase behavior in metered dose inhaler formulations. In: Hickey AJ (ed) Inhalation aerosols – physical and biological basis for therapy, 2nd edn. Informa Healthcare, New York, pp 347–371
40. Jones SA, Martin GP, Brown MB (2006) Stabilisation of deoxyribonuclease in hydrofluoroalkanes using miscible vinyl polymers. J Control Release 115:1–8
41. Kappl M, Butt HJ (2002) The colloidal probe technique and its application to adhesion force measurements. Part Part Syst Charact 19:129–143
42. Kiselev SB, Ely JF (2003) Generalized corresponding states model for bulk and interfacial properties in pure fluids and fluid mixtures. J Chem Phys 119:8645–8662
43. Kleinstreuer C, Zhang Z, Donohue JF (2008) Targeted drug-aerosol delivery in the human respiratory system. Annu Rev Biomed Eng 10:195–220
44. Kreyling WG, Semmler-Behnke M, Seitz J, Scymczak W, Wenk A, Mayer P, Takenaka S, Oberdorster G (2009) Size dependence of the translocation of inhaled iridium and carbon nanoparticle aggregates from the lung of rats to the blood and secondary target organs. Inhal Toxicol 21(Suppl 1):55–60

45. Kumari A, Yadav SK, Yadav SC (2010) Biodegradable polymeric nanoparticles based drug delivery systems. Colloids Surf B Biointerfaces 75(1):1–18
46. Lai SK, O'Hanlon DE, Harrold S, Man ST, Wang YY, Cone R, Hanes J (2007) Rapid transport of large polymeric nanoparticles in fresh undiluted human mucus. Proc Natl Acad Sci USA 104(5):1482–1487
47. Lai SK, Wang YY, Hanes J (2009) Mucus-penetrating nanoparticles for drug and gene delivery to mucosal tissues. Adv Drug Deliv Rev 61(2):158–171
48. Lavorini F, Fontana GA (2009) Targeting drugs to the airways: the role of spacer devices. Expert Opin Drug Deliv 6(1):91–102
49. Lewis D (2007) Metered-dose inhalers: actuators old and new. Expert Opin Drug Deliv 4(3):235–245
50. Liao YH, Brown MB, Jones SA, Nazir T, Martin GP (2005) The effects of polyvinyl alcohol on the in vitro stability and delivery of spray-dried protein particles from surfactant-free HFA 134a-based pressurised metered dose inhalers. Int J Pharm 304(1–2):29–39
51. McCallister JW, Moore WC (2008) Hydrofluoroalkane preparations of fluticasone propionate. Expert Rev Resp Med 2:433–442
52. McDonald KJ, Martin GP (2000) Transition to CFC-free metered dose inhalers – into the new millennium. Int J Pharm 201(1):89–107
53. Mitchell JP, Nagel MW (2009) Oral inhalation therapy: meeting the challenge of developing more patient-appropriate devices. Expert Rev Med Devices 6(2):147–155
54. Mogalian E, Myrdal PB (2007) Pharmaceutical solvents for pulmonary drug delivery. In: Augustijns P, Brewster M (eds) Solvent systems and their selection in pharmaceutics and biopharmaceutics, vol VI. Springer, New York, pp 427–441
55. Nicolini G, Scichilone N, Bizzi A, Papi A, Fabbri LM (2008) Beclomethasone/formoterol fixed combination for the management of asthma: patient considerations. Ther Clin Risk Manag 4(5):855–864
56. Norwood DL, Paskiet D, Ruberto M, Feinberg T, Schroeder A, Poochikian G, Wang Q, Deng TJ, DeGrazio F, Munos MK, Nagao LM (2008) Best practices for extractables and leachables in orally inhaled and nasal drug products: an overview of the PQRI recommendations. Pharm Res 25(4):727–739
57. Nyambura BK, Kellaway IW, Taylor KM (2009) Insulin nanoparticles: stability and aerosolization from pressurized metered dose inhalers. Int J Pharm 375(1–2):114–122
58. Nyambura BK, Kellaway IW, Taylor KM (2009) The processing of nanoparticles containing protein for suspension in hydrofluoroalkane propellants. Int J Pharm 372(1–2):140–146
59. O'Riordan TG, Smaldone GC (2008) Aerosol delivery systems. In: Kay AB, Kaplan AP, Bousquet J, Holt PG (eds) Allergy and allergic diseases, vol 1, 2nd edn. Blackwell Publishing, Maiden, pp 768–782
60. Overhoff KA, Engstrom JD, Chen B, Scherzer BD, Milner TE, Johnston KP, Williams RO III (2007) Novel ultra-rapid freezing particle engineering process for enhancement of dissolution rates of poorly water-soluble drugs. Eur J Pharm Biopharm 65(1):57–67
61. Overhoff KA, Johnston KP, Tam J, Engstrom J, Williams RO (2009) Use of thin film freezing to enable drug delivery: a review. J Drug Deliv Sci Technol 19(2):89–98
62. Oversteegen L (2008) Inhaled medicines: product differentiation by device. Innov Pharm Technol 28:62–65
63. Oversteegen L, Rovini H, Belsey MJ (2007) Respiratory drug market dynamics. Nat Rev Drug Discov 6(9):695–696
64. Panyam J, Labhasetwar V (2003) Biodegradable nanoparticles for drug and gene delivery to cells and tissue. Adv Drug Deliv Rev 55(3):329–347
65. Patel N, Marlow M, Lawrence MJ (2003) Fluorinated ionic surfactant microemulsions in hydrofluorocarbon 134a (HFC 134a). J Colloid Interface Sci 258(2):354
66. Patel N, Marlow M, Lawrence MJ (2003) Formation of fluorinated nonionic surfactant microemulsions in hydrofluorocarbon 134a (HFC 134a). J Colloid Interface Sci 258(2):345
67. Patton JS, Byron PR (2007) Inhaling medicines: delivering drugs to the body through the lungs. Nat Rev Drug Discov 6(1):67–74

68. Peguin RPS, da Rocha SRP (2008) Solvent–solute interactions in hydrofluoroalkane propellants. J Phys Chem B 112:8084–8094
69. Peguin RPS, Selvam P, da Rocha SRP (2006) Microscopic and thermodynamic properties of the HFA134a-water interface: atomistic computer simulations and tensiometry under pressure. Langmuir 22(21):8826–8830
70. Peguin RPS, Wu L, da Rocha SRP (2007) The Ester Group: how hydrofluoroalkane-philic is it? Langmuir 23(16):8291–8294
71. Peguin RP, Kamath G, Potoff JJ, da Rocha SR (2009) All-atom force field for the prediction of vapor-liquid equilibria and interfacial properties of HFA134a. J Phys Chem B 113(1):178–187
72. Pison U, Welte T, Giersig M, Groneberg DA (2006) Nanomedicine for respiratory diseases. Eur J Pharmacol 533(1–3):341–350
73. Prokopovich P, Theodossides S, Rahnejat H, Hodson D (2009) Friction in ultra-thin conjunction of valve seals of pressurised metered dose inhalers. Wear 268:845–852
74. Quinn EA, Forbes RT, Williams AC, Oliver MJ, McKenzie L, Purewal TS (1999) Protein conformational stability in the hydrofluoroalkane propellants tetrafluoroethane and heptafluoropropane analysed by Fourier transform Raman spectroscopy. Int J Pharm 186(1):31–41
75. Ridder KB, Davies-Cutting CJ, Kellaway IW (2005) Surfactant solubility and aggregate orientation in hydrofluoroalkanes. Int J Pharm 295(1–2):57–65
76. Rogueda P (2005) Novel hydrofluoroalkane suspension formulations for respiratory drug delivery. Expert Opin Drug Deliv 2(4):625–638
77. Rogueda P (2007) Pharmaceutical spray formulation comprising a hydrofluoroalkane and an acylated cyclodextrin. US Patent 2007/0104652 A1
78. Rogueda PG, Traini D (2007) The nanoscale in pulmonary delivery. Part 1: deposition, fate, toxicology and effects. Expert Opin Drug Deliv 4(6):595–606
79. Rogueda PGA, Traini D (2007) The nanoscale in pulmonary delivery. Part 2: formulation platforms. Expert Opin Drug Deliv 4(6):607–620
80. Rosenholm JB (2010) Solvent and surfactant induced interactions in drug dispersions. Colloid Surface Physicochem Eng Aspects 354:197–204
81. Rytting E, Nguyen J, Wang X, Kissel T (2008) Biodegradable polymeric nanocarriers for pulmonary drug delivery. Expert Opin Drug Deliv 5(6):629–639
82. Saiprasad S, Onyskiw P, Conti D, Wu L, da Rocha SRP (2010). DNase particle formulation for pressurized metered-dose inhalers. Manuscript in Preparation
83. Scheuch G, Siekmeier R (2007) Novel approaches to enhance pulmonary delivery of proteins and peptides. J Physiol Pharmacol 58(Suppl 5, Pt 2):615–625
84. Selvam P, Peguin RPS, Chokshi U, da Rocha SRP (2006) Surfactant design for the 1,1,1,2-tetrafluoroethane-water interface: ab initio calculations and in situ high-pressure tensiometry. Langmuir 22(21):8675–8683
85. Selvam P, Chokshi U, Gouch A, da Rocha SRP (2008) Ethoxylated copolymer surfactants for the hfa134a-water interface: interfacial activity, aggregation behavior and biomolecule. Soft Matter 4:357–366
86. Semmler-Behnke M, Takenaka S, Fertsch S, Wenk A, Seitz J, Mayer P, Oberdorster G, Kreyling WG (2007) Efficient elimination of inhaled nanoparticles from the alveolar region: evidence for interstitial uptake and subsequent reentrainment onto airways epithelium. Environ Health Perspect 115(5):728–733
87. Sham JOH, Yu Z, Finlay WH, Roa WH, Lobenberg R (2004) Formulation and characterization of spray-dried powders containing nanoparticles for aerosol delivery to the lung. Int J Pharm 269:457–467
88. Shekunov BY, Chattopadhyay P, Seitzinger J, Huff R (2006) Nanoparticles of poorly water-soluble drugs prepared by supercritical fluid extraction of emulsions. Pharm Res 23(1):196–204
89. Shekunov BY, Chattopadhyay P, Tong HH, Chow AH (2007) Particle size analysis in pharmaceutics: principles, methods and applications. Pharm Res 24(2):203–227
90. Sherwood JK, Alex S, Salama G, Obenauer-Kutner L, Huyck S, Berry J, Sequeira J, Brouet G, Marie C (2007) Particle size coarsening induced by valve silicone in a metered dose inhaler. Drug Dev Ind Pharm 33(2):155–162

91. Shoyele SA (2008) Engineering protein particles for pulmonary drug delivery. Methods Mol Biol 437:149–160
92. Shoyele SA, Cawthorne S (2006) Particle engineering techniques for inhaled biopharmaceuticals. Adv Drug Deliv Rev 58(9–10):1009–1029
93. Smyth HDC (2003) The influence of formulation variables on the performance of alternative propellant-driven metered dose inhalers. Adv Drug Deliv Rev 55(7):807–828
94. Smyth HD (2005) Propellant-driven metered-dose inhalers for pulmonary drug delivery. Expert Opin Drug Deliv 2(1):53–74
95. Soppimath KS, Aminabhavi TM, Kulkarni AR, Rudzinski WE (2001) Biodegradable polymeric nanoparticles as drug delivery devices. J Control Release 70(1–2):1–20
96. Steckel H, Wehle S (2004) A novel formulation technique for metered dose inhaler (MDI) suspensions. Int J Pharm 284(1–2):75–82
97. Stefely J (2002) Novel excipients for inhalation drug delivery: expanding the capability of the MDI. Drug Deliv Technol 2:1–8
98. Stein SW, Myrdal PB (2004) A theoretical and experimental analysis of formulation and device parameters affecting solution MDI size distributions. J Pharm Sci 93(8):2158–2175
99. Stein SW, Stefely J (2003) Reinventing metered dose inhalers: from poorly efficient CFC MDIs to highly efficient HFA MDIs. Drug Deliv Technol 3:46–51
100. Steytler DC, Thorpe M, Eastoe J, Dupont A, Heenan RK (2003) Microemulsion formation in 1,1,1,2-tetrafluoroethane (R134a). Langmuir 19(21):8715
101. Sung JC, Pulliam BL, Edwards DA (2007) Nanoparticles for drug delivery to the lungs. Trends Biotechnol 25(12):563–570
102. Tam JM, Engstrom JD, Ferrer D, Williams RO III, Johnston KP (2010) Templated open flocs of anisotropic particles for pulmonary delivery with pressurized metered dose inhalers. J Pharm Sci 99(7):3150–3165
103. Tang BC, Dawson M, Lai SK, Wang YY, Suk JS, Yang M, Zeitlin P, Boyle MP, Fu J, Hanes J (2009) Biodegradable polymer nanoparticles that rapidly penetrate the human mucus barrier. Proc Natl Acad Sci USA 106(46):19268–19273
104. Tarasova A, Burden F, Gasteiger J, Winkler DA (2010) Robust modelling of solubility in supercritical carbon dioxide using Bayesian methods. J Mol Graph Model 28(7):593–597
105. Terzano C (2001) Pressurized metered dose inhalers and add-on devices. Pulm Pharmacol Ther 14(5):351–366
106. Traini D, Rogueda P, Young P, Price R (2005) Surface energy and interparticle forces correlations in model pMDI formulations. Pharm Res 22(5):816–825
107. Traini D, Young PM, Price R, Rogueda P (2006) A novel apparatus for the determination of solubility in pressurized metered dose inhalers. Drug Dev Ind Pharm 32(10):1159–1163
108. Traini D, Young PM, Rogueda P, Price R (2006) Investigation into the influence of polymeric stabilizing excipients on inter-particulate forces in pressurised metered dose inhalers. Int J Pharm 320(1–2):58–63
109. Traini D, Young PM, Rogueda P, Price R (2007) In vitro investigation of drug particulates interactions and aerosol performance of pressurised metered dose inhalers. Pharm Res 24(1):125–135
110. Tsapis N, Bennett D, Jackson B, Weitz DA, Edwards DA (2002) Trojan particles: large porous carriers of nanoparticles for drug delivery. Proc Natl Acad Sci USA 99(19): 12001–12005
111. van Krevelen DW (1990) Properties of polymers. Their correlationwith chemical structure; their numerical estimation and prediction from additive group contributions, 3rd rev. ed.; Elsevier Scientific, Amsterdam, 1990
112. Vervaet C, Byron PR (1999) Drug-surfactant-propellant interactions in HFA-formulations. Int J Pharm 186(1):13–30
113. Videira MA, Botelho MF, Santos AC, Gouveia LF, de Lima JJ, Almeida AJ (2002) Lymphatic uptake of pulmonary delivered radiolabelled solid lipid nanoparticles. J Drug Target 10(8): 607–613
114. Wilby M (2005) pMDI performance – increasing performance consistency of pMDIs. Drug Deliv Technol 5:1–6

115. Williams RO III, Hu C (1997) A study of an epoxy aerosol can lining exposed to hydrofluoroalkane propellants. Eur J Pharm Biopharm 44(2):195–203
116. Williams RO III, Hu C (2000) Moisture uptake and its influence on pressurized metered-dose inhalers. Pharm Dev Technol 5(2):153–162
117. Williams RO III, Hu C (2001) Influence of water on the solubility of two steroid drugs in hydrofluoroalkane (HFA) propellants. Drug Dev Ind Pharm 27:71–79
118. Wright P (2006) Transition to CFC-free metered-dose inhalers in IPAC Submission to the Medical Technical Options Committee. pp 1–15
119. Wu L, da Rocha SRP (2007) Biocompatible and biodegradable copolymer stabilizers for hydrofluoroalkane dispersions: a colloidal probe microscopy investigation. Langmuir 23(24): 12104–12110
120. Wu L, da Rocha SRP (2008) Applications of the atomic force microscope in development of propellant-based inhalation formulations. KONA Powder Part J 26:106–128
121. Wu L, Al-Haydari M, da Rocha SRP (2007) Novel propellant-driven inhalation formulations: engineering polar drug particles with surface-trapped hydrofluoroalkane-philes. Eur J Pharm Sci 33(2):146–158
122. Wu L, Peguin RPS, da Rocha SRP (2007) Understanding solvation in hydrofluoroalkanes: ab initio calculations and chemical force microscopy. J Phys Chem B 111:8096–8104
123. Wu L, Peguin RPS, Selvam P, Chokshi U, da Rocha SRP (2007) Molecular scale behavior in alternative propellant-based inhaler formulations. In: Hickey AJ (ed) Inhalation aerosols – physical and biological basis for therapy, 2nd edn. Informa Healthcare, New York, pp 373–397
124. Wu L, Bharatwaj B, Panyam J, da Rocha SRP (2008) Core-shell particles for the dispersion of small polar drugs and biomolecules in hydrofluoroalkane propellants. Pharm Res 25(2): 289–301
125. Yamamoto H, Kuno Y, Sugimoto S, Takeuchi H, Kawashima Y (2005) Surface-modified PLGA nanosphere with chitosan improved pulmonary delivery of calcitonin by mucoadhesion and opening of the intercellular tight junctions. J Control Release 102(2):373–381
126. Yang W, Peters JI, Williams RO III (2008) Inhaled nanoparticles – a current review. Int J Pharm 356(1–2):239–247
127. Yang W, Tam J, Miller DA, Zhou J, McConville JT, Johnston KP, Williams RO III (2008) High bioavailability from nebulized itraconazole nanoparticle dispersions with biocompatible stabilizers. Int J Pharm 361(1–2):177–188
128. Young PM, Price R, Lewis D, Edge S, Traini D (2003) Under pressure: predicting pressurized metered dose inhaler interactions using the atomic force microscope. J Colloid Interface Sci 262(1):298–302
129. Zhao Y, Brown MB, Jones SA (2009) Engineering novel topical foams using hydrofluroalkane emulsions stabilised with pluronic surfactants. Eur J Pharm Sci 37(3–4):370–377
130. Zhao Y, Moddaresi M, Jones SA, Brown MB (2009) A dynamic topical hydrofluoroalkane foam to induce nanoparticle modification and drug release in situ. Eur J Pharm Biopharm 72(3):521–528
131. Zhao Y, Brown MB, Jones SA (2010) Pharmaceutical foams: are they the answer to the dilemma of topical nanoparticles? Nanomedicine 6(2):227–236
132. Zheng JY, Fulu MY, Lee DY, Barber TE, Adjei AL (2001) Pulmonary peptide delivery: effect of taste-masking excipients on leuprolide suspension metered-dose inhalers. Pharm Dev Technol 6(4):521–530

Chapter 9
Science and Technology of Dry Powder Inhalers

Timothy M. Crowder and Martin J. Donovan

Abstract The technology of dry powder inhalers has often surpassed our scientific understanding of the mechanisms of formulation, aerosol performance, and manufacture controls for these devices. However, with greater research effort being devoted to these systems, the science is beginning to drive the technological innovation. In this chapter, dry powder inhaler technology is discussed in relation to this growing body of scientific understanding. In addition, a section on the practical implications, most notably on manufacturing of these systems, is also included.

Keywords Design • DPIs • Dry powder inhalers • Manufacturing • Mechanism

9.1 Introduction

From a pragmatic viewpoint, the development of a novel dry powder inhaler (DPI) is centered on two objectives: creating a device that can efficiently entrain and disperse a powdered drug formulation for inhalation therapy, while utilizing a design and function sufficiently novel to avoid patent infringement. Accordingly, this has produced no shortage of prospective designs, prototypes, and devices incorporating a diverse array of powder dispersion strategies and de-agglomeration principles. Running the gamut from cyclones, baffles, and impellers, to the inclusion of external energy sources applying mechanical, electrical, and pneumatic forces to the powdered dose, this abundance reflects the carte blanche at an inventor's disposal when conceptualizing a new inhaler. However, while the harvest from the patent literature may be bountiful, very few patented DPI designs reach the market, as an array of factors including development cost, manufacturability, portability, and ease

T.M. Crowder (✉)
GlaxoSmithKline, Durham, NC, USA
e-mail: timothy.m.crowder@gsk.com

H.D.C. Smyth and A.J. Hickey (eds.), *Controlled Pulmonary Drug Delivery*, Advances in Delivery Science and Technology, DOI 10.1007/978-1-4419-9745-6_9,

of use distinguish between a design that is plausible and one that is practical. The objective of this chapter is to provide a survey of the plurality of dispersion strategies employed in DPIs, drawing on examples from both the scientific and patent literature, and is not intended to compare the relative merits of one specific device to another.

9.2 Pulmonary Drug Delivery

The overall performance of a therapeutic inhalation regimen is a complex interplay between the physiology and anatomy of the patient's pulmonary system and disease state, the physicochemical properties of the drug formulation, and the device employed to aerosolize and deliver the dose. With regards to the patient, proper training and instruction on the use of their prescribed inhaler, though often neglected, can produce a marked improvement in performance. However, apart from the performance boon provided through education, the anatomy of a patient's airways will dictate both the inspiratory force they will produce and the duration over which it is sustained, and the formulation and device must compensate for any shortcomings in the patient's inspiratory output.

Evolving from a common ancestor, the development of therapeutic inhalation devices has diverged into three distinct classes: nebulizers, pressurized metered-dose inhalers (pMDIs), and DPIs, with the latter being the most recent addition to the family [1]. Given their bulkiness, external power requirements, and time required for inhalation, nebulizers lack portability and are generally confined to the home or clinic, although hand-held nebulizers have been introduced [2]. In contrast, pMDIs employing a pressurized gas propellant to aerosolize the dose are highly portable and inexpensive. Inaugurated in the 1950s, pMDIs dominated the portable inhalation drug delivery market for many years, essentially running unopposed until the unveiling of the first DPI, the following decade (Spinhaler™, Aventis) [3]. Even then, nascent DPIs were characterized by low efficiency, and delivered a dose from a capsule that had to be manually loaded prior to each actuation; a sharp contrast to pMDIs and their 200 dose capacity. However, pMDIs were not without their drawbacks, and chief among them was their use of the ozone-depleting chlorofluorocarbons (CFCs) as propellants [4]. In 1987, members of the international community, recognizing the irreparable damage CFCs have on the environment, signed the Montreal Protocol pledging to phase out production of CFCs by 1996 [2]. In need of an alternative propellant, CFCs have been replaced by hydrofluoroalkanes (HFAs), although the transition has not proven seamless, as density and solubility differences between the propellants have led to reformulation issues. The high velocity of the emitted dose from a pMDI also requires coordination between patient inhalation and device actuation to avoid deposition in the throat, which many patients find difficult [5]. In addition, advanced therapeutics, including peptides, proteins, and gene vectors, exhibit poor stability at room temperature when formulated into the aqueous solutions delivered by pMDIs [6].

9.3 Dry Powder Inhalers

Developed as an alternative to the pMDIs, DPIs have their own unique advantages and limitations, and can be broadly categorized into either passive or active devices [3]. Passive DPIs are breath-actuated, deriving the energy for powder dispersion and aerosol formation solely from the patient's inspiratory maneuver, alleviating the coordination of actuation and inhalation that many patients find problematic with pMDIs [4]. However, patients can produce a wide range of flow rates, which is reflected in the high interpatient variability in the dose delivered from these devices. Additionally, some patients can only generate flow rates too low to produce an aerosol cloud, and treatment with DPIs can be ineffectual. In light of this, many DPI developers have concluded that the best strategy lies in decoupling powder dispersion from inhalation and, following the path of pMDIs, have designed active DPIs incorporating an assortment of auxiliary energy sources to aerosolize the dose.

DPIs can be further classified as either single dose or multidose devices, the latter being delineated into multiunit dose or reservoir inhalers. The single-dose inhalers represent the first generation of DPIs (Spinhaler™, Rotahaler™), although some devices in this category are still currently marketed (Aerolizer™, Handihaler™). These inhalers generally deliver the dose from a gelatin capsule inserted into the device prior to each actuation. The multiunit dose inhalers disperse individual doses pre-metered by the device manufacturer into blisters, while reservoir DPIs contain enough powder for multiple doses (typically 60–200) within the device, metering individual doses prior to actuation. Advantages of DPIs employing factory-metered doses include environmental protection of the powder and consistency of the dose relative to their reservoir counterparts. However, a drawback is the higher cost of these devices, stemming from the multicomponent designs and dedicated factory production lines [4]. Regardless of how the dose is provided, all DPIs contain a de-agglomeration principle to address the challenges inherent to producing aerosols from a dry powder.

9.4 DPI Formulations

To be effectively delivered into the lung, drug particles are generally required to fall in the size range between 1 and 5 μm; particle size reduction is typically performed through "top down" processes such as jet milling, which yields highly cohesive particles with ill-defined size distributions and morphologies [5]. Particle engineering, a "bottom up" approach to producing respirable size particles at the initial particle formation step, holds promise in addressing the drawbacks of attrition processes such as jet milling. However, few technologies have been industrialized at least in part due to the increased cost of goods for these nascent techniques [7]. The cohesive interactions between the powder particles arise from a combination of electrostatic, capillary, and van der Waals forces, although by allowing the particles time to dissipate excess electric charges, and in the absence of high relative humidity, it is the latter that are most important [8]. Individually weak, but collectively robust, van der Waals interactions are the dominant attractive forces in particles with diameters under 10 μm, exerting a "velcro effect" that impedes powder dispersion. To aid in the entrainment and de-agglomeration of these cohesive particles, dry powder formulations are generally binary, or interactive, blends, with the bulk, >95% (w/w), comprised of large inert carrier particles (50–100 μm), to improve the flowability and metering properties of the formulation [9]. In developing such an interactive blend, the interactions between the drug and carrier particles must be balanced, where they are strong enough to form a stable mixture, but sufficiently weak such that they are readily susceptible to dispersion when entrained. However, while the addition of the large carrier particles allows the powder to be readily entrained in a flow stream, the adhesive interactions between the drug and carrier hinders the formation of an aerosol comprised of primary drug particle sizes. Consequently, drug particles that fail to detach from the carriers are deposited in the throat and upper airways [10].

In lieu of developing inhalers that can enhance the de-agglomeration forces, an alternative approach is to reduce the cohesive and adhesive interactions within the powder through modifications to both drug and carrier particles [1]. For micronized drug, one strategy is to increase the diameter of the particle, thereby lowering the surface area-to-volume ratio, without compromising its aerodynamic diameter. The aerodynamic diameter of particle (d_a) governs how it behaves in a flow stream, and is related to its true diameter (d_p) and density (ρ_p) as:

$$d_a = d_p \sqrt{\frac{\rho_p}{\rho_o}}$$

where is ρ_o is the unit density, and ρ_p/ρ_o is the specific gravity of the particle [8]. From the above, it can be seen that an increase in the diameter of a particle can be balanced by a concomitant reduction in its density, an approach embodied by the large porous particles [11]. Additional developments in powder dispersion performance include smaller porous particles, pulmospheres, and the use of supercritical fluid technology. However, in comparison to unmodified drug, these particles can be costly to manufacture, requiring the use of spray drying or spray-freeze drying [1]. Moreover, given their increased volume, larger reservoir systems must be developed to accommodate the formulation, requiring the production of devices specific to the powder [1].

9.5 DPI Development

While it is the finished product that is unveiled to the public, the development a DPI requires an enormous temporal and financial investment, with a design initially rooted in a theoretical foundation, the prototypes refined through multiple rounds of experimentation and empirical observations, and the most optimized device emerging at the end. As inhalers are developed in a proprietary atmosphere, it is undeniably the inventors' prerogative to withhold the information of a DPI's formative stages from public view, especially if disclosure should benefit their competitors [5]. However, this also deprives us from gaining a more intrinsic understanding of the device components, and the extent to which they contribute to the performance of the inhaler. Why is a flow channel a certain diameter? What is the effect on device performance if the mouthpiece is lengthened, or widened? Unfortunately studies designed to address these questions are limited, but those that exist provide much information on DPI performance [12–16]. Additionally, the recent emergence of computational fluid dynamics (CFD) in device development, allowing the visualization of the spatial and temporal profiles of the flow stream, specifically in terms of the pressure and velocity distributions within the DPI during inhalation, offers an insight into the fundamental nature of device performance [17, 18].

While a wide variety of DPIs have been developed, their proliferation provides testament to the conspicuous absence of a device encompassing all the qualities of the ideal DPI. Each category of DPI mentioned above provides its own design challenges and opportunities. Due to the general simplicity and low production costs, passive devices are highly desirable, although by virtue of their function, they are prone to flow rate dependence and inconsistent dosing [6]. For an active delivery device employing an auxiliary powder de-agglomeration source, the challenge to the developer is incorporating a self-contained energy source that is reliable throughout the life of the inhaler, without compromising the compactness desired in a portable drug delivery device. This last statement is exemplified by the joint venture between Pfizer and Nektar Therapeutics to develop the insulin dry powder formulation, Exubera®. Over a decade in development, Exubera was launched with high expectations and much fanfare, only to be summarily scuttled a year later and remembered as a cautionary tale. The sinking of the inaugural insulin dry powder inhalation system has been extensively autopsied, the implications of its commercial failure thoroughly dissected, and while its disappointing market performance has been attributed to multiple sources, a general consensus is that the bulky design of the inhaler itself did nothing to advance its prospects [19].

Presented below is a brief overview of the powder de-agglomeration principles found in both commercial devices and the patent literature.

9.5.1 Turbulence Inducement

In its most basic form, a DPI is comprised of three regions: an air inlet port through which outside air enters the device during inhalation, a powder-holding chamber, and an outlet port delivering the dose to the patient. However, simply passing a flow stream across a static powder bed does not provide sufficient shear forces to effectively aerosolize the dose. When a particle is in a flow stream, it is subjected to two types of forces. The first of these are the body forces, such as gravity and electromagnetism, which act throughout the bulk of the particle and are described in terms of force per unit mass. The second category are the surface forces, generally provided as force per unit area and consisting of normal and shear, or tangential, stresses; it is the tangential stresses that are most important in powder de-agglomeration [20].

Turbulent flows are marked by highly irregular and rapid fluctuations of velocity in both time and space containing high energy eddies that continuously buffet the drug particles, subjecting them to shear stresses through accelerations in different directions [8]. When these accelerations generate forces of sufficient magnitude, the drug can detach from an aggregate particle or carrier. A frequently encountered mode of inducing turbulence in an inhaler is to supply spiraling channels for the dose as it exits the device; a design popularized by the Turbuhaler™ [21]. However, while effective at generating turbulence, spiral paths also increase the surface area of the flow channel, enhancing the amount of particle deposition within the device. An example of this can be seen in the Turbuhaler™, where over 20% of the dose may be retained in the mouthpiece. By comparison to the Diskhaler™, possessing a straight-channel mouthpiece with a much smaller surface area, approximately 5% of the nominal dose is lost [22].

In lieu of spiraling flow channels, many DPIs of more recent vintage contain tangential inlets opening into a cylindrical chamber to generate a high energy cyclone within the device. As an example, the NEXT™ DPI, a reservoir inhaler, contains a cyclone chamber composed of two non-concentric arcs (US Patent No. 7,107,988). CFD analysis was used to optimize the design and dimensions of the chamber, and the resulting geometry was shown to eliminate "dead spots" where drug deposition may have occurred [17]. An alternate example is found in the Conix™ inhaler, using a patented reverse cyclone technology. When the patient inhales, air is drawn into a cyclone chamber, establishing a vortex. As the flow stream travels down the cyclone, it encounters a blocked path at the bottom of the chamber, inducing the air flow to reverse direction and travel back up through a circular outlet [23].

9.5.2 Mechanical Forces

In contrast to providing swirl channels or cyclone chambers, other inhalers incorporate designs applying mechanical forces to the powder. An example is the Spiros™ inhaler, where a powder-laden flow stream is carried through a battery-driven impeller to form an aerosol. An alternative embodiment (US Patent No. 6,237,591) uses a turbine to drive the impeller, labeling it a passive DPI, as the turbine is propelled when the patient inhales.

Apart from impellers, mechanical forces are provided by low density beads contained within the dispersion chamber (US Patent No. 6,971,384). When the patient inhales, the lightweight beads are driven by the entrained flow, repeatedly colliding with each other and the walls of the inhaler. As the dose is carried through the chamber, the powder is de-agglomerated as it is caught between the bead–bead and bead–wall impactions. Additionally, powder can be dispersed by employing a spring-driven hammer to strike a blister containing the dose (US Patent No. 5,655,523) [24].

When compared to inhalers that rely on turbulence, mechanically driven devices are more complex, requiring additional moving parts that may be prone to failure. However, an alternate strategy to generate mechanical impaction forces, without overly increasing the complexity of a device, is through the use of baffles (US Patent No. 5,724,959). Located downstream of the powder-holding chamber, a narrow channel opens into a larger volume region containing an impactor plate, producing an abrupt change in the flow path. Due to their increased stopping distance, larger particles will be unable to follow the flow stream and navigate around the plate, colliding with the impactor and detaching drug from both aggregates and carriers. The width of the plate and the distance between it and the narrow channel opening alters the cut-off size of the particles that will be intercepted.

9.5.3 *Pneumatic Forces*

9.5.3.1 Compressed Gas

The most straightforward way to impart pneumatic forces to a powder bed is via a compressed gas source. However, while DPIs have been designed that utilize an external compressed gas source (US Patents, No. 5,875,776, and 5,775,320 as examples), more recent designs incorporate a method to compress air within the device through a manual pump. The most notable example is the Nektar Pulmonary Inhaler used to deliver Exubera (previously known as the Inhale DPI, Inhale Therapeutic Systems, US Patent No. 6,257,233). The DPI, which decouples device actuation from inhalation, is comprised of a lower pressurization chamber where the compressed air is generated, and an upper receiving chamber that holds the aerosolized powder following actuation. The air is compressed when the patient manually primes a handle coupled to a piston within the lower chamber. When actuated, the compressed air is discharged past the dose, drawing the powder into the upper chamber from where the patient subsequently inhales. Another example is the Vectura Aspirair™, employing a bolus of air that is manually compressed by the patient via a corkscrew-type manual pump, and discharged into the powder bed upon inhalation.

9.5.3.2 Vacuum

An alternative approach to pneumatically driven dispersion is provided in US Patent No. 6,138,673, illustrating a reservoir DPI that creates a vacuum within the body of the device. The inhaler body can be extended by twisting the top portion relative to the lower portion two revolutions, expanding the volume of the container from

550 to 750 mL, and creating a vacuum inside when the body is closed to the atmosphere. The device is manually actuated by the patient, inhaling while pressing a trigger that opens a valve within the body, exposing it to ambient air which enters the device through the dose holding chamber and aerosolizes the powder.

9.5.3.3 Synthetic Jetting Technology

The application of synthetic jets to powder dispersion is another means of applying pneumatic forces and is described in US Patent No. 7,334,577. Synthetic jetting may be produced in a chamber bounded on one end by a wave generating device and on the opposite end by a rigid wall with a small orifice [25]. When acoustic waves are emitted at high frequency and amplitude from the generator, a jet of air, directed outward from the chamber, is produced. This "synthetic" jet is comprised of vortical air puffs corresponding to the generator's frequency, which may be a piezoelectric element (discussed in detail below) or an electrodynamic transducer. When actuated, a dry powder dose located in the jet-producing chamber and resting above the generator is levitated and dispersed by the high frequency vibrations. As the aerosolized powder nears the orifice, it is expelled from the chamber by the synthetic jet and into a flow channel where it is carried to the patient.

9.5.4 Sustained Exposure to Flow Stream

One of the drawbacks commonly cited with passive DPIs is that their reliance on the patient's inspiratory flow rate can render them incapable of providing sufficient energy to effectively aerosolize and de-agglomerate the dose [26]. However, when considering that the duration of an inspiratory maneuver can last well beyond 3 s, and that the majority of the powder departs the inhaler within the first second of inhalation (well before the peak inspiratory flow rate is reached), it becomes clear that much of the energy available in the flow stream is never allowed the opportunity to interact with the dose [27]. Accordingly, the poor performance that plagues these devices cannot be solely attributed to an inadequate supply of energy, but also to an inefficient application of the available energy. To address this problem, inhalers have been developed incorporating designs that can lengthen the time through which the dose interacts with the flow stream, sustaining the energy transfer between the air and powder. Alternatively, the device can delay the exposure of the dose to the flow stream until a predetermined flow rate through the device has been attained, exposing the powder to the maximum energy levels of the flow.

9.5.4.1 Delayed Exposure

An example of this design incorporates a diverting flow channel in parallel with a secondary channel passing through the dosing chamber (US Patent No. 6,561,186).

The flow channels share a common occluding mechanism connected by a rotatable vane, such that when one channel is open, the other is closed. When the flow rate reaches a predetermined level, the vane closes, occluding the first channel, while simultaneously opening the second channel and allowing the flow access to the dose. In the Skyehaler™ (US Patent No. 6,182,655) a valve shield is moved within the inhaler in response to the suction generated from inhalation. When this force is sufficiently strong (>1.5 kPa) the valve shield opens a shutter, exposing the dose cavity to the flow stream. While these examples are from passive devices, the use of more complex air flow sensors is becoming a common feature in many active DPIs under development to automatically coordinate aerosol production with inhalation.

9.5.4.2 Air Classifier Technology

A vector quantity, the detachment forces acting on a drug particle adhered to a carrier is characterized by both a magnitude and a direction. When holding the magnitude constant, detachment forces exert their maximum effect when they act in a direction directly opposite to the adhering force [8]. However, due to the short duration of time to which a dose is subjected to separation forces in a typical DPI, coupled to the random nature of the detachment forces, only a small fraction of the drug particles will be correctly aligned to experience a force having a favorable combination of magnitude and direction sufficient to detach them from the carrier particle surface.

Air classifier technology was developed as a particle de-aggregation principle to sustain the interaction between the dose and the flow stream. A basic classifier is comprised of a cylindrical chamber, containing at least one tangential air inlet and a discharge outlet beginning at the center of one of the circular ends of the chamber [9]. Named for its ability to segregate, or classify, particles according to their size, it operates as a balance between centrifugal and drag forces. The drag force is proportional to the first power of the particle diameter and dominates for fine particles, while proportional to the cube of the particle diameter, larger carrier particles will experience a greater centrifugal force. When the powder is carried into the classifier through the tangential inlet, the large carrier particles, subjected to strong centrifugal forces, will be confined to the edge of the chamber, repeatedly impacting against the inner wall. As smaller drug particles are detached from the carriers through either drag forces or inertial forces arising from the carrier collisions, they will be pulled into the center of the chamber, under the discharge outlet, and removed from the classifier. Aggregated drug particles are broken up through collisions with both carrier particles and the classifier wall; the geometry of the classifier can be modified to alter the cut-off diameter of particle exiting the chamber [9, 28, 29].

While the major fraction of drug that will detach does so within the first half-second of inhalation, these are the particles that are subjected to the strongest removal forces, generally drug aggregates and large primary particles [30]. As the dose continues circumnavigating the classifier, a secondary population, primarily smaller drug particles located in higher sites on the carrier and less susceptible to removal forces, is further released; studies indicate that the amount of drug exiting the classifier over the time interval from 0.5 to 2 s following inhalation can exceed 50% of the amount released within the first half-second [31]. Moreover, low flow rates can be compensated by sustaining the detachment forces acting on the dose, and produce fine particle fractions comparable to higher flow rates. Examples of DPIs incorporating air classifiers as their de-agglomeration principle are provided by the Twincer™ and Novolizer™ [29, 32].

The Twincer™ is a single-dose device containing two air classifiers in parallel; a design that permits a large dose, up to 25 mg, for delivering antibiotics as treatment for cystic fibrosis [32]. Delivering the dose from a blister, the powder is entrained and divided between two separate channels, each feeding into a discreet classifier; two additional tangential air inlets per classifier provide the necessary turbulence to detach the drug from the carrier. That the carrier particles are retained within the classifier following actuation is of no consequence, as the inhaler is a low-cost device, formed from three layered, molded plastic plates, and is intended to be disposable following each use. In contrast, the Novolizer™ is a reservoir inhaler, and eliminating the deposition of drug and carrier within the classifier following inhalation becomes important [29]. To achieve this, the Novolizer™ employs an eight-sided classifier, where each opening is a tangential air inlet. In addition, the classifier walls are not isometric, consisting of four long sides to accelerate the carriers and four short sides to generate impactions, and distributed in an alternating pattern. Using this arrangement, the Novolizer™ can expel upwards 95% of the dose from the device following each actuation.

9.5.5 Vibration-Induced Dispersion

9.5.5.1 Capsule Vibrations

Though not always explicit, inhalers that employ capsules to deliver a pre-metered dose rely heavily on mechanical vibrations to facilitate powder de-agglomeration. An example of a DPI that induces rapid capsule oscillations is the Handihaler™. When a patient inhales through this device, the airflow enters the DPI opposite the mouthpiece, passing through a short, narrow inlet tube and abruptly opening into a larger volume capsule chamber [18]. As the flow stream emerges from the narrower passage into the larger, the flow decelerates, causing the boundary layer to separate from the inner walls of the inhaler and reattaching further upstream. This produces an annular region in the interim where fluid does not flow downstream, instead recirculating as a turbulent eddy resulting in a low pressure area [20]. The Handihaler exploits this phenomenon to rapidly vibrate the capsule during inhalation; the flow stream entering the larger chamber pushing the capsule forward while the low pressure region pulls its back. The capsule chamber was designed to be ample enough to allow the capsule room to oscillate, but sufficiently confined to prevent the capsule from tilting on its side and inadvertently occluding the inlet. As shown in studies with the Aerolizer™, the rapid capsule vibrations, coupled with the shear forces produced from passing the powder through the narrow perforations in the capsule wall, provides an effective means of powder de-aggregation [13].

9.5.5.2 Aeroelastic Vibrations

An alternative form of vibration-induced dispersion is a passive DPI (US Patent No. 11,713,180) that operates on the aerodynamic principle of "flutter." When an aeroelastic object is placed within the path of a flow stream, it begins to oscillate, the energy of which in turn feeds further oscillations, rapidly intensifying the amplitude and frequency of its vibration. When a powdered dose is placed on an aeroelastic film, it is effectively aerosolized, requiring very low threshold flow rates to induce flutter in the film [33].

9.5.5.3 Piezoelectric Driven Dispersion

The piezoelectric effect was initially observed in 1880 by the Curie brothers, who noticed that anisotropic crystals, i.e., crystals lacking a center of symmetry, emit an electrical signal when stressed. Conversely, the application of an electrical signal produces mechanical deformation in the crystal; when this electrical stimulus is supplied in the form of an oscillating potential, rapid crystal vibrations are generated [34]. Piezoelectric polymers have long found wide application in numerous fields, including as components of nebulizers for pulmonary drug delivery, and have recently been introduced as a dispersion mechanism in DPIs.

An example of the piezoelectric de-agglomeration principle is found in the Microdose inhaler (US Patent No. 6,026,809), where a drug-containing blister is brought into contact with a piezoelectric vibrator housed within the device. The DPI includes an air flow sensor that activates the piezoelectric element as the patient inhales, transferring mechanical energy to the blister and imparted into the dry powder formulation within to disperse the dose through the blister openings [35]. Additionally, the frequency at which the piezoelectric element operates is not arbitrary, but set to match the blister's resonance frequency. Briefly explained, the acoustic resonance of a system is its tendency to absorb more energy when oscillated at one of its own natural frequencies of vibration. By matching the frequency of the piezoelectric element to the resonance frequency of the blister, stronger oscillations can be generated without augmenting the effort required from the energy source.

An alternative application of the piezoelectric effect has been developed by Oriel Therapeutics, Inc. (US Patent No. 6,889,690), disclosing a multidose inhaler wherein the blisters containing the powder are comprised of a piezoelectric polymer material, incorporating the piezoelectric element into the blister itself. During inhalation, an electrical stimulus is provided to the blister, prompting the piezoelectric substrate to oscillate rapidly, vibrating the blister and ejecting the dose into the flow stream. Adjusting the shape of the blister can yield specific oscillation frequencies tailored to the flow characteristics of the formulation contained within; vibrating the blister to match the resonance frequency of the powder can optimize the dispersion potential [36]. Both of the above inhalers are active devices, requiring a self-contained power source to stimulate piezoelectric element.

9.5.6 *Electric and Magnetic Fields*

While less common, de-agglomeration principles capitalizing on the high charge-to-mass ratio of micronized powders are found in the patent literature. In one example (US Patent No. 6,089,227), a DPI is comprised of two separate chambers, the barrier between them containing a rotating cylinder with a portion of its perimeter located in each chamber. One chamber is a powder reservoir, and the other is in fluid communication with an air inlet and outlet. When an electric field is generated in the reservoir, the charged powder is attracted to the surface of the electrically neutral cylinder, which slowly rotates, exposing the adhered drug particles to a second electric field, opposite in polarity from the first, in the adjacent chamber. This prompts the drug to detach from the dosing drum and flock to the electrode producing the field, although as this detachment is coordinated with inhalation, the particles are carried to the patient.

US Patent No. 6,328,033 also discloses an inhaler that utilizes an electric field, albeit an oscillating one. In this device, the electric field is oscillated between the top and bottom of the powder-holding chamber, entraining the drug particles as they are rapidly drawn from one side to the other. The electric field is maintained while the patient inhales through the device, and the flow passing through the chamber conveys the dose downstream. An additional embodiment of this design relies on a magnetic field for de-agglomeration, wherein the dry powder formulation is coated onto particles having a magnetic core. Upon actuation, a rapidly oscillating magnetic field is applied between the upper and lower sections of the dose package, causing the larger magnetic particles to rapidly vibrate, expelling the powder from their surfaces where the flow stream carries the dose to the patient. The magnetic field is sustained throughout the duration of the inhalation maneuver, confining the magnetic particles within the device.

9.6 Device Resistance and Flow Rate

The role of resistance in DPIs, and the extent to which it influences device performance, is ambiguous. The equation relating flow rate (Q), device resistance (R), and the pressure drop (ΔP) across an inhaler was provided by Clark and Hollingsworth [37]:

$$\sqrt{\Delta \mathrm{P}} = \mathrm{QR}$$

For a given pressure drop, a higher flow rate will be generated through a lower resistance device. But how does device resistance translate into performance? When increasing the resistance in a DPI, either the air inlet is narrowed, or there is a constriction along the flow path, usually at the point where the fluid stream encounters the dose. As known from the flow continuity equation, reducing the cross sectional area increases the flow velocity, which in turn increases the kinetic energy carried by the flow stream; accordingly it seems intuitive that increased resistance would improve performance [20]. However, high resistance is not the only determinant of device performance. A comparison between the high resistance Pulvinal™ and the lower resistance Turbuhaler™, both passive inhalers, demonstrated better dispersion performance in the latter DPI [38]. Equally important is the method in which the kinetic energy from the flow rate is employed; in other words, the efficiency of the powder de-agglomeration principle. However, the importance of device resistance is being diminished through the introduction of inhalers that no long rely solely on a brief, almost instantaneous interaction between the flow stream and powder (ACT, cyclones), and active DPIs providing external energy sources.

Studies comparing low and high resistance passive inhalers generally conclude that while low resistance devices allow a much wider range of flow rates, the performance of high resistance inhalers exceeds that of its low resistance counterparts [39, 40]. However, high resistance devices show significant flow rate dependence, and the amount of delivered drug varies widely across the spectrum of generated flow rates. In contrast, low resistance devices provide a more consistent dose across a wider range of inspiratory efforts. Additionally, a significant patient population, especially children and the elderly, have difficulty generating a sufficient flow rate through the higher resistance devices [41].

Closely related to resistance, is the flow rate through the device. While higher flow rates improve the kinetic energy levels imparted to the dose, increasing the flow rate through an inhaler does not improve performance ad infinitum. As shown in studies with the Aerolizer™ there is a flow rate where the fine particle fraction is maximized and throat deposition is minimal; increasing the flow rate beyond this point hinders performance as a larger fraction of the dose will be deposited in the throat and mouth [14]. Accordingly, while high flow streams are desirable when encountering the dose, they are counterproductive if the velocity is not dampened as it exits the inhaler. One strategy to accomplish this is to widen the flow path in the mouthpiece, which significantly lowers the axial velocity of the flow stream as it exits the device, reducing throat and mouth deposition without comprising the turbulence levels [16]. Alternatively, other DPIs employ a coaxial sheath of air produced by flow bypass channels to form a buffer that surrounds the aerosol as it exits the device (Novolizer™), or incorporate baffles and curved paths to reduce flow velocity (Skyehaler™), although particle deposition may become problematic in these designs.

9.7 Meshes and Screens

Ubiquitous to many commercial DPIs is a mesh, or screen, situated between the powder-holding chamber and the inhaler exit, commonly tasked with performing the often overlooked, yet vital role of preventing the escape of the capsule, and capsule fragments, from the device during inhalation. However, it is also noted that meshes are integral components of inhalers that do not employ capsules, but rather disperse powder from blisters packs or reservoirs. The presence of the mesh can influence the flow field generated within an inhaler, both upstream and downstream of its location, and accordingly impact the overall performance of the device.

A mesh can be thought of as a distributed resistance that can effect a change in the flow direction of a fluid stream, coupled with a reduction in pressure [42], and can act as both a suppressor and generator of turbulence. In its role as a suppressor, the turbulence downstream of the mesh has been dampened in both scale and intensity subsequent to its passage through the mesh. In this case, the aim is obtaining a spatially uniform flow to both to avoid generating further turbulence, and to remove existing turbulence from the flow stream. Meshes can also serve to increase the turbulence downstream of their location up to 10% relative of the upstream value. In this case, the meshes are commonly coarse, with porosities exceeding 45% of the total area.

Reports in the literature examining the influence of meshes on inhaler performance are at odds with each other, casting meshes in the role of both spectator and active participant in affecting device performance. A study investigating the influence of a mesh on carrier particle dispersion concluded that no benefit to overall performance (as measured by the FPF_{total}) was obtained when powder was passed through a mesh, compared to the absence of the mesh [26]. Any small improvement in particle de-aggregation the mesh may have provided was offset by the drug retained on the mesh ($\approx$5% of the nominal dose), resulting in no significant difference in performance. In contrast, Coates et al., examining the effect of the grid inside the Aerolizer™ DPI, concluded that the grid does significantly affect overall inhaler performance. For these studies, dispersion performance of the Aerolizer™ outfitted with three different grids (the original grid geometry and two grids of increasing porosity) was compared, showing that the original aerolizer grid yielded no significant differences in the fine particle fraction of the emitted dose. This implies that the mesh had no effect on drug de-aggregation, as noted in Voss and Finlay, but rather exerted its influence through reducing the number of impactions between the mouthpiece and the inhaler, as a significantly higher fraction of drug deposits inside the inhaler mouthpiece with increasing grid voidage. In the Aerolizer™, the grid serves as a turbulence suppressor, straightening the flow by reducing the level of tangential flow generated in the device [12]. As the turbulence is suppressed downstream of the powder dispersion chamber, overall performance is unaffected. In reconciling these opposing views, the length of the inhaler downstream of the mesh becomes important, and the dispersion apparatus used by Voss and Finlay was not a commercial inhaler, but an experimental set up with an extended flow path downstream of the grid. In this case, the laminar flow created by the mesh may have degenerated into turbulence flow prior to exiting the device.

9.8 DPI Product Manufacturing

The remaining sections in this chapter focus upon considerations of DPI device design and, particularly, dispersion mechanisms utilized in the creation of a powder aerosol. This section describes in general terms the end-to-end manufacturing process involved to create a finished DPI drug product. By necessity, the precise manufacturing unit operations and process parameters will vary significantly according to the formulation, device type, and device design, therefore only generic unit operations and some trade-offs are presented here. The discussion also focuses on manufacturing considerations that should be made during product development.

Typically, inhalation drug formulations and devices are manufactured as separate supply chains brought together either at the point of filling, final assembly or, in the case of capsule DPI designs, in packaging. The overall value stream is often divided into primary, or drug substance, and secondary, or drug product unit operations.

Primary manufacture has as its goal, the production of an active drug substance (active pharmaceutical ingredients or APIs) with defined and controlled crystal form and habit and physicochemical properties. Key steps in primary manufacture from the perspective of a DPI product are the particle forming step and isolation of drug particles from solvents used in synthesis [43]. Traditionally, particles formed in primary manufacture are well above respirable size. Particle engineering for inhalation has as its goal, the production of respirable size particles at the particle forming step [44]. Larger particles can be isolated by filter drying or centrifugation. Particle engineering of respirable size particles may require more complex isolation techniques such as spray drying or supercritical fluid evaporation as filtration may be inefficient or may result in the formation of strong aggregates [45].

For traditional small molecule products, micronization is typically the unit operation bridging primary and secondary manufacture. A number of particle size reduction techniques and equipment are available with air jet milling being widely used to achieve attrition of particles down to respirable size. Air jet milling relies on particle–particle and particle–wall collisions to fracture particles. Air jet milling is a high energy process that can result in creation of surface amorphous content and high surface energy as well as triboelectrification [46]. To avoid downstream variability, control of these effects is often accomplished by relaxation through quarantine or active conditioning [47]. Scale-up from development must take into account process changes that may occur due to build up of product on the surfaces of the micronizer during extended production batch runs as compared to smaller batches produced during development.

As inhalation drug substances are typically highly potent and bioavailable due to their size, often with therapeutic doses in the low microgram range, personnel protection must be factored into the manufacturing process and isolation can add significantly to the capital cost. The most intensive exposure control is required at micronization and at dispensing since the product is aerosolized in its pure form at these steps. As dry products of small size, careful consideration of explosion control must also be taken [48].

For highly potent respirable compounds, process handling and device metering would be significantly complicated if the API were not blended with a bulking agent. Therefore, production of the majority of DPI products entails physical mixing of the active drug substance with one or more excipients to form a homogeneous powder blend. Typically, micronized API is mixed with a comparatively coarse carrier (e.g., lactose) [49].

The most common types of blenders used in pharmaceutical production are tumbling (low shear) and convective (high shear) blenders. With a low shear process, the presence of even mildly cohesive material can significantly complicate the motion patterns of the powder within the blender resulting in poor mixing characteristics. Due to the typically cohesive nature of DPI powders, they do not lend themselves easily to a low shear process [50]. The convective blender is composed of a stationary vessel and rotating impeller that moves and stirs the powder. The goal of all blending approaches is the production of a homogenous blend with adequate adhesion between the drug and excipient to avoid downstream segregation while still allowing drug and excipient particles to be separated during aerosolization [51]. An approach to the scale-up of process parameters from development to manufacturing is equivalent blend energy which is a function of the blending speed, time, and blend size [50]. Triboelectrification may occur in the blending process and impact product performance. The degree of charging will be a function of the input materials, process humidity, and process equipment materials of construction [52]. These effects should be understood as equipment contact area, environmental controls and cleaning regimens may change during scale-up from the laboratory to a production environment [53]. Product shelf life will often begin at the point of blending so it is desirable to minimize all subsequent process times in order to maximize the saleable lifetime of the product.

Blended formulations must subsequently be filled either into the DPI metering system or into intermediate packages. Foil blisters and capsules are typical factory-metered containment systems [4]. For multidose systems, significant trade-offs are made in the design of the containment system and blend strength where ease of handling and filling increase as the weight filled and size of container increases, but this comes at the expense of device size. Examples of powder metering principles used to fill DPI devices are vibration feeding, dosator, rotary drum, and immersion filling [54]. Compaction is utilized in many of these systems to increase fill weight control through elimination of voids but as compaction increases the energy required to aerosolize the dose also increases [55]. Filling systems will often be bespoke designs matched to the DPI dose package. During early product development, these systems are likely to also be in development. Therefore, scale-up will require the understanding of filling effects on product performance and the relationship between the development and full scale filling systems.

Assembly of the device and dose containment system may be complex depending on the number and interaction of components and the degree to which devices can be subassembled prior to the introduction of the drug containment. In any case, highly automated assembly is required for a high volume product. The device is likely to be the highest cost component of the drug product. Robust manufacturing process understanding and in-process testing of the unit operations prior to assembly is highly desirable in order to minimize the risk of batch failure after committing the filled product to completed devices.

Since many inhalation actives as well as lactose are hygroscopic, product performance may be impacted by storage conditions [56]. Moisture ingress protection is a consideration for these formulations and may be provided by the formulation containment materials, an overwrap or an included desiccant. Supply chain storage and in-use moisture protection must be considered also recognizing that many patients may store their inhalers in a high humidity location such as a bathroom [4].

The typical DPI device will be fabricated of plastic components with manufacturing outsourced to a third party with molding expertise. Device design is discussed in subsequent chapters but it is worth stressing that while very complex designs can be fabricated in principle, these may not be practical to manufacture with the required consistency at volume and at a price that can be supported by the patient or formulary. Materials of construction must be considered for their cost, potential for extractables or leachables, and impact on device performance due to factors such as electrostatic attraction of API to the plastic [57, 58]. Consideration should also be given to sourcing and availability of materials. For even a large volume DPI product, the market leverage in the overall materials market will be low. Suppliers may change or discontinue materials formulations. Security of supply and dual sourcing are important considerations particularly for performance critical and patient or drug contact materials. Device refilling carries complexities of hygiene and assurance of proper assembly so most DPI devices are disposable. This requirement dictates the value of componentry included in the DPI. For electronic device designs, waste electronics disposal regulations also must be considered [59].

A comprehensive review of manufacturing unit operations, process parameters, and materials of construction would be the subject of multiple specialty texts. This section has attempted to capture considerations and trade-offs common to many typical DPI products and to suggest early consideration of these during DPI product development.

9.9 Conclusion

A diverse array of technologies have been designed to address the challenges of effectively aerosolizing a dry powder formulation. Recent years have seen an increase in the number of active DPIs in development, although passive devices, with their low costs and simple designs, will undoubtedly remain an active area of inhaler research, as they are especially attractive as disposable platforms for dry powder vaccine delivery. However, while much effort has been devoted to improving the powder dispersion performance of inhalers, the ability of the device to optimize inhalation therapy must be placed into perspective, as no matter how cleverly designed and masterfully crafted an inhaler may be, it cannot overcome patient misuse or a poorly prepared formulation. Future developments in DPIs will benefit greatly from parallel advancements in patient education and powder formulation technology.

9.10 Appendix

9.10.1 Patents

US Patent No. 11,713,180
US Patent No. 6,889,690

US Patent No. 6,138,673
US Patent No. 7,107,988
US Patent No. 6,237,591
US Patent No. 6,971,384
US Patent No. 5,655,523
US Patent No. 5,724,959
US Patent No. 5,775,320
US Patent No. 6,257,233
US Patent No. 6,182,655
US Patent No. 5,087,710
US Patent No. 6,328,033
US Patent No. 5,875,776
US Patent No. 7,334,577
US Patent No. 6,561,186
US Patent No. 6,026,809
US Patent No. 6,089,227

References

1. Frijlink HW, de Boer AH (2005) Trends in the technology-driven development of new inhalation devices. Drug Discov Today Technol 2:47–57
2. Smyth HDC, Leach CL (2005) Alternative propellant aerosol delivery systems. Crit Rev Ther Drug Carrier Syst 22:493–534
3. Islam N, Gladki E (2008) Dry powder inhalers (DPIs) – a review of device reliability and innovation. Int J Pharm 360:1–11
4. Ashurst I, Malton A, Prime D, Sumby B (2000) Latest advances in the development of dry powder inhalers. Pharm Sci Technol Today 3:246–256
5. Crowder TM, Louey MD, Sethuraman VV, Smyth HDC, Hickey AJ (2001) An odyssey in inhaler formulation and design. Pharm Technol 7:99–107
6. Frijlink HW, De Boer AH (2004) Dry powder inhalers for pulmonary drug delivery. Expert Opin Drug Deliv 1:67–86
7. Chow A, Tong H, Chattopadhyay P, Shekunov B (2007) Particle engineering for pulmonary drug delivery. Pharm Res 24:411–437
8. Finlay WH (2001) The mechanics of inhaled pharmaceutical aerosols. Academic, London
9. de Boer A, Hagedoorn P, Gjaltema D, Goede J, Frijlink H (2003) Air classifier technology (ACT) in dry powder inhalation Part 1. Introduction of a novel force distribution concept (FDC) explaining the performance of a basic air classifier on adhesive mixtures. Int J Pharm 260:187–200
10. Hickey AJ (2004) Pharmaceutical Inhalation aerosol technology. Marcel Dekker, New York, NY
11. Edwards DA, Hanes J, Caponetti G, Hrkach J, Ben-Jebria A, Eskew ML, Mintzes J, Deaver D, Lotan N, Langer R (1997) Large porous particles for pulmonary drug delivery. Science 276: 1868–1872
12. Coates MS, Fletcher DF, Chan H-K, Raper JA (2004) Effect of design on the performance of a dry powder inhaler using computational fluid dynamics. Part 1: grid structure and mouthpiece length. J Pharm Sci 93:2863–2876
13. Coates MS, Fletcher DF, Chan H-K, Raper JA (2005) The role of capsule on the performance of a dry powder inhaler using computational and experimental analyses. Pharm Res 22: 923–932

14. Coates MS, Chan H-K, Fletcher DF, Raper JA (2005) Influence of air flow on the performance of a dry powder inhaler using computational and experimental analyses. Pharm Res 22:1445–1453
15. Coates MS, Chan H-K, Fletcher DF, Raper JA (2006) Effect of design on the performance of a dry powder inhaler using computational fluid dynamics. Part 2: air inlet size. J Pharm Sci 95: 1382–1392
16. Coates MS, Chan H-K, Fletcher DF, Chiou H (2007) Influence of mouthpiece geometry on the aerosol delivery performance of a dry powder inhaler. Pharm Res 24:1450–1456
17. Brambilla G, Cocconi D, Armanni A, Smith S, Lye E, Burge S (2006) Designing a novel dry powder inhaler: the NEXTTM DPI (Part 1). Respir Drug Deliv 2:553–555
18. Wachtel H, Ertunc O, Koksoy C, Delgado A (2008) Aerodynamic optimization of Handihaler and Respimat: the roles of computational fluid dynamics and flow visualization. Respir Drug Deliv 1(2008):165–174
19. Mack GS (2007) Pfizer dumps Exubera. Nat Biotech 25:1331–1332
20. Gupta V, Gupta SK (1984) Fluid mechanics and its applications. New Age International, New Delhi
21. Wetterlin K (1988) Turbuhaler: a new powder inhaler for administration of drugs to the airways. Pharm Res 5:506–508
22. de Boer AH, Gjaltema D, Hagedoorn P (1996) Inhalation characteristics and their effects on in vitro drug delivery from dry powder inhalers Part 2: effect of peak flow rate (PIFR) and inspiration time on the in vitro drug release from three different types of commercial dry powder inhalers. Int J Pharm 138:45–56
23. Needham M, Fradley G, Cocks P (2010) Investigating the efficiency of reverse cyclone technology for DPI drug delivery. Respir Drug Deliv 2:369–372
24. Stein S, Hodson D, Alband T, Sitz R, Robison T, Wang Z, Chiou H, Simons J, McNally R, Ganser J (2010) The 3M™ taper dry powder inhaler device. Respir Drug Deliv 2:377–380
25. Aydin M, Akouka H, Becker D, Merrill T, Reynolds E, Shukla R, Gumaste A, and Byron DA (2008) Application of synthetic jetting for pulmonary delivery of drug candidates. Respir Drug Deliv 2:635–640
26. Voss A, Finlay WH (2002) Deagglomeration of dry powder pharmaceutical aerosols. Int J Pharm 248:39–50
27. Bisgaard H, Klug B, Sumby BS, Burnell PKP (1998) Fine particle mass from the Diskus inhaler and Turbuhaler inhaler in children with asthma. Eur Respir J 11:1111–1115
28. de Boer A, Hagedoorn P, Gjaltema D, Goede J, Kussendrager K, Frijlink H (2003) Air classifier technology (ACT) in dry powder inhalation. Part 2. The effect of lactose carrier surface properties on the drug-to-carrier interaction in adhesive mixtures for inhalation. Int J Pharm 260:201–216
29. de Boer A, Hagedoorn P, Gjaltema D, Goede J, Frijlink H (2006) Air classifier technology (ACT) in dry powder inhalation. Part 3. Design and development of an air classifier family for the Novolizer® multi-dose dry powder inhaler. Int J Pharm 310:72–80
30. de Boer AH, Hagedoorn P, Gjaltema D, Lambregts D, Irngartinger M, Frijlink HW (2004) The mode of drug particle detachment from carrier crystals in an air classifier-based inhaler. Pharm Res 21:2167–2174
31. de Boer AH, Hagedoorn P, Gjaltema D, Lambregts D, Irngartinger M, Frijlink HW (2004) The rate of drug particle detachment from carrier crystals in an air classifier-based inhaler. Pharm Res 21:2158–2166
32. de Boer A, Hagedoorn P, Westerman E, Lebrun P, Heijerman H, Frijlink H (2006) Design and in vitro performance testing of multiple air classifier technology in a new disposable inhaler concept (Twincer®) for high powder doses. Eur J Pharm Sci 28:171–178
33. Selvam P, McNair D, Truman R, Smyth HDC (2010) A novel dry powder inhaler: effect of device design on dispersion performance. Int J Pharm 401:1–6
34. Eggins BR (2002) Chemical sensors and biosensors. Wiley, West Sussez
35. Brown B, Rasmussen J, Becker D, Friend DR (2004) A piezo-electronic inhaler for local and systemic applications. Drug Deliv Technol 4:90–93
36. Crowder T, Hickey A (2006) Powder specific active dispersion for generation of pharmaceutical aerosols. Int J Pharm 327:65–72

37. Clark AR, Hollingworth AM (1993) The relationship between powder inhaler resistance and peak inspiratory conditions in healthy volunteers – implications for in vitro testing. J Aerosol Med 6:99–110
38. Meakin BJ, Ganderton D, PANZA I, Ventura P (1998) The effect of flow rate on drug delivery from the pulvinal, a high-resistance dry powder inhaler. J Aerosol Med 11:143–152
39. Srichana T, Martin GP, Marriott C (1998) Dry powder inhalers: the influence of device resistance and powder formulation on drug and lactose deposition in vitro. Eur J Pharm Sci 7:73–80
40. Mendes P, Pinto J, Sousa J (2007) A non-dimensional functional relationship for the fine particle fraction produced by dry powder inhalers. J Aerosol Sci 38:612–624
41. Johannes HW, Nigel DD, Joyce MW, Sunalene GD, Peter NL (1999) Inhalation therapy in asthma: Nebulizer or pressurized metered-dose inhaler with holding chamber? In vivo comparison of lung deposition in children. J Pediatr 135:28–33
42. Laws EM, Livesey JL (1978) Flow through screens. Annu Rev Fluid Mech 10:247–266
43. Lu J, Rohani S (2009) Polymorphism and crystallization of active pharmaceutical ingredients (APIs). Curr Med Chem 16:884–905
44. Iacocca RG, Burcham CL, Hilden LR (2010) Particle engineering: a strategy for establishing drug substance physical property specifications during small molecule development. J Pharm Sci 99:51–75
45. Sollohub K, Cal K (2010) Spray drying technique: II. Current applications in pharmaceutical technology. J Pharm Sci 99:587–597
46. Ward GH, Schultz RK (1995) Process-induced crystallinity changes in albuterol sulfate and its effect on powder physical stability. Pharm Res 12:773–779
47. Perkins MC, Bunker M, James J, Rigby-Singleton S, Ledru J, Madden-Smith C, Luk S, Patel N, Roberts CJ (2009) Towards the understanding and prediction of material changes during micronisation using atomic force microscopy. Eur J Pharm Sci 38:1–8
48. Jacobson R (2005) Inert milling systems. Pharm Eng. 25
49. Podczeck F (1998) The relationship between physical properties of lactose monohydrate and the aerodynamic behaviour of adhered drug particles. Int J Pharm 160:119–130
50. Bridson RH, Robbins PT, Chen Y, Westerman D, Gillham CR, Roche TC, Seville JPK (2007) The effects of high shear blending on [alpha]-lactose monohydrate. Int J Pharm 339:84–90
51. Staniforth JN (1995) Performance-modifying influences in dry powder inhalation systems. Aerosol Sci Technol 22:346–353
52. Eilbeck J, Rowley G, Carter PA, Fletcher EJ (2000) Effect of contamination of pharmaceutical equipment on powder triboelectrification. Int J Pharm 195:7–11
53. Rowley G, Mackin LA (2003) The effect of moisture sorption on electrostatic charging of selected pharmaceutical excipient powders. Powder Technol 135–136:50–58
54. Crowder TM (2007) Precision powder metering utilizing fundamental powder flow characteristics. Powder Technol 173:217–223
55. Tuley R, Shrimpton J, Jones MD, Price R, Palmer M, Prime D (2008) Experimental observations of dry powder inhaler dose fluidisation. Int J Pharm 358:238–247
56. Maggi L, Bruni R, Conte U (1999) Influence of the moisture on the performance of a new dry powder inhaler. Int J Pharm 177:83–91
57. US Food and Drug Administration (1998) Guidance for industry: metered dose inhaler (MDI) and dry powder inhaler (DPI) drug products, chemistry, manufacturing and controls documentation. US Department of Health and Human Sciences, FDA Center for Drug Evaluation and Research (CDER)
58. Podczeck F (1998) Evaluation of the adhesion properties of salbutamol sulphate to inhaler materials. Pharm Res 15:806–808
59. Widmer R, Oswald-Krapf H, Sinha-Khetriwal D, Schnellmann M, Böni H (2005) Global perspectives on e-waste. Environ Impact Assess Rev 25:436–458

Chapter 10
Science and Technology of Nebulizers and Liquid-Based Aerosol Generators

Aileen Gibbons and Hugh D.C. Smyth

Abstract Nebulizers were originally developed for and remain useful for particular clinical circumstances, and proof of concept trials. Recently however, due to considerable improvements on dose delivery and portability, nebulizers may find broader application for different patient groups and disease states. In this chapter, traditional and newer nebulizer technologies are reviewed. In addition, practical aspects relating to formulation issues and performance of these different systems are presented. Increasing nebulizer efficiencies and the integration of new mechanisms of aerosolization are included. With the ongoing advancement of nebulizer systems, we should see improvements in clinical outcomes and meet the needs of novel targeted therapies with precision-based, reproducible, and efficient nebulizer devices.

Keywords Nebulizer • Aerosol • Jet nebulizer • Ultrasonic nebulizer • Metered dose liquid inhaler • Adaptive aerosol delivery

10.1 Introduction

The first atomized inhaler was developed in France in 1858 by Sales-Girons, using pressure from a manual pump to draw liquid through a reservoir. This was followed closely in 1864, by a steam driven nebulizer, the Siegle Steam spray inhaler [60]. Complete portability was introduced in the 1950s with the advent of pMDI devices, since then nebulization therapy has been reserved mainly for a select subset of patients who are either too ill or too young to use pMDI or DPI devices effectively.

A. Gibbons (✉)
College of Pharmacy, The University of Texas at Austin,
Mailstop A1920, Austin, TX 78712, USA
e-mail: agibbons@mail.utexas.edu

H.D.C. Smyth and A.J. Hickey (eds.), *Controlled Pulmonary Drug Delivery*,
Advances in Delivery Science and Technology, DOI 10.1007/978-1-4419-9745-6_10,

However, nebulizers play a critical role when used, they can deliver the large doses of therapeutic agent, such as antibiotics and mucolytics, necessary to treat severe infections in cystic fibrosis [32]. Such large doses would not be conveniently delivered via pMDI or DPI inhalers [25]. For example, Tobramycin used as a topical antibiotic for Pseudomonas endobronchial infections in cystic fibrosis is delivered in 300 mg unit doses [25, 33, 71]. Like Tobramycin, nebulization has been used to deliver a wide range of other locally acting medications such as gentamycin [35], albuterol [39], budesonide [1], saline [69], and more recently, dornase alpha (rhDNase) [34].

10.2 Nebulizer Uses, Advantages, and Limitations

The advantages of nebulizers set them apart from other classes of device. Unlike pMDIs, nebulizers do not require coordination of actuation and inhalation. Nebulization devices are quite intuitive to use, and as such, require little or no training in their use. Users of other devices, particularly pMDI devices, on the other hand, require initial training on the correct use of their inhaler, as well as regular retraining, to ensure optimal benefit is derived from their pMDI [65]. Also, unlike passive DPI devices that are flow-rate dependant, nebulizers require only tidal breathing to sufficiently entrain droplets into the lungs. The phasing out of CFC use due to the Montreal Protocol induced a surge in research based around replacing existing CFC-based pMDI devices with less ozone-depleting HFA-pMDIs and looking for DPI devices for another avenue of aerosol formulation to fill the gap [59]. Despite this resurgence in aerosol formulation and device enhancement, the advancement of nebulization devices was largely neglected. However, technical complexities commonly associated with formulating active pharmaceutical ingredients for pMDI and DPI devices, and the costs involved [64], has led to an increased interest toward the development and advancement of nebulizer devices.

When screening and developing new APIs for topical or systemic delivery to the lungs, nebulization systems allow quicker realization of the product from "bench to bedside." In the early stages of drug development, aerosolization through nebulization is the most straightforward route to realizing clinical outcome, primarily due to the ease with which a solution or suspension for nebulization can be formulated without the processing and formulation complexities inherent in pMDI and DPI development [64].

It is often the case that novel API's will first be tested in investigative clinical trials using a nebulizer device, before assessing the benefits, if any, of reformulating the drug for another device. Many investigational studies have been conducted using nebulizers in early clinical trials of novel therapeutics directed at lung delivery for local or systemic effect [7, 8, 26, 43].

Despite their importance in inhalation therapy, nebulizers have certain disadvantages associated with their use including: noise generation during operation (60 dB for jet nebulizers), bulky and cumbersome design, lengthy administration times, and lack of portability. In addition, performance efficiency of the available nebulizers is variable and depends on numerous factors, e.g., nebulizer gas flow for jet nebulizers, fill volume of the nebulizer reservoir, as well as the physicochemical attributes of the drug under nebulization, such as viscosity for ultrasonic nebulizers [58]. It can be surmised, therefore that accurate predictions cannot be made on the drug available for lung deposition, unless that particular device has been studied with the drug undergoing nebulization [25].

Increasingly, formulation difficulties, inefficient dose delivery, and progressively complex device designs incumbent in pMDIs and DPIs have led aerosol researchers to seek out more cost-effective and efficient device solutions. The early nebulization devices are an attractive alternative to pMDIs and DPIs for a simple and immediate in vitro and in vivo study of liquid-based APIs; however, recent developments of novel expensive APIs require a delivery device that will (a) minimize drug loss, (b) increase dosing efficiency, and (c) facilitate optimal targeting to the desired regions of the airways, for either local or systemic delivery. Early nebulization devices could not fulfill such criteria, but newer nebulization devices have addressed many of the issues traditionally associated with the early prototypes and are becoming an attractive route for aerosolization of novel therapeutics and optimizing the delivery of older therapeutics.

10.3 The Jet Nebulizer

The conventional jet nebulizer has been used for nearly a century and is a well-established alternative for pMDI and DPI devices in the treatment of a particular patient population where the use of alternative devices would be inhibitive, or who require large bolus doses of a therapeutic agent. Jet nebulization occurs when compressed air is forced through a narrow orifice within the nebulizer creating a negative pressure. This draws liquid upward through a feeder tube by the Bernoulli effect. Upon entering the air stream, the liquid is broken up by air turbulences within the jet, and also by impaction upon one or more baffles within the nebulizer. Small droplets leave the nebulizer in an aerosol cloud, while large droplets remain in the nebulizing chamber by way of inertial impaction or the baffle and will undergo renebulization [44]. This renebulization occurs to more than 90% of molecules nebulized [3, 15, 35]. Baffles facilitate the generation of a slow aerosol, which can be entrained into the inspiratory air flow and delivered to the lungs. Without this impaction mechanism, the velocity of the aerosol would be such that impaction in the throat would certainly occur [50]. This impaction of large droplets on baffles within the reservoir of the nebulizer as well as the recycling of liquid, help to control the size, velocity, and volume flow of the emerging droplets [50]. An example of a conventional jet nebulizer device is the Acorn II® (Marquest, Englewood, CO, USA) (Fig. 10.1).

Certain variable factors can affect the performance of the jet nebulizer, such as the physical properties of the liquid, most notably its viscosity, the volume of liquid present in the reservoir, as well as external factors like temperature and humidity. Manufacturer's specifications for particle size distribution are usually derived from nebulizing normal saline solution. However, it is has been shown that liquids of different viscosities can significantly affect the particle size of droplets produced. In a

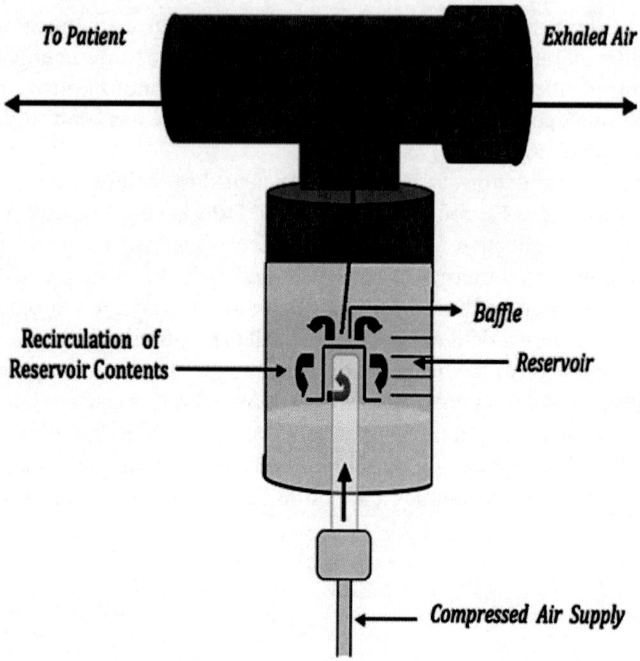

Fig. 10.1 A schematic of a conventional nebulizer, the Acorn II® Jet Nebulizer

study comparing the nebulization of Gentamicin solution and using four different jet nebulizers, to nebulized normal saline solution, the antibiotic solution was found to alter the particle size distribution. In addition, more than half of the drug remained in the nebulizer at the end of nebulization [47].

Generally, nebulizer devices will be operated until maximum output, i.e., until the reservoir runs dry, or the "dead volume" is reached and generation of an aerosol no longer occurs [12, 63]. Since jet nebulization relies on the interaction of the liquid and gas phase for aerosol generation, dramatically altering the proportion of one of these components will impact on the quality of the aerosol generated. This is often witnessed toward the end of aerosol generation when the liquid level falls below a certain threshold in the reservoir, causing an inconsistent spray (sputtering) [40]. Changes in solution temperature and humidity of the dilution air during nebulization can influence the size and solute concentration of the aerosol droplets produced [40]. Cooling and concentration effects of jet nebulizers on the reservoir liquid are well known [55]. Heat loss occurs due to the evaporation of the nebulizer solution to saturate the gas used to generate the aerosol as well as cooling due to adiabatic expansion of the generating gas [45]. This in turn leads to increase in solute concentration and potential for precipitation.

A major concern associated with the use of jet nebulizers for aerosolizing expensive therapeutics is the high potential for waste in these devices. This occurs mainly due to the large portion of liquid that remains unavailable for nebulization, i.e., the "dead volume," which can be as much as 2 mL. Liquid adhesion to reservoir surfaces is also a known issue for jet nebulizer output to the extent that physicians will routinely advise operators to "tap" the walls of the reservoir to detach the droplets [20]. For unvented jet nebulizers, drug is inhaled during inspiration and lost to the environment during expiration [11]. Loss of aerosol to the environment can lead not only to suboptimal dosing for the patient, but also involuntary exposure to care givers. In a study examining passive inhalation of Pentamidine by caregivers in a hospital, significant levels of Pentamidine were found in urine samples of the hospital staff [6]. Newer jet nebulizer designs have minimized such passive exposures by building a vent into the device which allows for maximized inspiratory delivery and minimized drug loss.

Jet nebulizer design has not significantly advanced over the years, although some of the principle shortcomings have been addressed. To reduce the accumulation of liquid droplets in the inner surfaces of the jet nebulizer reservoir, considerations have been made to alter the surface material used in their manufacture. Incorporating materials known to enhance wettability should reduce attachment of droplets to the device walls, allowing more liquid to become available for nebulization [50]. Reducing the internal surface area will also reduce the area available for attachment. Device dimensions have also been addressed to minimize the other issue of dead volume, and in so doing create aerosol-conserving reservoirs. For example, in the Pari IS-2 device an inverted cone is used to maximize the depth of the fluid close to the liquid feed jet [14, 37].

Implementation of breath-enhanced, or vented, nebulization in the newer jet nebulizer models, as mentioned previously, has improved aerosol output by entraining the patient's inspiratory flow into the device, and minimizing loss of particles that would otherwise be expired [13]. This also allows for increased rate of output with increasing inspiratory flow, with output falling back to baseline during expiration when no flow is entrained, e.g., Pari LC® Star (Pari GmbH, Starnberg, Germany) [38]. Other jet nebulizer devices use breath-actuation technology, designed to produce an aerosol only upon inspiration, allowing for better drug conservation than the breath-enhanced, e.g., the AeroEclipse II® BAN (Trudell Medical International, London, Canada). In a study comparing the rate and amount of deposition as well as consistency of dose delivered, of breath-enhanced and breath-actuated jet nebulizers, Leung et al demonstrated high respirable fractions for both nebulizers, the rate of deposition was greatest for the breath-enhanced Pari LC Star®, while the AeroEclipse® was the least wasteful of the two [38].

The choice of nebulizer device depends on the formulation to be nebulized. Jet nebulization has been known to disrupt liposome carriers [28, 53] and also to destabilize some macromolecular APIs [2]. In a study where the protein recombinant secretory leukocyte protease inhibitor (rSLPI) was encapsulated in liposomes, nebulization using the Acorn II® jet nebulizer caused disruption of the liposomes to the extent that $41.1 \pm 3.8\%$ of the encapsulated protein leaked into the continuous phase due to the nebulization process [28]. However, no degradation of the protein was observed due to jet nebulization in this instance [28]. Concerns regarding jet nebulization of proteins relates to the fact that more than 90% of the formulation is ultimately refluxed, which can cause denaturation of the protein due to repeated stress and exposure of proteins to air–water interfaces [51]. However, proteins have been nebulized successfully without detriment to the active moiety, e.g., rDNase [34]. Moreover, jet nebulization has been observed in some studies to be less likely than ultrasonic nebulizers to cause destabilization of proteins [62, 64].

10.4 The Ultrasonic Nebulizer

Jet nebulizer devices dominated the nebulizer market until the introduction of a novel and more portable device, the electronic ultrasonic wave nebulizer in 1964 [60]. Ultrasonic nebulizers incorporate a piezoelectric crystal that vibrates at high frequencies (1–3 MHz) to generate a fine slow-moving mist of droplets. Unlike the jet nebulizer, these devices do not require a supply of compressed air and so are more portable; however, they still require a source of electricity for power. Oscillation of the piezo crystal induces the formation of waves or fountains in the liquid reservoir that propagates to the surface. Droplet formation occurs by capillary wave formation and cavitation. Cavitation bubbles are formed at low frequencies while capillary wave formation occurs at higher frequencies. At low frequencies, the cavitation theory proposes that the piezo crystal vibrates the liquid thus forming cavitation bubbles. Cavitation bubbles are formed due to negative pressure produced by the sound waves which causes some vapor in the liquid to come out of the solution in the form of bubbles [23]. As the air bubbles rise to the surface the internal pressure of the bubbles equilibrates with the atmosphere, causing implosion of the bubbles. As the bubbles collapse, part of the liquid breaks free and forms droplets. At high frequencies, the capillary wave theory proposes that the piezo oscillations produce small waves that interfere to form peaks at the air–liquid interface where they become unstable, break free, and form droplets [4].

Ultrasonic nebulizers have proven invaluable when delivering aerosols to patients under mechanical ventilation, as the delivered volumes are much less than that required for jet nebulization, particularly for the newer generation of vibrating mesh ultrasonic nebulizers which deliver as little as 0.5 mL. In addition, significant changes to ventilation circuits that are required when jet nebulizers are used are not necessary when using the ultrasonic nebulizer devices [10]. Ultrasonic nebulizers overcome many of the major disadvantages associated with jet nebulizers. In terms of aiding patient compliance and acceptability, these devices are less bulky, less cumbersome, length of treatment time is reduced, and nebulization is relatively quiet. The issue of product waste is surmounted; dead volume is not an issue for these devices, and the volumes of liquid nebulized are comparatively small, and can be as low as 0.5 mL. Also, the surface area within the reservoir is minimal, and this prevents large amount of liquid depositing on the surface walls. Ultrasonic nebulizers generally have a higher output rate, but this may be achieved with the production of larger median droplet diameters in older models [30, 58].

As with many other inhalation devices, various device design and formulation factors can influence the mass and particle size produced from the ultrasonic systems, such as the viscosity, density, vapor pressure, and surface tension of the liquid, the frequency employed to activate the piezo crystal, the size and dimensions of the nebulizing chamber, and the presence or absence of baffles [30]. The effect of drug concentration and cooling, as is seen in jet nebulization, does not occur in ultrasonic nebulizers, although heating of the reservoir liquid has been shown to occur with temperatures rising up to 20°C above ambient temperature [15, 55, 58].

Due to the high frequencies employed in the operation of these devices and with that, the potential for heat generation, ultrasonic nebulization can degrade heat labile drugs, for example in the case of lactate dehydrogenase, which was rendered completely inactive after 20 min of nebulization using the DeVilbiss Aerosonic® nebulizer [52]. There is evidence to show, with respect to certain liposomal carrier-mediated delivery systems, that ultrasonic nebulizers are less disruptive to the carrier compared with jet nebulization [56]; however, older ultrasonic nebulizers are not very efficient at aerosolizing viscous formulations [42] and depending on the liposome composition, may disrupt liposomes causing the active to leak out into the continuous phase [9]. Aerosolization of suspensions may also prove problematic in conventional ultrasonic nebulizers. In a study where two formulations, a budesonide suspension and a Terbutaline solution, were nebulized in two different nebulizers, the Spira Elektro 4® jet nebulizer (Respiratory Care Centre, Hämeenlinna, Finland) and the Spira Ultra® (Respiratory Care Centre, Hämeenlinna, Finland) ultrasonic nebulizer, it was concluded that due to the poor output of budesonide suspension from the ultrasonic nebulizer relative to the jet nebulizer, that budesonide suspension should not be nebulized using conventional ultrasonic nebulizers [49].

Recent developments in new generation ultrasonic nebulizers provide an alternate strategy to improve the ultrasonic nebulizer's ability to aerosolize suspensions and viscous liquids more effectively by generating the aerosol mechanically using a vibrating mesh [18, 19], e.g., Aeroneb® (Aerogen Ltd, Galway, Ireland) (Fig. 10.2), and PARI eFLOW®. These new generation ultrasonic nebulizers enable high respirable fractions, low velocity aerosols by synchronizing a laser-bored mesh plate with the piezo crystal to produce fine mists of droplets [70]. The technology works by oscillations of the mesh plate, which in turn pumps fluid from a small volume

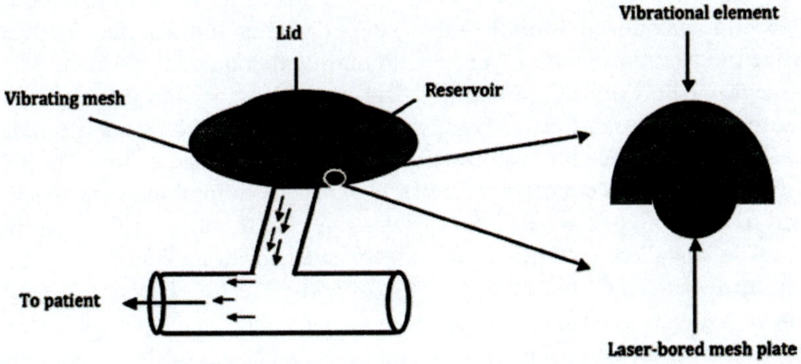

Fig. 10.2 Schematic of the Aeroneb Pro® vibrating mesh nebulizer, insert shows fine mesh plate with vibrational element (schematic adapted from Ghazanfari et al. [27])

reservoir through the tapered holes to generate droplets in the respirable size range. These nebulizers have been shown to effectively aerosolize sensitive classes of APIs and drug carriers such as recombinant proteins and liposomes [19, 70]. However, this class of nebulizer has also demonstrated a relationship between nebulization and formulation properties. Depending on the device used, the emitted dose may be lower, and the aerodynamic diameter of the aerosol droplet larger, than that obtained by jet nebulization [19, 27, 67]. The Omron Micro*Air*® VMT (Omron Healthcare, Bannockburn, IL, USA) generates droplets through passively vibrating mesh apertures. The ultrasonic vibrations in this device are passively conducted via a thin layer of liquid to the mesh plate. Due to the lower vibratory energy employed, formulation parameters such as liquid viscosity has a more profound effect on the operation of the vibration mechanism [27]. In a study by Ghazanfari et al., new generation ultrasonic devices were pushed to their limit in their ability to aerosolize viscous liquids. They demonstrated that at 2.75cP, the Omron® device ceased nebulization entirely, while the Aeroneb® continued, but with discontinuous aerosol production [27]. As with other aerosol devices, it is clear that cognizance of the physicochemical nature of the liquid to be aerosolized as well as early integration of the aerosol device is essential to effective drug development with ultrasonic nebulizers.

10.5 Optimizing Nebulizer Delivery

The cost to benefit ratio for employing add-on intelligent technology in nebulizer devices for aerosolization of inexpensive drugs would not be reasonable, as the developers would in essence price themselves out of the market. Currently, the area of delivering low cost medications is filled by jet nebulizers, which are considered the inexpensive mainstay of nebulization through their cost-effective housing of injection-molded plastic, which is either autoclavable or disposable. The recent introduction of “intelligent” nebulizers into the ever-expanding portfolio of nebulizer devices has consequently been born out of a need, not for optimizing delivery of preexisting nebulized formulations, but optimizing the accurate delivery of expensive therapeutic agents and those with narrow therapeutic windows.

As mentioned previously, earlier nebulization devices, such as jet nebulizers can be wasteful due to loss of drug arising from; environment losses upon exhalation; residual drug unavailable for nebulization in the reservoir “dead volume” or walls of the reservoir; as well as concentration effects of drug solutions and suspensions that can lead to clogging or the nebulizer systems over the duration of the nebulization process. The so-called “intelligent” nebulizers endeavor to maximize drug delivery and minimize drug loss. To achieve this, the Adaptive Aerosol Delivery AAD nebulizers have been designed to continuously adapt to changes in the patient’s breathing pattern, to deliver the aerosol at the most appropriate periods during patient inspiration when generation of the aerosol will be maximally delivered to the lungs. By monitoring and actively adapting to changes in patient’s breathing patterns, AAD devices enable more accurate and less variable delivery of drugs to the lungs [16].

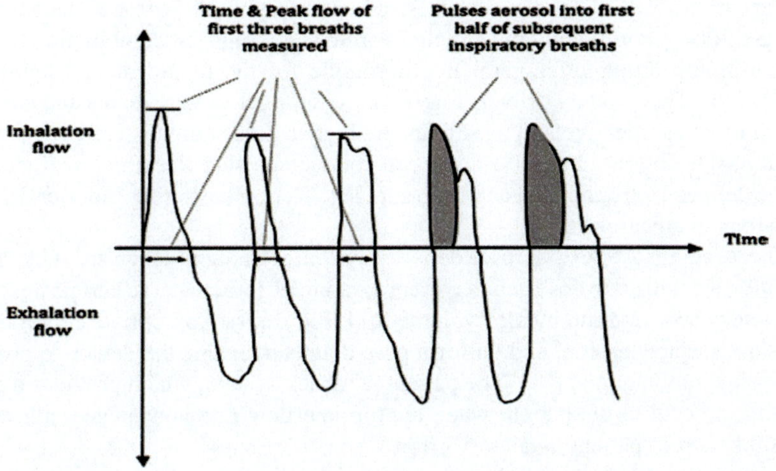

Fig. 10.3 Breath tracing showing how the Adaptive Aerosol Devices monitor and use the peak flow of the first three inhalations, to determine the duration of aerosol production needed to target the beginning of a breath. The device will then pulse droplets during the first half of subsequent inspirations

The I-neb™ AAD system (Philips Respironics, Murrysville, PA, USA) monitors the peak flow of the first three inhalations achieved by the patient and uses this information to determine the duration of aerosol production required to target the optimal time for delivery, i.e., the first half of an inspiratory breath (Fig. 10.3). This allows adequate time for the droplets to reach the site of deposition within the lungs. Any change in patient's inspiration can be monitored and used to recalculate subsequent aerosolization timing, if required. Integration of a patient's feedback interface within the device and the I-neb Insight® (web-based feedback system) further enhances adherence and compliance of patients to the treatment [17, 48]. The benefit of this technology over nonadaptive devices has been demonstrated [38]. One study compared the accurate dosing capability of an AAD device, the HaloLite® (Medic-Aid Ltd, West Sussex, UK), and breath-enhanced and breath-actuated devices, the Pari® LC Star and the AeroEclipse®, respectively. The HaloLite® AAD device was shown to deliver more accurate amounts of drug due to its adaptive aerosol delivery technology [38]. An in vitro study comparing the performance of three generations of AAD systems, the HaloLite®, I-neb®, and Prodose® (Philips Respironics, Murrysville, PA, USA), showed comparable aerodynamic profiles and output of nebulized iloprost [66].

10.6 Metered Dose Liquid Inhalers

Novel inhaler devices that incorporate microprocessor technology into aerosol generation have been in development since the 1990s, of these, Respimat® Soft Mist™ Inhaler (SMI) (Boehringer Ingelheim, Ingelhein, Germany) is the first to receive market approval [70]. The product Berodual® Respimat® SMI was launched in January 2004 incorporating the anticholinergic and β_2 agonist combination therapy of ipatropium bromide/fenoterol hydrobromide for the treatment of asthma and COPD [36]. This product delivers lower doses of the active ingredients and has been shown to be as effective and as safe as the higher doses administered via conventional pMDI [29]. Studies have also been conducted using Respimat SMI technology to deliver Tiotropium [5] and Budesonide [57], demonstrating superior delivery efficiency in all cases.

The Aradigm AerX™ aerosol delivery system (Aradigm, Hayward, CA, USA) provides a multi-unit dose inhaler preprogrammed to actuate at certain levels of inspiratory flow rate and inhaled volume [61] (Fig. 10.4). Appropriate conditions of pressure, surface tension, and uniform pore diameter, enable the device to produce a fine slow moving mist [61]. The patient feedback system, which provides a green light to indicate to the patient when appropriate flow rate is achieved, allows for maximization of patient inhalatory effort.

By avoiding the incorporation of propellants into the mechanism of the device, the metered dose liquid inhalers (MDLI) systems generate a slower moving aerosol, thus minimizing the potential for oropharyngeal impaction commonly associated with high velocity propulsion [70]. Comparison studies carried out on the AerX system and a microprocessor-controlled pMDI (SmartMist®) demonstrated this fact. The AerX system resulted in just 6.9% deposition in the oropharynx and stomach compared with 42% for the SmartMist® pMDI [22]. Moreover, the MDLI has obvious advantages from a formulatory point of view. The absence of a propellant in the MDLI device reduces the pMDI associated formulatory complications of compatibility, stability, and/or solubility issues of the API in a propellant-based system.

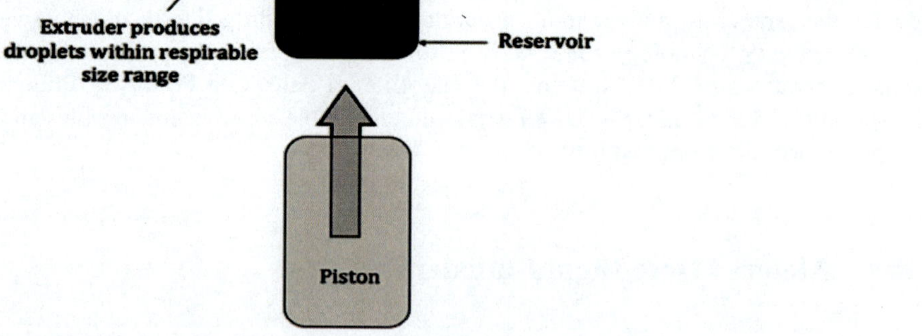

Fig. 10.4 A schematic of the Aradigm AerX™ Inhaler (adapted from Watts et al. [70])

The AerX® system has been tested with various compounds, for local targeting in the lungs, such as rhDNase [24], as well as for systemic delivery of compounds including insulin [21, 31, 46] and opioid analgesics morphine and fentanyl [41, 54]. In opioid analgesia, the AerX® system has been shown to be highly effective in its systemic delivery of inhaled morphine, with peak plasma concentrations occurring at just 2.7 min after administration [68]. A 12 week proof-of-concept trial was conducted on inhaled insulin using the AerX® insulin Diabetes Management System (iDMS) in collaboration with Novo Nordisk®. Relative to subcutaneous injection of insulin in type-2 diabetics, this study found no major hypoglycemic events between the two groups. In addition, there was no observed significant difference in patient's lung function after standard lung function tests were performed on the patient groups [31]. Unfortunately in early 2008, Novo Nordisk pulled the AerX iDMS product, quoting an apparent lack of commercial potential due to the recent failure of Pfizer Inc.'s Exubera® product. However, the recent collaboration between Dance Pharma® and Aerogen®, in January 2011, suggests that the elusive inhaled insulin market may still be within reach.

10.7 Conclusions

Nebulizers were once considered only for use in particular clinical circumstances, and proof of concept trials. Recently however, due to considerable improvements on dose delivery and portability, nebulizers are peaking interest as select candidates for accurate, reproducible, and efficient drug delivery [15, 60]. The case may be made that newer generations of nebulizer devices, particularly the AAD systems, have diverged from the well-established and affordable conventional nebulizers, to become increasingly more complex and expensive and thus impractical for utilization with standard therapies. A cost to benefit check should clearly be considered early on when evaluating the choice of nebulizer device for particular drug development. The expenditure may be warranted, when formulating specialized, expensive APIs, and those with a narrow therapeutic window. The recent rapid evolution of the nebulizer device has overcome the shortcomings of conventional nebulizers and broadened the possibilities for a range of new inhaled therapeutics. Novel therapeutics for local and systemic delivery are continuously under development. With the ongoing advancement of nebulizer systems we should see improvements in clinical outcomes and meet the needs of novel targeted therapies with precision-based, reproducible, and efficient nebulizer devices.

References

1. Agertoft L, Andersen A, Weibull E, Pedersen S (1999) Systemic availability and pharmacokinetics of nebulised budesonide in preschool children. Arch Dis Child 80:241–247
2. Albasarah YY, Somavarapu S, Taylor KM (2010) Stabilizing protein formulations during air-jet nebulization. Int J Pharm 402:140–145

3. Alderson PO, Secker-Walker RH, Stominger DB, Markham J, Hill RL (1974) Pulmonary deposition of aerosols in children with cystic fibrosis. J Pediatr 84:479–484
4. Arzhavitina A, Steckel H (2010) Surface active drugs significantly alter the drug output rate from medical nebulizers. Int J Pharm 384:128–136
5. Bateman ED, Tashkin D, Siafakas N, Dahl R, Towse L, Massey D, Pavia D, Zhong NS (2010) A one-year trial of tiotropium Respimat plus usual therapy in COPD patients. Respir Med 104:1460–1472
6. Beach JR, Campbell M, Andrews DJ (1999) Exposure of health care workers to pentamidine isethionate. Occup Med (Lond) 49:243–245
7. Boguniewicz M, Martin RJ, Martin D, Gibson U, Celniker A, Williams M, Leung DY (1995) The effects of nebulized recombinant interferon-gamma in asthmatic airways. J Allergy Clin Immunol 95:133–135
8. Borish LC, Nelson HS, Lanz MJ, Claussen L, Whitmore JB, Agosti JM, Garrison L (1999) Interleukin-4 receptor in moderate atopic asthma. A phase I/II randomized, placebo-controlled trial. Am J Respir Crit Care Med 160:1816–1823
9. Bridges P, Taylor KMG (1998) Nebulisers for the generation of liposomal aerosols. Int J Pharm 173:117–125
10. Cameron D, Clay M, Silverman M (1990) Evaluation of nebulizers for use in neonatal ventilator circuits. Crit Care Med 18:866–870
11. Coates AL, Allen PD, MacNeish CF, Ho SL, Lands LC (2001) Effect of size and disease on estimated deposition of drugs administered using jet nebulization in children with cystic fibrosis. Chest 119:1123–1130
12. Coates AL, Ho SL (1998) Drug administration by jet nebulization. Pediatr Pulmonol 26:412–423
13. Coates AL, MacNeish CF, Lands LC, Meisner D, Kelemen S, Vadas EB (1998) A comparison of the availability of tobramycin for inhalation from vented vs unvented nebulizers. Chest 113:951–956
14. Corcoran TE, Dauber JH, Chigier N, Iacono AT (2002) Improving drug delivery from medical nebulizers: the effects of increased nebulizer flow rates and reservoirs. J Aerosol Med 15:271–282
15. Dennis JH, Hendrick DJ (1992) Design characteristics for drug nebulizers. J Med Eng Technol 16:63–68
16. Denyer J, Dyche T (2010) The adaptive aerosol delivery (AAD) technology: past, present, and future. J Aerosol Med Pulm Drug Deliv 23(Suppl 1):S1–S10
17. Denyer J, Nikander K, Smith NJ (2004) Adaptive aerosol delivery (AAD) technology. Expert Opin Drug Deliv 1:165–176
18. Dhand R (2002) Nebulizers that use a vibrating mesh or plate with multiple apertures to generate aerosol. Respir Care 47:1406–1416, discussion 1416–1408
19. Elhissi AM, Faizi M, Naji WF, Gill HS, Taylor KM (2007) Physical stability and aerosol properties of liposomes delivered using an air-jet nebulizer and a novel micropump device with large mesh apertures. Int J Pharm 334:62–70
20. Everard ML, Evans M, Milner AD (1994) Is tapping jet nebulisers worthwhile? Arch Dis Child 70:538–539
21. Farr SJ, McElduff A, Mather LE, Okikawa J, Ward ME, Gonda I, Licko V, Rubsamen RM (2000) Pulmonary insulin administration using the AERx system: physiological and physicochemical factors influencing insulin effectiveness in healthy fasting subjects. Diabetes Technol Ther 2:185–197
22. Farr SJ, Warren SJ, Lloyd P, Okikawa JK, Schuster JA, Rowe AM, Rubsamen RM, Taylor G (2000) Comparison of in vitro and in vivo efficiencies of a novel unit-dose liquid aerosol generator and a pressurized metered dose inhaler. Int J Pharm 198:63–70
23. Flament MP, Leterme P, Gayot A (2001) Study of the technological parameters of ultrasonic nebulization. Drug Dev Ind Pharm 27:643–649
24. Geller D, Thipphawong J, Otulana B, Caplan D, Ericson D, Milgram L, Okikawa J, Quan J, Bowman CM (2003) Bolus inhalation of rhDNase with the AERx system in subjects with cystic fibrosis. J Aerosol Med 16:175–182
25. Geller DE (2005) Comparing clinical features of the nebulizer, metered-dose inhaler, and dry powder inhaler. Respir Care 50:1313–1321, discussion 1321–1312

26. Gessler T, Schmehl T, Hoeper MM, Rose F, Ghofrani HA, Olschewski H, Grimminger F, Seeger W (2001) Ultrasonic versus jet nebulization of iloprost in severe pulmonary hypertension. Eur Respir J 17:14–19
27. Ghazanfari T, Elhissi AM, Ding Z, Taylor KM (2007) The influence of fluid physicochemical properties on vibrating-mesh nebulization. Int J Pharm 339:103–111
28. Gibbons AM, McElvaney NG, Taggart CC, Cryan SA (2009) Delivery of rSLPI in a liposomal carrier for inhalation provides protection against cathepsin L degradation. J Microencapsul 26:513–522
29. Goldberg J, Freund E, Beckers B, Hinzmann R (2001) Improved delivery of fenoterol plus ipratropium bromide using Respimat compared with a conventional metered dose inhaler. Eur Respir J 17:225–232
30. Greenspan B (2007) Ultrasonic and electrohydrodynamic methods for aerosol generation. In: Hickey AJ (ed) Inhalation aerosols. physical and biological basis for therapy. Informa Healthcare, New York, pp 285–306
31. Hermansen K, Ronnemaa T, Petersen AH, Bellaire S, Adamson U (2004) Intensive therapy with inhaled insulin via the AERx insulin diabetes management system: a 12-week proof-of-concept trial in patients with type 2 diabetes. Diab Care 27:162–167
32. Hess DR (2000) Nebulizers: principles and performance. Respir Care 45:609–622
33. Ho SL, Coates AL (1999) Effect of dead volume on the efficiency and the cost to deliver medications in cystic fibrosis with four disposable nebulizers. Can Respir J 6:253–260
34. Hung JC, Hambleton G, Super M (1994) Jet nebulising systems for recombinant human DNase I. Arch Dis Child 71:558–559
35. Ilowite JS, Gorvoy JD, Smaldone GC (1987) Quantitative deposition of aerosolized gentamicin in cystic fibrosis. Am Rev Respir Dis 136:1445–1449
36. Kassner F, Hodder R, Bateman ED (2004) A review of ipratropium bromide/fenoterol hydrobromide (Berodual) delivered via Respimat Soft Mist Inhaler in patients with asthma and chronic obstructive pulmonary disease. Drugs 64:1671–1682
37. Knoch M, Wunderlich E, Geldner S (1994) A nebulizer system for highly reproducible aerosol delivery. J Aerosol Med 7:229–237
38. Leung K, Louca E, Coates AL (2004) Comparison of breath-enhanced to breath-actuated nebulizers for rate, consistency, and efficiency. Chest 126:1619–1627
39. MacNeish CF, Meisner D, Thibert R, Kelemen S, Vadas EB, Coates AL (1997) A comparison of pulmonary availability between Ventolin (albuterol) nebules and Ventolin (albuterol) respirator solution. Chest 111:204–208
40. Malone RA, Hollie MC, Glynn-Barnhart A, Nelson HS (1993) Optimal duration of nebulized albuterol therapy. Chest 104:1114–1118
41. Mather LE, Woodhouse A, Ward ME, Farr SJ, Rubsamen RA, Eltherington LG (1998) Pulmonary administration of aerosolised fentanyl: pharmacokinetic analysis of systemic delivery. Br J Clin Pharmacol 46:37–43
42. McCallion ON, Taylor KM, Thomas M, Taylor AJ (1995) Nebulization of fluids of different physicochemical properties with air-jet and ultrasonic nebulizers. Pharm Res 12:1682–1688
43. McElvaney NG, Doujaiji B, Moan MJ, Burnham MR, Wu MC, Crystal RG (1993) Pharmacokinetics of recombinant secretory leukoprotease inhibitor aerosolized to normals and individuals with cystic fibrosis. Am Rev Respir Dis 148:1056–1060
44. Mercer TT (1973) Production and characterization of aerosols. Arch Intern Med 131:39–50
45. Mercer TT (1981) Production of therapeutic aerosols; principles and techniques. Chest 80:813–818
46. Moses RG, Bartley P, Lunt H, O'Brien RC, Donnelly T, Gall MA, Vesterager A, Wollmer P, Roberts A (2009) Safety and efficacy of inhaled insulin (AERx iDMS) compared with subcutaneous insulin therapy in patients with type 1 diabetes: 1-year data from a randomized, parallel group trial. Diabet Med 26:260–267
47. Newman SP, Pellow PG, Clay MM, Clarke SW (1985) Evaluation of jet nebulisers for use with gentamicin solution. Thorax 40:671–676
48. Nikander K, Arheden L, Denyer J, Cobos N (2003) Parents' adherence with nebulizer treatment of their children when using an adaptive aerosol delivery (AAD) system. J Aerosol Med 16:273–281

49. Nikander K, Turpeinen P, Wollmer P (2009) The conventional ultrasonic nebulizer proved inefficient in nebulizing a suspension. J Aerosol Med 12:47–53
50. Niven R, Hickey AJ (2007) Part III: pharmaceutics and pharmaceutical technology. atomization and nebulizers. In: Hickey AJ (ed) Inhalation aerosols. physical and biological basis for therapy. Informa Healthcare, New York, pp 253–284
51. Niven RW (1995) Delivery of biotherapeutics by inhalation aerosol. Crit Rev Ther Drug Carrier Syst 12:151–231
52. Niven RW, Ip AY, Mittelman S, Prestrelski SJ, Arakawa T (1995) Some factors associated with the ultrasonic nebulization of proteins. Pharm Res 12:53–59
53. Niven RW, Speer M, Schreier H (1991) Nebulization of liposomes II. The effects of size and modeling of solute release profiles. Pharm Res 8:217–221
54. Otulana B, Okikawa J, Linn L, Morishige R, Thipphawong J (2004) Safety and pharmacokinetics of inhaled morphine delivered using the AERx system in patients with moderate-to-severe asthma. Int J Clin Pharmacol Ther 42:456–462
55. Phipps PR, Gonda I (1990) Droplets produced by medical nebulizers. Some factors affecting their size and solute concentration. Chest 97:1327–1332
56. Pillai R, Petrak K, Blezinger P, Deshpande D, Florack V, Freimark B, Padmabandu G, Rolland A (1998) Ultrasonic nebulization of cationic lipid-based gene delivery systems for airway administration. Pharm Res 15:1743–1747
57. Pitcairn G, Reader S, Pavia D, Newman S (2005) Deposition of corticosteroid aerosol in the human lung by Respimat Soft Mist inhaler compared to deposition by metered dose inhaler or by Turbuhaler dry powder inhaler. J Aerosol Med 18:264–272
58. Rau JL (2002) Design principles of liquid nebulization devices currently in use. Respir Care 47:1257–1275, discussion 1275–1258
59. Ross DL, Gabrio BJ (1999) Advances in metered dose inhaler technology with the development of a chlorofluorocarbon-free drug delivery system. J Aerosol Med 12:151–160
60. Sanders M (2007) Inhalation therapy: an historical review. Prim Care Respir J 16:71–81
61. Schuster J, Rubsamen R, Lloyd P, Lloyd J (1997) The AERX aerosol delivery system. Pharm Res 14:354–357
62. Shoyele SA, Slowey A (2006) Prospects of formulating proteins/peptides as aerosols for pulmonary drug delivery. Int J Pharm 314:1–8
63. Smith EC, Denyer J, Kendrick AH (1995) Comparison of twenty three nebulizer/compressor combinations for domiciliary use. Eur Respir J 8:1214–1221
64. Steckel H, Eskandar F, Witthohn K (2003) The effect of formulation variables on the stability of nebulized aviscumine. Int J Pharm 257:181–194
65. van Beerendonk I, Mesters I, Mudde AN, Tan TD (1998) Assessment of the inhalation technique in outpatients with asthma or chronic obstructive pulmonary disease using a metered-dose inhaler or dry powder device. J Asthma 35:273–279
66. Van Dyke RE, Nikander K (2007) Delivery of iloprost inhalation solution with the HaloLite, Prodose, and I-neb adaptive aerosol delivery systems: an in vitro study. Respir Care 52:184–190
67. Wagner A, Vorauer-Uhl K, Katinger H (2006) Nebulization of liposomal rh-Cu/Zn-SOD with a novel vibrating membrane nebulizer. J Liposome Res 16:113–125
68. Ward ME, Woodhouse A, Mather LE, Farr SJ, Okikawa JK, Lloyd P, Schuster JA, Rubsamen RM (1997) Morphine pharmacokinetics after pulmonary administration from a novel aerosol delivery system. Clin Pharmacol Ther 62:596–609
69. Wark PA, McDonald V, Jones AP (2005). Nebulised hypertonic saline for cystic fibrosis. Cochrane Database Syst Rev. (2):CD001506
70. Watts AB, McConville JT, Williams RO III (2008) Current therapies and technological advances in aqueous aerosol drug delivery. Drug Dev Ind Pharm 34:913–922
71. Weber A, Morlin G, Cohen M, Williams-Warren J, Ramsey B, Smith A (1997) Effect of nebulizer type and antibiotic concentration on device performance. Pediatr Pulmonol 23:249–260

Chapter 11
Excipients Utilized for Modifying Pulmonary Drug Release

Poonam Sheth and Paul B. Myrdal

Abstract Experimentally, there are a plethora of excipients used for pulmonary-administered agents and formulations. However, far fewer have been tested in humans. The limited use of excipients is indicative of the difficulty of selecting an excipient that will not limit aerosolization performance, is safe with little-to-no immunogenicity, and is effective in sustaining drug release. These issues and topics are discussed in this chapter with a particular focus on modified release.

Keywords Excipients • Inhalation • Sustained release

11.1 Introduction

Modified, namely sustained, release of drugs is especially important in clinical practice. Medications that are typically administered via the pulmonary route frequently need to be dosed multiple times a day, which increases the risk of nonadherence to drug therapies. The development and application of sustained release pulmonary drug products could greatly benefit patients, at large, as described throughout this book. It reasons that the rate of release of drug molecules from sustained release formulations will dictate the duration of activity: rapid release yields high absorption rate with relatively higher concentrations and a short duration of action while an overly slow release rate may result in nontherapeutic drug levels.

To modify the release of drugs, a variety of excipients can be used to either decrease the absorption rate or the clearance rate of drugs. Excipients are chemicals intentionally added to a drug formulation other than the active pharmaceutical ingredient. In this chapter, liposomes, polymers, cyclodextrins, and other excipients

P.B. Myrdal (✉)
University of Arizona, College of Pharmacy, 1703 East Mabel Street, Tucson, AZ 85721, USA
e-mail: myrdal@pharmacy.arizona.edu

H.D.C. Smyth and A.J. Hickey (eds.), *Controlled Pulmonary Drug Delivery*, Advances in Delivery Science and Technology, DOI 10.1007/978-1-4419-9745-6_11,

that have favorable in vitro or in vivo release rate profiles are detailed. Sustained release via pulmonary delivery can be demonstrated by the following findings.

1. Increased duration of drug that is present in bronchoalveolar lavage fluid.
2. Longer time to achieve maximum concentration (T_{max}).
3. Lower maximum concentration of drug (C_{max}) in serum.
4. Longer presence of biomarkers in bronchoalveolar lavage fluid.
5. Slower appearance of biomarker.
6. Lower maximum concentration of biomarker in serum.
7. Longer duration of pharmacodynamic activity.

The studies presented in this chapter will use these physiological findings to suggest the ability of given excipients to modify drug release.

11.2 Liposomes

11.2.1 What Are Liposomes?

As described in Chap. 14, liposomes are vesicles enclosing an aqueous environment with phospholipids that form bilayers or multilamellar structures. Since there are a variety of lipids that can be used, these excipients are versatile and can form liposomes with a wide range of structural properties. The chemical structures of phospholipids discussed in this chapter are listed in Fig. 11.1.

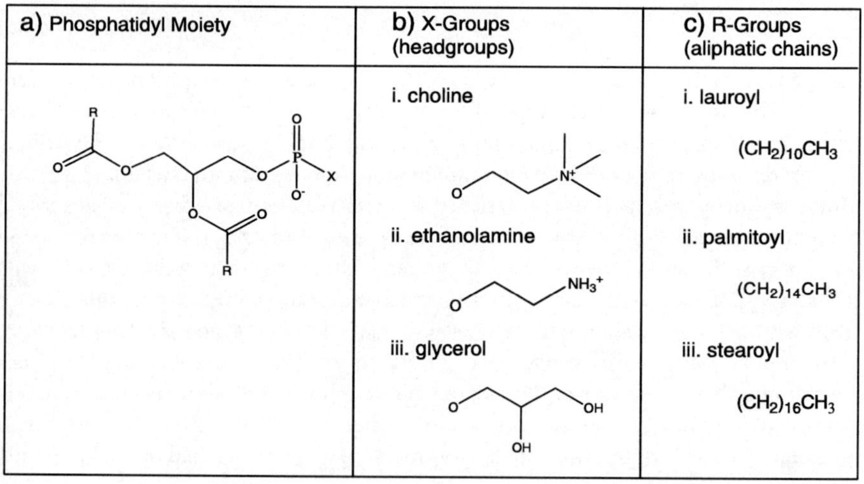

Fig. 11.1 General chemical structures of some phospholipids. Phospholipids used in liposomal formulations have a phosphatidyl moiety backbone with a specific head- and R-groups. At times, the R-group can simply be hydrogen atoms. The phospholipids discussed in this chapter include phosphatidylcholine (PC), dilauroylphosphatidylcholine (DLPC), dipalmitoylphosphatidylcholine (DPPC), distearoylphosphatidylcholine (DPPC), distearoylphosphatidylglycerol (DSPG), dipalmitoylphosphatidylethanolamine (DPPE), and distearoylphosphatidylethanolamine (DSPE)

In general, phospholipids are considered to have low toxicity since they have been found to be biologically inert, with little antigenicity or pyrogenicity [50]. There have been no accounts of adverse effects on lung function. Changes in forced expiratory volume in 1 s or forced vital capacity as a function of different doses of lipids in humans up to 1-h post-inhalation were not observed by Schreier et al. [66]. This suggests that liposomal inhalation formulations may not cause increased airway resistance or decreased lung compliance.

11.2.2 How Do Liposomes Function in Controlled Release?

A drug molecule encapsulated in a liposome must diffuse through at least one hydrophilic and lipophilic layers prior to release to the physiological environment for absorption by the body. The rate of release is modulated by all of the following: (a) liposome surface charge and size, (b) drug:lipid ratio, (c) the composition and chain length of the phospholipid, and (d) presence of cholesterol or other excipients that affect membrane fluidity [90]. Examples of these effects can be seen with dipalmitoyl phosphatidylglycerol (DPPG) imparting a negative charge which facilitates the inclusion of certain drug molecules into liposomes and cholesterol increasing the stability of liposomes [66]. Size also has an appreciable effect on a liposomal formulation's sustained release potential. If the individual liposome is less than 0.1 μm in diameter, it is less readily opsonized than its larger counterparts, thus having longer residence time in the lung [66]. However, vesicles that are exceedingly small are easily exhaled and not delivered to the alveoli. Hence, liposomes with a diameter of 50–200 nm are optimal for avoiding phagocytosis and trapping sufficient drug load and have relevant clinical applications [68]. The phase transition temperature (T_c) of lipids also affects drug release rate [68]. If the temperature of the environment is lower than the T_c, the lipid will be in a well-ordered, gel phase with increased stability than if the temperature of the environment is higher than the T_c. Using lipids with a T_c greater than physiological temperature will make the liposome less fluid and a better candidate for retaining the encapsulated drug for a longer period.

11.2.3 Examples of Common Liposomes Used

11.2.3.1 Comparing Liposomal and Solution Formulations

Several liposomal formulations have been evaluated for the potential to modify the release of drugs in the lung. These formulations are compared to solution formulations of the drug to establish the effect of the liposome on the release profile of a given drug. Table 11.1 provides a summary of studies that investigate the ability of drug-liposome encapsulation for modifying drug release. While the evidence in utilizing dilauroyl

Table 11.1 Summary of select studies that evaluate the pharmacokinetics of various liposome formulations for pulmonary drug delivery

Drug	Liposome	Subject	Results	References and Other Studies with Similar Results
Cyclosporin A	DLPC only	Mice; administered via intranasal instillation	Liposome carrier is retained up to 16.9 times longer than the $t_{1/2}$ of solution cyclosporine A in healthy lungs and up to 7.5 times longer in inflamed lungs	[5]
Isoniazid, rifampicin, and ethambutol	DPPC:drugs 1:1 (w/w)	In vitro release study utilizing diffusion model	Formulation was adequate for aerosolization. Nonencapsulated drug was released within 5–6 h, while encapsulated drug was released over 24 h without burst effect	[13]
Porcine-sodium and zinc insulins	7:2 DPPC:CH (w/w)	Rats; administered via intratracheal instillation	Increased absolute bioavailability with liposomal formulation; prolonged T_{max} and increase in AAC_{gluc} with liposomal formulation. Sustained hypoglycemic effect	[14, 17, 52]
Cisplatin	DPPC and CH	Humans; administered by aerosol	In a safety and pharmacokinetics study of SLIT™ cisplatin, found no dose limiting toxicity or nephrotoxicity, ototoxicity, and neurotoxicity. Data from Transave Inc. suggests that the formulation can sustain drug release over 50 h, while free cisplatin is released within 20 h	[49, 74, 83]

Detirelix	52:8.4:39.6 molar ratio of DSPC:DSPG:CH	Dogs; administered by intratracheal instillation	PD results consistent with sustained drug release associated with encapsulated drug. Drug that is adsorbed to the surface of the liposome has a K_a similar to that of the solution formulation. See increase in $t_{1/2}$ and MRT with liposomal formulation. Postmortem histopathology reveals minor degree of local inflammation of lungs with liposomal formulation	[6, 71]
Fentanyl	PC:CH at a ratio of 10:1 (w/w)	Human; drug administered by nebulization	Data suggests decrease in C_{max} to one-fourth the value and prolonged T_{max} with liposomal formulation compared to intravenous formulation. Plasma concentrations were sustained for longer duration with liposomal formulation than with intravenous formulation. No toxicity noted; patients complained of a greasy smell and taste associated with the nebulized formulation	[33, 67]

Abbreviations used in the table: half-life ($t_{1/2}$); area above the glucose remaining vs. time curve (AAC_{gluc}); sustained release lipid inhalation targeting (SLIT™); pharmacodynamic (PD); absorption rate (K_a); mean residence time (MRT); cholesterol (CH); for all other abbreviations, refer to the text above phosphatidylcholine (DLPC) for sustained release pulmonary formulations is limited, there are many studies that found that DLPC formulations did not pose safety concerns in mice or humans [36, 39, 78, 79]. In addition, it was found that DLPC formulations have the potential to specifically deliver medications to the lungs and limit systemic effects of drugs such as cyclosporine A [78]. As a naturally occurring phospholipid in lung surfactant, dipalmitoyl phosphatidylcholine (DPPC) has gained much attention in formulations delivered to the lung. It is speculated that DPPC liposomes have a minimal adverse effect profile. There are many promising studies that suggest that it is a good excipient to modulate sustained release of drugs when formulated with cholesterol [14, 52, 74]. Distearoyl phosphatidylcholine (DSPC) has been used in formulation of drugs that are delivered to the lung. Namely, two prominent examples include using DSPC liposomes intratracheally to deliver detirelix, a luteinizing hormone releasing-hormone antagonist, and triamcinolone acetonide phosphate, a corticosteroid [6, 71]. In both instances, the DSPC liposome also had distearoyl phosphatidylglycerol (DSPG), which imparted a negative surface charge. DSPG's negative charge has been associated with enhancing pulmonary residence times of terbutaline in guinea pigs and increased bioavailability of 6-carboxy-fluorescein [6]. Due to favorable liposomal characteristics seen in previously published literature, it is likely that the inclusion of DSPG will prove to be favorable in pulmonary delivery of detirelix and triamcinolone. Finally, there are several studies that have looked at the effects of various phosphatidylcholine (PC) formulations, including soy and egg PC (ePC) along with other excipients, to modulate the release rate of drug from liposomes [33, 67]. Based on these studies, it appears that there are many liposomal structures that can modulate the release rate of an encapsulated drug.

11.2.3.2 Comparing Liposomal Formulations to Each Other

Not all phospholipids function equally well, or even at all, for modifying release for every drug. Selecting an optimal liposomal composition largely depends on the interaction of the lipid with the drug molecule's properties. A study that examined the release rate of liposomally encapsulated ^{3}H-terbutaline in male guinea pigs by intratracheal administration demonstrates this relationship [23]. This experiment utilized formulations that contained varying combinations of DSPC, DPPC, DPPG, ePC, and partially hydrogenated egg PC (phPC) with or without cholesterol (see Table 11.2). The partial hydrogenation of phospholipids to an iodine value (IV) of 40 increased half-life ($t_{1/2}$) slightly, while unencapsulated drug and more complete hydrogenation of phospholipids to an IV of 1 produced a similar $t_{1/2}$ as that observed with complete saturation. Increasing the carbon chain length from 16, in DPPC, to 18, in DSPC, did not affect the kinetics of the drug. Moreover, the mere addition of cholesterol to the formulation increased $t_{1/2}$ by tenfold. Granted all these results are specific to ^{3}H-terbutaline, but the study presents how the composition of liposomes alters the sustained release profile and that liposomes prolong residence time of the bronchodilator in the lungs. In addition, this study demonstrates one way to test various liposomal formulations for in vivo sustained release effect.

Table 11.2 Composition, size, and pulmonary half-lives of liposome-encapsulated ^{3}H-terbutaline formulations administered by intratracheal instillation to guinea pigs

Lipid components	Molar ratio	Mean particle size (nm)	Half-life (h)
Unencapsulated	–	–	1.3
DPPC:DPPG	95:5	207	1.4
ePC:ePG:CH	55:5:40	238	4.8
phPC (IV40):ePG:CH	55:5:40	270	5.4
phPC (IV1):ePG:CH	55:5:40	270	16.3
DPPC:DPPG:CH	55:5:40	243	17.5
DSPC:DSPG:CH	55:5:40	231	17.9

Abbreviations used in the table: egg phosphatidylglycerol (ePG); cholesterol (CH); for all other abbreviations, refer to the text above. Adapted from Fielding and Abra [23]

Liposomes can also be formulated to render multiple mechanisms for sustained release. For instance Huang et al. used hydrogenated PC, cholesterol, and PEGylated-dipalmitoylphosphatidyl ethanolamine (PEG-DPPE) at molar ratios of 70:30 PC:cholesterol or 70:30:1 PC:cholesterol:PEG-DPPE in mice to deliver insulin via the pulmonary system [32]. Increasing the ratio of cholesterol to PC decreased liposomal fluidity, thereby increasing formulation stability. Furthermore, PEG-DPPE formed a three-dimensional steric hindrance around the liposome that increased the liposome stability and decreased the rate of insulin release from the liposome compared to the PC-cholesterol formulation. The data supports that both formulations have a sustained release effect beyond that conferred with the physical mixture of insulin and blank liposomes over a period of 6 h. The safety of these formulations was evaluated by studying the types of leukocytes and morphology of cells in alveolar lavage fluid; there were no signs of virulence, inflammation, or immunoreaction in the lungs. Similar results related to sustained release can be seen with in vitro release of beclomethasone dipropionate from PEG-distearoyl phosphatidylethanolamine (DSPE) micelles [27]. The use of PEGylation is further discussed in this chapter under conjugate excipients.

11.3 Polymers

11.3.1 What Are Polymers?

Polymers are macromolecules that are composed of repeating structures, which are covalently bonded. They are among the most investigated systems for controlling the release of drugs via various routes of administration. Polymers can be attached to or encapsulate drugs to prolong the effects of drugs. The most widely used natural polymers are albumin, chitosan, and hyaluronic acid, and the most frequently used synthetic polymers are poly(lactic-co-glycolic acid) (PLGA), poly(lactic acid) (PLA), and poly(ethylene glycol) (PEG) [50]. Of these polymers, PEG is frequently incorporated into sustained release formulations by covalently bonding PEG to the active drug to increase molecular weight, while all the other polymers mentioned above are generally used to encapsulate the drug particles. Polymers can encapsulate drugs in nanoparticles, microspheres, and large porous particles. Chapter 12 further differentiates these formulations and provides a more detailed description of polymers used for pulmonary drug delivery. This section is limited to polymers that modify drug release from the dosage form by encapsulating the agent; details about how PEG functions in controlling drug release can be found in Sect. 11.4.

11.3.2 How Do Polymers Function in Controlled Release?

Polymers can function in controlled release by three mechanisms that depend upon the properties of the polymer. Those mechanisms are by diffusion, swelling followed by diffusion, and degradation. Drug diffusion from a polymer system requires drug to pass through the polymeric matrix to the external environment [8, 50]. The mechanism of drug release dependent upon swelling of the polymeric matrix followed by drug diffusing is similar to diffusion-dependent drug release. Drug release by degradation occurs by hydrolysis or bulk erosion resulting in decreasing the molecular weight of the polymer until dissolution occurs. While these mechanisms are described briefly and discretely in this section, generally polymer erosion and channeling act in conjunction to alter the release of drugs.

11.3.3 Examples of Polymers Used for Sustained Release

Various polymers have been used in modifying drug release of pulmonary agents. These polymers are separated based on natural or synthetic origin and listed in Table 11.3. In addition, Table 11.4 presents a list of copolymers used in pulmonary delivery systems that have some ability to modify drug release. Pertaining studies are further discussed in Chap. 12 in this book.

11.4 Conjugate Excipients

11.4.1 What Are Conjugate Excipients?

Conjugation, in chemistry, is defined as a chemical compound that is formed by covalently joining two or more compounds. The examples of conjugation reviewed in this chapter involve dextrans (DEX) and poly(ethylene glycol) (PEG). Both of these are hydrophilic macromolecules. DEX are branched polysaccharides made up of glucose molecules. Typically, DEX are used therapeutically to expand plasma volume. PEGylation is the act of covalently attaching PEG chains to other molecules; they can be used to decrease immunogenicity and increase duration of action of the drug [90].

Table 11.3 Structures of natural and synthetic polymers used in modified release pulmonary delivery systems

Polymer	Structure	Modified release studies
Natural polymers		
Albumin	Any protein that is moderately soluble in salt solutions and experiences heat-driven coagulation	[47]
Carrageenan	Kappa-carrageenan 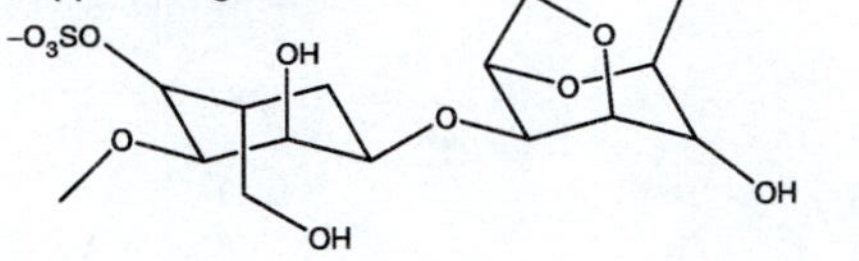	[86, 87]
	Iota-carrageenan 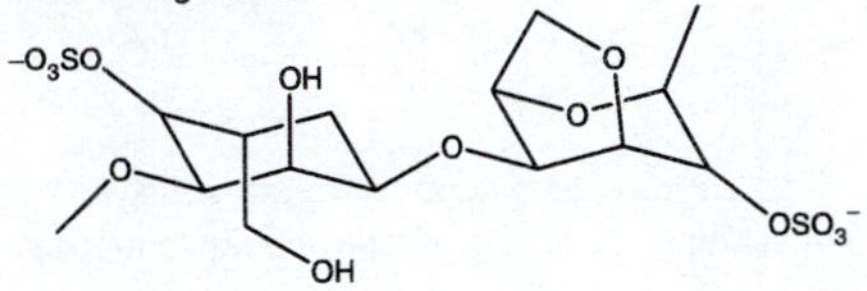	
	Lambda-carrageenan 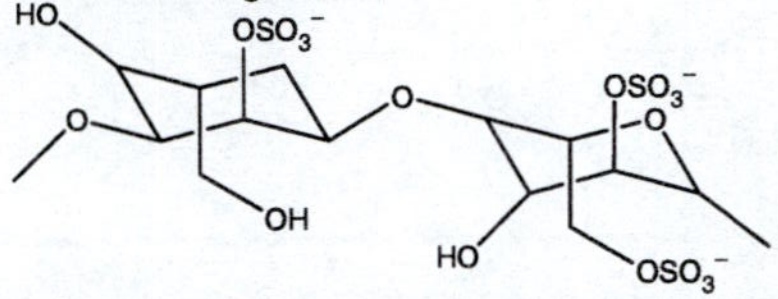	

(continued)

Table 11.3 (continued)

Polymer	Structure	Modified release studies
Chitosan	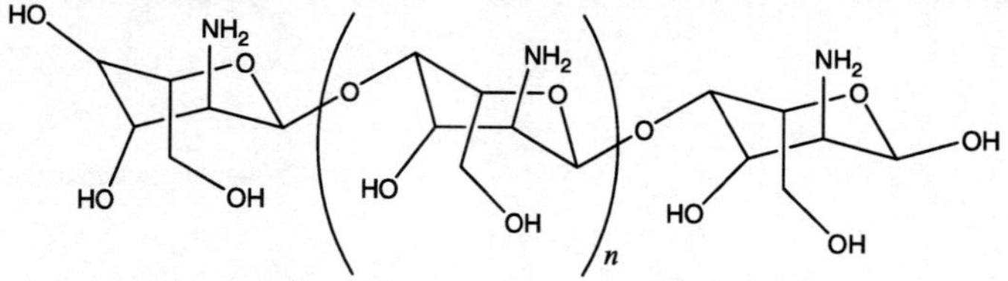	[43, 45, 46, 59, 75, 88]
Gelatin	Produced by partial hydrolysis of collagen and is composed of primarily glycine, proline, and hydroxyproline	[51, 54]
Hyaluronic acid	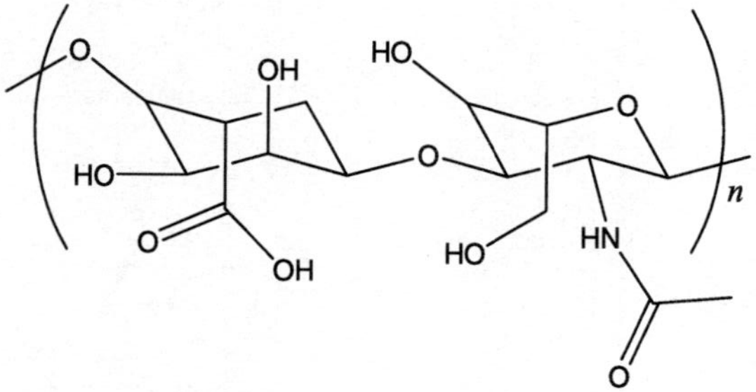	[55, 61, 73]
Synthetic polymers		
Poly(acrylic acid) and derivatives	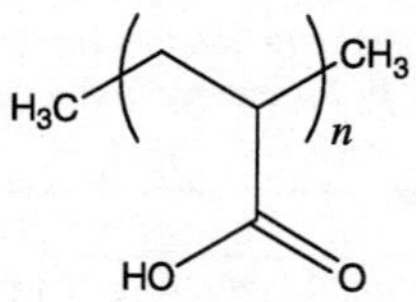	[2, 21, 91]

Poly(ethylene glycol) (PEG)	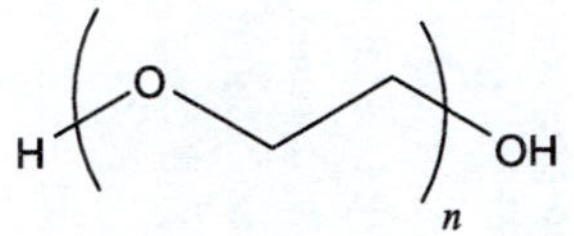	Utilized in combination or as a PEG conjugate. See Table 11.4
Poly(glycolic acid) (PGA)	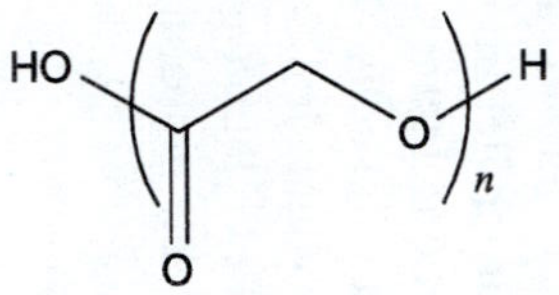	Utilized in combination. See Table 11.4
Poly(lactic acid) (PLA) Oligolactic acid (OLA)[a]	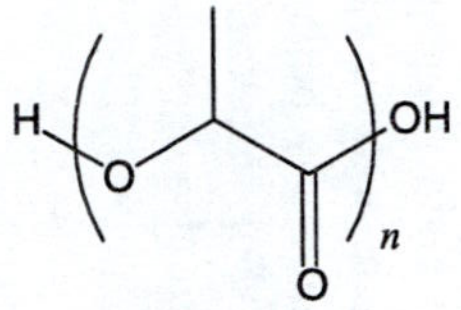	[18, 22, 40, 48, 57, 70]
Poly(vinyl alcohol) (PVA)	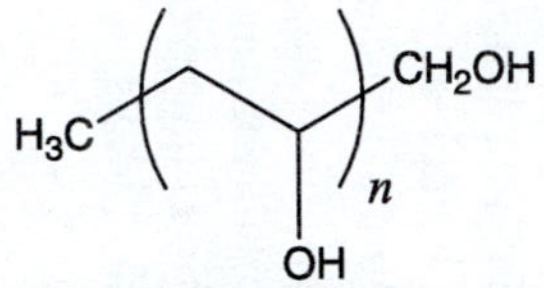	[1, 63, 64]

[a]Oligolactic acid contains a terminal acetate group

Table 11.4 Studies that present modified drug release for pulmonary delivery drugs using various copolymers

Copolymer	References
Poly(lactic acid) – poly(ethylene glycol) – poly(lactic acid) (PLA-PEG-PLA)	[62, 84]
Poly(γ-benzyl-L-glutamate) – poly(ethylene oxide) – poly(γ-benzyl-L-glutamate) (PBLG-PEO-PBLG)	[58]
Poly(ethylene glycol) – sebacic acid (PEG-SA)	[25, 26]
Poly(lactic-co-glycolic) acid (PLGA)	[4, 35, 37, 44, 56, 65, 72, 82, 89]

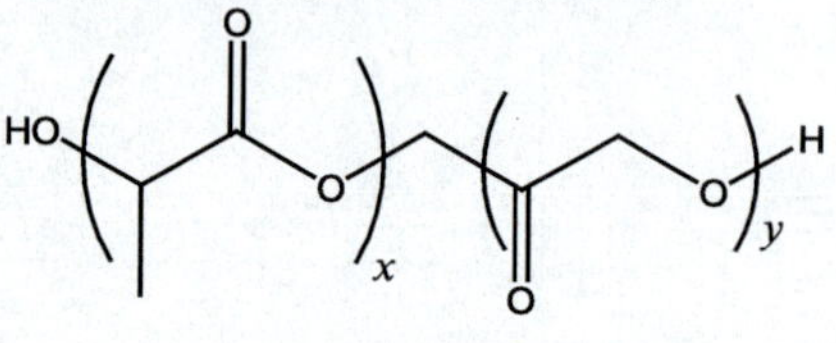

x = number of units of lactic acid
y = number of units of glycolic acid

The attachment of conjugates to a drug relies on modifying the drug without sterically hindering the active site [68]. Specific examples of these drug conjugates are discussed below.

11.4.2 *How Do Conjugate Excipients Function in Controlled Release?*

Drug conjugation with DEX or PEG appears to delay drug absorption and prevents drug clearance. The predominant limiting factor for hydrophilic drug absorption though airway epithelium is molecular weight: conjugations increase molecular weight of the drug, which limits diffusion through the epithelial cells [90]. In addition, high molecular weight drugs are generally bulky, which limits their renal excretion and increases circulation time. For instance, PEGylation is associated with increased circulatory time; however, it also decreases the probability of enzymatic degradation of proteins [24, 77].

11.4.3 *Examples of Excipient Conjugates*

11.4.3.1 Dextran

While DEX has been used as excipients and therapeutic agents by various routes of administration, there are a limited number of studies supporting the use of DEX in controlling drug release for pulmonary agents. A study by Williams and Taylor presents results of in vitro release of cromoglycate from a covalent complex of cromoglycic acid (CGA) and DEX using a series of hydrolysis experiments [81]. It was found that the CGA-DEX complex that contained 0.8% CGA (w/w) released cromoglycate with a $t_{1/2}$ of 39 min; however, with 40% CGA (w/w), the release $t_{1/2}$ was 290 min. It was determined that the $t_{1/2}$ of release is related to the degree of substitution (i.e., percentage DEX in the CGA formulation): increasing the drug load increased $t_{1/2}$ of release. Larsen et al. studied the release of benzoate esters with varying DEX content and found that DEX can modulate $t_{1/2}$, but it was independent of the degree of substitution [81].

11.4.3.2 PEGylation

PEGylation has been used as an excipient conjugate in a formulation to sustained drug release. Initially, PEGylation was evaluated as an excipient in injectable formulations. For instance, Ho et al. developed and studied the effect of intravenous PEGylated L-asparaginase in rabbits compared to the non-PEGylated enzyme [31]. They found that the $t_{1/2}$ of the PEGylated protein was 144 h and significantly longer than that of the non-PEGylated version ($t_{1/2}$ = 20 h). Similar results can be seen in the development of PEGylated insulin for pulmonary delivery. Monomethoxy-PEG-750-insulin has an in vitro $t_{1/2}$ which is twice as long compared to regular insulin when digested with chymotrypsin [9, 42]. This extensive $t_{1/2}$ rationalized subsequent in vivo testing of inhaled PEGylated insulin in rats [41]. When administered intratracheally, PEG-750-insulin suppressed glucose for 12 h, which is twice as long as that observed with non-PEGylated insulin. These results suggest that PEGylation of proteins increases the duration of pharmacotherapeutic effect.

PEGylation is frequently used in conjunction with other excipients that further sustain drug effects. For instance, doxorubicin, antisense oligonucleotides, and small interfering RNA were administered intratracheally to mice with orthotopic human lung cancer using PEGylated liposomes [28]. The liposomes were composed of ePC:cholesterol at a molar ratio of 51:44. Liposomes are readily eliminated from the blood by the reticuloendothelial system; however, PEGylation avoids this problem. In vivo, the inhaled PEGylated liposomes provided higher peak concentrations and lung retention times than intravenous administration of the formulation. Furthermore, intratracheal administration of calcium phosphate and PEGylated insulin has shown to increase the duration of pharmacodynamic effects in rats [29]. Insulin administered in this formulation had a longer $t_{1/2}$ and mean residence time (MRT) than intratracheal administration of insulin solution. Pharmacodynamic analysis present that the calcium phosphate-PEG formulation had a statistically significant larger area above the effect curve than that seen with insulin administration, suggesting that the pharmacodynamic effects are much greater with the calcium phosphate and PEGylated formulation than with insulin solution. Overall, the use of multiple excipients that prolong drug release is gaining much interest among pharmaceutical scientists.

11.5 Cyclodextrins

11.5.1 What Are Cyclodextrins?

Cyclodextrins (CD) are cyclic oligosaccharides with hydrophilic outer surfaces and hydrophobic central cavities. They form inclusion complexes with drugs, where the drug bonds noncovalently to the CD by partial or complete inclusion into the hydrophobic cavity (see Fig. 11.2). CD can be used to increase drug solubility and dissolution rate, decrease volatility, alter release rates, modify local irritation, and increase stability of drugs [20, 60].

There are three types of CDs used in pharmaceutical formulations: they are α-, β-, and γ-CD. The nomenclature is based on the number of glycopyranose units in a given CD ring; α-, β-, and γ-CD have six, seven, and eight units, respectively. Furthermore, the CD rings can be modified as seen in Fig. 11.3. In general, it has been found that drugs form favorable complexes with β-CD than α-CD, because the 6 Å cavity diameter of the β-CD better accommodates aromatic groups found on most drug molecules [20]. In addition to the difference in size of the three CD rings,

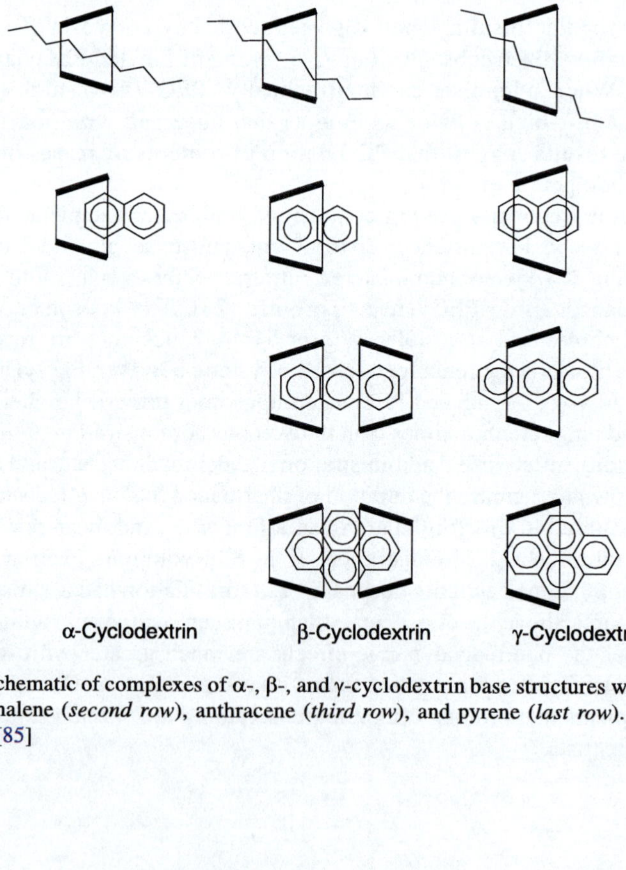

Fig. 11.2 Schematic of complexes of α-, β-, and γ-cyclodextrin base structures with decane (*top row*), naphthalene (*second row*), anthracene (*third row*), and pyrene (*last row*). Adapted from Yalkowksy [85]

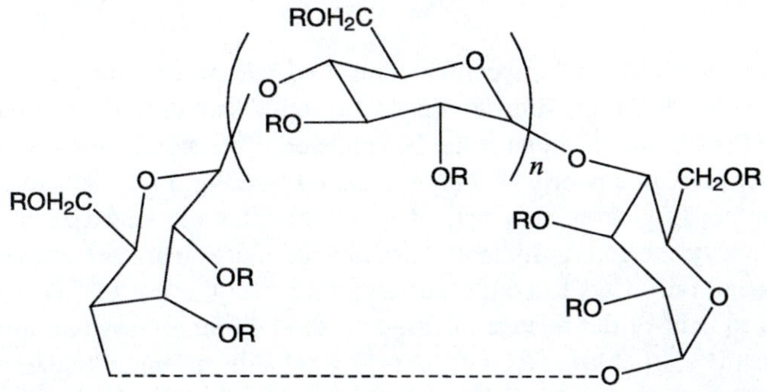

Cyclodextrin	R	*n*	Abbreviation
α-cyclodextrin	H	4	α-CD
β-cyclodextrin	H	5	β-CD
γ-cyclodextrin	H	6	γ-CD
Dimethyl-β-cyclodextrin	CH_3 or H	5	DM-β-CD
Methyl-β-cyclodextrin	CH_3 or H	5	M-β-CD
Random methyl-β-cyclodextrin	CH_3 or H	5	RM-β-CD
Hydroxyethyl-β-cyclodextrin	CH_2CH_2OH or H	5	HE-β-CD
Hydroxypropyl-β-cyclodextrin	$CH_2CHOHCH_3$ or H	5	HP-β-CD
Sulfobutylether-β-cyclodextrin	$(CH_2)_4SO_3Na$ or H	5	SBE-β-CD

Fig. 11.3 Chemical structures of various cyclodextrins. This is a general structure for cyclodextrin rings, with various R-groups and number of glycopyranose units. The abbreviations provided in this figure are used throughout the discussion of cyclodextrins within this chapter. Adapted from Rajewski and Stella [60]

there is also a difference in their water solubilities. Under ambient conditions, the aqueous solubilities of α-, β-, and γ-CDs are 13, 2, and 26% (w/w), respectively [20]. It is suggested that the decrease in the solubility of β-CD, despite a greater number of hydroxyl groups compared to that of α-CD, is due to the formation of internal hydrogen bond. There are multiple CDs that can be investigated for formulations. There are several FDA-approved products that contain CDs, including intravenous alprostadil (α-CD, Caverject®), oral and intravenous intraconazole (HP-β-CD, Sporanox®), and intramuscular ziprasidone maleate (SBE-β-CD, Geodon®).

The choice of which CD to include in a given formulation can be partly limited by its safety profile. CDs are eliminated renally without metabolism [60]. Generally, administration of high doses of CD is associated with nephrotoxicity. Specific modifications of CDs are also associated with specific toxicities [20]. For example, sulfated-CDs exhibit a heparin-like effect that increases blood clotting times, which greatly limits its applicability in drug delivery. At high concentrations, β-CD can cause hemolysis of erythrocytes. There are also accounts of damage to nasal mucosa with various CDs including DM-β-CD, M-β-CD, HP-β-CD, HE-β-CD, and SBE-β-CD, ranked from greatest to least potential for damage [20]. There is no record that CDs elicit an immune response in mammals.

11.5.2 How Do Cyclodextrins Function in Controlled Release?

The mechanism by which CDs confer controlled release of a drug may vary. One postulation is that CDs retard drug release through dissolution-limited mechanisms [7]. Slowed dissolution of drugs could be achieved if the association constant, K_{assoc}, is sufficiently high for a poorly water soluble drug (see Fig. 11.4). It is also suggested that CDs protect drugs from enzymatic degradation, thus allowing a sustained release effect [90]. Partly, the ambiguity in the mechanism of action arises from inconclusive results of using only CDs to confer sustained release; the use of CD with another mechanism to control the release of drugs in the pulmonary system appears to be more promising [50]. Alone, CDs do not offer reliable sustained release effects for all drugs or a particular class of drugs. However, as a function in polymer systems, CDs decrease drug diffusivity by increasing molecular weight with a complementary

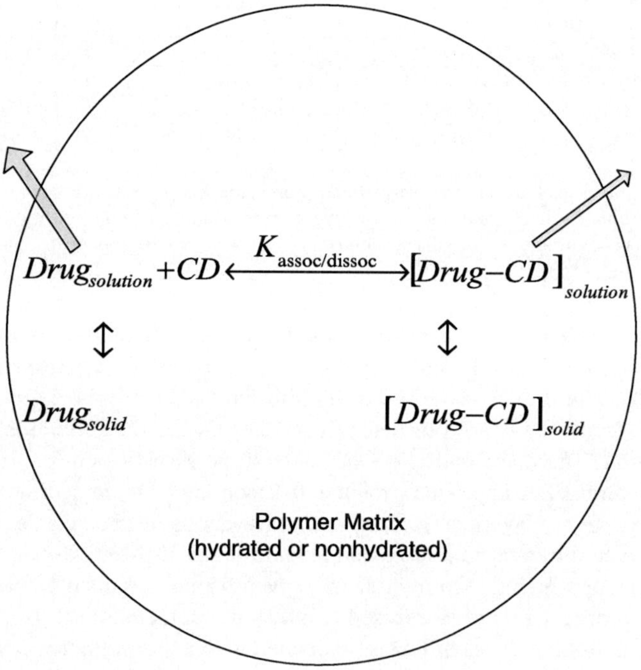

Fig. 11.4 Drug release mechanism from cyclodextrin-drug complex encapsulated in a polymer matrix. This figure presents the possible mechanistic dynamics of drug and drug-cyclodextrin (Drug-CD) complex release from a polymer matrix. At equilibrium between the association and dissociation of the complex, drug in solution and complexed drug can diffuse out of the polymeric matrix. However, if the free drug concentration is below the saturation point, less drug-cyclodextrin complexes will form and the total drug release from the polymeric microsphere will be less, thereby sustaining drug release. Adapted from Bibby et al. [7]

increase in complexation or cross-linking the polymer with the CD all of which may prolong the release of a drug from a dosage form [20]. Hence, the proceeding section reviews the use of CDs alone and with other excipients.

11.5.3 Evidence for the Use of Cyclodextrins in Controlled Release Formulations

11.5.3.1 Use of Cyclodextrins Only

There are no well-designed studies that present convincing evidence supporting the use of CDs alone in conferring sustained release of a given drug. Many of the studies present conflicting results, and it appears that the in vitro dissolution studies do not correlate well to the in vivo pharmacokinetics studies.

There are three notable studies that look at in vitro potential of using CDs in pulmonary delivery; a couple of these studies also look at short term in vivo toxicity [10, 34, 69]. In one study, CDs without drug were assessed for the feasibility of using them for pulmonary delivery [34]. It was found that HP-β-CD, γ-CD, and RM-β-CD can be successfully aerosolized and that the resulting droplet size is compatible with deposition in the respirable fraction of the lung. In this study, lung and kidney histology, bronchial responsiveness to methacholine, and blood urea levels appeared to be normal for all CDs, except for an increase in lymphocyte count in the bronchoalveolar lavage fluid associated with administration of γ-CD. A phase-solubility study found that the solubility of salbutamol varies with the type of CD used; greatest solubility of CD-complexed salbutamol was seen with DM-β-CD compared to other CDs (DM-β-CD > β-CD > γ-CD > α-CD) [10]. Larger stability constant values were found for DM-β-CD and β-CD indicating that salbutamol interacts more strongly with these CDs and that these CDs better accommodated salbutamol into the CD cavity. This study sets the precedence for the primary use of β-CD for salbutamol formulations. In an in vitro study, salbutamol was formulated as a dry powder with either DM-β-CD or γ-CD [69]. The formulation that contained γ-CD enhanced drug delivery to lower stages of the twin-stage impinger with 65% deposition compared to DM-β-CD (50%) or control formulation (40%). It was found that both formulations' drug release rates were fast: greater than 70% release in first 5 min and nearly all drug released within first half-hour. This data does not support sustained release properties of either of the CD studied. In addition, it was also found that DM-β-CD caused increase in blood urea nitrogen, creatinine, and hemolysis of red blood cells compared to γ-CD or control formulations in a short term in vivo toxicology study. While CDs can be successfully aerosolized, the in vitro data suggests that it cannot produce sustained release effects.

The in vivo findings relating to the efficacy of CD formulations in sustaining drug release for pulmonary delivered dosage forms remain ambiguous. A frequently cited study by Cabral Marques et al. presents the pharmacokinetics of three CDs and concludes that HP-β-CD is the only CD tested that had a decreased rate of absorption from rabbit lungs after intratracheal administration; the other two CDs studied were β-CD and DM-β-CD [11]. The mean absorption time (MAT) for HP-β-CD was 113 vs. 26 and 21 min for β-CD and DM-β-CD. The MAT values also exhibited differences between HP-β-CD and the other β-CDs that were consistent with controlled release kinetics. The T_{max} also varied greatly for HP-β-CD (113 vs. 30 min for β-CD and 22 min for DM-β-CD). Based on these pharmacokinetic results, the research group concluded that HP-β-CD has some potential as a drug carrier for sustained pulmonary delivery. Subsequently, Marques and colleagues utilized the information they gathered about HP-β-CD in formulating salbutamol with CDs and administering it to rabbits via intratracheal instillation [13]. Most of the pharmacokinetic parameters of HP-β-CD-complexed salbutamol suggest that there is some absorption rate-limitation effect, but the lack of a significant extended release profile indicates that while HP-β-CD has some potential in modifying the release of salbutamol, it may not be clinically substantial. Similar results were found in another study that complexed HP-β-CD with rolipram and testosterone for in vivo pulmonary delivery [80]. It was found through in vitro studies that HP-β-CD can decrease the rate of rolipram release by half; however, in vivo the $t_{1/2}$ of the drug was less than 60 s and the complexation with the CD did not alter rate of drug loss from the lungs. Since rolipram does not have high association strength with HP-β-CD, testosterone was also studied. Similar results were found with testosterone, suggesting that the rate of drug release does not solely depend on the association strength of the drug with the CD. Based on the studies conducted by Cabral Marques et al. and Wall et al., it is accurate to conclude that the physical chemical properties associated with the use of CD in sustained release formulation is not completely understood and that the characteristics of the drug influence these formulations to a great extent.

11.5.3.2 Use of Cyclodextrins in Concert with Other Excipients to Control Drug Release

CDs have been used with polymers to alter the release of medication with more promising results than formulations utilizing only CD inclusion complexes. First, it is important to appreciate the mechanisms by which CDs function in polymer delivery systems (see Fig. 11.4). One mechanism is diffusion-limited release of drug [7]. The solid drug exists within the hydrated polymer matrix, and equilibrium is established between free drug and CD complexes when the concentration of free drug is sufficiently above the saturation point. Thus, the presence of CD, in this case, does not decrease the concentration of free drug. Drug release, then, results from combined diffusion of free drug and drug-CD complexes, and the total drug release would be greater than that attained by the inclusion of drug into polymer microspheres. However, if the free drug present is at a concentration below the saturation point, then the addition of CD would decrease free drug concentration within the microsphere by complexation, which in turn would decrease drug diffusibility. Due to the decrease in the concentration gradient of free drug, there would be a decrease in total drug released compared to formulations with drug encapsulated in polymeric microspheres. Another mechanism by which polymers and CDs function in altering drug release is by functioning as channeling or wicking agents that promote erosion of the polymeric matrix. This hypothesis has been confirmed by microscopy, which presents erosion of the matrix by increased porosity. Moreover, both mechanisms can function simultaneously.

There are several examples of the use of PLGA and modified β-CD in promoting sustained release. In an in vitro study that sought to produce dry powder insulin formulation with PLGA and HP-β-CD, the CD was used as an absorption enhancer [76]. In formulations without HP-β-CD, a time lag of 3 days was seen prior to the start of insulin release. However, with the CD-PLGA formulation, an initial burst phase that released approximately 60% of the insulin was followed by a second slow release phase during which insulin was released from the microparticles up to 45 days. The release mechanism, proposed by the researchers, is that CD alters the diffusion of insulin from the polymer and also facilitates polymeric erosion. Similar results were seen in an in vivo study that encapsulated insulin and DM-β-CD complex with PLGA and administered to rats via intratracheal instillation [3]. It was found that the hypoglycemic response was efficient and prolonged compared to other formulations. Hence, PLGA may be used to confer sustained release of insulin complexed with CD via the pulmonary route.

Polymers, such as PLGA, are not the only excipient that can be combined with CDs to alter release profiles of various medications; liposomes can also be used. DSPC was utilized to deliver HP-β-CD complexes of dehydroepiandrosterone (DHEA), retinol, and dexamethasone by the intravenous route to rats [53]. Between 2 and 60 min postadministration, there was more DHEA released than retinol and dexamethasone; furthermore, the amount of dexamethasone released was the least among the three agents. The amount of drug found in plasma with the use of CD-liposome formulation was less than liposomally entrapped drug: for instance, for DHEA, 6.2% was released with liposome-only formulation while 4.9% was released with CD-liposome formulation. A similar trend as that seen for DHEA was seen for retinol and for dexamethasone. The proposed mechanism for this sustained release system is that drug displacement from the CD is followed by drug accommodation into the lipid bilayer of the liposome, which increases liposomal fluidity thus increasing permeability. Ultimately, these findings suggest that drug-CD delivery via liposomal systems can act as a means to control the release of a wide range of therapeutic agents. This study presents that this combination of liposome and CD has potential for sustaining the release of some drugs and may be explored via the pulmonary route.

11.6 Alternate Modalities to Sustain Drug Release

This section covers a class of miscellaneous excipients and modalities that control the release of drug based on limiting dissolution rate. In general, the formulations discussed are in the solid state and primarily rely on decreasing absorption rate of drugs by slowing dissolution of the drug from the dosage form [90]. Examples discussed in this section include amorphous sugars, microcrystals, and suspensions.

Amorphous sugars are produced when a suitable viscous solution of carbohydrates cools very rapidly to below its glass transition temperature (T_g), thereby not permitting sufficient time for the crystal lattice to form. For example, when sucrose is cooled slowly it results in crystalline sugar (i.e., hard candy), but when quickly cooled it can form cotton candy. Microcrystals are solid particles that comprise a collection of smaller crystalline drug particles, which can be seen under the microscope. Finally, suspensions are a heterogeneous mixture, where the particles suspended in the liquid form are solid and can be seen by the naked eye.

11.6.1 How Do These Modalities Function in Controlled Release?

As previously stated, modified release relies on the dissolution rate of the amorphous sugar, microcrystal, or solid particles in suspensions to control the release of drug from the dosage form and for absorption by the lung epithelium. Thus, properties of the excipient or drug that decrease dissolution rate at physiological and storage conditions can potentially be used for sustaining drug release.

One of the amorphous sugars commonly used to achieve controlled release is trehalose. Trehalose is a disaccharide with a high T_g of 118°C [30]. It is used to stabilize proteins as solid solutions in amorphous sugars, since it is a nonreducing sugar that is not reactive with peptides. Most sugars cannot be used for this purpose, because they are readily water-soluble or have a T_g significantly lower than typical drug storage conditions. It was determined that in a highly humid environment, such as within the lung, these microparticles absorb water that cause them to lose relaxation energy and crystallize [19]. The delay in crystallization depends on the amount of water present, and the release of the drug is temporally linked to the crystallization of the trehalose derivatives.

Microcrystals can be prepared by seed zone or conventional seeding methods to decrease the surface area to volume ratio of drugs, thus conferring sustained release. With reducing the surface area of protein crystals, decreased exposure to solvent is achieved, thus decreasing the overall solubilization rate. While this seems like an attractive mode for sustained drug release, large molecular weight and decreased flexibility of most proteins limit the application of microcrystallization [68].

Suspensions, similar to microcrystals, depend on the resistance of the solid to dissolve. In general, the greater the crystallinity, the slower the dissolution and corresponding absorption rate. Suspension formulations are generally compared to solution formulations of drugs. Solutions can be formed by dissolving salts of the drug in an appropriate solvent or through the use of a cosolvent.

11.6.2 *Examples of Alternate Modalities*

11.6.2.1 Amorphous Sugars

In the study described in Sect. 11.6.1 of this chapter relating to the release of drug from trehalose microparticles, sustained release of insulin was achieved [19]. Ester derivatives of trehalose were used to increase the lipophilicity of the microparticles, thus reducing the extent or rate of crystallization in hopes to achieve sustained release of insulin. It was found that insulin was detectable in rat plasma after 24 h post-intratracheal instillation at concentrations that are greater than that achieved with inhaled insulin solution. Based on these findings, it is evident that amorphous sugars, such as trehalose, can modify peptide release from an inhaled dosage form.

Chougule and colleagues compared spray dried tacrolimus with a liposomal formulation of tacrolimus with lactose, sorbitol, or trehalose in vitro and in vivo by intratracheal instillation in rats [15]. The liposome was composed of hydrogenated PC and CH at a molar ratio of 8:2. It was found that of the three liposomal formulations, the trehalose formulation was most favorable for forming low density particles with adequate flowability. The in vitro drug release profiles presented that 90% of the drug was released from the nonliposomal formulation within 6 h and from the liposomal formulations within 18 h. In vivo studies present that drug release was prolonged with the liposomal formulations (in order of drug release retardation: nonliposomal formulations sorbitol $<$ lactose $<$ trehalose). The terminal $t_{1/2}$ of tacrolimus was 3.9, 16, and 11 h from the nonliposomal, liposomal with trehalose, and liposomal with lactose formulations, respectively. The difference in half-lives for the various amorphous sugars used in the formulation suggests that the choice of carbohydrate can also alter drug release rate.

11.6.2.2 Microcrystals

Pulmonary delivery of microcrystal formulations of insulin has shown some success for both in vitro and in vivo studies [38]. Through in vitro analysis, it was found that the insulin microcrystals formed homogenous rhombohedral shapes with some rhombus forms without aggregates that are suitable for pulmonary delivery. In diabetic rats, it was found that intratracheal instillation of a microcrystal suspension reduced glucose levels and prolonged hypoglycemia over 13 h compared to insulin solution; the T_{max} of solution and suspension were 2 and 3 h, respectively. In this case, the zinc-insulin hexamer must dissociate from the microcrystal prior to being absorbed into the circulatory system via the alveolar epithelium. With the solution formulation, the dissociation step does not apply, thus the absorption rate is not limited by drug dissolution rate. Overall, a greater total reduction in blood glucose was seen in rats receiving the microcrystal formulation vs. the insulin solution (34.4 vs. 25.0%). In addition, it was found that with a challenge glucose bolus, the rats that received insulin solution experienced an increase in blood glucose to 90% of the baseline glucose value, whereas those that received the microcrystal formulation only experienced an increase to 50% of baseline glucose value. However, the difference was not clinically significant to demonstrate the feasibility of using microcrystals in sustaining insulin release in the lungs.

11.6.2.3 Suspensions

The use of suspensions in prolonging drug release can be presented through an in situ study of rat lungs that were assayed for unabsorbed xanthone derivatives from a sodium salt solution and suspension formulations [16]. It was found that the suspension was absorbed three to four times slower than the solution formulation. The $t_{1/2}$ of absorption of the solution was found to be 13 min and that of the suspension was 61.7 min for one of the two derivatives tested. For the second derivative, a two-stage absorption pattern was seen that comprised an initial absorption of the suspension which was slower than that of the solution, but the terminal absorption rates were similar with an absorption $t_{1/2}$ of 3 min. The derivative with the shorter absorption $t_{1/2}$ has an octanol-water partition coefficient greater than that of the other derivative (2,493 vs. 59). The differences between the two xanthone derivatives' octanol-water partition coefficient suggest that the increased lipid solubility increases the rate of absorption of the drug from suspension. The dissolution of the xanthone derivatives is limited by the formulation's chemical properties. Once dissolved in the lungs, xanthones are absorbed by diffusion across the lipoidal region of the lung, and the absorption of the organic acid is primarily controlled by lipid solubility of the unionized species of the drug.

11.7 Conclusion

As presented throughout this chapter, there are a plethora of excipients used for experimental formulations of pulmonary-administered agents. There are several formulations such as the liposomal cisplatin, sodium cromoglycate, and fentanyl that have been tested in humans by inhalation delivery systems. The limited use of such excipients is indicative of the difficulty of selecting an excipient that will not limit aerosolization performance, is safe with little-to-no immunogenicity, and is effective in sustaining drug release. More about the regulatory requirements can be found in later chapters of this book. As standard testing methodologies are developed and a better understanding of the chemical interaction of the drug and excipient is understood, it is likely that the list of potential excipients will grow and more inhaled drug products that are sustained release will become available.

References

1. Adi H, Young PM, Chan H, Salama R, Traini D (2010) Controlled release antibiotics for dry powder lung delivery. Drug Dev Ind Pharm 31:119–126
2. Alhusban FA, Seville PC (2009) Carbomer-modified spray-dried respirable powders for pulmonary delivery of salbutamol sulfate. J Microencapsul 26:444–455
3. Aguiar MMG, Rodrigues JM, Cunha AS (2004) Encapsulation of insulin-cyclodextrin complex in PLGA microspheres: a new approach for prolonged pulmonary insulin delivery. J Microencapsul 21:553–564
4. Arnold MM, Gorman EM, Schieber LJ, Munson EJ, Berkland C (2007) NanoCipro encapsulation in monodisperse large porous PLGA microparticle. J Control Release 121:100–109
5. Arppe J, Vidgren M, Waldrep JC (1998) Pulmonary pharmacokinetics of cyclosporine A liposomes. Int J Pharm 161:205–214
6. Bennett DB, Tyson E, Mah S, de Groot JS, Hegde SG, Terao S, Teitelbaum Z (1994) Sustained delivery of detirelix after pulmonary administration of liposomal formulations. J Control Release 32:27–35
7. Bibby DC, Davies NM, Tucker IG (2000) Mechanisms by which cyclodextrins modify drug release from polymeric drug delivery systems. Int J Pharm 197:1–11
8. Brannon-Peppas L (1997) Polymers in controlled drug delivery. Med Plast Biomater 4:34–44
9. Bueche B, Song Y, Lobo J, Leach C, Guo L, Viegas T, Bentley M, Kuo M (2003) PEGylated insulin for pulmonary delivery: characterization and formulation challenges. AAPS J 5:S1
10. Cabral Marques HM, Hadgraft J, Kellaway IW (1990) Studies of cyclodextrin inclusion complexes. I. The salbutamol-cyclodextrin complex as studied by phase solubility and DSC. Int J Pharm 63:259–266
11. Carbral Marques HM, Hadgraft J, Kellaway IW, Taylor G (1991) Studies of cyclodextrin inclusion complexes. IV. The pulmonary absorption of salbutamol from a complex with 2-hydroxypropyl-β-cyclodextrin in rabbits. Int J Pharm 77:303–307
12. Carbral Marques HM, Hadgraft J, Kellaway IW, Taylor G (1991) Studies of cyclodextrin inclusion complexes. III. The pulmonary absorption of β-, DM-β- and HP-β-cyclodextrins in rabbits. Int J Pharm 77:297–302
13. Chimote G, Banerjee R (2009) Evaluation of antitubercular drug-loaded surfactants as inhalable drug-delivery systems for pulmonary tuberculosis. J Biomed Mater Res 89A: 281–292
14. Chono S, Fukuchi R, Seki T, Morimoto K (2009) Aerosolized liposomes with' dipalmotoyl phosphatidylcholine enhance pulmonary insulin delivery. J Control Release 137:104–109
15. Chougule M, Padhi B, Misra A (2007) Nano-liposomal dry powder inhaler of tacrolimus: preparation, characterization, and pulmonary pharmacokinetics. Int J Nanomed 2: 675–688
16. Chowhan ZT, Amaro AA (1976) Pulmonary absorption studies utilizing in situ rat lung model: designing dosage regiment for bronchial delivery of new drug entities. J Pharm Sci 65:1669–1672
17. Cook RO, Pannu RK, Kellaway IW (2005) Novel sustained release microspheres for pulmonary drug delivery. J Control Release 104:79–90
18. Coowanitwong I, Arya V, Kulvanich P, Hochhaus G (2008) Slow release formulations of inhaled rifampin. AAPS J 10:342–348
19. Davidson IG, Langner EJ, Plowman SV, Blair JA (2003) Release mechanism of insulin encapsulated in trehalose ester derivative microparticles delivered via inhalation. Int J Pharm 254:211–222
20. Davis ME, Brewster ME (2004) Cyclodextrin-based pharmaceutics: past, present and future. Nat Rev 3:1023–1035
21. Donovan M, McGill S, Smyth H (2008) Swellable particles for sustained release drug delivery to the lungs. Respir Drug Deliv 3:857–860

22. El-Baseir MM, Kellaway IW (1998) Poly(L-lactic acid) microspheres for pulmonary drug delivery: release kinetics and aerosolization studies. Int J Pharm 175:135–145
23. Fielding RM, Abra RM (1992) Factors affecting the release rate of terbutaline from liposome formulations after intratracheal instillation in the guinea pig. Pharm Res 9:220–223
24. Fishburn CS (2008) The pharmacology of PEGylation: balancing PD with PK to generate novel therapeutics. J Pharm Sci 97:4167–4183
25. Fu J, Fiegel J, Hanes J (2004) Synthesis and characterization of PEG-based ether anhydride terpolymers: novel polymers for controlled drug release. Macromolecules 37:7174–7180
26. Fu J, Fiegel J, Krauland E, Hanes J (2002) New polymeric carriers for controlled drug delivery following inhalation or injection. Biomaterials 23:4425–4433
27. Gaber NN, Darwis Y, Peh K, Tan YT (2006) Characterization of polymeric micelles for pulmonary delivery of beclomethasone dipropionate. J Nanosci Nanotechnol 6:3095–3101
28. Garbuzenko OB, Saad M, Betigeri S, Zhang M, Vetcher AA, Soldatenkov VA, Reimer DC, Pozharov VP, Minko T (2009) Intratracheal versus intravenous liposomal delivery of siRNA, antisense oligonucleotides and anticancer drug. Pharm Res 26:382–394
29. Garcia-Contreras L, Morçöl T, Bell SJD, Hickey AJ (2003) Evaluation of novel particles as pulmonary delivery systems for insulin in rats. AAPS PharmSci 5(2):article 9
30. Hardy JG, Chadwick TS (2000) Sustained release drug delivery to the lungs: an option for the future. Clin Pharmacokinet 39:1–4
31. Ho DHW, Wang CY, Lin JR, Brown N, Newman RA, Krakoff IH (1987) Polyethylene glycol-L-asparaginase and L-asparaginase studies in rabbits. Drug Metab Dispos 16:27–29
32. Huang Y, Wang C (2006) Pulmonary delivery of insulin by liposomal carriers. J Control Release 113:9–14
33. Hung OR, Whynot SC, Varvel JR, Shafer SL, Mezei M (1995) Pharmacokinetics of inhaled liposome-encapsulated fentanyl. Anesthesiology 83:277–284
34. Ikeda Y, Kimura K, Hirayama F, Arima H, Uekama K (2000) Controlled release of a water-soluble drug, captopril, by combination of hydrophilic and hydrophobic cyclodextrin derivatives. J Control Release 66:271–280
35. Kawashima Y, Yamamoto H, Takeuchi H, Fujioka S, Hino T (1999) Pulmonary delivery of insulin with nebulized DL-lactide/glycolide copolymer (PLGA) nanospheres to prolong hypoglycemic effect. J Control Release 62:279–287
36. Knight V, Koshkina NV, Waldrep JC, Giovanella BC, Gilbert BE (1999) Anticancer exffect of 9-nitrocamptothecin liposome aerosol on human cancer xenografts in nude mice. Cancer Chemother Pharmacol 44:177–186
37. Koushik K, Dhanda DS, Cheruvu NPS, Kompella UB (2004) Pulmonary delivery of deslorelin: large-porous PLGA particles and HPβCD complexes. Pharm Res 21:1119–1126
38. Kwon J, Lee B, Lee J, Kim C (2004) Insulin microcrystal suspension as a long acting formulation for pulmonary delivery. Eur J Pharm Sci 22:107–116
39. Latimer P, Menchaca M, Snyder RM, Yu W, Gilbert BE, Sanders BG, Kline K (2009) Aerosol delivery of liposomal formulated paclitaxel and vitamin E analog reduces murine mammary tumor burden and metastases. Exp Biol Med 234:1244–1252
40. Leach CL, Hameister WM, Tomai MA, Hammerbeck DM, Stefely JS (2000) Oligolactic acid (OLA) biomatricies for sustained release of asthma therapeutics. Respir Drug Deliv VII 2:291–294
41. Leach CL, Patton JS, Perkins KM, Kuo M, Roberts MJ, Guo L, Michael B (2002) PEG-insulin delivered by the pulmonary route provides prolonged systemic activity compared with insulin alone. AAPS J 4(S1). Available from: http://www.aapsj.org/
42. Leach CL, Kuo M, Bueche B, Fishburn S, Viegas T, Bossard M, Guo L, Bentley MD, Hobbs CH, Cherrington AD, Patton JS (2004) Modifying the pulmonary absorption and retention of proteins through PEGylation. Respir Drug Deliv IX 1:69–77
43. Learoyd TP, Burrows JL, French E, Seville PC (2006) Leucine-modified PLGA-based respirable spray-dried powders for sustained drug delivery. Respir Drug Deliv 2:437–440
44. Learoyd TP, Burrows JL, French E, Seville PC (2006) Sustained drug delivery from chitosan-based respirable spray-dried powders. Respir Drug Deliv 2:441–443

45. Learoyd TP, Burrows JL, French E, Seville PC (2008) Chitosan-based spray-dried respirable powders for sustained delivery of terbutaline sulfate. Eur J Pharm Biopharm 68:224–234
46. Learoyd TP, Burrows JL, French E, Seville PC (2009) Sustained delivery by leucine modified chitosan spray-dried respirable powders. Int J Pharm 372:97–104
47. Li FQ, Hu JH, Lu B, Yao H, Zhang WG (2001) Ciprofloxacin-loaded bovine serum albumin microspheres: preparation and drug-release in vitro. J Microencapsul 18:825–829
48. Li Z, Stefely JS (2004) Aerosol sustained release formulations with oligolactic acid (OLA). Respir Drug Deliv IX 2:321–324
49. Li Z, Zhang Y, Wurtz W, Lee JK, Malinin VS, Durwas-Krishnan S, Meers P, Perkins WR (2008) Characterization of nebulized liposomal amikacin (Arikase™) as a function of droplet size. J Aerosol Med Pulm Drug Deliv 21(3):245–254
50. Louey MD, Garcia-Contreras L (2004) Controlled release products for respiratory delivery. Am Pharm Rev 7:82–87
51. Lu B, Zhang Q, Yang H (2003) Lung-targeting microspheres of carboplatin. Int J Pharm 265:1–11
52. Lui F, Shao Z, Kildsig DO, Mitra AM (1993) Pulmonary delivery of free and liposomal insulin. Pharm Res 10:228–232
53. McCormack B, Gregoriadis G (1996) Comparative studies of the fate of free and liposome-entrapped hydroxypropyl-β-cyclodextrin/drug complexes after intravenous injection into rats: implications in drug delivery. Biochim Biophys Acta 1291:237–244
54. Morimoto K, Katsumata H, Yabuta T, Iwanaga K, Kakemi M, Tabata Y, Ikada Y (2002) Gelatin microsphere as a pulmonary delivery system: evaluation of salmon calcitonin absorption. J Pharm Pharmacol 52:611–617
55. Morimoto K, Metsugi K, Katsumata H, Iwanaga K, Kakemi M (2001) Effects of low-viscosity sodium hyaluronate preparation on pulmonary absorption of rh-insulin in rats. Drug Dev Ind Pharm 27:365–371
56. Mu L, Feng SS (2001) Fabrication, characterization and in vitro release of paclitaxel (Taxol®) loaded poly (lactic-co-glycolic acid) microspheres prepared by pray drying technique with lipid/cholesterol emulsifiers. J Control Release 76:239–254
57. Müller RH, Maaβen S, Wayhers H, Specht F, Lucks JS (1996) Cyctotoxcity of magnetite-loaded polylactide, polylactide/glycolide particles and solid lipid nanoparticles. Int J Pharm 138:85–94
58. Nah J, Jeong Y, Cho C, Kim S (2000) Drug-delivery system based on core-shell type nanoparticles composed of poly(γ-benzyl-L-glutamate) and poly(ethylene oxide). J Appl Polym Sci 75:1115–1126
59. Naikwade S, Bajaj A (2009) Preparation and in vitro evaluation of budesonide spray dried microparticles for pulmonary delivery. Sci Pharm 77:419–441
60. Rajewski RA, Stella VJ (1996) Pharmaceutical applications of cyclodextrins. 2. In vivo drug delivery. J Pharm Sci 85:1142–1169
61. Rouse JJ, Whateley TL, Thomas M, Ecclecston GM (2007) Controlled drug delivery to the lungs: influence of hyaluronic acid solution conformation on its adsorption to hydrophobic drug particles. Int J Pharm 330:175–182
62. Ruan G, Feng S (2003) Preparation and characterization of poly(lactic acid) poly(ethylene glycol)-poly(lactic acid) (PLA-PEG-PLA) microspheres for controlled release of paclitaxel. Biomaterials 24:5037–5044
63. Salama RO, Traini D, Chan H, Sung A, Ammit AJ, Young PM (2009) Preparation and evaluation of controlled release microparticles for respiratory protein therapy. J Pharm Sci 98: 2709–2717
64. Salama RO, Traini D, Chan H, Young PM (2008) Preparation and characterisation of controlled release co-spray dried drug-polymer microparticles for inhalation 2: evaluation of in vitro release profiling methodologies for controlled release respiratory aerosols. Eur J Pharm Biopharm 70:145–152
65. Schliecker G, Schmidt G, Fusch S, Wombacher R, Kissel T (2003) Hydrolytic degradation of poly(lactide-co-glycolide) films: effect of oligomers on degradation rate and crystallinity. Int J Pharm 266:39–49

66. Schreier H, Gonzalez-Rothi RJ, Stecenko AA (1993) Pulmonary delivery of liposomes. J Control Release 24:209–223
67. Schreier H, McNicol KJ, Ausborn M, Soucy DM, Derendorf H, Stecenko AA, Gonzalez-Rothi RJ (1992) Pulmonary delivery of amikacin liposomes and acute liposome toxicity in the sheep. Int J Pharm 87:183–193
68. Shoyele SA (2008) Controlling the release of proteins/peptides via the pulmonary route. Drug Deliv Syst 437:141–148
69. Srichana T, Suedee R, Reanmongkol W (2001) Cyclodextrin as a potential drug carrier in salbutamol dry powder aerosols: the in vitro deposition and toxicity studies of the complexes. Respir Med 95:513–519
70. Stefely JS, Duan DC, Myrdal PB, Ross DL, Schultz DW, Leach CL (2000) Design and utility of a novel class of biocompatible excipients for HFA-based MDIs. Respir Drug Deliv VII 1:83–90
71. Suarez S, Gonzalez-Rothis RJ, Schreier H, Hochhaus G (1998) Effect of dose and release rate on pulmonary targeting of liposomal triamcinolone acetonide phosphate. Pharm Res 15:461–465
72. Sung JC, Padilla DJ, Garcia-Contreras L, VerBerkmoes JL, Durbin D, Peloquin CA, Elbert KJ, Hickey AJ, Edwards DA (2009) Formulation and pharmacokinetics of self-assembled rifampicin nanoparticle systems for pulmonary delivery. Pharm Res 26:1847–1855
73. Surendrakumar K, Martyn GP, Hodgers ECM, Jansen M, Blair JA (2003) Sustained release of insulin from sodium hyaluronate based dry powder formulations after pulmonary delivery of beagle dogs. J Control Release 91:385–394
74. Taylor KMG, Taylor G, Kellaway IW, Stevens J (1989) The influence of liposomal encapsulation on sodium cromoglycate pharmacokinetics in man. Pharm Res 6:633–636
75. Tahara K, Sakai T, Yamamoto H, Takeuchi H, Hirashima N, Kawashima Y (2009) Improved cellular update of chitosan-modified PLGA nanospheres by A549 cells. Int J Pharm 382:198–204
76. Ungaro F, De Rosa G, Miro A, Quaglia F, La Rotonda MI (2006) Cyclodextrins in the production of large porous particles: development of dry powders for the sustained release of insulin in the lungs. Eur J Pharm Sci 28:423–432
77. Veronese FM, Pasut G (2005) PEGylation, successful approach to drug delivery. Drug Discov Today 10:1451–1458
78. Waldrep JC, Arppe J, Jansa KA, Vidgren M (1998) Experimental pulmonary delivery of cyclosporine A by liposome aerosol. Int J Pharm 160:239–249
79. Waldrep JC, Gilbert BE, Knight CM, Black MB, Scherer PW, Knight V, Eschenbacher W (1997) Pulmonary delivery of beclomethasone liposome aerosol in volunteers: tolerance and safety. Chest 111:316–323
80. Wall DA, Marcello J, Pierdomenico D (1994) Administration as hydroxylpropyl-β-cyclodextrin complexes does not slow rates of pulmonary drug absorption in rat. STP Pharma Sci 4:63–68
81. Williams AS, Taylor G (1992) Synthesis, characterization and release of cromoglycate from dextran conjugates. Int J Pharm 83:233–239
82. Wischke C, Schwendeman SP (2008) Principles of encapsulating hydrophobic drugs in PLA/PLGA microparticles. Int J Pharm 364:298–327
83. Wittgen BPH, Kunst PWA, van der Born K, van Wijk AW, Perkins W, Filkiewicz FG, Perez-Soler R, Nicholson S, Peters GJ, Postmus PE (2007) Phase I study of aerosolized SLIT cisplatin in the treatment of patients with carcinoma of the lung. Clin Cancer Res 13:2414–2421 http://clincancerres.aacrjournals.org/content/13/8/2414.short
84. Wu L, da Rocha SRP (2007) Biocompatible and biodegradable copolymer stabilizers for hydrofluoroalkane dispersions: a colloidal probe microscopy investigation. Langmuir 23:12104–12110
85. Yalkowksy SH (1999) Solubility and solubilization in aqueous media. Oxford University Press, New York

86. Yamada K, Kamada N, Odomi M, Okada N, Nabe T, Fujita T, Kohno S, Yamamoto A (2005) Carrageenans can regulate the pulmonary absorption of antiasthmatic drugs and their retention in the rat lung tissues without any membrane damage. Int J Pharm 293:63–72
87. Yamada K (2007) Control of pulmonary absorption of drugs by various pharmaceutical excipients. Yakugaku Zasshi 127:631–641
88. Yamamoto H, Kuno Y, Sugimoto S, Takeuchi H, Kawashima Y (2005) Surface modified PLGA nanosphere with chitosan improved pulmonary delivery of calcitonin in mucoadhesion and opening of the intracellular tight space. J Control Release 102:373–381
89. Yang Y, Bajaj N, Xu P, Ohn K, Tsifansky MD, Yeo Y (2009) Development of highly porous large PLGA microparticles for pulmonary drug delivery. Biomaterials 30:1947–1953
90. Zeng XM, Martin GP, Marriott C (1995) The controlled delivery of drugs to the lung. Int J Pharm 124:149–164
91. Zhang Q, Shen Z, Nagai T (2001) Prolonged hypoglycemic effect of insulin-loaded polybutylcyanoacrylate nanoparticles after pulmonary administration to normal rats. Int J Pharm 218:75–80

Chapter 12
Polymers for Pulmonary Drug Delivery

Poonam Sheth and Paul B. Myrdal

Abstract One method by which to modify the release rate of a drug in the pulmonary tract, is through the use of polymeric systems. There are several natural and synthetic polymers that have been studied in vitro or in vivo in animals to assess pulmonary drug delivery and sustained release. This chapter focuses on some of the more common non-conjugated polymers utilized for inhalation drug delivery and draws examples from an array of pre-clinical findings.

Keywords Inhalation • Polymers • Sustained release

12.1 What Are Polymers?

Polymers are macromolecules that are composed of covalently bonded repeating structures. Polymers can be attached to or encapsulate drugs to prolong the effects of drugs. Albumin, chitosan, and hyaluronic acid are the most widely used natural polymers; while the most frequently used synthetic polymers are poly(lactic-co-glycolic acid) (PLGA), poly(lactic acid) (PLA), and poly(ethylene glycol) (PEG) [28]. Of these polymers, PEG is frequently incorporated into sustained release formulations by covalently bonding PEG to the active drug to increase molecular weight; details about how PEG functions in controlling drug release can be found in Chap. 11 [16, 46]. The other polymers mentioned above are generally used to encapsulate drug particles and are further discussed in this chapter. Chapter 11 also contains a table with the chemical structure or description of the polymers described in this chapter.

Polymers can be used to make formulations that contain nanoparticles, microparticles, and large porous particles (LPP) for pulmonary drug delivery. In context of

P.B. Myrdal (✉)
University of Arizona, College of Pharmacy, 1703 East Mabel Street, Tucson, AZ 85721, USA
e-mail: myrdal@pharmacy.arizona.edu

H.D.C. Smyth and A.J. Hickey (eds.), *Controlled Pulmonary Drug Delivery*, Advances in Delivery Science and Technology, DOI 10.1007/978-1-4419-9745-6_12,

inhalation drug delivery, particle size is an essential consideration to achieve pulmonary deposition. While the probability of lung deposition varies as a function of particle size, the most respirable particles have aerodynamic diameters between 0.5 and 5 μm. This can be characterized as spherical microparticles having geometric medians between 0.5 and 5 μm with a unit density. As density changes, the aerodynamic diameter will follow as a square root function of the density. For most polymers, the raw material densities will be greater than one (ca. 1.1–1.5 g/cm^3). Nanoparticles are small colloidal particles that have a geometric diameter of 1–1,000 nm [48]. Due to their small size, they can increase dissolution properties, enhance bioavailability of hydrophobic drugs, and deliver drugs intracellularly [5]. Disadvantages with nanoparticles include increased particle–particle interactions in the dosage form that decrease the functionality of aerosolization, and poor pulmonary drug delivery efficiency due to decreased inertia which does not permit deposition into the respirable fraction of the lung [48]. Decreased drug delivery efficiency occurs because individual nanoparticles do not follow classical deposition pathways: sedimentation takes greater time for smaller particles than larger particles and Brownian diffusion predominates. To overcome these issues, microparticle composites are frequently used. Microparticle composites are composed of nanoparticles, containing the drug, which are held together within a polymeric matrix; the nanoparticles may dissociate in vivo from the microparticle carrier, prior to drug dissolution and absorption. LPPs, having geometric diameters much greater than that of nanoparticles (5–30 μm for LPPs), also have low mass densities and as a result can have aerodynamic diameters less than 5 μm [12]. Since these particles have aerodynamic diameters comparable to smaller, nonporous particles, they can be deposited in the lungs with aerosolization because particle mass diameter dictates aerosol deposition. LPPs are advantageous for pulmonary drug delivery because they decrease the tendency of drug agglomeration compared to small, nonporous particles and can escape phagocytotic clearance due to increased particle size [11].

12.2 How Do Polymers Function in Controlled Release?

Polymers can function in controlled release by a variety of mechanisms that depend upon the unique properties of the polymer-drug system. As illustrated in Fig. 12.1, three main mechanisms of modifying drug release are drug diffusion, polymer swelling (followed by drug diffusion), and polymer degradation. Drug diffusion from a polymer system requires drug to pass through the polymeric matrix to the external environment [6]. In this system, the polymer is stable in the physiologic environment and does not degrade or swell during the process; the drug molecules pass through the pores of the polymer structure. Due to the nature of diffusion dependency, the rate of drug diffusion decreases over time since the drug particles progressively have a longer distance to travel, which requires a greater time. The rate of release is controlled by concentration and solubility of the drug and the type, porosity, and tortuosity of the polymer matrix.

The mechanism of drug release dependent upon swelling of the polymeric matrix is similar to diffusion-dependent drug release since it generally requires drug to diffuse

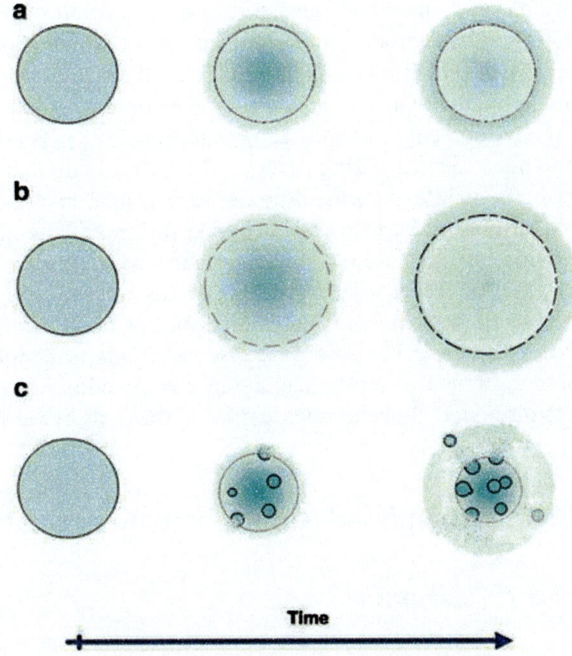

Fig. 12.1 Mechanisms of drug delivery via polymeric systems. The three basic mechanisms of drug release from polymeric drug delivery systems include (**a**) diffusion, (**b**) matrix swelling, and (**c**) erosion and/or degradation. The *shading* in this figure represents relative concentrations; darker the color the greater the concentration

Table 12.1 Physiologic environment changes which stimulate hydrogel swelling

Stimulus	Hydrogel	Mechanism
pH	Acidic or basic hydrogel	Change in pH leads to swelling
Ionic strength	Ionic hydrogel	Change in ionic strength due to change in concentration of ion within the gel results in swelling
Chemical species	Hydrogel with electron-accepting groups	Electron-donating compounds and formation of charge/transfer complexes result in swelling
Thermal	Thermoresponsive hydrogel (i.e., poly (*N*-isopropylacrylamide))	Change in temperature alters polymer– and water–polymer interactions which leads to swelling

Table adapted from Brannon-Peppas [6]

away from the polymer system as polymer swelling progresses. When polymers that utilize this mechanism are placed in physiological environments, they can absorb fluid from the surrounding environment to initiate swelling. This increase in aqueous solvent content within the formulation and subsequent increase in pore size of the polymer matrix facilitates drug diffusion, especially for water soluble molecules. Environmental changes such as a change in acidity, temperature, or ionic strength can trigger polymeric swelling (see Table 12.1).

Drug release by degradation generally occurs by hydrolysis or bulk erosion, which results in decreasing the molecular weight of the polymer until dissolution occurs. For sustained release, the drug particles are either coated with slowly dissolving material or incorporated into a slowly dissolving matrix. The rate of release of the drug is then controlled by the size of drug particles, core particle formulation, the type and thickness of the coating, or the matrix and the physiochemical environment where the drug is released [28]. There are two ways of degrading polymers to release drug particles: bulk degradation and surface erosion. Bulk degradation disintegrates the drug delivery system in a fairly uniform manner throughout the matrix. Surface erosion breaks down only the surface exposed to the solvent, resulting in a release rate that is proportional to the surface area of the drug delivery system [6]. Most microparticle systems release drug by bulk erosion; however, polymeric systems with polyanhydrides or polyorthoesters utilize surface erosion. While these mechanisms were discussed independent of each other, the polymeric drug systems frequently degrade in vivo by combining two or more of the modalities discussed above to release drug to the body for absorption.

12.3 Examples of Natural Polymers Used for Sustained Release

12.3.1 Albumin

Albumin refers to any protein that is moderately soluble in concentrated salt solutions and experiences heat coagulation. Albumin can be found in egg whites and in blood. Bovine serum albumin microspheres were used to develop a ciprofloxacin dry powder inhalation formulation [26]. The formulations tested include ciprofloxacin:albumin at molar ratios of 1:1, 1:2, and 1:4. It was found that the characteristics of the formulation made it possible to deliver the drug via the pulmonary route. Furthermore, drug release can be retarded by thermal denaturation of the ciprofloxacin-loaded microspheres. BSA microspheres can be stabilized by heat treatment, which induces the formation of cross-links that increase the hydrolytic resistance of the microspheres. With prolonging heat treatment and/or elevating temperature, the water solubility of the microspheres was decreased and the time for in vitro enzymatic degradation was prolonged. Overall, it was found that the release rate of ciprofloxacin from the microsphere formulations was slower than the dissolution rate of the pure drug.

12.3.2 Carrageenan

Carrageenans are naturally occurring sulfated polysaccharides that fill the void space in the cellulose backbone of certain species of red seaweed. There are three main commercial classes of carrageenans: kappa, iota, and lambda. Kappa is in the form of a rigid gel, while iota forms soft gels; lambda does not form a gel and is used as a thickening agent in dairy products. The difference between the three forms of carrageenan is mediated by the number and positions of ester sulfate groups on the repeating galactose units (see Table 11.3). Theophylline- and fluticasone-loaded carrageenans were studied in vivo in rats to determine drug release profile after pulmonary administration [57, 58]. In these studies, the effects of gelatin, iota-, and kappa-carrageenans and sodium alginate were determined on the absorption of drugs. The sodium alginate formulation decreased the maximum plasma concentration (C_{max}) values of the test drugs without prolonging the duration of detectable serum concentrations compared to control treatments. In addition, gelatin did not significantly affect pharmacokinetic parameters of theophylline or fluticasone. Theophylline release could be significantly regulated by 1% iota-carrageenan, which decreased C_{max} to one third of the control. Moreover, the 1% iota-carrageenan formulation prolonged the time to maximum plasma concentration (T_{max}) and doubled the area under the curve (AUC) of the theophylline plasma concentration-time plot. Fluticasone release, on the other hand, was best regulated with 0.5% kappa-carrageenan. This formulation prolonged T_{max}, decreased C_{max}, and nearly doubled AUC compared to control. All of the formulations, except 1% lambda-carrageenan, appeared to have the same toxicities as phosphate-buffered saline in terms of an increase in neutrophils, total protein, and lactose dehydrogenase in bronchoalveolar lavage fluid at 24 h postadministration. One percent lambda-carrageenan appears to elicit an immune and inflammatory response in rats. Overall, Yamada et al. present that carrageenans can be used to modify drug release and different classes may preferentially modify the release of a certain drug.

12.3.3 Chitosan

Chitosan, along with polymers such as hydroxypropylcellulose and poly(hydroxyethyl-methacrylate), is a mucoadhesive polymer that can prolong the residence time of drug carriers at the absorption site by preferential interactions with mucous membranes [36, 52]. Chitosan is a biocompatible and biodegradable polysaccharide derived by deacetylation of naturally occurring chitin [36]. With the use of mucoadhesive nanoparticles, one would expect to see improved drug absorption and potential controlled drug release [52]. In addition, these carriers may also protect the encapsulated drug from enzymatic degradation until the drug is released from the drug delivery system.

Chitosan microparticles and nanoparticles have been used in a number of in vitro studies to prolong drug release. In comparison to gelatin microparticles and LPPs, chitosan microparticles and LPPs exhibited favorable extended release of budesonide from a dry powder formulation [36]. With chitosan, drug release was extended over 4 h with high respirable fractions compared to only 10 min with gelatin. The half-life of drug release for each formulation was as follows: gelatin microsphere, 4 min; gelatin LPP, 2 min; chitosan microsphere, 27 min; and chitosan LPP, 56 min. By monitoring drug content, content uniformity, and in vitro drug release, it was determined that chitosan microparticles released drug by diffusion of the content through the matrix with zero-order kinetics. Formulations with chitosan along with the aerosolization enhancer, leucine, have also been able to modify drug release of terbutaline sulfate, salbutamol, and beclomethasone dipropionate [23–25]. Similar extended drug release results can be seen when chitosan is used to modify PLGA nanospheres [51, 59]. Chitosan-modified PLGA nanospheres were used to improve pulmonary delivery of elcatonin, a calcitonin derivative [59]. It was found that chitosan-modified PLGA nanospheres were eliminated more slowly and enhanced absorption of elcatonin from the lungs of guinea pigs than PLGA nanospheres. The elimination rate constant of the chitosan-modified PLGA formulation from the lung was approximately one third of the nonmodified PLGA nanospheres. In addition, serum calcium levels decreased to 80% of the baseline value and the effect was prolonged over 24 h with the administration of the chitosan-modified formulation, which is significantly longer than that of the PLGA nanosphere formulation. Yamamoto et al. suggest that the mucoadhesive property of chitosan prevented clearance of the nanosphere from the lungs. Furthermore, the results present that chitosan-modified PLGA reduce serum calcium levels by a greater extent during the initial phase than elcatonin solution or elcatonin-loaded PLGA nanospheres. Based on these findings, the authors suggest that chitosan functions in opening intracellular tight junctions, much like it does in the gastrointestinal system, to improve drug absorption. A recent study by the researchers that presented the effects of chitosan-modified PLGA interestingly shows that these nanospheres are preferentially absorbed by human lung adenocarcinoma cells in an energy-dependent manner by clathrin-mediated endocytosis [51]. Through the comparison of chitosan to gelatin and PLGA formulations, it is evident that chitosan has some potential for sustaining drug release. Chitosan and hyaluronic acid (another natural polymer) nanosphere formulation is capable of modifying heparin release; this indicates that combining polymers in formulation, much like the PLGA example above, can further modify drug release [37].

12.3.4 Gelatin

Gelatin, another natural polymer, is an irreversible hydrolyzed form of collagen. It is a widely used biodegradable coating with good biocompatibility and harmless metabolic products [29]. There are several in vitro accounts of the use of gelatin to decrease drug release rate. Carboplatin-loaded gelatin microspheres decreased the release rate of carboplatin; the half-life of carboplatin from gelatinous microspheres was 49.7 min and 92.04% of the drug load was released in the first 10 h, while 92.15% of pure drug was dissolved in the first half-hour [29]. The release kinetics of carboplatin from gelatin microspheres can be described as a biexponential model; initially, a mean of 40.12% of drug was released in the first half-hour followed by a sustained drug release. Morimoto et al. further present that the use of gelatin microspheres for sustaining drug release may be associated with the electrical charge on the microsphere [31]. In vitro drug release studies demonstrate that at pH 7.0, salmon calcitonin was released from the positively charged, basic gelatin microsphere (isoelectric point 9.0) in 2 h with a cumulative release of 85%; however, only 40% of drug load was released from the negatively charged, acidic gelatin formulation (isoelectric point 5.0) over 2 h. The acidic formulation's drug release leveled off at 2 h, indicating that the entire drug load was not released from the microsphere during the course of the study. It was found that the microspheres could be administered to rats via the pulmonary route and that both microsphere formulations conferred sustained pharmacologic effect compared to calcitonin solution when administered by intratracheal instillation. The calcitonin solution had decreased therapeutic effect on blood calcium levels and a shorter duration of effect compared to the microsphere formulations in rats. The in vitro data, supported by in vivo findings, presents the feasibility of administering gelatin microspheres by injection or inhalation to prolong drug effect.

12.3.5 Hyaluronic Acid

Hyaluronic acid (HA), also known as hyaluronan, is a naturally occurring polymer composed of D-glucuronic acid and *N*-acetyl-D-glucosamine linked by β-1, 4 and β-1, 3 glycosidic bonds. It is an amorphous material, which slowly dissolves in water to form a highly viscous solution, even at low concentrations [50]. There are several reasons why HA is an attractive polymer for delivery of medications to the lungs as listed below [39]:

1. It is endogenous to the pulmonary environment.
2. It plays a function in various inflammatory mediators and agglutination of alveolar macrophages; inhibits phagocytosis.
3. It is bioadhesive: high molecular weight hyaluronan anchors associated proteins and enzymes, thereby preventing their removal by the mucociliary escalator.

There are some studies that examine the use of HA polymers to deliver drugs to the lungs. Overall, these studies present that these polymers can be used to deliver a variety of drugs in a relatively safe and efficacious manner [32, 39, 50]. For instance, Surendrakumar et al. presented that hyaluronic acid can be used to extend the release of insulin when administered as a dry powder formulation intratracheally to canines [50]. The administration of pulmonary dry powder insulin was compared to the administration of insulin with HA alone, HA with hydroxypropyl cellulose and HA with free zinc ions. It was found that in comparison to the insulin-only formulation, the formulation with HA increased both mean residence time (MRT, mean of 91 vs. 29 min) and half-life (mean of 63 vs. 20 min). In addition, it was found that the HA hydroxypropyl cellulose increased MRT sevenfold, increased AUC/dose fivefold, and increased T_{max} threefold compared to insulin-only formulation. Similar findings were associated with the HA-zinc formulation: ninefold increase in MRT, 2.5-fold increase in AUC/dose, and a threefold increase in T_{max} in comparison to insulin-only formulation. Characteristic plasma controlled release pharmacokinetics are seen with HA with or without hydroxypropyl cellulose or zinc as evidenced by an increase in half-life and delayed time to reach maximum concentrations. Furthermore, hydroxypropyl cellulose has been shown to increase bioavailability and increase the protection of insulin against enzymatic peptidase activity prior to absorption. Zinc functions in a different manner to prolong the sustained release effect associated with HA; zinc ions complex with HA to induce a conformational change from a helical structure to a globular structure and decreases insulin's dissolution rate. Overall, it appears that HA is an attractive polymer for pulmonary drug delivery.

12.4 Examples of Synthetic Polymers Used for Sustained Release

12.4.1 *Poly(Lactic Acid)*

PLA, also known as polylactide, has been studied extensively through in vitro studies as drug carriers for various therapeutic agents. In an in vitro study of PLA, PLGA, and solid lipid nanoparticles with varying content of magnetite, an intravenous contrast agent for magnetic resonance imaging, it was found that the degradation velocity of the PLA formulation decreased with increase in molecular weight of the polymer [34]. Both polymer formulations of magnetite degraded by hydrolysis fairly slowly compared to the solid lipid nanoparticle formulation. Incubation of low molecular weight PLA formulation with human granulocyte presented that the formulation led to a distinct reduction in viability with an increase in magnetite concentration from 0.1 to 0.3%. This cytotoxicity was shown to be relatively low and was attributed to the effect of the contrast medium and not the excipients. PLA has been used for a variety of medical applications including sutures, medical staples, screws, dressing, and orthopedic implants [3, 38]. Several studies present the use of PLA, in vitro, for inhalation agents. One such study loaded PLA microspheres with either nedocromil sodium or beclomethasone dipropionate in the form of a dry powder suitable for aerosolization [13]. Nedocromil was released from PLA microspheres, in sink conditions, over 8 days with a burst effect that varied depending on particle size: for a 2.79 μm mass median diameter distribution, 27% of the drug load was immediately released and a greater release of drug during the first 5 min was seen with larger particles. For the beclomethasone formulation, the drug release was studied in nonsink conditions using 50% isopropanol dissolution medium. Sink conditions could not be used due to the low aqueous solubility of beclomethasone. Seventy-five percent of beclomethasone was released from the microspheres over the span of 120 h. Similar results were found with the in vitro and in vivo release of rifampicin from PLA particles in a pharmacokinetics study in rats; with 10% (w/w) PLA formulation, a decrease in C_{max} and an increase in T_{max} were noted [8]. The studies presented above demonstrate that PLA formulations degrade by bulk erosion via hydrolysis and can be used to deliver a variety of agents.

Fig. 12.2 Chemical structure of oligo(lactic acid)

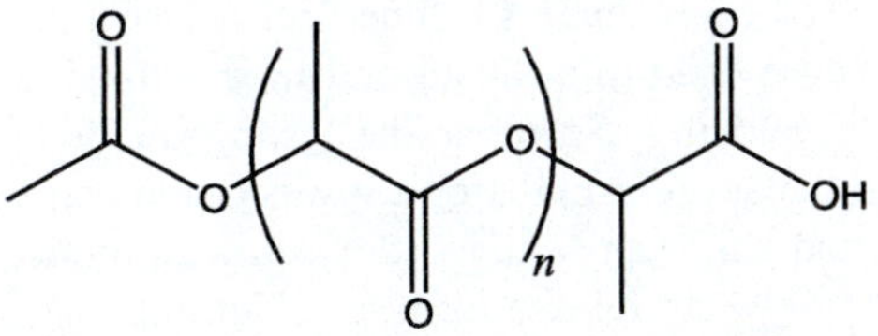

12.4.2 Oligo(Lactic Acid)

Recently, oligo(lactic acid) (OLA) polymers have emerged as potential excipients for sustained pulmonary drug release [21, 27]. OLAs, while similar to PLAs, have been differentiated through the modification of the terminal alcohol, most commonly through acetylation (see Fig. 12.2). In addition these OLA derivatives typically have shorter lengths, with lactic acid repeat units generally between 5 and 20. These attributes of OLAs yield excellent solubility in HFA 134a and 227 propellants. As a result, OLAs can be used in solution formulations contained in metered dose inhalers, affording a means by which to produce polymer microspheres in situ, post-actuation. This is a significant cost benefit as it eliminates the need for the manufacture and qualification of preformed drug encapsulated microspheres [27]. OLAs are absorbed and metabolized relatively rapidly which is beneficial with respect to lung accumulation; however, this can also limit sustained release. Leach et al. have offered evidence of OLAs providing sustained release of a steroid in a dog model. Modification of the head groups for the polymer system permits added diversity for controlled release properties. Li and Stefely illustrated through three undisclosed end groups attached to OLA that the end group functionality affects modified drug release. OLA modifications were even shown to provide sustained release for the water soluble albuterol, in vitro. In alignment with PLAs, OLAs have been found to be nontoxic in rats, guinea pigs, and dogs, with no bronchospastic evidence [21, 27]. In addition to sustained release, OLAs have been utilized as suspension and solubilizing excipients in MDIs.

12.4.3 Poly(Vinyl Alcohol)

Poly(vinyl alcohol) (PVA) is generally used to produce poly(vinyl butyral), a resin used for laminated safety glass for automobile windshields or as a lubricant in hard contact lens rewetting solutions. However, it is being researched as an excipient that can sustain release of drugs; there are several in vitro accounts of using PVA to modify release of small molecule and protein drugs. Research has shown that 1% PVA (w/w) can sustain the release of 5(6)-carboxyfluorescein when administered by inhalation to rats [1]. Additionally, Salama and colleagues evaluated the release of disodium cromoglycate from microparticles that were composed of various amounts of PVA using three different methods [43]. The methods most commonly used to evaluate sustained release of inhaled products are generally the same as those used to evaluate oral medications. Recently, much discourse is centered on developing a comparable testing method to evaluate the in vitro drug release of inhalation formulations. It was found that the USP Apparatus 2 (dissolution model) and USP Apparatus 4 (flow through model) could only differentiate disodium cromoglycate release from formulations that contained PVA and those that did not; it was not sufficiently sensitive to detect the difference in concentration of PVA in the microspheres [43]. There was no significant difference between the release rate studies of the two USP Apparatuses. The Franz-cell diffusion model, on the contrary, was able to detect difference in release rates of drug from various PVA microspheres. The diffusion model treats the inhalation formulation analogous to a diffusion process, where the formulation is merely wetted and the drug will diffuse out into the surrounding media. The Franz-cell model of diffusion presented that with 0% PVA, a mean of 98% of the drug was released in 60 min and with 90% PVA, a mean of 45% of the drug was released in 60 min. The authors conclude that the diffusion model best presents the sustained release for this inhalation formulation; however, this model can only be used for polymers that are known to control drug release by diffusion through the matrix. Salama et al. studied the release of bovine serum albumin from PVA microparticles using the Franz-cell model [41]. While increasing PVA content has a negative effect on aerosol performance in vitro, it was found that high PVA concentration in the formulation could effectively sustain protein release. The PVA particles were biocompatible and noncytotoxic; they exhibited physiochemical properties that suggested that the formulation was suitable for dry powder inhalation. Microparticles with 90% PVA (w/w) released 50% bovine serum albumin in the first 6 h, which is considerably longer than microparticles that did not contain PVA (100% of the drug load was released in an hour). Similar results as the previous two studies were seen with 10% ciprofloxacin and/or doxycycline loaded into PVA microparticles [1]. With either antibiotic, it was found that less than 50% of the drug load was released in vitro over a period of 6 h compared to greater than 90% released in a half-hour from microparticle formulations that did not contain PVA. The benefit of such sustained release is that the drugs, such as ciprofloxacin or doxycycline, can be delivered locally with increased local resident time, thus improving therapeutic efficacy. Through these studies, it is evident that PVA has definite ability to modify drug release by diffusion-limitation, but the in vivo effects of this polymer remain unknown.

12.4.4 Acrylic Acid Derivatives

There are various types of carbomers, which are high molecular weight polymers of acrylic acid that can be cross-linked with allylsucrose or allylpentaerythritol [2]. Traditionally, they have been used in pharmaceutical and cosmetic formulations for suspensions or increasing viscosity. Alhusban and Seville studied the use of carbomers 971P and 974P to prepare dry powder inhalation formulations of albuterol

20–50% (w/w) [2]. Carbomer 971P has a fishnet structure with a low level of cross-linking, allowing it to rapidly swell upon hydration; carbomer 974P, on the contrary, has a fuzzball-like structure with a high degree of cross-linking that does not permit it to swell as rapidly. It was found that the dissolution of albuterol from the control formulation released 100% of the drug within 45 min. In comparison, the carbomer formulations exhibited sustained drug release; 90% of the drug load was released in an average of 12.76 h with 50% carbomer 971P and 41.78 h with 50% carbomer 974P. In the dry state, albuterol was entrapped within a glassy core of the carbomer matrix; however, following hydration, a gelatinous layer was formed with drug dispersed in discrete microgels composed of individual polymers. The osmotic pressure within the hydrogel disrupts the structure, permitting albuterol to diffuse through the gel layer and enter the surrounding solution. Hence, the molecular structure of the polymer, as evidenced by the different rate of release of albuterol from the formulations, can affect the ability of the polymer to sustain drug release. Another acrylic derivative, polybutylcyanoacrylate, is also known to sustain drug release, in vitro [61]. Insulin release from such polymeric nanoparticles was modeled by a burst effect followed by a biexponential drug release function. In vivo testing in rats presented a minimum blood glucose level of 46.9, 30.4, and 13.6% of baseline values after pulmonary delivery of 5, 10, and 20 IU/kg, respectively. In addition, these doses had durations of effect for 4, 4, and 8 h, respectively. The duration of blood glucose level below 80% of baseline was much longer for the polybutylcyanoacrylate formulation than for insulin solutions, and a dose-dependent increase in duration of glucose level below 80% of baseline was seen for the insulin-loaded nanoparticles. Similar sustained release potential has been seen with the use of hydroxyterminated PEG-acrylate and dithiothreitol hydrogel loaded with *N*-acetyl cystine and rhodamine B [10]. In vitro and in vivo studies present that acrylic acid derivatives can be successfully used in sustaining release and pulmonary delivery of medications.

12.5 Examples of Copolymers Used in Sustained Release

12.5.1 Introduction to Copolymers

Copolymers, sometimes referred to as block polymers, are polymers consisting of two or more different monomers. The combinations of various monomers offer greater flexibility and large variety for pharmaceutical formulations.

12.5.2 Review of Poly(Lactic-Co-Glycolide Acid) (PLGA)

Poly(lactic-co-glycolide acid) is one of the most widely studied copolymers in pharmaceutics and is composed of the monomers of lactic (LA) and glycolic acids (GA). PLGA, along with its monomers, degrades by bulk erosion: PLGA is hydrolyzed to polylactides and polyglycolides, which are further broken down to lactic and glycolic acids, in vivo. Lactic and glycolic acids can enter the Kreb's cycle to be decomposed to carbon dioxide and water. In general, the polymer's properties, such as amount of water-uptake, degradation time, polymer end-groups, ratio of lactic to glycolic acids, and crystallinity of PLA, affect the ability for PLGA formulations to sustain drug release [55]. As long as 4–6 weeks duration of drug release can be achieved by 50:50 LA:GA with low to medium molecular weight polymers. Even slower release can be achieved by increasing the LA content in comparison to the GA content. For instance, incorporation of lactic acid oligomers into PLGA increases initial mass loss and extent of water absorption; however, it extends the lag phase until onset of polymer erosion, thus decreasing the overall rate of release of the drug [44]. As the example above suggests, PLGA drug release is triphasic. The burst release is associated with drug adhered to the surface of the formulation; the lag time depends on molecular weight and end-capping of the polymer. The lag time is followed by a slow or absent diffusion-controlled release that finally ends in drug release due to polymer erosion.

There are several studies that present the application of PLGA in modifying drug release for various drugs for pulmonary delivery. Hickey et al. provided preliminary studies on the use of PLGA microspheres with rifampicin for treating tuberculosis in a guinea pig model [17]. Sung and colleagues have investigated rifampicin-loaded PLGA as a dry powder porous particle formulation [49]. In vitro, the drug release could be characterized by an initial burst followed by the remainder of the load gradually releasing from the formulation over a span of 8 h. When administered to guinea pigs by intratracheal insufflation, rifampicin was detectable up to 6–8 h in the plasma and remained detectable in lung tissue and cells beyond 8 h. Moreover, the formulation was prepared as aggregates of porous nanoparticles, such that the aggregate once delivered into the body can deposit in the lungs as a dry form, thus acting as a Trojan delivery system for the nanoparticles. In all porous nanoparticle-aggregate particle (PNAP) formulations, the serum concentration of rifampicin remained elevated compared to intravenous, oral, and porous particle formulation with L-leucine. Half-life of the 80% drug load, PLGA PNAP was 3.79 ± 2.37 h and was longer than that of the porous particle formulation without PLGA (2.26 ± 1.63 h). There are several other studies that explore the in vitro utility of PLGA loaded with either rifampicin as treatment or tuberculosis antigen as a vaccine for tuberculosis by the pulmonary route [30, 45, 47]. Recently, Wang and coworkers have utilized PLGA microparticles loaded with muramyl dipeptide and/or trehalose debehenate to study administration of tuberculosis vaccine via the pulmonary route [55]. It was found that there were statistically elevated indicators of biochemical response in THP-1 cells, which may indicate toxicity to human alveolar macrophages. Such research highlights the need for corresponding in vivo experiments to evaluate cytotoxic potential of PLGA drug delivery systems.

Many drugs, other than antituberculosis agents, have been studied with PLGA. A significant difference between control and test formulation of insulin was seen with the PLGA nanosphere composition of 75:25 LA:GA [19]. An in vivo study of insulin-loaded PLGA nanospheres in guinea pigs following 20 min nebulization presented significant hypoglycemia that was prolonged over 48 h compared to only 6 h for intratracheal administration of insulin solution. Similar results are seen with deslorelin-loaded PLGA LPP administration to male rats [20]. At the end of 7 days postdose administration, the plasma concentration for rats that received the LPP formulation was 120 and 2.5-fold greater than that of those who received deslorelin powder or small conventional deslorelin-loaded PLGA particles, respectively. In addition to the studies described above, PLGA formulations have been used successfully, in vitro, to modify the release of beclomethasone dipropionate, doxorubicin, TAS-103 (a topoisomerase inhibitor), and ciprofloxacin [4, 22, 53, 60]. Overall, PLGA appears to be effective in sustaining the release of a variety of drugs including antibiotics, hormones, hormone agonists, and chemotherapeutics. In addition, there were no overt toxicities to healthy human lung macrophages noted with a microparticle PLGA formulation and no signs of pulmonary inflammation noted in mice [49].

As with other excipients described in this chapter and Chap. 11, PLGA can be combined with L-α-dipalmitoyl phosphatidylcholine (DPPC), cholesterol, PVA, or gelatin to further extend the release of paclitaxel from PLGA microspheres [33]. It was found that paclitaxel released most rapidly from PLGA-only formulations and reached 50% in vitro drug release in 110 days; the addition of PVA to the formulation resulted in a slightly slower release (41% in 110 days). Gelatin retarded drug release such that only 35% of the paclitaxel load was released in 110 days. Formulations with excipients in addition to PLGA increased the encapsulating efficacy of the PLGA microspheres. Without these emulsifiers, the encapsulating efficiency during the formulating process is 61.1%; however, with the addition of the emulsifiers the encapsulating efficiency increases to 63.1% with 2% PVA and to 84.1% with 2% DPPC. It is postulated that the drug release is retarded with the addition of these excipients to the PLGA microsphere by decreasing the particle porosity due to the strong interaction between the hydrophobic nature of the excipient and the drug. This example serves to present that the characteristic of the drug molecule and addition of excipients may affect the sustained release properties of the formulation.

12.5.3 Review of Miscellaneous Copolymers

The studies that relate to copolymers' potential to modify the release of drugs are limited to in vitro release-kinetics studies. One such study incorporated various amounts of paclitaxel into poly(lactic acid)-poly(ethylene glycol)-poly(lactic acid) (PLA-PEG-PLA) in hopes to improve paclitaxel release rate compared to paclitaxel-loaded PLGA [40]. Since paclitaxel is extremely hydrophobic and PLGA is also hydrophobic, the release of paclitaxel was extremely slow with only 50% of drug released within 3 months. The use of PLA-PEG-PLA copolymer was postulated to improve release of paclitaxel due to the incorporation of hydrophilic PEG into the hydrophobic PLA chain. The incorporation of hydrophilic PEG would, also, increase biocompatibility since biocompatibility is highly correlated with hydrophilicity of an exogenous surface. It was found that the PLA-PEG-PLA microsphere formulation resulted in faster drug release than PLGA, with approximately 49.6% sustained release of paclitaxel within 1 month. The increased drug release with the copolymer formulation can partially be explained by the increased porous morphology in addition to the incorporation of hydrophilic PEG: the addition of acetone into the organic solvent during formulation increased microsphere porosity. Higher drug load resulted in decreased drug release rate from the microspheres. Another study using PLA-PEG-PLA block polymers to deliver salbutamol base presented that the overall concentration, molecular weight, surfactant tail length, and ratio between PLA and PEG affected the drug's cohesiveness and aerosolizability [56].

Similar results were seen with clonazepam in an in vitro release-kinetics study utilizing copolymers as release modifiers [35]. The copolymer nanoparticle formulation was based on poly(γ-benzyl-L-glutamate) (PBLG) as the hydrophobic part and poly(ethylene oxide) (PEO) as the hydrophilic part to form the triblock of PBLG-PEO-PBLG. The rate of drug release decreased with increased drug content and was affected by the solvents used during the formulation process: it was found that the release rate of clonazepam was decreased with 1,4-dioxane and tetrahydrofuran compared to that with dimethylformamide, dimethylacetamide, and dimethyl sulfoxide. The release of drug was extended over 2–3 days and modeled by zero-order kinetics.

Copolymer formulations for sustained drug release can include polyether-anhydride formulations that contain hydrophobic sebacic acid (SA) and hydrophilic PEG monomers [15]. Fu et al. showed that an increase in PEG content was associated with an increase in degradation rate of the polymer, thus demonstrating that the in vitro release from such polymers is greatly associated with the relative PEG content. The researchers suggest that the change in degradation rate is associated with the increased hydrophilicity of the microsphere with increased concentration of PEG; increased hydrophilicity would lead to greater aqueous solubility of the dosage form. In an in vitro study evaluating the release of plasmid DNA from such microspheres, it was found that while the microspheres degraded within hours, plasmid DNA was released up to 6 days and that the DNA's interaction with the monomers influenced sustained release properties to a greater extent than that with the copolymer. Similar sustained release properties were seen with salbutamol acetonide loaded into solid lipid microparticles composed of Poloxamer 188, a block polymer of poly(propylene oxide) (PPO) and PEO, with glyceryl behenate [18]. While the in vitro data clearly suggests that sustained drug release can be achieved by various copolymer formulations, it may not translate to be true in vivo.

In addition to the polymers described above, PEG and SA have been used in a triblock formulation with 1,3-bis(carboxyphenoxy)propane (CPP) to produce drug carrier particles suitable for injection or inhalation [14]. It was found that CPP-SA copolymers erode from the surface and are hydrophobic. LPPs with PEG-CPP-SA could be efficiently aerosolized as a dry powder formulation into the deep lungs. The density of the LPP could be decreased by increasing PEG content. The PEG-CPP-SA LPPs were loaded with rhodamine B base. The increase in hydrophobicity by increasing SA or CPP content decreased drug release rate such that 10:90 PEG:SA released 60% of the agent in 4 h, SA-only polymers released 60% drug in 18 h, and 10:70:20 PEG:SA:CPP released 60% of the drug load in 2.5 days. Similar results were seen when PEG-CPP-SA LPPs were loaded with LacZ plasmid DNA [14]. Hydrophilic LPPs released the test agent at a greater rate than hydrophobic formulations.

12.6 Limitations and Safety Concerns

There are various natural and synthetic polymers studied for pulmonary drug delivery; however, safety concerns limit their applicability. The effect of polymers on cell viability and the potential for bioaccumulation has affected their clinical use. It was found that PLA and PLGA polymer formulations decrease viability of cells, increase airway inflammation, increase neutrophil count, and increase incidence of hemorrhage [7, 9]. Greater cell toxicity is seen with PLGA than with sodium hyaluronate and chitosan [42]. Other than cytotoxicity, long residence time of polymer-drug formulations may be associated with pulmonary accumulation due to the slow degradation of the system [7]. Such concerns need to be further investigated prior to conducting extensive clinical trials with inhaled polymeric drug delivery systems.

References

1. Adi H, Young PM, Chan H, Salama R, Traini D (2010) Controlled release antibiotics for dry powder lung delivery. Drug Dev Ind Pharm 36:119–126
2. Alhusban FA, Seville PC (2009) Carbomer-modified spray-dried respirable powders for pulmonary delivery of salbutamol sulfate. J Microencapsul 26:444–455
3. Amecke B, Bendix D, Entenmann G (1992) Resorbable polyesters: composition, properties, and applications. Clin Mater 10:47–50
4. Arnold MM, Gorman EM, Schieber LJ, Munson EJ, Berkland C (2007) NanoCipro encapsulation in monodisperse large porous PLGA microparticle. J Control Release 121:100–109
5. Bailey MM, Berkland CJ (2009) Nanoparticle formulations in pulmonary drug delivery. Med Res Rev 29:196–212
6. Brannon-Peppas L (1997) Polymers in controlled drug delivery. Med Plast Biomater 4:34–44
7. Cook RO, Pannu RK, Kellaway IW (2005) Novel sustained release microspheres for pulmonary drug delivery. J Control Release 104:79–90
8. Coowanitwong I, Arya V, Kulvanich P, Hochhaus G (2008) Slow release formulations of inhaled rifampin. AAPS J 10:342–348
9. Courrier HM, Butz N, Vandamme TF (2002) Pulmonary drug delivery systems: recent development and prospectives. Crit Rev Therap Drug Carrier Syst 19:425–498
10. Donovan M, McGill S, Smyth H (2008) Swellable particles for sustained release drug delivery to the lungs. Respir Drug Deliv 3:857–860
11. Edwards DA, Ben-Jebria A, Langer R (1998) Recent advances in pulmonary drug delivery using large, porous inhaled particles. J Appl Physiol 85:379–385
12. Edwards DA, Hanes J, Caponetti G, Hrkach K, Ben-Jebria A, Eskew ML, Mintzes J, Deaver D, Lotan N, Langer R (1997) Large porous particles for pulmonary drug delivery. Science 276:1868–1871

13. El-Baseir MM, Kellaway IW (1998) Poly(L-lactic acid) microspheres for pulmonary drug delivery: release kinetics and aerosolization studies. Int J Pharm 175:135–145
14. Fu J, Fiegel J, Hanes J (2004) Synthesis and characterization of PEG-based ether anhydride terpolymers: novel polymers for controlled drug release. Macromolecules 37:7174–7180
15. Fu J, Fiegel J, Krauland E, Hanes J (2002) New polymeric carriers for controlled drug delivery following inhalation or injection. Biomaterials 23:4425–4433
16. Garcia-Contreras L, Morçöl T, Bell SJD, Hickey AJ (2003) Evaluation of novel particles as pulmonary delivery systems for insulin in rats. AAPS Pharm Sci 5(2):article 9
17. Hickey AJ, Suarez S, Bhat M, O'Hara P, Lalor CB, Atkins K, Hopfer R, McMurry DN (1998) Efficacy of rifampicin-poly(lactide-co-glycolide) microspheres in treating tuberculosis. Respir Drug Deliv VI 1:201–209
18. Jaspart S, Bertholet P, Piel G, Dogné J, Delattre L, Evrard B (2007) Solid lipid microparticles as a sustained release system for pulmonary drug delivery. Eur J Pharm Biopharm 65: 47–56
19. Kawashima Y, Yamamoto H, Takeuchi H, Fujioka S, Hino T (1999) Pulmonary delivery of insulin with nebulized DL-lactide/glycolide copolymer (PLGA) nanospheres to prolong hypoglycemic effect. J Control Release 62:279–287
20. Koushik K, Dhanda DS, Cheruvu NPS, Kompella UB (2004) Pulmonary delivery of deslorelin: large-porous PLGA particles and HPβCD complexes. Pharm Res 21:1119–1126
21. Leach CL, Mameister WM, Tomai MA, Hammerbeck DM, Stefely JS (2000) Oligolactic acid (OLA) biomatricies for sustained release of asthma therapeutics. Respir Drug Deliv VII 2: 291–294
22. Learoyd TP, Burrows JL, French E, Seville PC (2006) Leucine-modified PLGA-based respirable spray-dried powders for sustained drug delivery. Respir Drug Deliv 2:437–440
23. Learoyd TP, Burrows JL, French E, Seville PC (2006) Sustained drug delivery from chitosan-based respirable spray-dried powders. Respir Drug Deliv 2:441–443
24. Learoyd TP, Burrows JL, French E, Seville PC (2008) Chitosan-based spray-dried respirable powders for sustained delivery of terbutaline sulfate. Eur J Pharm Biopharm 68:224–234
25. Learoyd TP, Burrows JL, French E, Seville PC (2009) Sustained delivery by leucine modified chitosan spray-dried respirable powders. Int J Pharm 372:97–104
26. Li FQ, Hu JH, Lu B, Yao H, Zhang WG (2001) Ciprofloxacin-loaded bovine serum albumin microspheres: Preparation and drug-release in vitro. J Microencapsul 18:825–829
27. Li Z, Stefely JS (2004) Aerosol sustained release formulations with oligolactic acid (OLA). Respir Drug Deliv IX 2:321–324
28. Louey MD, Garcia-Contreras L (2004) Controlled release products for respiratory delivery. Am Pharm Rev 7:82–87
29. Lu B, Zhang Q, Yang H (2003) Lung-targeting microspheres of carboplatin. Int J Pharm 265:1–11
30. Lu D, Garcia-Contreras L, Xu D, Kurtz SL, Liu J, Braunstein M, McMurray DN, Hickey AJ (2007) Poly (lactide-c-glycolide) microspheres in respirable sizes enhance an in vitro T cell response to *Mycobacterium tuberculosis* antigen 85B. Pharm Res 24:1834–1843
31. Morimoto K, Katsumata H, Yabuta T, Iwanaga K, Kakemi M, Tabata Y, Ikada Y (2002) Gelatin microsphere as a pulmonary delivery system: evaluation of salmon calcitonin absorption. J Pharm Pharmacol 52:611–617
32. Morimoto K, Metsugi K, Katsumata H, Iwanaga K, Kakemi M (2001) Effects of low-viscosity sodium hyaluronate preparation on pulmonary absorption of rh-insulin in rats. Drug Dev Ind Pharm 27:365–371
33. Mu L, Feng SS (2001) Fabrication, characterization and in vitro release of paclitaxel (Taxol®) loaded poly (lactic-co-glycolic acid) microspheres prepared by pray drying technique with lipid/cholesterol emulsifiers. J Control Release 76:239–254
34. Müller RH, Maaβen S, Wayhers H, Specht F, Lucks JS (1996) Cyctotoxcity of magnetite-loaded polylactide, polylactide/glycolide particles and solid lipid nanoparticles. Int J Pharm 138:85–94

35. Nah J, Jeong Y, Cho C, Kim S (2000) Drug-delivery system based on core-shell type nanoparticles composed of poly(γ-benzyl-L-glutamate) and poly(ethylene oxide). J Appl Polym Sci 75:1115–1126
36. Naikwade S, Bajaj A (2009) Preparation and in vitro evaluation of budesonide spray dried microparticles for pulmonary delivery. Sci Pharm 77:419–441
37. Oyarzun-Ampuero FA, Brea J, Loza MI, Torres D, Alonso MJ (2009) Chitosan hyaluronic acid nanoparticles loaded with heparin for the treatment of asthma. Int J Pharm 381:122–129
38. Pulapura S, Kohn J (1992) Trends in the development of bioresorbable polymers for medical applications. J Biomater Appl 6:215–250
39. Rouse JJ, Whateley TL, Thomas M, Ecclecston GM (2007) Controlled drug delivery to the lungs: Influence of hyaluronic acid solution conformation on its adsorption to hydrophobic drug particles. Int J Pharm 330:175–182
40. Ruan G, Feng S (2003) Preparation and characterization of poly(lactic acid) poly(ethylene glycol)-poly(lactic acid) (PLA-PEG-PLA) microspheres for controlled release of paclitaxel. Biomaterials 24:5037–5044
41. Salama RO, Traini D, Chan H, Sung A, Ammit AJ, Young PM (2009) Preparation and evaluation of controlled release microparticles for respiratory protein therapy. J Pharm Sci 98:2709–2717
42. Salama R, Traini D, Chan H, Young PM (2009) Recent advances in controlled release pulmonary therapy. Curr Drug Deliv 6:404–414
43. Salama RO, Traini D, Chan H, Young PM (2008) Preparation and characterisation of controlled release co-spray dried drug-polymer microparticles for inhalation 2: evaluation of in vitro release profiling methodologies for controlled release respiratory aerosols. Eur J Pharm Biopharm 70:145–152
44. Schliecker G, Schmidt G, Fusch S, Wombacher R, Kissel T (2003) Hydrolytic degradation of poly(lactide-co-glycolide) films: effect of oligomers on degradation rate and crystallinity. Int J Pharm 266:39–49
45. Shi S, Hickey AJ (2010) PLGA microparticles in respirable sizes enhance an in vitro T cell response to recombinant *Mycobacterium tuberculosis* antigen TB10.4-Ag58B. Pharm Res 27:350–360
46. Shoyele SA (2008) Controlling the release of proteins/peptides via the pulmonary route. Drug Deliv Syst 437:141–148
47. Suarez S, O'Hara P, Kazantseva M, Newcomer CE, Hopfer R, McMurray DN, Hickey AJ (2001) Respirable PLGA microspheres containing rifampicin for the treatment of tuberculosis: screening in an infectious disease model. Pharm Res 18:1315–1319
48. Sung JC, Pulliam BL, Edwards DA (2007) Nanoparticles for drug delivery to the lungs. Trends Biotechnol 25:563–570
49. Sung JC, Padilla DJ, Garcia-Contreras L, VerBerkmoes JL, Durbin D, Peloquin CA, Elbert KJ, Hickey AJ, Edwards DA (2009) Formulation and pharmacokinetics of self-assembled rifampicin nanoparticle systems for pulmonary delivery. Pharm Res 26:1847–1855
50. Surendrakumar K, Martyn GP, Hodgers ECM, Jansen M, Blair JA (2003) Sustained release of insulin from sodium hyaluronate based dry powder formulations after pulmonary delivery of beagle dogs. J Control Release 91:385–394
51. Tahara K, Sakai T, Yamamoto H, Takeuchi H, Hirashima N, Kawashima Y (2009) Improved cellular update of chitosan-modified PLGA nanospheres by A549 cells. Int J Pharm 382: 198–204
52. Takeuchi H, Yamamoto H, Kawashima Y (2001) Mucoadhesive nanoparticulate systems for peptide drug delivery. Adv Drug Deliv Rev 47:39–54
53. Tomoda K, Ohkoshi T, Hirota K, Sonavane GS, Nakajima T, Terada H, Komuro M, Kitazota K, Makino K (2009) Preparation and properties of inhalable nanocomposite particles for treatment of lung cancer. Colloids Surf B Biointerfaces 17:177–182
54. Wang C, Muttil P, Lu D, Beltran-Torres AA, Garcia-Contreras L, Hickey AJ (2009) Screening for potential adjuvants administered by the pulmonary route for tuberculosis vaccines. AAPS J 11:139147

55. Wischke C, Schwendeman SP (2008) Principles of encapsulating hydrophobic drugs in PLA/PLGA microparticles. Int J Pharm 364:298–327
56. Wu L, da Rocha SRP (2007) Biocompatible and biodegradable copolymer stabilizers for hydrofluoroalkane dispersions: a colloidal probe microscopy investigation. Langmuir 23: 12104–12110
57. Yamada K, Kamada N, Odomi M, Okada N, Nabe T, Fujita T, Kohno S, Yamamoto A (2005) Carrageenans can regulate the pulmonary absorption of antiasthmatic drugs and their retention in the rat lung tissues without any membrane damage. Int J Pharm 293:63–72
58. Yamada K (2007) Control of pulmonary absorption of drugs by various pharmaceutical excipients. Yakugaku Zasshi 127:631–641
59. Yamamoto H, Kuno Y, Sugimoto S, Takeuchi H, Kawashima Y (2005) Surface modified PLGA nanosphere with chitosan improved pulmonary delivery of calcitonin in mucoadhesion and opening of the intracellular tight space. J Control Release 102:373–381
60. Yang Y, Bajaj N, Xu P, Ohn K, Tsifansky MD, Yeo Y (2009) Development of highly porous large PLGA microparticles for pulmonary drug delivery. Biomaterials 30:1947–1953
61. Zhang Q, Shen Z, Nagai T (2001) Prolonged hypoglycemic effect of insulin-loaded polybutylcyanoacrylate nanoparticles after pulmonary administration to normal rats. Int J Pharm 218: 75–80

Chapter 13
Particle Engineering Technologies for Pulmonary Drug Delivery

Nashwa El-Gendy, Mark M. Bailey, and Cory Berkland

Abstract Particle engineering has seen many applications in the field of pulmonary drug delivery due to the intimate relationship between particle physicochemistry and aerosol product performance. In this chapter, the science behind established and emerging particle engineering technologies is reviewed. Fundamental principles of particle engineering will be introduced. Following a discussion of how aerosol delivery technologies integrate with particle engineering, a detailed review of particle synthesis methods is included. This encompasses micronization and other top–down methods, spraying methods, precipitation technologies, and other emerging processes that may advance the field of pulmonary drug delivery.

Keywords Aerosols • Nanotechnology • Particle engineering

13.1 Introduction

Advancing the field of pulmonary drug delivery requires continued development of particle engineering technologies. These technologies range from traditional processing techniques such as spray drying and milling to more advanced techniques such as encapsulating drugs within engineered microparticles with controllable geometries. Nanoparticle-based formulations have also emerged. Improvements in particle engineering could lead to greater control over aerodynamic particle size, which may lead to more accurate and precise pulmonary dosing. This could enable physicians to better treat local lung diseases such as asthma and chronic obstructive

C. Berkland(✉)
Department of Chemical and Petroleum Engineering,
Department of Pharmaceutical Chemistry, University of Kansas,
Lawrence, KS, USA
e-mail: berkland@ku.edu

H.D.C. Smyth and A.J. Hickey (eds.), *Controlled Pulmonary Drug Delivery*,
Advances in Delivery Science and Technology, DOI 10.1007/978-1-4419-9745-6_13,

pulmonary disease (COPD) as well as systemic diseases. Particle-engineering technologies that are commonly used in industry are reviewed. More advanced techniques that are under development are also discussed as a means to enhance the effectiveness of pulmonary drug formulations.

13.2 Pulmonary Physiology and Drug Delivery

The functions of the lungs are to facilitate gas exchange between the blood and the external environment and to maintain homeostatic systemic pH [1]. The conducting airways are composed of the trachea, which bifurcates into the bronchi. The bronchi branch into smaller bronchioles, and ultimately the terminal bronchi, which terminate with the alveolar sac. There are approximately 300 million alveoli in the lungs, with a combined surface area greater than 100 m^2, and with an alveolar epithelial layer as thin as 0.1 μm [2, 3]. This large surface area, combined with a thin barrier between the pulmonary lumen and the bloodstream, creates conditions that are well suited for efficient mass transfer, offering a potential conduit for systemic drug delivery.

The surface of the alveoli is composed primarily of type I pneumocytes, which share a basement membrane with the pulmonary capillaries. Type II pneumocytes are also present, which secrete lung surfactant, as well as alveolar macrophages, which engulf large particles [4]. The conducting airways are lined with ciliated cells, which are covered with a thin layer of mucus. These cilia beat in a coordinated, rhythmic fashion that constantly move the mucus layer toward the proximal airways, where it is either expectorated or swallowed [1]. This process, known as mucociliary clearance, presents a major physical barrier to pulmonary drug delivery [5].

Pulmonary drug formulations currently exist for many local lung diseases such as asthma, COPD, and cystic fibrosis [3]. Many other local lung diseases are prime candidates for inhalation therapy, including pulmonary hypertension and cancer. Treating these diseases locally is advantageous because the drug avoids first-pass metabolism and deposits directly at the disease site, thereby mitigating undesirable systemic side effects and facilitating higher local dose [1].

Recently, there have been many advances in systemic drug delivery through the lungs. Systemic delivery may ultimately require particles that are designed with suitable properties for delivery to the alveolar region. Once deposited, particles must dissolve or release their contents before they are engulfed and cleared by alveolar macrophages. Additionally, drug molecules, particularly macromolecules such as small peptides, proteins, and nucleic acids, are subject to enzymatic degradation in this region. After being released into solution, the drug must diffuse or actively transport through the epithelial barrier, basement membrane, and endothelium to access the blood. Systemic drug delivery through the lungs has been reported to have many advantages, including rapid onset of action and an increased bioavailability over oral dosage forms [1].

13.3 Particle Engineering Fundamentals

Mechanisms of particle deposition in the lungs are shown in Fig. 13.1. Inertial impaction typically occurs in the oropharyngeal region, where a large particle will impact the back of the throat as the direction of the air stream changes. Sedimentation often occurs in the bronchial region, where the air velocity diminishes and particles settle and deposit by gravity. Finally, diffusion occurs in the alveolar region, where particles with $d_a << 1$ μm may remain in the air due to Brownian motion [6].

The goal of particle engineering is to design particles with desirable attributes for optimal delivery to the lungs, as well as maximal bioavailability. These attributes include a narrow size distribution, drug stability, high dispersibility, and the ability to control drug release kinetics [7]. On a practical level, a desirable particle engineering process should reduce manufacturing complexity, lower costs, and minimize environmental impact.

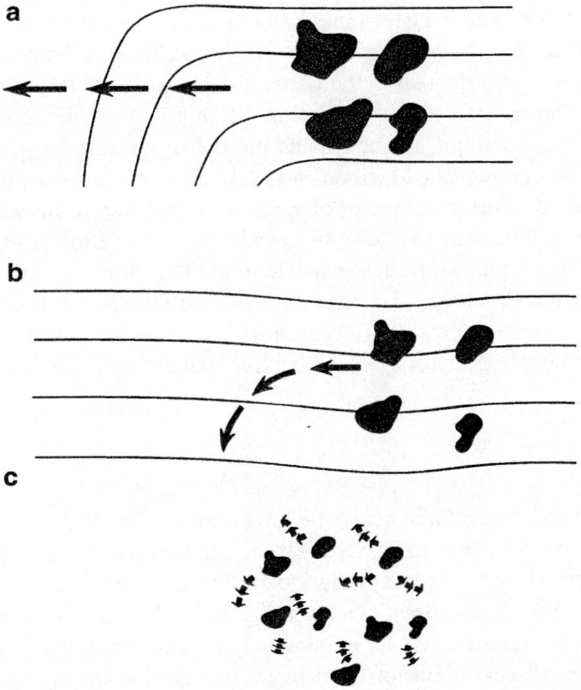

Fig. 13.1 Schematic showing mechanisms of particle deposition in the lungs. The *lines* represent streamlines of the inhaled air, and the *arrows* show the path taken by the particle. (**a**) Inertial impaction commonly occurs in the oropharyngeal region, where a large particle will impact the back of the throat and not be carried further by the air stream. (**b**) Sedimentation often occurs in the bronchial region, where the air velocity diminishes and particles settle by gravity and are deposited. (**c**) Diffusion is anticipated in the alveolar region, where very small particles ($d_a < 1$ μm) diffuse due to Brownian motion

The fundamental parameter used in defining how a particle will deposit in the respiratory system is the aerodynamic diameter (d_a). The aerodynamic diameter is defined as the diameter of a sphere of geometric diameter d_a and density ρ that reaches the same settling velocity as a nonspherical particle. The aerodynamic diameter is given below [1, 6–11]:

$$d_a = d_v \sqrt{\frac{\rho}{\chi \cdot \rho_o}}, \tag{13.1}$$

where d_v is the geometric diameter, ρ_o is the reference density (usually that of water), χ is the shape factor (1 for a sphere, but changes as shape deviates from spherical), and ρ is the particle density [7]. The aerodynamic diameter determines how far a particle will travel into the respiratory tract. For instance, large particles (d_a >10 μm) will deposit in the oropharyngeal region by means of inertial impaction, or will deposit via sedimentation in the oropharyngeal or bronchial region. At the other extreme, very small particles (d_a << 1 μm) are likely to reach the alveolar region, but may not settle [1, 12]. Particles with d_a between 1 and 5 μm have an increased probability of depositing in the terminal bronchial and alveolar regions. This particle size range is ideal for many indications in pulmonary drug delivery, since a portion of the material will still deposit in the upper airways as well [13].

Numerous forces affect the deposition of particles in the pulmonary tract, including gravity, the drag force of air, and inertial forces acting on the particle [8]. The aerodynamic diameter (d_a) given in (13.1) allows one to parameterize these forces so that the deposition behavior of particles with different shapes and densities can be compared. The shape factor (χ) given in (13.1) is defined as the ratio of the drag force acting on a nonspherical particle to the drag force acting on a sphere of the same volume and velocity. For aerosols, the inertial effects of air are negligible compared to viscosity effects; therefore, Stoke's law applies for the drag force [8]. Based on this, the shape factor (χ) can be expressed as:

$$\chi = \frac{F_a}{3 \cdot \pi \cdot \eta \cdot V \cdot d_v}, \tag{13.2}$$

where F_a is the resistance force acting on the particle, η is the dynamic viscosity of the fluid (air), and V is the particle velocity relative to that of air [8]. Empirically, the shape factor acts as a parameter that quantifies nonsphericity.

The aerodynamic diameters of aerosolized particles (specifically the mass median aerodynamic diameter, or MMAD) are typically measured using a cascade impactor (Fig. 13.2). The cascade impactor is an inertial separation device that functions by separating the particles into aerodynamic size fractions upon actuation of the inhaler or delivery device. The cascade impactor is divided into stages, where each stage has a different orifice size. Each stage acts like a filter that catches any particles that fail to make it through the holes. The orifice size of each stage is correlated to a particle diameter range via the Stoke's number, which is the ratio of the particle stopping distance to the characteristic length of the obstacle (in this case, the orifice size) [8]. The Stoke's number is:

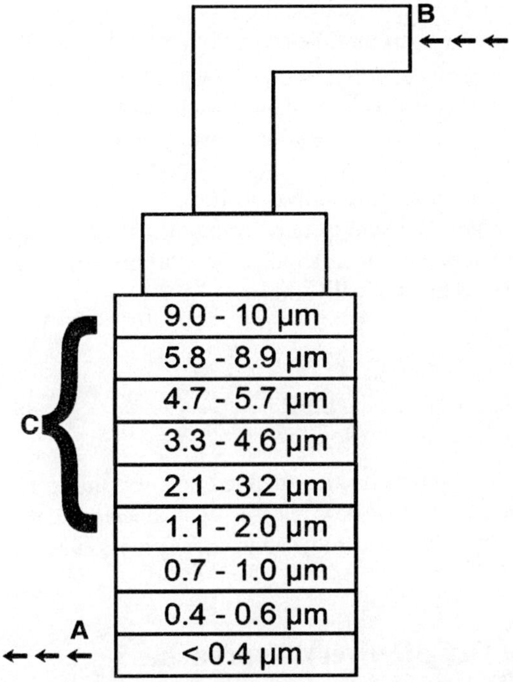

Fig. 13.2 Diagram of a cascade impactor, which is used to measure the mass mean aerodynamic diameter (MMAD) of particles by inertial separation. A vacuum is drawn at (**A**), and the inhaler device is attached at (**B**). Upon actuation of the inhaler, the dose is entrained through the different stages (**C**), where individual particles will either travel through the stage nozzle or deposit on the stage, depending on their Stokes number (i.e., aerodynamic diameter). Each stage represents a different aerodynamic diameter range, with the larger particles being deposited at the top and the smaller particles at the bottom. The mass of particles deposited on the stages is then determined and divided by the total dose to give the particle fraction for that particular size range

$$\mathrm{Stk} = \frac{d_{\mathrm{a}}^2 \cdot C_{\mathrm{c}} \cdot \rho \cdot U}{18 \cdot \eta}, \tag{13.3}$$

where d_{a} is the aerodynamic diameter of the particle, C_{c} is the Cunningham correction factor for slip flow (a correction to the drag coefficient for small particles), ρ is the density of the particle, U is the velocity of the particle at the hole exit, and η is the kinematic viscosity of air [8]. The Cunningham correction factor is [14]:

$$C_{\mathrm{C}} = 1 + \frac{2 \cdot \lambda}{d_{\mathrm{a}}} \cdot \left(A_1 + A_2 \cdot \exp\left(\frac{-A_3 \cdot d_{\mathrm{a}}}{\lambda} \right) \right), \tag{13.4}$$

where d_{a} is the particle aerodynamic diameter, λ is the mean free path of the particle, and A_n are experimentally determined coefficients dependent on the identity of the fluid [14]. The Stoke's number determines if a particle will move with the streamline as it flows around the obstacle, or if it will continue on its trajectory and deposit on the stage. If Stk >1, the particle will continue in a straight line (i.e., impact the stage) as the fluid moves around the obstacle, and if Stk <1, then the particle will follow the fluid streamline around the obstacle.

The fine particle fraction (FPF) of a pulmonary drug formulation is defined as the fraction of particles smaller than a certain aerodynamic diameter (typically ~5 μm) [15]. It is determined by measuring the mass of particles impacted within each stage of the cascade impactor representative of the "fine" particles and dividing by the total mass of the dose. The sum of mass fractions for all stages with characteristic aerodynamic diameters equal to or smaller than the specified aerodynamic diameter is equal to the FPF. Mathematically, this is (13.5):

$$\mathrm{FPF}_5 = \frac{\sum_{i=1}^{n} m_\mathrm{i}}{M}, \tag{13.5}$$

where FPF_5 is the fine particle fraction (defined as d_a less than 5 μm), m_i is the mass of stage i, M is the total mass of the dose released upon inhaler actuation, and n is the number of stages with characteristic aerodynamic diameters less than or equal to 5 μm.

13.4 Aerosol Drug Delivery Approaches

Formulation of inhaled aerosols is particularly challenging because it engages both the fabrication of an aerosol and the selection of a device for aerosol dispersion. Several types of inhalation devices are used for liquid solutions, suspensions, emulsions, or micronized dry powders formulated for pulmonary drug delivery. The most common types of inhalation devices are nebulizers, metered dose inhalers (MDIs), and dry powder inhalers (DPIs). Each Inhaler has unique strengths and weaknesses [16].

A pulmonary nebulizer machine or "nebulizer" is an electronic device that transforms liquid into a very fine mist, which is then inhaled deeply into the lungs. There are different types of nebulizers, although the most common are jet nebulizers (atomizers) and ultrasonic nebulizers. Jet nebulizers are connected by tubing to a compressed air source that causes air or oxygen to flow at high velocity through a liquid to turn it into an aerosol, which is then inhaled by the patient. Air jet nebulizers can generate both smaller particles having MMAD ~2–5 μm and coarse aerosols. Nebulizers, however, tend to deposit large amount of drug in the oropharyngeal cavity and often require long exposure times.

Ultrasonic nebulizers are electrically driven and do not require compressed gas. These devices use high frequency vibration caused by a piezoelectric crystal transducer to break up surface water into droplets. The resulting dense mist can penetrate smaller airways when inhaled by the patient. In particular, greater aerosol delivery can be achieved from a large capacity ultrasonic nebulizer when compared to a jet nebulizer during mechanical ventilation in vitro. Ultrasonic nebulizers have another advantage in that delivery is completed more rapidly than jet nebulizers [17].

Generally, nebulizers are suitable for drugs with high dose requiring little patient coordination or skill; however, treatment using nebulizers is time consuming, less efficient, and less convenient. For these reasons, nebulizers are typically limited to home or hospital use and are not convenient for chronic disease management.

An MDI is a pressurized, hand-held device with a metering valve. Upon actuation of the valve, it releases a measured dose of drug dissolved or suspended in liquefied propellants (typically hydrofluoroalkanes) with other excipients. Medication is delivered to the lungs as a short burst of aerosolized medicine that is inhaled by the patient. It has been reported that engineered particles composed of lipid-coated budesonide microcrystals may provide an acceptable alternative formulation technology for MDIs in new hydrofluoroalkane propellants [18]. MDIs have also been used for aerosol delivery of polar low molecular weight molecules and macromolecules (such as peptides) to the lungs by the solubilization of polar compounds into dimethyl ether (DME) and DME/propane utilizing lecithin inverse microemulsions. Water-in-propellant DME/lecithin and DME/propane/lecithin microemulsions were shown to produce aerosols with particle size distributions suitable for pulmonary delivery [19, 20]. MDIs also have some disadvantages such as oropharyngeal deposition of drugs due to the high velocity of propellants and the requirement of proper technique for correct use (coordinating the peak of inspiration with the actuation of the inhaler). In addition, implications that chlorofluorocarbons (CFCs) and now HFAs damaged the ozone layer drove down the development of MDIs [11].

A DPI is a breath-activated, nonpressurized device that delivers medication to the lungs in the form of a dry powder. DPI systems offer an efficient and environmentally friendly delivery system. These devices eliminate the need for propellants, and are less dependent on coordination of inhalation and device actuation. DPIs are generally formulated as one-phase, solid particle blends. Dry powders tend to reduce the rate of chemical degradation and the possibility of reaction with contact surfaces compared to wet formulations. Other advantages of dry powder systems include the potential for delivering a small or large mass of drug per breath, and their applicability to both soluble and insoluble drugs [21, 22].

Difficulties can arise in producing sufficient air flow through the device to entrain the drug and subsequently break up the powder into aerosol particles that are small enough to reach the lungs. Insufficient patient inhalation flow rates may lead to reduced dose delivery and incomplete deaggregation of the powder, resulting in unsatisfactory device performance. Furthermore, fine powder particles often have a tendency to stick to each other. These agglomerates can be difficult to break apart into breathable particles. Currently, devices with restricted flow and tortuous paths are used to disaggregate powders. A key to efficient aerosol delivery is the formulation of readily dispersible powders with small particle size by particle engineering. Advanced particles would enable the use of small, cheap, and user-friendly devices to generate efficient dispersion and dissemination of the drug aerosol, even at low inspiration flow rates common for compromised pediatric or geriatric patients [23].

DPIs can be categorized into three types based on design (Fig. 13.3). The first group comprises single unit dose systems (usually employing a capsule). The second type is a multiple unit dose dispensing system and the third class of DPI consists of

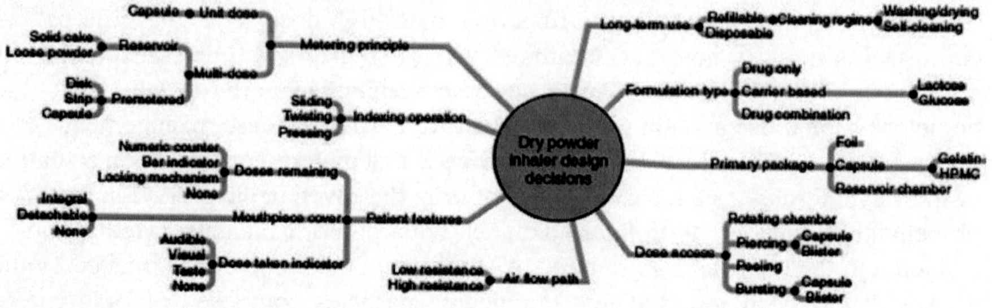

Fig. 13.3 Design decisions affecting the development of dry powder inhalers (Taylor et al. [24], used with permission)

devices that contain multiple doses. The use of a capsule as a unit container is convenient and often does not lead to a marked loss of drug through residual drug remaining in the capsule after activation. Moisture retention by gelatin capsules or the drug powder may cause a problem with stability of the drugs and ease of powder dispersion. Hence, the subsequent generation of multiple unit dose devices has emerged to solve such problems by protecting the drug from moisture until the point of administration. In some newer devices, the FPF is often not improved. The dose uniformity in some multiple dose systems can be poor and may therefore have to be filled with more drug than is needed to maintain dose uniformity [25, 26].

In general, formulators usually face many challenges that make designing an inhaled aerosol device quite difficult. An important challenge is the efficient production of drug particles that have a very small size from a device that can be used easily. Another challenge is that individuals tend to inhale in different ways and as a result, an inhaled aerosol is subjected to a variable air velocity and duration. A third challenge is the anatomical variation in the geometry of mouth–throat airway passages. The differing effects of device mouthpieces on mouth–throat geometry, their effect on the fluid mechanics in the mouth–throat, and the differences in the lung geometry between individuals are important considerations. Finally, different disease states can result in further differences in anatomy and breathing patters, which affect the spatial distribution of drug particles.

13.5 Particle Engineering Technologies for Pulmonary Delivery

The restrictions associated with MDI have encouraged the investigation of dry powder systems. Consequently, a variety of novel DPIs has emerged and a number of them have been introduced in the market. Currently, all marketed dry powder inhalation products are delivered from passive DPIs that depend on the patient's inspiratory effort as the only source of energy for particle dispersion. However, despite improvement in device design, the efficiency of DPIs is still poor mainly due to significant powder dispersion problems resulting in poor lung deposition and high dose variability. This has led to a further shift of focus toward engineering particle properties as a means of improving powder dispersion and ultimately DPI performance [27]. To obtain particles appropriately sized for inhalation, methods for size reduction have been required. Common techniques used in the pharmaceutical industry include milling, spray drying, precipitation from solution, and more recently nanoparticle-based formulations.

13.5.1 *Milling Technology*

Milling is the mechanical process of reducing large particles to a powder of micro- or nano-size (Fig. 13.4). Milling can be achieved by jet milling and ball milling on a wet or dry basis. Both techniques can produce particles less than 5 μm in diameter; however, the energy input for each of the techniques is different. The physical properties of powders may differ depending on the milling technique used, which can have a significant impact on aerosolization. Dry milling has been known to create partially amorphous materials with surface charge, possibly resulting in particle agglomeration. This problem can be eliminated by specialized milling methods such as fluid energy milling at elevated humidity and wet milling technology [28].

13.5.1.1 Fluid Energy Milling

Fluid energy or jet mills create particles under 25 μm through particle impaction in high velocity air. Compressed air produces a high-speed jet stream that passes the feed funnel and draws powder into a grinding chamber. Pulverizing nozzles are installed around the grinding chamber and inject additional high-speed air into the grinding chamber in a rotational direction. The centrifugal air flow accelerates

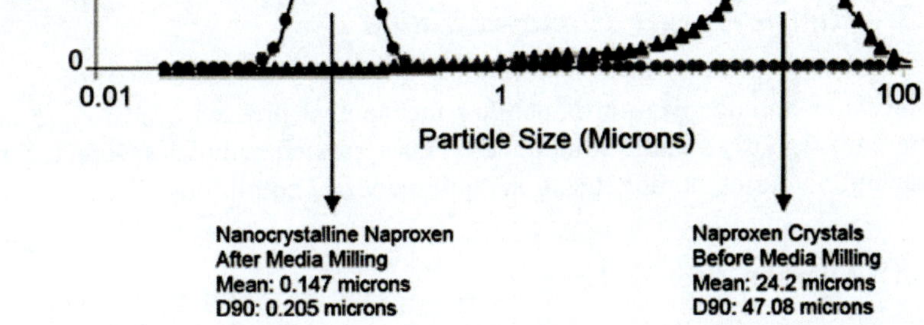

Fig. 13.4 The particle size distribution profile of naproxen crystals before and after milling. Before milling, the drug crystals had a mean particle size of 24.2 μm. After being processed for 30 min in a media mill, the mean particle size of the nanocrystalline dispersion was 0.147 μm with D_{50}=0.205 μm (adapted from Merisko-Liversidge et al. [30], used with permission)

particles and reduces particle size by particle-to-particle impaction. The air flow drives large particles toward the perimeter, but small particles move toward the center where they exit through the outlet [29].

For decreasing the amorphous content in powders produced by milling, fluid energy milling can be carried out at elevated humidity to enhance in situ crystallization. It has been reported that the milled products are mostly crystalline with particle size distributions similar to those produced by the conventional milling process. The system entails control of the relative humidity (e.g., 30–70%) of the milling chamber by humidifying the feed gas (e.g., by superheated steam to minimize condensation) used for milling the powder [30, 31].

Narrow size distributions of jet-milled and spray-dried mannitol particles have been produced in the respirable (1–10 μm) range. A previous report indicated that the particle size (d_{50}) and particle morphology differed between the jet-milled and spray-dried particles, whereas similarities were noted in other physicochemical properties. Significant differences were observed between the aerosol dispersion of the angular jet-milled particles and spherical spray-dried particles. MMAD for both technologies ranged from 3.8 to 6 μm. Higher MMADs were obtained with increasing d_{50} for both jet-milled and spray-dried particles, which was attributed to the ratio between aerodynamic and geometric diameters. Spray-dried particles produced higher values of FPF compared with jet-milled particles, but differences between jet-milled and spray-dried particles were not observed in emitted dose (ED), MMAD, or geometric standard deviation (GSD) [32].

13.5.1.2 Wet Milling Technology

Wet milling is an aqueous milling process capable of reducing the particle size to nanoparticles [33]. Fine-ball mills are popular because of their simplicity and scalability. Since the particles are produced in water, amorphous regions in the particles may undergo recrystallization. Thus, the wet-milled powder is anticipated to be crystalline and more stable to moisture than powders produced by dry milling [34]. Nanoparticle-based aerosols produced by this technique are reviewed in Sect. 5.7.

13.5.2 High-Pressure Homogenization

High-pressure homogenization is another mechanical process used to reduce the particle size of poorly soluble compounds. This approach includes several technologies depending on the homogenization equipment and conditions.

13.5.2.1 Microfluidizer Technology

Particles can be produced by a high-shear process using jet-steam homogenizers, such as microfluidizers. Particle collision, shear forces, and cavitation forces are created as a result of a frontal collision of two fluid streams under pressures up to

1,700 bar, which can lead to particle size reduction to the nanometer range. Surfactants are typically necessary to stabilize the desired particle size. In many cases, 50–100 passes are needed for sufficient particle size reduction. This principle has been applied by SkyePharma, Inc. for its IDD-P™ technology [35].

13.5.2.2 Piston-Gap Homogenization

Particle engineering in this process is based on cavitation forces created in high-pressure homogenizers. The suspension of a drug and surfactant is forced through a very small gap of the homogenizer under pressure of approximately 1,500 bar, usually using 3–20 homogenization cycles. Due to the small gap, the high streaming velocity of the suspension and the increased dynamic fluid pressure (according to Bernoulli's law) are compensated by the reduction in the static pressure on the fluid below the boiling point of water at room temperature. Consequently, water starts boiling at room temperature and gas bubbles are formed that collapse when the fluid leaves the homogenization gap. These cause cavitation-induced shock waves, thereby shearing the particles [36, 37]. Budesonide nanosuspensions have been successfully formulated using a piston-gap homogenizer depending on the processing pressure, and number of cycles applied [38, 39].

DissoCubes™ technology is another example of this process developed by Müller et al. in 1994, which was recently released as a patent owned by SkyePharm PLC. Enhanced dissolution behavior and the cubic shape of the resulting nanocrystals were reported [40]. Scholer et al. prepared atovaquone nanosuspensions using DissoCubes™ technology with a mean diameter of 279 nm $\pm$7 and mean polydispersity index of 0.18 $\pm$0.001 [41].

NanoPure® technology is another approach using the piston-gap homogenizer, owned and developed by PharmaSol GmbH in Berlin. This technology exploits dispersion media with a low vapor pressure and optional homogenization at low temperatures. The cavitation in the homogenization gap is reported to be very little or nonexistent. The remaining shear forces, particle collisions, and turbulence are enough to attain particle size reduction. The possible low temperatures while homogenizing permit the processing of thermally labile drugs. It is feasible to apply the process to nonaqueous media to protect drugs from hydrolysis [42]. NanoEdge™ technology, owned by Baxter Company, uses a precipitation step with subsequent attrition by applying high shear and/or thermal energy [43]. Nanopure XP technology, owned by PharmaSol GmbH, is a combination method using a pretreatment step followed by homogenization to create particles below 100 nm. It allows the extension of the performance of Nanopure technology to very hard and crystalline materials [44].

Homogenization processes are advantageously used to engineer particles for many reasons. High-pressure homogenizers are available with various capacities ranging from a few hundred to a few thousand liters per hour, allowing large-scale production of a drug delivery system. Drug crystallinity has been reported to be changed by high-pressure homogenization processes. A portion of poorly water soluble drugs in the particles was amorphous in some cases, while the drugs were wholly amorphous in other cases. In addition, homogenization may be exploited after rapid precipitation to further decrease particle size. A possible disadvantage of high-pressure homogenization processes includes the amorphous fraction produced in some cases, since this can be a challenge with respect to pharmaceutical stability [38].

13.5.3 Spray Drying Technology

Spray drying was the breakthrough technology that was explored in the 1940s as a promising alternative method of fabricating nanoparticles and microparticles with desirable flow and dispersion characteristics for pulmonary administration, perhaps, without the need for coarse carriers. Spray drying has been shown to be a scalable, robust, and pharmaceutically viable process for powder production. The first pharmaceutical product targeting systemic treatment through the pulmonary delivery route was inhaled insulin Exubera™. This dry powder was formulated by a spray drying process to engineer homogeneous particles containing an accurate amount of drug and excipients [45].

Anti-asthmatic drugs such as salbutamol sulfate, terbutaline sulfate, isoprenaline sulfate, and sodium cromoglycate have been formulated using this technology. In addition to drug production, spray drying has been used to produce carrier particles. For example, spherical lactose carrier particles containing budesonide have been prepared by this method. Spray drying is not limited to aqueous solutions. Nonaqueous systems have also been used to prepare porous particles suitable for aerosol delivery [46–48]. This technology has attracted much attention to respiratory drug delivery as a result of its potential application to the formulation of therapeutic proteins and peptides for systemic delivery via the lungs. There have been numerous studies illustrating the possibility of engineering protein particles with the desired physical characteristics for inhalation. In general, the MMAD of the particles generated by this technique is in the range of 1.5–4.0 μm, as observed for insulin, parathyroid hormone, human calcitonin, α-1 antitrypsin, β-interferon [49–51], influenza vaccine [52] as well as nucleic acid drugs, DNA/lipid vectors, viral vectors, and vesicle structures [53].

A typical spray drying process consists of three steps: (a) spraying of dissolved drug and excipient solution through an atomization nozzle; (b) drying of sprayed droplets at elevated temperatures; and (c) separation of dried product from the air (Fig. 13.5). For each of these operating steps, a wide variety of process designs are available, depending on specific needs or applications. The particle size of the dried particles is directly affected by the atomizer performance. There are various types of atomizers used in industrial spray drying applications, depending on the droplet size required. These include rotary atomizers, pressure nozzles, two-fluid nozzles, ultrasonic atomizers, and more recently four-fluid nozzles with in-line mixing. The later has been developed for the production of composite particles [28, 54].

Although the drying air temperature can be relatively high (>100°C), the actual temperature of the evaporating droplets is considerably lower due to cooling by the

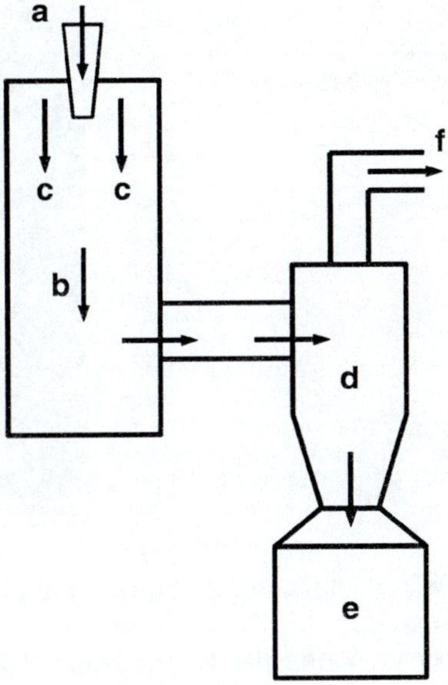

Fig. 13.5 Schematic diagram of the spray drying process. Drug and excipient solution enter the drying chamber (**b**) through the nozzle (**a**), where it is carried by a hot air flow and the solvent is evaporated (**c**). The droplets that form then enter a cyclone separation chamber (**d**), where the solvent is evaporated, particles are collected (**e**), and exhaust is expelled (**f**) (adapted from Chan et al. [28])

latent heat of vaporization. Thus, thermal degradation of the active ingredient may not be as great of a concern as it first appears. The system can also be tailored for better product recovery and larger scale production, especially for thermally sensitive materials such as proteins and peptides. For example, by replacing the bag-filter unit with a vacuum system, the drying airflow resistance can be reduced, thus allowing the protein product to be dried at a lower inlet air temperature. Design of high-efficiency cyclone separator is also essential for this technology to be economically acceptable on an industrial scale [55].

Spray drying technology has been used for fabrication of low-density particles. Particles of low density (ρ) are advantageous for aerosol drug delivery because of their relatively large volume diameter (d_v) and small aerodynamic diameter (d_a) leading to better dispersibility and more efficient deep lung deposition. FPF reaching 65–95% has been reported [50]. Furthermore, large porous particles produced in this manner were proposed to escape from the natural phagocytic clearance in the lungs, which may enhance bioavailability. Currently, low-density particles can be engineered by the PulmoSphere™ process, involving spraying an emulsion of fluorocarbon in water stabilized by phospholipid, where the drug is dissolved or dispersed in the external aqueous phase containing excipients such as hydroxyl ethyl starch. The fluorocarbon acts as a blowing agent at high temperature to produce foam-like porous or hollow structures with a tapped density <0.1 g/cm^3 (Fig. 13.6) [56].

PulmoSpheres™ were originally developed as ultrasound contrast media [57] and have been administered to the lung by liquid instillation. [58] They have been used for pulmonary delivery of several small molecules and peptides [59, 60] and have been recommended for vaccination applications [61], modulation of immune responses,

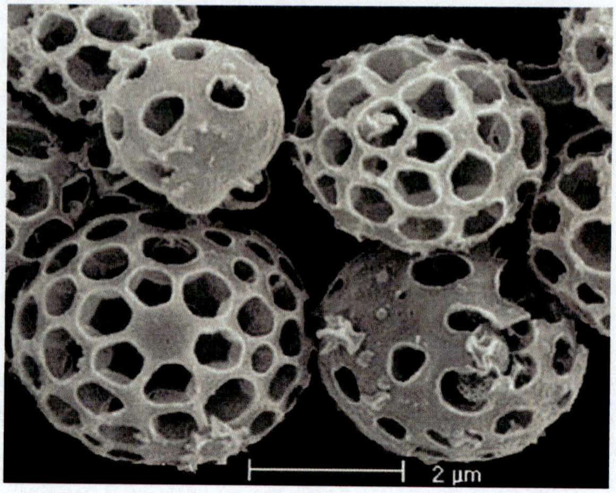

Fig. 13.6 PulmoSpheres™ solid foam particles (Vehring [66], used with permission)

[62] or immunoglobulin delivery [63]. The efficacy of these advanced particles in DPIs [59] or MDIs has been shown to exceed that of conventional dosage forms. It has been reported that the pulmonary delivery of tobramycin using PulmoSpheres™ was 9 times as efficient as delivery by a nebulizer [64]. A study on budesonide reported a twofold efficiency advantage relative to a conventional powder formulation [59].

Alternatively, a particulate form incorporating nanoparticles into micron-scale hollow or porous structures has been engineered to overcome the issues of storing and delivering nanoparticle suspensions to the lungs. This process was developed by Tsapis et al. and is referred to as "porous nanoparticle-aggregate particles" (PNAPs) [65]. These porous particles can be prepared by cospray drying of drugs, such as insulin, albuterol, or estradiol dissolved in an ethanolic solution, with simple sugars and/or polysaccharides (e.g., lactose) and additives such as human serum albumin and dipalmitoyl phosphatidylcholine (DPPC) [66, 67]. During spray drying, such mixtures may produce walled hollow particles, which collapse into a crumpled paper-like capsule structure. Similar to prior spray drying approaches, the size, density, and morphology of the resulting particles can all be controlled by varying the concentrations of DPPC and lactose [67]. The matrix of the carrier microparticle can consist of only nanoparticles or additional inert pharmaceutical excipients, such as amino acids, sugars, or phospholipids. Upon exposure to water the matrix of the PNAPs dissolves and readily releases the nanoparticles.

The formation of PNAPs during the drying process can be characterized by the Péclet number (*Pé*), a dimensionless mass transport number that compares the times for droplet drying and nanoparticle diffusion, and is defined by the equation 6:

$$Pé = \frac{R^2}{\tau_d D}, \tag{13.6}$$

where R is the radius of the droplet, t_d is the time required for a droplet to dry, and D is the solute (or nanoparticle) diffusion coefficient. A porous particle is formed when the droplet dries quickly and solutes or nanoparticles diffuse slowly and thus stay at the surface of the drying droplet without enough time to diffuse to the center ($Pé$ >1). The solutes and nanoparticles accumulated at the evaporating front of the droplet form a shell that dries to become the hollow microparticle. The nanoparticles are held together in this form by both physical forces (e.g., van der Waals force) and the continuous matrix of excipients. PNAP characteristics can be controlled by manipulating variables such as nanoparticle size, nanoparticle composition, excipients used to form the PNAP matrix, solution solids concentration, and nanoparticle concentration in the final particle. The drug release from PNAP can be influenced by the drug concentration in both the nanoparticles and carrier particles [11].

13.5.3.1 Spray Freeze Drying

Spray freeze drying (SFD) has been explored for pharmaceutical applications since the early 1990s. This process involves spraying an aqueous drug solution via a two-fluid or an ultrasonic nozzle into a spray chamber filled with a cryogenic liquid (liquid nitrogen or refrigerant such as CFC or fluorocarbon) followed by ventilation or lyophilization to obtain dry, free-flowing micronized powders. The spraying process can be performed directly into (spray freezing into liquid; SFL) or above the surface of the cryogenic liquid, depending on the position of the nozzle [68, 69]. Atomization into the nitrogen vapor above the liquid gas leads to gradual agglomeration and solidification of the solution droplets that pass through the vapor phase and then settle to the surface of the cryogenic liquid. As a result, broad particle size distributions and nonmicronized dry powders may result. To circumvent these problems, spray freezing into liquid (SFL) was developed and this process was patented by The University of Texas at Austin in 2001 [70] and commercialized by Dow Chemical Company. The SFL process can achieve intense atomization and rapid freezing rates. The ultra-rapid freezing rates avoid the phase separation of solutes within the feed solution and induce formation of amorphous structures with high surface areas, enhanced wetting, and significantly enhanced dissolution rates [68]. It has been reported that the SFL process can use organic solvents such as acetonitrile as the solution source solvent in addition to an aqueous solution or water/organic cosolvent such as water and tetrahydrofuran (THF). The SFL acetonitrile system offers several advantages, including increasing the drug loading in the SFL feed solution and decreasing the drying time for lyophilization following the SFL process. A recent study also demonstrated that rapidly dissolving SFL micronized powders have been produced by SFL with an organic solvent mixture [71].

Compared to spray drying, SFD produces light and porous particles with enhanced aerosol performance and the production yield is almost 100%. The method has been applied to prepare rhDNase and anti-IgE antibody [72] particles for inhalation and can also be used for anti-asthmatic compounds. However, SFD methods are not as well established and widely utilized as spray drying due to higher complexity, scale-up, and expense. Despite these encumbrances, with the introduction of new large-scale spraying techniques and enhancing drying cycles, SFD has become a viable alternative to spray drying that is not limited to aqueous solutions [71].

13.5.3.2 Controlled Evaporation of Droplets

Like spray drying, controlled evaporation of droplets is a single-step continuous process; however, this method achieves better control over the temperature and residence time of droplets. In this process, the drug solution is atomized into a carrier gas for drying using an ultrasonic nebulizer. The droplets suspended in a carrier gas are then fed into a tubular flow reactor housed in a constant temperature oven for evaporation. Since the feed rate and temperature are adjustable, the temperature and residence time of the droplets can be controlled. The method has the potential to control particle morphology and drug polymorphic form and has been used to produce beclomethasone dipropionate particles [73].

13.5.3.3 Evaporation of Low-Boiling-Point Solutions

Evaporation of low-boiling-point solutions entails simply the dissolution of the active pharmaceutical ingredient in a low-boiling-point organic solvent, followed by atomization and evaporation of the resulting droplets to produce dry particles. This concept is somewhat similar to both spray drying and supercritical fluid (SCF) expansion [28].

13.5.4 Precipitation Technology

Precipitation technology is one of the oldest methods for the production of microparticles and nanoparticles. The first application of this technique was known as hydrosol technology that was developed by List and Sucker and owned by Sandoz (now Novartis) [74]. In general, the drug is dissolved in a solvent and subsequently added to an antisolvent, leading to the precipitation of finely dispersed drug particles.

13.5.4.1 Direct Controlled Crystallization

Controlled crystallization of hydrophobic drugs in the respirable size range can be obtained by an antisolvent precipitation technique using growth-retarding stabilizing additives such as hydroxypropylmethylcellulose (HPMC). It has been reported that precipitated drug particles such as budesonide, predonsiolone, fluticasone, and disodium cromoglycate have a higher FPF than jet-milled samples. These particles often have lower amorphous content than that of the mechanically treated micronized materials. This technology has been used for engineering Zinc-free insulin crystals in the inhalable size range of 0.2–5 μm. These crystals were found to be more stable than powders of essentially the same composition prepared by spray drying, freeze drying, vacuum drying, or oven drying [75–77].

Fig. 13.7 SEM micrographs of various budesonide particles: (**a**) prepared using an acetone–water system; (**b**) prepared using a methanol–water system; (**c**) prepared by the methanol–water system in the presence of HPC (0.07%, g/mL); (**d**) prepared using a methanol–water system in the presence of Tyloxapol (0.16%, g/mL)

An antisolvent precipitation approach for preparing ultra-fine budesonide particles suitable for dry powder inhalation produced particles with controlled shape. Crystals of rectangular shape were obtained from an acetone–water system while elliptical particles were obtained from a methanol–water system. Both were no more than 5 μm in length and 300 nm in thickness, and formed agglomerates. When stabilizers such as hydroxypropyl cellulose and tyloxapol were present, uniform spindle or ellipsoidal particles less than 5 μm in diameter could be prepared (Fig. 13.7). All the prepared particles had the same crystalline structure as the commercial budesonide product, as assessed by X-ray diffraction. Higher FPF (~69%) values were observed from the agglomerates formed from flaky particles compared to spindle or ellipsoidal particles [78].

13.5.4.2 Direct Crystallization of Spherical Agglomerates

Another approach used for pulmonary formulations involves antisolvent precipitation of drug solution in a water-miscible organic solvent, followed by addition of a bridging solvent, which is immiscible or partially miscible with water. For instance, incorporation of ethyl acetate into the water/acetone crystallization medium led to the fabrication of spherical agglomerates (200–300 μm) composed of primary crystals in the respirable range (MMAD = 1.3–2.7 μm) [7]. Deagglomeration was achieved by mixing the agglomerated particles with a lactose carrier for 2 min or more, and the attached primary particles could then be easily separated from the lactose during inhalation. Spherical crystallization can also be attained by quenching a hot organic or aqueous organic solution of the drug with a cold organic or aqueous organic solvent. The quench solvent should be miscible with the drug solvent, for example, to produce spherical accretions of salmeterol xinafoate microcrystals. The resulting agglomerates were free-flowing, friable, and readily micronizable to a material suitable for inhalation delivery [7].

The challenge in all precipitation processes is that the drug needs to be soluble in at least one solvent, which creates problems for newly developed drugs that are insoluble in both aqueous and mild organic media. Another problem associated with this technology is that the formed nanoparticles need to be stabilized to avoid growth into micrometer crystals. As a result, it can be difficult to reproducibly generate particles in the micron range for aerosol delivery [79–81].

13.5.4.3 Sono-Crystallization

Ultrasound has been applied to control the precipitation process. The particle size can be altered by the ultrasound-induced mixing and the effect of cavitation on supersaturation and nucleation, which may provide more uniform particles. Fluticasone propionate and salmeterol xinafoate are anti-asthmatic drugs successfully prepared using this technique. Furthermore, lowering the mean particle size can be achieved by adding high concentrations of growth-retarding or stabilizing excipients [28].

13.5.4.4 Microprecipitation by Opposing Liquid Jets

Excessive turbulence and intense mixing during precipitation can be achieved using a jet of drug solution opposing a jet of antisolvent coming through two opposing nozzles mounted in a small chamber. The crucial process parameters include the speed of the liquid jets and concentration of the drug solution. A high jet stream speed or a high drug concentration was found to give finer particles but at higher residual solvent level. The volume ratio of drug solution to antisolvent is also anticipated to influence the precipitation process [28].

13.5.4.5 Evaporative Precipitation into Aqueous Solution Process (EPAS)

The EPAS process uses rapid phase separation to nucleate and grow nanoparticles and microparticles of poorly water soluble drugs. The drug is first dissolved in a low boiling liquid organic solvent. Then, this solution is pumped through a tube where it is heated under pressure to a temperature above the solvent's normal boiling point and then sprayed through a fine atomizing nozzle into a heated aqueous solution. The nozzle is immersed into the aqueous solution to ensure that the nucleating drug particles contact the nonsolvent with hydrophilic surfactant rapidly, thus, inhibiting crystallization and growth of the drug particles. The stable aqueous drug suspension is then dried by one of several techniques including ultra rapid freezing in conjunction with lyophilization or spray drying. The rapid evaporation of the heated organic solution in EPAS seems to result in fast nucleation leading to amorphous nanoparticle suspensions [82].

13.5.5 Supercritical Fluid (SCF) Technology

An SCF processes offer an alternative approach for engineering microparticles and nanoparticles directly from drug solutions. SCFs have gaseous and liquid properties at the temperature and pressure above their critical point (T_c, critical temperature; P_c, critical pressure; Fig. 13.8). CO_2 is the most commonly used SCF for pharmaceutical applications because it has a low T_c (31.2°C) and P_c (7.4 MPa), is nonflammable, nontoxic, and inexpensive. SCF can serve as a solvent or an antisolvent [83, 84]. Early SCF processes demonstrated limited ability to engineer particles in the respirable size range; however, SCF technologies are rapidly developing [85–93].

Rapid expansion of supercritical solutions (RESS) atomizes a high-pressure solution of the drug substance in an SCF into a low-pressure vessel, where it undergoes expansion through an orifice to create extremely high supersaturation. This phenomenon can drive homogeneous nucleation and subsequent particle formation in the precipitation unit. This process is limited only for substances with high SCF solubility ($\geq 10^{-3}$ kg/kg). Fluids such as DME may be necessary when processing polar molecules; however, DME is flammable and difficult to recycle. RESS can be optimized to achieve a relatively narrow particle size distribution and can also be

Fig. 13.8 Typical phase diagram showing the supercritical fluid region

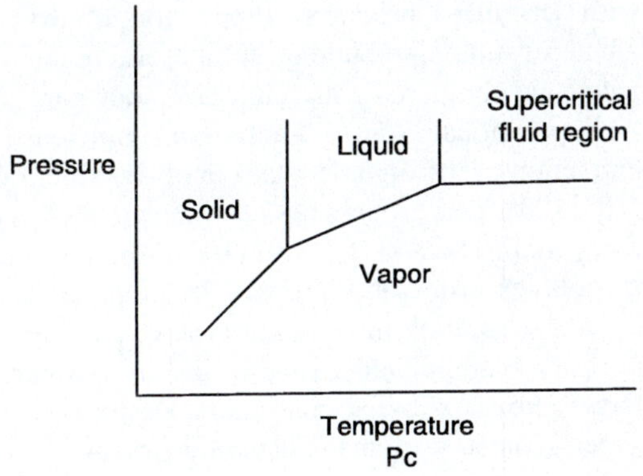

used for the coating of micron-sized particles with CO_2-soluble polymers or waxes. Also, the addition of an organic solvent miscible with CO_2 can improve the solubility of the drug; however, efficient removal of this solvent is required after expansion [94].

Johnston et al. developed a process based on Rapid expansion from Supercritical to Aqueous Solution (RESAS) to produce stable nanosuspensions of poorly water soluble drugs. This process includes the induction of rapid nucleation of the SCF-dissolved drugs in the presence of surface modifying agents resulting in particle formation with a desirable size distribution in a very short time [95, 96]. Another modification is Rapid Expansion of Liquefied-Gas Solution and Homogenization (RELGSH), which was a combination of RESAS with high-pressure homogenization developed by Pace et al. to create a stable and high-payload drug delivery system. The microparticle or nanoparticle suspensions were formed by rapid expansion into an aqueous medium of a compressed solution of the compound. Surface modifiers were added in a liquefied gas, and the resulting aqueous suspension was homogenized with a high-pressure homogenizer [97, 98].

In other SCF processes, referred to as Supercritical Anti-Solvent (SAS), the fundamental particle formation mechanism is based on rapid precipitation when a drug solution is brought into contact with an SCF. For instance, Solution-Enhanced Dispersion by Supercritical Fluids (SEDS) involves rapid dispersion and mixing of the drug solution with the SCF (typically in a coaxial spray arrangement) and extraction of the solvent into the SCF [7]. SEDS technology is able to produce crystalline powders of small molecules with a low amount of residual solvents and may offer direct fabrication of respirable drug particles. Additional advantages include the ability of supercritical solvents to be competently separated by decompression from organic cosolvents and solid powders, facilitating a single-step, clean and solvent recyclable process [83, 99].

As most of the drugs have very limited or no solubility in CO_2, these SAS processes have attracted much attention because they offer a direct means for engineering micron-sized dry powders, such as budesonide and fluticasone [100, 101]. A number of pure and coated (with lecithin) particulate products that are suitable for formulation into pMDIs have been obtained with this technology. Coated particles produced with SAS processes showed a significant increase in FPF when compared with jet-milled products. Production of dry powders of salmeterol xinafoate, albuterol sulfate, terbutaline sulfate, and fenoterol hydrobromide have been investigated and compared with micronized powders [7].

Supercritical Fluid Extraction of Emulsions (SFEE) is a novel process based on extraction of the organic phase in oil-in-water or multiple emulsions using SC CO_2 [102, 103]. This process has been shown to yield dramatically different crystal when compared to SAS (Fig. 13.9) [102]. Another method developed by Kompella et al. involves exposure of microparticles prepared by an emulsion solvent evaporation to SC CO_2, resulting in a similar type of particle as prepared by SFEE, with particle porosity manipulated by adjusting the suspension depressurization rate. PLGA large-porous particles containing Deslorelin, a gonadotropin-releasing hormone super-agonist, were made using this method [104].

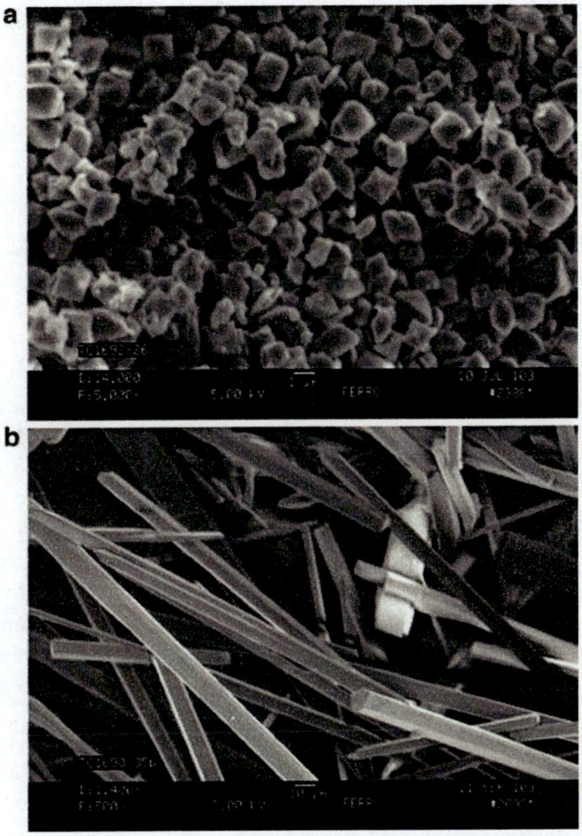

Fig. 13.9 (**a**) Morphology of griseofulvin (GF) crystals produced using the SFEE method (volume-weighted diameter (VMD) = 978 nm, number-weighted mean diameter (NMD) = 784 nm) in comparison to (**b**) crystals produced using supercritical antisolvent precipitation under the same conditions of pressure, temperature, and solution flow rate

13.5.6 Porous Microparticle Fabrication Techniques

An interesting design motif often employed in engineering pulmonary drug formulations is that of large, porous microparticles. Porous particles may be ideal for pulmonary drug delivery because of their low density and large surface area. Most commonly, large porous particles have been produced by spray drying as discussed earlier; however, multiple emulsion techniques, which includes a porogen in the organic phase, have also been reported [105, 106]. In general, water with a surfactant or another pore-forming agent is then emulsified into a volatile organic phase (W/O emulsion) containing the matrix-forming excipient (with or without porogen). This emulsion is then emulsified again into another water phase containing a surfactant (W/O/W emulsion). As the organic phase evaporates, the encapsulant (e.g., polymer) hardens into microparticles. The porogen can then be washed away or dissolved, revealing a porous microparticle structure that can have very diverse geometries depending on the polymer and porogen used (Table 13.1) [105–107].

Table 13.1 Porous microparticle fabrication techniques

Drug	Encapsulant	Porogen	Technique	Size (MMAD) (μm)	Reference
BSA (model drug)	PLGA	Cyclodextrin derivative	W/O/W emulsion	3.0	[105]
Plasmid DNA	PLGA	None	W/O/W emulsion	3.8	[106]
Ciprofloxacin	PLGA	Canola oil or silicon oil	Precision particle fabrication	4.0	[107]

MMAD mas median aerodynamic diameter

Although multiple emulsion techniques have generated promising results, they do not offer great control over the variability in the particle size distribution [108]. To overcome this limitation, several studies have aimed to develop a Precision Particle Fabrication technique that sprays a drug and porogen-containing polymer solution through a nozzle under acoustic excitation to produce uniform droplets [107–109]. In one study, the authors used either canola oil or silicon oil as a porogen. They found that the canola oil formed particles with a porous, web-like interior, while silicon oil formed particles with a solid exterior and hollow interior (Fig. 13.10) [107]. Particle morphology could be controlled by selecting oil porogens that were partially miscible or completely phase separated from matrix-forming polymer. The diffusive transport of the porogen (oil droplets) compared to the boundary velocity of the decreasing droplet radius (*Pé* number), was also a factor dictating particle appearance.

13.5.7 Nanoparticle-Based Formulations

In addition to the aerodynamic properties of particles, it is advantageous to consider how particle engineering will affect the rate of drug release. For instance, nanoparticle-based formulations (defined here as $d_v < 1$ μm) are gaining considerable interest because of their enhanced release kinetics. Nanoparticle drug formulations possess very unique physical properties due to their small size [1]. For instance, the saturation solubility of drugs formulated as very small nanoparticles can be greater than that of bulk drug, as described by:

$$c(r) = c_{\infty} \cdot \exp\left(\frac{2 \cdot \gamma \cdot V_{\mathrm{m}}}{r \cdot R \cdot T}\right) \qquad (13.7)$$

where γ is the interfacial surface tension between the particle surface and the surrounding medium, V_{m} is the molar volume of the drug, r is the radius of the particle, R is the gas constant, and T is the absolute temperature [110]. In addition to enhanced thermodynamic properties like saturation solubility, nanoparticles also

Fig. 13.10 Scanning electron micrographs of large porous particles using various porogens, as demonstrated by Arnold, et al. (**a**) Surface morphology and (**b**) interior morphology of particles prepared with canola oil as a porogen. (**c**) Surface morphology and (**d**) interior morphology of particles prepared using silicon oil as a porogen. The scale bars represent 10 μm (courtesy Arnold et al. [107])

demonstrate enhanced mass transfer properties due to their large total surface area and high surface curvature [6]. This principle is generically captured by the Noyes-Whitney equation [102].

$$\frac{\mathrm{d}m}{(m\mathrm{d}t)} = \frac{k_\mathrm{d} \cdot k_\mathrm{s}}{k_\mathrm{d} + k_\mathrm{s}} \cdot s \cdot (c_0 - c) \tag{13.8}$$

where $\mathrm{d}m/(m\,\mathrm{d}t)$ is the fraction of mass dissolved in time, s is the surface area, c is the concentration of drug in the bulk, and c_0 is the saturation solubility of the drug. The coefficients k_d and k_s are the diffusion and the surface kinetic components of the mass transfer coefficient [7, 102]. Typically, k_d is dependent on the stirring regime of the system, and k_s is dependent on the solution thermodynamics [102].

One approach that has been investigated as a means to deliver nanoparticles to the lungs without compromising their aerodynamic properties is to engineer microparticle–nanoparticle composite particles. Particles less than 1 μm in aerodynamic diameter generally reach the alveolar region but may fail to deposit due to diffusional effects that prevent them from settling in the alveolar lumen. Developing particles that have the properties currently desired for optimal deposition in the lungs (1 μm<d_a<5 μm), but retain the enhanced pharmacokinetic properties of nanoparticles has been a challenge [1].

PNAPS represent one possible approach; however, nanoparticles have also been incorporated into micron-sized carrier particles using a technique such as spray drying or SFD [111, 112]. Typically, nanoparticles are synthesized using a precipitation technique. After purification, the particles are mixed into an aqueous solution containing dissolved lactose (or other appropriate matrix material), which is then spray dried to produce nanoparticle-containing lactose particles [111, 112]. Presumably, after inhalation and deposition into the alveolar region of the lungs, the lactose will dissolve, leaving behind their therapeutically active nanoparticle cargo. This technique has been investigated as a means to formulate the antineoplastic agent doxorubicin [111] and human insulin [113].

Another technique used to synthesize nanoparticles for pulmonary delivery is wet milling nanotechnology, which is capable of reducing particle size to smaller than 400 nm. This approach has been led by Elan Pharmaceuticals [33]. Milling is often conducted using yttrium-stabilized zirconium oxide beads or highly cross-linked plastic beads, such as hard polystyrene, or glass beads, ranging from 0.05 to 3 mm, in a media mill. Impact and shear forces between the moving balls grind suspended solid particles. Mixing intensity and the number of contact points are the main factors that determine the size-milled particles. Grinding media of 1 mm or less in size is preferred to be effective in attrition [71, 114]. Milling time can range from minutes to days depending on the energy input and the brittleness of the drug particles. Milling can be conducted under refrigeration to minimize the thermal degradation of the active ingredient. Osatrander and coworkers reported budesonide milled to particle size of 166 nm, which was then spray dried to produce powder with a mean particle size of 1.35 μm. A surface modifier (such as polyvinyl pyrrolidone (PVP), lecithin, and/or cellulose derivatives) was added during or after milling to prevent agglomeration of the nanoparticles. This milled budesonide was used for nebulized delivery to the lungs [115].

Another possible method to overcome the challenge of imbuing nanoparticles with the aerodynamics of small microparticles is to engineer a cluster of drug nanoparticles that can be aerosolized and efficiently delivered to the lungs [116]. Recently, El-Gendy et al. have reported a NanoCluster SN™ process that is being commercialized by Savara Pharmaceuticals [9, 117]. Typically, nanoparticles may first be synthesized using a precipitation or other appropriate technique. Pure nanoparticle suspensions of hydrophobic drugs naturally agglomerate through via water exclusion or 'hydrophobic' interactions if stabilizers (e.g., surfactants) are not included. More stable nanoparticle suspensions may be flocculated by disrupting the colloidal stability of the nanoparticles. For example, electrostatic nanoparticle stability may be interrupted by salting out using an amino acid or sodium chloride [116]. NanoCluster-SN™ assembly of nanoparticles in solution has been investigated as a means to formulate nifedipine [10], human insulin [6], a combination of paclitaxel and cisplatin [117], and budesonide among others (Fig. 13.11) [9].

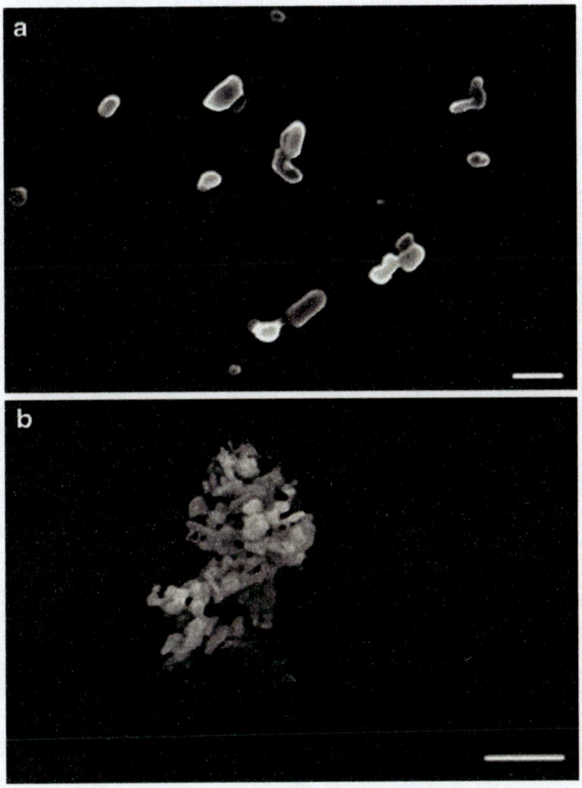

Fig. 13.11 Scanning electron micrographs of nifedipine nanoparticle clusters engineered using the NanoCluster-SN™ technology platform. Panel (**a**) shows nifedipine nanoparticles prior to flocculation NanoClusters. The scale bar represents 1 μm. Panel (**b**) shows nanoclusters after flocculation. The scale bar represents 10 μm. Cascade impaction measurements indicated that the flocculates for this formulation had a mass median aerodynamic diameter (MMAD) of 1.4 ± 0.1 μm (courtesy Plumley et al. [10])

13.6 Summary

Current research in pulmonary drug formulations has shown that many diverse technologies have been used to manufacture drug-containing particles for pulmonary delivery. These technologies range from traditional techniques such as milling and spray drying to the more recent SCF techniques and nanotechnology. Technologies for engineering particle shape, density, and size will continue to be honed to enhance the effectiveness of pulmonary drug formulations. Continued development of simple engineering approaches may enable more drugs to be delivered through this route for local treatment of lung diseases or systemic therapy.

References

1. Bailey MM, Berkland CJ (2009) Nanoparticle formulations in pulmonary drug delivery. Med Res Rev 29(1):196–212
2. Brain JD (2007) Inhalation, deposition, and fate of insulin and other therapeutic proteins. Diab Technol Ther 9:S4–S15

3. Patton JS, Byron PR (2007) Inhaling medicines: delivering drugs to the body through the lungs. Nat Rev Drug Discov 6(1):67–74
4. Patton JS (1996) Mechanisms of macromolecule absorption by the lungs. Adv Drug Deliv Rev 19(1):3–36
5. Asgharian B, Hofmann W, Miller FJ (2001) Mucociliary clearance of insoluble particles from the tracheobronchial airways of the human lung. J Aerosol Sci 32(6):817–832
6. Bailey MM, Gorman EM, Munson EJ, Berkland C (2008) Pure insulin nanoparticle agglomerates for pulmonary delivery. Langmuir 24(23):13614–13620
7. Chow AHL, Tong HHY, Chattopadhyay P, Shekunov BY (2007) Particle engineering for pulmonary drug delivery. Pharm Res 24(3):411–437
8. de Boer AH, Gjaltema D, Hagedoorn P, Frijlink HW (2002) Characterization of inhalation aerosols: a critical evaluation of cascade impactor analysis and laser diffraction technique. Int J Pharm 249(1–2):219–231
9. El-Gendy N, Gorman EM, Munson EJ, Berkland C (2009) Budesonide nanoparticle agglomerates as dry powder aerosols with rapid dissolution. J Pharm Sci 98(8):2731–2746
10. Plumley C, Gorman EM, El-Gendy N, Bybee CR, Munson EJ, Berkland C (2009) Nifedipine nanoparticle agglomeration as a dry powder aerosol formulation strategy. Int J Pharm 369(1–2):136–143
11. Sung JC, Pulliam BL, Edwards DA (2007) Nanoparticles for drug delivery to the lungs. Trends Biotechnol 25(12):563–570
12. Rogueda PG, Traini D (2007) The nanoscale in pulmonary delivery. Part 1: deposition, fate, toxicology and effects. Expert Opin Drug Deliv 4(6):595–606
13. Shekunov BY, Chattopadhyay P, Tong HHY, Chow AHL (2007) Particle size analysis in pharmaceutics: principles, methods and applications. Pharm Res 24(2):203–227
14. Cunningham E (1910) On the velocity of steady fall of spherical particles through fluid medium. Proc R Soc Lond 83(563):357–365
15. Hickey AJ, Martonen TB, Yang Y (1996) Theoretical relationship of lung deposition to the fine particle fraction of inhalation aerosols. Pharm Acta Helv 71(3):185–190
16. Telko MJ, Hickey AJ (2005) Dry powder inhaler formulation. Respir Care 50(9):1209–1227
17. Harvey CJ, O'Doherty MJ, Page CJ, Thomas SH, Nunan TO, Treacher DF (1997) Comparison of jet and ultrasonic nebulizer pulmonary aerosol deposition during mechanical ventilation. Eur Respir J 10(4):905–909
18. Tarara TE, Hartman MS, Gill H, Kennedy AA, Weers JG (2004) Characterization of suspension-based metered dose inhaler formulations composed of spray-dried budesonide microcrystals dispersed in HFA-134a. Pharm Res 21(9):1607–1614
19. Sommerville ML, Cain JB, Johnson CS, Hickey AJ (2000) Lecithin in-verse microemulsions for the pulmonary delivery of polar compounds utilizing dimethylether and propane as propellants. Pharm Dev Technol 7(3):273–288
20. Sommerville ML, Cain JB, Johnson CS, Rypacek F, Hickey AJ (2002) Lecithin microemulsions in dimethylether and propane for the generation of pharmaceutical aerosols containing polar solutes. Pharm Dev Technol 7(3):273–288
21. Chougule MB, Padhi BK, Jinturkar KA, Misra A (2007) Development of dry powder inhalers. Recent Pat Drug Deliv Formul 1(1):11–21
22. Ashurst II, Malton A, Prime D, Sumby B (2000) Latest advances in the development of dry powder inhalers. Pharm Sci Technol 3(7):246–256
23. Lalor CB, Hickey AJ (1997) Generation and characterization of aerosols for drug delivery to the lung. In: Akwete Lex Adjei, Pramod KG (ed) Inhalation delivery of peptides and proteins, vol. 107. p 235–276
24. Taylor A, Gustafsson P (2005) Do all dry powder inhalers show the same pharmaceutical performance? Int J Clin Pract 59(149):7–12
25. Srichana T, Martin GP, Marriott C (1998) Dry powder inhalers: the influence of device resistance and powder formulation on drug and lactose deposition in vitro. Eur J Pharm Sci 7(1):73–80
26. Chrystyn H (2007) The Diskus: a review of its position among dry powder inhaler devices. Int J Clin Pract 61(6):1022–1036

27. Lobo JM, Schiavone H, Palakodaty S, York P, Clark A, Tzannis ST (2005) SCF-engineered powders for delivery of budesonide from passive DPI devices. J Pharm Sci 94(10):2276–2288
28. Chan HK, Chew NY (2003) Novel alternative methods for the delivery of drugs for the treatment of asthma. Adv Drug Deliv Rev 55(7):793–805
29. Austin LG, Trass O (1997) Size reduction of solids crushing and grinding equipment. In: Fayed ME, Otten L (eds) Handbook of powder science & technology. Chapman & Hall, New York, pp 586–634
30. Merisko-Liversidge E, McGurk SL, Liversidge GG (2004) Insulin nanoparticles: a novel formulation approach for poorly water soluble Zn-insulin. Pharm Res 21(9):1545–1553
31. Müller RH, Bohm BHL, Grau MJ (2000) A formulation approach for poorly soluble and poorly bioavailable drugs. In: Wise DL (ed) Hand book of pharmaceutical controlled release technology., pp 345–358
32. Louey MD, Van Oort M, Hickey AJ (2004) Aerosol dispersion of respirable particles in narrow size distributions produced by jet-milling and spray-drying techniques. Pharm Res 21(7):1200–1206
33. Liversidge GG, Cundy KC, Bishop JF, Czekai DA (1992) Surface Modified Drug Nanoparticles, US Patent 5,145,684
34. Müller RH, Möschwitzer J, Bushrab FN (2006) Manufacturing of Nanoparticles by milling and homogenization technique. In: Gupta RB, Kompella UB (eds) Nanoparticle technology for drug delivery., pp 21-–51
35. Bruno RP, McIlwrick R (1999) Microfluidizer processor technology for high performance particle size reduction, mixing and dispersion. Eur J Pharm Biopharm 56:29–36
36. Rabinow BE (2004) Nanosuspensions in drug delivery. Nat Rev Drug Discov 3(9):785–796
36. Zhang D, Tan T, Gao L, Zhao W, Wang P (2007) Preparation of azithromycin nanosuspensions by high pressure homogenization and its physicochemical characteristics studies. Drug Dev Ind Pharm 33(5):569–575
38. Muller RH, Jacobs C, Kayser O (2001) Nanosuspensions as particulate drug formulations in therapy. Rationale for development and what we can expect for the future. Adv Drug Deliv Rev 47(1):3–19
39. Muller RH, Becker R, Kruss B, Peters K (1999) Pharmaceutical Nanosuspensions for Medicament Administration as Systems with Increased Saturation Solubility and Rate of Solution. US Patent 5,858,410
40. Müller RH, Jacobs C, Kayser O (2003) DissoCubes – a novel formulation for poorly soluble and poorly bioavailable drugs. In: Rathbone MJ, Hadgraft J, Roberts MS (eds) Modified-release drug delivery systems. Marcel Dekker, New York, pp 135–49
41. Scholer N, Krause K, Kayser O, Muller RH, Borner K, Hahn H, Liesenfeld O (2001) Atovaquone nanosuspensions show excellent therapeutic effect in a new murine model of reactivated toxoplasmosis. Antimicrob Agents Chemother 45(6):1771–1779
42. Bushrab NF, Müller RH (2003) Nanocrystals of poorly soluble drugs for oral administration. J New Drugs 5:20–2
43. Kipp JE, Wong JCT, Doty MJ, Rebbeck CL (2003) Microprecipitation method for preparing submicron suspensions. US Patent 6607784. B2 USA
44. Müller RH, Möschwitzer J (2005) Method and apparatus for the production of ultrafine particles and coating of such particles. DE 10 2005 053862.2 Application, Germany
45. Atkins PJ (2005) Dry powder inhalers: an overview. Respir Care 50:1304–1312
46. Vidgren MT, Vidgren PA, Paronen TP (1987) Comparison of physical and inhalation properties of spray-dried and mechanically micronized disodium cromoglycate. Int J Pharm 35:139–144
47. Forrester RB, Boardman TD (1986) Inhalation pharmaceuticals, US Patent 4590206
48. Chawla A, Taylor KMG, Newton JM, Johnson MCR (1994) Production of spray dried salbutamol sulfate for use in dry powder aerosol formulation. Int J Pharm 108:233–240
49. Platz RM, Patton JS, Foster L, Eljamal M (2003) Spray drying of macromolecules to produce inhalable dry powders. U.S. Patent 6,582,728
50. Platz RM, Patton JS, Foster L, Eljamal M (2004) Compositions and methods for the pulmonary delivery of aerosolized macromolecules. US Patent 6,797,258
51. Platz RM, Patton JS, Foster L, Eljamal M (2005) Composition for pulmonary administration comprising a drug and a hydrophobic amino acid. United States Patent. 6,921,527

52. Smith DJ, Bot S, Dellamary L, Bot A (2003) Evaluation of novel aerosol formulations designed for mucosal vaccination against influenza virus. Vaccine 21:2805–2812
53. Eljamal M, Patton JS, Foster L, Platz RM (1999) Compositions and methods for nucleic acid delivery to the lung. United States Patent. 5,994,314
54. Pan PJ, Chou CL, Chiou HJ, Ma HL, Lee HC, Chan RC (2003) Extracorporeal shock wave therapy for chronic calcific tendinitis of the shoulders: a functional and sonographic study. Arch Phys Med Rehabil 84(7):988–993
55. Maa YF, Nguyen PA, Sit K, Hsu CC (1998) Spray-drying performance of a bench-top spray dryer for protein aerosol powder preparation. Biotechnol Bioeng 60:301–309
56. Tarara T, Weers J, Dellamary L (2000) Engineered powders for inhalation. Respir Drug Deliv Proc 7:413–416
57. Trevino LA, Schutt EG, Klein DH, Tarara TE, Weers JG, Kabalnov A (1998) Stabilized Gas Emulsion Containing Phospholipid for Ultrasound Contrast Enhancement, Alliance Pharmaceutical Corp. USA Patent 5,798,091
58. Smith DJ, Gambone LM, Tarara T, Meays DR, Dellamary LA, Woods CM, Weers J (2001) Liquid dose pulmonary instillation of gentamicin PulmoSpheres formulations: tissue distribution and pharmacokinetics in rabbits. Pharm Res 18(11):1556–1561
59. Duddu SP, Sisk SA, Walter YH, Tarara TE, Trimble KR, Clark AR, Eldon MA, Elton RC, Pickford M, Hirst PH, Newman SP, Weers JG (2002) Improved lung delivery from a passive dry powder inhaler using an Engineered PulmoSphere powder. Pharm Res 19(5):689–95
60. Hirst PH, Pitcairn GR, Weers JG, Tarara TE, Clark AR, Dellamary LA, Hall G, Shorr J, Newman SP (2002) In vivo lung deposition of hollow porous particles from a pressurized metered dose inhaler. Pharm Res 19(3):258–264
61. Bot AI, Smith DJ, Bot S, Dellamary L, Tarara TE, Harders S, Phillips W, Weers JG, Woods CM (2001) Receptor-mediated targeting of spray-dried lipid particles coformulated with immunoglobulin and loaded with a prototype vaccine. Pharm Res 18(7):971–979
62. Bot AI, Tarara TE, Smith DJ, Bot SR, Woods CM, Weers JG (2000) Novel lipid-based hollow-porous microparticles as a platform for immunoglobulin delivery to the respiratory tract. Pharm Res 17(3):275–283
63. Dellamary L, Smith DJ, Bloom A, Bot S, Guo GR, Deshmuk H, Costello M, Bot A (2004) Rational design of solid aerosols for immunoglobulin delivery by modulation of aerodynamic and release characteristics. J Control Release 95(3):489–500
64. Newhouse MT, Hirst PH, Duddu SP, Walter YH, Tarara TE, Clark AR, Weers JG (2003) Inhalation of a dry powder tobramycin PulmoSphere formulation in healthy volunteers. Chest 124(1):360–366
65. Tsapis N, Bennett D, Jackson B, Weitz DA, Edwards DA (2002) Trojan particles: large porous carriers of nanoparticles for drug delivery. Proc Natl Acad Sci USA 99(19):12001–12005
66. Vehring R (2008) Pharmaceutical particle engineering via spray drying. Pharm Res 25(5):999–1022
67. Vanbever R, Pliquett UF, Preat V, Weaver JC (1999) Comparison of the effects of short, high-voltage and long, medium-voltage pulses on skin electrical and transport properties. J Control Release 60(1):35–47
68. Rogers TL, Johnston KP, Williams RO 3rd (2001) Solution-based particle formation of pharmaceutical powders by supercritical or compressed fluid CO2 and cryogenic spray-freezing technologies. Drug Dev Ind Pharm 27(10):1003–1015
69. Yu Z, Rogers TL, Hu J, Johnston KP, Williams RO 3rd (2002) Preparation and characterization of microparticles containing peptide produced by a novel process: spray freezing into liquid. Eur J Pharm Biopharm 54(2):221–228
70. Williams RO, Hu J, Rogers TL, Barron MK, Young TJ, Yu Z, Johnston KP (2003) Process for Production of Nanoparticles and Microparticles by Spray Freezing into Liquid. U.S. Patent 20030041602
71. Hu J, Johnston KP, Williams RO 3rd (2004) Nanoparticle engineering processes for enhancing the dissolution rates of poorly water soluble drugs. Drug Dev Ind Pharm 30(3):233–245
72. Soppimath KS, Aminabhavi TM, Kulkarni AR, Rudzinski WE (2001) Biodegradable polymeric nanoparticles as drug delivery devices. J Control Release 70(1–2):1–20

73. Watanabe W, Ahonen P, Kauppinen E, Jarvinen R, Brown D, Jokiniemi J, Muttonen W (2002) Novel method for the synthesis of inhalable multicomponent drug powders with controlled morphology and size. In: Dalby RN, Byron PR, Peart J, Farr SJ (eds) Proceedings of respiratory drug delivery VIII, vol 2. Davis Horwood International, North Carolina, pp 795–797
74. List M, Sucker H (1992) Pharmaceutical colloid hydrosols for injection. GB patent 2200048
75. Steckel H, Rasenack N, Villax P, Muller BW (2003) In vitro characterization of jet-milled and in-situ-micronized fluticasone-17-propionate. Int J Pharm 258(1–2):65–75
76. Steckel H, Rasenack N, Muller BW (2003) In-situ-micronization of disodium cromoglycate for pulmonary delivery. Eur J Pharm Biopharm 55(2):173–180
77. Rasenack N, Steckel H, Muller BW (2003) Micronization of anti-inflammatory drugs for pulmonary delivery by a controlled crystallization process. J Pharm Sci 92(1):35–44
78. Hu T, Zhao H, Jiang L, Le Y, Chen JF, Yun J (2008) Engineering pharmaceutical fine particles of budesonide for dry powder inhalation (DPI). Ind Eng Chem Res 47:9623–9627
79. Baldyga J, Henczka M, Shekunov BY (2004) Fluid dynamics, mass-transfer and particle formation in supercritical fluids. In: York P, Kompella UB, Shekunov BY (eds) Supercritical fluid technology for drug product development. Marcel Dekker, New York, pp 91–157
80. Rasenack N, Muller BW (2004) Micron-size drug particles: common and novel micronization techniques. Pharm Dev Technol 9(1):1–13
81. Hu T, Chiou H, Chan H, Chen J, Yun J (2008) Preparation of inhalable salbutamol sulphate using reactive high gravity controlled precipitation. J Pharm Sci 97(2):944–949
82. Chen X, Young TJ, Sarkari M, Williams RO III, Johnston KP (2002) Preparation of cyclosporine A nanoparticles by evaporative precipitation into aqueous solution. Int J Pharm 242(1–2):3–14
83. Rehman M, Shekunov BY, York P, Lechuga-Ballesteros D, Miller DP, Tan T, Colthorpe P (2004) Optimisation of powders for pulmonary delivery using supercritical fluid technology. Eur J Pharm Sci 22(1):1–17
84. Vemavarapu C, Mollan MJ, Lodaya M, Needham TE (2005) Design and process aspects of laboratory scale SCF particle formation systems. Int J Pharm 292(1–2):1–16
85. Lindsay AD, Omilinsky BA (1992) Method of preparing mixtures of active ingredients and excipients using liquid carbon dioxide. United States Patent 5169433
86. Matson DW, Fulton JL, Petersen RC, Smith RD (1987) Rapid expansion of supercritical fluid solutions: Solute formation of powders, thin films and fibres. Ind Eng Chem Res 26:2298–2306
87. Weidner E, Knez Z, Novak Z (1995) Process for preparing particles or powders. International Patent Publication WO 95/21688
88. Gallagher PM, Coffey MP, Krukonis VJ, Klasutis N (1989) Gas anti-solvent recrystallization: new process to recrystallize compounds insoluble in supercritical fluids. In: Johnston KP, Penninger J (eds) Supercritical fluid science and technology. ACS Symposium Series 406. American Chemical Society, Washington DC. pp 334–354
89. Bleich J, Muller BW, Wabmus W (1993) Aerosol solventextraction system: a newmicroparticle production technique. Int J Pharm 97:111–117
90. Bodmeier R, Wang H, Dixon DJ, Mawson S, Johnston KP (1995) Polymeric microspheres prepared by spraying into compressed carbon dioxide. Pharm Res 12(8):1211–1217
91. Bertucco A, Pallado P, Benedetti L (1996) Formation of biocompatible polymer microspheres for controlled drug delivery by a supercritical antisolvent technique. Process Technol Proc High Press Chem Eng 12:217–222
92. York P, Hanna M (1995) Salmeterol Xinafoate with controlled particle size. International Patent Publication WO 95/01324
93. Gupta RB, Chattopadhyay P (2002) Method of forming nanoparticles and microparticles of controllable size using supercritical fluids and ultrasound. US Patent Application US20020000681 A1
94. Perrut M, Clavier J (2003) Supercritical fluid formulation: process choice and scale-up. Ind Eng Chem Res 42:6375–6383
95. Young TJ, Johnston KP, Mishima K, Tanaka H (1999) Encapsulation of lysozyme in a biodegradable polymer by precipitation with a vapor-over-liquid antisolvent. J Pharm Sci 88(6): 640–650

96. Young TJ, Mawson S, Johnston KP, Henriksen IB, Pace GW, Mishra AK (2000) Rapid expansion from supercritical to aqueous solution to produce submicron suspensions of water-insoluble drugs. Biotechnol Prog 16(3):402–407
97. Pace SN, Pace GW, Parikh IM (1999) A.K., Novel injectable formulations of insoluble drugs. Pharm Technol 3:116–134
98. Pace GW, Vachon MG, Mishra AK, Henrikson IB, Krukonis V (2001) Processes to GenerateSubmicron Particles of Water-Insoluble Compounds. U.S. Patent 6,177,103
99. Palakodaty S, York P, Pritchard J (1998) Supercritical fluid processing of materials from aqueous solutions: the application of SEDS to lactose as a model substance. Pharm Res 15(12):1835–1843
100. Steckel H, Thies J, Müller BW (1997) Micronizing of steroids for pulmonary delivery by supercritical carbon dioxide. Int J Pharm 152:99–110
101. Steckel H, Muller BW (1998) Metered-dose inhaler formulations with beclomethasone-17,21-dipropionate using the ozone friendly propellant R 134a. Eur J Pharm Biopharm 46(1):77–83
102. Shekunov BY, Chattopadhyay P, Seitzinger J, Huff R (2006) Nanoparticles of poorly water-soluble drugs prepared by supercritical fluid extraction of emulsions. Pharm Res 23(1):196–204
103. Chattopadhyay P, Huff R, Shekunov BY (2006) Drug encapsulation using supercritical fluid extraction of emulsions. J Pharm Sci 95(3):667–679
104. Koushik K, Kompella UB (2004) Particle and device engineering for inhalation drug delivery. Drug Del Technol 4:40–50
105. Kwona MJ, Baea JH, Kima JJ, Nab K, Lee ES (2007) Long acting porous microparticle for pulmonary protein delivery. Int J Pharm 333(1–2):5–9
106. Mohamed F, van der Walle CF (2006) PLGA microcapsules with novel dimpled surfaces for pulmonary delivery of DNA. Int J Pharm 311(1–2):97–107
107. Arnold MM, Gonnan EM, Schieber LJ, Munson EJ, Berkland C (2007) NanoCipro encapsulation in monodisperse large porous PLGA microparticles. J Control Release 121(1–2):100–109
108. Berkland C, Kim KK, Pack DW (2001) Fabrication of PLG microspheres with precisely controlled and monodisperse size distributions. J Control Release 73(1):59–74
109. Berkland C, King M, Cox A, Kim K, Pack DW (2002) Precise control of PLG microsphere size provides enhanced control of drug release rate. J Control Release 82(1):137–147
110. Lindfors L, Skantze P, Skantze U, Rasmusson M, Zackrisson A, Olsson U (2006) Amorphous drug nanosuspensions. 1. Inhibition of Ostwald ripening. Langmuir 22(3):906–910
111. Azarmi S, Tao X, Chen H, Wang ZL, Finlay WH, Lobenberg R, Roa WH (2006) Formulation and cytotoxicity of doxorubicin nanoparticles carried by dry powder aerosol particles. Int J Pharm 319(1–2):155–161
112. Sham JOH, Zhang Y, Finlay WH, Roa WH, Lobenberg R (2004) Formulation and characterization of spray-dried powders containing nanoparticles for aerosol delivery to the lung. Int J Pharm 269(2):457–467
113. Grenha A, Seijo B, Remunan-Lopez C (2005) Microencapsulated chitosan nanoparticles for lung protein delivery. Eur J Pharm Sci 25(4–5):427–437
114. Junghanns JU, Muller RH (2008) Nanocrystal technology, drug delivery and clinical applications. Int J Nanomedicine 3(3):295–309
115. Ostrander KD, Hovey DC, Knapp DA, Parry-Billings M (2000) Potential delivery advantages of spray-dried Nanocrystals colloidal budesonide with the Clickhaler®. In: Dalby RN, Byron PR, Peart J, Farr SJ (eds) Proceedings of respiratory drug delivery VII. Serentec Press, North Carolina, pp 447–449
116. Shi LJ, Plumley CJ, Berkland C (2007) Biodegradable nanoparticle flocculates for dry powder aerosol formulation. Langmuir 23(22):10897–10901
117. El-Gendy N, Berkland C (2009) Combination chemotherapeutic dry powder aerosols via controlled nanoparticle agglomeration. Pharm Res 26(7):1752–1763

Chapter 14
Liposomes for Pulmonary Drug Delivery

Janani Swaminathan and Carsten Ehrhardt

Abstract Liposomes have been widely used in pulmonary drug delivery for multiple applications including solubilization, sustained release, cellular and intracellular targeting, minimization of toxicity, and facilitation of absorption. In this chapter, formulation aspects, aerosolization, and an extensive overview of the use of pulmonary drug delivery of liposomes for disease and drug classes are provided. Specifically, this chapter examines liposomes from in vitro work to clinical studies.

Keywords Controlled release • Drug targeting • Gene delivery • Nebulization

14.1 Liposomes

Liposomes are concentric bilayered vesicles in which an aqueous volume is entirely enclosed by a membranous lipid bilayer composed mainly of natural or synthetic phospholipids. The structure of a liposome vesicle is shown in Fig. 14.1. Liposomes are the commonly used lipid-based drug delivery systems, other lipid carriers being solid lipid nanoparticles (SLN), oily suspensions, submicron lipid emulsions, lipid implants, lipid microtubules, and lipid microspheres [1].

14.1.1 Classification of Liposomes

Liposomes can be classified based on various parameters as shown in Table 14.1

(I) According to their size and lamellarity [2] (Liposome classification based on size and lamellarity is explained in the Table 14.2 and is represented in Fig. 14.2)

C. Ehrhardt (✉)
School of Pharmacy and Pharmaceutical Sciences, Trinity College Dublin, Panoz Institute, Dublin, Ireland
e-mail: ehrhardc@tcd.ie

H.D.C. Smyth and A.J. Hickey (eds.), *Controlled Pulmonary Drug Delivery*, Advances in Delivery Science and Technology, DOI 10.1007/978-1-4419-9745-6_14,

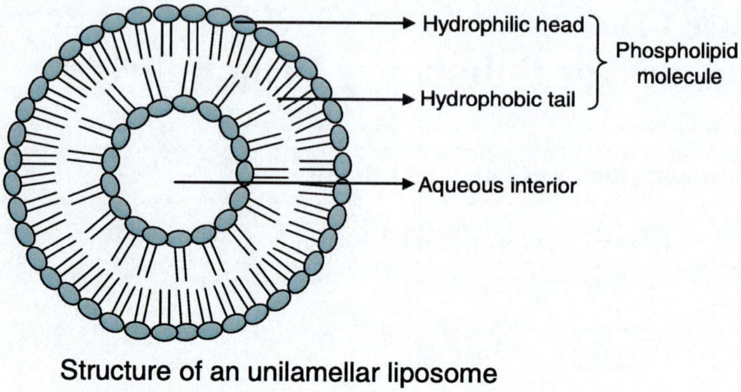

Fig. 14.1 Liposome structure

Table 14.1 Classification of liposomes

(I)	According to size and lamellarity
	1. Unilamellar vesicles
	2. Oligolamellar vesicles
	3. Multilamellar vesicles
	4. Multivesicular vesicles
(II)	According to method of preparation
	1. Liposomes prepared by solvent-free procedures
	(a) Dry-film or thin-film hydration
	(b) Lipid partitioning of hydrophobic peptides
	(c) Lipid surfactants
	(d) Polymorphic lipid
	(e) Freeze–thaw liposomes
	(f) Freeze-dehydration liposomes
	(g) Osmotically dried vesicles
	2. Liposomes prepared by solvent free procedures
	(a) Reversed-phase evaporation technique
	(b) Stable plurilamellar vesicle procedure
	(c) Ethanol injection technique
(III)	According to surface properties
	1. Conventional liposomes
	2. Sterically stabilized or stealth liposomes
(IV)	According to application
	1. Diagnostic liposomes
	2. Therapeutic liposomes

(II) According to the method of their preparation [2–5] (e.g., preparation by solvent-free or solvent-containing procedures)
(III) According to their surface properties [6–9] (e.g., conventional liposomes, sterically stabilized liposomes, or immunoliposomes)
(IV) According to their application [10, 11] (e.g., diagnostic liposomes or therapeutic liposomes)

Table 14.2 Classification of liposomes according to size and lamellarity

Vesicle type	Abbreviation	Diameter size	Number of lipid bilayers
1. Unilamellar vesicles	UV	20 nm–1 μm	One
(a) Small unilamellar vesicle	SUV	20–100 nm	One
(b) Large unilamellar vesicle	LUV	More than 100 nm	One
(c) Giant unilamellar vesicle	GUV	More than 1 μm	One
2. Oligolamellar Vesicle	OLV	0.1–1.0 μm	Approximately 5
3. Multilamellar Vesicle	MLV	More than 0.5 μm	5–25
4. Multivesicular Vesicle	MV	More than 1 μm	Multicompartmental structure

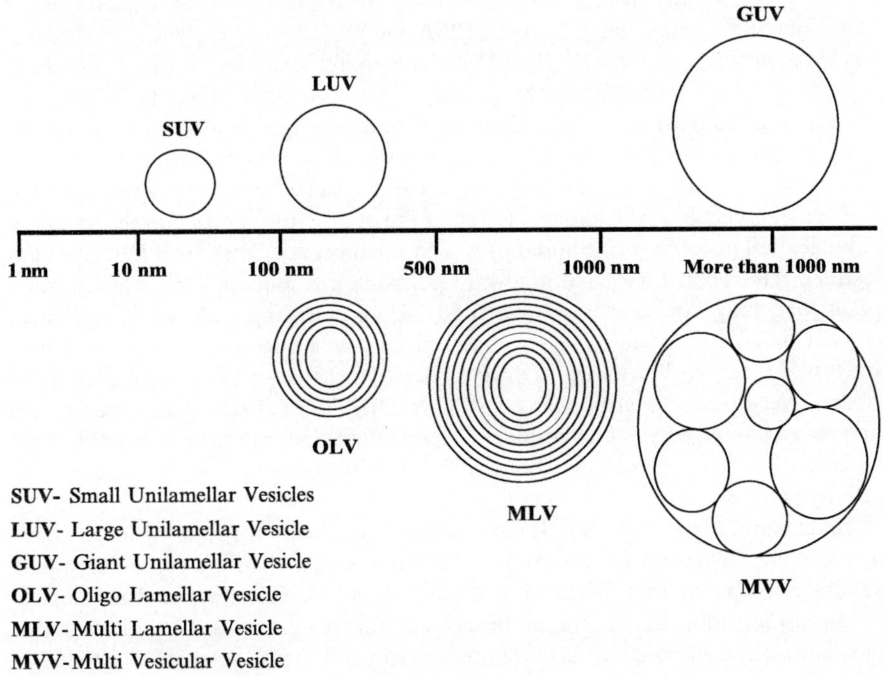

Fig. 14.2 Classification of liposomes according to size and lamellarity

14.2 Liposomes for Pulmonary Drug Delivery

In spite of availability of other advanced drug delivery systems such as microparticles and nanoparticles, liposomes are considered to be promising candidates for pulmonary drug delivery [12] due to the following reasons:

1. Ability to solubilize poorly soluble drugs
2. Capacity to provide a reservoir for sustained release, prolonging local and systemic therapeutic levels

3. Facilitation of intracellular delivery of drugs especially to alveolar macrophages
4. Avoidance of local irritation of lung tissue
5. Ability to target specific cell populations using surface-bound ligands or antibodies
6. Potential to be absorbed across the epithelium intact to reach the systemic circulation

Another reason for their choice is the chemical similarity of liposome formulations with lung surfactant. Human pulmonary surfactant is a mixture of lipids and proteins, of which the lipids account for nearly 20% of total tissue dry weight. The lipid pool comprises about 80% phospholipids, principally dipalmitoylphosphatidylcholine (DPPC), and 8% neutral lipids (cholesterol and free fatty acids). Other phospholipid components of surfactant are phosphatidyl glycerol (PG) and phosphatidyl inositol (PI). The four surfactant proteins (SPs), SP-A, SP-B, SP-C, and SP-D, account for 12% of the surfactant content. Of these, SP-A and SP-D are large, glycosylated water-soluble molecules, whereas SP-B and SP-C are highly hydrophobic small peptides.

Some early studies performed by various scientists suggest the safety and efficacy of liposomes as pulmonary drug delivery systems. Geiger et al. [13] reported the rapid uptake of radiolabelled DPPC liposomes in phosphate-buffered saline by alveolar type I and II cells and slower uptake of the same by macrophages. Oyarzun et al. [14]; Oyarzun and Clements [15] and Pre et al. [16] reported on the safety of radiolabelled liposomes composed of D- and L-isomers of DPPC and DPPC:DPPG instilled into rabbit lungs. No untoward pulmonary responses (i.e., arterial blood gases, pH, lung appearance, lavage fluid lactate dehydrogenase activity, protein content, and cell consistency) were observed following liposome instillation, which suggested safety of liposomes as drug carriers. It was clear from the results of the studies that liposomes comprising DPPC or DPPC:DPPG when delivered to the lung associate rapidly with the alveolar surfactant and enter the surfactant pool without disturbing normal metabolic activity or stimulating alveolar macrophage activity [17, 18].

Juliano and Mc Cullough [19] demonstrated that the half-life of cytosine arabinoside encapsulated in liposomes was 12 times longer than the same compound administered in solution. Mizushima et al. [20] successfully prepared aerosols of lipid-prostaglandin E1, a potent bronchodilator with irritating properties that precludes it use in man. He also demonstrated the effectiveness of the aerosol in protecting guinea pigs from histamine-induced dyspnea [17].

14.3 Factors to Be Considered in Formulating Liposomes for Pulmonary Delivery

14.3.1 Particle Size and Surface Charge

One of the most important formulation parameter for pulmonary delivery is the particle size. Owing to the well-defined branching architecture of the tracheobronchial tree leading to progressive narrowing of its airways at succeeding bifurcations, size of liposomal vesicle becomes a critical property in formulation development. Larger particles (4–7 μm) tend to deposit in the airways, while particles between 1 and 3 μm and those in submicron range (<1 μm) reach the lower airways and deeper lung. When delivering aerosolized liposomes, the liposomal particles become a part of the aerosol droplet, and pulmonary deposition is considered a function of aerosol droplet size rather than of liposomal vesicle size, which was elegantly demonstrated by Farr et al. [17, 21, 22].

Surface charge refers to the net charge of the liposomal surface (ζ-potential). The ζ-potential of liposomes can be amended by incorporating charged lipids (positive and negative) in the formulation. Liposomes bearing net positive or negative charge are known to accumulate to a greater extent in the lung in comparison to uncharged/neutral vesicles [17]. However, even liposomes prepared from neutral lipids possess nonzero ζ-potentials over a wide range of ionic strengths. This is due to the fact that zwitterionic head group of neutral lipids absorbs anions or cations, leading to slightly negative or positive ζ-potentials [17, 23].

14.3.2 *Rigidity and Stability of Liposomes in Bronchoalveolar Lavage Fluid (BAL)*

The rigidity of liposomal membranes constitutes yet another important formulation property, especially, during nebulization of liposomes or assessment of liposomal stability in BAL. Rigidity can be brought about by the use of rigid phospholipids or incorporation of cholesterol. The factor is also known to have an influence on the leakage of encapsulated drugs particularly when the temperatures during nebulization are higher than the phase transition temperature (Tc) of the lipid mixture. Stability in BAL is also enhanced with increasing transition temperature of the lipid mixture.

It was observed by Niven and coworkers that nebulization of liposomes above Tc of the phospholipids resulted in increased release of encapsulated carboxyfluorescein. Bridges et al. [24] studied the influence of rigidity using air-jet and ultrasonic nebulizers. They concluded a greater disruption of fluid egg-PC liposomes and increased resistance of DPPC:Chol liposomes with both nebulizer types [24]. Zaru et al. [25] reported better encapsulation and retention of ripampicin after nebulization with rigid liposomes in comparison to fluid ones. Moreover, the retention was found better with increasing rigidity or higher Tc. In other words DSPC:Chol liposomes were more resistant to nebulization compared to DPPC:Chol and PC:Chol liposomes [25]. Shek et al. [22] demonstrated increased stability of liposomes in BAL with increasing Tc in the order DMPC<DPPC<DSPC encapsulating glutathione. Incorporation of cholesterol had a positive effect toward membrane stability in DPPC and DMPC liposomes, whereas it caused membrane fluidization and increased glutathione release from DSPC liposomes [22].

14.3.3 Liposomal Dry-Powder Inhalers

Conventional DPIs are formulated either as loose agglomerates of micronized drug particles with particle size less than 5 μm or as carrier–drug mixtures of micronized drug particles adhered onto surface of large lactose carriers (see Chap. 9 of this book for more). However, current trends in DPI formulation include the preparation of liposomal suspensions, which are then converted into dry liposomal powders after the addition of cryoprotectants, carriers, additives, and antiadherents (alone or in combination) by lyophilization, spray drying, spray freeze-drying, or supercritical fluid technology. To the liposomal dry powder obtained coarse carriers, lubricants and antiadherents are added and sieved to obtain the final liposomal dry-powder formulation.

Recent studies have suggested the success of DPI technology for pulmonary delivery of liposomal drugs. Bi et al. [26] demonstrated the hypoglycemic ability of spray-freeze-dried insulin-loaded liposomes in rats delivered by DPI. Insulin liposomes were prepared by reversed-phase evaporation technique, and the liposomal suspension thus obtained was spray-freeze-dried. On administration to rats by intratracheal instillation, insulin-loaded liposomes reduced the blood glucose level in rats. This effect lasted markedly longer in comparison to that obtained by treatment with insulin solution [26].

Shahiwala et al. [27] reported the effectiveness of therapy with levonorgesterol in comparison to oral administration of the same. Liposomes of levonorgesterol were prepared by reversed-phase evaporation technique, and the liposomal suspension was subjected to freeze-drying. On intratracheal instillation, the bioavailability of the formulations decreased in the following order: liposomal formulation of the drug > physical mixture of drug and liposomal constituents > plain drug formulation > oral administration of the drug. Moreover, the orally administered formulation showed a high initial C_{max} followed by subtherapeutic concentrations, whereas formulations after pulmonary administration showed a lesser initial C_{max} followed by maintenance of effective plasma concentration up to 60 h [27].

The biggest disadvantage of liposomal dry-powder formulation is leakage of encapsulated drug during lyophilization and jet milling. Desai et al. [28] suggested a novel approach to overcome these deleterious effects; using lactose as cryoprotectant during lyophilization, they proved 90% drug retention in liposomes after reconstitution, which was reduced to 40% on jet milling. They then tried preparation of a phospholipid-based dry powder formulation that does not require a lyophilization step. This was made possible by blending phospholipids, drug, and cryoprotectant in powder form and micronizing them. The micronized powder was stored at room temperature (in contrast to liposomal suspensions that need to be refrigerated), which on reconstitution gave MLVs [28].

14.3.4 Nebulization of Liposomes

Nebulization is most common approach in pulmonary delivery of liposomes (see Chap. 10 of this book for more detail on nebulizers).

Several studies have been performed on the mentioned nebulizers to assess their efficiency and compatibility in liposomal drug delivery. Factors to be considered during nebulization include the aerosol droplet size, rate of aerosol output, temperature of the drug solution reservoir, viscosity of the drug solution, lipid concentration, rigidity of liposome composition, and PEGylation-related effects.

Steckel et al. [29] reported the factors affecting aerosol performance during nebulization with jet (Pariboy) and ultrasonic (Multisonic) nebulizers. They published the series of changes occurring to the drug solution in the reservoir over 10 min of nebulization time. The Pariboy device was characterized by an initial drastic decrease in temperature, leading to decreased saturated vapor pressure, increased viscosity and aerosol droplet size. Later, the particle size decreased due to increased solute concentration in the reservoir and decrease in surface tension. In the multisonic nebulizer, they observed a drastic increase in temperature initially, leading to increase in saturated vapor pressure, decrease in viscosity, increased solute concentration, reduced surface tension, and reduced aerosol droplet size. Later, with decrease in temperature, the particle size increased [29].

Bridges et al. [24, 30] demonstrated the factors influencing jet nebulization of liposomes using three different nebulizers. An increase in aerosol droplet size and viscosity was noted with an increase in lipid concentration from 5 to 80 mg/ml, which was more pronounced with Pari (Pari LC) and Cirrus, but not Sidestream. The aerosol output in general decreased with increasing lipid concentrations, which was, however, the least with the Sidestream. They reported a relationship between the mean liposome size and residual lipid concentration in the nebulization reservoir. An increase in mean liposome size beyond 2.5 μm led to increased retention of the same in the chamber due to reduced ability to be included in the aerosol. This group also reported an inverse relationship between the size of air-jet orifice and aerosol droplet size, which may be explained as an increase in size of air-jet orifice, leading to reduced pressure of driving gas and shear energy for droplet formulation contributing to increased droplet size [24, 30].

Elhissi et al. [31] compared liposome nebulization via jet (Pari LC plus) and vibrating mesh (Aeroneb Pro) nebulizers. Size of liposomes reduced following nebulization with both nebulizers; however, the reduction was more pronounced with the Pari, which is thought to damage the liposomes to a greater extent. Entrapped drug losses accounted for 88% with the Pari and 44% for the Aeroneb device. It was also concluded that extrusion of liposomes through 1-μm membrane filter enhanced stability to nebulization and reduced drug loss. The Aeroneb device had shorter nebulization time and produced higher mass output rates compared to the Pari, whereas the latter one produced smaller droplets, higher mass, and drug outputs [31].

Leung et al. [32] studied the stability of liposomes of different composition (Egg PC and Egg PC:Chol 1:1) to ultrasonic nebulization using a Medix ultrasonic nebulizer. They published a linear relationship between lipid concentration and viscosity. They also reported less marked size reduction of liposomal vesicles in the nebulizer reservoir on inclusion of 50 mol% cholesterol. Cholesterol imparts rigidity to the bilayer, making it more resistant to disruptive forces during nebulization [32].

Recently, Gaspar et al. [33] have assessed the potential of a vibrating mesh nebulizer (Aeroneb Pro) and an intratracheal nebulization catheter (AeroProbe) to aerosolize liposomal suspensions. A small size reduction of nonextruded liposomes following nebulization was reported, which was not observed with extruded formulations: These data confirm Elhissi's results (2007). Other physicochemical parameters such as ζ-potential and phospholipid content did not change significantly following nebulization. Hence, both nebulizers were deemed suitable for pulmonary delivery of liposomal formulations with the additional advantage of AeroProbe, that is, allows targeted delivery into select lung regions [33].

Rau et al. [34] compared the performance of five nebulizer models of three types-constant output (Misty-Neb and Sidestream), breath-enhanced (Pari LCD) and breath-actuated (Circulaire and AeroEclipse) in terms of inhaled drug, exhaled drug, drug lost to deposition in the apparatus and drug left in dose bottle. Drug inhaled or deposited in the breathing simulator was highest with the AeroEclipse (38.7 ± 1.3%) and lowest with the Pari LCD (15.2 ± 4.2%). Drug exhaled or lost to ambient to air was least with the AeroEclipse (6.6 ± 3.3%) and most with the Misty-Neb (26.8 ± 0.7%). Percentage of drug lost due to deposition in the apparatus was highest with the Circulaire (75.8 ± 0.5%) and lowest with the AeroEclipse (51 ± 2.1%). The duration of nebulization was shortest with the Circulaire and longest with the AeroEclipse [34].

Leung et al. [35] investigated the rate, consistency, and efficiency of drug delivery with breath-enhanced (Pari LC Star), breath-actuated (Halolite), and breath-enhanced breath-actuated (AeroEclipse) nebulizers. The nebulizers were tested for their in vitro performance and the respiratory pattern of seven cystic fibrosis patients. The rate of drug deposition was highest for the Pari LC star and lowest for the Halolite device. The rate of deposition was independent of the respiratory pattern for the Pari LC Star and the AeroEclipse, but was dependent on respiratory rate for the Halolite. The AeroEclipse had least amount of drug wastage. They finally concluded that the Pari LC Star would be the nebulizer of choice to achieve minimum treatment time, that the AeroEclipse would be preferred to minimize drug wastage, and that the Halolite would be the best to accurately deliver a specific drug dose [35].

Nerbrink et al. [36] demonstrated the effect of humidity in entrained air on constant output (Micro-Neb) and breath-enhanced (Pari LC+) nebulizers. Their results indicated an increase in aerosol size with increasing rH with the Micro-Neb, whereas the Pari LC+ did not show any change in aerosol droplet size. The results could be explained as follows: In the Micro-Neb, the entrained air mixes with small amount of cool high-humidity aerosol in the T-tube, during which the temperature of entrained air decreases and that of aerosol-laden air increases. Evaporation is thought to occur, the extent of which depends on temperature difference between entrained air and aerosol laden air. If the rH of the ambient air is high, indicating a low water holding capacity, the aerosol droplets absorb the humidity and grow in size and vice versa. By contrast, the Pari LC+ does not have a T-tube allowing the mixing of entrained air and aerosol. The entrained air, however, is brought in contact with aerosol at the aerosol generation point, where humidification occurs. No change in aerosol droplet size is noticed even with low rH of entrained air due to

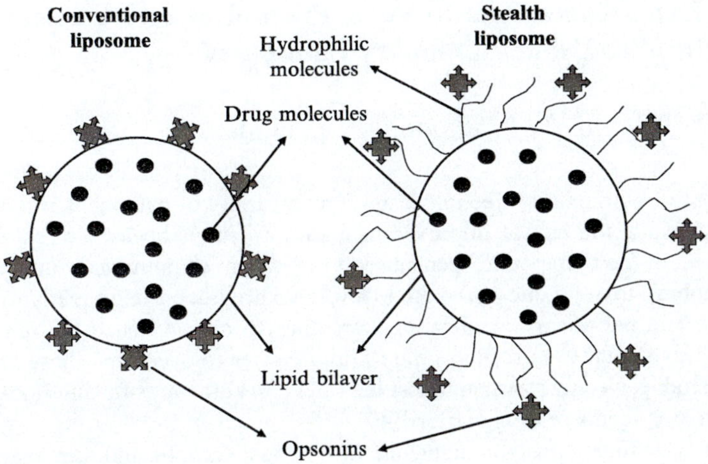

Fig. 14.3 STEALTH liposome

vast surface area and massive production of droplets which at any time ensures a high water-vapor potential [36].

Addition of PEG molecules provides stealth effect to liposomes prolonging their circulation time (Fig. 14.3) [37]. When considering pulmonary delivery, stability of PEGylated liposomes to nebulization becomes an important assessment factor. Anabousi et al. [70] demonstrated the effect of nebulization on liposomes and their stability in lung surfactant. PEGylated and non-PEGylated liposomes of different rigidity (DSPC and DPPC) were used. On nebulization through air-jet (Pari LC star) and ultrasonic (Aeroneb Rro) nebulizers, PEGylated liposomes showed marginal size and zeta potential changes, in comparison to non-PEGylated liposomes. When stability was assessed in commercial lung surfactant (Alveofact), PEGylated preparations (DSPC > DPPC) were more stable than non-PEGylated ones. The effect of temperature on stability in Alveofact was also determined, and the stability was better at 4°C than at 37°C, based on drug retention data [38].

In a related study Kleemann et al. [39] reported the stability of rigid PEGylated and non-PEGylated to nebulization using three different nebulizers: vibrating mesh (Aeroneb Pro), ultrasonic (Optineb), and air-jet (Pari LC). A significant decrease in hydrodynamic diameter was observed for those nebulized through the Pari LC and Optineb, whereas no change was observed on nebulization through the Aeroneb, suggesting less damage caused by the nebulizer. Drug retention post nebulization, was about 70% for non-PEGylated preparation and between 39 and 47% for the PEGylated liposomes [39].

14.4 Experiences from In Vitro, Preclinical, and Clinical Studies Using Pulmonary Liposomes

14.4.1 Treatment of Pulmonary Infections

Treatment of pulmonary infections caused by a variety of pathogens is a challenging task despite the fact of improved technology. This is because of solubility of drug (most of the therapeutic agents used in treatment of pulmonary infections are hydrophobic), toxicity due to the drug (insoluble drug deposits in the lung causing toxicity), and pulmonary localization (targeting specific areas of the lung) problems. It was thought to use liposome as drug carrier to overcome these problems. Several studies were performed to test the safety and efficacy of using liposomes as drug delivery agents or drug carriers [40].

Pulmonary tuberculosis remains one of the most complicated to manage forms of bacterial infection in the lung and the use of liposomes for pulmonary delivery of antitubercular drugs provides an alternative therapy for the disease. The classical treatment protocols are long-lasting; drugs reach the mycobacteria-infected macrophages in low amounts and do not persist long enough to develop the desired antimycobacterial effect [41, 42]. Chimote and Banerjee [43] demonstrated preparation of DPPC liposomes with antitubercular drugs, isoniazid (INH), rifampicin (RFM), and ethambutol (ETH). Entrapment efficiencies were $30.04 \pm 2.05\%$, $18.85 \pm 2.92\%$, and $61.47 \pm 3.32\%$ for INH, RFM, and ETH, respectively. In vitro deposition using a twin impinger showed $12.06 \pm 1.87\%$ of INH, $43.30 \pm 0.87\%$ of RFM, and $22.07 \pm 2.02\%$ of ETH deposited in the alveolar chamber upon nebulization for a minute using a jet nebulizer (Micelfluss F400). The formulation was stable even after a month when stored at 4°C [43].

Ciprofloxacin is a potent broad-spectrum antibiotic with good activity against most gram-negative bacteria and gram-positive cocci. Owing to failure of oral and iv therapy in treating infections of the lung (caused by *M. tuberculosis*, *M. avium*, *M. intracellulare*, *H. influenza*, *P. aeruginosa*, and *N. meningitidis*), liposome-encapsulated ciprofloxacin was tested. Conley et al. [44] reported efficacious treatment of *Francisella tularensis* infection in mice using ciprofloxacin encapsulated PC:Chol liposomes delivered by jet nebulization. Twelve commercially available nebulizers used in the study were A1800 (ARS Vital Aire), DVB7427 and DVB5601 (Devilbiss), Microcirrus (DHD Medical Products), Hosp3753 and Hosp952 (Hospitak), HudTU, HudUD2, and HudMM (Hudson RCI), Int1112220E (Intertech), Marq (Marquest Medical Products), and PurRD Raindrop (Puritan-Bennett). All mice treated with aerosolized liposome-encapsulated ciprofloxacin survived the infection caused with 10 times the 50% lethal dose of *F. tularensis*, while those treated with free nonencapsulated ciprofloxacin and control mice succumbed to the infection [44]. In a related study, Wong et al. [45] noticed significantly higher levels of drug deposition and prolonged drug retention in the lower respiratory tract after aerosol administration of liposome-encapsulated ciprofloxacin in comparison to intravenous injection of the same. They also concluded that ciprofloxacin given in free form was ineffective [45].

Enviroxime is an antiviral agent that inhibits the replication of all rhinoviruses and enteroviruses at concentrations as low as 10–100 ng/ml. Marked hydrophobicity and water insolubility of this compound make it a good candidate for liposome encapsulation. Gilbert et al. [46] succeeded in encapsulating enviroxime in liposomes (LE) and generated aerosols of the same using a Puritan-Bennett nebulizer. A study was conducted with five human volunteers who were exposed to LE for 1 h. One hour post treatment, a large quantity of enviroxime was still present in nasal wash, and no side effects were noted following LE administration [46]. Wyde et al. [47] tested the antirhinovirus activity and toxicity of LE in a cell culture model and its administration as small-particle aerosol in mice. Free enviroxime and LE had equivalent efficacies; however, LE was 10-fold safer or 50-fold less toxic than free enviroxime. Moreover, free enviroxime could not be delivered by small-particle aerosol due to its hydrophobic nature. In vivo studies with mice revealed significant amounts of the drug in lungs and nose, 20 min post treatment [47]. Knight et al. also reported the undiminished efficacy and reduced cellular toxicity on using LE, in comparison to free enviroxime [48].

Amphotericin B is an antifungal agent used in the treatment of systemic fungal infection caused by *Cryptococcus* and *Candida* species. Gavalda et al. [49] studied the efficacy of nebulized liposomal Amphotericin B (AmB) in an immune-suppressed murine model of invasive pulmonary aspergillosis. Mice treated with liposomal AmB showed improved survival rate when compared to the control group [49]. Previously, Gilbert et al. [50] had demonstrated the efficacy of small-particle aerosols containing liposomal AmB against *Candida* in mice. After 2-h, twice weekly administration for 1, 2, or 3 weeks, the mean time and percent of survival increased [50]. Ruijgork et al. [51] compared the efficacy of pulmonary aspergillosis treatment using aerosolized nonliposomal AmB (Fungizone) and liposomal AmB (AmBisome). They concluded similar efficacy with both preparations in all aspects [51].

Gilbert et al. [52] reported the suitability of treatment of Cryptococcus infections in the lung and brain using aerosols of AmB liposomes (AmB-L). Following intranasal inoculation of *Cryptococcus*, pulmonary infection is contracted immediately which spreads to the brain within 2–3 weeks. A single 2-h AmB-L nebulization was able to afford good reduction in the number of Cryptococcus organisms, when administered 24 h, 7 or 14 days post inoculation. The same therapy was effective even 21 days post inoculation, which showed an increase in the survival of infected animals [52]. Clark et al. [53] formulated AmB lipid complexes (AmB-LC) comprising AmB and phospholipids such as DMPC and DMPG and evaluated the toxicity and efficacy in mice. The 50% lethal dose for AmB-LC was 40 mg/kg, while it was 3 mg/kg for Fungizone. AmB-LC were efficacious against systemic *Candida* infections also in immunocompromised animals infected with *Candida*.

Gondal et al. [54] compared the efficacy of systemic *Candida* infection in mice with free AmB and AmB-L. Untreated mice had a median survival of 7 days with all mice dead by 12 days. A single dose of free AmB and AmB-L resulted in about 38% and 88–100% survival rates by day 42, respectively. With multiple dose regimens of free AmB and AmB-L, survival rate was about 38% and greater than 90% by day 42, confirming better treatment of the infection with AmB-L [54].

Ruijgrok et al. [55] compared the efficacy of treatment of pulmonary aspergillosis with aerosolized AmB desoxycholate and aerosolized liposomal AmB. Liposomal AmB significantly prolonged the survival rate of mice. AmB desoxycholate inhibited surfactant function in dose-dependent fashion, whereas liposomal AmB had no detrimental effect on surface activity of surfactant [55].

14.4.2 Cancer

The lung is the common site of primary and metastatic malignant diseases. Mainly due to the late diagnosis, lung cancer has poor survival rates and treatment options are generally reduced to surgical procedures. Because of the limited success of systemic chemotherapy, inhalation treatment of malignancies of the lung has been investigated in recent years [56]. One of the options is administration of cytokines as aerosol. However, administration of free cytokines comes with major limitations such as short half-lives and solubility problems. As a result, studies were performed to estimate the feasibility and efficacy of liposome-encapsulated interleukin administration in animal models [57, 58].

Anderson et al. [57] formulated liposomes encapsulating interleukin-2 (IL-2) and determined its efficacy in mice with pulmonary micrometastases. When treated once daily with liposome-encapsulated IL-2 intrathoracically, the results were significantly better in terms of longer survival and reduced number of pulmonary metastases, when compared with free IL-2 via the intrathoracic route [57]. Khanna et al. [58] performed similar studies with dogs to compare the immunological activation of inhaled free and liposomal IL-2. No toxicity was observed with either formulation; however, the BAL leukocyte cell count increased significantly following liposomal IL-2 inhalation [58]. The same group then studied the systemic toxicity of IL-2 liposomes in dogs to conclude on the report of objective regression of naturally occurring metastases after 1 month of nebulized IL-2 therapy [59]. They concluded that pet dogs with pulmonary metastases and primary lung carcinomas accepted the inhalation treatment with nebulized IL-2 liposomes. Moreover, the treatment was safe and nontoxic [59]. Ten et al. [60] conducted studies in humans with primary immune deficiency and hepatitis C to assess the patient acceptance, safety, toxicity, and immune effects of nebulized IL-2 liposomes. Aerosolized IL-2 liposomes were well tolerated, and no changes were observed in pulmonary function and chest X-ray [60]. These studies suggest the efficiency of aerosolized liposome-encapsulated IL-2 in human patients.

Studies on liposome encapsulation of water-insoluble anticancer drugs such as cisplatin and paclitaxel were also carried out. Wittgen et al. [61] investigated the safety and pharmacokinetics of aerosolized SLIT (Sustained-release Lipid Inhalation Targeting) cisplatin in 18 patients with lung carcinoma. Cisplatin SLIT was well tolerated, and safety data demonstrated no hematological toxicity, nephrotoxicity, ototoxicity, or neurotoxicity, which are typical side effects limiting the use of this drug in free form. Best overall response was stable disease in 13 patients, while 4 had progressive disease and 1 patient received only one course, suggesting the feasibility and safety of cisplatin SLIT therapy [61]. Koshkina et al. [62] evaluated the pulmonary pharmacokinetics and therapeutic efficacy of nebulized liposome-encapsulated paclitaxel in murine renal carcinoma pulmonary metastases model. The most effective treatment was obtained after inhalation of the drug for 30 min, three days per week, confirmed by significant reduction of lung weights and reduced number of tumor foci on lung surfaces [62].

Camptothecin is a plant alkaloid possessing anticancer properties in vitro. Two of its derivatives, 9-nitrocamptothecin (9-NC), which is water insoluble, and 9-nitro-20s-camptothecin, which is water soluble, have been studied to assess their anticancer properties in vivo. Knight et al. [63] tested the anticancer properties of 9-NC against human breast, colon, and lung cancer xenografts in nude mice when administered as liposome aerosols. DLPC-9-NC liposomes when administered, greatly reduced tumor growth, and tumors were undetectable after several weeks with colon tumors being least responsive [63]. Koshkina et al. [64] used liposome-encapsulated 9-NC (L-9NC) to assess the efficacy of treatment against melanoma and osteosarcoma lung metastases in mice. Mice were treated from the day after iv injection of melanoma cells with L-9NC for 1 h, 5 days/week for 3 weeks, which led to reduced lung weights and reduced tumor foci. The same was the case with L-9NC aerosol therapy initiated on the ninth week after iv tumor injection that was continued for 8–10 weeks [64].

Knight et al. [65] demonstrated the importance of 9-NC in treatment of human cancer subcutaneous xenografts and pulmonary cancer metastases in mice. With subcutaneous tumors, aerosol of L-9NC showed significant improvement, which was not evident with oral 9-NC, while intramuscular L-9NC in slightly higher doses than the aerosol showed detectable anticancer activity. L-9NC was also effective against cancer metastases [65]. Zhang et al. [66] designed a study to investigate the in vitro release and in vivo tissue distribution and damage to the lungs by 9-NC liposomes. In vitro studies were conducted in pH 7.4 PBS. Following pulmonary delivery, mean residence time of L-9NC was 3.4 times longer than 9-NC solution. L-9NC, in addition to causing less lung damage, showed sustained release characteristics of 9-NC [66]. Gilbert et al. [67] evaluated the potential toxicity of L-9NC in an 8 week subacute toxicity study in 14 dogs. Full-face exposures for 60 min were conducted 5 consecutive days a week for 8 weeks. L-9NC exposed animals showed no toxicity and so did the control group treated with empty liposomes [67]. Verschraegen et al. [68] evaluated the feasibility and safety of aerosol administration of L-9NC in human volunteers. Patients with primary or metastatic lung cancer received aerosolized L-9NC five consecutive days/week for 1, 2, 4 or 6 weeks followed by 2 weeks of rest. Aerosol administration proved safe and feasible with 9-NC detectable in plasma shortly after administration. A recommended dose for phase II studies of 13.3 μg/kg/day was reported [68].

Most human tumors overexpress receptors for growth factors and peptide hormones which are being increasingly studied and utilized as means for selective targeting of anticancer agents. Transferrin receptors (TfR) are one such example [69]. Anabousi et al. [70] assessed the uptake levels and cytotoxicity of transferrin-conjugated liposomes containing doxorubicin in vitro. Cells of (lung) cancerous origin showed a higher TfR expression than healthy alveolar epithelial cells, leading to enhanced uptake and increased levels of cytotoxicity. Liposome uptake by cancerous cells was temperature dependent and could be inhibited by excess free Tf. They concluded that Tf-conjugated liposomes were good candidates to deliver cytotoxic drugs such as doxorubicin to sites of lung cancer by inhalation [70].

14.4.3 Macrophage Targeting

Alveolar macrophages (AM) play an important role in primary defense mechanism. On activation, they produce a variety of inflammatory by-products such as tumor necrosis factor (TNF)-α, monocyte chemoattractant protein (MCP)-1, and macrophage inflammatory protein (MIP)-2. Ischemia-Reperfusion (I/R) is an unavoidable inflammatory response, following lung transplantation, which is said to be caused by cells in the donor lung rather than circulating leucocytes from the recipient. Increased TNF-α levels following an inflammation can be prevented by anti-TNF-α antibody pretreatment. It is hypothesized that TNF-α deficiency offers protection from I/R.

Studies were performed to prove the advantage of macrophage depletion following transplantation. Zhao et al. [71] demonstrated that depletion of alveolar macrophages will reduce cytokine/chemokine expression and lung injury after I/R in buffer-perfused, isolated mouse lung model. Following treatment with liposome clodronate, which was used to bring about AM depletion, cytokine/chemokine production reduced significantly, which was not the case in control treated with buffer [71]. However, Nakamura et al. [72] reported the aggravation of I/R injury, following clodronate liposome treatment in Lewi's rats. Two groups of rats were treated with clodronate-liposomes in Hank's Buffered Salt Solution (HBSS) and control (treated with HBSS only) intratracheally. After 3 days, the lungs were subjected to I/R when it was noted that the clodronate-liposome group showed severe lung injury within 10 min, leading to termination of experiment, while the control group could be studied for 60 min. They also concluded that clodronate significantly decreased dynamic airway compliance/increased airway resistance and increased concentrations of polymorphonuclear neutrophils and MIP-2 [72]. Leemans et al. [73] demonstrated the depletion of AM by liposome-encapsulated dichloromethylene diphosphonate (AM-), during pulmonary tuberculosis is protective. The AM- group displayed reduced outgrowth of bacteria and less granuloma formation that suggested they were protected against lethality, which was not seen in control (AM+) [73].

Liposome size and charge have an influence on macrophage uptake as reported by some researchers. Chono et al. [74] studied the influence of particle size of ciprofloxacin liposomes on AM uptake. They prepared liposomes of five different particle sizes (100, 200, 400, 1,000, and 2,000 nm) with hydrogenated soy bean phosphatidyl choline (HSPC), cholesterol, and dicetyl phosphate (DCP). Delivery efficiency of ciprofloxacin to rat AM increased with increase in particle size up to 1,000 nm, over which there was no significant change. Ciprofloxacin concentrations in rat AM until

24 h post administration were higher than the minimum inhibitory concentration with 1,000-nm liposomes. [74] Chono et al. [75] then reported the influence of liposome size and mannose modification on AM uptake. Five different liposome sizes, as mentioned in previous study, were considered. Uptake of nonmodified liposomes increased with increase in size over the range 100–2,000 nm following pulmonary administration. Uptake of mannose modified liposomes was always greater than nonmodified ones [75]. Lambros et al. [76] evaluated the impact of charge (neutral, negative, and positive) on disposition of AmB liposomes. Neutral and positive liposomes followed biexponential kinetics, while negative liposomes showed monoexponential clearance. The α and β half-lives for positive liposomes were 1.3 and 15.1 days, respectively, whereas for the neutral ones the α and β half-lives were 2.3 and 22 days. For the negative liposomes the half-life was 4.5 days [76].

Liposomes can be coated with AM specific ligands such as *O*-palmitoyl mannan (OPM) and *O*-palmitoyl pullulan (OPP) to improve the efficiency of targeting. Vyas et al. [77] compared the efficiency of nonmodified and mannose-modified liposomes in pulmonary delivery of AmB-loaded liposomes. They noticed preferential accumulation of mannose-modified liposomes in AM compared to nonmodified ones. Mannose-modified liposomes, in addition to rapid attainment of high drug concentration in lungs, could maintain the same the same over prolonged period of time [77]. Vyas et al. [78] studied the AM disposition of nonmodified and mannosylated liposomes of rifampicin. Percent viability of *M. smegmatis* following administration was 7–11% in ligand-anchored aerosols, 31.6% in rifampicin-loaded liposomal aerosol and 45.7% in plain-drug aerosol, suggesting the efficacy of liposome-anchored aerosol in treatment [78]. Chono et al. [79] investigated the antibacterial effects following the pulmonary delivery of liposomal ciprofloxacin to alveolar macrophages in rats with lipopolysaccharide-induced pneumonia. The results showed a higher concentration of ciprofloxacin in alveolar macrophages and lung epithelial lining fluid while lower concentrations of the same in plasma over time when administered as a liposomal formulation, suggesting better treatment efficacy [79].

14.4.4 Pulmonary Arterial Hypertension (PAH)

Iloprost is a prostacyclin analog that is used in the treatment of PAH that requires to be dosed several times a day (about 6–9). Kleeman et al. [39] investigated the ability of liposomes in providing a sustained release formulation of iloprost to reduce its inhalation frequency. Liposomes of varying surface charges were formulated and their stability to aerosolization was assessed using three different nebulizers (air-jet, ultrasonic, and vibrating mesh). Results suggested that liposomes composed of DPPC and Chol showed better stability (70% drug encapsulation post nebulization) compared to the liposomes formulated with DPPC, DPPE-PEG, and Chol (20–50% drug encapsulation post nebulization). They concluded that the vibrating mesh nebulizer offered significant advantages over the others in terms of production of small aerosol droplets, high output, and lowest deleterious physical influence on all investigated liposomal combinations [39].

14.4.5 Peptide and Protein Delivery

Pulmonary delivery of peptides and proteins is still a viable alternative to injection [26]. Formulation for pulmonary insulin delivery was the aim of many research groups due to the number of patients affected by diabetes and the frequency of insulin injections to be taken by them. Liu et al. [80] worked on the pulmonary delivery of dimeric and hexameric insulin-loaded liposomes. They reported that both insulins were absorbed into systemic circulation producing a significant hypoglycemic response. However, the initial hypoglycemic response from hexameric insulin was slower than from dimeric insulin [80]. Huang et al. [81] demonstrated an optimal encapsulation of insulin into liposomes could be achieved by the detergent dialysing method, and that insulin was stable in solution form. Animal (Balb/c mice) studies showed an effective reduction in plasma glucose level with reduction in extra pulmonary side-effects on administration of aerosolized insulin-encapsulated liposomes [81]. Chono et al. [82] developed aerosolized DPPC liposomes encapsulating insulin. They reported the enhanced pulmonary delivery of insulin to rat lungs with DPPC liposomes, which was not the case with DSPC, DMPC, DLPC and DOPC liposomes [82].

Bi et al. [26] designed a dry powder inhalation of insulin loaded liposomes by the spray freeze drying method. Sucrose was found to be the best cryoprotectant in an 1:6 ratio of lipid:sucrose. After intratracheal instillation in rats, the formulation showed a successful hypoglycemic effect with long-lasting low blood glucose level and a relative bioavailability of 38.38% in the group treated with 8 IU/kg of insulin [26]. Karathanasis et al. [83] developed an aerosol insulin carrier based on agglomerated vesicle technology. The carrier consisted of insulin-loaded liposomes cross-linked via chemical bridges cleavable by cysteine. Contact with cysteine led to disruption of liposomal walls and rapid release of encapsulated insulin. Studies in rats showed a rapid decrease in glucose levels while delivery of cysteine triggered a further drop of glucose implying acceleration of insulin release from the liposomal carrier [83].

Hajos et al. [84] achieved encapsulation of vasoactive intestinal peptide (VIP) into liposomes for treatment of lung diseases such as asthma and pulmonary hypertension. After nebulization, particle size and encapsulation efficiency were unaltered. When tested in an ex vivo model, sustained release of VIP and extended pharmacological effect were obtained with liposomal VIP in comparison to free VIP [84]. Stark et al. [85] developed a liposomal formulation of VIP and proved its potential in terms of protection of VIP from enzymatic cleavage. They assessed the stability of VIP in bronchoalveolar lavage fluid (BALF) by high-performance liquid chromatography and mass spectroscopy and reported better stability of liposome-encapsulated VIP in comparison to free VIP. Fluorescence spectroscopy studies revealed that the tight association of VIP with lipid membrane was minimally affected following incubation in BALF. Finally, the biological activity of VIP remained intact following incubation in BALF [85]. Rubinstein et al. [86] compared the effectiveness of VIP self-associated with sterically stabilized liposomes (VIP-SSL) following intratracheal and subcutaneous administrations in spontaneously hypertensive hamsters.

Maximal effect of mean arterial pressure (MAP) normalization was obtained in 10–20 min following administration that lasted for 6 h following administration by either route. This MAP reduction achieved with VIP-SSL group was better than the control group [86].

14.4.6 Nucleotide Delivery

Nucleotide delivery by inhalation holds promise for the treatment of pathological conditions such as lung cancer, cystic fibrosis (CF), and asthma. Aerosolized gene or RNAi therapy provides a direct noninvasive access to the lung regions with the ability to deliver the payload directly to the target site. Nebulization of naked DNA or siRNA molecules results in rapid destruction of such large and fragile molecules. Thus, complexation and encapsulation in cationic phospholipids is considered a method of choice in pulmonary nucleic-acid delivery [87, 88].

Schwarz et al. [89] compared the transfection ability of liposome-encapsulated DNA post nebulization in A549 lung cells using different lipid combinations. They concluded that samples made from DMRIE {*N*-(2-hydroxyethyl)-*N*,*N*-dimethyl-2,3-bis(tetradecytoxy)-1-propaniminium bromide} and DOPE {Dioleyl Phosphatidyl Ethanolamine} were more stable to nebulization with a Puritan-Bennett 1600 device, compared to those made with DOSPA {2,3-dioleyloxy-*N*-[2(sperminecarboxamido)ethyl]-*N*,*N*-dimethyl-1-propanaminium trifluroacetate} by assessing the transfection ability of the preparations in vitro [89].

Selective gene inhibition by antisense oligodeoxynucleotide (AS-ODN) or by small interference RNA (siRNA) therapeutics promises the treatment of diseases that cannot be cured by conventional drugs. However, antisense therapy is hindered due to poor stability in physiological fluids and limited intracellular uptake. Li et al. [90] achieved targeted delivery of AS-ODN and siRNA into lung cancer cells. They mixed AS-ODN or siRNA with carrier DNA that was complexed with protamine, a highly positively charged peptide, and the resulting particles were coated with cationic liposomes of DOTAP:Chol (1:1 molar ratio) to obtain liposome-polycation-DNA nanoparticles. Ligand targeting and steric stabilization was made possible by attachment of DSPE-PEG molecule. They concluded that ligand-targeted and sterically stabilized nanoparticles provide selective delivery of AS-ODN to lung cancer cells [90].

The expression of a number of genes is altered in cancer, and modifying the expression of these genes is an attractive method of cancer treatment. Thus, siRNA therapies can be considered as potential cancer treatments. Several studies have been performed in this field using both viral and nonviral vectors [91].

Kleemann et al. [92] compared the structure and stability of four different polyethyleneimine (PEI) DNA polyplexes following air-jet (Pari LC star) and ultrasonic (Optineb) nebulization. The four PEI-DNA polyplexes investigated were branched PEI (bPEI), linear PEI (linPEI), polyethylene glycol grafted PEI (PEGPEI), and biodegradable PEI (bioPEI). Characterization tests that went into structure and stability determination were morphology of DNA polyplexes by atomic force microscopy and physicochemical properties such as particle size and zeta potential before and after nebulization. Results revealed a greater alteration of polyplex structure following air-jet nebulization in comparison to ultrasonic nebulization. Moreover, PEGPEI polyplexes showed no change in shape and size compared to their counterparts, suggesting maximum stability of the same [92].

14.5 Concluding Remarks

Over the last 20 years, liposomes have been investigated for pulmonary drug delivery. The main drive for research in this field comes with the advantages that liposomes provide as carriers; the first and foremost reason being their chemical similarity to lung surfactant. Others include the ability to solubilize poorly soluble drugs, capacity to provide sustained release reservoir, facilitation of alveolar macrophage delivery, and avoidance of local irritation of lung tissue. The downsides, however, are mostly concerning stability of the formulations in terms of shelf life and during aerosolization.

Nebulization has been the most common approach for the pulmonary delivery of liposomes. Issues that prevented widespread use of nebulizers include long nebulization cycles and loss or degradation of drug with increase in temperature in the nebulization chamber. These problems were faced with air-jet and early-generation ultrasonic nebulizers and have been overcome with the advance of novel vibrating-mesh, breath-actuated, and breath-enhanced devices. Some researchers are also looking into liposomal dry-powder inhalers as an alternative to nebulizers. This has not been explored to the fullest yet.

Advances with respect to formulation technology include surface modifications to target specific cells such as cancer cells and macrophages, or to increase physical stability and prevent endocytosis.

The future will show if it is possible to design liposomes that can cross the epithelial air–blood barrier uncompromised and serve as systematically circulating depot formulation. However, the ultimate success in the field leading to transfer of formulations from bench to bedside would depend not only on the formulation technology but also on the nebulizer design and characteristics.

References

1. Rawat M, Singh D, Saraf S, Saraf S (2008) Lipid carriers: a versatile vehicle for proteins and peptides. Yakugaku Zasshi 128(2):269–280
2. Samad A, Sultana Y, Aqil M (2007) Liposomal drug delivery systems: an update review. Curr Drug Deliv 4:297–305
3. Weiner AL (1994) Liposomes for protein delivery: selecting manufacture and developmental processes. Immunomethods 4:201–209
4. Guo J, Ping Q, Zhang L (2000) Transdermal delivery of insulin in mice by using lecithin vesicles as carrier. Drug Deliv 7:113–116

5. Lenk RP, Janoff AS, Ostro MJ (1985) Novel multilayered lipid vesicles: Comparison of physical characteristics of multilamellar liposomes and stable plurilamellar vesicles. Biochemistry 24:2833–2842
6. Storm G, Crommelin DJA (1998) Liposomes: quo vadis? PSTT 1:19–31
7. Bangham AD (1972) Model membranes. Chem Phys Lipid 8:386–392
8. Bangham AD, Standish MM, Watkins JC, Weissmann G (1967) The diffusion of ions from a phospholipid model membrane system. Protoplasma 63:183–187
9. Immordino ML, Dosio F, Cattel L (2006) Stealth liposomes: review of the basic science, rationale and clinical applications, existing and potential. Int J Nanomed 3:297–315
10. Seltzer SE (1989) The role of liposomes in diagnostic imaging. Radiology 171:19–21
11. Rhodes BA (1976) Liposomes and vesicles: a new class of radiopharmaceuticals? J Nucl Med 17:1102–1103
12. Gaspar MM, Bakowsky U, Ehrhardt C (2008) Inhaled liposomes – current strategies and future challenges. J Biomed Nanotechnol 4:1–13
13. Geiger K, Gallagher ML, Hedley-White J (1975) Cellular distribution and clearence of aerosolized dipalmitoyl lecithin. J Appl Physiol 39:759–766
14. Oyarzun MJ, Clements JA, Baritusso A (1980) Ventilation enhances pulmonary alveolar clearance of radioactive dipalmitoyl phosphatidylcholine in liposomes. Am Rev Respir Dis 121:709–721
15. Oyarzun MJ, Clements J (1977) Ventilatory and cholinergic control of pulmonary surfactant in the rabbit. J Appl Physiol 43:39–45
16. Pre J, Perret G, Bladier D (1983) Lecithin content estimate of human alveolar lining layer, comparison with mouse, rat and rabbit. Comp Biochem Physiol 76:393–395
17. Mihalko PJ, Schreier H, Abra RM (1988) Liposomes: a pulmonary perspective. In: Gregoriadis G (ed) Liposomes as drug carriers. Wiley, NewYork, pp 679–694
18. Goerke J (1998) Pulmonary surfactant: functions and molecular composition. Biochim Biophys Acta 1408:79–89
19. Juliano RL, McCullough HN (1980) Controlled delivery of an anti-tumor drug, localized action of liposome encapsulated cytosine arabinoside administered via the respiratory system. J Pharmacol Exp Ther 214:381–387
20. Mizushima Y, Hoshi K, Aihara H, Kurachi M (1983) Inhibition of bronchoconstriction by aerosol of a lipid emulsion containing prostaglandin E1. J Pharm Pharmacol 35:397
21. Farr SJ, Kellaway IW, Parry-Jones D, Woolfrey SG (1985) The clearance of 99m Tc labelled liposomes from the lungs of volunteers. Proceed Intern Symp Control Rel Bioact Mater 12:219–220
22. Shek PN, Suntres ZE, Brooks JI (1994) Liposomes in pulmonary applications: physicochemical considerations, pulmonary distribution and antioxidant delivery. J Drug Target 2: 431–442
23. Ulrich AS (2002) Biophysical aspects of using liposomes as delivery vehicles. Biosci Rep 22:129–150
24. Bridges PA, Taylor KMG (1998) Nebulizers for the generation of liposomal aerosols. Int J Pharm 173:117–125
25. Zaru M, Mourtas S, Klepetsanis P, Fadda AM, Antimisiaris SG (2007) Liposomes for drug delivery to the lungs by nebulisation. Eur J Pharm Biopharm 67:655–666
26. Bi R, Shao W, Wang Q, Zhang N (2008) Spray-freeze-dried dry powder inhalation of insulin-loaded liposomes for enhanced pulmonary delivery. J Drug Target 16:639–648
27. Shahiwala A, Misra A (2004) Pulmonary absorption of liposomal levonorgesterol. AAPS PharmSciTech 5:1–5
28. Desai TR, Wong JP, Hancock REW, Finlay WH (2002) A novel approach to the pulmonary delivery of liposomes in dry powder form to eliminate the deleterious effects of milling. J Pharm Sci 91:482–491
29. Steckel H, Eskandar F (2003) Factors affecting aerosol performance during nebulisation with jet and ultrasonic nebulisers. Eur J Pharm Sci 19:443–455
30. Bridges PA, Taylor KMG (2000) An investigation of some of the factors influencing the jet nebulisation of liposomes. Int J Pharm 204:69–79

31. Elhissi AMA, Faizi M, Naji WF, Gill HS, And Taylor KMG (2007) Physical stability and aerosol properties of liposomes delivered using an air-jet nebuliser and a novel micropump device with large mesh apertures. Int J Pharm 334:62–70
32. Leung KKM, Bridges PA, Taylor KMG (1996) The stability of liposomes to ultrasonic nebulisation. Int J Pharm 145:95–102
33. Gaspar MM, Gobbo O, Ehrhardt C (2009) Generation of liposome aerosols with the Aeroneb Pro and the AeroProbe nebulisers. J Liposome Res. doi: DOI: 10.1080/08982100903085150
34. Rau JL, Ari A, Restrepo RD (2004) Performance comparison of nebuliser designs: constant-output, breath-enhanced and dosimetric. Respir Care 49:174–179
35. Leung K, Louca E, Coates AL (2004) Comparison of breath enhanced to breath actuated nebulisers for rate, consistency and efficiency. Chest 126:1619–1627
36. Nerbrink OL, Pagels J, Pieron CA, Dennis JH (2003) Effect of humidity on constant output and breath enhanced nebuliser designs when tested in the EN 13544-1 EC standard. Aerosol Sci Technol 37:282–292
37. Moghimi SM, Szebini J (2003) Stealth liposomes and long circulating nanoparticles: critical issues in pharmacokinetics, opsonisation and protein binding properties. Prog Lipid Res 42:463–478
38. Anabousi S, Kleemann E, Bakowsky U, Kissel T, Schmehl T, Gessler T, Seeger W, Lehr CM, Ehrhardt C (2006) Effect of PEGylation on the stability of liposomes during nebulisation and in lung surfactant. J Nanosci Nanotechnol 6:3010–3016
39. Kleemann E, Schmehl T, Gessler T, Bakowsky U, Kissel T, Seeger W (2007) Iloprost-containing liposomes for aerosol application in pulmonary arterial hypertension: Formulation aspects and stability. Pharm Res 24:277–287
40. Gilbert BE (1996) Liposomal aerosols in the management of pulmonary infections. J Aerosol Med 9:111–122
41. Pandey R, Khuller GK (2005) Antitubercular inhaled therapy: opportunities, progress and challenges. J Antimicrob Chemother 55:430–435
42. Gaspar MM, Cruz A, Fraga AG, Castro AG, Cruz ME, Pedrosa J (2008) Developments on drug delivery systems for the treatment of mycobacterial infections. Curr Top Med Chem 8:579–591
43. Chinmote G, Banerjee R (2008) Evaluation of antitubercular drug-loaded surfactants as inhalable drug-delivery systems for pulmonary tuberculosis. J Biomed Mater Res A 89:281–292
44. Conley J, Yang H, Wilson T, Blasetti K, Di Ninno V, Schnell G, Wong JP (1997) Aerosol delivery of liposome-encapsulated ciprofloxacin: aerosol characterisation and efficacy against Francisella tularensis infection in mice. Antimicrob Agents Chemother 41:1288–1292
45. Wong JP, Yang H, Blasetti KL, Schnell G, Conley J, Schofield LN (2003) Liposome delivery of ciprofloxacin against Francisella tularensis infection. J Control Release 92:265–273
46. Gilbert BE, Six HR, Wilson SZ, Wyde PR, Knight V (1988) Small particle aerosols of enviroxime-containing liposomes. Antivir Res 9:355–365
47. Wyde PR, Six HR, Wilson SZ, Gilbert BE, Knight V (1988) Activity against rhinoviruses, toxicity and delivery in aerosol of enviroxime in liposomes. Antimicrob Agents Chemother 32:890–895
48. Knight V, Gilbert B (1988) Antiviral therapy with small particle aerosols. Eur J Clin Microbiol Infect Dis 7:721–731
49. Gavalda J, Martin M, Lopez P, Gomis X, Ramirez J, Rodriguez D, Len O, Puigfel Y, Ruiz I, Pahissa A (2005) Efficacy of nebulised liposomal amphotericin B in treatment of experimental pulmonary aspergillosis. Antimicrob Agents Chemother 49:3028–3030
50. Gilbert BE, Wyde PR, Lopez-Berestein G, Wilson SZ (1994) Aerosolised amphotericin B-liposomes for treatment of systemic candida infections in mice. Antimicrob Agents Chemother 38:356–359
51. Ruijgrok EJ, Vulto AG, Van Etten EW (2000) Aerosol delivery of amphotericin B desoxycholate (Fungizone) and liposomal amphotericin B (AmBisome): aerosol characteristics and in-vivo amphotericin B deposition in rats. J Pharm Pharmacol 52:619–627
52. Gilbert BE, Wyde PR, Wilson SZ (1992) Aerosolised liposomal amphotericin B for treatment of pulmonary and systemic Cryptococcus neoformans infections in mice. Antimicrob Agents Chemother 36:1466–1471

53. Clark JM, Whitney RR, Olsen SJ, George RJ, Swerdel MR, Kunselman L, Bonner DP (1991) Amphotericin B lipid complex therapy of experimental fungal infections in mice. Antimicrob Agents Chemother 35:615–621
54. Gondal JA, Swartz RP, Rahman A (1989) Therapeutic evaluation of free and liposome encapsulated amphotericin B in the treatment of systemic candidiasis in mice. Antimicrob Agents Chemother 33:1544–1548
55. Ruijgrok EJ, Vulto AG, Van Etten EW (2001) Efficacy of aerosolised amphotericin B desoxycholate and liposomal amphotericin B in the treatment of invasive pulmonary aspergillosis in severely immunocompromised rats. J Antimicrob Chemother 48:89–95
56. Rao RD, Markovic SN, Anderson PM (2003) Aerosol therapy for malignancy involving the lungs. Curr Cancer Drug Targets 3:239–250
57. Anderson PM, Katsanis E, Leonard AS, Schow D, Loeffler CM, Goldstein MB, Ochoa AC (1990) Increased local antitumor effects of interleukin 2 liposomes in mice with MCA-106 sarcoma pulmonary metastases. Cancer Res 50:1853–1856
58. Khanna C, Hasz DE, Klausner JS, Anderson PM (1996) Aerosol delivery of interleukin 2 liposomes is non-toxic and biologically effective: canine studies. Clin Cancer Res 2: 721–734
59. Khanna C, Anderson PM, Hasz DE, Katsanis E, Neville M, Klausner JS (1997) Interleukin-2 liposome inhalation therapy is safe and effective for dogs with spontaneous pulmonary metastases. Cancer 79:1409–1421
60. Ten RM, Anderson PM, Zein NN, Temesgen Z, Clawson ML, And Weiss W (2002) Interleukin-2 liposomes for primary immune deficiency using the aerosol route. Int Immunopharmacol 2:333–344
61. Wittgen BP, Kunst PW, van der Born K, Peters GJ, Perez-Soler R, Metzheiser B, Nicholson S, Perkins W, Pilkiewicz F, Postmus PE (2006) Phase I study of aerosolised SLIT cisplatin in the treatment of patients with carcinoma of the lung. J Clin Oncol 24:7131
62. Koshkina NV, Waldrep JC, Roberts LE, Golunski E, Melton S, Knight V (2001) Paclitaxel liposome aerosol treatment induces inhibition of pulmonary metastases in murine renal carcinoma model. Clin Cancer Res 7:3258–3262
63. Knight V, Koshkina NV, Waldrep JC, Giovanella BC, Gilbert BE (1999) Anticancer effect of 9-nitrocamptothecin liposome aerosol on human cancer xenografts in nude mice. Cancer Chemother Pharmacol 44:177–186
64. Koshkina NV, Kleinerman ES, Waidrep C, Jia SF, Worth LL, Gilbert BE, Knight V (2000) 9-nitrocamptothecin liposome aerosol treatment of melanoma and osteosarcoma lung metastases in mice. Clin Cancer Res 6:2876–2880
65. Knight V, Kleinerman ES, Waldrep JC, Giovanella BC, Gilbert BE, Koshkina NV (2000) 9-nitrocamptothecin liposome treatment of human cancer subcutaneous xenografts and pulmonary cancer metastases in mice. Ann N Y Acad Sci 922:151–163
66. Zhang LJ, Xing B, Wu J, Xu B, Fang XL (2008) Biodistribution in mice and severity of damage in rat lungs following pulmonary delivery of 9-nitrocamptothecin liposomes. Pulm Pharmacol Ther 21:239–246
67. Gilbert BE, Seryshey A, Knight V, Brayton C (2002) 9-nitrocamptothecin liposome aerosol: lack of subacute toxicity in dogs. Inhal Toxicol 14:185–197
68. Verschraegen CF, Gilbert BE, Loyer E, Huaringa A, Walsh G, Newman RA, Knight V (2004) Clinical evaluation of the delivery and safety of aerosolised liposomal 9-nitro-20(s)-camptothecin in patients with advanced pulmonary malignancies. Clin Cancer Res 10:2319–2326
69. Widera A, Norouziyan F, Shen WC (2003) Mechanisms of TfR-mediated transcytosis and sorting in epithelial cells and applications toward drug delivery. Adv Drug Deliv Rev 55:1439–1466
70. Anabousi S, Bakowsky U, Schneider M, Huwer H, Lehr CM, Ehrhardt C (2006) In vitro assessment of transferrin-conjugated liposomes as drug delivery systems for inhalation therapy of lung cancer. Eur J Pharm Sci 29:367–374
71. Zhao M, Fernandez LG, Doctor A, Sharma AK, Zarbock A, Tribble CG, Kron IL, Laubach VE (2006) Alveolar macrophage activation is a key initiation signal for acute lung ischemia reperfusion injury. Am J Physiol Lung Cell Mol Physiol 291:L1018–L1026

72. Nakamura T, Abu-Dahab R, Menger MD, Schafer U, Vollmar B, Wada H, Lehr CM, Schafers HJ (2005) Depletion of alveolar macrophages by clodronate-liposomes aggravates ischemia-reperfusion injury of the lung. J Heart Lung Transplant 24:38–45
73. Leemans JC, Juffermans NP, Florquin S, van Rooijen N, Vervoordeldonk MJ, Verbon A, van Deventer SJ, van der Poll T (2001) Depletion of alveolar macrophages exerts protective effects in pulmonary tuberculosis in mice. J Immunol 166:4604–4611
74. Chono S, Tanino T, Seki T, Morimoto K (2006) Influence of particle size on drug delivery to rat alveolar macrophages following pulmonary administration of ciprofloxacin incorporated into liposomes. J Drug Target 14:557–566
75. Chono S, Tanino T, Seki T, Morimoto K (2007) Uptake characteristics of liposomes by rat alveolar macrophages: influence of particle size and surface mannose modification. J Pharm Pharmacol 59:75–80
76. Lambros MP, Bourne DWA, Abbas SA, Johnson DL (1997) Disposition of aerosolised liposomal amphotericin B. J Pharm Sci 86:1066–1069
77. Vyas SP, Quraishi S, Gupta S, Jaganathan KS (2005) Aerosolised liposome-based delivery of amphotericin B to alveolar macrophages. Int J Pharm 296:12–25
78. Vyas SP, Kannan ME, Jain S, Mishra V, Singh P (2004) Design of liposomal aerosols for improved delivery of rifampicin to alveolar macrophages. Int J Pharm 269:37–49
79. Chono S, Tanino T, Seki T, Morimoto K (2008) Efficient drug delivery to alveolar macrophages and lung epithelial lining fluid following pulmonary administration of liposomal ciprofloxacin in rats with pneumonia and estimation of its antibacterial effects. Drug Dev Ind Pharm 34(10):1090–1096
80. Liu FY, Shao Z, Kildsig DO, Mitra AK (1993) Pulmonary delivery of free and liposomal insulin. Pharm Res 10:228–232
81. Huang YY, Wang CH (2006) Pulmonary delivery of insulin by liposomal carriers. J Control Release 113:9–14
82. Chono S, Fukuchi R, Seki T, Morimoto K (2009) Aerosolised liposomes with dipalmitoyl phosphatidylcholine enhance pulmonary insulin delivery. J Control Release 137:104–109
83. Karathanasis E, Bhavane R, Annapragada AV (2006) Triggered release of inhaled insulin from the agglomerated vesicles: pharmacodynamic studies in rats. J Control Release 113(2): 117–127
84. Hajos F, Stark B, Hensler S, Prassl R, Mosgoeller W (2008) Inhalable liposomal formulation of vasoactive intestinal peptide. Int J Pharm 357:286–294
85. Stark B, Andreae F, Mosgoeller W, Edetsberger M, Gaubitzer E, Koehler G, Prassl R (2008) Liposomal vasoactive intestinal peptide for lung application: protection from proteolytic degradation. Eur J Pharm Biopharm 70(1):153–164
86. Rubinstein I, Ikezaki H, Onyüksel H (2006) Intratracheal and subcutaneous liposomal VIP normalizes arterial pressure in spontaneously hypertensive hamsters. Int J Pharm 316(1–2): 144–147
87. Laube BL (2005) The expanding role of aerosols in systemic drug delivery, gene therapy and vaccination. Respir Care 50:1161–1176
88. Griesenbach U, Geddes DM, Alton EW (2006) Gene therapy progress and prospects: cystic fibrosis. Gene Ther 13:1061–1067
89. Schwarz LA, Johnson JL, Black M, Cheng SH, Hogan ME, Waldrep JC (1996) Delivery of DNA-cationic liposome complexes by small-particle aerosol. Hum Gene Ther 10:731–741
90. Li SD, Huang L (2006) Targeted delivery of antisense oligodeoxynucleotide and small interference RNA into lung cancer cells. Mol Pharm 3:579–588
91. Durcan N, Murphy C, Cryan SA (2008) Inhalable siRNA: potential as a therapeutic agent in the lungs. Mol Pharm 5(4):559–566
92. Kleemann E, Dailey LA, Abdelhady HG, Gessler T, Schmehl T, Roberts CJ, Davies MC, Seeger W, Kissel T (2004) Modified polyethylenimines as non-viral gene delivery systems for aerosol gene therapy: investigations of the complex structure and stability during air-jet and ultrasonic nebulization. J Control Release 100(3):437–450

Chapter 15
Nanoparticles for Pulmonary Delivery

Alan B. Watts and Robert O. Williams III

Abstract This chapter aims to provide a rational for the use of nanoparticles in pulmonary delivery as well as an overview of strategies and physiological implications of nanoparticle delivery to the lungs. Formulation aspects of nanoparticle systems in the form of liquid dispersions and inhaled dry powders are also reviewed. The chapter also addresses the expanse of lung toxicology research surrounding the inhalation of nanoparticulate pollutants present in the environment. The observations of the adsorption, distribution, metabolism, and excretion (ADME) of these insoluble nanoparticles may lead to a greater understanding of biological interactions of inhaled drug nanoparticles and their ultimate site of action.

Keywords Inhalation • Lung deposition • Nanoparticles • Nanotechnology

15.1 Introduction

The ability to create, organize, and detect nanoparticles of both organic and inorganic material has had an exceptional impact on recent advances made in electronics, sensors, building materials, imaging, and medicine. The breadth of applications of nanoscience can be attributed to the capability for manipulation of chemical, physical, or biological properties of nearly any material using nanoengineering so that it is better tailored for its intended use. Both theoretical and experimental data have shown that many materials behave differently on the very small scale, giving researchers the ability to engineer materials with more desirable properties.

R.O. Williams III (✉)
Division of Pharmaceutics, College of Pharmacy, University of Texas at Austin, Mailstop A1920, Austin, TX 78712, USA
e-mail: williro@mail.utexas.edu

H.D.C. Smyth and A.J. Hickey (eds.), *Controlled Pulmonary Drug Delivery*, Advances in Delivery Science and Technology, DOI 10.1007/978-1-4419-9745-6_15,

In this chapter, the application of nanoscience to improve the pulmonary delivery of therapeutic molecules will be reviewed. While many formulation technologies such as liposomes, cyclodextrins, micelles, and dendimers fall under the nanoscience umbrella due to their small size, these are typically thought of as separate formulation technologies and will not be discussed in this chapter. Furthermore, functionalized insoluble nanoparticles (i.e., PEG functionalized particles) have seen little investigation for pulmonary delivery and will also not be included. This chapter aims to provide a rationale for the use of nanoparticles in pulmonary delivery as well as an overview of strategies and physiological implications of nanoparticle delivery to the lungs in the form of liquid dispersions and inhaled dry powders. Additionally, the author felt it appropriate to address the expanse of lung toxicology research surrounding the inhalation of nanoparticulate pollutants present in the environment. While therapeutic drug nanoparticles do not share the same toxicology with "polluting" nanoparticles, observation of the adsorption, distribution, metabolism, and excretion (ADME) of these insoluble nanoparticles may lead to a greater understanding of biological interactions of inhaled drug nanoparticles and their ultimate site of action. Moreover, continued research in nanoparticle toxicology may lead to more stringent regulatory guidelines for development of nanoparticle for human use.

15.2 Nanoparticles in Drug Delivery

Because of the multidisciplinary nature of nanoscience, there is often overlap between fields and even discrepancy over the definition of the field itself. The most commonly used definition in research and industry, and the one accepted by the International Standards Organization (ISO), defines a nanoparticle as being between 1 and 100 nm in all three dimensions [40]. Many governments and environmental groups, claiming that particles larger than 100 nm may still have health consequences, have chosen to define nanoparticles as being up to 200 or 300 nm [92]. Alternatively, the Food and Drug Administration (FDA) does not recommend applying a strict definition to "nanomaterial" due to broad context, complexity, and variety of fields involved [93]. For the purposes of pharmaceutical science, the definition of nanomaterial is undefined; however, many authors recommend the inclusion of particles measuring under 1,000 nm in each dimension [47, 51]. It could be that the definition is more inclusive in this case because particles less than 1 μm often demonstrate enhanced permeability and absorption when present in a biological system. Additionally, material properties such as aggregation, dissolution, and solubility begin to change once a pharmaceutical material is reduced below the 1 μm threshold, so it seems appropriate that the definition of "nanomaterial" includes a wider range of particle diameters in pharmaceutical research than in other fields.

Nanoscience became relevant in pharmaceutical research approximately 20 years ago when drug formulation scientists began investigating new methods to further reduce the size of drug particles and were able to produce stable particles in the nano-range. High surface area and increased dissolution pressure of these formulations resulted in more rapid dissolution and significantly higher solubility in

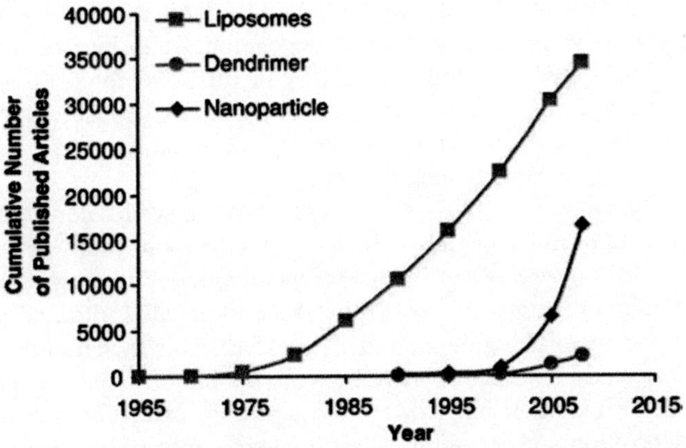

Fig. 15.1 Publications in nanomedicine Since 1965 (reprinted by permission, from [112])

laboratory testing. This basic application of nanoscience within the pharmaceutical industry has lead to marketed products with improved efficacy compared to their micronized predecessors (e.g., Rapamune®, Megace® ES) and holds potential to improve other pipeline drugs that would not have otherwise been considered due to low solubility and/or permeability. Marketed success combined with the growing number of drug molecules exhibiting poor solubility have resulted in an exceptionally rapid growth rate of publications addressing this topic (Fig. 15.1).

15.2.1 Pulmonary Delivery of Nanoparticles

While nanoparticle formulations for oral and intravenous use have proven successful in the laboratory, clinic, and marketplace, the use of inhaled nanoparticle therapeutics has been limited. Currently, no inhaled nanoparticle formulation has been approved for public use; however, successful clinical studies have been completed (e.g., nanobudesonide) and a variety of novel technologies and formulation approaches have been investigated in the laboratory. Given that many nanoformulations for inhalation have not reached the clinical stage, the bulk of this chapter will focus on investigation of novel approaches to nanoparticle delivery in vitro and in preclinical models.

Compared to oral and intravenous formulations, a simple method to achieve nanoparticle delivery to the lungs is not obvious. Nanoparticles in their native form are delivered to the lungs in a very inefficient and unpredictable manner. Upon aerosolization, interparticulate forces between nanoparticles predominate over inertial separation, resulting in aggregates of uncontrolled size. Dry nanomaterial is historically difficult to fill or meter due to particle cohesion resulting in poor flow properties. Likewise, cohesive aggregates are difficult to disperse into respirable aerosols and cannot be readily deaggregated with normal flow rates seen in passive dry powder inhalers. In addition to problems aerosolizing neat nanomaterial, studies have shown that up to 80% of inhaled particles below 1 μm are exhaled due to the lack of inertial and gravitational forces needed for deposition. Deposition of nanoparticle in the lung is time dependant and occurs due to diffusive forces associated with Brownian motion. Compared to inertial and sedimentary deposition, which are the mechanisms by which micronized particles deposit in the lungs, diffusion is a very inefficient means of achieving lung deposition.

Interestingly, as particle size continues to decrease well below 1 μm, the efficiency of lung deposition begins to increase due to more rapid diffusion. In the ultrafine particle range (less than 100 nm) computer modeling shows that deposition in the lower airway begins to rise so that 50% of inhaled ultrafine particles are deposited in the alveolar region [6, 109] (Fig. 15.2). This model of deposition was confirmed in a study in humans where between 40 and 60% of 40 nm inhaled particles were deposited in the lungs. Moreover, a strong correlation was seen between total lung deposition and breathing pattern, where deposition increased as inhalation volume and breath duration increased (Fig. 15.3). This computer model and clinical study both support the notion that Brownian motion predominates at these very small size ranges, resulting in more rapid diffusion than seen in larger nanoparticles. In a subsequent clinical study, it was found that patients with airway inflammation and narrowing due to asthma had higher deposition rates of ultrafine particles than those with healthy lungs [7]. Factors that might contribute to heightened deposition in these populations include an increased alveolar volume and airway turbulence often accompanying asthmatic conditions. While it has been proven that effective nanoparticle deposition can take place using ultrafine powders, deposition variability due to breathing pattern, expensive production processes, and difficulties handling ultrafine material have prevented it from being a feasible method for drug delivery of nanoparticulates to the lungs.

The undesirable material properties for handling and aerosolization of nanoparticles may lead one overlook the therapeutic benefits of nanoparticle delivery that have been demonstrated by other methods of delivery. It is quite true that neat particles smaller than 1 μm do not efficiently deposit on the pulmonary epithelium

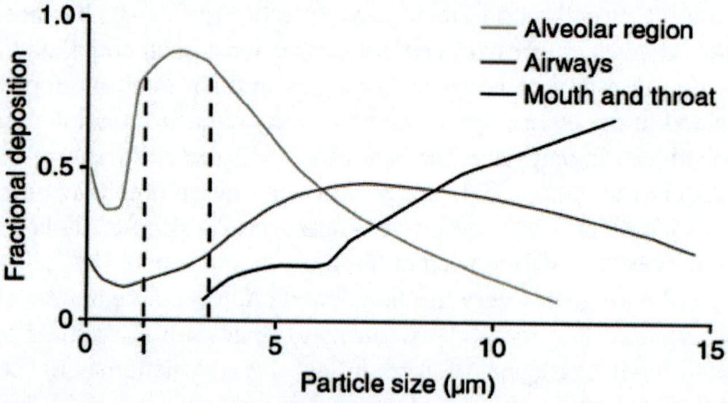

Fig. 15.2 Particle deposition in areas of the lung as a function of particle size (reprinted by permission from Yang et al. [109], originally from Byron [6])

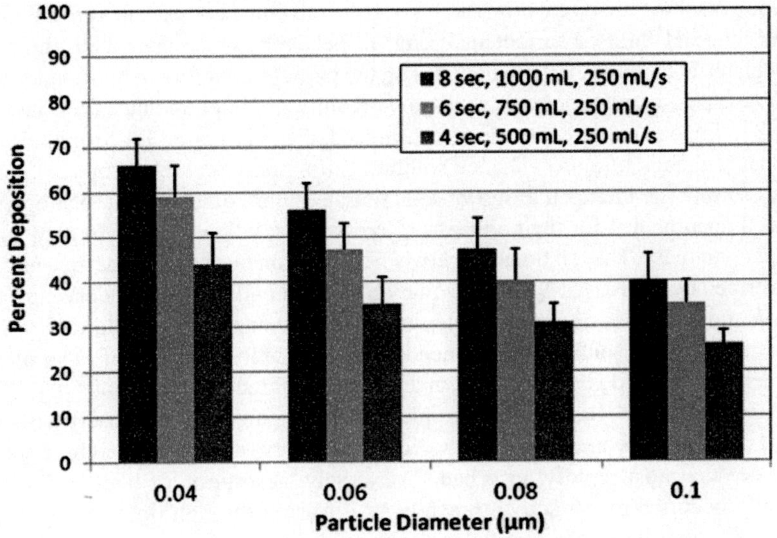

Fig. 15.3 Particle deposition as a function of nanoparticle size and breathing pattern (respiratory time, tidal volume, respiratory flow rate) (modified from Jaques et al. [43])

and would prove extremely difficult to develop into a full scale manufacturing process; however, several formulation strategies may be employed to enable drug delivery. To overcome this fundamental hurdle, pharmaceutical scientists have investigated the use of both liquid and solid carriers as well as particle engineering to improve aerosolization and provide appropriate aerodynamic diameters for deep lung deposition. In this chapter, research and development of nanoparticulate drugs for nebulization and dry powder inhalation will be discussed at length. Although discrete nanoparticles are not well suited for inhalation, many creative approaches have been taken to tailor nanoparticles for inhalable dosage forms, allowing the benefits of nanoparticle drug delivery to be imparted through inhalation.

15.2.2 Absorption of Drug Nanoparticles in the Lung

Many drugs intended for pulmonary drug delivery are formulated as aqueous aerosols where the drug is in solution and immediately available for absorption once deposited on the lung epithelial surface. Unless dissolved drug precipitates upon contact with pulmonary fluid, the overall bioavailability of nebulized solutions only relies on the extent of epithelial absorption of the molecular drug. In all other pulmonary formulations, a drug particle reaches the lungs and must both dissociate and be absorbed (although not always in that order) before it can impart its therapeutic effect.

Upon reaching the upper bronchioles or lower alveoli, nanoparticles are submersed in the mucosal fluid or surfactant lining, respectively. Once deposited, the fate of particulates in the lungs largely depends on the particle solubility in pulmonary fluids. In some cases, nanoparticles may dissolve readily in the pulmonary fluid layer and absorb to impart their therapeutic effect. Indeed, the very reason for formulation as a nanoparticulate is often to enhance solubility of a poorly soluble drug (e.g., budesonide) so that the therapeutic response is initiated more quickly [50]. Nanoparticles are well documented for their ability to increase dissolution rate due to high surface area per unit mass. The relationship between dissolution rate and surface area of a solid is described by the Noyes-Whitney equation. Additionally, increased curvature at the liquid/solid interface of nanoparticles increases the resulting dissolution pressure, and ultimately, drug solubility. This phenomenon is described in terms of vapor pressure between a liquid and gas in the Kelvin and Ostwald-Freudlich equations [61, 75].

After dissolution, the rate of absorption of a drug molecule is closely associated with its lipophilicity and molecular weight, where low molecular weight, lipophilic drugs are the most rapidly absorbed. Conversely, absorption of hydrophilic drugs typically occurs through active transport or diffusion through tight junctions [69]. For slowly dissolving drug particles deposited on the lung epithelium, the normal lung clearance mechanisms typically facilitate particle removal. Particles ranging from approximately 0.5–10 μm in diameter are phagocytized by alveolar macrophages, cleared from the airway, and transported to dendritic cells and lung draining lymph nodes [48]. In this regard, nanoparticles play an important role in the avoidance of the lung's normal clearance mechanisms in the alveolar region. Due to their ability to avoid macrophage uptake, many researchers have used nanoparticles for sustained release applications by incorporating a slow eroding material that releases loaded drug over a matter of hours or days. Other insoluble drug nanoparticles may transfect across the epithelial cell layer through endocytosis via alveolar caveolae and entersystemic circulation as an intact particle [30].

As a result of the interest in drug targeting technologies for intravenous administration, the fate of systemic nanoparticles has been thoroughly investigated. Studying the biodistribution of insoluble nanoparticles may give a clearer picture of the distribution on very slowly dissolving nanoparticles that have been administered to the lungs and are able to permeate into circulation. Characteristics of neat insoluble nanoparticles such as particle size, ionic charge, shape, and aggregation state all may determine the targets and degrees of nanoparticle accumulation. Researchers have found that very subtle changes in particle size can greatly affect the volume of distribution of a given nanomaterial. Gold nanoparticles, a substrate commonly used for PEGylation and drug targeting for intravenous administration, have shown that distribution into various organs can vary widely depending on particle size. After intravenous administration, gold nanoparticles of 10 nm were found to be distributed throughout the body, showing accumulation in the lungs and other major organs, while 50 nm particles were only present in the liver and spleen [15]. Similarly, absorption and distribution of nanoparticles reaching the body through inhalation are also directly controlled by particle size [52, 58, 60]. While drug nanoparticles typically dissociate in the lungs and do not reach systemic circulation, highly insoluble nanoparticles resulting from combustion of fossil fuels have been shown to reside in the lung for prolonged periods and also transfect unaltered into the circulatory system. The concerns and potential implications of inhaled insoluble nanoparticles will be discussed later in this chapter.

15.3 Pulmonary Delivery Strategies for Nanoparticulates

Although neat nanoparticles are able to achieve lung deposition to varying degrees depending on breathing rate and airway turbulence, they do so inefficiently, relying of Brownian diffusion for drug delivery to the pulmonary mucosa. On the other hand, delivery of drugs to the lungs requires highly efficient and reproducible particle deposition in order to provide reliable therapy and prove compliant with regulatory requirements. Consequently, nanoparticles intended for pulmonary deposition must be either incorporated in a carrier or aggregated so that their bulk size is within the required 1–5 μm aerodynamic diameter necessary for deep lung deposition. There are multiple strategies that have been used to accomplish this including formation of liquid dispersions, solid dispersions, or particle aggregates. Each method possesses different benefits and weaknesses and must be evaluated based on the active ingredient, mechanism of aerosolization, and therapeutic requirements of the specific indication.

15.3.1 Nebulized Dispersions

A clear delivery strategy to enable the deposition of nanoparticles in the deep lung is through nebulization of an aqueous colloidal dispersion or suspension. For drugs that are unsuitable for pulmonary delivery as an aqueous solution, nebulization of disperse nanoparticulates presents many advantages. First, when compared to micronized drug dispersions, nanoparticles can be distributed more uniformly throughout 1–5 μm droplets [66], allowing for greater dose uniformity and more effective deep lung deposition due the increased drug presence in droplets less than 3 μm. When compared to a microparticle suspension, studies have shown that improved dose uniformity and higher average delivered dose can be achieved [104]. Second, due to their near-solution rheology, nebulized nanoparticle dispersions have been shown to have less of a negative influence on nebulizer function and aerosol droplet size when compared to microparticles [14]. Researchers have investigated nebulization of nanoparticles produced by milling, antisolvent precipitation, rapid freezing, and supercritical fluid production techniques, many of which require the incorporation of a stabilizing polymer/surfactant to limit particle growth and/or aggregation.

A key criterion to consider when delivering an aqueous dispersion of therapeutic nanoparticles to the lungs is the choice of stabilizing material. Many synthetic surfactants and stabilizers that are normally used in oral and intravenous applications are not well suited for the lungs, particularly in high concentrations. As a result of their toxicity and immunogenicity in the lungs as well as their tendency to foam during nebulization, surfactants normally used in pharmaceutical formulations (e.g., sodium lauryl sulfate) should be used sparingly or not at all in nebulized dispersions. Consequently, nanodispersions intended for the lungs often use small amounts of natural surface active agents to stabilize a nanodispersion, or alternatively, design formulations that are intended for redispersion before nebulization. A potential advantage of dry formulations intended for dispersion in the clinic is that aqueous stability is only needed for the minutes between redispersion and nebulization, meaning minimal amounts of surfactant is necessary. Conversely, formulations for reconstitution typically require the assistance of a health professional, present an inconvenience to medical staff, and could result in contamination.

15.3.1.1 Stabilized Nanodispersions

Surfactant stabilized colloidal dispersions for nebulization have been developed to facilitate nanoparticulate drug delivery to the lungs. Incorporation of either synthetic (Tweens, Spans, poloxamers) or natural surfactants (dipalmitoyl phosphatidylcholine (DPPC), lecithins, leucines) is necessary to prevent aggregation of dispersed particulates in these formulations and insure colloidal stability. Researchers in inhalation have utilized this formulation approach and have identified numerous aerosolization and physiological advantages.

Recently, a nanodispersion of solubility-enhancing amorphous cyclosporine was produced and proven to be stable in aqueous media, avoiding both aggregation and crystallization after 10 months at 4°C. A controlled precipitation technique stabilized with polysorbate 80 yielded particles with a mean diameter of 300 nm, allowing for effective nebulization to ICR mice. When compared to ethanol-based nebulized solutions, nanoparticulate cyclosporine facilitated higher lung tissue levels while minimizing systemic absorption [88]. As is the case with aerosolized cyclosporine, reduction of systemic levels can be particularly important for localized pulmonary therapy where the administered moiety is known to have many untoward side effects [39].

Budesonide, a glucocorticoid with poor water solubility, has also proven to benefit from formulation as a nanoparticulate. Using a lab-scale wet milling technique where budesonide was dispersed for milling in a Tween 80 stabilized solution, Yang et al. [109] produced a monodisperse suspension containing particles less than 500 nm in diameter [108]. Pharmacokinetic profiles in a rodent model showed levels produced by the intratracheally instilled nanosuspension comparable to those produced by intratracheal and intravenous solution. Only a slight delay in absorption was seen after administration of the nanosuspension due to the time needed for particle dissolution in lung fluid. While this study demonstrated the "solution-like" properties of nanoparticulate budesonide, the potential irritating effects of Tween 80 on the pulmonary mucosa may need to be considered before initiation of clinical trials.

Addressing this issue, budesonide has also been developed as a nanosuspension for nebulization using nonimmunogenic surfactants [41]. By inclusion of 0.5% w/w lecithin and 0.2% w/w Tyloxapol in a piston gap homogenization process, a stable nanodisperse formulation containing 1.0% w/w budesonide was produced. A detailed explanation of the diminution mechanism is provided in a review by Gao et al. [27]. Enhanced solubility due to increased dissolution pressure, more homogenous drug dispersion within aerosol droplets, and prolonged residence time at the pulmonary epithelium were among the advantages that might be expected by the reduction in particle size. Compared to micronized dispersions, increased drug homogeneity within a nanodispersion allows for improved deep lung deposition [27] and shortened nebulization time [54]. The safety and pharmacokinetics of nanodisperse budesonide, or nanobudesonide, was evaluated in 16 healthy volunteers and compared with the marketed budesonide dispersion for nebulization, Pulmicort Respules® (AstraZeneca, Wilmington, DE). Plasma levels resulting from inhalation of nanobudesonide indicated dose proportional therapeutic concentrations and more rapid drug absorption compared to Pulmicort Respules [50]. Additionally, no clinically significant adverse effects were observed after delivery of nanobudesonide or placebo containing only suspension stabilizers. These findings support the hypothesis of increased solubility and absorption offered by nanoparticulate formulations and provides clinical evidence of the potential benefits offered by inhalation of nanoparticles. Continued clinical studies are being pursued for use of submicron budesonide to treat pediatric asthma by MAP Pharmaceuticals (Mountain View, CA).

Both air-jet and vibrating mesh nebulizers were evaluated with a nanoparticulate buparvaquone formulation made by high pressure homogenization. Synthetic surfactants polxamer 188 and polyvinyl alcohol were used in equal parts to the drug to provide suspension stability; although some nebulization methods produced particle aggregation. Study results made clear that particle aggregation within the reservoir and aerosolized droplets may vary based on nebulizer selection [33].

Improvements in aerosol performance have also been observed after cascade impaction testing of nanoparticulate beclomethasone propionate produced by milling. Nanoparticles produced in early studies through ball milling were stabilized with poly(vinyl alcohol) and proved to nebulize effectively [104]. More recently, tyloxapol was used to stabilize particles at a mean size of 164 nm after 4 days of milling. When tested for comparison to the marketed pMDI formulation Vanceril® (Key Pharmaceuticals, Kenilworth, NJ), nebulized beclomethasone nanoparticles delivered by an Omron ultrasonic nebulizer produced up to twice the respirable fraction (36 vs. 72%) [66]. An additional benefit of the nebulization of beclomethasone is that it is compliant with the Kyoto protocol, which has restricted the use of ozone damaging propellants used in pMDIs [26]. Nanoscale beclomethasone dipropionate has also been produced by a controlled precipitation technique where solvent combinations and mechanical stirring were optimized so that no stabilizing agent was required. In comparison to micronized drug, smaller particle size (850 nm) and more narrow particle size range (200–1,200 nm) showed that this technique could be effective, even in the absence of crystal growth inhibiting surfactants [98].

While a high potency formulation with no surfactant content is desirable, there was no indication of the stability of neat drug in dispersion and redispersability was not demonstrated. Significantly, smaller nanoparticles can be produced by combining and optimizing stabilizer concentration, mixing energy, and antisolvent feed rate [55]. Itraconzole particles less than 300 nm were created by optimizing these variables and incorporating concentrations of poloxamer 407 well below the critical micelle concentration. Findings showed that increased stabilizer concentrations and cooled process temperatures reduced surface tension resulting in more rapid nucleation and inhibited crystal growth.

15.3.1.2 Nanoscale Powders for Redispersion

As an alternative to employing a stabilized colloidal dispersion for nebulization, other researchers have reduced potential instability by creating a formulation intended for dispersion before administration. An added benefit of reduction or elimination of the need for stabilizing surfactants can also be realized through this approach. Growing interest in the use of nanoparticles as carriers for targeted and biopharmaceutical therapy has also led to the use of formulations for redispersion, given that many of the therapeutics and polymeric carriers are subject to hydrolytic degradation and are not stable as dispersions for prolonged periods. When investigating the feasibility of nebulizing powders reconstituted from lyoprotective agents such as lactose, mannitol, glucose, and sucrose matrices, researchers demonstrated that stability is preserved without significantly altering particle size [67].

A rapid freezing technique has been applied to create nanostructured formulations intended for nebulization after redispersion in aqueous media. These amorphous nanodispersions are capable of providing enhanced solubility to poorly water soluble drugs, allowing for heightened bioavailability in bronchoalveolar fluid. As a treatment and/or prophylactic measure against invasive fungal infections caused by *Aspergillus* spp., researchers have developed an inhaled itraconazole formulation that has shown the ability to supersaturate simulated biological media. Formulations of drug and surfactant were rapidly frozen using a spray freezing into liquid (SFL) cryogenic technique to produce amorphous drug particles in the nanoparticulate range. After reconstitution and nebulization of the aqueous dispersion to healthy mice, it was hypothesized that rapid wetting and dissolution occurs, allowing for itraconazole lung concentrations necessary for inhibition of spore germination [57]. When compared to the marketed oral itraconazole therapy, Sporanox®, inhaled itraconazole produced by SFL showed tenfold greater lung concentrations in mice while still maintaining serum concentrations above the minimum lethal concentration for *Aspergillus fumigatus* [96]. A similar amorphous itraconazole formulation was produced in a study by Yang et al. [109]; however, the biocompatible excipients, mannitol and lecithin, were included in order to produce a formulation with increased solubility and limited pulmonary toxicity. Over a 12 month storage period, the stability of the solubility-enhancing amorphous drug formulation was confirmed by powder X-ray diffraction [110].

Other studies incorporating thin film freezing (TFF) technology have been used to provide pulmonary deposition of immunosuppressive formulations containing tacrolimus. Amorphous and crystalline tacrolimus nanoparticles were investigated as a dispersion in aqueous media and delivered to mice via nebulization through a nose-only dosing chamber. In order to avoid surfactant-induced pulmonary irritation, this formulation was produced as a dry powder for redispersion, thus avoiding the requirement of colloid stabilizing surfactants. Both formulations resulted in similar bioavailability in vivo; however, amorphous tacrolimus resulted in more rapid systemic drug absorption, as evident through a decrease in t_{max} and an increase in C_{max} [83]. Evidence of supersaturation in simulated lung fluid during dissolution testing would lead one to believe that the enhanced solubility of this material is the main factor behind the more rapid drug absorption seen after administration of amorphous tacrolimus. An indication where inhaled tacrolimus may prove valuable is in immunosuppressive maintenance therapy for lung transplant recipients. In a lung transplanted rodent model, a tacrolimus dispersion for nebulization provided lung concentrations consistent with those shown to prevent allograft rejection [100]. It is suspected that lipophilic interactions between tacrolimus and lung surfactant led to low systemic drug levels in this model; systemic concentration was 50-fold less than concentrations measured in lung tissue. TFF technology for production of tacrolimus formulations for dispersion demonstrated the capability to effectively disperse in aqueous media so that aerosolization could occur and deep lung delivery could be achieved. Furthermore, combination of this technology with a lipophilic drug resulted in high lung concentrations while limiting resulting blood levels, thus avoiding therapy-limiting systemic side effects often associated with tacrolimus.

15.3.1.3 Drug Encapsulated Nanodispersions

In cases where the therapeutic molecule may be poorly soluble, subject to hydrolytic or enzymatic degradation, or intended for controlled release, drugs are often encapsulated in a carrier material. An added advantage is the ability to surface-modify these particles to impart targeting and/or stealth particle properties.

Tumor-targeting nanoparticles have been investigated for nebulization with the goal of achieving high concentrations of chemotherapeutic agents in the lungs while reducing systemic exposure. Preferential accumulation of targeted nanoparticles was observed in the lungs of tumor-induced nude mice after exposure to nebulized aerosols containing surface-modified gelatin nanoparticles. In this study, gelatin nanoparticles were produced by an antisolvent precipitation technique, cross-linked with glutaraldehyde to reduce solubility, and conjugated with biotinylated epidermal growth factor (EGF) to improve drug targeting. Findings proved that these particles effectively accumulated in lung tissue while showing no signs of toxicity [91]. More recently, targeting gelatin nanoparticles were loaded with the chemotherapeutic agent cisplatin and proven efficacious in a lung cancer model. As seen in Fig. 15.4, mice showed high lung concentrations of cisplatin after chemotherapeutic aerosol administration and significant reduction in tumor volume [90]. In this same

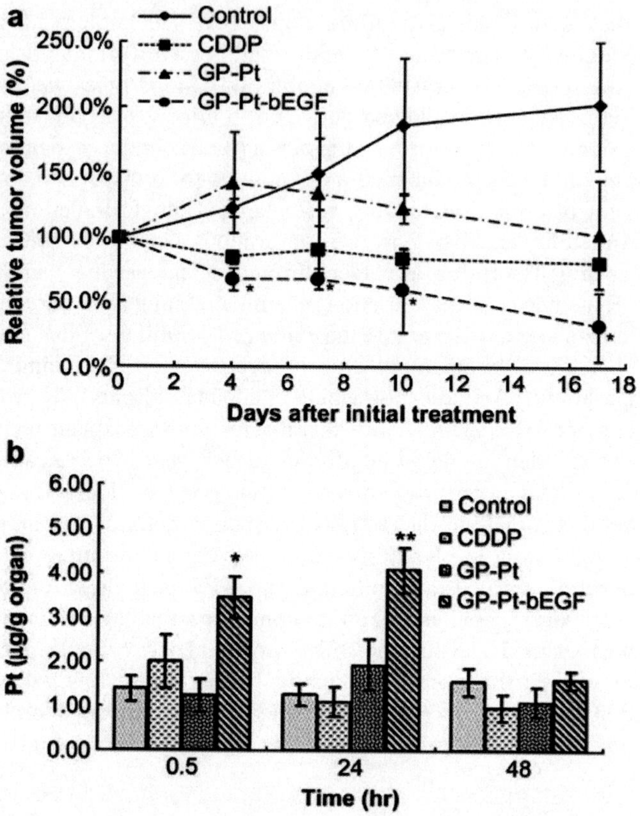

Fig. 15.4 Change in (**a**) tumor volume and (**b**) lung drug concentration in mice after pulmonary administration of a chemotherapeutic treatment. *Asterisk(s)* denote significant difference (*$P<0.05$; **$P<0.01$) from control. CDDP, free cisplatin; *GP-Pt* cisplatin in gelatin nanoparticles; *GP-Pt-bEGF* cisplatin in biotinylated-EGF gelatin nanoparticles (reprinted by permission from Tseng et al. [90])

model, nontargeted cisplatin caused marked weight loss in dosed mice, which is indicative of systemic toxicity caused by high blood concentrations. These results seem to indicate effective localization of cisplatin in the lungs by incorporation in novel targeting nanoparticles. In addition to gelatin-based nanoparticles, particles prepared with human serum albumin have been investigated, specifically as a material suitable for gene therapy [5].

Manipulation of drug release characteristics through slow erosion of a nonimmunogenic polymer has also been achieved using pulmonary administration of encapsulated drugs. Poly(lactic-*co*-glycolic acid) (PLGA) is a commonly researched bioerodible polymer that has been used in multiple studies of sustained release formulation for oral, intravenous, nasal, and pulmonary delivery. One such formulation intended for pulmonary delivery via nebulization used peptide loaded PLGA nanoparticles that had been surface-modified with chitosan, a known absorption enhancer [106]. In addition to increasing particle mucosal retention over nonmodified nanoparticles, the presence of chitosan was shown to enhance the absorption across the pulmonary epithelium. It is suggested that the tight junction-opening effect of chitosan observed in intestinal and nasal mucus membranes may also occur in the lungs. Other carrier materials suitable for prolonged release applications have also been investigated for delivery via nebulization that may prove to be more suitable for pulmonary products given that they erode more rapidly than PLGA. A novel PLGA derivative studied by Dailey et al. showed the ability to form nanoparticles below 200 nm in a surfactant-free process that erode more rapidly than PLGA nanoparticles [13]. Further evaluation of inflammatory response in BALB/c mice after administration of modified PLGA demonstrated that theoretical dosing levels produce a similar inflammatory response to isotonic glucose [12]. Another novel branched polyester capable of rapid erosion, DEAPA, was used to make drug loaded nanoparticles during a solvent displacement process. During in vitro testing, a burst release of over 90% of the drug occurred in the first 50 min of release testing, showing that a therapeutic bolus is available more rapidly than would be expected from PLGA particles. Furthermore, aerosol nebulization was unaffected by the dispersion of nanoparticulates, likely because of their solution-like behavior [3].

Cation-induced gelification of alginate was used to encapsulate antitubercular agents and was subsequently dispersed in normal saline for administration to an animal model. By providing a single-dose controlled release formulation to the lungs, more localized pulmonary concentrations were observed in addition to therapeutic levels in the liver and spleen. Controlled release of drugs administered pulmonarily shows potential to circumvent noncompliance issues commonly associated with daily dosing antitubercular medications [42] and avoid the metabolic pathways of the reticuloendothelial system [111]. Controlled release formulations delivered to the lungs would not be possible using micron-sized particles because of their tendency to be captured by phagocytic clearance mechanisms.

15.3.1.4 Lipid Nanoparticles

As will be discussed later in this chapter, pulmonary and systemic toxicity induced by insoluble nanomaterials is a major concern when considering a drug formulation intended for inhalation. While it has been proven that some biodegradable polymeric materials do not induce significant toxicity, such as PLGA [12], many researchers have opted for natural occurring excipients when formulating drugs for lung delivery. Liposomal drug delivery to the lungs attracts many researchers and clinicians because the delivery components contain many of the same saturated and unsaturated hydrocarbons found in human tissues and fluids. Significant clinical studies have been performed and are currently underway investigating liposomal delivery to the lungs [4, 17, 77]. Although liposomal delivery systems technically are of nanoparticulate dimensions (<200 nm), they are often classified as their own drug delivery category, and are consequently discussed in another chapter of this book.

Alternatively, solid lipid nanoparticles (SLN) represent a lipid-based drug delivery strategy that combines the biocompatibility of liposomes with the stability and production efficiency of polymeric particles. Furthermore, the solid and robust surface present in these formulations allows for surface modifications such as ligand targeting [46] and PEGylation [28]. SLN, typically made by homogenization or emulsion precipitation, have been used to deliver drugs as a solid molecular dispersion or as a drug encapsulating lipid shell. Normally, the lipophilic/hydrophilic nature of the drug and production technique will determine whether drug is encapsulated within a lipid shell or present as a solid solution in the lipid matrix. Clozapine, a very lipophilic drug used to treat schizophrenia, was found to dissolve with the triglyceride matrix, allowing for the production of drug loaded SLN by hot homogenization [97]. The enzymatic protection imparted by the lipid carrier showed potential to protect clozapine from an extensive first pass effect, and in turn, increase the its therapeutic potency.

Similar to aforementioned polymeric nanoparticles, SLN have also been investigated as a means to provide controlled release of chemotherapeutic pharmaceuticals for treatment of tuberculosis. Rifampicin, isoniazid, and pyrazinamide were incorporated into separate lipid nanoparticles by an emulsion precipitation technique [68, 36]. When nebulized to healthy and tuberculosis-infected guinea pigs, SLN showed dramatic improvement of bioavailability and prolonged residence in the lungs, liver, and spleen when compared to intravenous, oral, and unencapsulated pulmonary drug administration.

A novel method of SLN preparation was developed, characterized, and investigated with a metered liquid aerosol generating device [8]. An emulsion containing one of two poorly soluble drugs, indomethacin or ketoprofen, was dissolved along with a lipid base in chloroform and stabilized in an aqueous external phase with lecithin. After homogenization, chloroform was extracted by supercritical fluid, leaving lipid-based particles well under 100 nm in diameter. As a result of the small size for the dispersed particles, the emitted dose of this nanosuspension proved comparable to that of a solution. Moreover, the fine particle dose as a percentage of the initial loaded dose was 39 and 40% for indomethacin, and ketoprofen SLN formulations, respectively. In an efficacy model, nebulization of SLN once a week was shown to be therapeutically equivalent to 46 daily doses of oral therapy [68].

Another variation of the lipid carrier nanoparticle strategy, lipid nanocapsules, are described as a vector hybrid of liposomal and polymeric nanoencapsulated particles [34]. This delivery platform is embodied by drug solubilized within a solid lipid core, which will become liquid at biorelevant temperatures, surrounded by a rigid lecithin membrane. These particles can range from 25 to 100 nm and have proven suitable for aqueous dispersion and nebulization [38], particularly with vibratory mesh nebulization technology. In lipid nanocapsules containing the chemotherapeutic agent paclitaxel, aerosol production showed no reduction of in vivo efficacy against NCI-H460 human lung cancer cells when compared to non-nebulized nanoparticle dispersions, proving the nebulization process does not destroy the encapsulated formulation.

15.3.2 Dry Powders for Inhalation

Many researchers have demonstrated that the delivery of nanoparticles to the lungs is possible through incorporation in a carrier. Dispersion of nanoparticles into aqueous media for nebulization has proven to be a suitable carrier in a range of preclinical and animal studies; however, nanodispersions in liquid carriers are not without drawbacks. First, many dispersions stabilized as a suspension are likely to experience chemical and physical instability due to hydrolysis, particle settling, or aggregation. Additionally, these suspensions require the use of stabilizers that must be proven safe and nontoxic in the lungs before human use. Nanoparticles for redispersion offer greater stability, but typically require a clinician for administration and consequently would not be convenient for everyday treatment.

A viable alternative to aqueous nebulization of nanoparticles is found in dry powder aerosol delivery. Formulation of inhaled pharmaceuticals as dry powders insures greater chemical and physical stability when compared to dispersions for nebulization. This benefit is mainly due to the more limited molecular mobility of a dry system, in turn reducing the molecular and interfacial interactions that lead to degradation or agglomeration, respectively. Additionally, medication inhaled in the solid form can often be given as a single dose in one breath, reducing treatment time and improving patient compliance.

15.3.2.1 Nanodisperse Microspheres

As a formulation technique for oral delivery, microencapsulation is often used as a means to manipulate drug release, prevent enzymatic degradation, mask taste, or improve bioavailability. For pulmonary delivery, microencapsulation may be used for these purposes; however, the predominant reason for implementation of this strategy is to form particles with the appropriate aerodynamic diameter for deep lung deposition.

A common approach to produce powders suitable for inhalation is spray drying. This production method is often used to make respirable neat drug particles or drug dispersions in a stabilizing matrix. Researchers have also explored this method to produce nanoparticles dispersed within carrier microparticles [80]. In a study investigating the feasibility of spray drying as technique to incorporate nanoparticles into carrier microparticles, gelatin and polycyanoacrylate nanoparticles were dispersed in a lactose matrix microparticle and characterized by dissolution, fluorescent labeling, and cascade impaction. Resulting data showed that after redispersion in aqueous media, nanoparticles remained un-aggregated, meaning the benefits of nanoparticle size would likely be present in vivo. Additionally, impaction testing showed that respirable lactose matrix particles could produce fine particle fractions ranging from 38 to 42%. This approach to deliver nanoparticulates to the lungs could be particularly beneficial for peptide and protein based pharmaceuticals due to the absence of water in the final formulation and the relatively low processing temperatures (40–45°C). As a continuation of this research, Azarmi et al. have shown that this method of drug delivery to the lungs can be beneficial in targeting lung cancer cells [2]. Doxorubicin, used as a model chemotherapeutic agent, was formulated in poly(butylcyanocrylate) nanoparticles intended for delivery via encapsulation in lactose carrier particles made by spray-freeze drying. In vitro uptake into H460 and A549 lung cancer cell lines showed that nanoparticulate doxorubicin more effectively targeted cells than free drug, potentially due to more readily occurring endocytosis rather than passive diffusion.

An interesting variation of spray drying incorporating two simultaneously operating liquid spray nozzles was recently used to create drug nanoparticle loaded microparticles for inhalation. This approach has proven helpful in the dispersion of poorly water soluble drug nanoparticles in a water soluble carrier material, where a common solvent is not available. Pranlukast hemihydrates, a leukotriene antagonist for treatment of asthma, was spray dried by this method to produce nanodisperse drug within a mannitol microparticle capable of reaching the deep lung [59]. Upon reconstitution in aqueous media, drug nanoparticles were determined to range between 100 and 500 nm in diameter. Improved bioavailability after oral administration was demonstrated in Sprague Dawley rats, where the total absorption of the spray dried formulation was approximately 100 times that of the bulk drug. Improvements in bioavailability were attributed to enhanced solubility and absorption of nanoparticulate pranlukast.

For delivery of therapeutic molecules that are easily degraded by production and/or physiological conditions, nanoparticle encapsulation can serve as an effective means to preserve formulation potency. In the lungs, macrophage uptake, esterases, proteases, and various epithelial metabolic pathways represent substantial challenges for delivery of highly degradable macromolecules such as peptides, proteins, or nucleic acids. Nanoencapsulation offers a solution to in vivo degradation by allowing uptake of the therapeutic substance without exposure to denaturing environment. Furthermore, dispersion of these nanoparticles in a solid microparticle matrix may provide a more suitable alternative to nebulization of an aqueous dispersion due to the additional stability imparted by the absence of water.

Delivery of small interfering RNA (siRNA) to lungs seems to be well suited for nanoencapsulation in a dry powder carrier because of the potential for enhanced stability and increased absorption. By interfering with, or "switching off," disease causing genes, siRNA are able to treat pulmonary diseases such as cystic fibrosis, respiratory syncytial virus, and severe acute respiratory syndrome. To enable local delivery to the lungs, researchers have used polymers such as PLGA to create nanoencapsulated formulations capable of avoiding macrophage clearance.

To effectively reach the deep lung, spray dried dispersions of nanoparticles in carbohydrates and sugars like lactose, mannitol, and trehalose have been investigated. In some cases, optimization of spray drying conditions and excipient concentrations has produced nanoparticle concentrations making up over 50% (w/w) of the spray dried particle without causing substantial degradation of the thermally liable nucleic acid [44].

The slow eroding capabilities of PLGA in biological fluid has lead to its use in many sustained release formulations intended for oral and intravenous administration [74, 103]. Likewise, researchers have investigated its use for sustained release within the lungs. As mentioned previously, researchers demonstrated that not only is PLGA safe for human use in oral and intravenous drug products, but there is also reason to believe it is safe for inhalation [12]. In addition to identification of a suitable slow eroding polymer, sustained release in the lungs is also complicated by the body's natural clearance mechanisms: the mucociliary escalator and phagocytosis. These mechanisms have the ability to clear deposited particulates within a matter of hours, thereby removing formulated particles from their intended site of delivery. In the deep lung, phagocytic clearance predominates; however, macrophage uptake and clearance is limited to particles ranging from 0.5 to 10 μm in diameter. Consequently, formulation of large microparticles or small nanoparticles capable of deposition in the deep lung may result in increased residence time, and thus, prolonged drug release to the pulmonary epithelial layer. To evade macrophage detection and phagocytosis, PLGA nanoparticles containing insulin have been formulated and shown to provide significantly prolonged therapeutic effect after intratraceal administration in Wister rats when compared to insulin solution [105]. Alternatively, ciprofloxacin dispersed in a large porous PLGA matrix was produced and shown to provide sustained drug release over several days in laboratory testing [1]. Highly localized treatment of pulmonary infection associated with cystic fibrosis may be possible through development of formulations such as these. Porous or low density pulmonary formulations characterized by a large geometric diameter were initially published in a work by Edwards et al. [20] and present as a novel means to achieve deep lung deposition and clearance evasion. These and similar formulations incorporating aggregated nanoparticles and nanostructured particles will be discussed briefly in this chapter; however, detailed discussions of this subject will be addressed in chapters focused on particle engineering techniques.

While PLGA offers a safe and effective means of macromolecular encapsulation, slow erosion rates may hinder rapid therapy and acidic microenvironments produced may lead to protein denaturation. As an alternative to PLGA encapsulation, other encapsulating polymers such as degradable polyesters and naturally occurring polymers have gained interest. For rapid release of siRNA occurring over approximately 3 h, Nguyen et al. used a solvent displacement method to produce modified PVA/PLGA based nanoparticles less than 200 nm in diameter [63]. To investigate the potential of naturally occurring polymers for nanoencapsulation, macromolecule-containing chitosan nanoparticles have been investigated for pulmonary delivery via spray dried mannitol/lactose microparticle carriers [29]. A high loading capacity of the nanoparticles (75–80%) resulted in burst release of the model protein in vitro from both discrete nanoparticles and microparticle encapsulated nanoparticles. Equivalent time and magnitude of release of both discrete and encapsulated nanoparticles suggested that no release altering effects were imparted by inclusion in a microparticle carrier matrix. Interestingly, findings showed that higher nanoparticle loading improved the spherical morphology of spray dried particles. The authors attributed this phenomenon to dispersed solid nanoparticles that acted as a latticework for the formation of a spherical morphology during the spray drying process. Naturally occurring polymers have also been investigated for sustained release of nanoparticles in the lungs. Cook et al. investigated glyceryl behenate, tripalmitin, and hydrogenated palm oil as controlled release excipients, demonstrating that sustained release of hydrophilic nanoparticles with minimal burst release is possible with more biocompatible excipients [11].

In order to insure effective nanoparticle deaggregation after dissolution of the microparticle matrix carrier, Ely et al. have investigated the use of an active release mechanism for spray dried microparticles [23]. By the incorporating various concentrations of sodium carbonate and citric acid in the carrier matrix, an effervescent microparticle was produced that was suitable for inhalation. Upon exposure to moisture, sodium carbonate reacted to form carbon dioxide which produced additional dissolution energy within the matrix to deaggregate nanoparticles. When a similar lactose nanoparticle matrix was evaluated for dissolution, resulting aggregated nanoparticle diameters were significantly larger than those produced by formulations using effervescent technology. Through the application of effervescent excipients, actively deaggregated formulations could produce more discrete nanoparticles available for absorption, ultimately enhancing drug bioavailability.

15.3.2.2 Aggregated Nanoparticles

As mentioned previously, deposition of discrete aerosolized nanoparticles in the lower airways can be difficult due to the negligible effect of inertial and sedimentation forces needed for impaction. An interesting and effective strategy to increase the effects of these forces on nanoparticulate dry aerosols is to form low density aggregates of multiple nanoparticles. These nanoaggregates typically have low densities (often less than 0.1 g/cm^3) and may take the shape of hollow spheres [32, 89], spherical agglomerates [37], nonspherical flocculates [56, 71, 82], or aggregated plates [53, 76]. A variety of production techniques have been modified or developed to facilitate manufacture of these nanoaggregates including spray drying, salt flocculation, and rapid freezing processes.

Spray drying is often used for production of pharmaceuticals for inhalation and has been used to form nanoparticle aggregates as both hollow and solid spheres. Formation of hollow spheres composed of aggregated nanoparticles was initially proposed for pulmonary delivery by Tsapis et al. [89]. Their studies showed hollow microparticles composed of discrete aggregated nanoparticles which could be formed by spray drying nanoparticle dispersions in a rapidly evaporating liquid carrier. The ratio between liquid carrier evaporation and nanoparticulate diffusion in that carrier is often described in a Peclet number, where the greater the Peclet number, the more likely a hollow or collapsed spray dried product will be produced. After liquid evaporation, nanoparticles are left imbedded in surfactants or other excipients present, or in some cases, held in place without excipients through Van der Waals forces. In this way, spray drying can be used to create aggregated nanoparticles for inhalation containing only neat drug. Other groups have also evaluated this technique, investigating caviots such as the influence of nanoparticle size and chemical nature, excipients, and processing conditions on hollow particle structure and density [31, 32].

In addition to production of hollow spheres, spray drying has also been used to produce spherical aggregates of nanoparticles. After formation of nanoparticles using a novel high gravity controlled precipitation technique [10], salbutamol nanoparticles were spray dried to produce loose high surface area aggregates of approximately 2 μm in diameter. Using the Aerolizer® (Novartis, Basel, Switzerland), the fine particle fraction (emitted) was 71%, likely due to the unique morphology and high surface area of these spray dried aggregates [37].

Process technologies other than spray drying have also been used to control aggregation, or flocculation, of nanoparticles. Manipulation of the electrostatic repulsion/attraction of individual particles has been demonstrated as another effective means to self-associate nanoparticles for pulmonary delivery. Flocculation of polymeric nanoparticles was successfully controlled using charged polymeric PLGA nanoparticles in a high homogenization process [81]. Positive and negative charges were imparted in PLGA nanoparticles through the incorporation of polyvinylamine and poly(ethylene-maleic anhydride), respectively. After liquid removal, flocculated particles had a density of approximately 0.1 g/cm^3 and demonstrated aerodynamic diameters within the respirable range.

Highly potent drug flocculates have also been produced through manipulation of particle electrostatics; however, in this case, disruption of stabilizing charges was used to control flocculation. Polymeric stabilizers and surfactants are often used to produce stable colloidal solutions, particularly during the manufacture and storage of nanoparticulate drug formulations [102]. Steric hindrance and electrostatic forces effectively prevent particle growth and aggregation in these formulations. In a novel variation of this concept, Plumley et al. have used sodium chloride to disrupt the nanoparticle stabilizing effects of stearic acid on nifedipine nanoparticles [71]. The resulting nanoaggregates characterized by cascade impaction and laser diffraction testing demonstrated a mean aerodynamic diameter of 1.4 μm and a mean geometric diameter of approximately 10 μm, respectively. The large difference between the aerodynamic and geometric diameters was not unexpected due to the low bulk density (0.07 g/cm^3) of this aggregated material. Similar formulations have been investigated using leucine to destabilize and aggregate anticancer agents, paclitaxel and cisplatin [21]. Flocculation for production of respirable drug nanoaggregates is currently being investigated by Savara Pharmaceuticals (Austin, TX) for delivery of high potency formulations to the lungs. In addition to producing low density particles suitable for inhalation, salt flocculation has also been demonstrated as an effective method for the extraction of nanoparticles from colloidal dispersion, while maintaining nanoparticle morphology [56]. For collection of nanoparticles intended to be delivered in their amorphous state, salt flocculation followed by filtration allows for readily redispersible, low density powder.

A platform for accurate delivery of drug to the deep lung using self-assembled submicron plates has been developed by Mannkind Corporation (Valencia, CA). In the Technosphere® platform, aggregated plates composed of crystalline fumaryl diketopiperazine (FDKP) serve as a high surface area substrate for the adsorption of an active pharmaceutical ingredient. Much like enteric coated dosage forms intended for oral delivery, FDKP contains carboxylic acid groups that will ionize and rapidly dissolve in the neutral pH of pulmonary fluid. Solubility of the drug layered substrate combined with high surface area of aggregated crystalline plates shows promise for the use of this platform for rapid delivery of drugs for systemic indications. Initial clinical studies have been conducted to determine the feasibility of Technosphere® for effective delivery of salmon calcitonin, parathyroid hormone, and human insulin to systemic circulation via the lungs [53]. Promising clinical data has been collected after an open-label, randomized, cross-over study was conducted in 16 nonsmoking volunteers with type 2 diabetes. Pharmacokinetic, pharmacodynamic, and safety data was evaluated after administration of a 48 unit nominal dose of Technosphere® insulin and compared with that of normal subcutaneous injections of human insulin. Pulmonary administration of human insulin provided a significantly higher C_{max} 15 min postadministration than did subcutaneous injections; although, the doses given resulted in a nearly equivalent bioavailability, as determined by the systemic AUC. Consequently, blood glucose resulting from a standardized meal was 48% lower in subjects receiving inhaled insulin rather than subcutaneous insulin [76]. As demonstrated by the delivery of inhaled insulin, the nearly instantaneous partitioning of drug into systemic circulation via the lungs may prove to be a more effective method of treating disease exacerbation or acute symptoms of diabetes mellitus and other systemic diseases.

15.3.2.3 Other Engineered Particles

Although they will not be described in detail in this chapter, it should be mentioned that other advanced formulation technologies for inhalation have been thoroughly researched and, in some cases, have proven successful in clinical trials. While not composed of discrete nanoparticles, many low density microparticles with nanostructured morphologies have been produced by novel adaptations of spray drying and demonstrate exceptional aerosol performance. The resulting powders for aerosolization provide highly efficiency delivery due to their narrow size distribution and ease of deaggregation. Two such formulation technologies have proven quite successful in early laboratory testing and have continued on in clinical trials.

PulmoSphere® technology, developed by Inhale Therapeutic Systems (San Carlos, CA) and now owned by Novartis (Basel, Switzerland), uses a liquid emulsion feedstock to produce hollow, porous spray dried spheres for use in DPIs and pMDIs [16, 18]. The porous nature of the spray dried spheres is achieved through fluorocarbon evaporation during the drying process. Gamma scintigraphy studies in human subjects have been performed to evaluate the deposition of PulmoSphere® after administration with a pMDI and have demonstrated roughly twice the lung deposition when compared to a normal pMDI formulation [35]. Further clinical studies focused on administration of PulmoSpheres® as dry powder for inhalation. Tobramycin was formulated as a dry powder using this technology and was shown to give sevenfold more efficient lung deposition than nebulized tobramycin in 14 healthy volunteers [62]. Correspondingly, both Cmax and serum AUC of PulmoSphere® tobramycin was twice that of inhaled nebulized tobramycin. Ciprofloxacin has also demonstrated effective deposition via PulmonSphere® in healthy males with minimal systemic exposure [85].

Another inhalation platform incorporating low density particles manufactured by spray drying is AIR® technology (Alkemers Inc., Cambridge, MA). Through production of a collapsed, high surface area microparticle, this technology also produces powders with high mean geometric diameter and low aerodynamic diameter. The ability of these particles to avoid the normal clearance mechanisms of the lungs has also been demonstrated in studies investigating the sustained delivery of insulin [20]. Subsequent spray drying studies have shown that control over particle characteristics can be accomplished by changing spray drying parameters and excipient concentration [95]. Currently, phase II clinical trials are underway for evaluation of an anticolinergic, trospium chloride, using this formulation technology.

More recently, a new variety of nanostructured particles have been produced using a rapid freezing technology. This platform produces a low density pharmaceutical matrix that may be redispersed for nebulization [57, 83], dispersed in HFA for delivery via a pMDI [24], or aerosolized via DPI [101]. When dispersed in HFA and aerosolized, the HFA actively templates the dispersed low density (approximately 0.1 g/cm^3) flocculates into an appropriate geometric diameter for inhalation. For aerosolization with a passive DPI, an adequate shearing force was necessary to fracture friable matrices into respirable microparticles [101]. Findings showed that a passive DPI operating at 4 kPa provided sufficient energy for in situ production of a highly respirable, low density matrix particle. This technology is currently being developed by Enavail LLC (Austin, TX) to produced inhaled pharmaceuticals with increased bioavailability.

15.4 Nanoparticle Toxicology

While nanomedicine is an exciting new field showing potential for many therapeutic and diagnostic breakthroughs, mounting concerns have been voiced by regulatory agencies and research groups over insoluble nanoparticles and their effects on biological systems. Indeed, the very same characteristics of high surface area and increased bioavailability that makes nanosized drugs attractive for pharmaceutical applications, also translates into toxicity concerns for nonbiodegradable materials, particularly upon inhalation. Increased surface area of nanomaterials results in increased immunobiological reactivity, which can be further exacerbated by their prolonged tissue retention due to the ability to evade normal physiological clearance mechanisms. Additionally, particles categorized as "ultrafine," having a diameter less than 100 nm, can easily navigate through tight junctions and intracellular space, allowing them to readily reach systemic circulation after deposition in the lungs.

Due to the potential for confusion or overlap between nanoparticulates for drug delivery and insoluble nanoparticles used/produced in industry, it is important to understand the increasing concern surrounding the toxicity of both ultrafine combustion-derived nanoparticles (CDNP) and advanced nanoengineered particles present in industrial environments [79]. Investigations into the tolerability and pulmonary safety of biodegradable materials commonly used in drug delivery are also discussed and differentiated from nonbiodegradable inhaled particulates.

15.4.1 Combustion-Derived Particles

Much attention and growing concern has been focused on nanoparticles that reach the lung unintentionally due to respirable environmental pollutants. While the adverse effects of industrial air pollution have been known since the beginning of industrialized society, only now are researchers determining the dangers of inhaled particulate matter, specifically particles in the nanoscale. The majority of toxicology research conducted on inhaled particles is focused on CDNP that are encountered everyday, and play an import role in respiratory and cardiovascular health. Just one cubic centimeter of urban air contains approximately 10^7 particles with a diameter less than 300 nm, many of which are carbon particulates produced as byproducts from combustion engines [58]. These inhaled toxics include carbon black and diesel particulates and are typically characterized by size as coarse (>2.5 μm), fine (between 2.5 and 0.1 μm), and ultrafine (<0.1 μm). Environmental toxicologists now understand that ultrafine combustion-derived particles, which are by definition in the nano-range, are responsible for the majority of the untoward effects associated with particulate inhalation [64, 78]. In a clinical study, 27 asthmatics exposed to aerosolized particles with varying levels of ultrafines proved to have a more decreased peak expiratory flow after inhalation of increased levels of ultrafine particles [70]. Inhalation of CDNP has been proven to cause exacerbations of preexisting conditions such as asthma and COPD as well as increasing pulmonary inflammation in health subjects. Pulmonary epithelial cells and leukocytes (e.g., monocytes, macrophages) can be significantly disrupted by ultrafine particulates resulting in inhibition of phagocytosis and uncontrolled inflammation. When investigating the local toxicological effects of inhaled carbon black Printex 90 particles, scientist found that cytochrome P450 1B1 was downregulated up to 130-fold [22], which would severely limit the ability of the lungs to fight opportunistic infection. The oxidative stress imparted by combustion-derived particulates results in proliferation of pro-inflammatory cytokines IL-6, IL-8, and TNF-α, ultimately leading to airway constriction [107].

Additionally, disruption of the lung's natural defense mechanisms through CDNP has also been associated with tumor formation and malignancy.

Unfortunately, toxicity of inhaled ultrafines is not limited to pulmonary toxicity. Many particles smaller than 100 nm have the capability to circumvent the epithelial cell barrier of the lungs, thus entering systemic circulation fully intact. Translocation of inhaled CDNP to circulation has been shown to significantly increase systemic inflammation, which can lead to atherosclerosis, thrombosis, and infarction [86]. Accumulation of these nanoparticles in highly vascular tissues, specifically those of the reticuloendothelial system, may also lead to organ-specific toxicity. Within systemic circulation, disruption of the normal function of the vascular endothelial has also been shown to occur after CDNP exposure, most likely due to endogenous fibrinolysis [19].

15.4.2 Nanoengineered Particles

Recently, new engineering practices in nanotechnology have been developed to improve a host of everyday products: electrical circuits, composite materials, sensors, medical devices, and drug delivery devices. By manipulation of these materials on the small scale, the material properties of the substance, such as electrical conductance, material strength, or drug loading capacity, may be tailored to suit the needs of their application. Most nanoengineered materials are constructed from elements that render them nonbiodegradable, enabling them to maintain structure and function in nearly all natural environments. Unfortunately, the robust nature of these particles combined with their ability to evade normal biological elimination mechanisms have lead to growing concerns surrounding the occupational health of individuals exposed to these materials. Development of advanced nanoengineered particles such as carbon nanotubes, quantum dots, gold nanoparticles, and fullerenes have enabled advances in a broad array of fields; however, recent investigations in inhaled nanoparticle toxicology have revealed potential health consequences of these particles if not properly contained [73].

Carbon nanotubes are insoluble cylindrical particles made up of one or multiple layers of carbon atoms. Although proven to be quite useful to impart mechanical strength or electrical conductance to a material, these fiber-like particles have been proven to elicit inflammation in the lungs [99] as well as thrombosis and DNA damage systemically [49]. Moreover, the high aspect ratio and agglomeration tendency of these particles allow them to reach the lung periphery when inhaled, not unlike asbestos. Studies in animal models have demonstrated a potential link between inflammation and granuloma in the abdominal cavity of mice given intraperitoneal injections of carbon nanotubes. The pathology of the carbon nanotube induced inflammatory response in mesothelial lining of the murine chest is noted as having a striking similarity to the immunological response to asbestos [72, 87]. While further studies must be conducted to confirm carcinogenicity in humans [45], these studies have fed the growing concern that nanomaterial have many beneficial applications; however, like asbestos, could present serious health risks.

Other nanoscale particulates have also demonstrated toxic properties in the lungs and major tissues that are cause for concern and further investigation. Fullerens, hollow hexagnal structures containing approximately 60–80 carbon atoms, have many applications as catalysts, lubricants, and cosmetics; however, they have been associated with cytotoxic and DNA damaging effects, particularly when exposed to light [73]. In a gold nanoparticle study conducted in a alveolar type-II cell line, it was demonstrated that contaminants, such as sodium citrate, on the surface of gold nanoparticles could have cytotoxic effects and increase release of lactate dehydrogenase [94]. Similarly, Titanium oxide nanoparticles have proven to induce lung injury by activation of inflammatory pathways in mice [9]; and more recently caused tumor conversion to malignancy in a benign tumor mouse model [65].

Due to their highly insoluble and nonbiodegradable properties, engineered nanoparticles have the ability to reside in biological systems for prolonged periods of time. While composed of material that is typically nontoxic in normal size ranges, the combination of surface area, rapid tissue distribution, and prolonged residence time of nanoengineered particles has shown potential to cause many adverse effects in animal models and humans.

15.4.3 Regulatory Stance

Concerns in fields of environmental and occupational health combined with the growing catalog of literature covering nanopharmaceuticals have lead to new initiatives by the FDA to better regulate products containing nanomaterial. In 2006, the FDA assigned a Nanotechnology Task Force the duty of creating guidances to assist in the design of studies for physicochemical and biological characterization, specific to nanomaterials. Particular interest has been dedicated to the standardization of testing to determine the ADME for nanopharmaceuticals and the identification of the basic material properties that influence them [112]. To address regulatory policy, the task force is also investigating issues of identification of nanomaterial, the agency's range of authority, labeling of products containing nanomaterial, and potential overlap with environmental policy [25]. The FDA is still in the early stages of developing polices to regulate this rapidly expanding area of drug delivery, and are currently investigating nanoparticle formulations on a case by case basis.

15.4.4 Bioerodable Nanoparticles

As a consequence of the extensive research surrounding environmental toxicology and the unsettling realization of the toxicity of CDNP, the notion of inhaling nanoparticles for therapeutic purposes may seem counterproductive. Many researchers are investigating nanoparticles for delivery to the lungs based on the same principles that make them so toxic in the cases mentioned above. In short, nanoparticles can increase the bioreactivity and biodistribution of a given substance. If that substance is a known pharmaceutical agent with a detailed safety profile, reformulation as a nanoparticulate may allow for enhanced potency and increased distribution in targeted tissues.

The root cause of nanoparticle toxicity in the lungs is insolubility. Unlike combustion-derived and metallic particles, drug particles that are soluble, erodible, or biodegradable will often not induce an immunological response, unless the material is fundamentally toxic on the molecular level. Indeed, current dry powder inhalers marketed to treat asthma, COPD, and influenza all deliver microparticles to the lungs and must undergo dissolution to impart a therapeutic effect. As microparticles dissolve they must pass through the nanoparticulate range before completely dissociating. This common-sense approach provides a logical rationale that particle size alone cannot impart toxicity.

In order to evaluate the inflammatory potential of a common nanomaterial for drug delivery to the lungs, Dailey et al. investigated pulmonary delivery of biodegradable and nonbiodegradable polymeric nanoparticles in mice [12]. Theoretical doses of either PLGA or polystyrene were administered to anesthetized BALB/c mice by intratracheal instillation and compared with the inflammatory effect of a glucose solution control. To characterize the inflammatory response, multiple signaling and response factors were quantified in murine bronchoalveolar lavage fluid including ingress of inflammatory white blood cells, specifically eosinophils, neutrophils, and basophils. Figure 15.5a shows that at 24 h after instillation of two different doses of polystyrene nanoparticles, inflammatory cell (polymorphonucleocyte) infiltration proved to be significantly higher than the control group. Alternatively, bronchoalveolar lavage cell counts of PLGA dosed mice showed no significant increase in inflammatory cells. After 14 days, polystyrene nanoparticles in the 75 nm size range proved to persist in induction of heightened inflammatory cell levels, suggesting that these particles are impervious to normal pulmonary clearance mechanisms (Fig. 15.5b). This finding supports the conclusions by environmental toxicologists that insoluble ultrafines can avoid, and in some cases inhibit, alveolar macrophage clearance [22].

When evaluated for their potential for inflammatory action upon deposition on the pulmonary mucosa, both gelatin and human serum albumin based nanoparticles proved to be nonimmunogenic. After administration of albumin and gelatin particles, lactate dehydrogenase and interleukin 8 levels in cell cultured pulmonary epithelial cells proved to be similar to those found in controls. Conversely, polycylcyanoacrylate nanoparticles showed heightened toxicity dependant on the length of its alkyl sidechain [5]. Although unproven in large clinical trials, these studies suggest that biodegradable and immunologically inert materials delivered as nanoparticles to the lungs will not have adverse effects on lung physiology or function.

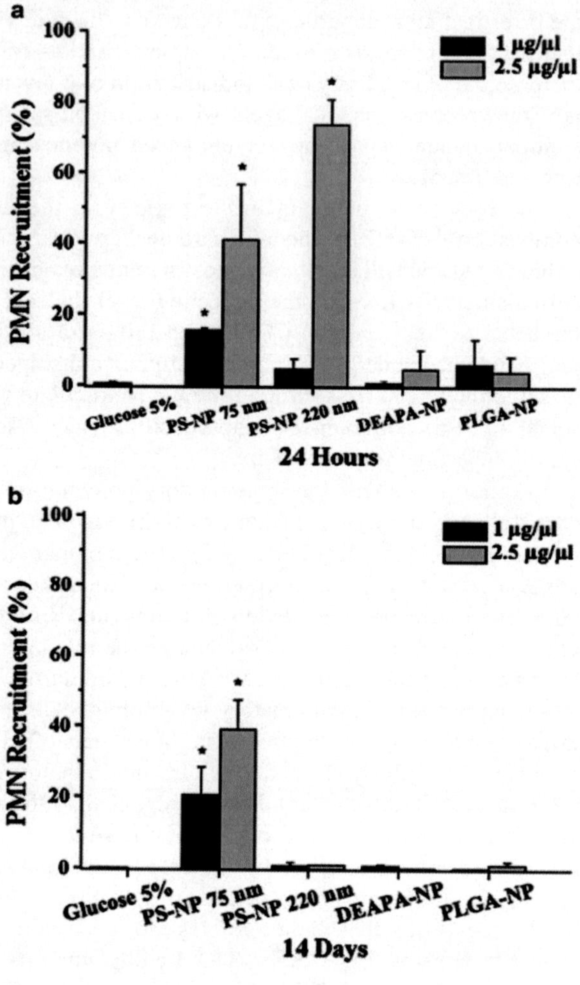

Fig. 15.5 Inflammatory cell recruitment at (**a**) 24 h and (**b**) 14 days after particle instillation. *Asterisks* denotes significant difference from the glucose control ($P<0.05$). *PMN* polymorphonucleocyte; *PS-NP* polystyrene nanoparticles; *DEAPA-NP* poly(lactic-co-glycolic acid) derivative nanoparticles; *PLGA-NP* poly(lactic-*co*-glycolic acid) nanoparticles (reprinted by permission from Dailey et al. [12])

15.5 Conclusion

Nanoparticle inhalation as a means to deliver pharmaceuticals to the lungs shows great potential to enhance therapy for pulmonary and systemic indications by increasing solubility, improving permeability, and allowing for sustained release of deposited drug to the pulmonary epithelium and systemic circulation. While a substantial amount of laboratory research has been conducted in the field of drug nanoparticles, there is a clear lack of nanoparticle-specific inhalation models and clinical trials to support benefits of nanoparticle delivery seen in vitro.

Most research involving ADME of inhaled nanoparticles has not been to determine efficacy or PK/PD, but has been to evaluate the potential toxicity of nontherapeutic nanoparticles not intended for inhalation. Unfortunately, this has lead to a public misconception of the therapeutic potential of inhaled nanoparticles for the purposes of drug delivery. Recently, additional attention has been dawn to the potentially deadly effects of inhaled nanoparticles in the workplace due to the tragic death of two Chinese factory workers [84]. Even though toxicologists disagree in this case as to whether these deaths were caused by nanoparticle inhalation or toxic fume inhalation, some claim these to be the first incidences of death due to occupational exposure to nanoparticles. While cases such as these highlight the growing need for proper industrial regulations to limit unwanted nanoparticle exposure, these findings should not be used to draw inference on the safety of the intentional delivery of therapeutic drug nanoparticles to the lung.

A variety of technologies have been applied to effectively manufacture and deliver nanoparticles to the lung through nebulization and dry powder inhalation. Future studies will need to prove the stability and scalability of these production methods as well as determine the influence of drug nanoparticles (as compared micronized particles) on the bioavailability and ADME of drugs and excipients in animals and humans. While the successful application of nanotechnology to pharmaceuticals has thus far been limited to oral and intravenous drug formulations, the body of research surrounding nanoparticles for pulmonary delivery suggests that this technology could provide improved therapy for inhaled drug products as well.

References

1. Arnold MM, Gorman EM, Schieber LJ, Munson EJ, Berkland C (2007) NanoCipro encapsulation in monodisperse large porous PLGA microparticles. J Control Release 121:100–109
2. Azarmi S, Roa WH, Löbenberg R (2008) Targeted delivery of nanoparticles for the treatment of lung diseases. Adv Drug Deliv Rev 60:863–875
3. Beck-Broichsitter M, Gauss J, Packhaeuser C, Lahnstein K, Schmehl T, Seeger W, Kissel T, Gessler T (2009) Pulmonary drug delivery with aerosolizable nanoparticles in an ex vivo lung model. Int J Pharm 367:169–178
4. Behr J, Zimmermann G, Baumgartner R, Leuchte H, Neurohr C, Brand P, Herpich C, Sommerer K, Seitz J, Menges G, Tillmanns S, Keller M (2009) Lung deposition of a liposomal cyclosporine A inhalation solution in patients after lung transplantation. J Aerosol Med Pulm Drug Deliv 22:121–129
5. Brzoska M, Langer K, Coester C, Loitsch S, Wagner TOF, Mallinckrodt CV (2004) Incorporation of biodegradable nanoparticles into human airway epithelium cells–in vitro study of the suitability as a vehicle for drug or gene delivery in pulmonary diseases. Biochem Biophys Res Commun 318:562–570
6. Byron PR (1986) Prediction of drug residence times in regions of the human respiratory tract following aerosol inhalation. J Pharm Sci 75:433–438
7. Chalupa DC, Morrow PE, Oberdoster G, Utell MJ, Frampton MW (2004) Ultrafine particle deposition in subjects with asthma. Environ Health Perspect 112:879–882
8. Chattopadhyay P, Shekunov BY, Yim D, Cipolla D, Boyd B, Farr S (2007) Production of solid lipid nanoparticles suspensions using supercritical fluid extraction of emulsions (SFEE) for pulmonary delivery using the AERx system. Adv Drug Deliv Rev 59:444–453

9. Chen H-W, Su S-F, Chien C-T, Lin W-H, Yu S-L, Chou C-C, Chen JJW, Yang P-C (2006) Titanium dioxide nanoparticles induce emphysema-like lung injury in mice. FASEB J 20:2393–2395
10. Chen J-F, Zhou M-Y, Shao L, Wang Y-Y, Yun J, Chew NYK, Chan H-K (2004) Feasibility of preparing nanodrugs by high-gravity reactive precipitation Int. J Pharm 269:267–274
11. Cook RO, Pannu RK, Kellaway IW (2005) Novel sustained release microspheres for pulmonary drug delivery. J Control Release 104:79–90
12. Dailey LA, Jekel N, Fink L, Gessler T, Schmehl T, Wittmar M, Kissel T, Seeger W (2006) Investigation of the proinflammatory potential of biodegradable nanoparticle drug delivery systems in the lung. Toxicol Appl Pharmacol 215:100–108
13. Dailey LA, Kleemann E, Wittmar M, Gessler T, Schmehl T, Roberts C, Seeger W, Kissel T (2003) Surfactant-free, biodegradable nanoparticles for aerosol therapy based on the branched polyesters. DEAPA-PVAL-g-PLGA. Pharm Res 20:2011–2020
14. Dailey LA, Schmehl T, Gessler T, Wittmar M, Grimminger F, Seeger W, Kissel T (2003) Nebulization of biodegradable nanoparticles: impact of nebulizer technology and nanoparticle characteristics on aerosol features. J Control Release 86:131–144
15. De Jong WH, Borm PJA (2008) Drug delivery and nanoparticles: applications and hazards. Int J Nanomed 3:133–149
16. Dellamary LA, Tarara TE, Smith DJ, Woelk CH, Adractas A, Costello ML, Gill H, Weers JG (2000) Hollow porous particles in metered dose inhalers. Pharm Res 17:168–174
17. Drew RH, Dodds Ashley E, Benjamin DKJ, Davis RD, Palmer SM, Perfect JR (2004) Comparative safety of amphotericin B lipid complex and amphotericin B deoxycholate as aerosolized antifungal prophylaxis in lung-transplant recipients. Transplantation 77:232–237
18. Duddu SP, Sisk SA, Walter YH, Tarra TE, Trimble KR, Clark AR, Eldon MA, Elton RC, Pickford M, Hirst PH, Newman SP, Weers JG (2002) Improved lung delivery from a passive dry powder inhaler using an engineered pulmosphere powder. Pharm Res 19:689–695
19. Duffin R, Mills NL, Donaldson K (2007) Nanoparticles – a thoracic toxicology perspective. Yonsei Med J 48:561–572
20. Edwards DA, Hanes J, Caponetti G, Hrkach J, Ben-Jebria A, Eskew ML, Mintzes J, Deaver D, Lotan N, Langer R (1997) Large porous particles for pulmonary drug delivery. Science 276:1868
21. El-Gendy N, Berkland C (2009) Combination chemotherapeutic dry powder aerosols via controlled nanoparticle agglomeration. Pharm Res 26:1752–1763
22. Elder A, Gelein R, Silva V, Feikert T, Opanashuk L, Carter J, Potter R, Maynard A, Ito Y, Finkelstein J (2006) Translocation of inhaled ultrafine manganese oxide particles to the central nervous system. Environ Health Perspect 114:1172–1178
23. Ely L, Roa W, Finlay WH, Löbenberg R (2007) Effervescent dry powder for respiratory drug delivery. Eur J Pharm Biopharm 65:346–353
24. Engstrom J, Tam J, Miller M, Williams R, Johnston K (2009) Templated open flocs of nanorods for enhanced pulmonary delivery with pressurized metered dose inhalers. Pharm Res 26:101–117
25. FDA (2007) Nanotechnology: a report on the U.S. Food and Drug Administration Nanotechnology Task Force, 25 July 2007. Rockville
26. Forte R, Dibble C (1999) The role of international environmental agreements in metered dose inhaler technology changes: pMDI in transition: CFC to HFA formulations. J Allergy Clin Immunol 104:S216–S270
27. Gao L, Zhang D, Chen M (2008) Drug nanocrystals for the formulation of poorly soluble drugs and its application as a potential drug delivery system. J Nanopart Res 10:845–862
28. Garcia-Fuentes M, Prego C, Torres D, Alonso MJ (2005) A comparative study of the potential of solid triglyceride nanostructures coated with chitosan or poly(ethylene glycol) as carriers for oral calcitonin delivery. Eur J Pharm Sci 25:133–143
29. Grenha A, Seigo B, Remunan-Lopez C (2005) Microencapsulated chitosan nanoparticles for lung protein delivery. Eur J Pharm Biopharm 25:417–437

30. Gumbleton M (2001) Caveolae as potential macromolecule trafficking compartments within alveolar epithelium. Adv Drug Deliv Rev 49:281–300
31. Hadinoto K, Phanapavudhikul P, Kewu Z, Tan RBH (2006) Novel formulation of large hollow nanoparticles aggregates as potential carriers in inhaled delivery of nanoparticulate drugs. Ind Eng Chem Res 145:3697–3706
32. Hadinoto K, Phanapavudhikul P, Kewu Z, Tan RBH (2007) Dry powder aerosol delivery of large hollow nanoparticulate aggregates as prospective carriers of nanoparticulate drugs: effects of phospholipids. Int J Pharm 333:187–198
33. Hernández-Trejo N, Kayser O, Steckel H, Müller RH (2005) Characterization of nebulized buparvaquone nanosuspensions – effect of nebulization technology. J Drug Target 13: 499–507
34. Heurtault B, Saulnier P, Pech B, Proust J-E, Benoit J-P (2002) A novel phase inversion-based process for the preparation of lipid nanocarriers. Pharm Res 19:875–880
35. Hirst PH, Pitcairn GR, Weers JG, Tarara TE, Clark AR, Dellamary LA, Hall G, Shorr J, Newman SP (2002) In vivo lung deposition of hollow porous particles from a pressurized metered dose inhaler. Pharm Res 19:258–264
36. Hu FQ, Yuan H, Zhang HH, Fang M (2002) Preparation of solid lipid nanoparticles with clobetasol propionate by a novel solvent diffusion method in aqueous system and physicochemical characterization. Int J Pharm 239:121–128
37. Hu T, Chiou H, Chan H-K, Chen J-F, Yun J (2008) Preparation of inhalable salbutamol sulphate using reactive high gravity controlled precipitation. J Pharm Sci 97:944–949
38. Hureaux J, Lagarce F, Gagnadoux F, Vecellio L, Clavreul A, Roger E, Kempf M, Racineux J-L, Diot P, Benoit J-P, Urban T (2009) Lipid nanocapsules: ready-to-use nanovectors for the aerosol delivery of paclitaxel. Eur J Pharm Biopharm 73:239–246
39. Iacono AT, Johnson BA, Grgurich WF, Youssef JG, Corcoran TE, Seiler DA, Dauber JH, Smaldone GC, Zeevi A, Yousem SA, Fung JJ, Burckart GJ, McCurry KR, Griffith BP (2006) A randomized trial of inhaled cyclosporine in lung-transplant recipients. N Engl J Med 354:141–150
40. ISO/TS-27687 (2008) ISO Standards, 2008. Nanotechnologies – terminology and definitions for nanoobjects – nanoparticle, nanofibre, and nanoplate, International Organization of Standardization
41. Jacobs C, Müller RH (2002) Production and characterization of a budesonide nanosuspension for pulmonary administration. Pharm Res 19:189–194
42. Janakan N, Senevirantne R (2008) Factors contributing to medication noncompliance of newly diagnosed smear-positive pulmonary turberculosis patients in the district of Colombo, Sri Lanka. Asia Pac J Public Health 20:214–223
43. Jaques PA, Kim CS (2000) Measurement of total lung deposition of inhaled ultrafine particles in healthy men and women Inhal Tox 12:715–731
44. Jensen DMK, Cun D, Maltesen MJ, Frokjaer S, Nielsen HM, Foged C (2010) Spray drying of siRNA-containing PLGA nanoparticles intended for inhalation. J Control Release 142(1):138–145
45. Kane AB, Hurt RH (2008) Nanotoxicology: the asbestos analogy revisited. Nat Nanotechnol 3:378–379
46. Kaur IP, Bhandari R, Bhandari S, Kakkar V (2008) Potential of solid lipid nanoparticles in brain targeting. J Control Release 127:97–109
47. Keck CM (2010) Particle size analysis of nanocrystals: improved analysis method. Int J Pharm 390(1):3–12
48. Kirby AC, Coles MC, Kaye PM (2009) Alveolar macrophages transport pathogens to lung draining lymph nodes. J Immunol 183:1983–1989
49. Kisin ER, Murray AR, Keane MJ, Shi X-C, Schwegler-Berry D, Gorelik O, Arepalli S, Castranova V, Wallace WE, Kagan VE, Shvedova AA (2007) Single-walled carbon nanotubes: geno- and cytotoxic effects in lung fibroblast V79 Cells. J Toxicol Environ Health A 70:2071–2079

50. Kraft WK, Steiger B, Beussink D, Quiring JN, Fitzgerald N, Greenberg HE, Waldman SA (2004) The pharmacokinetics of nebulized nanocrystal budesonide suspension in healthy volunteers. J Clin Pharmacol 44:67–72
51. Kreuter J (2001) Nanoparticulate systems for brain delivery of drugs. Adv Drug Deliv Rev 47:65–81
52. Kwon J-T, Hwang S-K, Jin H, Kim D-S, Minai-Tehrani A, Yoon H-J, Choi M, Yoon T-J, Han D-Y, Kang Y-W, Yoon B-I, Lee J-K, Cho M-H (2008) Body distribution of inhaled fluorescent magnetic nanoparticles in the mice. J Occup Health 50:1–6
53. Leone-Bay A, Grant M (2006) Technosphere technology: a platform for inhaled protein therapeutics. ONdrugDELIVERY, 8–11
54. Luangvala MM, Castillo MC, Bosco AP (2009) A novel submicron formulation of nebulized budesonide significantly decreases the nebulization time from a variety of commercially available jet nebulizers. J Allergy Clin Immunol 123:S81
55. Matteucci ME, Hotze MA, Johnston KP, Williams RO III (2006) Drug nanoparticles by antisolvent precipitation: mixing energy versus surfactant stabilization. Langmuir 22: 8951–8959
56. Matteucci ME, Paguio JC, Miller MA, Williams RO, Johnston KP (2008) Flocculated amorphous nanoparticles for highly supersaturated solutions. Pharm Res 25:2477–2487
57. McConville J, Overhoff K, Sinswat P, Vaughn J, Frei B, Burgess D, Talbert R, Peters J, Johnston K, Williams R (2006) Targeted high lung concentrations of itraconazole using nebulized dispersions in a murine model. Pharm Res 23:901–911
58. Medina C, Santos-Martinez MJ, Radomski A, Corrigan OI, Radomski MW (2007) Nanoparticles: pharmacological and toxicological significance. Br J Pharmacol 150:552–558
59. Mizoe T, Ozeki T, Okada H (2007) Preparation of drug nanoparticle-containing microparticles using a 4-fluid nozzle spray drier for oral, pulmonary, and injection dosage forms. J Control Release 122:10–15
60. Mossman BT, Brom PJ, Castranova V, Costa DL, Donaldson K, Kleeberger SR (2007) Mechanisms of action of inhaled fibers, particles and nanoparticles in lung and cardiovascular diseases. Part Fibre Toxicol 4:1–10
61. Müller RH, Jacobs C, Kayser O (2001) Nanosuspensions as particulate drug formulations in therapy: rationale for development and what we can expect for the future. Adv Drug Deliv Rev 47:3–19
62. Newhouse MT, Hirst PH, Duddu SP, Walter YH, Tarara TE, Clark AR, Weers JG (2003) Inhalation of a dry powder tobramycin pulmosphere formulation in healthy volunteers. Chest 124:360–366
63. Nguyen J, Steele TWJ, Merkel O, Reul R, Kissel T (2008) Fast degrading polyesters as siRNA nano-carriers for pulmonary gene therapy. J Control Release 132:243–251
64. Oberdörster G (2000) Pulmonary effects of inhaled ultrafine particles. Int Arch Occup Environ Health 74:1–8
65. Onuma K, Sato Y, Ogawara S, Shirasawa N, Kobayashi M, Yoshitake J, Yoshimura T, Iigo M, Fujii J, Okada F (2009) Nano-scaled particles of titanium dioxide convert benign mouse fibrosarcoma cells into aggressive tumor cells. Am J Pathol 175:2171–2183
66. Ostrander KD, Bosch HW, Bondanza DM (1999) An in-vitro assessment of a nanocrystal(TM) beclomethasone dipropionate colloidal dispersion via ultrasonic nebulization. Eur J Pharm Biopharm 48:207–215
67. Packhaeuser C, Lahnstein K, Sitterberg J, Schmehl T, Gessler T, Bakowsky U, Seeger W, Kissel T (2009) Stabilization of aerosolizable nano-carriers by freeze-drying. Pharm Res 26:129–138
68. Pandey R, Khuller GK (2005) Solid lipid particle-based inhalable sustained drug delivery system against experimental tuberculosis. Tuberculosis 85:227–234
69. Patton JS, Fishburn CS, Weers JG (2004) The lungs as a portal of entry for systemic drug delivery. Proc Am Thorac Soc 1:338–344
70. Peters A, Wichmann H, Tuch T, Heinrich J, Heyder J (1997) Respiratory effects are associated with the number of ultrafine particles. Am J Respir Crit Care Med 155:1376–1383

71. Plumley C, Gorman EM, El-Gendy N, Bybee CR, Munson EJ, Berkland C (2009) Nifedipine nanoparticle agglomeration as a dry powder aerosol formulation strategy. Int J Pharm 369:136–143
72. Poland CA, Duffin R, Kinloch I, Maynard A, Wallace WAH, Seaton A, Stone V, Brown S, MacNee W, Donaldson K (2008) Carbon nanotubes introduced into the abdominal cavity of mice show asbestos-like pathogenicity in a pilot study. Nat Nanotechnol 3:423–428
73. Powell MC, Kanarack MS (2006) Nanomaterial health effects-part 1: background and current knowledge. Wis Med J 105:16–21
74. Putney SD, Burke PA (1998) Improving protein therapeutics with sustained-release formulations. Nat Biotechnol 16:153–157
75. Rabinow BE (2004) Nanosuspensions in drug delivery. Nat Rev Drug Discov 3:785–796
76. Rave K, Heise T, Pfutzner A, Boss AH (2007) Coverage of postprandial blood glucose excursions with inhaled technosphere insulin in comparison to subcutaneously injected regular human insulin in subjects with type 2 diabetes. Diabetes Care 30:2307–2308
77. Rijnders BJ, Cornelissen JJ, Slobbe L, Becker MJ, Doorduijn JK, Hop WCJ, Ruijgrok EJ, Löwenberg B, Vulto A, Lugtenburg PJ, de Marie S (2008) Aerosolized liposomal amphotericin b for the prevention of invasive pulmonary Aspergillosis during prolonged neutropenia: a randomized, placebo controlled trial. Clin Infect Dis 46:1401–1408
78. Seaton A, Godden D, MacnNee W, Donaldson K (1995) Particulate air pollution and acute health effects. Lancet 345:176–178
79. Seaton A, Tran L, Aitken R, Donaldson K (2010) Nanoparticles, human health hazard and regulation. J Roy Soc Interface 7:S119–S129
80. Sham JOH, Zhang Y, Finlay WH, Roa WH, Löbenberg R (2004) Formulation and characterization of spray-dried powders containing nanoparticles for aerosol delivery to the lung. Int J Pharm 269:457–467
81. Shi L, Plumley C, Berkland C (2007) Biodegradable nanoparticle flocculates for dry powder aerosol formulation. Langmuir 23:10897–10901
82. Shi W, Liu T, Xie L, Wang S (2005) FK506 in a biodegradable glycolide-co-clatide-co-caprolactone polymer for prolongation of corneal allograft survival. Curr Eye Res 30:969–976
83. Sinswat P, Overhoff KA, McConville JT, Johnston KP, Williams RO III (2008) Nebulization of nanoparticulate amorphous or crystalline tacrolimus – single-dose pharmacokinetics study in mice. Eur J Pharm Biopharm 69:1057–1066
84. Song Y, Li X, Du X (2009) Exposure to nanoparticles is related to pleural effusion, pulmonary fibrosis and granuloma. Eur Respir J 34:559–567
85. Stass H, Baumann-Noss S, Delesen H, Nagelschmitz J, Willmann S, Edginton A, Staab D (2008) Pharmacokinetics of ciprofloxacin PulmoSphere® inhalational powder. J Cyst Fibros 7:S26
86. Suwa T, Hogg JC, Quinlan KB, Ohgami A, Vincent R, Eeden SFV (2002) Particulate air pollution induces progression of atherosclerosis. J Am Coll Cardiol 39:935–942
87. Takagi A, Hirose A, Nishimura T, Fukumori N, Ogata A, Ohashi N, Kitajima S, Kanno J (2008) Induction of mesothelioma in p53+/− mouse by intraperitoneal application of multi-wall carbon nanotube. J Toxicol Sci 33:105–116
88. Tam JM, McConville JT, Robert O, Williams I, Johnston KP (2008) Amorphous cyclosporin nanodispersions for enhanced pulmonary deposition and dissolution. J Pharm Sci 97:4915–4933
89. Tsapis N, Bennett D, Jackson B, Weitz DA, Edwards DA (2002) Trojan particles: Large porous carriers of nanoparticles for drug delivery. Proc Natl Acad Sci USA 99:12001–12005
90. Tseng C-L, Su W-Y, Yen K-C, Yang K-C, Lin F-H (2009) The use of biotinylated-EGF-modified gelatin nanoparticle carrier to enhance cisplatin accumulation in cancerous lungs via inhalation. Biomaterials 30:3476–3485
91. Tseng C, Wu SY, Wang W, Peng C, Lin F, Lin C, Young T, Shieh M (2008) Targeting efficiency and biodistribution of biotinylated-EGF-conjugated gelatin nanoparticles administered via aerosol delivery in nude mice with lung cancer. Biomaterials 29:3014–3022
92. U.K.DERFA (2006) UK voluntary reporting scheme for engineered nanoscale materials. Chemicals and Nanotechnologies Division, Defra

93. U.S. FDA (2007) Nanotechnology: a report of the U.S. Food and Drug Administration Nanotechnology Task Force. 25 July 2007 http://www.fda.gov/ScienceResearch/SpecialTopics/Nanotechnology/NanotechnologyTaskForceReport2007/default.htm
94. Uboldi C, Bonacchi D, Lorenzi G, Hermanns MI, Pohl C, Baldi G, Unger RE, Kirkpatrick CJ (2009) Gold nanoparticles induce cytotoxicity in alveolar type-II cell lines A549 and NCIH44I. Part Fibre Toxicol 6:1–12
95. Vanbever R, Mintzes JD, Wang J, Nice J, Chen D, Batycky R, Langer R, Edwards DA (1999) Formulation and physical characterization of large porous particles for inhalation. Pharm Res 16:1735–1742
96. Vaughn JM, McConville JT, Burgess D, Peters JI, Johnston KP, Talbert RL, Williams RO III (2006) Single dose and multiple dose studies of itraconazole nanoparticles. Eur J Pharm Biopharm 63:95–102
97. Venkateswarlu V, Manjunath K (2004) Preparation, characterization and in vitro release kinetics of clozapine solid lipid nanoparticles. J Control Release 95:627–638
98. Wang Z, Chen J-F, Le Y, Shen Z-G, Yun J (2007) Preparation of ultrafine beclomethasone dipropionate drug powder by antisolvent precipitation. Ind Eng Chem Res 46:4839–4845
99. Warheit DB, Laurence BR, Reed KL, Roach DH, Reynolds AM, Webb TR (2004) Comparative pulmonary toxicity assessment of single-wall carbon nanotubes in rats. Toxicol Sci 77:117–125
100. Watts AB, Cline AM, Saad AR, Johnson SB, Peters JI, Williams RO (2010) Characterization and pharmacokinetic analysis of tacrolimus dispersion for nebulization in a transplanted rodent model. Int J Pharm 384:46–52
101. Watts AB, Son Y-J, McConville JT, Williams RO (2009) Nanostructured tacrolimus produced by ultra-rapid freezing for dry powder inhalation. AAPS Annual Meeting, Los Angeles
102. Watts AB, Williams RO (2010) Formulation and production strategies for enhancing bioavailability of poorly absorbed drugs. In: Rogge MC, Taft DR (eds) Drugs and the pharmaceutical sciences: preclinical drug development. Informa Healthcare, New York, pp 161–195
103. Weiwei Z, Takahisa M, You Han B, Ramesh R, Alfred KC, Steven EK (2006) Development of a sustained-release system for perivascular delivery of dipyridamole. J Biomed Mater Res B Appl Biomater 77B:135–143
104. Wiedmann TS, DeCastro L, Wood RW (1997) Nebulization of nanocrystals: production of a respirable solid-in-liquid-in-air colloidal dispersion. Pharm Res 14:112–116
105. Yamamoto H, Hoshina W, Kurashima H, Takeuchi H, Kawashima Y, Yokoyama T, Tsujimoto H (2007) Engineering of poly(DL-lactic-*co*-glycolic acid) nanocomposite particles for dry powder inhalation dosage forms of insulin with spray-fluidized bed granulating system. Adv Powder Technol 18:215–228
106. Yamamoto H, Kuno Y, Sugimoto S, Takeuchi H, Kawashima Y (2005) Surface-modified PLGA nanosphere with chitosan improved pulmonary delivery of calcitonin by mucoadhesion and opening of the intercellular tight junctions. J Control Release 102:373–381
107. Yang H-M, Ma JYC, Castranova V, Ma JKH (1997) Effects of diesel exhaust particles on the release of interleukin-1 and tumor necrosis factor-alpha from rat alveolar macrophages. Exp Lung Res 23:269–284
108. Yang JZ, Young AL, Chiang P, Thurston A, Pretzer DK (2008) Fluticasone and budesonide nanosuspensions for pulmonary delivery: preparation, characterization, and pharmacokinetic studies. J Pharm Sci 97(11):4869–4878
109. Yang W, Peters JI, Williams RO III (2008) Inhaled nanoparticles – a current review. Int J Pharm 356:239–247
110. Yang W, Tam J, Miller DA, Zhou J, McConville JT, Johnston KP, Williams RO III (2008) High bioavailability from nebulized itraconazole nanoparticle dispersions with biocompatible stabilizers. Int J Pharm 361:177–188
111. Zahoor A, Sadhna S, Khuller GK (2005) Inhalable alginate nanoparticles as antitubercular drug carriers against experimental tuberculosis. Int J Antimicrob Agents 26:298–303
112. Zolnik BS, Sadrieh N (2009) Regulatory perspective on the importance of ADME assessment of nanoscale material containing drugs. Adv Drug Deliv Rev 61:422–427

Chapter 16
Pulmonary Delivery of Plasmid DNA for Disease Prevention and Therapy

Simon Heuking and Gerrit Borchard

Abstract For gene delivery to the lung, the challenges are high, but successful treatment of cystic fibrosis or achieving immunity against the global infectious diseases provide an allure that cannot be ignored. This chapter summarizes and reviews nonviral DNA delivery for both gene therapy and DNA vaccination in the lung. Aerosolization of DNA is evaluated, and the stability during this process is discussed. Carriers for DNA are then discussed including lipoplexes and polyplexes, with a particular focus on systems that achieve good transfection and minimize potential toxicity. Then principles of DNA vaccination are introduced, and the advantages of pulmonary vaccination are discussed. Finally, the transport of plasmid DNA vaccines into the lungs is reviewed.

Keywords DNA delivery • DNA vaccination • Gene therapy • Non-viral gene delivery • Pulmonary route of application

16.1 Introduction

As such, the concept of gene delivery to the lung for therapeutic as well as preventive purposes appears to be straight forward: relatively easy application meeting with patient compliance, direct accessibility of the target tissue, low enzymatic activity (compared to the oral route of application) and absence of pH gradient. However, as with all "simple" concepts, the devil is in the details. In this case, as somebody put it correctly in the three challenges afflicting all efforts in drug development: *Delivery*, *Delivery*, and *Delivery*. The first delivery aspect is to deposit an aerosol DNA bearing particles at the correct site in the lung, maintaining DNA

G. Borchard (✉)
School of Pharmaceutical Sciences, University of Geneva, University of Lausanne, 30, quai Ernest Ansermet, 1211 Geneva, Switzerland
e-mail: gerrit.borchard@unige.ch

H.D.C. Smyth and A.J. Hickey (eds.), *Controlled Pulmonary Drug Delivery*, Advances in Delivery Science and Technology, DOI 10.1007/978-1-4419-9745-6_16,

integrity in lieu of shear forces necessary to produce aerosol droplet sizes suitable for inhalation. The second is to enable the uptake of DNA carrier systems by, if possible, the "correct" target cells. The last aspect is related to obtaining successful *transfection*, i.e., the successful expression and processing of the protein encoded in the DNA delivered. For all three of these stages, technologies have been developed over several decades. In this chapter, we would like to focus on supplying the reader with an overview over some aspects of – nonviral – DNA delivery for both gene therapy and DNA vaccination in the lung.

16.2 Aerosolization of pDNA

The first report of pulmonary delivery of aerosolized DNA, complexed with cationic liposomes, was described by Stribling et al. [63]. Using DOTMA (*N*-[1-(2,3)-dioleyl-oxy)propyl]-*N,N,N*-triethylammonium), in combination with the fusigenic lipid DOPE (dioleoyl-phosphatidylethanolamine) to form complexes with a pDNA encoding for chloramphenicol acetyl transferase (CAT), successful transfection of pulmonary tissues in mice was achieved in the absence of toxic side effects. In this case, the aerosolization was achieved by nebulization in what turned out to be a rather inefficient process. Application of aerosol had to be maintained for a total duration of 4–5 h, aerosolizing 6 mg of total DNA, of which only about 10 μg was finally deposited in the lung. However, this study sparked the interest in aerosol delivery of pDNA for a number of genetic diseases, including cystic fibrosis (CF) [2, 50], lung cancer [27], and inflammatory afflictions of the lung, such as asthma [2].

From these first trials, it became also clear that the nebulization of a large and relatively fragile molecule poses a challenge. Especially seen the forces necessary to produce droplet sizes suitable for inhalation (about 2 μm), shear stress occurring during nebulization is to be minimized in order to maintain structural integrity of the DNA delivered [15]. A recent paper by Lentz et al. [41] investigated the effect of the strain rate on the integrity of plasmid DNA (pDNA) aerosolized from different delivery systems (ultrasonic, vibrating mesh and jet nebulizer, electrostatic spray). The strain rate γ is defined as the derivative of velocity (d*u*) in correlation to linear distance traveled (d*y*) [56]:

$$\gamma = \mathrm{d}u / \mathrm{d}y.$$

The study revealed the lowest destabilizing effect for pDNA aerosolized from electrostatic sprays, and the highest for jet nebulizers. In spite of a relatively low strain rate, also vibrating mesh nebulizers exerted a destabilizing effect on plasmid DNA, possibly due to the interaction of the molecule with the vibrating grid [41]. As was already shown, DNA degradation in ultrasonic nebulizers was due to cavitation, i.e., the collapse of air bubbles creating shock waves, which can damage the DNA [42]. In contrast to aerosolized "naked" DNA, complexation with positively charged molecular entities, condensing and compacting DNA, has been shown to lead to a stabilization of the DNA delivered, besides resulting in a potential increase in transfection efficiency observed [57].

16.3 Lipoplexes and Polyplexes

In nonviral delivery, DNA is usually condensed by electrostatic interaction with either cationic lipids to form so-called lipoplexes or cationic polymers (polyplexes). Examples for the former, in addition to the above mentioned DOTMA/DOPE liposomes, are 3-[*N*-[(*N′*,*N′*-dimethylamino)ethane]carbamoyl]cholesterol (DC-cholesterol), used for the application of a plasmid expressing for the cystic fibrosis transmembrane conductance regulator (CFTR) protein in mice [1]. Since then, a number of lipoplex formulations have been developed and successfully tested in animals [26, 49].

In clinical trials, however, lipoplex-mediated transfection of pulmonary tissues has suffered from comparably low efficiency [6], as well as proinflammatory side effects [58]. The recent use of an improved lipid, the Genzyme lipid 67 (GL-67), called the "gold-standard" [25] in pulmonary gene delivery, has resulted in an increased transient transfection and expression of a reporter gene after instillation in mouse lungs [40]. The lipid, consisting of a spermine headgroup covalently attached to cholesterol "anchor," used in a liposomal formulation including dioleoylphosphatidyl ethanolamine (DOPE), dimyristoylphosphatidyl ethanolamine (DMPE), and polyethylene glycol (PEG) has also been tested in clinical studies. In CF patients, treatment resulted in significant, though partial, restoration of the chloride channel activity disturbed in CF by the expression of the CFTR protein [2, 3].

Another cationic lipid, guanidinium-cholesterol-bis-guanidinium-tren-cholesterol (BGTC), showed relatively high transfection efficiency in vitro and in vivo when used in combination with DOPE [16]. Applied as an aerosol to mice, expression of the reporter gene CAT was however lower than for polyplexes prepared with polyethyleneimine (PEI), applied by the same route [22]. In addition, cytokine (TNF-alpha, IL-1 beta) levels measured in the bronchoalveolar fluid (BALF) after exposure were higher for BGTC-based systems than for the PEI polyplexes.

Inflammatory side effects in lipoplex-mediated gene delivery in the lung have not been attributed to the cationic lipids themselves, but more to the presence of (and exposure to) CpG sequences absent in mammalian, but present in bacterial plasmid DNA. These sequences have been identified as ligands of the Toll-like receptor (TLR) 9 [62]. TLR belongs to a group of pathogenic pattern recognition receptors that are a part of the innate immune system [30]. TLR ligands have therefore recently been introduced as novel adjuvants to boost the mucosal immune response following pulmonary application of vaccines (see below).

Polymers have long been used for condensing DNA, forming condensates ("polyplexes") held together by electrostatic forces between the negatively charged DNA and positively charged polymers [36]. This condensation both reduces the size of the DNA delivered, and masks its negative charge, both factors making DNA amenable

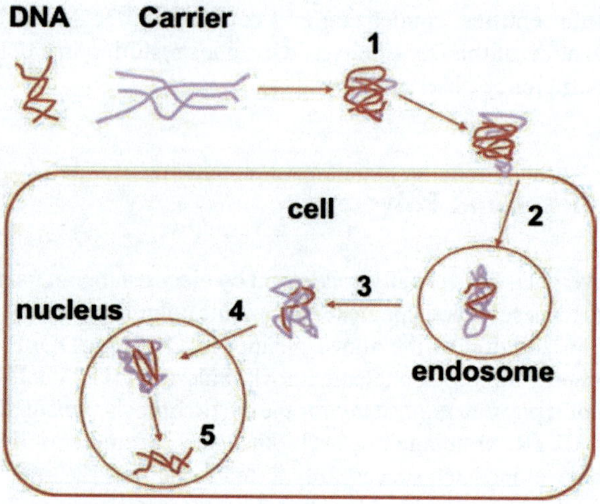

Fig. 16.1 Concept of polyplex-mediated transfection: *1*: formation of polyplex by electrostatic condensation of DNA and polymer; *2*: polyplex endocytosis and endosomal uptake; *3*: "endosomal escape"; *4*: transport through cytoplasm and nuclear localization; *5*: intracellular dissociation of plasmid DNA from polymer

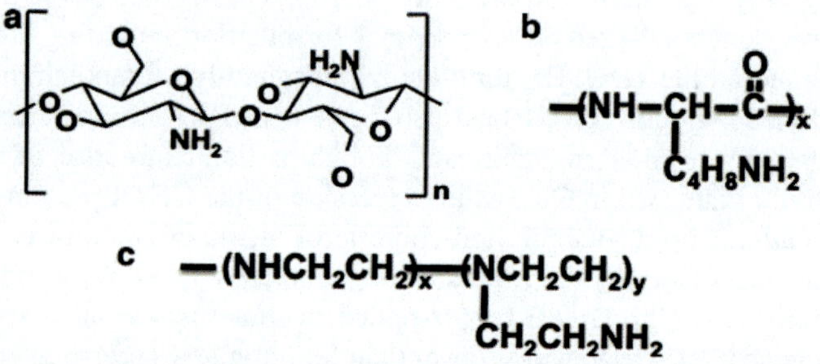

Fig. 16.2 Molecular structures of (**a**) chitosan, (**b**) poly-L-lysine (PLL), and (**c**) polyethyleneimine (PEI)

for endocytosis by target cells. Cationic polymers, especially PEI, may also assist in the "endosomal escape," i.e., the release of intact plasmid DNA from the endosomal compartment [12], as depicted in Fig. 16.1. In this case, the polymer acts as a "proton sponge," by virtue of protonation of its multiple amino groups. This, in turn, buffers the pH in the endosomal compartment, causing the endosomal membrane to rupture and release the polyplex into the cytoplasm [53]. In addition, condensation with polymers was shown to stabilize DNA against enzymatic degradation [13].

Examples of widely used polymers in nonviral gene delivery are poly(L-lysine) (PLL) [39, 43], PEI [12], poly(2-(dimethylaminoi)ethylmethacrylate) (pDMAEMA) [67], as well as biodegradable polysaccharides such as chitosan and its derivatives [11, 48]. Among these, PEI, PLL, and chitosan (Fig. 16.2) and respective derivatives have been shown to successfully transfect lung epithelial cells in vitro and in vivo.

PEI shows a high density of primary, secondary, and tertiary amine function in its structure (Fig. 16.2), and is therefore suited to condense DNA plasmids and act as a proton sponge for lysosomal escape and facilitated intracellular trafficking of the vector. PEI has therefore been considered as a nonviral delivery system in gene therapy of CF [20], a genetic disease caused by the aberration of a chloride channel protein, the CFTR protein, expressed in mucosal epithelia in the intestinal tract and the lung [46]. As the cause of fatality is linked to the pulmonary effects of CF, gene therapy through application of the correcting gene by aerosolization was attempted early on. In this regard, PEI was shown to protect DNA against degradation during the aerosolization process [17], to lead to a higher transfection rates compared to most cationic lipids [16], with transfection using PEI polyplexes not inhibited by lung surfactant [19]. However, PEI is significantly cytotoxic, which appears to depend on the density of the positive charges as well as on the molecular weight of the polymer [33], and the degree of molecular branching [70]. Cytotoxicity of PEI and aggregation tendency of PEI/DNA complexes could greatly be reduced by grafting PEI with PEG, with reduction in cytotoxicity being dependent on the degree of PEG grafting, and not on the molecular weight of PEG [52, 55].

In a recent study [65], branched PEI was compared to a newly developed copolymer from oligo(ethylene glycol) methyl ether methacrylate (OEGMA) and poly *N,N*-dimethylaminoethylmethacrylate (pDMAEMA) in terms of toxicity and transfection efficiency after intratracheal application in Balb/c mice. The copolymer could prevent polyplex aggregation at high concentrations in isotonic solutions, rendering formulations useful for in vivo application. Gene expression in vivo apparently improved by sevenfold when compared to branched PEI.

Condensation of plasmid DNA by poly-L-lysine (PLL) appears to be dependent on the interaction of the primary amine groups of the polymer with negatively charged DNA. For the formation of stable polyplexes, the number of amine groups available appears to be of crucial importance. Polyplexes formed with PLL below a molecular weight of 3 kDa were shown to be unstable [38]. On the other hand, polyplexes prepared with high molecular weight PLL tended to form aggregates in isotonic solutions [43] and exerted considerable cytotoxicity [14]. To overcome these challenges, PLL was grafted with PEG, resulting in an A–B-type block copolymer [71]. Polyplexes formed with these copolymers maintained their stability in serum, and can be applied in suspensions at a concentration exceeding 12 mg/ml DNA in isotonic saline [44]. Successively, PEG-PLL polymers were employed in preclinical studies in mice, transfecting 60–75% of epithelial cells lining the bronchial tract after a single intratracheal application [75]. Based on these favorable data, a first clinical trial was initiated [35], testing the application of a plasmid expressing for the CFTR protein and condensed with PEGylated PLL (CK_{30}) in a total of 12 CF patients in a placebo-controlled, double-blind, dose-escalation intranasal study. Primary endpoints – safety and tolerability – were met, and the secondary endpoint – gene transfer efficiency – was assessed as well. A partial to complete response was seen in eight subjects, and correction persisted for 6 days, for 1 patient even up to 4 weeks after single nasal application of the polyplex formulation.

Chitosan is a linear, cationic polysaccharide consisting of randomly distributed β-(1-4)-linked D-glucosamine (deacetylated unit) and *N*-acetyl-D-glucosamine (acetylated unit). It is industrially produced by alkaline deacetylation of chitin, which is the structural element in the exoskeleton of crustaceans [61]. The degree of deacetylation of chitosan ranges from 40 to 98%. Different molecular weights (range of >100,000 to <2,000 Da) and viscosity grades are on the market. The primary amino function affords the possibility of easily altering the chemical modification of chitosan. The pK_a value of this amino group was determined to be around 6.5 and does not vary significantly, even for different degrees of *N*-acetylation. Hence, chitosan is positively charged and soluble in acidic solutions, whereas unmodified chitosan is insoluble at physiological pH values (7.2–7.4). Chitosan has been considered to serve as a vector for gene delivery [60] due to biocompatibility and biodegradability [5]. A study dedicated to unravel the relationship between the structure of chitosan, i.e., its molecular weight and degree of deacetylation, its toxicity in vitro, and transfection efficiency after intratracheal administration in mice was performed recently [34]. It was shown that the percentage of deacetylation must be greater than 65% to achieve stable complexes with plasmid DNA, and achieve transgene expression. In terms of acute toxicity, measured by incubation with 293 cells, an IC50 value of 630 mg/ml was measured for ultrapure chitosan (UPC) of a deacetylation degree of 83%, in comparison to an IC50 value of just 75 μg/ml for PEI of a molecular weight of 800 kDa [24]. After intratracheal instillation, polyplexes were mostly found in the airways of the central lung. PEI transfection was seen to be more effective and longer lasting than with UPC, which itself was comparable to cationic lipids. It was considered to be safe for pulmonary application, provided chitosan is used in its ultrapure form [31].

16.4 Principles of Plasmid DNA Vaccination

With an increasing knowledge of the immune system and its molecular answers to infections, it becomes more and more obvious that antibody inducing vaccines might not be the appropriate solution to intracellular infections (such as tuberculosis, influenza, hepatitis, or HIV/AIDS) requiring for their prevention a potent cytotoxic T lymphocyte response (CTL). Plasmid deoxyribonucleic acid (DNA) vaccination might be an answer to that due to its hallmark of inducing a strong CTL response in orchestration with CD4+ T helper cells (cellular immunity) as well as its generation of antibodies (humoral immunity).

In general, a plasmid DNA vaccine consists of a bacterial plasmid vector into which a gene is inserted encoding for one or more antigenic protein(s). When compared to gene therapy, plasmid DNA vaccination is supposed to be already efficient at a relatively low level of gene expression. Plasmid DNA vaccines are produced in bacteria (such as *Escherichia coli*) and after purification injected into the host [29].

Wolff et al. [72] reported for the first time that an injection of plasmid DNA (encoding for the bacterial enzyme beta-galactosidase) into muscle cells can result in lasting protein expressions, even after 2 months of injection. This technique of in situ expression of protein(s) was then applied for eliciting an antibody response by Tang et al. [64]. They were the first to demonstrate that an injection of plasmid DNA by gene gun is able to elicit an immune response in vivo against the delivered human growth hormone. Further studies in mice and chicken injected with influenza plasmid DNA demonstrated protection against following viral challenge [21, 66]. Starting from these days, several plasmid DNA vaccines were found to provide protective immunity in various animal models [37]. In comparison to current protein vaccines, plasmid DNA vaccination exhibits many advantages [9, 37], such as:

1. Favoring a cellular immune response, which is most desired for the prophylaxis against intracellular pathogens. This cell-mediated immunity is developed thanks to an induction of major histocompatibility complex (MHC) I-restricted CTLs, also killer T cells termed. CTLs are a result of in situ DNA transfection, subsequent intracellular processing of polypeptides, followed by a presentation of the corresponding epitope(s) to antigen-presenting cells (APC).
2. Allowing the genetic construction of multiple antigens of choice included into the same vector, which in turn leads to transcription of different antigenic proteins (epitopes) in situ by one single vaccination.
3. Possessing an intrinsic adjuvant, unmethylated 5′-deoxycytidine-phosphate-guanosine (CpG)-motifs, which are stimulating the innate immune system via the TLR-9. Activation of TLR-9 leads to a favorable T helper cell type 1 (Th1) biased immune response with induction of proinflammatory cytokines.
4. Prolonging the expression of the antigenic protein(s) enables a continuous stimulation of the immune response.
5. Being easily produced, up-scaled and stored at higher temperatures without causing loss of activity. These properties render DNA vaccines economically very attractive and are certainly an advantage for their transport, especially to countries in the southern hemisphere.
6. DNA vaccination is considered as safe immunization for immunocompromised hosts as it does not induce vector immunity (in contrast to viral or bacterial delivery systems) and therefore can be used for repeated boosting.

However, in order to successfully transfect host cells, plasmid DNAs have in general to overcome a couple of extra and intracellular barriers before desired antigen(s) can be expressed (see Fig. 16.3).

In addition to these barriers, potential safety issues raised [9, 37] and have to be addressed at preclinical and clinical level:

- Integration of plasmid DNA into the host genome
- Immunological tolerance to the encoded antigen
- Development of anti-DNA antibodies or autoimmunity disorders against DNA
- Development of antibiotic resistance

Despite promising results of plasmid DNA immunizations in preclinical trials [28, 37, 66], studies in nonhuman primates and humans have failed so far in achieving protective immunity [32, 68, 69]. Consequently, amelioration strategies

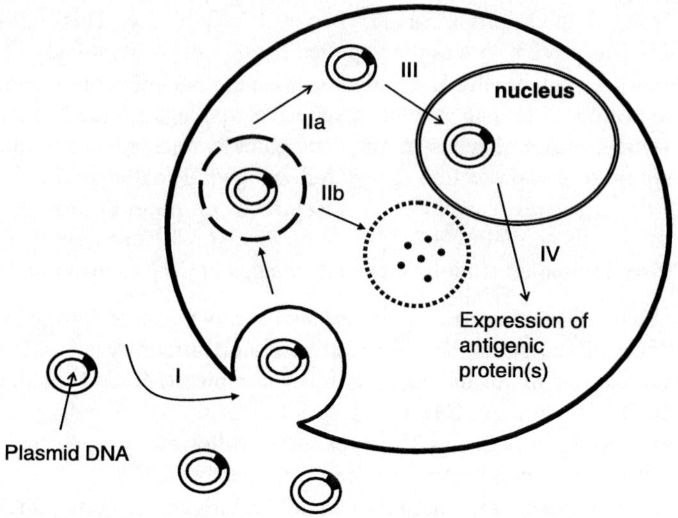

Fig. 16.3 Potential barriers for the transport of plasmid DNA into the nucleus. (*I*) Plasmid DNA has to withstand enzymatic degradation by DNases (extra and intracellularly). (*IIa*) Endocytosed plasmid DNA needs to escape from the endosomal compartment; otherwise, it will be degraded within the lysosome due to an acidic pH and presence of digestive enzymes (*IIb*). (*III*) In order to enter the nucleus, plasmid DNA has to travel through the cytosol by diffusion. (*IV*) Transcription and subsequent translation steps for antigenic protein(s) have to be taken

of DNA vaccination were exploited, ranging from plasmid optimization over coformulation with adjuvants to changing the route of administration to, e.g., pulmonary vaccination.

16.4.1 Advantages of Pulmonary Vaccination

Most known pathogens invade the human body through mucosal epithelia (such as nasal, oral, and pulmonary). Hereby, airborne bacterial and viral infections in the lung tract (e.g., influenza, measles, and tuberculosis) are major reason for high rate of deaths per annum [74]. In fighting against such lung infections, pulmonary delivery of vaccines mimics the natural way of infection and might therefore be an appropriate way for their prevention. In animal models, aerosol delivery involves intratracheal instillation and insufflation or the use of exposure chambers, whereas for clinical trials a delivery device is required. Microsized particles (1–5-μm in size) are generated by dry-powder inhalers and aerosols from liquid-suspended particles by nebulizers, which are then delivered into the respiratory tract [47]. In history, aerosol vaccination was applied in human subjects for more than a century and includes aerosol vaccines against anthrax, plaque, tularemia, smallpox, tetanus, and botulism [23]. When compared to common parenteral immunizations, the following benefits of aerosolized vaccines are matter of discussions:

1. Delivery of vaccines into the respiratory tract can trigger the secretion of local IgA antibodies, which are in turn capable of crossing epithelia and preventing further entrance of pathogens.
2. The particular noninvasive nature of antigen delivery into the lungs circumvents the common use of needles and syringes, which are the main cause for unsafe injections (e.g., needle-stick injuries). According to WHO sources [51], those unsafe injections cause 8.2 million cases hepatitis B, 2.3 million cases of hepatitis C, and 0.1 million cases of HIV/AIDS around the world.
3. The application of pulmonary dry-powder vaccines could stop the common imperative of an intact cold chain for storage, what is mandatory for conventional vaccines.
4. For the administration of vaccines by using inhalers, no specially trained medical personnel will be required.

In addition, two independent studies in infants underline the potential of aerosol vaccination and will be highlighted in more details. First, 4,327 schoolchildren (age 5–14) in South Africa received a measles vaccines (either Schwarz or Edmonston–Zagreb (EZ) vaccine) by aerosol or by subcutaneous (s.c.) administration. After 1 month of treatment, aerosol vaccination with the EZ strain caused seroconversion of 326 (84.7%) schoolchildren, which was superior to 257 (78.8%) who received the subcutaneous EZ vaccine or 176 (62.2%) who received subcutaneous Schwarz vaccine. The authors stated that this method of aerosol vaccination might also be suited for mass campaigns in pediatric populations [18].

In a second study, a measles vaccine (EZ) or measles-rubella (Edmonston–Zagreb with RA27/3) was administered to Mexican schoolchildren via inhalation or injection. Interestingly, titers of neutralizing antibodies for the aerosol group were around fourfold greater (52–64%) in comparison to the group, which received the injected vaccine (4–23%). Moreover, fewer side effects were observed after aerosol than injection administration of vaccines [7]. Besides, Wong-Chew et al. [73] reported that a measles aerosol vaccine in 9-month-old Mexican infants stimulated strong cellular immunity as measured by the Th1 cytokine interferon-gamma (IFN-γ).

Considering reported findings, pulmonary delivery of antigens holds certainly promise as an immunogenic and safe way of vaccination for the prevention of airborne pathogens. Moreover, it is an exceedingly attractive approach for developing countries in the south, where not always cold chain, correct syringe disposal, and trained personal can be guaranteed.

16.4.2 Transport of Plasmid DNA Vaccines into the Lungs

Pulmonary plasmid DNA vaccination is a rather new and promising concept of vaccination and might enable in the future immunizations against intracellular pulmonary pathogens, such as *Mycobacterium tuberculosis*, respiratory syncytial

Table 16.1 Overview of pulmonary DNA vaccination studies

Encoded protein	Delivery route	Delivery system	References
Hepatitis B surface antigen	Intratracheal (instillation)	None	[45]
Eight epitopes of *M. tuberculosis*	Intratracheal (microsprayer)	Chitosan NP	[8]
Hemagglutinin	Aerosol (nebulizer)	PEI NP	[54]
Rv1733c	Intratracheal (microsprayer)	PGLA-PEI NP	[10]

PLGA poly(D,L-lactide-co-gylcolide); *PEI* polyethylenimine; *NP* nanoparticulates; *RSV* respiratory syncytial virus

virus (RSV), and severe acute respiratory syndrome coronavirus (SARS) [9]. Until now, most in vivo studies focused mainly on gene therapy to the lung [4] and expectations raised that the most efficient and safe pulmonary gene delivery systems will find their application also for transport of plasmid DNA vaccines.

Regarding the delivery of plasmid DNA into the lung, in addition to common intracellular barriers for gene therapy (see Fig. 16.1), further extracellular obstacles have to be anticipated and overcome, e.g., withstanding shear forces during aerosolization and crossing the respiratory mucus layer, which is covering conducting airways or the liquid layer in the alveoli [59].

So far, pulmonary delivery of plasmid DNA vaccines was only reported for a very few antigens (see Table 16.1) by different research groups and will be discussed in the following.

Lombry et al. [45] studied the immunogenicity of two protein antigens, ovalbumin (OVA) and hepatitis B surface antigen (HBsAg) and their encoding plasmid DNAs, pOVA and pHBsAg, after intratracheal instillation or injection into mice. Pulmonary immunizations induced equipotent cellular and humoral immune responses when related to injected vaccines. Interestingly, antigen and plasmid DNA immunizations favored a Th2 and a Th1 response, respectively.

Moreover, Bivas-Benita et al. [8] encapsulated a plasmid DNA vaccine encoding eight T-cell epitopes of *M. tuberculosis* into chitosan nanoparticles and applied them intratracheally into HLA-A2 transgenic mice. DNA nanoparticles elicited maturation of dendritic cells (DC) and stimulated an increased secretion of IFN-γ cytokines in comparison to pulmonary delivery of the plasmid DNA in solution or via intramuscular immunization.

The same group evaluated the in vitro and in vivo immunogenicity of a plasmid DNA vaccine encoding for the latency antigen Rv1733c of *M. tuberculosis* [10]. pRv1733c was formulated in poly(D,L-lactide-*co*-gylcolide) (PLGA)-PEI nanoparticles. DNA nanoparticles matured human DCs and induced secretion of two Th1 cytokines, TNF-α and IL-12, to a similar extent as the positive control lipopolysaccharide (LPS). Priming of mice with pRv1733c nanoparticles and boosting after 3 weeks with Rv1733c protein enabled superior levels of T cell proliferation as intramuscular immunization controls. In addition, the same trend was observed for IFN-γ secretions, where aerosol delivery of nanoparticulates in conjunction with the protein boost triggered the highest release of IFN-γ.

So far, the only study of pulmonary DNA vaccination demonstrating next to immunogenicity also protective efficacy was reported by Orson et al. [54]. An influenza antigen, hemagglutinin (HA; from viral strain A/PR8/34), expressing plasmid DNA (pHA) was incorporated in PEI particles and aerosolized into mice. When compared to intravenous delivery of the same HA plasmid in macroaggregated albumin (MAA)-PEI particles, less virus neutralizing antibodies were found after 2 weeks postimmunization. However, when plasmids encoding the cytokines IL-12 and granulocyte-macrophage colony stimulating factor (GM-CSF) were co-aerosolized in pHA PEI particles, a significant increase in neutralizing titer was remarked together with protection against subsequent influenza challenge.

16.5 Conclusions

Fallbacks – sometimes fatal – have sent gene therapy from the bedside back to the bench. In this scenario, the application of nonviral gene delivery systems appears as an advantageous alternative to viral carriers. However, the dilemma remains that nonviral delivery systems are either toxic and therefore not safe for human use, or much less efficient in transfecting target cells than their viral counterparts. The same holds true for genetic vaccines: while quite successful in animal studies, DNA vaccine trials in human subjects fail to reach their endpoints of establishing protective immunity.

On top of this, the administration site of the lung is a difficult one. Though relatively accessible by inhalation, toxicity and thus safety concerns play a major role when designing clinical trials for pulmonary vaccine delivery, although safety for measles vaccines applied by inhalation was shown. The introduction of novel excipients, such as ultrapure chitosan or modified PEI, will prove difficult in this context. However, research must continue to develop novel gene therapy and DNA vaccine carrier systems, especially for the lung. The challenges are high, but successful gene therapy of CF and achieving immunity against the global killer tuberculosis are too big benefits to be missed.

References

1. Alton EW, Middleton PG, Caplen NJ, Smith SN, Steel DM, Munkonge FM, Jeffery PK, Geddes DM, Hart SL, Williamson R (1993) Non-invasive liposome-mediated gene delivery can correct the ion transport defect in cystic fibrosis mutant mice. Nat Genet 5:135–142
2. Alton EW, Griesenbach U, Geddes DU (1999) Gene therapy for asthma: inspired research or unnecessary effort? Gene Ther 6:155–156
3. Alton EWFW, Stern M, Farley R, Jaffe A, Chadwick SL, Phillips J, Davies J, Smith SN, Browning J, Davies MG, Hodson ME, Durham SR, Li D, Jeffery PK, Scallan M, Balfour R, Eastman SJ, Cheng SH, Smith AE, Meeker D, Geddes DM (1999) Cationic lipid-mediated *CFTR* gene transfer to the lungs and nose of patients with cystic fibrosis: a double-blind placebo-controlled trial. Lancet 353:947–954

4. Aneja MK, Geiger JP, Himmel A, Rudolph C (2009) Targeted gene delivery to the lung. Expert Opin Drug Deliv 6:567–583
5. Arai K, Kinumaki T, Fujita T (1968) Toxicity of chitosan. Bull Tokai Reg Fish Lab 43:89–94
6. Audouy SAL, de Leij LFMH, Hoekstra D, Molema G (2002) In vivo characteristics of cationic liposomes as delivery vectors for gene therapy. Pharm Res 19:1599–1605
7. Bennett JV, Fernandez de Castro J, Valdespino-Gomez JL, Garcia-Garcia Mde L, Islas-Romero R, Echaniz-Aviles G, Jimenez-Corona A, Sepulveda-Amor J (2002) Aerosolized measles and measles-rubella vaccines induce better measles antibody booster responses than injected vaccines: randomized trials in Mexican school children. Bull World Health Organ 80:806–812
8. Bivas-Benita M, van Meijgaarden KE, Franken KL, Junginger HE, Borchard G, Ottenhoff TH, Geluk A (2004) Pulmonary delivery of chitosan-DNA nanoparticles enhances the immunogenicity of a DNA vaccine encoding HLA-A*0201-restricted T-cell epitopes of *Mycobacterium tuberculosis*. Vaccine 22:1609–1615
9. Bivas-Benita M, Ottenhoff TH, Junginger HE, Borchard G (2005) Pulmonary DNA vaccination: concepts, possibilities and perspectives. J Control Release 107:1–29
10. Bivas-Benita M, Lin MY, Bal SM, van Meijgaarden KE, Franken KL, Friggen AH, Junginger HE, Borchard G, Klein MR, Ottenhoff TH (2009) Pulmonary delivery of DNA encoding *Mycobacterium tuberculosis* latency antigen Rv1733c associated to PLGA-PEI nanoparticles enhances T cell responses in a DNA prime/protein boost vaccination regimen in mice. Vaccine 27:4010–4017
11. Borchard G (2001) Chitosan in gene delivery. Adv Drug Deliv Rev 52:145–150
12. Boussif F, Lezoualc'h F, Zanta MA, Mergny MD, Scherman D, Demeneix B, Behr JP (1995) A versatile vectore for gene and oligonucleotide transfer into cells in culture and in vivo: polyethyleneimine. Proc Natl Acad Sci USA 92:7297–7301
13. Capan Y, Woo BH, Gebrekidan S, Ahmed S, DeLuca P (1999) Stability of poly(L-lysine)-complexed plasmid DNA during mechanical stress and DNase I treatment. Pharm Dev Technol 4:491–498
14. Choi YH, Liu F, Kim JS, Choi YK, Park JS, Kim SW (1998) Polyethylene glycol-grafted poly-L-lysine as polymeric gene carrier. J Control Release 54:39–48
15. Crook K, McLachlan G, Stevenson BJ, Porteous DJ (1996) Plasmid DNA molecules complexed with cationic liposomes are protected from degradation by nucleases and shearing by aerosolisation. Gene Ther 3(83):4–839
16. Densmore CL, Giddings TH, Waldrep JC, Kinsey BM, Knight V (1999) Gene transfer by guanidinium-cholesterol:dioleoylphosphatidyl-ethanolamine liposome-DNA complexes in aerosol. J Gene Med 1:251–264
17. Densmore CL, Orson FM, Xu B, Kinsey BM, Waldrep JC, Hua P, Bhogal B, Knight V (2000) Aerosol delivery of robust polyethyleneimine-DNA complexes for gene therapy and genetic immunization. Mol Ther 1:180–188
18. Dilraj A, Cutts FT, de Castro JF, Wheeler JG, Brown D, Roth C, Coovadia HM, Bennett JV (2000) Response to different measles vaccine strains given by aerosol and subcutaneous routes to schoolchildren: a randomised trial. Lancet 355:798–803
19. Ernst N, Ulrichskötter S, Schmalix WA, Rädler J, Galneder R, Mayer E, Gersting S, Plank C, Reinhardt D, Rosenecker J (1999) Interaction of liposomal and polycationic transfection complexes with pulmonary surfactant. J Gene Med 1:331–340
20. Ferrari S, Pettenazzo A, Garbati N, Zacchello F, Behr J-P, Scarpa M (1999) Polyethyleneimine shows properties of interest for cystic fibrosis gene therapy. Biochim Biophys Acta 1447:219–225
21. Fynan EF, Robinson HL, Webster RG (1993) Use of DNA encoding influenza hemagglutinin as an avian influenza vaccine. DNA Cell Biol 12:785–789
22. Gautam A, Densmore CL, Waldrep JC (2001) Pulmonary cytokine responses associated with PEI-DNA aerosol gene therapy. Gene Ther 8:254–257
23. Giudice EL, Campbell JD (2006) Needle-free vaccine delivery. Adv Drug Deliv Rev 58:68–89
24. Godbey WT, Wu KK, Mikos AG (1999) Size matters: molecular weight affects the efficiency of poly(ethyleneimine) as a gene delivery vehicle. J Biomed Mater Res 45:268–275

25. Griesenbach U, Kitson C, Garcia SE, Farley R, Singh C, Somerton L, Painter H, Smith RL, Gill DR, Hyde SC, Chow Y-H, Hu J, Gray M, Edbrooke M, Ogilvie V, MacGregor G, Scheule RK, Cheng SH, Caplen NJ, Alton EWFW (2006) Inefficient cationic lipid-mediated siRNA and antisense oligonucleotide transfer to airway epithelial cells in vivo. Respir Res 7:26–41
26. Guillaume C, Delepine P, Droal C, Montier T, Tymen G, Claude F (2001) Aerosolization of cationic lipid–DNA complexes: Lipoplex characterization and optimization of aerosol delivery conditions. Biochem Biophys Res Commun 286:464–471
27. Hege KM, Carbone DP (2003) Lung cancer vaccines and gene therapy. Lung Cancer 41:103–113
28. Huygen K, Content J, Denis O, Montgomery DL, Yawman AM, Deck RR, DeWitt CM, Orme IM, Baldwin S, D'Souza C, Drowart A, Lozes E, Vandenbussche P, Van Vooren JP, Liu MA, Ulmer JB (1996) Immunogenicity and protective efficacy of a tuberculosis DNA vaccine. Nat Med 2:893–898
29. Huygen K (2005) Plasmid DNA vaccination. Microbes Infect 7:932–938
30. Iwasaki A, Medzhitov R (2005) Toll-like receptor control of the adaptive immune responses. Nat Immunol 5:987–995
31. Illum L (1998) Chitosan and its use as a pharmaceutical excipient. Pharm Res 15:1332–1339
32. Jaoko W, Nakwagala FN, Anzala O, Manyonyi GO, Birungi J, Nanvubya A, Bashir F, Bhatt K, Ogutu H, Wakasiaka S, Matu L, Waruingi W, Odada J, Oyaro M, Indangasi J, Ndinya-Achola J, Konde C, Mugisha E, Fast P, Schmidt C, Gilmour J, Tarragona T, Smith C, Barin B, Dally L, Johnson B, Muluubya A, Nielsen L, Hayes P, Boaz M, Hughes P, Hanke T, McMichael A, Bwayo J, Kaleebu P (2008) Safety and immunogenicity of recombinant low-dosage HIV-1 A vaccine candidates vectored by plasmid pTHr DNA or modified vaccinia virus Ankara (MVA) in humans in East Africa. Vaccine 26:2788–2795
33. Jeong JH, Song SH, Lim DW, Lee H, Park TG (2001) DNA transfection using linear poly(ethyleneimine) prepared by controlled acid hydrolysis of poly(2-ethyl-2-oxazoline). J Control Release 73:391–399
34. Köpping-Höggård M, Tubulekas I, Guan H, Edwards K, Nilsson M, Vårum KM, Artursson P (2001) Chitosan as a nonviral gene delivery system. Structure-property relationships and characteristics compared with polyethyleneimine in vitro and after lung administration in vivo. Gene Ther 8:1108–1121
35. Konstan MW, Davis PB, Wagener JS, Hilliard KA, Stern RC, Milgram LJH, Kowalczyk TH, Hyatt SL, Fink TL, Gedeon CR, Oette SM, Payne JM, Muhammad O, Ziady AG, Moen RC, Cooper MJ (2004) Compacted DNA nanoparticles adminstered to the nasal mucosa of cystic fibrosis subjects are safe and demonstrate partial to complete cystic fibrosis transmembrane regulator reconstitution. Hum Gene Ther 15:1–15
36. Kundu PP, Sharma V (2008) Synthetic polymeric vectors in gene therapy. Curr Opin Solid State Mater Sci 12:89–102
37. Kutzler MA, Weiner DB (2008) DNA vaccines: ready for prime time? Nat Rev Genet 9:776–788
38. Kwoh DY, Coffin CC, Lollo CP, Jovenal J, Banszczyk MG, Mullen P, Phillips A, Amini A, Fabrycki J, Bartholomew RM, Brostoff SW, Carlo DJ (1999) Stabilization of poly-L-lysine/DNA polyplexes for in vivo gene delivery to the liver. Biochim Biophys Acta 1444:171–190
39. Laemmli UK (1975) Characterization of DNA condensates induced by poly(ethylene oxide) and polylysine. Proc Natl Acad Sci USA 72:4288–4292
40. Lee R, Marshall J, Siegel C, Jiang C, Yew NS, Nichols MR, Nietupski JB, Ziegler RJ, Lane MB, Wang KX, Wan NC, Scheule RK, Harris DJ, Smith AE, Cheng SH (1996) Detailed analysis of structures and formulations of cationic lipids for efficient gene transfer to the lung. Hum Gene Ther 7:1701–1717
41. Lentz KY, Anchordoquy TJ, Lengsfeld CS (2006) Rationale for the selection of an aerosol delivery system for gene delivery. J Aerosol Med 19:372–384
42. Lentz KY, Anchordoquy TJ, Lengsfeld CS (2006) DNA acts as a nucleation site for transient cavitation in the ultrasonic nebulizer. J Pharm Sci 95:607–619
43. Liu G, Molas M, Grossmann GA, Pasumarthy M, Perales JC, Cooper MJ, Hanson RW (2001) Biological properties of poly-L-lysine-DNA complexes generated by cooperative binding of the polycation. J Biol Chem 276:34379–34387

44. Liu G, Li D, Pasumarthy MK, Kowalczyk TH, Gedeon C, Hyatt S, Payne JM, Miller TJ, Brunovskis P, Moen RC, Hanson RW, Cooper MJ (2003) Nanoparticles of compacted DNA transfect post-mitotic cells. J Biol Chem 278:32578–32586
45. Lombry C, Marteleur A, Arras M, Lison D, Louahed J, Renauld JC, Préatand V, Vanbever R (2004) Local and systemic immune responses to intratracheal instillation of antigen and DNA vaccines in mice. Pharm Res 21:127–135
46. Lommatzsch ST, Aris R (2009) Genetics of cystic fibrosis. Semin Respir Crit Care Med 30:531–538
47. Lu D, Hickey AJ (2007) Pulmonary vaccine delivery. Expert Rev Vaccines 6:213–226
48. Mao S, Sun W, Kissel T (2010) Chitosan-based formulations for delivrey of DNA and siRNA. Adv Drug Deliv Rev 62:12–27
49. Marshall J, Nietupski JB, Lee ER, Siegel CS, Rafter PW, Rudginsky SA, Chang CD, Eastman SJ, Harris DJ, Scheule RK, Cheng SH (2000) Cationic lipid structure and formulation considerations for optimal gene transfection of the lung. J Drug Target 7:453–469
50. McLachlan G, Davidson DJ, Stevenson BJ, Dickinson P, Davidson-Smith H, Dorin JR, Porteus DJ (1995) Evaluation in vitro and in vivo of cationic liposome-expression construct complexes for cystic fibrosis gene therapy. Gene Ther 2:614–622
51. Miller MA, Pisani E (1999) The cost of unsafe injections. Bull World Health Organ 77:808–811
52. Mishra S, Webster P, Davis ME (2004) PEGylation significantly affects cellular uptake and intracellular trafficking of non-viral gene delivery particles. Eur J Cell Biol 83:97–111
53. Nel AE, Mädler L, Velegol D, Xia T, Hoek EMV, Somasundaran P, Klaessig F, Castranova V, Thompson M (2009) Understanding biophysicochemical interactions at the nano-bio interface. Nat Mater 8:543–557
54. Orson FM, Kinsey BM, Densmore CL, Nguyen T, Wu Y, Mbawuike IN, Wyde PR (2006) Protection against influenza infection by cytokine-enhanced aerosol genetic immunization. J Gene Med 8:488–497
55. Petersen H, Fechner PM, Martin AL, Kunath K, Stolnik S, Roberts CJ, Fischer D, Davies MC, Kissel T (2002) Polyethyleneimine-graft-poly(ethylene glycol) copolymers: influence of copolymer block structure on DNA complexation and biological activities as gene delivery system. Bioconjug Chem 13:845–854
56. Ram A, Kadim A (1970) Shear degradation of polymer solutions. J Appl Polym Sci 14:2145–2156
57. Rudolph C, Schillinger U, Ortiz A, Plank C, Jauernik J, Rosenecker J (2004) Aerosol delivery of DNA-loaded nanoparticles: factors influencing particle stability and gene transfer efficiency. IEEE Conference on Nanotechnology, Minich, pp 249–251
58. Ruiz FE, Clancy JP, Perricone MA, Bebok Z, Hong JS, Cheng SH, Meeker DP, Young KR, Schoumacher RA, Weatherly MR, Wing L, Morris JE, Sindel L, Rosenberg M, Van Ginkel FW, Mcghee JR, Kelly D, Lyrene RK, Sorscher EJ (2001) A clinical inflammatory syndrome attributable to aerosolized lipid–DNA administration in cystic fibrosis. Hum Gene Ther 12:751–761
59. Sanders N, Rudolph C, Braeckmans K, De Smedt SC, Demeester J (2009) Extracellular barriers in respiratory gene therapy. Adv Drug Deliv Rev 61:115–127
60. Saranya N, Moorthi A, Saravanan S, Pandima Devi N, Selvamurugan N (2011) Chitosan and its derivatives for gene delivery. Int J Biol Macromol 48:234–238
61. Senel S, McClure SJ (2004) Potential applications of chitosan in veterinary medicine. Adv Drug Deliv Rev 56:1467–1480
62. Shimosato T, Fujimoto M, Tohno M, Sato T, Tateo M, Otani H, Kitazawa H (2010) CpG oligodeoxynucleotides induce strong up-regulation of interleukin 33 via Toll-like receptor 9. Biochem Biophys Res Commum 394:81–86
63. Stribling R, Brunette E, Liggitt D, Gaensler K, Debs R (1992) Aerosol gene delivery in vivo. Proc Natl Acad Sci USA 89:11277–11281
64. Tang DC, DeVit M, Johnston SA (1992) Genetic immunization is a simple method for eliciting an immune response. Nature 356:152–154

65. Üzgün S, Akdemir Ö, Hasenpusch G, Mauksch C, Golas MM, Sander B, Stark H, Imker R, Lutz J-F, Rudolph C (2010) Characterization of tailor-made copolymers of oligo(ethylene glycol) methyl ether methacrylate and N, N-dimethylaminoethyl methacrylate as nonviral gene transfer agents: influence of macromolecular structure on gene vector particle properties and transfection efficiency. Biomacromolecules 11:39–50
66. Ulmer JB, Donnelly JJ, Parker SE, Rhodes GH, Felgner PL, Dwarki VJ, Gromkowski SH, Deck RR, DeWitt CM, Friedman A (1993) Heterologous protection against influenza by injection of DNA encoding a viral protein. Science 259:1745–1749
67. van de Wetering P, Cherng JY, Talsma H, Crommelin DJ, Hennink WE (1998) 2-(Dimethylamino) ethyl methacrylate based (co)polymers as gene transfer agents. J Control Release 53:145–153
68. Wang R, Doolan DL, Le TP, Hedstrom RC, Coonan KM, Charoenvit Y, Jones TR, Hobart P, Margalith M, Ng J, Weiss WR, Sedegah M, de Taisne C, Norman JA, Hoffman SL (1998) Induction of antigen-specific cytotoxic T lymphocytes in humans by a malaria DNA vaccine. Science 16:476–480
69. Wang R, Epstein J, Baraceros FM, Gorak EJ, Charoenvit Y, Carucci DJ, Hedstrom RC, Rahardjo N, Gay T, Hobart P, Stout R, Jones TR, Richie TL, Parker SE, Doolan DL, Norman J, Hoffman SL (2001) Induction of CD4(+) T cell-dependent CD8(+) type 1 responses in humans by a malaria DNA vaccine. Proc Natl Acad Sci USA 98:10817–10822
70. Wightmann L, Kircheis R, Rossler V, Carotta S, Ruzicka R, Kursa M, Wagner E (2001) Different behavior of branched and linear polyethyleneimine for gene delivery in vitro and in vivo. J Gene Med 3:362–372
71. Wolfert MA, Schacht EH, Toncheva V, Ulbrich K, Nazarova O, Seymour LW (1996) Characterization of vectors for gene therapy formed by self-assembly of DNA with synthetic block co-polymers. Hum Gene Ther 7:2123–2133
72. Wolff JA, Malone RW, Williams P, Chong W, Acsadi G, Jani A, Felgner PL (1990) Direct gene transfer into mouse muscle in vivo. Science 23:1465–1468
73. Wong-Chew RM, Islas-Romero R, García-García Mde L, Beeler JA, Audet S, Santos-Preciado JL, Gans H, Lew-Yasukawa L, Maldonado YA, Arvin AM, Valdespino-Gómez JL (2006) Immunogenicity of aerosol measles vaccine given as the primary measles immunization to nine-month-old Mexican children. Vaccine 24:683–690
74. Woodland DL, Randall TD (2004) Anatomical features of anti-viral immunity in the respiratory tract. Semin Immunol 16:163–170
75. Ziady AG, Gedeon CR, Miller T, Quan W, Payne JM, Hyatt SL, Fink TL, Muhammad O, Oette S, Kowalczyk T, Pasumarthy MK, Moen RC, Cooper MJ, Davis PB (2003) Transfection of airway epithelium by stable PEGylated poly-L-lysine DNA nanoparticles in vivo. Mol Ther 8:936–947

Chapter 17
In Vitro Performance Testing for Pulmonary Drug Delivery

Yoen-Ju Son, Jolyon P. Mitchell, and Jason T. McConville

Abstract This chapter provides a detailed review of in vitro testing methods for inhalation products. Specifically, the current compendial methods for pulmonary drug delivery are presented, discussion of cascade impactor use and simplification, determination of aerosol electrostatics, static characterization of particles and powders, solubility screening, and a review of research leading to improved dissolution studies for these products.

Keywords Aerodynamic particle size • Aerosol • Aerosol electrostatics • Cascade impactor • Dissolution • Interparticulate interaction • Solubility

17.1 Compendial Methods

Multistage cascade impactors (CIs) are defined in the US and European pharmacopeias (Table 17.1) for the measurement of inhaler aerosol aerodynamic particle size distribution (APSD) for the following reasons [57]:

1. They operate on the principle of inertial size fractionation, making it possible to determine aerodynamic diameter directly.
2. The collected particles are also available for quantification of drug mass by appropriate analytical techniques, which excludes the possibility of excipient-only particles biasing the size distribution.

CIs are used both in the development and subsequent quality control (QC) of marketed inhaler products and the qualification of the clinical batches together with the

---

J.T. McConville (✉)
College of Pharmacy, University of Texas at Austin, Austin, TX, USA
e-mail: jtmcconville@mail.utexas.edu

H.D.C. Smyth and A.J. Hickey (eds.), *Controlled Pulmonary Drug Delivery*, Advances in Delivery Science and Technology, DOI 10.1007/978-1-4419-9745-6_17,

Table 17.1 Identification of apparatuses for APSD measurement in the US and European Pharmacopeias (from Mitchell and Dalby [55], with permission)

Impactor	US Pharmacopeia	European pharmacopeia[a]
Glass "Twin Impinger" (TI)	Not used	Apparatus A for pMDIs, DPIs, and nebulizers
Andersen 8-stage Without pre-separator (ACI)	Apparatus 1 for pMDIs	Apparatus D for pMDIs
Marple Miller Model 160 (MMI)	Apparatus 2 for DPIs	Not used
Andersen 8-stage With pre-separator (ACI)	Apparatus 3 for DPIs	Apparatus D for DPIs
Multistage liquid impinger (MSLI)	Apparatus 4 for DPIs	Apparatus C for pMDIs and DPIs
Next generation pharmaceutical impactor (NGI)	Apparatus 5 for DPIs Apparatus 6 for pMDIs	Apparatus E for pMDIs and DPIs

[a]Apparatus B was the single-stage "Metal Impinger" that was withdrawn from the European Pharmacopeia in 2005 (Supplement 5.1)

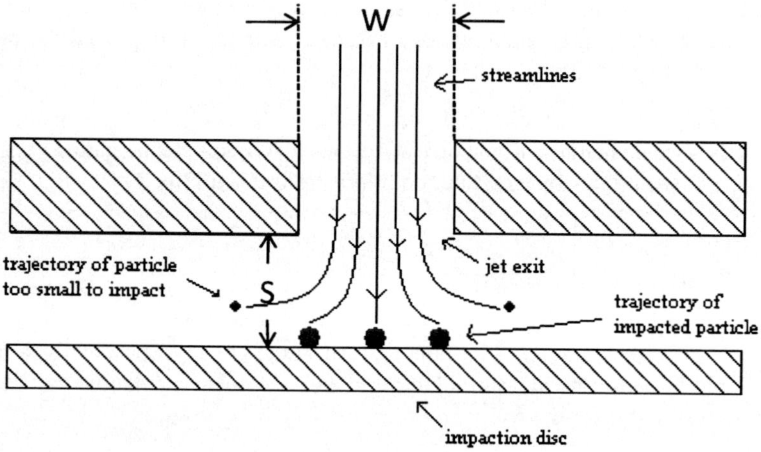

Fig. 17.1 Schematic of a single-stage inertial impactor

assessment of add-on devices affecting the size-properties of the emitted aerosol, such as spacers and holding chambers that are used with pMDIs [55].

In its simplest form, a single-stage impactor comprises a nozzle plate containing one or more circular or slot-shaped jets (nozzles) of diameter (W) located at a fixed distance (S) from a flat collection surface that is usually but not necessarily horizontal (Fig. 17.1). Incoming particles are size classified on the basis of their differing inertia, which reflects the resistance to a change in direction of the flow. The laminar flow streamlines diverge on approach to the collection surface. Particles with greater inertia will tend to cross these streamlines more readily to impact on the substrate, whereas finer particles with less inertia follow the streamlines and remain airborne as they pass the obstruction. Liquid impingers are a variant in which the particles are captured in a liquid

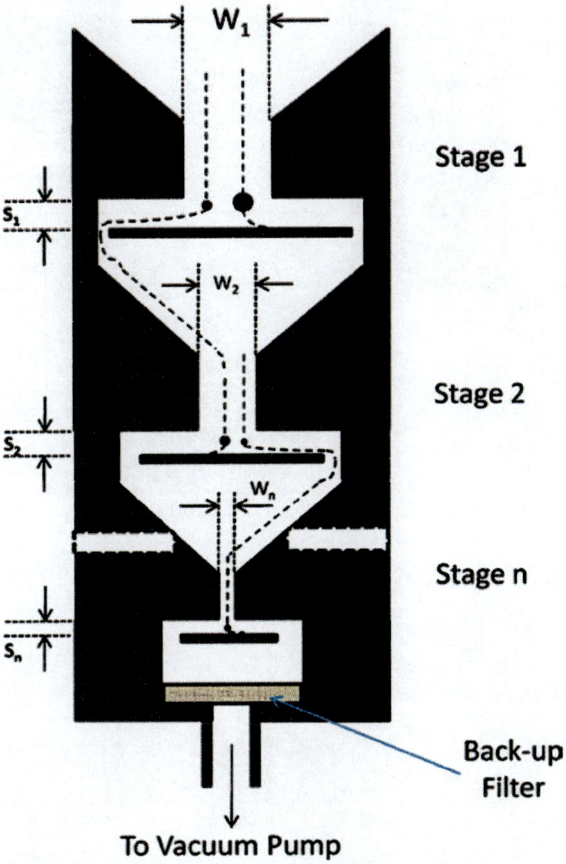

Fig. 17.2 Schematic of a multistage cascade impactor (CI)

medium rather than on a solid substrate, an advantage in that particle bounce and re-entrainment is avoided, and the liquid can be the eluent medium for recovery and assay of active pharmaceutical ingredient(s) (APIs). The CI (Fig. 17.2) is constructed by coupling several stages (typically seven or more) together with progressively finer and/or reduced number of jets such that particle velocity and hence inertia is progressively increased. In this way, the device fractionates the incoming aerosol into particles covering discrete and well-defined size ranges. The theory of cascade impaction is well described by Marple and Liu [49] and Rader and Marple [77].

The Andersen cascade impactor (ACI) (Fig. 17.3a) is currently still the most widely used impactor for inhaler aerosol testing, although the NGI (Fig. 17.3b) is becoming more widely used as familiarity with its capability grows. Both CIs can operate in the flow rate range of interest for pMDI and DPI testing (30–100 L/min), but the ACI requires the removal of the lowermost stages (6 and/or 7) and insertion of new stages at the upper end (stages −1 and −2) for use at flow rates of 60 and 100 L/min [65]. On the other hand, the NGI does not require stage replacement [53] and can operate over the entire range with excellent size resolution [51]. It can even be used without its pre-separator and a back-up filter substituted for the multiorifice collector [93] for

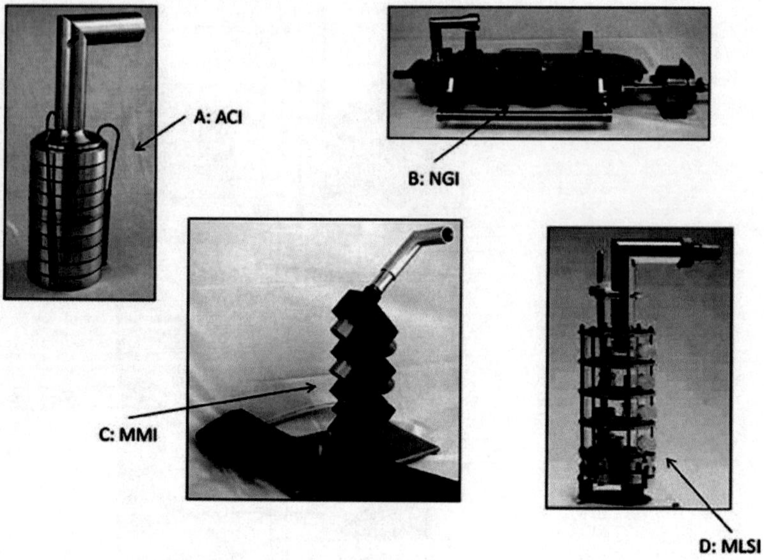

Fig. 17.3 Compendial CIs

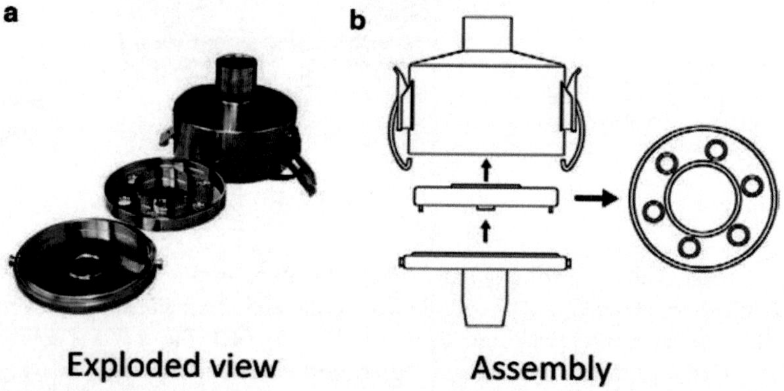

Fig. 17.4 NGI pre-separator

testing aerosols from nebulizers at the recommended flow rate of 15 L/min [52]. Other CIs in use are the Marple-Miller 5-stage design that uses collection cups, similar in concept to the NGI (Fig. 17.3c) and which is also available in a low flow version (4.9 and 12 L/min) for testing inhalers intended for pediatric use [69], as well as in the standard versions that operate at 30 and 60 L/min for routine testing of pMDIs and DPIs [50]. The 4-stage multistage liquid impinger (MSLI, Fig. 17.3d) is the last widely used CI for inhaler aerosol evaluations, being a popular choice where collection of the active substance in the eluent for subsequent assay is possible [5].

The compendial apparatuses are normally used with the Ph.Eur./USP induction port (throat), which has well-defined entry and exit profiles and comprises a right-angle bend intended as a crude simulation of the human oropharynx (see each apparatus in Fig. 17.3). A pre-separator (Fig. 17.4) may also be inserted between the induction port and entry to the CI with the purpose of preventing entry of excessively coarse particulate, such as carrier particles from certain DPIs, into the CI where they would otherwise bounce and become re-entrained in the flow due to their extremely high inertia even at the relatively low flow velocities in the upper stages of the CI.

17.2 Considerations with Cascade Impactor Use

Although the underlying principle for the CI method is simple to understand, its practical implementation is technically challenging because it is labor-intensive, requiring a high degree of manual skill. Christopher et al. have developed a comprehensive guide on the considerations that should be taken when developing a CI-based method for inhaler testing [15]. Factors that have to be taken into account in method development include the following:

(a) Collection solvent (impingers) or recovery solvent (dry impactor systems).
(b) Quantitation lower limit for the APIs.
(c) Use of collection surface coating (almost always necessary with dry impactor systems).
(d) Recovery techniques for the APIs.
(e) Use of a pre-separator (important for many DPIs, in particular those based on the drug-carrier particle delivery principle).
(f) Cleaning procedure for the CI.
(g) Electrostatic charge (very important if the CI collection surfaces are electrically insulating or isolated).
(h) Environmental factors (barometric pressure, temperature, relative humidity).
(i) Use of a back-up filter or the microorifice collector in the case of the NGI (Note that the MOC may not be fully effective if the formulation is predominantly composed of particles finer than about 1 μm aerodynamic diameter).

17.3 Simplifying the Cascade Impactor Method

Simplification of the CI measurement process is attractive, not only to save time per measurement but also to reduce the intrinsic variability associated with the process [10]. The Efficient Data Analysis concept was developed out of the recognition that the CI is not an analogue of the respiratory tract [56]. Thus, for product development and QC (Fig. 17.5a), two mutually independent metrics, one relating to the so-called impactor sized mass (ISM), which is the sum of small (SPM) and large particle mass (LPM), and the other being the ratio of LPM/SPM, have both been shown to be highly sensitive to small changes in both the amplitude (area under the curve) and location (size range) of differential mass-weighted APSDs from a variety

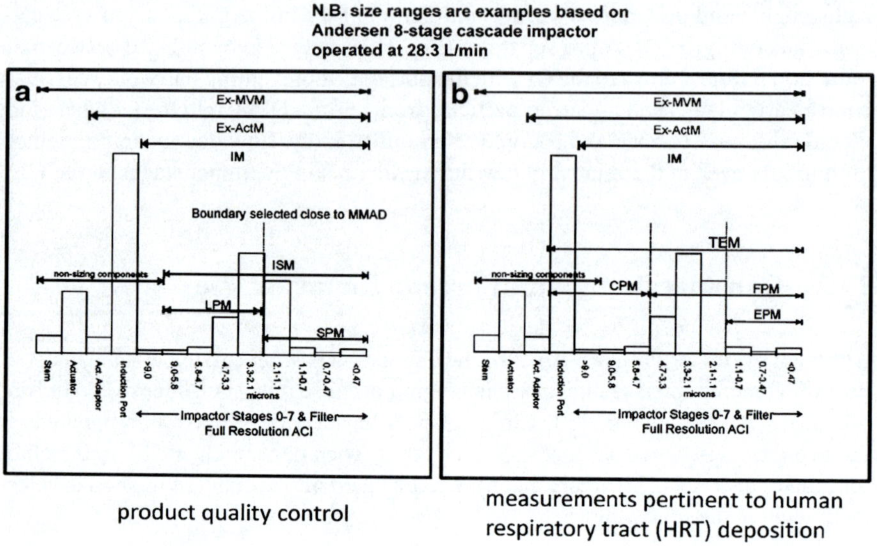

Fig. 17.5 EDA metrics for the ACI and their particular uses

of inhaler types determined by different full resolution cascade impaction systems [91]. The boundary between SPM and LPM is ideally chosen to coincide with the MMAD of the aerosol, but importantly, Tougas et al. showed that it can be displaced by a significant amount from this value without much deterioration in the correlation between APSD change and the magnitude of this ratio. The same group also showed that these simplified metrics, by being independent of each other, have better diagnostic capability for changes in APSD than stage groupings of data from full resolution systems, an important advantage for the LDA approach.

Measures of coarse (CPM), fine (FPM), and extra-fine particle (EPM) mass (Fig. 17.5b) may be sufficient to define the size-related behavior of inhaled particles in the human respiratory tract (HRT) [27] when seeking to provide data that are more clinically pertinent. Here the impactor mass collected by the entire CI including the initial stage (IM), which in the ACI has an undefined upper size bound, is more pertinent. When IM is added to the mass collecting in the induction port, this quantity defines the total aerosol mass entering the system, analogous to the mass that is inhaled by a patient.

The abbreviated impactor measurement (AIM) concept has recently been re-evaluated as a possible solution to enable EDA-based metrics to be determined directly [58]. Earlier work by Van Oort and Roberts pointed the way forward in terms of setting out a hierarchy of reduced stack ACI measurements [95], supported by the full resolution system (Fig. 17.6). As well as greatly reducing the labor-intensive methodology associated with full resolution CI APSD measurements [15], abbreviated impactor determinations have the potential to improve product batch disposition decision-making by allowing more samples from the lot (coverage) to be

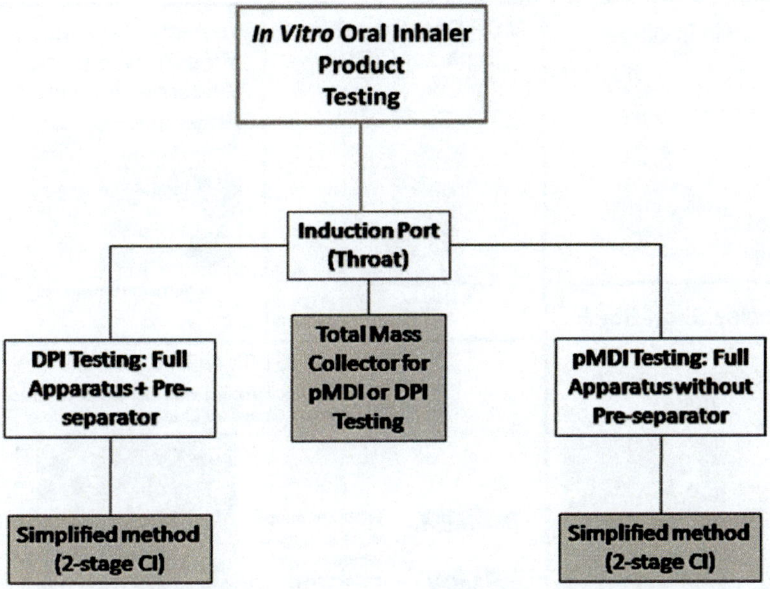

Fig. 17.6 Early proposal for AIM concept hierarchy in inhaler testing

evaluated in a given time period [91]. AIM-based methods also overcome the problem that with a typical inhaler-generated aerosol, several stages of full resolution CIs capture little or no API, reducing the overall precision of the method [10]. Furthermore, there is the potential that measurements may be made with the clinical dose (typically 1–2 actuations of the inhaler), rather than after 5 or more actuations, as is typical with full resolution systems to acquire sufficient drug deposits on stages collecting particles at the periphery of the APSD to permit acceptably accurate and precise assay of the API collected at these locations [15]. Additionally, by reducing the number of manipulations required to make a measurement, AIM-based methods should decrease the chances of operator-related errors. Furthermore, the AIM approach is more amenable to (semi) automation than multistage impactors. Finally, the use of less solvent for API recovery and quantitation is more environmentally friendly in the context of the emerging discipline of green chemistry.

It is important to understand that the AIM concept [60] is not confined to one particular configuration of impactor. Instead, many options exist (Fig. 17.7) from the Twin Impinger [34] to reduced versions of the ACI [47, 95] and the NGI [84]. The chosen abbreviated system must provide data that are compatible with the equivalent measures obtainable from the full resolution CI. In studies with pMDI-generated aerosols [58, 59], the following was demonstrated:

1. Precautions should be taken to minimize nonideal behavior, such as internal losses and particle re-entrainment by the use of tacky (adhesive) surfaces on the collection plates for each stage.

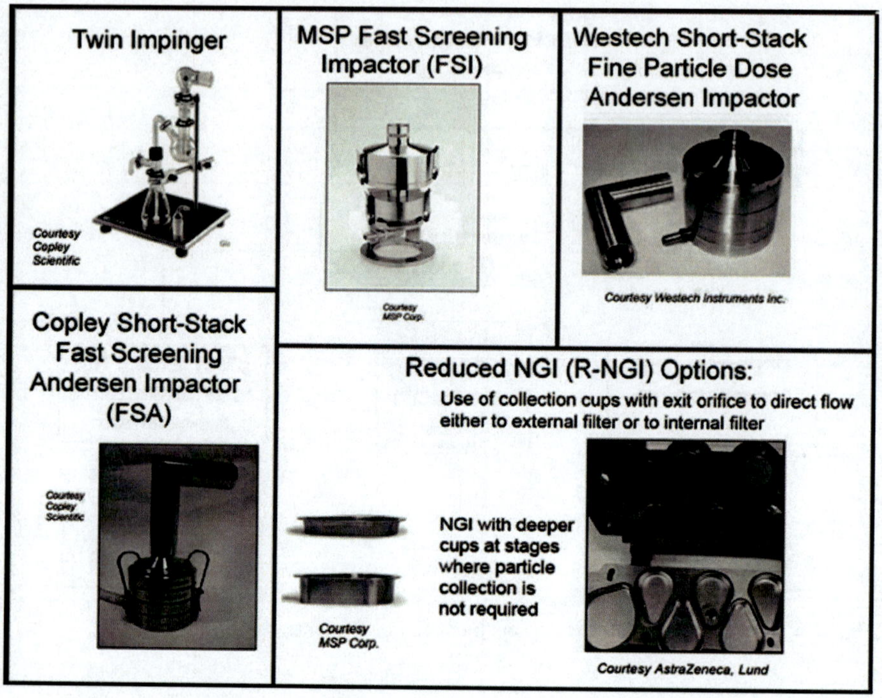

Fig. 17.7 Various options for AIM-based measurements

2. If the formulation contains volatiles such as ethanol, the dead space before the initial size-fractionation stage should be as similar as possible to that of the full resolution impactor.

More recently, a rigorous comparison was undertaken between two AIM-based CIs, one developed for product quality control testing (AIM-QC system) and the other for potential use in measurements pertinent to particle deposition in the lungs (AIM-HRT system), as well as the full resolution ACI (Fig. 17.8) with an HFA-pMDI-based product. This designed experiment demonstrated equivalent precision for all pertinent metrics (IM, ISM, LPM/SPM, and FPM) (Fig. 17.9), once the precautions listed above were taken [62]. However, EPM determined by this configuration was found to be about 8% greater than the equivalent metric obtained by the ACI. This source of bias was overcome by using a surfactant-saturated glass microfiber filter on the lower stage of this CI (Fig. 17.10) to prevent the divergent flow above the plate stripping away the surfactant when applied as a coating [61].

These experiments have demonstrated the practicality of AIM-based systems as substitutes for the full resolution ACI. In the future, it is likely that similar comparisons will be undertaken with other CI systems, in particular the NGI, which is now in widespread use [21]. Modified versions of the Twin Impinger having cut-point sizes closer to the range of interest (1.0–5.0 μm) at its operating flow rate of 60 L/min

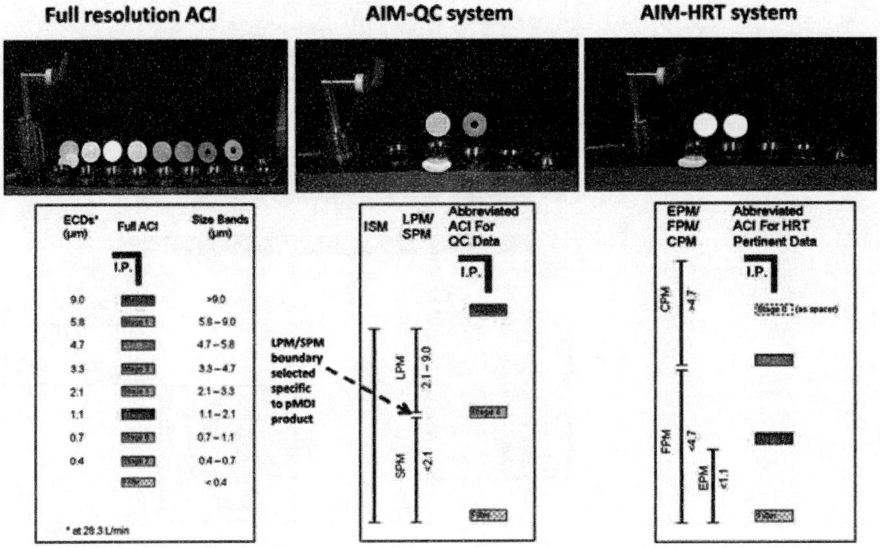

Fig. 17.8 Comparison between AIM-based CIs and full resolution ACI

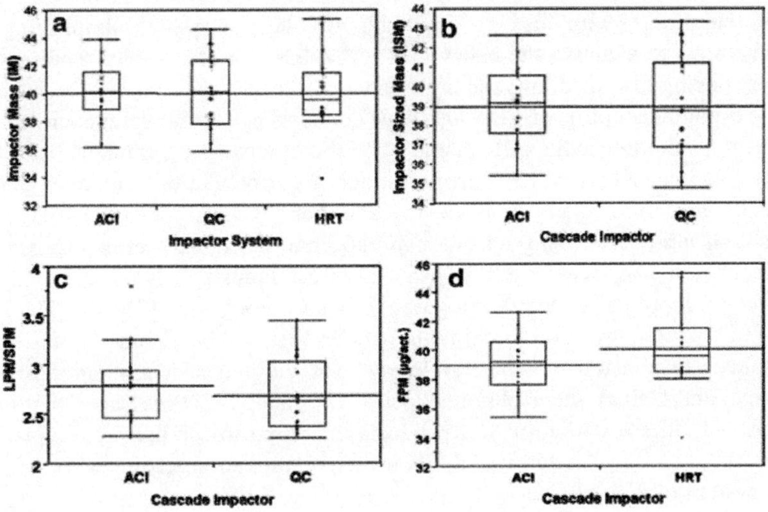

Fig. 17.9 Comparisons between AIM- and full resolution ACI-determined metrics for an HFA-albuterol product

may be developed if adequate precision is possible to control the finer jet diameter that would be required for the upper stage. This type of inertial fractionator avoids the problem of particle bounce by capturing the particles collecting on each stage in liquid medium that can also become the eluent for subsequent assay of the API.

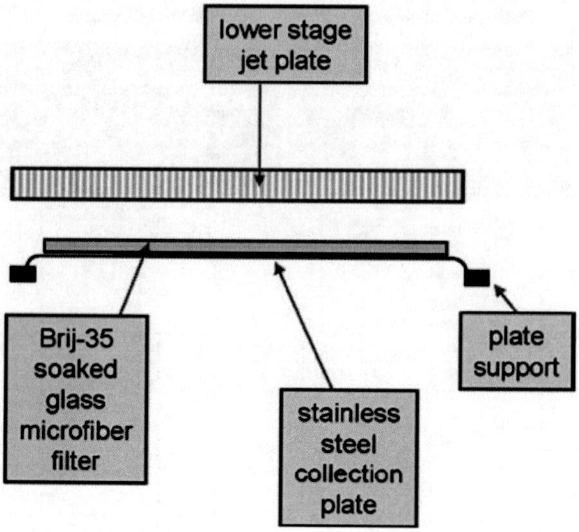

Fig. 17.10 Use of a surfactant-saturated filter to eliminate particle bounce in an AIM-HRT CI

Further work will also be required to investigate DPI-related issues, but the variety of different designs with this class of inhaler will likely necessitate simplifying such investigations to examine the behavior of AIM-based systems with representative inhalers having low, medium, and high flow resistances.

The eventual adoption of LDA and AIM concepts by regulatory agencies and the pharmacopeial compendia will depend upon the quality of experimental and theoretical evidence that can be developed in support of abbreviation. Although there are currently no formal plans in place for adoption, the European Pharmaceutical Aerosol Group (EPAG) and the International Pharmaceutical Aerosol Consortium on Regulation and Science (IPAC-RS), as cross-industry scientific groups, are involved in coordinating the development of such information [91, 98]. The intention is to lay the foundation for stimulus articles that will be intended to inform the committees responsible for the development of the pertinent monographs in the European and United States pharmacopeias. Ultimately, it is envisaged that manufacturers of either existing or new products intending to use these concepts in its development and production will do their own qualification studies with the AIM-based system of their choice.

17.4 Aerosol Electrostatics

Triboelectrification is the charge transfer process, which occurs during powder handling when sliding and frictional forces are involved [7]. In pulmonary drug delivery, triboelectric charge generated during powder handling, mixing, and aerosolization of formulations has been identified as an important property for understanding. The performance of inhalation products, such as powder attachment and detachment, as well as the lung deposition of active drug particles is quite dependent on the electric charge of the aerosol cloud. Powders used in dry powder inhalers (DPIs) or pressurized metered dose inhalers (pMDIs) optimally have a very fine particle size distribution (<5 μm) and generally very strong interparticulate forces, as they are mostly a mixture of fine respirable-sized particles and coarse carrier materials, or propellants respectively. It has been also reported that static electricity generated during powder handling and aerosolization improves the deposition of drug particles in the alveolar region and can be used in targeting the aerosol to the desired location of the respiratory tract [7, 8]. However, there is difficultly in evaluating the influences of electrostatic force on the aerosolization performance and the particle deposition to the target sites in vitro due to the extreme complexity and dynamic nature of aerosol systems. This is in part due to the fact that particles having different surface roughness and size distributions charge heterogeneously upon contacting each other, and environmental conditions such as humidity and temperature may not be constant during aerosolization.

The Faraday cage is a common method for measuring charge in aerosol droplets and particles [11, 14, 64, 71, 89]. The Faraday cage functions as a hollow conductor consisting of an electronically grounded outer conducting layer to eliminate ambient induction of current flow in the cage by external fields and an inner conducting layer connected to an electrometer to measure the charge induced on the particles. The detected current data in the particle mass are plotted against time, and then converted into a charge measurement by integration of the area under the plotted curve. The measured charge can be analyzed in two ways: net charge for total mass added to the cage and specific charge (charge-to-mass ratio) which is very useful to compare the charge between different test compounds.

However, the use of the Faraday method is very limited in measurement of the dynamic properties of aerosol clouds since the Faraday cage measures only the total charge of the entire formulation bulk. Several dynamic measurement systems have been introduced to make up for shortfalls of Faraday cage method as summarized in Table 17.2. The current measurement systems can be largely divided into two classes according to data collection: a direct (real-time) method using electrometer attached

Table 17.2 Various features of dynamic systems to measure aerosol electrostatic properties

Device	Theory	Aerodynamic separation	Measurement	Applicability
ELPI™	Faraday cage	ELPI (10–30 L/min, without pre-separator)	Direct measurement	pMDI DPI with modification
eNGI	Faraday cage	NGI (~100 L/min with/ without pre-separator)	Direct measurement	pMDI DPI
TSI	–	TSI (30–90 L/min)	Direct measurement	DPI
E-SPART	Laser Doppler velocimeter	–	Indirect measurement	pMDI DPI

to each impactor stage, and an indirect method using an electrical single-particle relaxation time (E-SPART) analyzer based on the motion of single-particle with respect to the oscillating electric field.

17.4.1 Direct Measurement

The electrical low pressure impactor (ELPI™) has been widely applied to measure the electrostatic charge on aerosol particles in the range from 0.028 to 10 μm [29, 36, 44, 48, 89, 101]. Originally, the concept of ELPI was introduced for real-time measurement of the particle size distribution of aerosol by means of detecting the current signal from each impactor stage containing collected particles using an electrometer. The operation principle of the ELPI involves applying charges to the aerosol particles using a corona charger, followed by introducing the charged aerosol into the 13-stage impactor at flow rate range of 10–30 mL/min. Each impactor stage functions as a Faraday cage to detect the total electrostatic charge of the deposited particles collected on each stage. Each stage is electrically insulated from each other, and a multichannel electrometer attached to each impactor stage measures the current induced by charged particles. The detected charge level is directly related to the particle size distribution since the signal detected from each impactor stage can be converted to the specific charge (or charge-to-mass ratio) for given particle cut-off sizes [44].

For pharmaceutical applications, the ELPI device requires modification to measure charges on the pharmaceutical aerosol particles since the conventional ELPI device was originally designed for environmental and industrial applications, such as air QC. The schematic diagram of the ELPI for pharmaceutical applications is shown in Fig. 17.11. The corona charger is either removed from the instrument or turned off while analyzing to detect innate charge on the particles. Aerosol samples are introduced through the USP throat using an adaptor to simulate the particle classification procedure of pharmaceutical impactors. It has been successfully demonstrated that the impact of propellants, excipients, the number of actuation, and environmental conditions on the charge profile of particles actuated from various MDI devices can be measured by using a modified ELPI device [29, 44, 101]. However, the ELPI has some limitations in evaluating DPI formulations, which mostly contain large carrier particles and are generated at higher flow rates than MDI formulations. Unlike the pharmaceutical impactors, the ELPI device does not have a pre-separator which is usually required to remove large carrier particles prior to particle classification, and the maximum operation flow rate is only 30 L/min as indicated in Table 17.2. To overcome these obstacles, a modification of the ELPI to incorporate a pre-separator with an ACI unit has been achieved [89, 104], and modification using a twin stage impinger (TSI) has also been demonstrated [89, 104]. Although several trials to utilize the ELPI for use with DPIs have been made, there still remain issues with respect to the manipulation of relevant flow rates, since the ELPI is not a device developed specifically for evaluating pharmaceutical aerosols.

The concept of the electrical next generation impactor (eNGI), using commercial pharmaceutical impactor, was first introduced by Hoe et al. [37, 38].

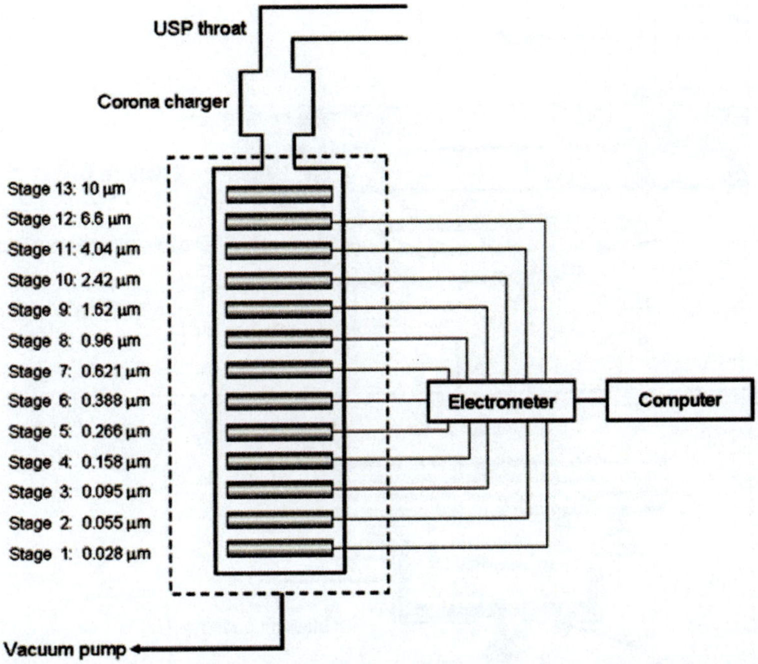

Fig. 17.11 Schematic of the electrical low-pressure impactor (ELPI™)

The basic measurement concept of the eNGI is very similar to that of ELPI, in that an aerosol inertial particle size classification apparatus was modified to act as a Faraday cage. As shown in schematic diagrams of eNGI (Fig. 17.12), the modification of the NGI was achieved by coating of each dose-collection plate to insulate each other and a multichannel electrometer was attached to each collection plate to display total charge/stage profiles [37, 38]. The eNGI device demonstrates better performances in versatility and accuracy than current ELPI-based devices, in particular, it offers a pre-separator option to remove large carrier particles, such as lactose, from a dry powder mixture. It is also capable of simultaneous comparison of mass and charge of aerosols at relevant pharmaceutical flow rates. Mass median aerodynamic diameter (MMAD) values obtained from dispersion studies for two DPI formulations, Pulmicort and Bricanyl, showed that data were statistically different in the ELPI for both DPI formulations when compared with those of NGI and eNGI [37]. In particular, the mass deposition in upper stage of ELPI (stage 13) and eNGI (stage 1) was shown to be different (Fig. 17.13a) for DPIs, which was quite different from the data from the eNGI, however, there was a similar capability in particle size distribution to the ELPI for pMDIs (Fig. 17.13b) [37, 38]. These results indicate that an effectiveness in removal of large carrier particles has a great influence for evaluation of particle size distributions, this can be an important factor in understanding an electrostatic charge dependency relative to the particle size distribution of aerosol dissipated from a given DPI device [37].

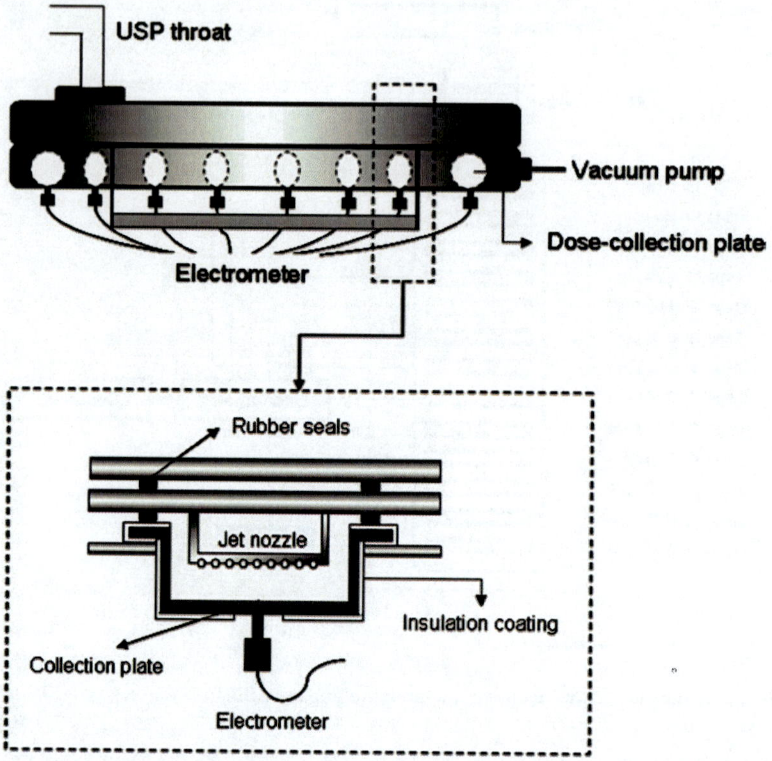

Fig. 17.12 Schematic of the electrical next generation impactor (eNGI)

17.4.2 Indirect Measurement

The E-SPART analyzer which is based on the principle of laser Doppler velocimeter (LDV) for particle sizing is a major indirect charge measurement system (Table 17.2). In this measurement system, particles are introduced by the actuation of aerosol generating devices into the aerosol settling chamber, which is followed by the sampling of particles by the E-SPART analyzer at a certain flow rate for detecting the oscillatory motion of particles. Aerodynamic particle size and charge are calculated from (1) the phase lag between the electric drive and the particle velocity and (2) the amplitude ratio of the particle motion to that of the electric drive (Fig. 17.14) [74, 78]. The E-SPART system has the major advantage of offering the independent measurement of particle size distributions of positive, negative, and neutral particles unlike the ELPI and the eNGI methods, which provide the average charge value at different particle size ranges [73, 74, 78]. Additionally, a dynamic aerosol classification using a pharmaceutical impactor is not required due to the unique detection mechanism of the E-SPART method, which is a single particle-based measurement

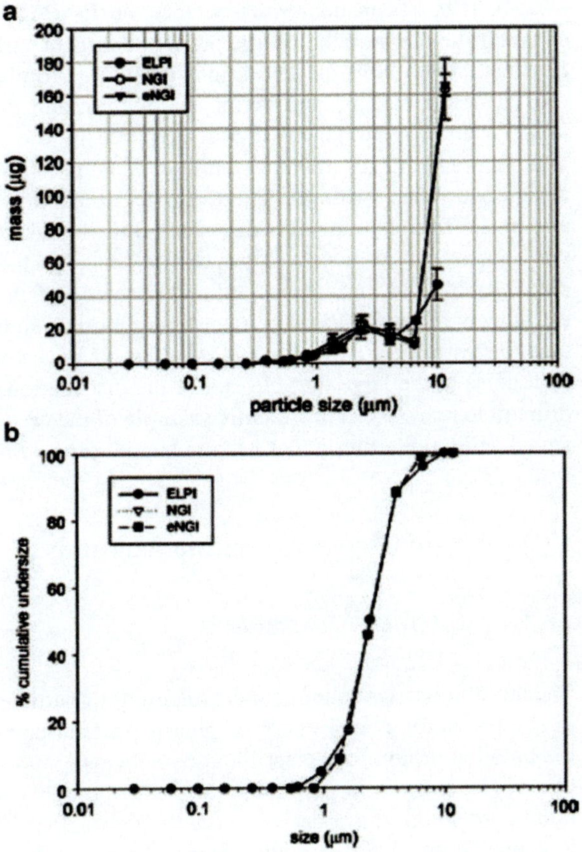

Fig. 17.13 Cumulative mass-weighted aerodynamic particle size distribution profiles for Pulmicort® (**a**) and Ventolin® (**b**) using the ELPI, NGI, and eNGI size fractionators at a flow rate of 30 L/min (Hoe et al. [37, 38] with permission)

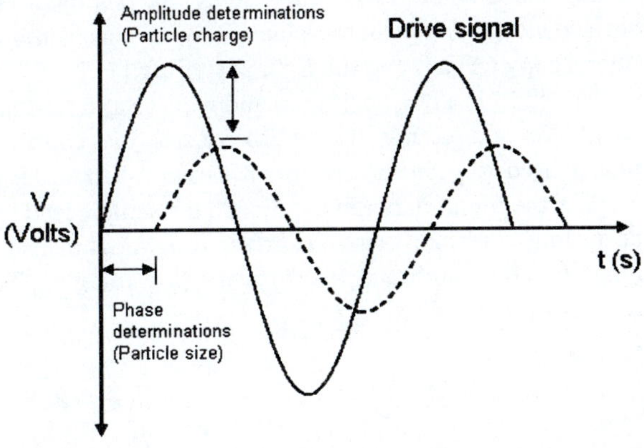

Fig. 17.14 Particle oscillatory motion in response to an alternative electric drive (this graph is modified from Saini et al. [78], with permission)

system. It has been demonstrated that the E-SPART measurement system has a capability to measure the charge profile of aerosol particles at different particle size distributions without particle impaction [78]. From the results, it was found that most aerosols had bipolar charge distribution and the individual charges showed higher values, although net charge-acquired by the aerosol was small, indicating that it has advantages in understanding the nature of materials used for formulation design over other methods. The E-SPART method is being applied to several DPI and pMDI formulations, however, care must be taken when this system is applied since many factors which may have influence on the result must be clearly validated, for instance, influences of the degree of particle separation inside the aerosol settling chamber and the flow rate during particle sampling. It is important to note that this instrument derived number, rather than mass-weighted PSDs, so that statistical noise at the large particle end of the distribution can be a problem. Also, it is difficult to know if a representative sample of the aerosol has been taken, due to the signal validation criteria that need to be met for a valid measurement.

17.5 Static Characterization Methods

17.5.1 Surface Morphology

The morphology (roughness or corrugation) of particles plays a very important role in aerosolization performance of pharmaceutical powders. It has been shown that the broad adhesion force distributions between particles are attributed to the morphology of surface, in particular, the surface roughness [2, 12, 40]. A considerable number of studies have been explored to correlate the surface structure of particles with powder flow and aerosolization performance, and the interest keeps growing as various engineered particles having unique structural features have been introduced in the market. Commonly, scanning electron microscopy is widely being applied to analyze the surface morphology of particles, however, this method is only limited to visualize the particle structure, and cannot provide numerical values for the degree of difference since it is a qualitative analysis method [12]. The quantification of surface morphology can be achieved by various techniques, such as the atomic force microscopy (AFM) and the light scattering [1, 2, 13, 35].

The AFM is a most common and well-established method to evaluate the surface roughness of a particle. The AFM consists of a cantilever with a sharp tip at its end that is used to scan the specimen surface. When the tip is brought into a sample surface, the forces between the tip and the sample lead to a deflection of the cantilever according to Hooke's law. In surface roughness measurement, the scanned image is quantified by image analysis software, post imaging. The software makes use of the relationship described by (17.1):

$$\mathrm{RMS}(R_{\mathrm{q}}) = \sqrt{\frac{(Z_i - \bar{Z})^2}{N}}, \tag{17.1}$$

where the root mean square roughness, RMS(R_q), is a quantifiable value to represent the degree of surface roughness. Z_i is the height at the position i (μm), $\overline{Z}$ is the average height (μm) of the Z values in the sample (Z-axis representing topographical height), and N is the number of data points sampled within the given area. The AFM is a very powerful technique that offers a high-resolution image scan while demonstrating a resolution at the nanometer scale. In a study conducted by Islam et al., the surface roughness of various lactose carriers was quantified by AFM imaging, and the results showed significant roughness, even though the SEM images of those particles seemed to be smooth [40]. It has also been verified quantitatively by AFM imaging that spray-dried BSA particles having a high degree of surface corrugation have a better aerosolization efficiency than that of smooth particles [2]. In this study, the fine particle fraction (FPF) values for particles having different degree of surface corrugation were plotted as a function of RMS value, and a linear correlation between FPF and RMS was found [2].

The degree of surface corrugation can also be expressed by surface fractal dimension (Ds) determined by a light scattering method [13, 86, 87]. The determination utilizes the Rayleigh-Gans-Debye (RGD) scattering theory that explains the structure of each single particle in an aggregate [87]. The value of Ds is obtained from slope of the scattering curve (Fig. 17.15) of intensity of scattered light ($I(q)$) plotted against scattering momentum (q) on a log–log scale in accordance with (17.2), below:

$$I(q) \propto q^{-6+\mathrm{Ds}}. \tag{17.2}$$

To predict the Ds value of individual particles accurately, the fractal region which is shown as a linear line in Fig. 17.15 has to be higher than 1/particle diameter [87]. The Ds value varies from 2 for a perfectly smooth surface, to 3

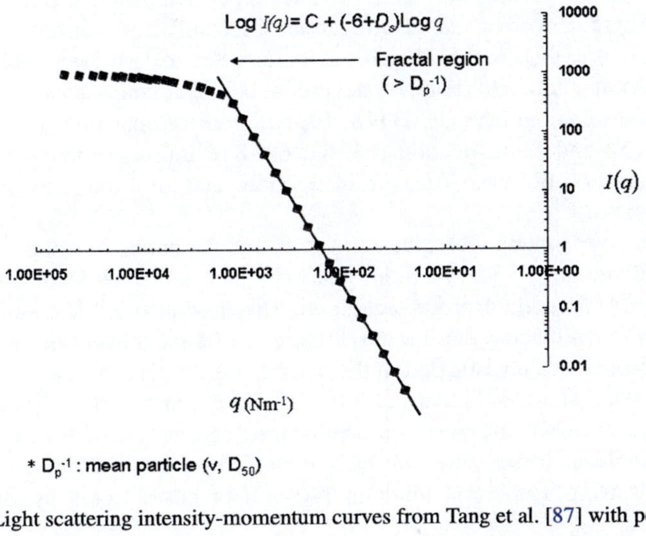

Fig. 17.15 Light scattering intensity-momentum curves from Tang et al. [87] with permission

for a very rough surface. In the measurement of spray-dried BSA particles, the Ds values for a relatively smooth particle and the most corrugated particle are 2.06 and 2.41, respectively and the powder dispersion FPF was found to improve with corrugated particles, as previously reported in other results using the AFM methods [13]. However, there are some limitations in dry powder applications since particles need to be suspended in the liquid media to avoid the multiple scattering associated with aggregated particles. Dry particles could possibly dissolve in the suspending medium which could change the surface morphology during sample preparation, thus appropriate media selection has to be achieved prior analysis.

17.5.2 Interparticulate Interaction in Dry Powder Formulations

Interparticulate forces, adhesive and cohesive force, in the dry powder mixture are a result of a complex combination of physical forces including Van der Waals forces, capillary forces, and electrostatic forces. The contribution of each force to the particle–particle adhesion is dependent on various factors, such as surface morphology, crystallinity, environmental conditions, and intrinsic material properties. For instance, at high humidity conditions the electrostatic charge between particles is generally reduced by the increasing capillary force, and under low relative humidity (i.e., <50% RH), capillary forces become less significant, and electrostatic force increases [103]. The balance of these forces is very critical in the blending, particle de-aggregation, and aerosolization properties of drug particles. In particular with carrier-based DPI formulations, the adhesion force between drug particles and large carrier particles is a predominant factor in determining the dose that is actually delivered to the lung, since only drug particles separated from the carrier can reach to the deep lung. However, the quantification of the adhesion force and the specific role of those forces on aerosolization behavior have not, to date, been fully addressed. Although adhesion forces in various systems have been determined by several methods, inverse gas chromatography (ICG) [18, 19, 90], particle detachment rate measurement [24, 25], and centrifugation [16, 67, 68, 83], those methods only offer an indirect estimation of interparticulate interactions and are limited to the measurement of bulk properties.

The direct measurement of adhesion forces and the adhesive–cohesive balances between individual colloidal particles and any given substrate become capable by utilizing the AFM colloid probe technique. The modification of commonly used AFM probes by replacing the fine tip at the end of a cantilever with a very small colloidal particle was investigated in the early 1990s to measure the force between a planar surface and a colloidal particle [28]. In pharmaceutical applications, in particular, pharmaceutical aerosol applications, a single drug particle can be mounted onto the cantilever to measure interparticulate forces. The particle attachment is achieved by using a micromanipulator [40, 93] or custom-built systems having

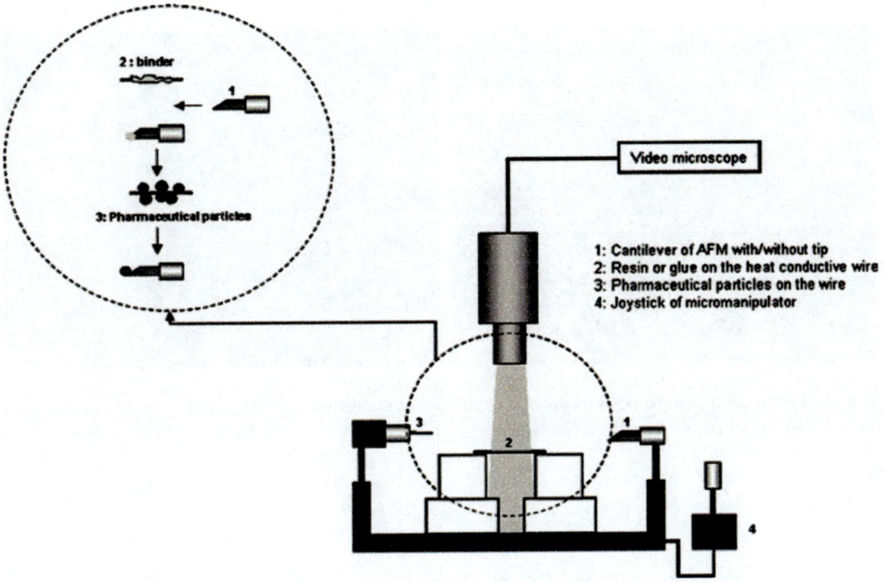

Fig. 17.16 Schematic of colloid probe preparation steps (this figure is modified from Tsukada et al. [93])

equivalent functions [2, 75, 100, 102]. The preparation steps are as follows (a schematic diagram is shown in Fig. 17.16):

1. Setting an AFM cantilever on one of the arms.
2. Adding a binder (glue or resin) on one wire and pharmaceutical particles on the other wire.
3. Contacting an apex of cantilever to the binder on the wire. (It is important to transfer only a small amount of the binder onto the cantilever to prevent embedding of the particle into the wet binder).
4. Contacting the cantilever to the particle on the wire. (Only one single particle must be attached onto the cantilever).

The whole process is performed using video microscopy as an aid to observation. Two different types of colloid probe, particle mounted on the cantilever tip, or the tipless cantilever, may be prepared as shown in Fig. 17.17a, b [93, 100]. A commercially available AFM cantilever is commonly used for preparing the colloid probes, but custom-made tipless cantilever prepared from a single wafer was recently introduced to minimize possible variation in the spring constant that might occur due to the thickness variation of the tip [100–102]. The spring constant of the fabricated cantilevers has to be determined by the attachment of known masses to the cantilever, since the spring constant is a critical factor in calculation of the adhesion force [17, 75].

The measurement of adhesion force can be achieved by moving the tip vertically using a piezoelectric translator, advancing and retreating from the sample plate as

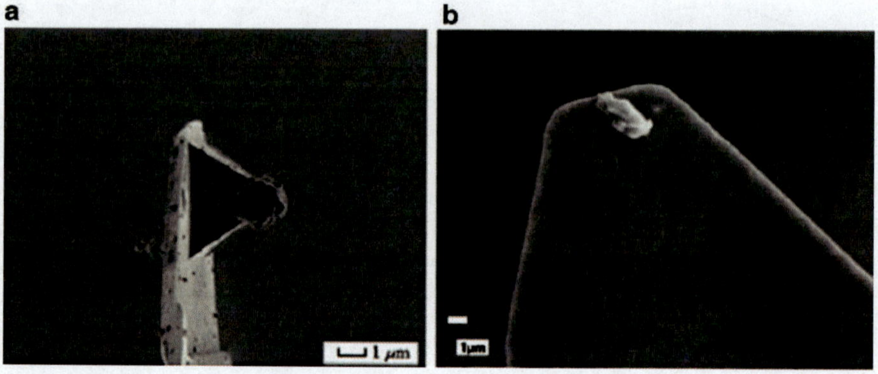

Fig. 17.17 SEM image of colloid probes where particle mounted on the cantilever tip (**a**) [93] and tipless cantilever (**b**) [99] with permission

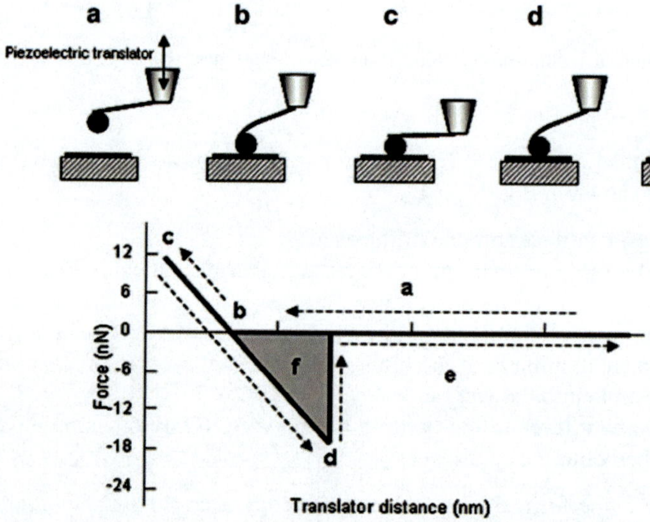

Fig. 17.18 Ideal force curve between an AFM cantilever drug probe and substrate: (**a**) approach of probe with no cantilever deflection, (**b**) contact of probe with no applied force, (**c**) applied force to the probe (**d**) negative cantilever deflection due to probe adhesion, (**e**) retraction of probe with no cantilever deflection, and (**f**) separation energy (this figure was modified from Young et al. [100])

shown in Fig. 17.18. The relative deflection of the cantilever is recorded as the tip approaches, contacts with, and moves away from a sample surface. The result of the force measurement is a plot of the cantilever deflection (Z_c) vs. the height position of piezoelectric translator (Z). The cantilever deflection is easily converted to a force measurement, as shown in (17.3), by multiplying it with the spring constant of the cantilever (*K*) to produce a force-distance curve (Fig. 17.18):

$$F = KZ_c. \tag{17.3}$$

The energy of separation between two test materials can also be calculated by integrating the area within the adhesion portion of a force-distance curve (Fig. 17.18). The use of force-distance curve offers an easy estimation of interparticulate forces for various test parameters, such as the influence of humidity [100, 102], the surface roughness of particles [2, 40], type of carrier materials [42, 43], and the structural influences of the colloid particle attached on the cantilever on the drug–drug interaction [9, 93]. These series of studies provide a variety of useful information that play a pivotal role in predicting blending, particle segregation, de-aggregation of fine particle, powder dispersion, and optimal environmental conditions. The separation energy of salbutamol (albuterol) sulfate (SS), for instance, has been shown to increase as environmental humidity was increased from 15% RH to 75% RH due to capillary forces that may be induced by the multilayer absorption of water onto the particle surface [100]. In contrast to the SS, an increase in humidity resulted in decreased separation force for triamcinolone acetonide (TAA) due to the elimination of dominant electrostatic forces at lower relative humidity [100–102]. As shown in above studies, the interparticulate forces are very drug specific, thus the screening of such factors related to the powder dispersion at a preformulation stage could make a useful contribution in the optimization of dry powder formulations.

Another useful technique, which makes use of the AFM colloid probe, is a cohesive–adhesive balance (CAB) measurement. The CAB enables the quantification of cohesive and adhesive forces without normalization of the probe-substrate contact area through the determination [9, 39]. Although the AFM colloid probe technique is very useful to measure the interparticulate forces, there is a lack of uniformity in contact area between the colloid probe and the substrate of second chemical, resulting in large variations in results, due to the irregular roughness of both colloid particle and the substrate surface. In order to overcome these limitations, the adhesion force and cohesion force balance between two materials can be measured, instead of the separation energy, by plotting the cohesive forces vs. the adhesive forces obtained with a series of colloid probes, the CAB of the interacting drug particles can be quantified by linear regression analysis of the respective slope [9]. In this measurement, the variation in the surface roughness of probes and substrates can be ignored, since the work of adhesion and contact angle are only dependent on the physical characteristics of materials. Moreover, the CAB measurement allows direct comparison of the cohesive and adhesive tendency of an API and carrier materials, since the CAB plot provides a specific cohesive/adhesive ratio that describes the cohesion of the colloidal probe material relative to its adhesion to the substrate of the second material. If the CAB ratio is less than 1, this is indicative of a situation where the adhesion force is greater than the cohesion force; when the CAB ratio is greater than 1, a system where the cohesion force is greater than adhesion force is present. The CAB technique has proven to be a powerful tool in understanding the blend uniformity, drug particle segregation, and powder dispersion properties in the carrier-based dry powder systems [9, 42, 43].

17.5.3 *Drug Solubility Screening for Propellants*

A variety of APIs have been formulated into pressurized metered-dose inhaler (pMDI) systems that contain supercritical propellants, mainly hydrofluoroalkanes (HFAs), to achieve a high-energy atomization of therapeutic APIs. The pMDI formulations are largely divided into suspension or solution-based systems. In solution-based pMDI formulations, the API is dissolved with nonvolatile cosolvents, such as ethanol, while in the suspension system micronized API particles are suspended in the propellant. In both types of formulation, the understanding of the solvation behavior of an API in various vehicles (propellants or the mixture of propellants and cosolvents) is very important for the development of stable pMDI formulations since solubilized drugs are easily subjected to chemical degradation, and crystal growth that can lead to a reduced deposition of drug into the lung [22, 76]. However, it is technically very difficult to determine drug solubility in a pressurized suspension formulation without perturbation of equilibrium established in the liquid, as the propellants are gases at the room temperature. The cooling of propellants could reduce the drug solubility and without cooling, gas-stated propellant will evaporate very quickly at room temperature.

The first in situ method for determining drug solubility in volatile propellant systems was introduced by Dalby et al. [22]. A device consisting of a closed gas-tight system with a donor container containing formulations, an empty receiving container, and a filtering unit positioned between two donor and receiving containers (Fig. 17.19a). Upon actuation of donor container, positioned on the top of the device, propellant and any dissolved drug pass through the filtering unit and are collected in the receiver. The weight of collected propellant is measured by subtracting the preweighed value of the receiver compartment from the weight postsample collection, followed by opening the receiver to release propellant, and subsequent collection of the dissolved drug. In this study, it was demonstrated how

Fig. 17.19 Schematic of the filtration unit (**a**), and the online reversed-phase HPLC system (**b**) for in situ drug solubility study in propellants (from Dalby et al. [22] with permission)

drug solubility in the propellant can result in Oswald ripening, a thermodynamically spontaneous process that occurs by dissolution of small particles in the carrier media to become energetically favorable large particles [22]. This in situ measurement system has been modified by Williams et al. [97] and Traini et al. [92] and successfully applied to evaluate the solubility of various APIs.

An online reversed-phase HPLC system developed by Gupta et al. [32, 33] is an another powerful technique to measure drug solubility. This method offers a one-step process, so that the sampling of dissolved drug and analysis can be achieved by direct sample injection. This online technique enables the quantification of low levels of analytes by eliminating several sample transfer and dilution steps. The instrumentation set-up consists of an HPLC system coupled with a manual sample injector instead of an auto-sampler, and a pMDI device connected with a filtration and injection assembly (Fig. 17.19b). Direct injection of dissolved drug into the HPLC system can be achieved by inserting the needle of the pMDI vial into the HPLC sample injector before pMDI actuation. The online system has several advantages compared to conventional methods, such as being less labor intensive, and having better detection sensitivity and reproducibility [33].

17.6 In Vitro Dissolution Studies

17.6.1 Existing Test Methods and Issues Associated with Current Methods

Dissolution testing allows one to examine the drug release behavior of pharmaceutical dosage forms in vitro in order to differentiate formulation types, and perhaps gives an estimate of a dissolution behavior in vivo. Dissolution testing is routinely used in QC studies such as: batch-to-batch consistency, stability, and detection of manufacturing deviations. Several USP general chapters (i.e., <711>, <724>, <1088>, and <1092>) play a key role in these evaluations as the gold standard tests. Those tests are routinely successfully applied to estimate in vitro dissolution behaviors of solid or semi-solid dosage forms [94]. However, there is no such universally accepted method to estimate the dissolution behavior of inhaled active ingredients, although many dissolution methods for testing aerosols have been investigated [31]. Designing a standardized method applicable to the lung is not an easy task, as the lung has several unique features which are difficult to replicate in vitro, such as the extremely small amount of aqueous fluid, and the presence of endogenous lung surfactants [31, 70].

For inhalation products, one of the most important steps with in vitro performance testing is characterization of the delivery of a given API from a specified delivery device using a pharmaceutical impactor/impinger to estimate actual dose delivered to the target site of the lung. Only a fraction of the API emitted from standard delivery devices is usually delivered to the target site of the deep lung since most inhaler products are a mixture of fine API particles and course carrier particles or propellants [45]. Thus, an ideal dissolution test procedure for inhalation formulations would involve particle classification followed by an evaluation of the dissolution behavior for the classified drug particles that may deposit at various sites in the respiratory tract. Additionally, a stagnant dissolution system, rather than well-stirred system, would be required to estimate lung dissolution as the volume of lung fluid is approximately 10–20 mL/100 m^2 [70].

Experimental difficulties exist in dose collection due to very fine powder and their electrostatic characteristics [31]. Therefore, to date most dissolution procedures on powders have been performed in the absence of aerodynamic classification, whereby formulations may have been directly dispersed into an apparatus II dissolution tester [4], or placed directly into a modified basket to prevent drug particles from escaping directly into the dissolution medium [41, 46]. However, those methods using a commercially available dissolution apparatus have following limitations:

1. Formulations intended for pulmonary delivery are hard to disperse homogeneously into the vessel/basket.
2. Dispersed particles stick on to the vessel wall and/or paddle/basket during such dissolution tests.
3. Floating powders may be inadvertently collected during the sampling procedure.
4. Existing paddle/basket apparatus provides well-stirred environment.

In an attempt to make up for some of the shortfalls of this type of testing using commercially available dissolution systems, several diffusion systems, a custom-built diffusion cell [20], a twin-stage impinger (TSI) [54], and a dissolution cell [3, 79], have been investigated. However, no single in vitro test system has yet emerged as the ideal choice for performing dissolution measurements for inhalation formulations.

17.6.2 New Dissolution Approaches

17.6.2.1 Flow-Through Cell

Several attempts to adapt a USP flow-though cell system (Apparatus 4) for evaluating in vitro dissolution behaviors of aerosol products have been made, since the flow-through apparatus offers specific sample cells associated with filter system that may hold powder and granular dosage forms inside the cell [6, 80]. In this system, the dissolution media is supplied by pump force, and the release/dissolution profiles of loaded drug inside the cell can be estimated from the amount of drug released from the flow-through cell. It has been demonstrated that the use of flow-though cell system could possibly differentiate the variables in dissolution patterns of several inhalation formulations by Taylor et al. [88]. However, there still remain issues with dose collection onto the flow-through cell and the hydrodynamic condition of the system in mimicking the dissolution behavior of delivered drug particles to the target lung site. In particular, the powder presentation inside the cell undoubtedly has a great influence on the overall release/dissolution pattern as the surface area, and the wettability of powder bed varies according to the loading method.

Aerodynamic classification to collect aerosol particles for dissolution study has been attempted by Davies and Feddah [23]. In this method, dose collection from DPI and pMDI devices onto a glass filter was achieved by aerodynamic classification using Anderson Mark II Cascade Impactor. Following dose collection, the glass filter containing classified particles was placed inside the modified flow-through cell, which was then subjected to dissolution testing. The system consists of a reservoir and an HPLC pump for supplying a dissolution media and a flow-through cell. The impacts of dissolution media, flow rate and surfactant on the dissolution of corticosteroids, budesonide, fluticasone propionate, and TAA, were evaluated using this modified device. The presented dissolution data in this article well reflected all the variables in test procedure, and indicated that this method had a discriminatory capability. Moreover, the procedure employed a relatively lower flow rate (0.7 mL/min) than that commonly used in the USP apparatus (4–16 mL/min), which could potentially create more stagnant conditions for dissolution.

17.6.2.2 Membrane Holder

The concept of using a membrane holder to assess the dissolution profiles of aerodynamically separated drug particles was recently introduced by Son and McConville [82]. This membrane holder was designed to be easily incorporated into the commercially available dissolution tester. The membrane holder acts as a sinker device with a diffusion layer at its surface to enable dissolution testing. In these studies, dose collection was achieved after aerodynamic size-fractionation directly on to polycarbonate membrane into the NGI collection cup of each stage, as shown in Fig. 17.20. Following aerodynamic separation, the classified drug particles on their respective membranes were sandwiched underneath another identical membrane (presoaked with dissolution medium). The drug now contained within two membranes is clamped into the holder and then placed into a dissolution vessel. The dissolution profiles for a model drug (hydrocortisone) were successively estimated by the amount of drug released from the membrane holder. It was found that there was significant difference between the bulk formulation and an aerodynamically classified formulation in the dissolution profile [82]. However, in this dissolution setup, a modification of NGI was required to collect dispersed particles on the membrane, and the entire classified dose collected on the membrane cannot be used for dissolution testing due to a substantial limitation imparted by the prototype frame holder size.

A subsequent modified membrane holder has been introduced by Son et al. [81]. It was designed specifically to be incorporated into the Next Generation Impactor (NGI) for better dose collection performance, than a previously reported prototype membrane holder [81]. The features of both a prototype and a modified holder are

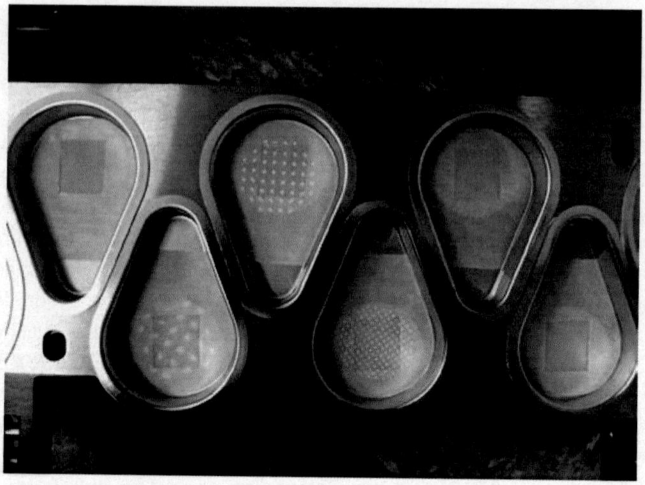

Fig. 17.20 Aerodynamic separation into the modified next generation impactor (NGI) for dose collection (from Son and McConville [82] with permission)

Table 17.3 Features of prototype and modified membrane holder for dissolution study

	Prototype membrane holder	Modified membrane holder
Housing material	Polyethylene	Stainless steel
Diffusion area	10 cm^2	18 cm^2
Membrane exposure	Two sides	One side
Membrane type	Polycarbonate (0.05 μm pore)	Polycarbonate (0.05 μm pore)
Dissolution apparatus incorporated	Commercial dissolution tester (paddle apparatus for small vol. vessel)	USP apparatus 2
Maximum dissolution volume hold (mL)	100	900
Optimized agitating speed (rpm)	50	75
Test material	Micronized hydrocortisone (HS) Lactose monohydrate/HS mixture	Ventolin® HFA Pulmicort™ Flexhaler
Aerodynamic separation	On the modified NGI dose-plate	On the impaction insert

compared in Table 17.3. Two main components of the dissolution setup include USP dissolution apparatus 2 and a newly designed membrane holder. The membrane holder assembly (Fig. 17.21a) consists of an NGI dissolution cup (a) which contains a removable impaction inset (b), a securing ring (c), two sealing 0-rings, and a polycarbonate (PC) membrane to function as a highly porous diffusional powder retaining layer.

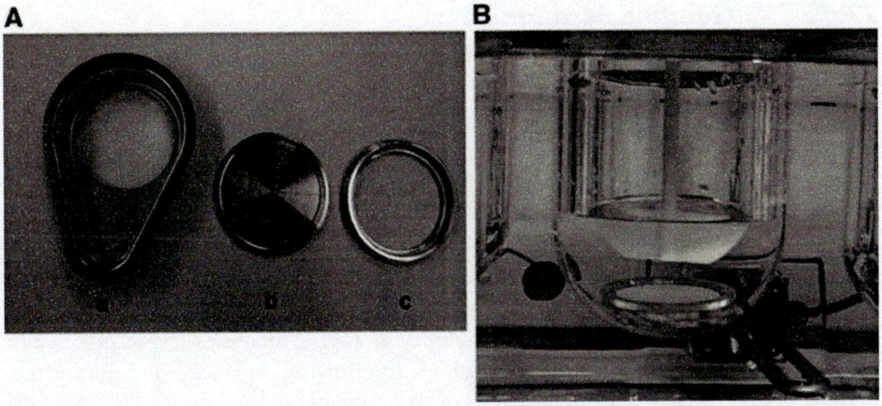

Fig. 17.21 Schematic diagram of the dissolution apparatus. Component (**A**) membrane holder assembly: (**a**) an NGI dissolution cup, (**b**) an impaction insert and (**c**) a securing ring, and component (**B**): dissolution station (from [81], with permission)

The dissolution test using membrane holder is conducted by the following procedures:

1. The NGI dissolution cup is placed in the NGI cup tray.
2. Dose collection is achieved by aerodynamic separation of given formulation.
3. The impaction insert is removed from the NGI dissolution cup.
4. A presoaked membrane was placed onto the top and sealed in place with the securing ring.
5. The sealed membrane holder was placed into each dissolution vessel (Fig. 17.21b).

The mechanism of this dissolution method can be explained by dissolution diffusion-controlled drug release from the membrane holder. During the dissolution process, the dispersed drug within the membrane holder undergoes dissolution, as dissolution media migrates through the pores on the membrane surface and the dissolved drug then releases out to the bulk media by diffusion [82]. The membrane holder may impart optimally stagnant conditions, in order for test compounds to be appropriate for comparing dissolution behaviors of formulations intended for lung administration. This membrane holder was designed to have a very thin liquid layer between the membrane, and the liquid layer was maintained to dissolve drug particle by continuous exchange of dissolution media through the pores by agitation. In other words, a dynamic equilibrium is rapidly established for solute exchange, following the commencement of a given dissolution test with the membrane holder. The dissolution profiles obtained in this study were shown to differ substantially between those from drugs having dissimilar hydrophobicity as well as the aerosols having different APSDs. Notably, this method can be used to examine the dissolution behaviors of inhalation dosage forms with similar particle distribution by selecting drug particles accumulated in same collection plate. It would be anticipated that the dissolution of particles having a similar size distributions would provide more interformulation discrimination.

17.6.3 Dissolution Medium

Two types of simulated lung fluid (SLF), simulated interstitial lung fluid (SILF), and simulated surfactant lung fluid (SSLF) have been widely applied to in vitro dissolution studies due to their similarity in composition to actual lung fluid [3, 23, 63]. The composition of the SILF and the SSLF is shown in Table 17.4. The simulate lung interstitial fluid (SLF) was developed by Moss [63]. The use of SSLF that is prepared by adding 0.02% of DPPC to SILF was suggested by Dennis et al. since DPPC is the primary surface-active component found in all mammalian lungs [30]. The SSLF has been used to predict the solubility and solubilization process of inhalation formulations which have very low aqueous solubility as DPPC increases the wettability of hydrophobic drugs, and prevent aggregation to allow an enhanced dissolution rate [72, 82, 96]. However, SLF media may not be preferable for the routine QC study for inhalation products due to its low buffering effect. In particular, the use of SLF media would not be recommended for evaluating inhalation dosage forms that show pH-dependency or sustained-release manner in dissolution profile. It has been reported that the pH of SLF media significantly increased up to 8.7 without CO_2 bubbling [63]. Additionally, in the membrane-based dissolution apparatus, it was found that the SSLF media may not be freely diffuse and reach the drug particles through the pore on the membrane surface as DPPC forms liposomal aggregates that may have a larger particle size than the membrane pore size in aqueous media [82]. For these reasons, recently, alternative biological buffer systems having similar pH range to that of the target lung site, such as phosphate buffer and PBS, have been applied for evaluating newly developed sustained release formulations [20, 46, 66, 82, 85]. The standardized dissolution media has not been developed yet because the media selection for

Table 17.4 Compositions of actual lung fluid, and two simulated lung fluids, SILF and SSLF

Ions	Actual (meq)	SILF (meq)	SSLF (meq)
Calcium	5.0	5.0	5.0
Magnesium	2.0	2.0	2.0
Potassium	4.0	4.0	4.0
Sodium	145.0	145.0	145.0
Total cations	156.0	156.0	156.0
Bicarbonate	31.0	31.0	31.0
Chloride	114.0	114.0	114.0
Citrate	–	1.0	1.0
Acetate	7.0	7.0	7.0
Phosphate	2.0	2.0	2.0
Sulfate	1.0	1.0	1.0
Protein	1.0	–	–
DPPC	–	–	200 μg/mL
Total anions	156.0	156.0	156.0
pH	7.3	7.3	7.3

Table is modified from Dennis et al. [26]

lung dissolution is quite dependent on the test apparatus, the chemical/physical properties of APIs as described above. Thus, careful and thorough examinations and validation in the selection of media must be made, especially if one seeks to establish valid in vitro–in vivo correlations.

References

1. Adi H, Traini D, Chan HK, Young PM (2008) The influence of drug morphology on the aerosolisation efficiency of dry powder inhaler formulations. J Pharm Sci 97:2780–2788
2. Adi S, Adi H, Tang P, Traini D, Chan HK, Young PM (2008) Micro-particle corrugation, adhesion and inhalation aerosol efficiency. Eur J Pharm Sci 35:12–18
3. Ansoborlo E, Guilmette RA, Hoover MD, Chazel V, Houpert P, Henge-Napoli MH (1998) Application of in vitro dissolution tests to different uranium compounds and comparison with in vivo data. Radiat Prot Dosim 79:33–37
4. Asada M, Takahashi H, Okamoto H, Tanino H, Danjo K (2004) Theophylline particle design using chitosan by the spray drying. Int J Pharm 270:167–174
5. Asking L, Olsson B (1997) Calibration at different flow rates of a multistage liquid impinger. Aerosol Sci Technol 27:39–49
6. Azarmi S, Roa W, Lobenberg R (2007) Current perspectives in dissolution testing of conventional and novel dosage forms. Int J Pharm 328:12–21
7. Bailey AG (1993) Charging of solids and powders. J Electrostat 30:167–180
8. Balachandran W, Machowski W, Gaura E, Hudson C (1997) Control of drug aerosol in human airways using electrostatic forces. J Electrostat 40–1:579–584
9. Begat P, Morton DAV, Staniforth JN, Price R (2004) The cohesive-adhesive balances in dry powder inhaler formulations I: direct quantification by atomic force microscopy. Pharm Res 21:1591–1597
10. Bonam M, Christopher D, Cipolla D, Donovan B, Goodwin D, Holmes S, Lyapustina S, Mitchell J, Nichols S, Pettersson G, Quale C, Rao N, Singh D, Tougas T, Van Oort M, Walther B, Wyka B (2008) Minimizing variability of cascade impaction measurements in inhalers and nebulizers. AAPS PharmSciTech 9(2):404–413
11. Byron PR, Peart J, Staniforth JN (1997) Aerosol electrostatics.1. Properties of fine powders before and after aerosolization by dry powder inhalers. Pharm Res 14:698–705
12. Chew NYK, Chan HK (2001) Use of solid corrugated particles to enhance powder aerosol performance. Pharm Res 18:1570–1577
13. Chew NYK, Tang P, Chan HK, Raper JA (2005) How much particle surface corrugation is sufficient to improve aerosol performance of powders? Pharm Res 22:148–152
14. Chow KT, Zhu K, Tan RBH, Heng PWS (2008) Investigation of electrostatic behavior of a lactose carrier for dry powder inhalers. Pharm Res 25:2822–2834
15. Christopher D, Curry P, Doub B, Furnkranz K, Lavery M, Lin K, Lyapustina S, Mitchell J, Rogers B, Strickland H, Tougas T, Tsong Y, Wyka B (2003) Considerations for the development and practice of cascade impaction testing, including a mass balance failure investigation tree. J Aerosol Med 16:235–247
16. Clarke MJ, Peart J, Cagnani S, Byron PR (2002) Adhesion of powders for inhalation: an evaluation of drug detachment from surfaces following deposition from aerosol streams. Pharm Res 19:322–329
17. Cleveland JP, Manne S, Bocek D, Hansma PK (1993) A nondestructive method for determining the spring constant of cantilevers for scanning force microscopy. Rev Sci Instrum 64:403–405
18. Cline D, Dalby R (2002) Predicting the quality of powders for inhalation from surface energy and area. Pharm Res 19:1274–1277
19. Columbano A, Buckton G, Wikeley P (2003) Characterisation of surface modified salbutamol sulphate-alkylpolyglycoside miciroparticles prepared by spray drying. Int J Pharm 253:61–70

20. Cook RO, Pannu RK, Kellaway IW (2005) Novel sustained release microspheres for pulmonary drug delivery. J Control Release 104:79–90
21. Copley M (2007) Assessing the impact of the NGI. Pharmaceutical Manufacturing and Packaging Sourcer, Autumn
22. Dalby RN, Phillips EM, Byron PR (1991) Determination of drug solubility in aerosol propellants. Pharm Res 8:1206–1209
23. Davies NM, Feddah MIR (2003) A novel method for assessing dissolution of aerosol inhaler products. Int J Pharm 255:175–187
24. de Boer AH, Hagedoorn P, Gjaltema D, Lambregts D, Irngartinger M, Frijlink HW (2004) The mode of drug particle detachment from carrier crystals in an air classifier-based inhaler. Pharm Res 21:2167–2174
25. de Boer AH, Hagedoorn P, Gjaltema D, Lambregts D, Irngartinger M, Frijlink HW (2004) The rate of drug particle detachment from carrier crystals in an air classifier-based inhaler. Pharm Res 21:2158–2166
26. Dennis NA, Blauser HM, Kent JE (1982) Dissolution fractions and half-times of single source yellowcake in simulated lung fluids. Health Phys 42:469–477
27. Dolovich MB, Mitchell JP (2004) Canadian Standards Association standard CAN/CSA/Z264.1-02:2002: a new voluntary standard for spacers and holding chambers used with pressurized metered-dose inhalers. Can Respir J 11:489–495
28. Ducker WA, Senden TJ, Pashley RM (1991) Direct measurement of colloidal forces using an atomic force microscope. Nature 353:239–241
29. Glover W, Chan HK (2004) Electrostatic charge characterization of pharmaceutical aerosols using electrical low-pressure impaction (ELPI). J Aerosol Sci 35:755–764
30. Goerke J (1998) Pulmonary surfactant: functions and molecular composition. Biochim Biophys Acta 1408:79–89
31. Grey VA, Hickey AJ, Balmer P, Davies NM, Dunbar C, Foster TS, Olsson BL, Sakagami M, Shah VP, Smurthwaite MJ, Veranth JM, Zaidi K (2008) The Inhalation Ad Hoc Advisory Panel for the USP performance tests of inhalation dosage forms. Pharmacopeial Forum 34:1068–1074
32. Gupta A, Myrdal PB (2004) On-line high-performance liquid chromatography method for analyte quantitation from pressurized metered dose inhalers. J Chromatogr A 1033:101–106
33. Gupta A, Myrdal PB (2005) A comparison of two methods to determine the solubility of compounds in aerosol propellants. Int J Pharm 292:201–209
34. Hallworth GW, Westmoreland DG (1987) The twin impinger – a simple device for assessing the delivery of drugs from metered dose pressurized aerosol inhalers. J Pharm Pharmacol 39:966–972
35. Hickey AJ, Mansour HM, Telko MJ, Xu Z, Smyth HDC, Mulder T, McLean R, Langridge J, Papadopoulos D (2007) Physical characterization of component particles included in dry powder inhalers. I. Strategy review and static characteristics. J Pharm Sci 96:1282–1301
36. Hickey AJ, Mansour HM, Telko MJ, Xu Z, Smyth HDC, Mulder T, McLean R, Langridge J, Papadopoulos D (2007) Physical characterization of component particles included in dry powder inhalers. II. Dynamic characteristics. J Pharm Sci 96:1302–1319
37. Hoe S, Traini D, Chan HK, Young PM (2009) Measuring charge and mass distributions in dry powder inhalers using the electrical next generation impactor (eNGI). Eur J Pharm Sci 38:88–94
38. Hoe S, Young PM, Chan HK, Traini D (2009) Introduction of the electrical next generation impactor (eNGI) and investigation of its capabilities for the study of pressurized metered dose inhalers. Pharm Res 26:431–437
39. Hooton JC, Jones MD, Price R (2006) Predicting the behavior of novel sugar carriers for dry powder inhaler formulations via the use of a cohesive-adhesive force balance approach. J Pharm Sci 95:1288–1297
40. Islam N, Stewart P, Larson I, Hartley P (2005) Surface roughness contribution to the adhesion force distribution of salmeterol xinafoate on lactose carriers by atomic force microscopy. J Pharm Sci 94:1500–1511
41. Jaspart S, Bertholet P, Piel G, Dogne JM, Delattre L, Evrard B (2007) Solid lipid microparticles as a sustained release system for pulmonary drug delivery. Eur J Pharm Biopharm 65:47–56

42. Jones MD, Harris H, Hooton JC, Shur J, King GS, Mathoulin CA, Nichol K, Smith TL, Dawson ML, Ferrie AR, Price R (2008) An investigation into the relationship between carrier-based dry powder inhalation performance and formulation cohesive-adhesive force balances. Eur J Pharm Biopharm 69:496–507
43. Jones MD, Hooton JC, Dawson ML, Ferrie AR, Price R (2008) An investigation into the dispersion mechanisms of ternary dry powder inhaler formulations by the quantification of interparticulate forces. Pharm Res 25:337–348
44. Kwok PCL, Glover W, Chan HK (2005) Electrostatic charge characteristics of aerosols produced from metered dose inhalers. J Pharm Sci 94:2789–2799
45. Labiris NR, Dolovich MB (2003) Pulmonary drug delivery. Part I: physiological factors affecting therapeutic effectiveness of aerosolized medications. Br J Clin Pharmacol 56:588–599
46. Learoyd TP, Burrows JL, French E, Seville PC (2008) Chitosan-based spray-dried respirable powders for sustained delivery of terbutaline sulfate. Eur J Pharm Biopharm 68:224–234
47. Lundbäck H, Wiktorsson B (2006) High throughput inhaler testing I: Fine particle dose. In: Dalby RN, Byron PR, Peart J, Suman JD, Farr SJ (eds) Respiratory Drug Delivery. Davis Healthcare International Publishing LLC, River Grove, Illinois, USA, pp 467–469
48. Marjamaki M, Keskinen J, Chen DR, Pui DYH (2000) Performance evaluation of the electrical low-pressure impactor (ELPI). J Aerosol Sci 31:249–261
49. Marple VA, Liu BYH (1974) Characteristics of laminar-jet impactors. Environ Sci Technol 8:648–654
50. Marple VA, Olson BA, Miller NC (1995) A low-loss cascade impactor with stage collection cups – calibration and pharmaceutical inhaler applications. Aerosol Sci Technol 22:124–134
51. Marple VA, Olson BA, Santhanakrishnan K, Mitchell JP, Murray SC, Hudson-Curtis BL (2003) Next generation pharmaceutical impactor (a new impactor for pharmaceutical inhaler testing). Part II: archival calibration. J Aerosol Med 16:301–324
52. Marple VA, Olson BA, Santhanakrishnan K, Roberts DL, Mitchell JP, Hudson-Curtis BL (2004) Next generation pharmaceutical impactor: a new impactor for pharmaceutical inhaler testing. Part III. Extension of archival calibration to 15 L/min. J Aerosol Med 17:335–343
53. Marple VA, Roberts DL, Romay FJ, Miller NC, Truman KG, Holroyd MJ, Mitchell JP, Hochrainer D (2003) Next generation pharmaceutical impactor (a new impactor for pharmaceutical inhaler testing). Part I: design. J Aerosol Med 16:283–299
54. McConville JT, Patel N, Ditchburn N, Tobyn MJ, Staniforth JN, Woodcock P (2000) Use of a novel modified TSI for the evaluation of controlled-release aerosol formulations. I. Drug Dev Ind Pharm 26:1191–1198
55. Mitchell J, Dalby R (2006) Characterization of Aerosol Performance. Chapter 5 in Pulmonary Drug Delivery – Basics, Applications and Opportunities for Small Molecules and Bio-Pharmaceuticals. Ed. K. Bechtold-Peters and H. Lüssen. Editio Cantor Verlag, Aulendorf, Germany, pp. 282–305
56. Mitchell JP, Dunbar C (2005) The interpretation of data from cascade impactors. J Aerosol Med 18:439–451
57. Mitchell JP, Nagel MW (2003) Cascade impactors for the size characterization of aerosols from medical inhalers: their uses and limitations. J Aerosol Med 16:341–377
58. Mitchell JP, Nagel MW, Avvakoumova V, MacKay H, Ali R (2009) The abbreviated impactor measurement (AIM) concept: part 1-influence of particle bounce and re-entrainment-evaluation with a "dry" pressurized metered dose inhaler (pMDI)-based formulation. AAPS PharmSciTech 10:243–251
59. Mitchell JP, Nagel MW, Avvakoumova V, MacKay H, Ali R (2009) The abbreviated impactor measurement (AIM) concept: part ii-influence of evaporation of a volatile component-evaluation with a "droplet-producing" pressurized metered dose inhaler (pMDI)-based formulation containing ethanol as co-solvent. AAPS PharmSciTech 10:252–257
60. Mitchell JP, Nagel MW, Copley M (2009) The abbreviated impactor measurement concept: a potentially faster and more precise way to assess quality of inhaled drug products. Inhalation 3:26–30

61. Mitchell JP, Nagel MW, Doyle CC, Ali RS, Avvakoumova VI, Christopher JD, Quiroz J, Strickland H, Tougas T, Lyapustina S (2010) Relative precision of inhaler aerodynamic particle size distribution (APSD) metrics by full resolution and abbreviated Andersen cascade impactors (ACIs). Part2: investigation of bias in extra-fine mass fraction with AIM-HRT impactor. AAPS PharmSciTech 11(3):1115–1118
62. Mitchell JP, Nagel MW, Doyle CC, Ali RS, Avvakoumova VI, Christopher JD, Quiroz J, Strickland H, Tougas T, Lyapustina S (2010) Relative precision of inhaler aerodynamic particle size distribution (APSD) metrics by full resolution and abbreviated Andersen cascade impactors (ACIs). Part 1: designed experiment. AAPS PharmSciTechnol 11(2):843–851
63. Moss OR (1979) Simulants of lung interstitial fluid. Health Phys 36:447–448
64. Murtomaa M, Mellin V, Harjunen P, Lankinen T, Laine E, Lehto VP (2004) Effect of particle morphology on the triboelectrification in dry powder inhalers. Int J Pharm 282:107–114
65. Nichols SC, Brown DR, Smurthwaite M (1998) New concept for the variable flow rate Andersen cascade impactor and calibration data. J Aerosol Med 11:S133–S138
66. O'Hara P, Hickey AJ (2000) Respirable PLGA microspheres containing rifampicin for the treatment of tuberculosis: manufacture and characterization. Pharm Res 17:955–961
67. Okada J, Matsuda Y, Fukumori Y (1969) Measurement of adhesive force of pharmaceutical powders by centrifugal method. I. Yakugaku Zasshi 89:1539–1544
68. Okada J, Matsuda Y, Fukumori Y (1971) Measurement of adhesive force of pharmaceutical powders by centrifugal method. 2. Relation between surface condition of substrates and separation force-particle residue curves. Yakugaku Zasshi 91:1207.
69. Olson BA, Marple VA, Mitchell JP, Nagel MW (1998) Development and calibration of a low-flow version of the Marple-Miller impactor (MMI (TM)). Aerosol Sci Technol 29:307–314
70. Patton JS (1996) Mechanisms of macromolecule absorption by the lungs. Adv Drug Deliv Rev 19:3–36
71. Peart J, Staniforth JN, Meakin BJ (1995) Electrostatic charge interactions in pharmaceutical dry powder aerosols. Electrostatics 1995(143):271–274
72. Pham S, Wiedmann TS (2001) Dissolution of aerosol particles of budesonide in Survanta (TM), a model lung surfactant. J Pharm Sci 90:98–104
73. Philip VA, Mehta RC, DeLuca PP, Mazumder MK (1997) E-SPART analysis of poly(D, L-lactide-co-glycolide) microspheres formulated for dry powder aerosols. Part Sci Technol 15:303–316
74. Philip VA, Mehta RC, Mazumder MK, DeLuca PP (1997) Effect of surface treatment on the respirable fractions of PLGA microspheres formulated for dry powder inhalers. Int J Pharm 151:165–174
75. Preuss M, Butt HJ (1998) Direct measurement of particle-bubble interactions in aqueous electrolyte: dependence on surfactant. Langmuir 14:3164–3174
76. Pritchard JN (2001) The influence of lung deposition on clinical response. J Aerosol Med 14:S19–S26
77. Rader DJ, Marple VA (1985) Effect of ultra-Stokesian drag and particle interception on impaction characteristics. Aerosol Sci Technol 4:141–156
78. Saini D, Biris AS, Srirama PK, Mazumder MK (2007) Particle size and charge distribution analysis of pharmaceutical aerosols generated by inhalers. Pharm Dev Technol 12:35–41
79. Sdraulig S, Franich R, Tinker RA, Solomon S, O'Brien R, Johnston PN (2008) In vitro dissolution studies of uranium bearing material in simulated lung fluid. J Environ Radioact 99:527–538
80. Siewert M, Dressman J, Brown CK, Shah VP (2003) FIP/AAPS guidelines to dissolution/in vitro release testing of novel/special dosage forms. AAPS PharmSciTech 4:E7
81. Son Y-J, Horng M, Copley M, McConville JT (2010) Optimization of an in vitro dissolution test method for inhalation formulations. Dissolution Technologies, 17(2), May 6–13
82. Son Y-J, McConville JT (2009) Development of a standardized dissolution test method for inhaled pharmaceutical formulations. Int J Pharm 382(1–2):15–22
83. Staniforth JN, Rees JE, Lai FK, Hersey JA (1981) Determination of interparticulate forces in ordered powder mixes. J Pharm Pharmacol 33:485–490

84. Stobbs B, McAulay E, Bogard H, Monsallier E (2009) Evaluation of the fast screening impactor for determining fine particle fraction of dry powder inhalers. Drug Delivery to the Lung 20, Edinburgh 158–161
85. Sung JC, Padilla DJ, Garcia-Contreras L, VerBerkmoes JL, Durbin D, Peloquin CA, Elbert KJ, Hickey AJ, Edwards DA (2009) Formulation and pharmacokinetics of self-assembled rifampicin nanoparticle systems for pulmonary delivery. Pharm Res 26:1847–1855
86. Tang P, Chan HK, Raper JA (2009) Validation of computation method to predict aerodynamic diameter of particles with rough surface. Powder Technol 192:74–84
87. Tang P, Chew NYK, Chan HK, Raper JA (2003) Limitation of determination of surface fractal dimension using N-2 adsorption isotherms and modified Frenkel-Halsey-Hill theory. Langmuir 19:2632–2638
88. Taylor MK, Hickey AJ, VanOort M (2006) Manufacture, characterization, and pharmacodynamic evaluation of engineered ipratropium bromide particles. Pharm Dev Technol 11:321–336
89. Telko MJ, Kujanpaa J, Hickey AJ (2007) Investigation of triboelectric charging in dry powder inhalers using electrical low pressure impactor (ELPI™). Int J Pharm 336:352–360
90. Thielmann F, Burnett DJ, Heng JYY (2007) Determination of the surface energy distributions of different processed lactose. Drug Dev Ind Pharm 33:1240–1253
91. Tougas TP, Christopher D, Mitchell JP, Strickland H, Wyka B, Van Oort M, Lyapustina S (2009) Improved quality control metrics for cascade impaction measurements of orally inhaled drug products (OIPs). AAPS PharmSciTech 10:1276–1285
92. Traini D, Young PM (2006) A novel apparatus for the determination of solubility in pressurized metered dose inhalers. Drug Dev Ind Pharm 32:1159–1163
93. Tsukada M, Irie R, Yonemochi Y, Noda R, Kamiya H, Watanabe W, Kauppinen EI (2004) Adhesion force measurement of a DPI size pharmaceutical particle by colloid probe atomic force microscopy. Powder Technol 141:262–269
94. USP 32- NF27, R., MD (2009) US Pharmacopeial Convention, Inc.
95. Van Oort M, Roberts W (1996) Variable flow-variable stage-variable volume strategy for cascade impaction testing of inhalation aerosols. Respiratory Drug Delivery 1:418–420
96. Wiedmann TS, Bhatia R, Wattenberg LW (2000) Drug solubilization in lung surfactant. J Control Release 65:43–47
97. Williams RO, Rogers TL, Liu J (1999) Study of solubility of steroids in hydrofluoroalkane propellants. Drug Dev Ind Pharm 25:1227–1234
98. Wyka B, Tougas T, Mitchell J, Strickland H, Christopher D, Lyapustina S (2008) Comparison of two approaches for treating cascade impaction mass balance measurements. J Aerosol Med 21:155–156
99. Young PM, Price R, Lewis D, Edge S, Traini D (2003) Under pressure: predicting pressurized metered dose inhaler interactions using the atomic force microscope. J Colloid Interface Sci 262:298–302
100. Young PM, Price R, Tobyn MJ, Buttrum M, Dey F (2003) Investigation into the effect of humidity on drug-drug interactions using the atomic force microscope. J Pharm Sci 92:815–822
101. Young PM, Sung A, Traini D, Kwok P, Chiou H, Chan HK (2007) Influence of humidity on the electrostatic charge and aerosol performance of dry powder inhaler carrier based systems. Pharm Res 24:963–970
102. Young PM, Tobyn MJ, Price R, Buttrum M, Dey F (2006) The use of colloid probe microscopy to predict aerosolization performance in dry powder inhalers: AFM and in vitro correlation. J Pharm Sci 95:1800–1809
103. Zeng XM, Martin G, Marriot C (2001) Particulate interactions in dry powder formulations for inhalation. Taylor and Francis, New York
104. Zhu KW, Ng WK, Shen SC, Tan RBH, Heng PWS (2008) Design of a device for simultaneous particle size and electrostatic charge measurement of inhalation drugs. Pharm Res 25:2488–2496

Chapter 18
In Vitro Cell Culture Models for Evaluating Controlled Release Pulmonary Drug Delivery

Stephen T. Buckley, Kwang-Jin Kim, and Carsten Ehrhardt

Abstract A variety of cell culture systems for modeling the pulmonary system have been developed. They offer the potential to study various cell biology-related questions in the lung field. In this chapter primary cell cultures, continuous disease models, and coculture models are discussed. The use of these models in biopharmaceutical research is then reviewed along with discussion on drug permeability, transporters, aerosol deposition studies, and in vitro–in vivo correlation studies. Notwithstanding the difficulties and challenges surrounding various aspects of in vitro respiratory epithelial models, mechanistic studies of pulmonary drug delivery using various tracheobronchial and alveolar mucosae are expected to provide us with a wealth of information in the coming years that will lead to newer and more efficient methods to treat lung-specific diseases using controlled and targeted approaches.

Keywords Continuous cell line • Drug disposition • Drug transporter • In vitro model • Primary culture

18.1 Introduction

18.1.1 Primary Cell Cultures

Primary cells are thought to offer the most accurate in vitro representation of the respiratory epithelium in its native form. In particular, they typically exhibit characteristics which more closely resemble those observed in vivo. This is especially important in the case of alveolar epithelium, where currently there are no cell lines which reliably mimic type I or type II pneumocytes. Utilization of respiratory

C. Ehrhardt (✉)
School of Pharmacy and Pharmaceutical Sciences, Trinity College Dublin, Panoz Institute, Dublin, Ireland
e-mail: ehrhardc@tcd.ie

H.D.C. Smyth and A.J. Hickey (eds.), *Controlled Pulmonary Drug Delivery*, Advances in Delivery Science and Technology, DOI 10.1007/978-1-4419-9745-6_18,

epithelial cells for primary culture entails their fresh isolation from lung tissues. As a consequence, primary cultures are associated with a number of limitations. For example, availability of airway tissues is often restricted, in particular those derived from humans. Moreover, cell yields may be poor while donor-to-donor variation can reduce the usefulness of these cultures. In contrast to cell lines, primary cultures are also limited by rather short lifespans. As a result, the use of primary cell cultures is associated with high costs and can be time-consuming. In addition, generating tight monolayers with high transepithelial electrical resistance (TEER), both of which are essential for drug absorption studies, can be difficult to achieve and often require great expertise in primary culture settings.

18.1.1.1 Primary Cell Cultures of Tracheobronchial Epithelia

Over the last 30 years, protocols for the isolation and culture of primary tracheobronchial epithelial cells obtained from lungs of many species have been developed. They include primary cultures of airway epithelial cells of the mouse [1], hamster [2], guinea pig [3], rat [4], ferret [5], rabbit [6, 7], dog [8], pig [9], cow [10], horse [11], and human [12–14]. Most of the protocols result in well-differentiated epithelial cells with mixed phenotypes. In general, these primary cell cultures are good for drug absorption studies for the first couple of subcultures after which they lose their ability to form tight junctions and their capacity to generate high TEER is lost. Indeed, successive passages of human tracheal epithelial cells have been shown to exhibit a linear drop in short-circuit current (I_{sc}) [15]. Using Ussing chamber techniques, these subcultures were shown to exhibit minimal rates of active Na^+ and Cl^- transport by passage 3 and beyond passage 4 the cells failed to generate any active ion transport. These data suggest that it is important to develop airway epithelial cell lines that retain the ability to differentiate, form tight junctions, and maintain ion channel/pump activities when grown in vitro [15].

Ready-to-use culture systems of human tracheobronchial cell layers exhibiting well-differentiated ciliated and goblet cell phenotypes are commercially available (Epiairway™ system, Mattek Corporation, Ashland, MA, USA) [16]. Despite Epiairway™ being marketed for use in drug delivery studies, little data have been reported to date. This may be reflective of low uptake and/or due to cost or usage in industry that is subject to confidentiality. Overall, primary culture is less convenient and economical than the use of cell lines. Consequently, primary cultured tracheobronchial cell layers have not been widely used for biopharmaceutical purposes [17].

18.1.1.2 Primary Cell Cultures of Alveolar Epithelia

Given the scarcity of suitable alveolar epithelial cell (AEC) lines that form functional tight junctions, the majority of in vitro studies of alveolar epithelial function (e.g., solute transport and metabolism) use primary cultures of AEC. Primary mammalian

AEC culture techniques involve isolation, purification, and cultivation of ATII cells obtained from either resected (human) lung tissues or isolated perfused (rodent) lungs. When plated on permeable supports or plastics under appropriate culture conditions, these ATII cells acquire features of type I cell-like phenotype and morphology [18–20]. Isolation of ATI pneumocytes from rat lungs has been recently described [21–23]. However, development of confluent ATI cell monolayers exhibiting electrically tight characteristics has yet to be reported. Of note, unlike many other cells in primary culture, AECs typically exhibit a very limited proliferative capacity and therefore are not suitable for passaging. Consequently, each data set requires a fresh preparation of cells resulting in a marked increase in costs. Moreover, it necessitates a reliable normalization scheme of data from each set of cell preparations.

Owing to the lack of availability of human lung tissues and ethical issues pertaining to the use of human tissues, most lung permeability studies have been based on isolation and culture of cells from animals, including mouse [24], rat [25], rabbit [26], and pig [27]. Since evidence for species difference between human and rodents might be more significant than once assumed [28], confirmation using primary human pneumocyte cultures has been performed for various aspects [29–31].

18.1.2 Continuous Cell Cultures

The use of continuous cell cultures in biopharmaceutical studies is widespread. They include those derived from tumors and transformed cell lines. Importantly, in contrast to primary cell cultures, they are immortal. The principal benefit of cell lines is their convenience of use. Culturing and maintenance of respiratory epithelial cell lines can be performed using standard reagents and protocols. Moreover, the yield associated with each passage of a particular cell line is typically high. Furthermore, they offer greater flexibility, with cells capable of being frozen and stored. In terms of cost, they are considerably more economical to use. This, together with their good reproducibility, makes them a particularly attractive option for use in high-throughput screening studies.

However, the use of cell lines is also associated with a number of disadvantages. In the case of transformants, they often lose differentiated properties with increasing passage number. In particular, their lack of capacity to form tight junctions can diminish their utility in transport studies [32]. In the case of tumor-derived cells, they exhibit excessive proliferative capacity, in addition to other abnormalities (e.g., mucus and surfactant content) in comparison to normal healthy cells [33].

18.1.2.1 Tracheobronchial Epithelial Cell Lines

In contrast to gastrointestinal in vitro testing where Caco-2 cells have emerged as the gold standard, there is no such consensus to date on the preferred cell line(s) for modeling the bronchial epithelium in vitro. Several detailed protocols for culture, maintenance, growth, and permeability assessment of tracheobronchial epithelial cell lines have been published in recent years [17], with the most commonly used systems being the Calu-3 and 16HBE14o– cell lines. Additionally, BEAS-2B are frequently used for studies of metabolism and the interaction of cells with xenobiotics, and the use of the NuLi-1 cell line has recently been explored [34, 35].

Calu-3 (American Type Culture Collection; ATCC HTB-55) is a human bronchial epithelial cell line derived from an adenocarcinoma of the lung [36]. This cell line has been shown to exhibit serous cell properties and forms confluent monolayers of mixed cell phenotypes, including ciliated and secretory cell types [37], but the cilia are formed very irregularly and seem to disappear with increasing passage number (unpublished observations CE). Calu-3 cells have shown utility as a model to examine transport [38–40] and metabolism in human bronchial epithelial cells for many therapeutic compounds [41]. Furthermore, they have been used in a number of particle–cell interaction studies [42–44].

Another continuous bronchial epithelial cell line, 16HBE14o–, was generated by transformation of normal bronchial epithelial cells obtained from a 1-year-old male heart–lung transplant patient. Transformation was accomplished with SV40 large T antigen using the replication defective pSVori– plasmid [45]. 16HBE14o– cells can be obtained from Dieter C. Gruenert, Ph.D., at the California Pacific Medical Center. The noncommercial availability of the cell line might be one of the reasons why they have been less widely used than Calu-3. 16HBE14o– cells have a nonserous, nonciliated phenotype and are generally round in shape and smaller in size than Calu-3. When grown under liquid-covered culture (LCC) conditions, 16HBE14o– form confluent, polarized cell layers with functional tight junctions and the expression of several drug transport proteins [33, 46]. By contrast, air-interfaced culture (AIC) conditions lead to cell layers of less desirable phenotypic and morphological traits. Since most epithelial cells, that are normally located at an air interface in vivo, function optimally under AIC conditions for cultivation, the superior performance of the 16HBE14o– cell line under LCC conditions is an exception for which the exact reasons or mechanism are currently unknown. Like Calu-3, the 16HBE14o– cell line has been utilized for many drug absorption and particle–cell interaction studies [47–49].

The BEAS-2B cell line was derived from normal human epithelial cells that were immortalized using the hybrid virus of adenovirus 12 and Simian virus 40 [50]. BEAS-2B is available from the ATCC (CRL-9609) and has been popular in studies of airway epithelial cell structure and function, including phenotyping and mechanistic investigation of cytokine regulation [51]. BEAS-2B cells have also been used to evaluate responses to challenges such as tobacco smoke [52], environmental particles [53, 54], and hyperoxia [55]. However, as it is challenging to generate TEER values higher than 100 ohm cm^2 using BEAS-2B cells [56], they have not been used much in drug permeability studies, but have frequently been utilized to investigate the expression and activity of drug metabolizing enzymes [57, 58].

A relatively new cell line, that has not been characterized to date for its use in biopharmaceutics, is based on primary airway epithelial cells infected with retroviruses expressing hTERT and HPV-16 E6/E7 (NuLi-1) [15]. NuLi-1 cells were cultured on plastic up to passage 30. When grown on collagen-coated, semipermeable membranes

(Millicell-PCF), NuLi-1 TEER decreased by ~50% over the 30 passages from 685 ± 31 to 389 ± 21 ohm cm^2. The TEER of NuLi-1 is similar to that observed with the primary bronchial cultures of 532 ± 147 ohm cm^2. Thus, NuLi-1 cells can form an electrically tight airway epithelial barrier that mimics active and passive ion transport properties of primary human bronchial epithelial cells [15].

18.1.2.2 Alveolar Epithelial Cell Lines

While a number of immortalized cell lines emanating from different cell types of the airway (i.e., tracheobronchial) epithelium of lungs from various mammalian species are available, reliable and continuously growing cell lines that possess AEC morphology and phenotype are not reported to date. Most studies use cell lines of alveolar epithelial origin for drug absorption studies, while the observations are hard or meaningless to extend to human situation.

Probably, the most frequently used alveolar epithelial model is the A549 cell line (American Type Culture Collection, ATCC CL-185); continuously growing cells derived from a human pulmonary adenocarcinoma that have some morphologic and biochemical features of the human pulmonary alveolar type II cell in situ [59]. A549 cells contain multilamellar cytoplasmic inclusion bodies, like those typically found in human lung ATII cells, although these hallmarks disappear as culture time increases. At early and late passage levels, the cells synthesize lecithin with a high percentage of disaturated fatty acids utilizing the cytidine diphosphocholine pathway [59]. The cell line has been utilized for many biopharmaceutical studies, albeit A549 cells lack the ability to form tight monolayers of polarized cells, due to the inability to form functional tight junctions [60–63]. Despite their obvious unsuitability, due to their lack of good TEER, some investigators utilized A549 cell layers in drug absorption studies [64, 65], leading to meaningless conclusions. The absorption data reported by these groups were not able to clearly show differences in transport rates between proteins and peptides of various sizes where the published permeability values were 2–4 orders of magnitude higher than those obtained in tight monolayer systems (e.g., human alveolar epithelial cell [hAEpC] monolayers) using identical compounds [61]. Moreover, the reported A549 TEER of ~600 ohm cm^2 by one group [64] may be in gross error. Their data indicate that large hydrophilic molecules are translocated across A549 cell layers at rates approaching near-free diffusion limits. In other words, such large fluxes probably took place via large gaps between A549 cells, contradicting to the very large TEER they reported. It should be noted that TEER of A549 cells is usually in the range of 20–60 ohm cm^2, which makes much more sense, as it is known that no functional tight junctions are present in A549 cell layers. Notwithstanding, A549 cells might be a useful model in other areas of biopharmaceutical research, including metabolism studies or cytotoxicity studies for various substances [62, 66, 67].

Other cell lines of an alveolar epithelial origin that are reported to date include R3/1 and L-2 (rat), MLE-12 and 15 (mouse), and H441 and TT1 (human). Of these, NCI-H441 cell line (ATCC HTB-174), emanated from a human lung adenocarcinoma, has been described to have characteristics of both ATII [68, 69] and bronchiolar (i.e., Clara) epithelial cells [70, 71]. Studies can thus be designed around to characterize metabolism and transport properties of these two particular cell types. There is an emerging evidence that H441 cells are capable to form monolayers of polarized cells, exhibiting moderately tight transepithelial with active Na^+ absorption [72, 73], although the cell line has not been used for drug absorption experiments so far.

Wikenheiser and coworkers generated a series of continuous AEC lines (MLE-7, -12, and -15) from transgenic mice harboring the SV40 large T antigen under the control of the human SP-C promoter region. These MLE cell lines maintained morphological and functional characteristics of distal respiratory epithelial cells normally lost after isolation and primary culture [74], which are consistent with those seen in nonciliated bronchiolar and ATII epithelial cells. However, morphological and functional characteristics associated with an individual cell type do not always appear to coexist in a clonal cell line. For example, MLE-12 cells express SP-C mRNA (i.e., indicative of ATII cells in the adult mouse), but other characteristics (e.g., SP-A mRNA expression or presence of lamellar bodies) are missing in MLE-12 cells. The heterogeneity of cellular markers in MLE cells may be related to the immortalization of ATII cells at various stages of lung development. Alternatively, MLE cell lines might represent distinct subtypes of distal respiratory epithelial cells. Culture conditions may also be an important determinant of expression of various cellular markers in MLE cells. When grown under AIC on hollow fibers, MLE-15 cells contain numerous lamellar bodies and secrete both SP-A and SP-B [75]. Currently, only MLE-12 can be obtained from the American Type Culture Collection (ATCC CRL-2110).

L-2 cells (ATCC HTB-149) have been isolated by clonal culture techniques from the adult rat lung. These cells appear to retain differentiated functions that are present in ATII cells of intact rat lungs. L-2 cells are diploid, epithelial cells. They contain osmiophilic lamellar bodies in their cytoplasm and synthesize lecithin by the same de novo pathways as in intact lungs [76]. L-2 cells are capable of forming confluent monolayers, although they do not exhibit as high a TEER as ATII cells in primary culture do [77]. Moreover, they possess ion transport and single channel characteristics indistinguishable from primary cultures of ATII cells, together with comparable amiloride-sensitive transepithelial current [77]. To date, L2 cells have not been systematically investigated regarding their suitability as a model for drug absorption studies.

The rat cell line R3/1 was established from cells obtained from bronchoalveolar tissues of fetal Wistar rats at 20 days of gestation. This cell line displays a phenotype with several characteristic features of ATI cells. R3/1 cells were analyzed to show a positive expression for both mRNA and protein for markers related to the ATI cell type (T1α, ICAM-1, connexin-43, and caveolin-1 and -2) [78]. However, it has been shown very recently that R3/1 cells are unable to form confluent monolayers due to a lack of expression of several tight junction proteins, such as occludin [79]. This excludes R3/1 cells from being used as an in vitro model for studies of alveolar absorption. In spite of this feature, given that their expression pattern of proteolytic enzymes is similar to that of type I-like cells of the rat alveolar epithelium in primary culture, the cell line may be suitable for studying stability of inhaled and endogenous proteins [79].

Most recently, a new alveolar type I-like cell line, transformed type 1 (TT1), has been established [80]. They are an immortalized cell line produced by retroviral transduction of primary human ATII cells with the catalytic subunit of human telomerase (hTERT) and a temperature sensitive mutant of Simian virus 40 large antigen (U19tsA58 LT). They exhibit a flattened and thin cell morphology, which is typical of type I cells. It has also been shown that TT1 cells contain endosomal vesicles within the cytoplasm and invaginations at the cell membrane. Importantly, they express type I cell markers such as caveolin-1, while lacking type II cell markers such as TTF-1, pro-SPC, and alkaline phosphatase [80]. They show positive staining for pan-cytokeratin, confirming their epithelial phenotype. However, they exhibit poor barrier properties. Staining for ZO-1 in TT1 cells is weak and discontinuous consistent with a TEER of ~55 ohm cm^2 and flu-Na P_{app} of $\sim 6\times 10^{-6}$ cm/s [81]. Consequently, they are unsuitable for modeling drug transport and epithelial barrier function in vitro.

18.1.3 Disease Models

Cystic fibrosis (CF) is an autosomal recessive disease and is one of the lethal genetic diseases. There are approximately 30,000 patients in the USA and Europe [82, 83]. CF is caused by mutation in the CFTR gene located at chromosome 7 and is associated mostly with defective chloride transport in airway epithelial cells. Lung pathology in CF includes abnormally low to nil chloride transport, increased mucus viscosity, reduced mucociliary clearance rates, recurrent infections, chronic inflammations, and airway damage. Because of the significance of the disease, the most visible gene therapy schemes under development are inhalable regimens (containing complementary DNA of CFTR) to treat CF by restoration of CFTR function in the airways of the lung.

In vitro models based on CF airway epithelium have been used widely to better understand the CF pathophysiology, to have a tool to study alterations in airway permeability, and to assess the efficiency of gene vectors. Utilizing approaches similar to those for cultivation of cells from healthy tissues, several protocols have been developed to culture airway epithelial cells harvested from CF patients [84, 85]. The development of immortalized cell lines with a CF phenotype has been a significant benefit for investigators in the CF field, alleviating the major limitation inherent in primary cell culture models, i.e., the very limited availability of suitable CF tissues [32, 86]. A number of these CF cell lines have been reported to show the ability to form polarized cell layers with the necessary bioelectrical tightness that makes them potentially useful for drug permeability studies. Examples of human immortalized CF airway epithelial cell lines are NCF3 [87], CFT1 [88], CFBE41o– [89], and CuFi-3 and -4 [15]. It should be noted that only the CFBE41o– and CuFi cell lines have been characterized to validate their long-term stability as reliable in vitro models.

18.1.4 Coculture Models

Up until recently, the vast majority of investigations have utilized cell lines or primary cultures in the form of monocultures. Monocultures are easy and convenient to establish and offer good viability of cells. However, the lung is composed of a complex collection of >40 different cell types. While the inherent intricacies of the lung cannot be totally recapitulated by artificial cell cultures, cocultures composed of two to three cell types ensure that cell–cell interactions which are important in determining and preserving phenotype and normal cell function are maintained.

In addition to the translocation of drug particles via endocytic pathways, emerging evidence suggests that nonendocytic mechanisms may also be of importance. Indeed, studies by Rothen-Rutishauser et al. [90, 91] using a triple coculture model of the air–blood barrier of the respiratory tract by culturing epithelial cells (A549 cells) on filter inserts together with placing human blood monocyte-derived macrophages and dendritic cells on the apical and basal sides, respectively, investigated interaction of these three different cells during the uptake of polystyrene particles from the apical side.

Recently, cocultivation of HBEC with the human lung fibroblast cell line Wi-38 cultured under AIC resulted in differentiation of HBEC into the three main bronchial cell phenotypes, namely basal, mucus-producing, and ciliary cells [92]. Importantly, the cilia in the surface of epithelial cells formed exhibited a beating frequency of 14–19 Hz which is comparable to that observed in vivo. Together, this model offers potential for use in investigating both acute and chronic effects following pulmonary exposure to pharmacological agents. Of note, monocultures demonstrated far less marked differentiation. Interestingly, cocultivation of 16HBE14o– cells with Wi-38 failed to initiate their differentiation.

In the pulmonary context, drug therapy is frequently required under pathophysiological conditions characterized by acute inflammation of the alveolo-capillary barrier. Consequently, models which mimic an impaired barrier could be helpful in predicting effects of drugs on alveolar epithelium exhibiting altered barrier function. To this end, Hermanns and colleagues [93] have developed a coculture model incorporating the epithelial cell line H441 together with primary isolated human pulmonary microvascular endothelial cells (HPMECs) or the endothelial cell line ISO-HAS-1 on opposite sides of a permeable filter insert in the absence and presence of proinflammatory stimuli. An inflamed alveolo-capillary barrier with impaired barrier function was generated when proinflammatory stimuli (e.g., tumor necrosis factor-alpha and interferon-gamma) were given, characterized by diminished TEER and enhanced transport of sodium fluorescein. The nanocarrier polyethyleneimine (PEI), complexed with Oregon Green (OG) or pDNA encoding enhanced green fluorescent protein (EGFP) gene, was used to evaluate cell uptake and gene transfer respectively. Interestingly, in contrast to monocultures, virtually no PEI–OG or EGFP expression was detected in cocultures. Moreover, transfection efficiency of an inflamed alveolo-capillary barrier was markedly lower than obtained with nonpolarized monolayers of H441 cells. Together, this suggests that polarized epithelial cells may impede uptake.

18.2 Use of In Vitro Models of Respiratory Epithelial Barriers in Biopharmaceutical Research

18.2.1 Drug Absorption Studies Using Tracheobronchial Epithelial Cells

Despite the fact that the alveolar epithelium with its extremely thin barrier (<0.5 μm) and vast surface area (>100 m^2) is regarded as the most favorable site from which to achieve delivery of therapeutic molecules into the systemic circulation following inhalation, several cell lines of the tracheobronchial origin have been employed to gain mechanistic information on drug transport and metabolism in the lung. This can be explained, in part, by the relative ease and low costs by which continuously growing tracheobronchial cell lines can be maintained, compared to primary cultures of AEC. The airways also constitute the target for therapeutic compounds for asthma and COPD and, depending upon the site of deposition and physicochemical properties of an inhaled compound, may contribute significantly to the pulmonary absorption of inhaled therapeutics.

18.2.1.1 Drug Permeability

The permeability of a large number of compounds has been measured in airway epithelial cell culture models, and the results of these studies have been reviewed [17], with dependence on molecular size and lipophilicity being seen in the Calu-3 and 16HBE14o– cell lines. Observations in these cell lines were similar to those made in primary rabbit tracheobronchial cell cultures. The use of airway epithelial cell models in pulmonary biopharmaceutical research in general appears to be well justified, since permeability (P_{app}) values observed in such in vitro cell culture models appear to be predictive of absorption from intact lungs [35]. Drug delivery applications such as absorption enhancement and the effects of inhaled formulation excipients on drug permeability have also been investigated using these two cell lines [17].

18.2.1.2 Active Transport Mechanisms in Tracheobronchial Epithelial Cells

ABC-Transporters

P-glycoprotein (P-gp) expression and functionality have been determined in several lung cell culture models. Studies of cultures of primary human bronchial epithelial cells have revealed the presence of P-gp [94, 95]. Interestingly, expression of P-gp appears to be time-dependent with mRNA levels increasing from day 7 to day

21 [96]. In agreement with this, P-gp-mediated transport of digoxin was apparent in cultures after 21 days, but not after 14 days. Functional studies suggested a basolateral localization of P-gp [95]. In 16HBE14o– cells, expression of P-gp has been shown at both mRNA and protein levels [46, 97, 98]. In contrast to normal human bronchial epithelial cells, immunocytochemical staining indicates that it is localized at the apical membrane in this cell line [46]. Functional activity was measured with Rh123 and the transport was inhibited by the P-gp inhibitor, verapamil [46]. For Calu-3 cells, published data are somewhat conflicting, in that some groups observed net secretion of the P-gp substrate [95, 99, 100], while others observed P-gp-mediated absorption [101], and another report says an undetectable level of P-gp expression [102]. Such inconsistencies may also be explained in part by the use of P-gp substrates and inhibitors which are nonspecific. Indeed, net secretion of Rh123 across Calu-3 cells may be explained by additional multidrug resistance-associated protein 1 (MRP1) activity in these cells (most likely with a basolateral localization), because Rh123 is also an MRP1 substrate [103]. Moreover, it appears that different Calu-3 clones may have been utilized in these studies, whereby the cell line may yield different phenotypes dependent upon the culture conditions used by investigators. Illustrating this, studies by Madlova and colleagues [95] suggested that functional expression of P-gp occurred only in Calu-3 monolayers at passages over 50 grown on cell culture inserts for 21 days. However, this hypothesis was contradicted in findings by Brillault et al. [98] who detected P-gp in Calu-3 cell layers at passages 22–30 grown for just 15 days and showed P-gp-mediated transport of moxifloxacin. In CFBE41o– CF airway epithelial cells, P-gp was expressed at both message and protein levels, although it lacked functionality in the efflux or secretion of Rh123 [89]. Whether this is a result of the CF phenotype or a finding specific to the cell line is unclear.

MRP1 protein has been shown to be expressed in freshly isolated human bronchial epithelial cells, where the efflux of carboxy-dichlorofluorescein (an MRP1 substrate) was inhibited by MK571, a specific MRP inhibitor [94]. Both 16HBE14o– and Calu-3 cell lines have also been reported to express MRP1 basolaterally, which is thought to contribute to the net absorption of certain substrates (e.g., monochlorobimane GSH conjugate) [103, 104]. Analysis of human respiratory epithelial cell culture models by PCR for ABC transporters revealed high levels of MRP1, MRP3, MRP5, and MRP7 and moderate levels of MRP2 and MRP6 [97]. While transcripts for MRP4 and MRP8 were detected in human bronchial epithelial cells, they were either absent or very weakly expressed in both Calu-3 and 16HBE14o– cell lines [97]. In 16HBE14o– cells, immunohistochemical staining for MRP1 was strong, while that for MRP4 was weak [105]. BCRP transcripts were detected in all human bronchial epithelial cell cultures [97]. While reports are conflicting as to mRNA levels in Calu-3 [97, 106], protein expression appears relatively high with immunocytochemical staining indicating apical localization. 16HBE14o– cells exhibit abundant expression [97]. Moreover, its high expression in 16HBE14o– was confirmed by strong immunostaining [105]. A summary of drug transporter expression in in vitro cell culture models of the respiratory epithelium is given in Table 18.1.

Table 18.1 Summary of drug transporter expression in in vitro cell culture models of the respiratory epithelium

Protein name	*In vitro* model	Cellular localization	References
ABC transporters			
Pgp/MDR1	hBEpC; Calu-3 (inconclusive data); 16HBE14o–; CFBE41o-; A549; rat ATII; rat type I-like; human ATII; human type I-like	Apical/ basolateral (inconclusive data)	[46, 89, 94–102, 107, 108]
MRP1	hBEpC; Calu-3; 16HBE14o–; BEAS-2B; A549; rat ATII; rat type I-like; human ATII; human type I-like	Basolateral	[94, 97, 100, 103, 104]
MRP2	hBEpC; Calu-3; 16HBE14o–; BEAS-2B; A549; human ATII; human type I-like		[97]
MRP3	hBEpC; Calu-3; 16HBE14o–; BEAS-2B; A549; human ATII; human type I-like		[97]
MRP4	hBEpC; Calu-3; A549; human ATII; human type I-like		[97]
MRP5	hBEpC; Calu-3; 16HBE14o–; BEAS-2B; A549; human ATII; human type I-like		[97]
MRP6	hBEpC; Calu-3; 16HBE14o–; BEAS-2B; A549; human ATII; human type I-like		[97]
MRP7	hBEpC; Calu-3; 16HBE14o–; BEAS-2B; A549; human ATII; human type I-like		[97]
MRP8	hBEpC; Calu-3; human ATII; human type I-like		[97]
BCRP	hBEpC; Calu-3; 16HBE14o–; BEAS-2B; A549; human ATII; human type I-like	Apical	[97, 105, 106]
ABCA1	MLE-12; A549, human ATII, rat ATII, mouse		[109]
ABCA3	A549; MLE-15; human ATII rat ATII; murine ATII		[110, 111]
SLC transporters			
GLUT1	Rat ATII		[112]
GLUT2	H441	Apical & basolateral	[113]
GLUT4	Rat ATII; H441		[112, 113]
GLUT5	Rat ATII		[112]
SGLT1	Rat ATII		[112]
OCT1	hBEpC; Calu-3; 16HBE14o–; BEAS-2B; MLE-12; A549; rat ATII; murine ATII		[97, 114–117]
OCT2	Calu-3 (inconclusive data); 16HBE14o– (inconclusive data); MLE-12; rat ATII (inconclusive data); murine ATII		[97, 114–116, 118]
OCT3	hBEpC; Calu-3; 16HBE14o–(inconclusive data); A549; rat ATII		[97, 114, 115, 117, 118]
OCTN1	hBEpC (inconclusive data); Calu-3; 16HBE14o–; BEAS-2B; A549	Apical	[97, 115, 117]
OCTN2	hBEpC; Calu-3; 16HBE14o–; BEAS-2B; A549	Apical	[97, 115, 117]
PEPT1	Calu-3; 16HBE14o–; BEAS-2B; A549	Apical	[97, 119]

(continued)

Table 18.1 (continued)

Protein name	*In vitro* model	Cellular localization	References
PEPT2	hBEpC; Calu-3; 16HBE14o–; BEAS-2B; A549; L2; rat ATII; murine ATII		[97, 119–121]
LAT-1	Rat ATII		[121]
ATB^{0+}	Rat ATII; murine ATI		[122, 123]
SVCT1	Rat ATII	Apical	[124]
SVCT2	Rat ATII	Apical	[124]
OAT1	No		[97]
OAT2	No		[97]
OAT3	No		[97]
OAT4	Calu-3; 16HBE14o–; BEAS-2B; A549		[97]
SLCO transporters			
OATP1A2	hBEpC; Calu-3; BEAS-2B; A549		[97]
OATP1B1	Calu-3; A549		[97]
OATP1B3	Calu3; BEAS-2B; A549		[97]
OATP1C1	Calu3; A549		[97]
OATP2A1	No		[97]
OATP2B1	Calu3; A549		[97, 125]
OATP3A1	hBEpC; Calu-3; 16HBE14o–; BEAS-2B; A549		[97]
OATP4A1	hBEpC; Calu-3; 16HBE14o–; BEAS-2B; A549		[97]
OATP5A1	Calu-3; BEAS-2B; A549		[97]
Lung resistance protein			
LRP	hBEpC; Calu-3; 16HBE14o–; CFBE41o-; A549		[46, 89, 126–129]

SLC-Transporters

Investigations into organic cation transporter (OCT) mRNA transcripts have revealed different expression patterns with an apparent dependence on culture conditions. All five OCTs were detected in Calu-3 and 16HBE14o– cell lines when grown in cell culture flasks at high passage numbers [114]. However in a separate study, RT-PCR analysis of both bronchial cell lines and primary bronchial cells showed an absence of OCT2 [97]. Furthermore, OCT3 transcripts were not detected in 16HBE14o–, while primary bronchial cells did not exhibit a signal for OCTN1 [97]. When grown at an AIC, primary human bronchial epithelial cells exhibited high levels of OCTN1 and OCTN2, which was localized at the apical membrane [115]. However, they expressed only low amount of OCT1-3 under these conditions [115]. OCT1 and OCT2 may mediate acetylcholine release and reabsorption from epithelial cells [130] and involvement of OCTN1 and OCTN2-mediated drug absorption from the airways in a pH-dependent manner has been demonstrated [115]. With respect to carrier-mediated cationic drug absorption in functional (i.e., electrically tight) cell layer cultures, only limited information on drug absorption profiles is available. Net absorption of salbutamol (albuterol) across both Calu-3 and 16HBE14o– cell layers has been shown [114]. A further study revealed that positively charged β_2-agonists such as salbutamol and formoterol are capable of modulating the activity of OCTs [115].

The presence of a high-affinity peptide transporter PEPT2 in bronchial epithelium has been demonstrated by measuring the uptake of oligopeptides into bronchial cells following intratracheal instillation [131]. The functional expression of PEPT2 in primary cultured human lung cells isolated from the trachea, bronchi, and bronchioles of multiple donor subjects has also been demonstrated [132]. While expression of PEPT2 transcripts has been shown in both primary human bronchial epithelial cells and bronchial cell lines, expression of PEPT1 appears to be restricted to Calu-3 and 16HBE14o–, with no signal detected in primary human bronchial epithelial cells [97]. In Calu-3 cells grown at an AIC both PEPT1 and the PEPT2*1 variant were detected [119]. Using Western blot, expression of PEPT1 was confirmed and shown to localize on the apical cell membrane [119]. RT-PCR analysis of the gene for cationic amino acid transporters expressed in Calu-3 cells detected amplicons of the expected size for *SLC7A1* (coding for hCAT1 transporter), *SLC7A2* (for transcripts hCAT2B and hCAT2A), *SLC7A4, SLC7A6* (for y+LAT2), *SLC7A7* (for y+LAT1), *SLC3A2* (for 4F2hc/CD98), and *SLC6A14* (for ATB^{0+}) [120]. In H441 cells, the glucose transporter GLUT-2 was found to be localized at both apical and basolateral membranes [113].

Organic anion transporters (OAT) have been characterized for their ability to mediate the absorption and elimination of endogenous and exogenous organic anions. In human bronchial epithelial cell culture models, OAT1, OAT2, and OAT3 transcripts were absent. In the case of OAT4, mRNA was detected only in Calu-3 and 16HBE14o– cells [97].

Lung Resistance Protein (LRP)

Lung resistance protein (LRP) is expressed in normal human bronchial epithelium as well as in other tissues potentially exposed to toxins [126, 127]. LRP was reported to be expressed in both Calu-3 cells [128] and 16HBE14o– model [46]. However, the role of LRP in different cellular locations is not clear to date.

Organic Anion Transporting Polypeptides (OATP)

OATPs are capable of transporting diverse classes of endogenous solutes and xenobiotic compounds. Expression of this family of transporters has been shown to vary considerably between different in vitro models of the bronchial epithelia. Calu-3 cells appear to express all OATP transporters except for OATP1A2 [97]. By contrast, 16HBE14o– cells showed transcripts for just OATP3A1 and OATP4A1 [97]. Similarly, normal bronchial cells exhibited expression for OATP3A1 and OATP4A1, in addition to OATP1A2 [97].

18.2.1.3 In Vitro–In Vivo Correlation Using Tracheobronchial Epithelial Cells

Data are emerging slowly on the use of in vitro respiratory epithelial models as predictive screens for drugs administered via pulmonary routes [39, 48, 133, 134]. Although excellent predictive power has been achieved, the experimental conditions are far from the reality where a patient would be administered with the drug as an aerosol cloud, since a significant proportion of absorption from the lung is likely to occur from the alveolar region.

18.2.2 Drug Absorption Studies Using Alveolar Epithelial cells

With the lung emerging as an attractive noninvasive delivery route for systemic absorption of biopharmaceuticals and fast acting conventional drugs, it appears pivotal to gather knowledge about modes of transport and the fate of these compounds at the distal blood–air barrier, i.e., the alveolo-capillary barrier. Although an in vitro–in vivo correlation (IVIVC) using AEC monolayers could not be established definitively at present time, various in vitro models can help elucidate underlying transport mechanisms of drug molecules, as well as their metabolic stability in the lung. Therefore, the available in vitro systems need to be carefully characterized and validated prior to using them in the context of biopharmaceutical research.

Absorptive processes have first been studied in the context of ion and fluid clearance from the lung in health and disease (e.g., edema), which were conducted largely using monolayers of rat pneumocytes [e.g., 135]. Following studies of active and passive ion movement across alveolar epithelium using the in vitro model, translocation of plasma proteins (which are commonly found in the alveolar lumen, e.g., albumin, immunoglobulin G, and transferrin) was investigated [136]. Since rat-based monolayers have been considered the gold standard in elucidation of AEC physiology/biology for several decades, it is not surprising that the same well-established primary rat pneumocyte monolayer model was also employed for drug delivery studies [137–139]. However, as stated above, newer models based on different animal species (which might more closely resemble human, e.g., derived from porcine lung [27]) were developed and a number of laboratories also started to use human AEC monolayers for drug absorption studies [63, 140].

Effros and Mason showed an apparent inverse relationship between molecular weights and rates of in vivo lung clearance of hydrophilic molecules of different sizes, both in log scales [141]. A similar relationship has also been found in two in vitro alveolar epithelial models of primary rat [142] and human [63] monolayers. By contrast, FITC-dextran permeabilities observed with A549 cells [64] are over two orders of magnitude greater than those found using primary AEC monolayers [17].

Active transport mechanisms by which some compounds traverse the distal lung blood–air barrier are more likely to provide new insights and improved strategies for pulmonary delivery of drugs into the systemic circulation, and targeting drugs to lung parenchymal cells [143]. The expression pattern of ATP-binding cassette (ABC)-transporters in both the healthy and pathological lung has recently been reviewed by van der Deen et al. [144]. Several transporters have been identified in AEC, which might have an impact on pulmonary bioavailability of drugs, especially lipophilic drugs (e.g., cytostatics). P-glycoprotein (P-gp, MDR1) is probably the best characterized efflux pump in the body. In 2001, Lehmann and coworkers reported the presence of P-gp and multidrug resistance-associated protein 1 (MRP1) in rat ATII cells in primary culture [94]. These findings were later confirmed using both human and rat lung tissue sections as well as ATI cell-like monolayers by immunohistochemistry and efflux studies using vinblastine and rhodamine 123 as respective substrates for P-gp and MRP1 [107, 108]. More recently, it has been shown that MRP-1 is located at the basolateral surface of primary cultures of both rat ATI-like and ATII cells [145]. Moreover, indomethacin, an inhibitor of MRP-1, increased the basolateral to apical transport of fluorescein across rat ATII cell monolayers [145]. In A549 cells, LRP has been detected at the nuclear envelope and in the cytoplasm [129].

ABCA1 and ABCA3 gene transcripts were found in cell lines (e.g., A549) of human lung origin, in freshly isolated ATII cells of human, rat, and mouse, as well as different tissues of rat, but the highest expression of ABCA1 and ABCA3 was observed in murine ATII cells [109, 110]. Molecular and biochemical studies show that ABCA1 mediates basolateral surfactant efflux, while ABCA3 is found in the limiting membrane of lamellar bodies [111]. Both transporters are members of a subfamily of ABC-transporters that are predominantly known to be involved in the regulation of lipid transport and membrane trafficking. Indeed, rat ATI-like cells have been shown to release lipid to apoA-I protein, and produce particles resembling high-density lipoprotein, suggesting a potentially important role for ABCA1 in maintaining lipid homeostasis in the lung [146]. It should also be noted that there are some conflicting reports regarding the activity and localization of ABC-transporters in mammalian AEC [144].

Other solute transporters (SLC) are also expressed in the alveolus. The mRNA transcripts of glucose transporters, GLUT1, GLUT4, GLUT5 and SGLT1, all have been detected in freshly isolated rat ATII cells by reverse transcription-polymerase chain reaction (RT-PCR) and western blot analyses. However, a gradual loss of sodium-glucose co-transport activity and transcription during transdifferentiation of rat ATII cells towards the type I-like cell phenotype in culture has been reported [112].

Efficient gas exchange and other functions (e.g., active Na^+ absorption) in alveoli are thought to be tied to transepithelial clearance of luminal proteins, albeit little is known about mechanisms of alveolar epithelial protein transport. Protein degradation followed by transport of resultant peptides and amino acids may play an important role in this process as well. Groneberg reported the expression and activity of the high-affinity proton-coupled peptide transporter, PEPT2, in rat and murine ATII cells [147].The neutral and cationic amino acid transporter, ATB^{0+}, is expressed in mouse ATI cells in situ [122]. Recent studies using rat ATII cells suggest that this transporter may contribute towards recovering cationic and neutral amino acids from alveolar luminal fluid [123]. The important role of peptide and amino acid removal was further supported by the finding of Rothenberg et al. that the arginine transport protein, cationic amino acid transporter 2 (CAT2), has a critical role in regulating lung inflammatory responses, in that CAT2-deficient mice revealed spontaneous inflammation in the lung [148]. More recently it has been reported that the rat AEC line L2 expresses PEPT2 and L-type amino acid transporter 1 (LAT-1) [121]. Furthermore, it appears that LAT plays an important role in mediating the uptake of nitric oxide by type II AECs [121].

Vitamin C (VC, L-ascorbic acid) is known to be essential for many enzymatic reactions. Sodium-dependent VC transporters (SVCT), SVCT1 and SVCT2, were recently identified and reported to be localized in the apical cell membrane of AECs in the lung of adult rats. These results suggest that SVCT proteins could transport the reduced form of VC from the airway/alveolar surface liquid into respiratory epithelial cells [124].

Primary rat and mouse ATII cells exhibit pH-sensitive, reversible transport of choline, a process blocked pharmacologically with phenoxybenzamine, an inhibitor of OCTs. RT-PCR products for choline transporters, OCT1 and OCT2, were detected in murine AEC, with only OCT2 protein being robustly expressed [116]. In 1999, Shen and coworkers published evidence that rabbit AEC in primary culture show a concentration-dependent net absorption of organic cations [149], although the molecular identity of the OCT involved was lacking. More recently, Ishiguro and colleagues [118] identified OCT3 protein expression in A549 cells and primary rat ATII cells. Work by Wang et al. [117] revealed OCT3, OCTN1 and OCTN2 protein expression in the A549 cell line. At the mRNA level, A549 cells have also been shown to express transcripts for OCT3, OCTN1 and OCTN2, although OCTN2 expression was markedly lower than either OCT3 or OCTN1 [97]. Of note, our laboratories published data on net absorption of salbutamol across hAEpC monolayers [150]. Since salbutamol carries a positive charge at physiological pH, involvement of OCT family member(s) in the alveolar epithelial absorption of salbutamol can be surmised. Recently, OATP2B1 has been shown to be expressed in A549 cells, contributing to the uptake of amiodarone [125].

Investigations utilizing tight monolayers of either rat or human AEC showed that clearance of a number of endogenous proteins (e.g., immunoglobulin G, albumin and transferrin) occurs in a cell energy-dependent manner [140, 151–153]. These observations strongly suggest that transcytotic processes (e.g., caveolae-mediated or via clathrin-coated pits) may be responsible for removal of these endogenous proteins from alveolar lining fluid. Most recently, investigations by Ikehata and colleagues [154] suggest that albumin is internalized into type II and type I-like rat AECs via clathrin-mediated endocytic pathway. Moreover, it appears that type II cells exhibit a much higher uptake, despite occupying a much smaller surface area of the lung [154]. Intriguingly, these proteins were net absorbed as intact molecules with variable degrees of degradation [136]. Although there appears to be no significant species differences in absorption rates for small lipophilic drugs, some interspecies differences are noted with water-soluble drugs absorbed from distal airspaces of in vivo mammalian lungs [155]. These species differences have not been systematically studied yet. However, some marked differences were reported for protein absorption rates found in primary human vs. rat AEC monolayers [140]. Absorption rates for most peptides and proteins in primary rat vs. human AEC monolayers, however, remain largely unknown.

18.2.3 Particle Impingement Studies

In vitro characterization of particle–cell interactions is an essential component of the evaluation of novel pharmaceutical aerosols for pulmonary delivery. Currently, in the majority of in vitro models, a suspension or solution of particles is applied to cells under submersion. Given that the airways are lined with just a few micrometers of viscous mucous or surfactant, cell monolayers covered in several millimeters of physiological buffers represent a poor resemblance to this. Until recently, few efforts have been made to accurately recreate the deposition of aerosols which occurs in vivo. As a result, little is known of the importance of the interaction between particles and cell layers in vivo. A number of groups have attempted to develop test systems which incorporate both an aerosol classification device and biological barrier. In doing so, such systems combine studies for deposition and absorption – crucial steps in the administration of an aerosol drug.

Studies by Fiegel et al. [42] utilized an Astra-type liquid impinger, under which a filter insert consisting of Calu-3 cells grown under either AIC or LCC was placed. Whilst AIC grown cells showed no change in barrier properties following aerosol administration, monolayers grown under LCC exhibited significant decreases in TEER and fluorescein-sodium transport. This difference was attributed to the dense mucous coating produced by AIC grown cells. Similar models have also been developed including a glass impinger together with 16HBE14o– cells [156] and an Andersen cascade impactor (ACI) in combination with Calu-3 cells [157].

Using a twin-stage impinger (TSI) incorporating a Calu-3 cell layer grown at AIC, Grainger and colleagues [134] investigated the permeability of large molecular weight hydrophilic solutes. Whilst the rate of transport of FITC-labeled dextran particles across the cell layer after deposition and dissolution of the particles in the cell surface secretions was ~20-fold higher than those applied as a solution, the P_{app} value was found to be similar for both. Notably, transport of FITC-dextran particles with molecular weights of between 4 and 70kDa showed excellent correlation with in vivo canine pulmonary clearance after intratracheal instillation of dextrans of similar molecular weights.

The importance of the deposition step in administration of aerosols was illustrated in investigations by Bur et al. [158]. Employing a multistage liquid impinger (MSLI) integrated with an inverted cell layer of Calu-3, the deposition and absorption of three dry powder formulations of salbutamol sulfate and budenoside were evaluated. In the absence of size separation by the MSLI the absorption rates from the aerosol formulations differed, but exhibited correlation with the size of the carrier lactose particles. However, when the aerosols were administered in conjunction with the MSLI, thus ensuring separation of the drug from its carrier, all three formulations showed identical absorption rates, verifying their bioequivalence.

Whilst the use of such cell culture systems is in its infancy, their benefit is clearly demonstrated. They enable methods of application of the drug which are of greater relevance, both physiologically and pharmaceutically. In doing so, they offering the possibility of improved characterization of the biopharmaceutics of aerosol formulations.

18.3 Concluding Remarks

A variety of cell culture systems for modeling the tracheobronchial epithelium have been developed and are currently available to the scientific community. These include primary cultures and cell lines of human and animal origins, in addition to airway cells which exhibit characteristics of lung disease such as CF. In contrast, models of the alveolar epithelial barrier are limited to primary cultures, with all currently available cell lines unable to recapitulate established in vivo properties of AECs. Moreover, coculture models incorporating respiratory epithelial cells together with other cell types (e.g., endothelial cells, macrophages, dendritic cells) continue to be developed further and require refinement. Given their greater complexity, they offer the potential to study various cell biology-related questions in the lung field (including drug metabolism and modulation of properties of lung air–blood barrier).

For permeability screening purposes, the good barrier properties of epithelial cells in vitro is essential and only cell lines and primary culture systems that form functional tight junctions in culture are suitable. For passively transported compounds, it appears that IVIVC with absorption from the rat lung can be obtained with cell lines (particularly, 16HBE14o– and Calu-3 cells) as well as primary cultures. Nontransport studies (e.g., metabolism of certain classes of xenobiotics) may be performed using cell line culture models (e.g., A549) for alveolar epithelium. However, it is important to note that there will always be limitations when modeling a complex organ such as the lung. For example, none of the currently available in vitro models for drug transport studies take into account the dynamic change of surface area caused by the inhalation-exhalation cycle. Pneumocytes cultured on flexible and permeable supports that form a tight barrier will be useful in that regard.

Notwithstanding the difficulties and challenges surrounding various aspects of in vitro respiratory epithelial models, mechanistic studies of pulmonary drug delivery using various tracheobronchial and alveolar mucosae are expected to provide us with a wealth of information in the coming years that will lead to (1) newer and more efficient methods to treat lung-specific diseases using targeting approaches and (2) help improve the bioavailability of those therapeutics that yield very poor absorption via other routes.

Acknowledgments STB is funded by an IRCSET Government of Ireland Postgraduate Scholarship in Science, Engineering and Technology. This work has been partly funded in part by a Strategic Research Cluster grant (07/SRC/B1154) under the National Development Plan co-funded by EU Structural Funds and Science Foundation Ireland (CE) and by Hastings Foundation, Whittier Foundation, and research grants ES017034, ES018782, HL038621, and HL064365 from the National Institutes of Health (KJK).

References

1. Oreffo VI, Morgan A, Richards RJ (1990) Isolation of Clara cells from the mouse lung. Environ Health Perspect 85:51–64
2. Kaufman DG (1976) Biochemical studies of isolated hamster tracheal epithelium. Environ Health Perspect 16:99–110
3. Robison TW, Dorio RJ, Kim KJ (1993) Formation of tight monolayers of guinea pig airway epithelial cells cultured in an air-interface: bioelectric properties. Biotechniques 15(3):468–473
4. Suda T, Sato A, Sugiura W, Chida K (1995) Induction of MHC class II antigens on rat bronchial epithelial cells by interferon-gamma and its effect on antigen presentation. Lung 173(2):127–137
5. Chung Y, Kercsmar CM, Davis PB (1991) Ferret tracheal epithelial cells grown in vitro are resistant to lethal injury by activated neutrophils. Am J Respir Cell Mol Biol 5(2):125–132
6. Liedtke CM (1988) Differentiated properties of rabbit tracheal epithelial cells in primary culture. Am J Physiol 255(6 pt 1):C760–C770
7. Mathias NR, Kim KJ, Robison TW, Lee VH (1995) Development and characterization of rabbit tracheal epithelial cell monolayer models for drug transport studies. Pharm Res 12(10):1499–1505
8. Welsh MJ (1985) Ion transport by primary cultures of canine tracheal epithelium: methodology, morphology, and electrophysiology. J Membr Biol 88(2):149–163
9. Black PN, Ghatei MA, Takahashi K, Bretherton-Watt D, Krausz T, Dollery CT, Bloom SR (1989) Formation of endothelin by cultured airway epithelial cells. FEBS Lett 255(1):129–132
10. Sisson JH, Tuma DJ, Rennard SI (1991) Acetaldehyde-mediated cilia dysfunction in bovine bronchial epithelial cells. Am J Physiol 260(2 pt 1):L29–L36
11. Sime A, McKellar Q, Nolan A (1997) Method for the growth of equine airway epithelial cells in culture. Res Vet Sci 62(1):30–33
12. Masui T, Wakefield LM, Lechner JF, LaVeck MA, Sporn MB, Harris CC (1986) Type beta transforming growth factor is the primary differentiation-inducing serum factor for normal human bronchial epithelial cells. Proc Natl Acad Sci USA 83(8):2438–2442
13. de Jong PM, van Sterkenburg MA, Kempenaar JA, Dijkman JH, Ponec M (1993) Serial culturing of human bronchial epithelial cells derived from biopsies. In Vitro Cell Dev Biol Anim 29A(5):379–387
14. Galietta LJ, Lantero S, Gazzolo A, Sacco O, Romano L, Rossi GA, Zegarra-Moran O (1998) An improved method to obtain highly differentiated monolayers of human bronchial epithelial cells. In Vitro Cell Dev Biol Anim 34(6):478–481
15. Zabner J, Karp P, Seiler M, Phillips SL, Mitchell CJ, Saavedra M, Welsh M, Klingelhutz AJ (2003) Development of cystic fibrosis and noncystic fibrosis airway cell lines. Am J Physiol 284(5):L844–L854
16. Chemuturi NV, Hayden P, Klausner M, Donovan MD (2005) Comparison of human tracheal/bronchial epithelial cell culture and bovine nasal respiratory explants for nasal drug transport studies. J Pharm Sci 94(9):1976–1985
17. Forbes B, Ehrhardt C (2005) Human respiratory epithelial cell culture for drug delivery applications. Eur J Pharm Biopharm 60(2):193–205
18. Demling N, Ehrhardt C, Kasper M, Laue M, Knels L, Rieber EP (2006) Promotion of cell adherence and spreading: a novel function of RAGE, the highly selective differentiation marker of human alveolar epithelial type I cells. Cell Tissue Res 323(3):475–488
19. Fuchs S, Hollins AJ, Laue M, Schaefer UF, Roemer K, Gumbleton M, Lehr CM (2003) Differentiation of human alveolar epithelial cells in primary culture – morphological characterisation and expression of caveolin-1 and surfactant protein-C. Cell Tissue Res 311:31–45
20. Danto SI, Shannon JM, Borok Z, Zabski SM, Crandall ED (1995) Reversible transdifferentiation of alveolar epithelial cells. Am J Respir Cell Mol Biol 12(5):497–502

21. Borok Z, Liebler JM, Lubman RL, Foster MJ, Zhou B, Li X, Zabski SM, Kim KJ, Crandall ED (2002) Na transport proteins are expressed by rat alveolar epithelial type I cells. Am J Physiol 282(4):L599–L608
22. Johnson MD, Widdicombe JH, Allen L, Barbry P, Dobbs LG (2002) Alveolar epithelial type I cells contain transport proteins and transport sodium, supporting an active role for type I cells in regulation of lung liquid homeostasis. Proc Natl Acad Sci USA 99(4):1966–1971
23. Chen J, Chen Z, Narasaraju T, Jin N, Liu L (2004) Isolation of highly pure alveolar epithelial type I and type II cells from rat lungs. Lab Invest 84(6):727–735 [Erratum in: Lab Invest 85(9):1181 (2005)]
24. Corti M, Brody AR, Harrison JH (1996) Isolation and primary culture of murine alveolar type II cells. Am J Respir Cell Mol Biol 14:309–315
25. Goodman BE, Crandall ED (1982) Dome formation in primary cultured monolayers of alveolar epithelial cells. Am J Physiol 243:C96–C100
26. Shen J, Elbert KJ, Yamashita F, Lehr CM, Kim KJ, Lee VH (1999) Organic cation transport in rabbit alveolar epithelial cell monolayers. Pharm Res 16(8):1280–1287
27. Steimer A, Franke H, Haltner-Ukomado E, Laue M, Ehrhardt C, Lehr CM (2007) Monolayers of porcine alveolar epithelial cells in primary culture as an in vitro model for drug absorption studies. Eur J Pharm Biopharm 66(3):372–382
28. King LS, Agre P (2001) Man is not a rodent: aquaporins in the airways. Am J Respir Cell Mol Biol 24(3):221–223
29. Wang J, Edeen K, Manzer R, Chang Y, Wang S, Chen X, Funk CJ, Cosgrove GP, Fang X, Mason RJ (2007) Differentiated human alveolar epithelial cells and reversibility of their phenotype in vitro. Am J Respir Cell Mol Biol 36(6):661–668
30. Bingle L, Bull TB, Fox B, Guz A, Richards RJ, Tetley TD (1990) Type II pneumocytes in mixed cell culture of human lung: a light and electron microscopic study. Environ Health Perspect 85:71–80
31. Ehrhardt C, Kim KJ, Lehr CM (2005) Isolation and culture of human alveolar epithelial cells. Methods Mol Med 107:207–216
32. Gruenert DC, Finkbeiner WE, Widdicombe JH (1995) Culture and transformation of human airway epithelial cells. Am J Physiol 268(3 pt 1):L347–L360
33. Ehrhardt C, Kneuer C, Fiegel J, Hanes J, Schaefer UF, Kim KJ, Lehr CM (2002) Influence of apical fluid volume on the development of functional intercellular junctions in the human epithelial cell line 16HBE14o–: implications for the use of this cell line as an in vitro model for bronchial drug absorption studies. Cell Tissue Res 308(3):391–400
34. Forbes B (2000) Human airway epithelial cell lines for in vitro drug transport and metabolism studies. Pharm Sci Technol Today 3(1):18–27
35. Sakagami M (2006) In vivo, in vitro and ex vivo models to assess pulmonary absorption and disposition of inhaled therapeutics for systemic delivery. Adv Drug Deliv Rev 58(9–10):1030–1060
36. Fogh J, Trempe G (1975) New human tumor cell lines. In: Fogh J (ed) Human tumor cells in vitro. Plenum, New York, pp 115–159
37. Shen BQ, Finkbeiner WE, Wine JJ, Mrsny RJ, Widdicombe JH (1994) Calu-3: a human airway epithelial cell line that shows cAMP-dependent Cl- secretion. Am J Physiol 266(5 pt 1): L493–L501
38. Foster KA, Avery ML, Yazdanian M, Audus KL (2000) Characterization of the Calu-3 cell line as a tool to screen pulmonary drug delivery. Int J Pharm 208(1–2):1–11
39. Mathia NR, Timoszyk J, Stetsko PI, Megill JR, Smith RL, Wall DA (2002) Permeability characteristics of calu-3 human bronchial epithelial cells: in vitro-in vivo correlation to predict lung absorption in rats. J Drug Target 10(1):31–40
40. Grainger CI, Greenwell LL, Lockley DJ, Martin GP, Forbes B (2006) Culture of Calu-3 cells at the air interface provides a representative model of the airway epithelial barrier. Pharm Res 23(7):1482–1490
41. Florea BI, Cassara ML, Junginger HE, Borchard G (2003) Drug transport and metabolism characteristics of the human airway epithelial cell line Calu-3. J Control Release 87(1–3):131–138

42. Fiegel J, Ehrhardt C, Schaefer UF, Lehr CM, Hanes J (2003) Large porous particle impingement on lung epithelial cell monolayers–toward improved particle characterization in the lung. Pharm Res 20(5):788–796
43. Cooney D, Kazantseva M, Hickey AJ (2004) Development of a size-dependent aerosol deposition model utilising human airway epithelial cells for evaluating aerosol drug delivery. Altern Lab Anim 32(6):581–590
44. Amidi M, Romeijn SG, Borchard G, Junginger HE, Hennink WE, Jiskoot W (2006) Preparation and characterization of protein-loaded N-trimethyl chitosan nanoparticles as nasal delivery system. J Control Release 111(1–2):107–116
45. Cozens AL, Yezzi MJ, Kunzelmann K, Ohrui T, Chin L, Eng K, Finkbeiner KE, Widdicombe JH, Gruenert DC (1994) CFTR expression and chloride secretion in polarized immortal human bronchial epithelial cells. Am J Respir Cell Mol Biol 10(1):38–47
46. Ehrhardt C, Kneuer C, Laue M, Schaefer UF, Kim KJ, Lehr CM (2003) 16HBE14o– human bronchial epithelial cell layers express P-glycoprotein, lung resistance-related protein, and caveolin-1. Pharm Res 20(4):545–551
47. Forbes B, Lim S, Martin GP, Brown MB (2002) An in vitro technique for evaluating inhaled nasal delivery systems. STP Pharma Sci 12:75–79
48. Manford F, Tronde A, Jeppsson AB, Patel N, Johansson F, Forbes B (2005) Drug permeability in 16HBE14o– airway cell layers correlates with absorption from the isolated perfused rat lung. Eur J Pharm Sci 26(5):414–420
49. Kudsiova L, Lawrence MJ (2008) A comparison of the effect of chitosan and chitosan-coated vesicles on monolayer integrity and permeability across Caco-2 and 16HBE14o– cells. J Pharm Sci 97(9):3998–4010
50. Reddel RR, Ke Y, Gerwin BI, McMenamin MG, Lechner JF, Su RT, Brash DE, Park JB, Rhim JS, Harris CC (1988) Transformation of human bronchial epithelial cells by infection with SV40 or adenovirus-12 SV40 hybrid virus, or transfection via strontium phosphate coprecipitation with a plasmid containing SV40 early region genes. Cancer Res 48(7):1904–1909
51. Atsuta J, Sterbinsky SA, Plitt J, Schwiebert LM, Bochner BS, Schleimer RP (1997) Phenotyping and cytokine regulation of the BEAS-2B human bronchial epithelial cell: demonstration of inducible expression of the adhesion molecules VCAM-1 and ICAM-1. Am J Respir Cell Mol Biol 17(5):571–582
52. Sun W, Wu R, Last JA (1995) Effects of exposure to environmental tobacco smoke on a human tracheobronchial epithelial cell line. Toxicology 100(1–3):163–174
53. Steerenberg PA, Zonnenberg JA, Dormans JA, Joon PN, Wouters IM, van Bree L, Scheepers PT, Van Loveren H (1998) Diesel exhaust particles induced release of interleukin 6 and 8 by (primed) human bronchial epithelial cells (BEAS 2B) in vitro. Exp Lung Res 24(1):85–100
54. Veranth JM, Kaser EG, Veranth MM, Koch M, Yost GS (2007) ytokine responses of human lung cells (BEAS-2B) treated with micron-sized and nanoparticles of metal oxides compared to soil dusts. Part Fibre Toxicol 4(1):2
55. Odoms K, Shanley TP, Wong HR (2004) Short-term modulation of interleukin-1beta signaling by hyperoxia: uncoupling of IkappaB kinase activation and NF-kappaB-dependent gene expression. Am J Physiol 286(3):L554–L562
56. Noah TL, Yankaskas JR, Carson JL, Gambling TM, Cazares LH, McKinnon KP, Devlin RB (1995) Tight junctions and mucin mRNA in BEAS-2B cells. In Vitro Cell Dev Biol Anim 31(10):738–740
57. Eaton EA, Walle UK, Wilson HM, Aberg G, Walle T (1996) Stereoselective sulphate conjugation of salbutamol by human lung and bronchial epithelial cells. Br J Clin Pharmacol 41(3):201–206
58. Proud D, Subauste MC, Ward PE (1994) Glucocorticoids do not alter peptidase expression on a human bronchial epithelial cell line. Am J Respir Cell Mol Biol 11(1):57–65
59. Lieber M, Smith B, Szakal A, Nelson-Rees W, Todaro G (1976) A continuous tumor-cell line from a human lung carcinoma with properties of type II alveolar epithelial cells. Int J Cancer 17(1):62–70

60. Kim KJ, Borok Z, Crandall ED (2001) A useful in vitro model for transport studies of alveolar epithelial barrier. Pharm Res 18:253–255
61. Forbes B, Ehrhardt C (2005) Human respiratory epithelial cell culture for drug delivery applications. Eur J Pharm Biopharm 60(2):193–205
62. Foster KA, Oster CG, Mayer MM, Avery ML, Audus KL (1998) Characterization of the A549 cell line as a type II pulmonary epithelial cell model for drug metabolism. Exp Cell Res 243(2):359–366
63. Elbert KJ, Schäfer UF, Schäfers HJ, Kim KJ, Lee VHL, Lehr CM (1999) Monolayers of human alveolar epithelial cells in primary culture for pulmonary absorption and transport studies. Pharm Res 16(5):601–608
64. Kobayashi S, Kondo S, Juni K (1995) Permeability of peptides and proteins in human cultured alveolar A549 cell monolayer. Pharm Res 12(8):1115–1119
65. Wang Z, Zhang Q (2004) Transport of proteins and peptides across human cultured alveolar A549 cell monolayer. Int J Pharm 269(2):451–456
66. Forbes B, Wilson CG, Gumbleton M (1999) Temporal dependence of ectopeptidase expression in alveolar epithelial cell culture: implications for study of peptide absorption. Int J Pharm 180(2):225–234
67. Anabousi S, Bakowsky U, Schneider M, Huwer H, Lehr CM, Ehrhardt C (2006) In vitro assessment of transferrin-conjugated liposomes as drug delivery systems for inhalation therapy of lung cancer. Eur J Pharm Sci 29(5):367–374
68. Duncan JE, Whitsett JA, Horowitz AD (1997) Pulmonary surfactant inhibits cationic liposome-mediated gene delivery to respiratory epithelial cells in vitro. Hum Gene Ther 8(4):431–438
69. Rehan VK, Torday JS, Peleg S, Gennaro L, Vouros P, Padbury J, Rao DS, Reddy GS (2002) 1Alpha,25-dihydroxy-3-epi-vitamin D3, a natural metabolite of 1alpha,25-dihydroxy vitamin D3: production and biological activity studies in pulmonary alveolar type II cells. Mol Genet Metab 76(1):46–56
70. Newton DA, Rao KM, Dluhy RA, Baatz JE (2006) Hemoglobin is expressed by alveolar epithelial cells. J Biol Chem 281(9):5668–5676
71. Zhang L, Whitsett JA, Stripp BR (1997) Regulation of Clara cell secretory protein gene transcription by thyroid transcription factor-1. Biochim Biophys Acta 1350(3):359–367
72. Shlyonsky V, Goolaerts A, Van Beneden R, Sariban-Sohraby S (2005) Differentiation of epithelial Na+ channel function. An in vitro model. J Biol Chem 280(25):24181–24187
73. Woollhead AM, Baines DL (2006) Forskolin-induced cell shrinkage and apical translocation of functional enhanced green fluorescent protein-human alphaENaC in H441 lung epithelial cell monolayers. J Biol Chem 281(8):5158–5168
74. Wikenheiser KA, Vorbroker DK, Rice WR, Clark JC, Bachurski CJ, Oie HK, Whitsett JA (1993) Production of immortalized distal respiratory epithelial cell lines from surfactant protein C/simian virus 40 large tumor antigen transgenic mice. Proc Natl Acad Sci USA 90(23):11029–11033
75. Grek CL, Newton DA, Qiu Y, Wen X, Spyropoulos DD, Baatz JE (2009) Characterization of alveolar epithelial cells cultured in semipermeable hollow fibers. Exp Lung Res 35(2): 155–174
76. Douglas WH, Kaighn ME (1974) Clonal isolation of differentiated rat lung cells. In Vitro 10(3–4):230–237
77. Helms MN, Chen XJ, Ramosevac S, Eaton DC, Jain L (2006) Dopamine regulation of amiloride-sensitive sodium channels in lung cells. Am J Physiol Lung Cell Mol Physiol 290(4):L710–L722
78. Koslowski R, Barth K, Augstein A, Tschernig T, Bargsten G, Aufderheide M, Kasper M (2004) A new rat type I-like alveolar epithelial cell line R3/1: bleomycin effects on caveolin expression. Histochem Cell Biol 121(6):509–519
79. Horalkova L, Radziwon A, Endter S, Andersen R, Koslowski RMW, Dolezal P, Ehrhardt C (2009) Characterisation of the R3/1 cell line as an alveolar epithelial cell model for drug disposition studies. Eur J Pharm Sci 36(4–5):444–450
80. Kemp SJ, Thorley AJ, Gorelik J, Seckl MJ, O'Hare MJ, Arcaro A, Korchev Y, Goldstraw P, Tetley TD (2008) Immortalization of human alveolar epithelial cells to investigate nanoparticle uptake. Am J Respir Cell Mol Biol 39(5):591–597

81. van den Bogaard EH, Dailey LA, Thorley AJ, Tetley TD, Forbes B (2009) Inflammatory response and barrier properties of a new alveolar type 1-like cell line (TT1). Pharm Res 26(5):1172–1180
82. Laube BL (2005) The expanding role of aerosols in systemic drug delivery, gene therapy, and vaccination. Respir Care 50(9):1161–1176
83. Boucher RC (2004) New concepts of the pathogenesis of cystic fibrosis lung. Eur Respir J 23(1):146–158
84. Stutts MJ, Cotton CU, Yankaskas JR, Cheng E, Knowles MR, Gatzy JT, Boucher RC (1985) Chloride uptake into cultured airway epithelial cells from cystic fibrosis patients and normal individuals. Proc Natl Acad Sci USA 82(19):6677–6681
85. Gruenert DC, Basbaum CB, Widdicombe JH (1990) Long-term culture of normal and cystic fibrosis epithelial cells grown under serum-free conditions. In Vitro Cell Dev Biol 26(4):411–418
86. Gruenert DC, Willems M, Cassiman JJ, Frizzell RA (2004) Established cell lines used in cystic fibrosis research. J Cyst Fibros 3(suppl 2):191–196
87. Scholte BJ, Kansen M, Hoogeveen AT, Willemse R, Rhim JS, van der Kamp AW, Bijman J (1989) Immortalization of nasal polyp epithelial cells from cystic fibrosis patients. Exp Cell Res 182(2):559–571
88. Olsen JC, Johnson LG, Stutts MJ, Sarkadi B, Yankaskas JR, Swanstrom R, Boucher RC (1992) Correction of the apical membrane chloride permeability defect in polarized cystic fibrosis airway epithelia following retroviral-mediated gene transfer. Hum Gene Ther 3(3):253–266
89. Ehrhardt C, Collnot EM, Baldes C, Becker U, Laue M, Kim KJ, Lehr CM (2006) Towards an in vitro model of cystic fibrosis small airway epithelium: characterisation of the human bronchial epithelial cell line CFBE41o-. Cell Tissue Res 323(3):405–415
90. Rothen-Rutishauser BM, Kiama SG, Gehr P (2005) A three-dimensional cellular model of the human respiratory tract to study the interaction with particles. Am J Respir Cell Mol Biol 32(4):281–289
91. Rothen-Rutishauser B, Mühlfeld C, Blank F, Musso C, Gehr P (2007) Translocation of particles and inflammatory responses after exposure to fine particles and nanoparticles in an epithelial airway model. Part Fibre Toxicol 25:4–9
92. Pohl C, Hermanns MI, Uboldi C, Bock M, Fuchs S, Dei-Anang J, Mayer E, Kehe K, Kummer W, Kirkpatrick CJ (2009) Barrier functions and paracellular integrity in human cell culture models of the proximal respiratory unit. Eur J Pharm Biopharm 72(2):339–349
93. Hermanns MI, Kasper J, Dubruel P, Pohl C, Uboldi C, Vermeersch V, Fuchs S, Unger RE, Kirkpatrick CJ (2010) An impaired alveolar-capillary barrier in vitro: effect of proinflammatory cytokines and consequences on nanocarrier interaction. J R Soc Interface 7(suppl 1):S41–S54
94. Lehmann T, Kohler C, Weidauer E, Taege C, Foth H (2001) Expression of MRP1 and related transporters in human lung cells in culture. Toxicology 167(1):59–72
95. Madlova M, Bosquillon C, Asker D, Dolezal P, Forbes B (2009) In-vitro respiratory drug absorption models possess nominal functional P-glycoprotein activity. J Pharm Pharmacol 61(3):293–301
96. Lin H, Li H, Cho HJ, Bian S, Roh HJ, Lee MK, Kim JS, Chung SJ, Shim CK, Kim DD (2007) Air-liquid interface (ALI) culture of human bronchial epithelial cell monolayers as an in vitro model for airway drug transport studies. J Pharm Sci 96(2):341–350
97. Endter S, Francombe D, Ehrhardt C, Gumbleton M (2009) RT-PCR analysis of ABC, SLC and SLCO drug transporters in human lung epithelial cell models. J Pharm Pharmacol 61(5):583–591
98. Brillault J, De Castro WV, Harnois T, Kitzis A, Olivier JC, Couet W (2009) P-glycoprotein-mediated transport of moxifloxacin in a Calu-3 lung epithelial cell model. Antimicrob Agents Chemother 53(4):1457–1462
99. Hamilton KO, Backstrom G, Yazdanian MA, Audus KL (2001) P-glycoprotein efflux pump expression and activity in Calu-3 cells. J Pharm Sci 90(5):647–658
100. Patel J, Pal D, Vangal V, Gandhi M, Mitra AL (2002) Transport of HIV-protease inhibitors across 1 alpha,25di-hydroxy vitamin D3-treated Calu-3 cell monolayers: modulation of P-glycoprotein activity. Pharm Res 19(11):1696–1703

101. Florea BI, van der Sandt IC, Schrier SM, Kooiman K, Deryckere K, de Boer AG, Junginger HE, Borchard G (2001) Evidence of P-glycoprotein mediated apical to basolateral transport of flunisolide in human broncho-tracheal epithelial cells (Calu-3). Br J Pharmacol 134(7):1555–1563
102. Yen WC, Corpuz MR, Prudente RY, Cooke TA, Bissonnette RP, Negro-Vilar A, Lamph WW (2004) A selective retinoid X receptor agonist bexarotene (Targretin) prevents and overcomes acquired paclitaxel (Taxol) resistance in human non-small cell lung cancer. Clin Cancer Res 10(24):8656–8664
103. Hamilton KO, Topp E, Makagiansar I, Siahaan T, Yazdanian M, Audus KL (2001) Multidrug resistance-associated protein-1 functional activity in Calu-3 cells. J Pharmacol Exp Ther 298(3):1199–1205
104. van der Deen M, Timmer-Bosscha H, Timens W, Postma DS, de Vries EG (2004) Effect of cigarette smoke extract on MRP1 function in bronchial epithelial cells. Proc Am Assoc Cancer Res 45:A508
105. van der Deen M, de Vries EG, Visserman H, Zandbergen W, Postma DS, Timens W, Timmer-Bosscha H (2007) Cigarette smoke extract affects functional activity of MRP1 in bronchial epithelial cells. J Biochem Mol Toxicol 21(5):243–251
106. Paturi DK, Kwatra D, Ananthula HK, Pal D, Mitra AK (2010) Identification and functional characterization of breast cancer resistance protein in human bronchial epithelial cells (Calu-3). Int J Pharm 384(1–2):32–38
107. Campbell L, Abulrob AN, Kandalaft LE, Plummer S, Hollins AJ, Gibbs A, Gumbleton M (2003) Constitutive expression of p-glycoprotein in normal lung alveolar epithelium and functionality in primary alveolar epithelial cultures. J Pharmacol Exp Ther 304(1):441–452
108. Endter S, Becker U, Daum N, Huwer H, Lehr CM, Gumbleton M, Ehrhardt C (2007) P-glycoprotein (MDR1) functional activity in human alveolar epithelial cell monolayers. Cell Tissue Res 328(1):77–84
109. Agassandian M, Mathur SN, Zhou J, Field FJ, Mallampalli RK (2004) Oxysterols trigger ABCA1-mediated basolateral surfactant efflux. Am J Respir Cell Mol Biol 31(2):227–233
110. Mulugeta S, Gray JM, Notarfrancesco KL, Gonzales LW, Koval M, Feinstein SI, Ballard PL, Fisher AB, Shuman H (2002) Identification of LBM180, a lamellar body limiting membrane protein of alveolar type II cells, as the ABC transporter protein ABCA3. J Biol Chem 277(25):22147–22155
111. Matsuzaki Y, Besnard V, Clark JC, Xu Y, Wert SE, Ikegami M, Whitsett JA (2008) STAT3 regulates ABCA3 expression and influences lamellar body formation in alveolar type II cells. Am J Respir Cell Mol Biol 38(5):551–558
112. Mamchaoui K, Makhloufi Y, Saumon G (2002) Glucose transporter gene expression in freshly isolated and cultured rat pneumocytes. Acta Physiol Scand 175(1):19–24
113. Kalsi KK, Baker EH, Medina RA, Rice S, Wood DM, Ratoff JC, Philips BJ, Baines DL (2008) Apical and basolateral localisation of GLUT2 transporters in human lung epithelial cells. Pflugers Arch 5:991–1003
114. Ehrhardt C, Kneuer C, Bies C, Lehr CM, Kim KJ, Bakowsky U (2005) Salbutamol is actively absorbed across human bronchial epithelial cell layers. Pulm Pharmacol Ther 18(3):165–170
115. Horvath G, Schmid N, Fragoso MA, Schmid A, Conner GE, Salathe M, Wanner A (2007) Epithelial organic cation transporters ensure pH-dependent drug absorption in the airway. Am J Respir Cell Mol Biol 36(1):53–60
116. Miakotina OL, Agassandian M, Shi L, Look DC, Mallampalli RK (2005) Adenovirus stimulates choline efflux by increasing expression of organic cation transporter-2. Am J Physiol 288(1):L93–L102
117. Wang T, Li J, Chen F, Zhao Y, He X, Wan D, Gu J (2007) Choline transporters in human lung adenocarcinoma: expression and functional implications. Acta Biochim Biophys Sin (Shanghai) 39(9):668–674
118. Ishiguro N, Oyabu M, Sato T, Maeda T, Minami H, Tamai I (2008) Decreased biosynthesis of lung surfactant constituent phosphatidylcholine due to inhibition of choline transporter by gefitinib in lung alveolar cells. Pharm Res 25(2):417–427

119. Søndergaard HB, Brodin B, Nielsen CU (2008) hPEPT1 is responsible for uptake and transport of Gly-Sar in the human bronchial airway epithelial cell-line Calu-3. Pflugers Arch 456(3):611–622
120. Rotoli BM, Bussolati O, Sala R, Gazzola GC, Dall'Asta V (2005) The transport of cationic amino acids in human airway cells: expression of system y+L activity and transepithelial delivery of NOS inhibitors. FASEB J 19(7):810–812
121. Granillo OM, Brahmajothi MV, Li S, Whorton AR, Mason SN, McMahon TJ, Auten RL (2008) Pulmonary alveolar epithelial uptake of S-nitrosothiols is regulated by L-type amino acid transporter. Am J Physiol Lung Cell Mol Physiol 295(1):L38–L43
122. Sloan JL, Grubb BR, Mager S (2003) Expression of the amino acid transporter ATB 0+ in lung: possible role in luminal protein removal. Am J Physiol 284(1):L39–L49
123. Uchiyama T, Fujita T, Gukasyan HJ, Kim KJ, Borok Z, Crandall ED, Lee VH (2008) Functional characterization and cloning of amino acid transporter B(0,+) (ATB(0,+)) in primary cultured rat pneumocytes. J Cell Physiol 214(3):645–654
124. Jin SN, Mun GH, Lee JH, Oh CS, Kim J, Chung YH, Kang JS, Kim JG, Hwang DH, Hwang YI, Shin DH, Lee WJ (2005) Immunohistochemical study on the distribution of sodium-dependent vitamin C transporters in the respiratory system of adult rat. Microsc Res Tech 68(6):360–367
125. Seki S, Kobayashi M, Itagaki S, Hirano T, Iseki K (2009) Contribution of organic anion transporting polypeptide OATP2B1 to amiodarone accumulation in lung epithelial cells. Biochim Biophys Acta 1788(5):911–917
126. Izquierdo MA, Scheffer GL, Flens MJ, Giaccone G, Broxterman HJ, Meijer CJ, van der Valk P, Scheper RJ (1996) Broad distribution of the multidrug resistance-related vault lung resistance protein in normal human tissues and tumors. Am J Pathol 148(3):877–887
127. Bouhamyia L, Chantot-Bastaraud S, Zaidi S, Roynard P, Prengel C, Bernaudin JF, Fleury-Feith J (2007) Immunolocalization and cell expression of lung resistance-related protein (LRP) in normal and tumoral human respiratory cells. J Histochem Cytochem 55(8):773–782
128. Berger W, Elbling L, Micksche M (2000) Expression of the major vault protein LRP in human non-small-cell lung cancer cells: activation by short-term exposure to antineoplastic drugs. Int J Cancer 88(2):293–300
129. Meschini S, Marra M, Calcabrini A, Monti E, Gariboldi M, Dolfini E, Arancia G (2002) Role of the lung resistance-related protein (LRP) in the drug sensitivity of cultured tumor cells. Toxicol In Vitro 16(4):389–398
130. Lips KS, Volk C, Schmitt BM, Pfeil U, Arndt P, Miska D, Ermert L, Kummer W, Koepsell H (2005) Polyspecific cation transporters mediate luminal release of acetylcholine from bronchial epithelium. Am J Respir Cell Mol Biol 33(1):79–88
131. Groneberg DA, Fischer A, Chung KF, Daniel H (2004) Molecular mechanisms of pulmonary peptidomimetic drug and peptide transport. Am J Respir Cell Mol Biol 30(3):251–260
132. Bahadduri PM, D'Souza VM, Pinsonneault JK, Sadee W, Bao S, Knoell DL, Swaan PW (2005) Functional characterization of the peptide transporter PEPT2 in primary cultures of human upper airway epithelium. Am J Respir Cell Mol Biol 32(4):319–325
133. Florea BI, Thanou M, Junginger HE, Borchard G (2006) Enhancement of bronchial octreotide absorption by chitosan and N-trimethyl chitosan shows linear in vitro/in vivo correlation. J Control Release 110(2):353–361
134. Grainger CI, Greenwell LL, Martin GP, Forbes B (2009) The permeability of large molecular weight solutes following particle delivery to air-interfaced cells that model the respiratory mucosa. Eur J Pharm Biopharm 71(2):318–324
135. Cheek JM, Kim KJ, Crandall ED (1989) Tight monolayers of rat alveolar epithelial cells: bioelectric properties and active sodium transport. Am J Physiol 256(3 pt 1):C688–C693
136. Kim KJ, Malik AB (2003) Protein transport across the lung epithelial barrier. Am J Physiol 284(2):L247–L259
137. Morimoto K, Yamahara H, Lee VH, Kim KJ (1993) Dipeptide transport across rat alveolar epithelial cell monolayers. Pharm Res 10(11):1668–1674
138. Matsukawa Y, Yamahara H, Lee VH, Crandall ED, Kim KJ (1996) Horseradish peroxidase transport across rat alveolar epithelial cell monolayers. Pharm Res 13(9):1331–1335

139. Widera A, Kim KJ, Crandall ED, Shen WC (2003) Transcytosis of GCSF-transferrin across rat alveolar epithelial cell monolayers. Pharm Res 20(8):1231–1238
140. Bur M, Huwer H, Lehr CM, Hagen N, Guldbrandt M, Kim KJ, Ehrhardt C (2006) Assessment of transport rates of proteins and peptides across primary human alveolar epithelial cell monolayers. Eur J Pharm Sci 28(3):196–203
141. Effros RM, Mason GR (1983) Measurements of pulmonary epithelial permeability in vivo. Am Rev Respir Dis 127(5 pt 2):S59–S65
142. Matsukawa Y, Lee VH, Crandall ED, Kim KJ (1997) Size-dependent dextran transport across rat alveolar epithelial cell monolayers. J Pharm Sci 86(3):305–309
143. Bosquillon C (2010) Drug transporters in the lung-do they play a role in the biopharmaceutics of inhaled drugs? J Pharm Sci 99:2240–2255
144. van der Deen M, de Vries EG, Timens W, Scheper RJ, Timmer-Bosscha H, Postma DS (2005) ATP-binding cassette (ABC) transporters in normal and pathological lung. Respir Res 6:59
145. Patel LN, Uchiyama T, Kim KJ, Borok Z, Crandall ED, Shen WC, Lee VH (2008) Molecular and functional expression of multidrug resistance-associated protein-1 in primary cultured rat alveolar epithelial cells. J Pharm Sci 97(6):2340–2349
146. Bates SR, Tao JQ, Yu KJ, Borok Z, Crandall ED, Collins HL, Rothblat GH (2008) Expression and biological activity of ABCA1 in alveolar epithelial cells. Am J Respir Cell Mol Biol 38(3):283–292
147. Groneberg DA, Paul H, Welte T (2006) Novel strategies of aerosolic pharmacotherapy. Exp Toxicol Pathol 57(suppl 2):49–53
148. Rothenberg ME, Doepker MP, Lewkowich IP, Chiaramonte MG, Stringer KF, Finkelman FD, MacLeod CL, Ellies LG, Zimmermann N (2006) Cationic amino acid transporter 2 regulates inflammatory homeostasis in the lung. Proc Natl Acad Sci USA 103(40):14895–14900
149. Shen J, Elbert KJ, Yamashita F, Lehr CM, Kim KJ, Lee VH (1999) Organic cation transport in rabbit alveolar epithelial cell monolayers. Pharm Res 16(8):1280–1287
150. Ehrhardt C, Kneuer C, Baldes C, Kim KJ, Lehr CM (2006) Albuterol is net absorbed across human alveolar epithelial cell monolayers. Proc Am Thor Soc 3:A761
151. Matsukawa Y, Yamahara H, Yamashita F, Lee VH, Crandall ED, Kim KJ (2000) Rates of protein transport across rat alveolar epithelial cell monolayers. J Drug Target 7(5):335–342
152. Kim KJ, Matsukawa Y, Yamahara H, Kalra VK, Lee VH, Crandall ED (2003) Absorption of intact albumin across rat alveolar epithelial cell monolayers. Am J Physiol 284(3):L458–L465
153. Kim KJ, Fandy TE, Lee VH, Ann DK, Borok Z, Crandall ED (2004) Net absorption of IgG via FcRn-mediated transcytosis across rat alveolar epithelial cell monolayers. Am J Physiol 287(3):L616–L622
154. Ikehata M, Yumoto R, Nakamura K, Nagai J, Takano M (2008) Comparison of albumin uptake in rat alveolar type II and type I-like epithelial cells in primary culture. Pharm Res 25(4):913–922
155. Schanker LS (1978) Drug absorption from the lung. Biochem Pharmacol 27(4):381–385
156. Lim ST, Forbes B, Martin GP, Brown MB (2001) In vivo and in vitro characterization of novel microparticulates based on hyaluronan and chitosan hydroglutamate. AAPS PharmSciTech 2(4):20
157. Cooney D, Kazantseva M, Hickey AJ (2004) Development of a size-dependent aerosol deposition model utilising human airway epithelial cells for evaluating aerosol drug delivery. Altern Lab Anim 32(6):581–590
158. Bur M, Rothen-Rutishauser B, Huwer H, Lehr CM (2009) A novel cell compatible impingement system to study in vitro drug absorption from dry powder aerosol formulations. Eur J Pharm Biopharm 72(2):350–357

Chapter 19
In Vivo Animal Models for Controlled-Release Pulmonary Drug Delivery

Lucila Garcia-Contreras

Abstract Prediction of drug delivery performance in the human lung is most often based on observations in commonly used laboratory animals. Even though cell cultures are a useful and inexpensive tool, they are limited to the study of single or a few processes. The complexity and numerous biological interactions of a drug and delivery system can be accurately evaluated only in vivo. The anatomical and physiological differences between species should be carefully considered and are reviewed in this chapter. In addition, the methods of aerosol delivery and dose calculation are presented. Then, techniques and methods to assess drug delivery performance in these in vivo models are reviewed. Lastly, animal models of disease are discussed. These factors are crucial for the correct interpretation of the results of studies and the subsequent extrapolation to humans.

Keywords Animal models • In vivo testing • Pulmonary disease models • Pharmacokinetics • Methods of aerosol administration • Aerosol dose • Pulmonary drug and vaccine delivery

19.1 Introduction

Pulmonary delivery of drugs has been employed mainly for the treatment of local respiratory diseases such as asthma and chronic obstructive pulmonary disease (COPD). In the last two decades, the research focus on pulmonary delivery has broadened to include a wider range of potential applications including the treatment of local infections and

L. Garcia-Contreras (✉)
Department of Pharmaceutical Sciences, College of Pharmacy, The University of Oklahoma Health Sciences Center, 1110 N. Stonewall Avenue, Oklahoma City, OK 73126, USA
e-mail: lucila-garcia-contreras@ouhsc.edu

H.D.C. Smyth and A.J. Hickey (eds.), *Controlled Pulmonary Drug Delivery*, Advances in Delivery Science and Technology, DOI 10.1007/978-1-4419-9745-6_19,

delivery of molecules for systemic action. The large absorptive surface area of the lung, thin alveolar epithelial barrier, and extensive vasculature make the pulmonary route the most promising noninvasive option for small and large molecules such as protein and peptide therapeutic compounds [1, 2]. Inhalation products are currently administered 2–4 times a day. Benefits of controlled-release products include reduced dosage frequency, extended duration of action, reduced incidence of side effects, and improved compliance resulting in better management of therapy and disease control [3].

The diversity of drugs and controlled-release products being proposed for pulmonary delivery and their applications pose a challenge in terms of the evaluation of their biological and pharmacological action. In order to obtain a clinical response from inhaled products, the drug should deposit in the lower airways, be released from the dosage form, penetrate to the site of action or into the general circulation prior to clearance or metabolism.

The site of drug deposition in the lungs depends on the particle size and size distribution of the drug particles or droplets, the inhaler device and formulation, breathing patterns of the subject, and airway geometry [4]. The duration of drug release is restricted to the residence time in the vicinity of the site of action for locally acting drugs or the site of absorption for systemically acting drugs. Aerosolized drugs can be cleared from the respiratory tract by different mechanisms depending on where the droplets/particles are deposited. The rate at which a drug is cleared is a function of the dynamic interaction of several factors including the rate of mucociliary clearance, the site of deposition, the physical state of the drug (solid or liquid), drug-release rate from the carrier, and physicochemical properties of the drug such as molecular weight, partition coefficient, and charge [5]. Particles deposited in the conducting airways are generally cleared by the mucociliary escalator, whereas particles deposited in the alveolar region are phagocytosed by macrophages [6]. Alveolar macrophages clear particles from the alveolar region by transport along the alveolar surface to the mucociliary escalator, translocation to the tracheobronchial lymph, or internal enzymatic degradation [5, 7, 8]. In addition, drugs reaching the lung endothelium may be subject to accumulation and/or metabolism by enzymes present in the lumen surface of the endothelium [9]. For systemic delivery of drugs, the principal barrier to absorption is the pulmonary epithelia, but for local therapy, absorption through this barrier is the major clearance mechanism of soluble drugs. The proposed routes of absorption include transport via membrane pores, via the intercellular tight junctions by passive diffusion, vesicular transport, active transport, and drainage to the lymphatics [10–14].

The complexity of the processes described above and their interactions can be accurately evaluated only *in vivo*. Even though cell cultures are a useful and inexpensive tool, they are limited to the study of single processes including mechanisms of drug transport in the lungs, the effect of permeation enhancers, formulation excipients or delivery systems and to assessment of the potential toxicity of a drug or excipients/drug delivery system [15]. *Ex vivo* tissue models such as the isolated perfused lung model (IPRL) are employed in pulmonary biopharmaceutics research to study the disposition kinetics of compounds delivered by the pulmonary route or mechanisms of drug transport [16–18]. This *ex vivo* model offers a number of advantages over isolated cell culture models because of the maintenance of the structure and functionality of the tissue. IPRL models have been shown to be kinetically predictive of *in vivo* outcomes for macromolecules, but they appear to be less predictive for small molecules. This has been attributed to the absence of the tracheobronchial circulation in this model. Small molecules are generally absorbed *in vivo* across the thick tracheobronchial epithelia, but large molecules are not absorbed; therefore, the lack of circulation in this region in IPRL makes modeling their absorption difficult. Other limitations of the IPRL include short viable periods of 2–3 h for the tissue and the high level of training and expertise required to set up the model [18]. Therefore, to evaluate properly the performance of controlled-release products for pulmonary delivery, animal models are essential. The anatomical and physiological differences between species available for testing should be carefully considered, as well as the method of aerosol delivery. These factors are crucial for the correct interpretation of the results of studies and the subsequent extrapolation to humans.

19.2 Animal Models for Pulmonary Drug Delivery

The evaluation of factors influencing the pulmonary delivery of drugs for local or systemic action in animal models allows the manipulation of the different variables involved in these complex processes that otherwise may not be easy or feasible in human subjects. Pulmonary drug delivery studies have employed different species including mice, rats, hamsters, guinea pigs, rabbits, dogs, sheep, monkeys, and non-human primates. Initial studies are often conducted in small rodents such as mice, rats, and guinea pigs because an assortment of dosing techniques that require a small amount of the drug can be employed, terminal procedures can be easily performed, and large numbers of these animals can be used for statistical validity. The use of larger species such as rabbits, dogs, sheep, and primates is considered when the study design is more complex or requires more sophisticated maneuvers, or when the administration of larger doses of drug or collection of larger amounts of body fluids for longer periods of time are required. Regulatory agencies and cost may limit the number of animals employed in these studies; therefore, this may influence the statistical analysis of the results [19].

Detailed reviews of the in vitro, ex vivo, and in vivo models available to assess pulmonary absorption and the disposition of inhaled therapeutics for local and systemic delivery have been published in the past 4 years by Sakagami [18], Cryan et al. [19] and Fernandes [20]. Sakagami [18] and Fernandes [20] reviewed the three types of models, but ex vivo models are reviewed in more detail by Sakagami [18]. Cryan et al. [19] focused more on animal models and methods of aerosol administration. The present chapter outlines the anatomical and physiological differences to be considered in the available animal models with respect to all processes involved in the delivery, deposition, and final disposition of drugs delivered by the pulmonary route. Technical aspects of dose calculation and their impact on the result of the study are reviewed, and some models of disease relevant to pulmonary delivery are presented.

19.2.1 *Factors to Be Considered in the Selection of an Animal Model*

The appropriate selection of an animal model is crucial when the quantitative description of airways may be used in mathematical models for predicting efficiencies of deposition and clearance of inhaled drugs. The structure and organization of cells and tissues in the lung pertaining to particular species are an important consideration in the selection of an animal model when the results are to be extrapolated to humans. The relative rates of aerosol deposition and clearance will determine the amount of drug present in the different regions of the respiratory tract that will be available for absorption or local pharmacological action; consequently, the species selected for testing will influence the resulting dose to the lungs. The anatomic and physiological characteristics of the lungs of many mammalian species have been previously described, and key differences are discussed below.

19.2.1.1 Considerations on Respiratory Mechanics

The route of breathing and the respiratory mechanics are important considerations in the selection of an animal model for pulmonary delivery and are presented in Table 19.1. Humans, monkeys, and dogs breathe by both nose and mouth, while most laboratory animals are obligate nose breathers. This has an impact in the amount of aerosol that the animal is able to inhale by passive inhalation in a particular study, which may result in the underestimation of the dose when extrapolating the results to humans

Respiratory parameters are presented in Table 19.1. The same parameter terminology is used for lung volumes in man as for other mammals, but the mechanics that define most of them is different [24]. Total lung capacity and residual volume are maximum and minimum volumes, achieved in cooperative humans by voluntary maximum effort, but in animals by application of external pressures whose magnitude and timing are not standardized.

Table 19.1 Respiratory parameters in mammals

Species	Route of breathing	Respiratory rate (breaths/min)	Tidal volume (mL)	Minute volume (L)
Mouse	Nose	163	0.15	0.023
Rat	Nose	85	1.5	0.10
Guinea pig	Nose	90	1.8	0.16
Rabbit	Nose	46	21.0	1.07
Dog	Nose/mouth	18	320.0	5.2
Monkey (rhesus)	Nose/mouth	40	21.0	0.86
Human (male)	Nose/mouth	12–20	400–616	5–8

Modified from Fernandes and Vanbever [20], Brown [21], Chaffee [22], and Sheldon [23]

Marked variations are also present in the lung volume and breathing frequency of different species (Table 19.1). Thus, if different species inhale the same aerosol for the same length of time they would not receive identical doses. In addition to respiratory parameters, variations in ventilation, collection efficiency, lung anatomy, and clearance mechanisms among species outlined below will also influence the doses of therapeutic aerosols achieved in a particular animal model.

19.2.1.2 Anatomical and Morphological Considerations for Large Airways

In overall shape, the human lung tends to be more spherical than the lungs of other species. The implication of this shape is reflected in the relatively symmetrical airway branching scheme at all levels in the human compared with the long tapering monopodial airways with small lateral branches characteristic of all the other species. The airway structures for different species are compared in Table 19.2 [25].

Airway branching, tracheal length-to-diameter (L/D) ratios, and the angle in airway bifurcation differ considerably among the compared species. Several studies have indicated that the differences in airway branching patterns, and the resulting effects on airflow across species can contribute to differences in regional deposition in the lungs [26, 27]. However, other factors such as differences in breathing patterns, obligate nose-breathing, and the physics of the aerosols delivered may contribute to

Table 19.2 Comparative airway structure: from replica casts

Species	Body mass (kg)	Airway branching	Trachea L/D (cm)	Tracheal bronchus	Major airway bifurcations	Airway L/D (ratio)
Rat	0.3	Strongly monopodial	2.30/0.26	Rare	Very sharp and very high throughout the lung	1.5
Hamster	0.14	Strongly monopodial	2.40/0.26	Absent	Very sharp	1.2
Guinea pig	1	Monopodial	5.7/0.4	Absent	Very sharp and high	1.7
Rabbit	4.5	Strongly monopodial	6.0/0.5	Absent	Sharp	1.9
Beagle dog	10	Strongly monopodial	17.0/1.6	Absent	Blunt tracheal bifurcation	1.3
Rhesus monkey	2	Monopodial	3.0/0.3	Absent	Mixed blunt and sharp	2.6
Human	80	Relatively symmetric	12.0/2.0	Extremely rare	Sharp for about the first ten generations; relatively blunt thereafter	2.2

Modified from Phalen and Oldham [25]

the observed effect [28]. It is probable that the way that humans breathe (oral/nasal) and their airway anatomy may lead to a larger amount of upper bronchial airway aerosol deposition and to greater deposition on localized surfaces near airway bifurcations compared to smaller mammals [29].

Staufer proposed that when the probability of deposition for inhaled aerosols was calculated by dimensional analysis it should be the same for different animals in the case of deposition controlled by sedimentation or turbulence mechanisms, but dimensional analysis should scale as (body weight)$^{0.1}$ for a diffusion-dominated deposition [30]. By contrast, McMahon et al. suggested that diffusion-dominated deposition is independent of body weight [31]. Studies in six different animal models showed that when the total amount of aerosol deposited was divided by body weight, the smaller animals had received more particles/g than the larger ones, but the collection efficiencies (fraction of inhaled aerosol deposited) for both the lungs alone and the lung, nose pharynx, and airways combined were substantially independent of body size.

In terms of the number of lobes, the left human lung is divided into two lobes, the superior and inferior, and the right lung is divided into three lobes. The right lungs from laboratory mammals including nonhuman primates are divided into four lobes. The left lungs from mice and rats are not divided, but those from larger mammals such as guinea pigs and rabbits are divided [32].

Table 19.3 shows the morphologic features of pleuras and bronchioles across different species [24]. The pulmonary or visceral pleura of large animals tend to be thick, whereas that of the smaller animals tends to be thin. The biological significance of thick or thin pulmonary pleuras appears to relate to collateral ventilation between adjacent lobules. Thick pleuras seen in the ox, pig, and some sheep prevent collateral

Table 19.3 Comparative lung biology: morphologic features of pleuras and bronchioles

Species	Pleuras	Bronchioles (nonrespiratory, nonalveolarized)	Structures in which terminal bronchiole ends	Bronchioles (respiratory)
Mouse, rat, hamster, guinea pig, rabbit	Thin	Several generations	Alveolar ducts or very short respiratory bronchioles	Absent or a single short generation
Dog, cat	Thin	Fewer generations	Respiratory bronchioles	Several generations; typical distal airway
Ox, sheep, pig	Thick	Several generations	Alveolar ducts or very short respiratory bronchioles	Absent or single short generation
Macaque monkey	Thick	Fewer generations, commonly only one	Respiratory bronchioles	Several generations; typical distal airway
Humans	Thick	Several generations	Respiratory bronchioles	Several generations

Modified from Tyler et al. [32]

ventilation between adjacent lobules, whereas dogs have excellent collateral ventilation because of their thin pleuras. Thus, comparatively simple species differences in morphologic aspects may relate to basic mechanisms of lung function [32].

The distal airways of the human lung consist of several generations of nonrespiratory bronchioles, ending in the terminal bronchiole. This is followed by about three generations of respiratory bronchioles and an additional three generations of alveolar ducts. Distal airways of animals differ from those of humans and from proximal airways. Cats, dogs, and macaque monkeys, for example, have fewer generations of nonrespiratory bronchioles than other animals or humans (Table 19.3). Respiratory bronchioles are poorly developed in hamsters, rabbits, and guinea pigs may have a single short generation or none [32].

19.2.1.3 Anatomical and Morphological Considerations at the Cellular Level

Table 19.4 shows the morphometric characteristics of cells in the alveolar region of the lungs of different species [33]. Turnover times for several cell populations in the lung are reported to be significantly different between species [34]. In addition, major differences in number, diameter, volume, and functionality have been reported between human, monkey, and rats/hamsters alveolar macrophages [35–38], which may have implications for effective clearance.

The different rates of clearance among species can also influence retention and, therefore, the total dose to the lungs. Brain and Mensah [39] demonstrated that *in situ* pulmonary macrophages of hamster cleared insoluble gold particles faster than rats, rabbits, and mice. Alveolar macrophages have retention half times of 50–80

Table 19.4 Morphometric parameters in the alveolar region of normal mammalian lungs

	Sprague-Dawley rat	Dog	Baboon	Human
Body weight (kg)	0.36±0.01	16±3	29±3	79±4
Lung volume (mL)	10.55±0.37	1,322±64	2,393±100	4,341±284
Total number of cells/lung (10^9)	0.89±0.04	114±13	99±9	230±25
Total lung cells (%)				
Alveolar type I	8.9±0.9	12.5±1.7	11.8±0.6	8.3±0.6
Alveolar type II	14.2±0.7	11.8±0.06	7.7±1.0	15.9±0.8
Endothelial	42.2±1.1	45.7±0.8	36.3±2.4	30.2±2.4
Interstitial	27.7±1.8	26.6±0.7	41.8±2.7	36.1±1.0
Macrophage	3.0±0.3	3.4±0.6	2.3±0.7	9.4±2.2
Average cell surface area (μm^2)				
Alveolar type I	5,320±694	3,794±487	4,004±383	5,098±659
Alveolar type II	123±20	107±15	285±85	183±14
Endothelial	1,105±72	1,137±127	1,040±209	1,353±67

Modified from Crapo et al. [33]

days in rats and about 10 times longer in humans [8]. Nguyen et al. [40] reported varying degrees of efficiency in the bactericidal activity of alveolar macrophages from humans, rabbits, rats, and hamsters. Rabbit alveolar macrophages avidly ingest and kill *Staphylococcus aureus*, whereas alveolar macrophages from rats ingest, but are inefficient at killing these bacteria [41].

Secretory cells of surface epithelium consist of mucous goblet-like cells from the nasal to the bronchial epithelium in most mammalian species including humans. In the pathogen-free rat, however, surface secretory cells in the thorax are mainly of the serous type, except for goblet cells found near the carina [42]. In addition, the chemical composition of the secretions stored in mucous goblet cells differs between species: in rats it is a nonsulfated, sialylated muco-substance, whereas in dogs it is sulfated, and in humans it varies with the region in the respiratory tract between these two types. This may have implications when selecting a model in which mucous secretions are important, such as for cystic fibrosis.

The Clara cell has been identified as the source of the secretory lining material for the bronchioles and as the site within the lung for the metabolism of xenobiotics via the cytochrome P-450 mono-oxygenase system. Plopper reports that Clara cells of humans and nonhuman primates do not share the ultrastructural features described for laboratory mammals [43]. In other species – dog, cat, and ferret – the nonciliated bronchiolar cell does not resemble that seen in either laboratory mammals or primates. Cytochrome P-450 monoxygenase activity has been demonstrated in mouse, rat, hamster, rabbit, and pig. No data are available regarding the activity in other species such as humans, guinea pig, dog, cat, and sheep.

Human basophils and mast cells are the objective of extensive study in several disease states, but they are of particular interest for asthma and other states involving inflammatory processes. As a consequence of antigenic activation, they release preformed mediators such as histamine from secretory granules within the cells. Therefore, differences in morphological and physiological characteristics among animal species need to be taken into account when performing studies involving these cells.

Dvorak et al. defined ultrastructural criteria useful in distinguishing basophils from mast cells in humans, guinea pigs, and mice [44]. Mature human basophils are 5–7 μm in diameter, with round cytoplasmic granules containing electron-dense particles. Mature guinea pig basophils are similar in size to those of humans and mice. They appear round with irregularly distributed blunt surface projections, but can assume motile configurations when appropriately stimulated. A mature mouse basophil has a segmented nucleus, oval homogeneously electron-dense cytoplasmic granules, and cytoplasmic glycogen particles. Human basophils can express two patterns of mediator release [45] by degranulation that differ completely from that in the guinea pig, and nothing appears to be known about the morphologic aspects of degranulation in mouse basophils.

Human mast cells generally are much larger (9–12 μm) than basophils, although occasional cells are as small as 6–7 μm. They have granules that are more numerous, smaller, and more variable than those of basophils. Guinea pig mast cells can be distinguished from basophils by criteria similar to those useful in humans.

Mouse mast cells lack cytoplasmic glycogen particles, have little Golgi apparatus or rough endoplasmic reticulum, and their granules are often larger than those of basophils.

19.2.1.4 Considerations on Lung Circulation

The pulmonary tree in humans, monkey, dog, and mouse is smoothly tapering so that elastic pulmonary arteries blend imperceptibly and without irregularity of wall thickness into muscular pulmonary arteries. This contrasts with rabbit, rat, and guinea pig where the elastic segment is thin-walled, and branching is associated with abrupt changes in lumen size and wall thickness. Similar differences are found in the muscular pulmonary arteries (Table 19.5). The medial thickness of muscular pulmonary arteries is similar in humans and monkeys, but in most other animals is significantly larger.

There is a pronounced variation between the species in the course, structure, and composition of the pulmonary veins (Table 19.5). In humans, the pulmonary veins are located at the periphery of the lobules in the interlobular fibrous septae, whereas in the mouse or sheep they are along the bronchial tree. In terms of composition, the pulmonary veins are muscular in guinea pigs and rats, while dogs, monkeys, and rabbits have pulmonary veins with thin fibrous walls like humans. Variations in localization, structure, and composition of pulmonary arteries and veins among different species may have an impact on the extent and absorption time of a drug delivered by the pulmonary route.

The pulmonary lymphatic vessels provide a unidirectional drainage system designed to maintain homeostasis of the interstitial areas by transporting excess tissue fluids (lymph) back to the blood stream. Leak and Jamuar reported differences in the quantity of pulmonary lymphatics in both pleural (superficial) and deep vessels [47]. Large animals have thick pleura with an extensive network of lymphatics

Table 19.5 Pulmonary vascular characteristics in mammals

	Muscular pulmonary arteries		
Species	Diameter range (μm)	Medial thickness (%)	Pulmonary veins
Mouse	20–300	4.4–11.7	Follow bronchial tree; cardiac muscle in media
Rat	20–350	2.6–26.0	Pads of smooth muscle; cardiac muscle in media
Guinea pig	–	4.3 ± 0.2	Thin, fibromuscular wall
Rabbit	40–150	3.8–10.6	Thin, mainly fibrous wall
Beagle dog	30–120	3.9–11.7	Thin, mainly fibrous wall
Sheep	30–200	5.2–11.8	Follow bronchial tree; thin muscular wall
Monkey	50–160	2.1–5.2	Thin, mainly fibrous wall
Humans	100–1,000	2.0–6.9	Thin, mainly fibrous wall

Modified from Kay [46]

throughout the connective tissue. Conversely, small animals such as the rat and rabbit have thin pleura with a sparse distribution of lymphatics. These differences between species may influence the results of studies involving drug absorption and disposition in and from the lung.

19.2.1.5 Considerations on Drug Absorption from the Lungs

The alveolar region in the lungs is considered to be the optimum site for systemic drug absorption, but the drug can be absorbed in other regions of the lung to a lesser extent. The mechanisms of absorption in the lungs and the factors affecting this process have been previously described in detail [14]. Depending on molecular weight and chemical structure of the drug, as well as physical state and excipients of the formulation, the routes of drug absorption across the airway epithelium can include passive and active transport mechanisms involving paracellular and transcellular transport, pore formation, vesicular transport, and drainage into the lymphatics.

The major barriers to transport are the alveolar epithelial cells and not the underlying endothelial cells [48]. Even though the alveolar epithelium is highly permeable to water, gases, and lipophilic substances, the permeability of large, hydrophilic substances, such as proteins is limited [49]. For protein and peptide drugs, this may increase the probability of rapid degradation in the lysosomal system. All the drug-metabolizing enzymes found in the liver are also found in the lung in smaller amounts [50]; however, they are present in much lower concentrations in the lungs than in the gastrointestinal tract [51]. The degree of drug metabolism appears to be drug-type dependent with small peptides being hydrolyzed rapidly, while for small molecules metabolism is minimal. A certain degree of metabolic clearance of insulin has been demonstrated in recent kinetic studies that would appear to impact on its <10% bioavailability in both human and animals alongside mucociliary clearance [17].

Differences in the rates of absorption have been observed among different species for hydrophilic drugs [52]. These compounds were absorbed 5 times faster in the mouse than in rat, and 2.5 times more slowly in rabbits than in rats. Conversely, lipophilic drugs were absorbed at the same rate in all species.

19.2.1.6 Variations in Biological Response

The innate responsiveness of the analogous cell, tissue, or organ may also differ among species, which may result in variations in the activation, degradation, excretion, or mechanism of action of a determined compound in each species. For instance, isoniazid, an antituberculosis drug, is well tolerated in monkeys at doses of 100 mg/kg, whereas in dogs, doses of 20 mg/kg produce convulsion, respiratory failure, and death. The reason is that monkeys can inactivate isoniazid by acetylation to some extent, while dogs do not have that capability [24].

Several examples of variation in biological responses across species can be found in the field of toxicology such as those to the rodenticide alpha-naphthylthiourea

Table 19.6 Acute toxicity (LD_{50}) of alpha-naphthylthiourea and paraquat in various species [39, 53, 54]

	ANTU		Paraquat	
Species	Route of administration	LD_{50} (mg/kg)	Route of administration	LD_{50} (mg/kg)
Mouse	Intraperitoneal	56	Oral	196
Rat	Intraperitoneal	8	Intraperitoneal	17
Guinea pig	Intraperitoneal	350	Intraperitoneal	3
Rabbit	Intraperitoneal	400	Dermal	280
Dog	Intraperitoneal	16	Oral	38
Monkey	Intraperitoneal	4,250	Oral	50

(ANTU) and the widely used herbicide paraquat. ANTU, like other lung toxic thioureas, produces a fibrin-rich pulmonary edema frequently accompanied by massive pleural effusions. Paraquat is highly toxic to mammals and humans when ingested, potentially leading to acute respiratory distress syndrome. The LD_{50} for ANTU varies by more than two orders of magnitude when rodents, rabbits, dogs, and monkeys are compared (Table 19.6). By contrast, the range of paraquat toxicity among these species is only about tenfold [39].

Thus, it is conceivable that a varying degree of biological responses may be obtained in different animal models when administering therapeutic agents by the pulmonary route.

19.3 Dose Calculation for Pulmonary Delivery Studies

Aerosol delivery to spontaneously breathing animals would be the most relevant method of drug delivery if the results of a study are to be extrapolated to therapy in humans. However, there are substantial limitations to conducting these studies in animals for comparison to humans. In addition to possible differences in the route of breathing, and cutoff diameter in the nose of laboratory animals [55], awake animals are difficult to train to follow particular breathing maneuvers. It is also rather complicated to deliver known quantities of a drug on a ventilatory air supply. Consequently, the dose delivered to animals cannot be easily determined.

There are numerous studies of aerosol delivery to animals in the existing literature, but the "dose" considered varies from study to study. For example, some studies may report the efficiency of delivery in terms of the delivered dose, without considering the fraction of the dose deposited in the periphery of the lung, which is important for locally acting drugs, or the absorbed dose that will determine the efficacy of drugs intended for systemic action. Thus, interpretation of results in these studies and extrapolation to humans should be made with caution. Figure 19.1 illustrates the relative magnitude of the drug amounts relating to the delivered, deposited, and absorbed dose compared to the total amount of drug that is placed in the device for delivery as well as some examples of techniques employed to determine each of these doses. The accuracy to determine each dose will impact on the accuracy of the results obtained.

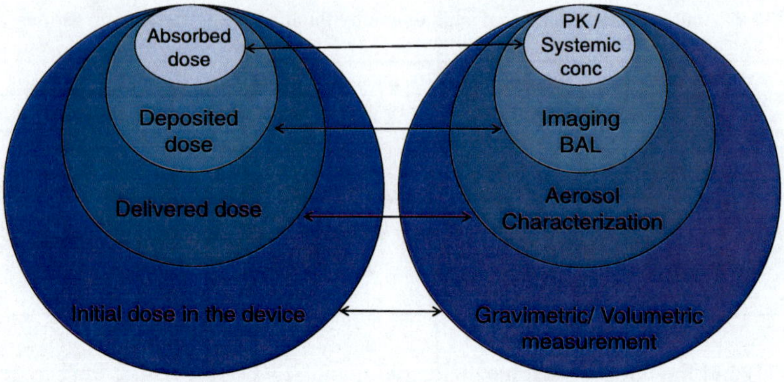

Fig. 19.1 Relative magnitude of the amount of drug (mass or volume) that could be measured as the different "doses" that may be considered in pulmonary delivery studies and examples of techniques to quantify them (the reciprocity is indicated by *arrows*) (modified from Cryan et al. [19])

19.3.1 Pulmonary Administration of Drugs in Animal Models

In pulmonary delivery studies, after selection of the appropriate animal model, the next challenge is to achieve a large delivered dose to maximize the deposited and absorbed doses. The efficiency of a delivered dose is heavily influenced by the method of aerosol administration. Drugs can be delivered to the lungs of animals by passive inhalation or direct administration methods that circumvent oropharyngeal and/or nasopharyngeal deposition. Accurate aerosol doses can be delivered to conscious animals by passive inhalation by complex methods used in the field of inhalation toxicology [56], but their use to deliver drugs may not be practical due to the requirement of large quantities of drug, the complexity of the equipment, and algorithms used to calculate the dose. When accurate doses are required and the amount of drug is limited, direct administration methods are a good option, since a finite amount of material can be delivered to animals. However, since the aerosol is "forced" into the lungs, the site of lung deposition may be different than with aerosol systems depending more directly on droplet or particle size generated by the device. Both methods of administration are described below.

19.3.1.1 Passive Inhalation

Aerosols can be generated from solutions or suspensions of drugs and dry-powder formulations. Jet nebulizers, ultrasonic nebulizers, and metered dose inhalers (MDIs) are commonly used to generate aerosols from liquid formulations, whereas dry-powder inhalers, the sonic sifter, and dry-powder generators such as the Wright dust feed and fluidized-bed generators have been employed to aerosolize powders [57]. The latter ones require tenths of grams of powders and are mainly employed in inhalation toxicology studies. However, while contract research organizations may use all forms of aerosol delivery devices, most published preclinical studies for drugs conducted with passive inhalation methods have employed nebulization to deliver drugs into an exposure chamber. A major limitation for the delivery of particles insoluble in biologically friendly media by nebulization is the concentration that can be achieved in the suspension. Nebulization of highly concentrated suspensions may result in particle precipitation in the nebulizer and the delivery of diluted aerosols. An alternative to nebulization of insoluble particles is to generate aerosols from dry powders. A fluidized-bed generator was considered to deliver drugs to positive-ventilated beagle dogs [58]. However, a large amount of drug was required, and the aerosol was still diluted by the introduction of compressed air, making this system impractical for daily delivery or multiple doses of a drug. The dry-powder dispersion chamber is a new system designed for the delivery of multiple doses of powders to small laboratory animals [59]. The apparatus is similar to a nose-only exposure chamber, with the advantages that the volume of the chamber is smaller and no air is introduced for the dispersion of powders, thus generating a standing cloud of highly concentrated drug powders. This chamber has been used in disposition and efficacy studies to deliver capreomycin and the investigational drug PA-824 for the treatment of tuberculosis in guinea pigs [59, 60].

Aerosolized drugs can be delivered to conscious animals by passive inhalation in exposure chambers that can be for whole body, head-only, and nose-only [56]. There are advantages and disadvantages inherent to each chamber. Whole-body inhalation chambers have the capability to deliver aerosol to unrestrained animals, but the probability of drug absorption by other routes including oral and percutaneous makes the assessment of the actual inhaled dose difficult. Significant interindividual variation might be anticipated as animals will vary in size, lung capacity, and breathing parameters. Examples of whole-body exposure chambers include those from CH technologies, Inc. (Westwood, NJ) (Fig. 19.2a) and the plethysmograph by EMMS (Hants, UK) (Fig. 19.2b).

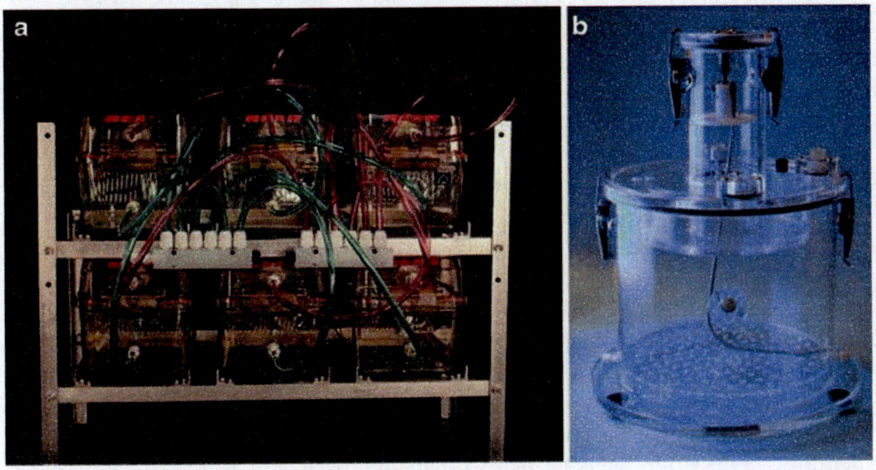

Fig. 19.2 Examples of whole-body exposure chambers by (**a**) CH technologies, Inc. and (**b**) EMMS (reproduced with permission of CH Technologies, Inc., and EMMS)

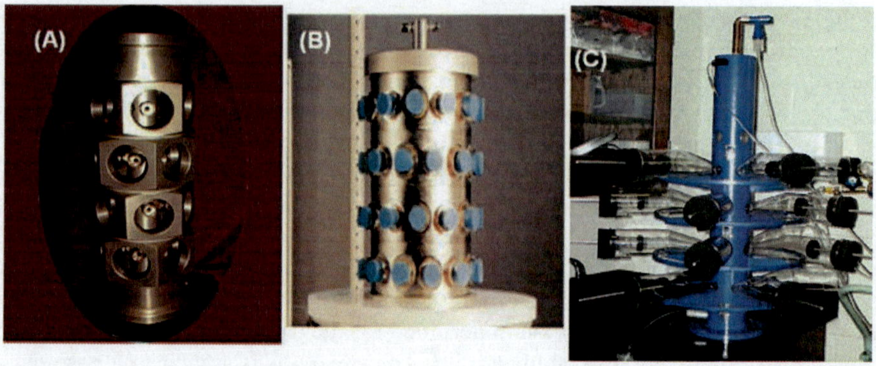

Fig. 19.3 Examples of nose-only exposure chambers by (**a**) Intox Products, LLC (**b**) CH technologies, Inc., and (**c**) ADG Developments, Ltd. (reproduced with permission of CH Technologies, Inc., and Intox Products, LLC)

Nose-only and head-only exposure chambers are generally circular or rectangular devices with animal holders in the outside part of the chamber and a central duct where the aerosol circulates. They are usually made of clear materials to monitor the animals [57]. These chambers eliminate or reduce the probability of drug exposure by multiple routes, but animals may experience stress related to the restraint necessary for these studies, and tight seals for the face or neck of the animal may be required [56]. Head-only exposure chambers such as the "Head dome" chamber are employed in studies involving large animals or human volunteers [61]. Commercially available nose-only exposure chambers can deliver aerosols to 4–48 animals at the same time and include those manufactured by CH Technologies, Inc. (Westwood, NJ), Intox Products, LLC (Moriarty, NM), TSE GmbH (Badhamburg, Germany), and ADG Developments, LTD (Herts, UK), among many others (Fig. 19.3).

The exposure chamber manufactured by ADG Developments (Fig. 19.3c) is designed to deliver aerosols to approximately twenty rodents. This chamber has been employed to deliver aerosols from rifampicin solutions and biodegradable particle suspensions for the treatment of tuberculosis [62, 63]. Independent evaluation of the dose delivered to each animal is strongly recommended, since inter-animal variability may play a role in the ultimate dose delivered. This could be addressed by inferring the dose from plasma concentrations in pharmacokinetic (PK) studies and/or using in vitro methods to estimate the dose at a given sampling flow rate.

The determination of the dose delivered to each animal in passive inhalation studies is complex because of variables related to the size, anatomy, and breathing characteristics of the animal, as well as variables pertaining to drug formulation, the aerosol-generating device, and the efficiency of the exposure chamber. Furthermore, the way that delivered dose was expressed varied from laboratory to laboratory, leading to confusion when comparing results performed in different laboratories. Therefore, in 2004 the Association of Inhalation Toxicologists (AIT) held a panel discussion to address the controversial issue of "How best to calculate and express dose in animal aerosol inhalation toxicology studies?" A document was published outlining, among other things, the guidelines on how to calculate and report the delivered dose and how to calculate respiratory minute volume (RMV), one of the most important factors in the calculation of delivered dose [64].

The AIT recommended that the dose should be reported as the delivered dose and be calculated according to the formula:

$$\mathrm{DD} = \frac{\mathrm{C} \times \mathrm{RMV} \times \mathrm{D} \times \mathrm{IF}}{\mathrm{BW}},$$

where DD is the delivered dose (mg/kg); *C* is the concentration of substance in air (mg/L); RMV is the respiratory minute volume of the animal or the volume of air inhaled in 1 min (L/min); *D* is the duration of exposure (min); BW is the bodyweight of the animal (kg); and IF is the proportion by weight of particles that are inhalable by the test species (respirable fraction). They suggested that the inclusion of IF was not essential provided that the aerosol had reasonable fraction in the respirable range for the intended species, but if IF was included, the way in which it was determined should be clearly stated. Regarding the RMV for mice, rats, dogs, and cynomolgus monkeys, it was recommended that it should be calculated according to the formula:

$$\mathrm{RMV} = 0.608 \times \mathrm{BW}^{0.852},$$

where RMV and BW should be expressed in L/min and kg, respectively. The AIT also suggested that, if the deposited dose was presented as supplementary information, the way in which it was calculated should be clearly stated [64].

19.3.1.2 Direct Administration

Direct administration methods include intratracheal liquid instillation, spray instillation, and dry-powder insufflation, but only the last two methods deliver drug in a form proxy to true aerosol. The devices employed for direct administration are shown in Fig. 19.4. In general, they involve visualization of the trachea of the animal with the help of a laryngoscope to place a thin stainless-steel tube in the trachea, near the carina, to administer drugs. The advantages of direct administration methods are

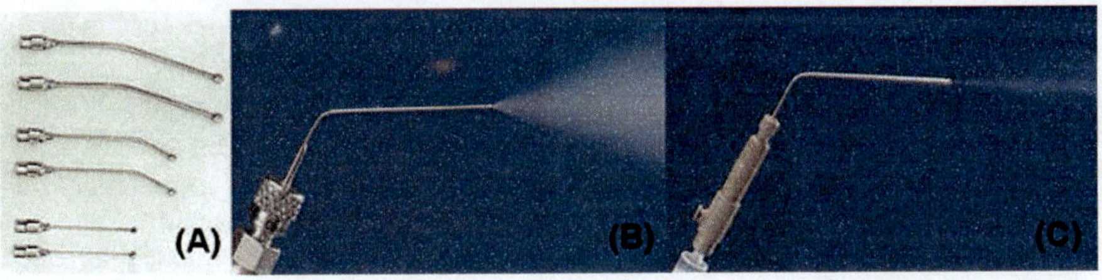

Fig. 19.4 Devices employed for direct intratracheal administration: (**a**) oral gavage needles (**b**) the MicroSprayer®, and (**c**) the Dry Powder Insufflator™ (reproduced with permission of Penn-Century, Inc., Philadelphia, PA)

that they circumvent oropharyngeal or nasopharyngeal deposition, the dose delivered can be accurately measured, and a range of dose sizes can be delivered by these methods. Administration of drugs by these methods is performed under light sedation or anesthesia; thus, depending on the animal and its health status, the use of these methods for multiple or consecutive dosing may not be recommendable. In addition, the procedure to insert the device may require training and practice, since multiple administrations may cause local irritation.

Direct administration devices have been employed in pulmonary delivery studies with animals to administer a variety of therapeutic compounds including antibiotics [62, 63, 65–67], antiproteases [68–70], bronchodilators [71], chemotherapy [72, 73], gene therapy [74, 75], lung surfactant [76], proteins and peptides [77–79], and vaccines [80, 81].

Oral gavage needles (Fig. 19.4a) are generally employed to deliver drugs by intratracheal liquid instillation in the form of a liquid bolus. Consequently, the drug is delivered to a limited area of the respiratory tract, mostly to the large airways, and it may not reach the alveolar region. In addition, if the amount of liquid delivered is large, it may cause significant respiratory distress in the animal, and/or a portion of the dose may be coughed up or swallowed. The MicroSprayer® (Penn Century, Philadelphia, PA) is employed for spray instillation of drug solutions or suspensions (Fig. 19.4b). The atomizer in the tip of a long, thin, stainless-steel tube generates a plume of liquid aerosol that can be deposited in the airways and deep lung. The droplet size of the aerosol produced by the MicroSprayer® is influenced by the physicochemical characteristics, such as viscosity of the solutions/suspension to be delivered. Determination of droplet size before animal studies are conducted is strongly recommended, but for suspensions this is crucial, to ensure that the suspended particles are smaller than the droplet size for an efficient aerosolization of the formulation. The efficiency of delivery using intratracheal liquid instillation and spray instillation was compared by delivering suspensions of insulin encapsulated in calcium phosphate particles to the lungs of rats, and the results are shown in Fig. 19.5. Insulin was absorbed to a greater extent when delivered by spray instillation resulting in glucose reduction for a longer period of time [77]. This was explained by differences in dose distribution. Liquid instillation delivered insulin to a smaller area, whereas finer droplets produced by the spray instillator may have delivered to a larger area, likely including the alveoli where absorption may be greater.

The Dry Powder Insufflator™ (Penn Century, Philadelphia, PA) has a small chamber that can be filled manually with a small amount of dry-powder formulations (Fig. 19.4c). The powder is then dispersed by applying small "puffs" of air to the device using an empty plastic syringe, and the procedure may be repeated a few more times to accommodate larger doses of powder. The dose of drug powder delivered to the lungs can be determined by weighing the device before and after administration. Particle size and distribution of dry powders delivered by insufflation have been reported to be essentially unaffected by passage through the device, as measured by a variety of techniques [82–87]. However, the proportions of powder deposited in the different lung lobes are influenced by the positioning of the stainless-steel tube of the device in the trachea of the animal during powder administration [88].

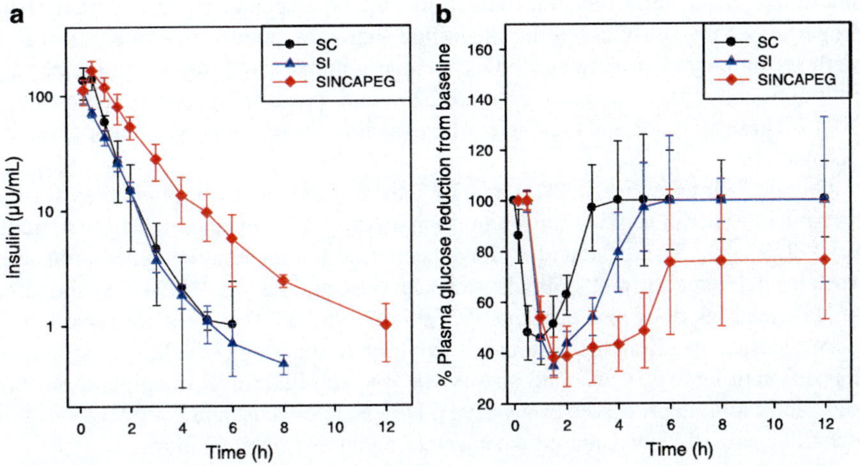

Fig. 19.5 (**a**) Insulin serum concentration vs. time and (**b**) percentage of plasma glucose reduction from baseline vs. time curves after administration of equivalent doses of insulin solution subcutaneously (SC) or intratracheal spray instillation (SI) and insulin CAP-PEG particles by intratracheal spray instillation (SINCAPEG) (modified from Garcia-Contreras et al. [77])

An assortment of dry-powder formulations of different sizes have been delivered by the Dry Powder Insufflator™ including micronized powders [89], nanoparticles [87], microparticles [63, 66, 67], and large porous particles [65, 85].

19.3.2 Methods to Assess Drug Deposition

The amount of drug deposited in the lung and its distribution are key parameters in the evaluation of the formulation and performance of devices for pharmaceutical aerosols.

In particular, when the amount of inhaled drug is very small and cannot be quantified by the lack of sensitivity of analytical methods, lung deposition data can act as a surrogate for clinical response, such as with inhaled anti-asthma drugs [90]. For accurate estimation of drug deposition into the lungs, the time between the exposure and the determination of the dose deposited should be kept to a minimum. If the measurement is delayed, deposition may be underestimated because much of the drug would have been removed by mucociliary clearance, absorption, and/or drug metabolism.

In experimental animals, noninvasive and invasive methods can be used to measure drug deposition. The main advantage with modern techniques used by noninvasive methods in animal studies is the possibility to follow the distribution, absorption, and elimination processes in vivo and the relative time course of each following drug administration. Radiolabeled (^{14}C, ^{3}H, ^{99m}Tc) drugs can easily be located and quantified in the lung, and the detection method depends on the characteristic properties of the substance to be quantified (half-life, energy, and dose). Imaging methods employed in animal studies include gamma scintigraphy, single-photon emission computed tomography (SPECT), and positron emission tomography (PET). These methods and their scientific principles have been reviewed in detail by Dolovich [91].

Gamma scintigraphy, SPECT, and PET have been successfully adapted for the imaging of experimental animals in pulmonary drug delivery and inhalation toxicology [72, 92–99]. These techniques have been used to image radiolabeled drugs and particles delivered to small animals such as mouse and rat [72, 92, 94, 98] and large animals such as dogs and baboons [93, 95–97, 99]. PET images are regarded as more accurate, detailing the regional distribution of the drug in the lungs, estimating deposition of large airway/small airways in vivo and functional imaging of physiologic and biologic processes in vivo [91]. However, the cost and complexity of this technique should be considered when selecting an imaging technique.

Real time in vivo fluorescence imaging is a new emerging technology that is currently used to collect temporal and spatial data from the same animal over multiple time points. By using drugs labeled with fluorophores such as small indocarbocyanine dyes, this technique can be employed in pulmonary delivery studies to allow real-time collection of data with significantly decreased animal numbers.

Lastly, when imaging techniques cannot be used to measure drug deposition, more invasive techniques such as bronchoalveolar lavage (BAL) can be employed for insoluble, inert particles. This technique is discussed in more detail in the next section.

19.3.3 Methods to Assess Absorption and Disposition of Drugs After Pulmonary Delivery

Once the drug has been deposited in the lungs, it can act locally or be absorbed into systemic circulation to have a systemic effect. Assuming that drug concentrations will be such that can be determined by current analytical techniques, the following methods can be employed to assess their absorption and disposition.

19.3.3.1 Determination of Local and Systemic Drug Concentrations

Quantification of drug concentrations in the epithelial lining fluid and lung tissue will give the necessary information to assess the efficacy of treatment for a drug intended for local action. This is not always possible, since the procedure is terminal for most animal models and large numbers of animals are required, making these studies costly and complicated by regulatory issues.

The amount of drug in the epithelial lining fluid can be estimated by analyzing drug concentrations in BAL fluid. BAL is the sampling of the lower respiratory tracts by instillation of sterile saline and subsequent aspiration of the fluid [100]. Lung cells, soluble proteins, lipids, and other chemical constituents from the epithelial surface of the lung can be sampled with this technique. In large animals such as dogs and nonhuman primates, BAL can be performed in vivo under mild sedation in a procedure similar to that performed in human subjects. A flexible fiberoptic bronchoscope is employed to pass as far as possible into the right middle lobe or left upper lobe of the lung. Normal saline solution is introduced and aspirated. Then, the aspirated fluid is collected and analyzed. If needed, BAL can also be performed in vivo for small laboratory animals [101], but for rodents, BAL is usually performed on excised lungs. A catheter inserted into the trachea is used to instill the saline solution. Typically, lavage volumes are approximately half of the total lung capacity of the section of the lung lavaged, but they can vary for small laboratory animals. Generally, two to four lavages are performed to evaluate drug concentration in BAL. In order to assess the dilution that may have occurred by addition of saline, the concentration of urea in the BAL can be measured as an endogenous marker [102] to determine drug concentrations accurately.

Determination of drug concentrations in tissue would likely correlate with drug action for locally acting drugs but may also provide information on drug metabolism in lung tissue, as reported for insulin [16]. Quantification of the drug in lung tissue may be difficult due to complex extraction procedures that may compromise the integrity of the drug.

Drug concentrations in plasma and serum are required to assess the pharmacological activity of drugs intended for systemic action. The blood collection sites are determined based on the volume and frequency of sampling required. For studies that require less frequent sampling in small animals, blood can be obtained from the tail vein or artery, saphenous vein, and the dorsal metatarsal vein [103]. In rabbits, the central ear artery and marginal ear veins can also be used [103]. For studies that require frequent blood sampling in small rodents, such as for PK studies, a cannula may be implanted in the jugular vein for a short period of time (usually less than a week) with a relatively complex surgical procedure. In dogs and other large animals, blood is usually drawn from a superficial implanted catheter from the femoral, cephalic or jugular veins [104, 105], and direct collection from the jugular vein has been also used in few studies [106–109].

19.3.3.2 Pharmacokinetics Models for the Disposition of Drugs in the Lungs

After pulmonary delivery of drugs, PK parameters characterizing drug disposition are influenced by the dynamic interaction of different factors such as the site of deposition, clearance mechanisms, drug formulation (solid or liquid), dissolution rate, mechanism of absorption, and drug metabolism [110]. Mechanisms of drug absorption in the lungs have been reviewed extensively elsewhere [10–14, 111, 112], so they will not be discussed here. As described in previous sections, most of these factors differ among species; thus, different disposition patterns should be expected in the different animal species that may influence the PKs of drugs delivered by the pulmonary route. PK parameters after aerosol administration of a drug can be calculated using compartmental or noncompartmental methods, and physiologically based PK models.

For compartmental analysis, the body is modeled as a system of compartments that are not physiologically or anatomically relevant. Intercompartmental transfer constants assume instantaneous drug distribution, and the rates of transfer between compartments and of drug elimination from a compartment are assumed to follow first-order or linear kinetics. The criteria to determine the best fitting model to a specific set of data are well defined [113]. The systemic disposition of most inhaled drugs is best characterized by one or two compartment models with central elimination, but the disposition of a few inhaled drugs fits better to a three-compartment model [110, 114].

When there is no appropriate compartmental model, noncompartmental analysis can be used. This method is not mechanistic and simply considers the body as a central homogeneous space, where drug input and elimination occur [114]. Drug is sampled in the central space, and the change of drug concentration over time is considered as a statistical distribution curve [115]. Calculation of PK parameters is based on the estimation of the area under the drug concentration–time profile.

Physiologically based pharmacokinetic (PBPK) incorporate physiologically and biochemically relevant information into the mathematical model describing the disposition of a given drug [116]. In general, a PBPK consists of different compartments with each representing a particular organ. The organs of interest are selected based on different factors such as the route of administration of the drug, extent of body distribution, binding characteristics, elimination pathways, and pharmacodynamic (PD) activity. Distribution rates are determined by the rate of blood flow to organs/tissues of distributional importance and partitioning between each organ/tissue and blood. Distribution volumes are based on actual size of organ/tissue and the partitioning of the drug between each organ/tissue and blood. Potential advantages of PBPK modeling include the ability to describe the time course of drug disposition in any organ/tissue of interest and the possibility of straightforward scaling of PK behavior across species [117].

Two models have been proposed to address lung residence time of soluble aerosols. In the first one, Byron [118] developed a mathematical method to determine residence times in the various regions of the respiratory tract as well as the maximum deposition in these regions based on optimization of modes of inhalation and particle size. The limitations of this model were that it did not differentiate between unreleased and released drug and that it did not separate absorption rate from dissolution rate. Subsequently, Gonda [119] proposed a compartmental model that simulated the effects of release rate and multiple dosing on the duration of "free drug levels" in the respiratory tract. This model also accounted for the possibility of excipient accumulation during chronic administration as a function of drug-release rate.

19.3.3.3 Methods to Calculate the Rate of Absorption and Pulmonary Bioavailability

The classical method to determine drug absorption is to measure the disappearance of drugs from the lungs or their appearance in systemic circulation. The rate of absorption is then calculated after curve fitting of semilogarithmic plots of the % remaining in the lung vs. time or drug concentration over time [120]. When a drug is delivered by passive inhalation, estimation of the pulmonary rate of absorption is complicated by different situations such as gastrointestinal absorption of the drug swallowed, but when a drug is delivered directly to the lungs this calculation is straightforward.

The rate and extent of drug absorption can be estimated by the Wagner–Nelson, the Loo–Riegelman, and the observational methods [113–115]. The first two are based on curve fitting by compartmental analysis but have limitations related to the number of compartments in the model [121, 122]. The observational method is based on noncompartmental analysis and is very simple to use [113, 114]. The maximum concentration (C_{max}) and the time to achieve that concentration (T_{max}) are obtained directly from the drug concentration vs. time plots and are used to calculate the rate of absorption. C_{max} depends on both the rate and extent of absorption whereas T_{max} takes discrete values and may be subject to reliability problems because its frequency is not normally distributed. Alternative measures include C_{max}/AUC_{∞}, C_{max}/T_{max}, C_{max}/AUC_{max} [114].

The contribution of several factors such as the physicochemical properties of the drug, the physical state of the formulation (solid or liquid), and possible interactions with excipients should be considered when interpreting the constant of absorption calculated for a drug delivered by the pulmonary route, since they can influence this parameter. For example, the constant of absorption may be underestimated if the rate of drug absorption is limited by poor drug solubility in the lung environment. Slow drug absorption may in turn influence the rate of elimination resulting in a phenomenon identified as flip-flop kinetics [115]. This PK behavior is typical of controlled-release formulations and has been observed for tobramycin [123] and insulin [17]. Likewise, estimation of the rate of drug absorption from the lungs simply on the basis of T_{max} is not recommended, since a short systemic half life can influence the magnitude of T_{max}.

If drug concentrations in plasma are such that can be determined by current analytical techniques, and data after IV administration of that drug is available, pulmonary bioavailability can be calculated using AUCs after IV and pulmonary administration using the following equation:

$$F' = \frac{AUC_{0\to\infty\,lung}}{AUC_{0\to\infty\,IV}} * \frac{Dose_{IV}}{Dose_{lung}} * 100.$$

It is imperative to use the accurate pulmonary dose in this calculation to avoid overestimation of bioavailability. Therefore, if bioavailability is to be determined in a particular study, direct methods of administration should be employed. Interpretation of bioavailability values from animal studies should be made with caution, since depending on the species selected, this parameter could be overestimated or underestimated in humans. Pulmonary bioavailability of insulin has been underestimated in rats and overestimated in rabbits and monkeys when compared to that of humans [124]. Although it is likely that factors such as differences in airway morphology and clearance mechanisms may have contributed to differences in insulin bioavailability in these species, the dosing techniques employed in these studies may have played a larger role. Bioavailability may have been underestimated in rats because intratracheal instillation was employed for animal dosing in that study [125], most likely covering less than 5% of the lung surface. By contrast, bioavailability may have been overestimated in studies with rabbits and monkeys where insulin was nebulized [126, 127], since a larger surface area is reached by this administration method.

The assessment of the bioavailability of drugs after pulmonary administration may be also limited by analytical issues related to the low drug concentrations in lung tissue and plasma. In these cases, an indirect method such as imaging (discussed in Sect. 19.3.2) and measurement of drug in urine (for β2-agonists) can be used to estimate drug bioavailability.

Lastly, when there is no other method to quantify the parent drug, the magnitude of its effects (pharmacodynamics) can be quantified to assess drug absorption. The decrease in glucose levels after administration of insulin formulations has been employed to assess the disposition of insulin formulations after pulmonary administration [77, 79]. The reduction of bacterial burden in affected organs after aerosol administration has also been used to assess the pulmonary disposition of antitubercular drugs [59, 60, 63, 66, 67].

19.4 Animal Models of Pulmonary Disease

Another important consideration in the selection of an animal model to study the pulmonary delivery of drugs is the availability of disease models to conduct preclinical evaluations of the efficacy of drugs and formulations and their relevance to humans in terms of the elements of disease. The feasibility of inducing a particular disease in a determined animal model, the susceptibility of the animal to the disease, and the cost of inducing and maintaining the disease in humane conditions are of paramount importance. Table 19.7 shows some examples of existing animal models of pulmonary diseases and their experimental treatments. The description of the method to induce the specific disease is beyond the scope of the present chapter, but a detailed description of the methods can be found in the respective references.

Table 19.7 Examples of animal models of pulmonary diseases and their treatments

Animal model	Disease	Drug/formulation	Administration method	References
Mouse	Asthma	Methacholine	Nebulizer (DeVilbiss UltraNeb®)	[128]
	Cancer	Liposomal paclitaxel and vitamin E	Nebulizer (Aerotech II)	[129]
		Carboxyl-terminal modulated protein	Nebulizer	[130]
	Cystic fibrosis (*Pseudomonas* infection)	MP-376 (levofloxacin inhalation solution)	MicroSprayer®	[131]
	Diabetes	Insulin	Ultrasonic nebulizer DeVilbiss Pro-UltraNeb 099HD	[132]
	Infectious disease: general	Vaccines and drugs	Dry Powder Insufflator®	[133]
	Infectious disease: tetanus	Tetanus toxoid in mucoadhesive particles	Intranasal	[134]
	Infectious disease: tularemia	Liposomal ciprofloxacin	Jet nebulizer PurRD	[135]
	Airway inflammation	IL-4 antisense oligonucleotide	Intranasal	[136]
Rat	Cystic fibrosis (*Pseudomonas* infection)	Ciprofloxacin in polyethylene glycol-coated liposomes	IV	[137]
	Diabetes	Insulin	Dry Powder Insufflator®	[138, 139]
	Pulmonary hypertension	ONU-1301MS (sustained-release prostacyclin analog)	SC	[140]
	Pneumonia	PEG-coated ciprofloxacin liposomes	IV	[141]
	Pulmonary fibrosis	VEGF	Intratracheal administration	[142]
	Lung transplant	Cyclosporine	Aerosol generator	[143]
Hamster	Cancer	5-fluorouracil lipid-coated nanoparticles	1.7 MHz ultrasonic driver with custom made baffles	[144]
Guinea pig	Tuberculosis	Rifampicin	Dry Powder Insufflator®	[62, 65–67]
			Nebulizer Acorn II	[63, 67]
		Capreomycin	Dry Powder Insufflator® and dry powder dispersion chamber	[59]

(continued)

Table 19.7 (continued)

Animal model	Disease	Drug/formulation	Administration method	References
Rabbit	Asthma	EPI-2010 (respirable antisense oligonucleotide)	Pari LCStar nebulizer	[145]
	Acute lung injury	Recombinant human Clara cell secretory protein	Intratracheal instillation	[146]
Dog	Asthma/COPD	Tiotropium bromide	Respimat soft mist inhaler	[147]
	Cancer (osteosarcoma)	Gemcitabine	Minimate compressor and nebulizer	[148]
	Cancer	Cisplatin	Aeroprobe INC	[149]
	Migraine	Naratriptan	Inhalation	[150]
	Hormonal therapy	Growth hormone-releasing factor, GRF TH9507	Dry Powder Insufflator®	[151]
		Decapeptide detirelix	Intratracheal instillation and UltraVent nebulizer	[152]
Pig	Ventilation	Gadopentetate dimeglumine	MiniHeart jet nebulizer	[153]
Sheep	Smoke inhalation injury	Alpha-tocopherol and vitamin E	DeVilBiss Pro-UltraNeb 099HD	[154]
	Lung injury by infection	Perfluorochemical gentamicin	Intratracheal instillation	[155]
Monkey	Cancer	Gemcitabine	Atomisor NL9M jet nebulizer	[95]
	Gene therapy	Adenoviral vectors with CFTR	MicroSprayer®	[74, 75]

A number of disease models have been developed in mice because physiological variables such as genes can be easily manipulated and large numbers can be used in studies at a relatively low cost. The delivery of large doses of aerosol to mice is difficult to achieve in short periods of time and the amount of biological fluids that can be collected is small. Nevertheless, mice have been frequently used to investigate drug delivery in disease states such as cystic fibrosis [131, 156] and lung cancer [73, 129, 130].

Guinea pigs and rats are often used for the study of drug delivery to the lungs because they are good models for a number of respiratory disease states. Studies of the comparative biology of the guinea pig have revealed that their respiratory tract physiology is remarkable similar to that of humans, particularly the response of the lung to inflammatory stimuli as well as the dermal response to both acute and chronic inflammatory mediators [157, 158]. For that reason, guinea pigs are considered as a good model of bronchoconstriction/bronchodilation to evaluate drugs employed in asthma treatment. The guinea pig is also a good model for infectious diseases in humans because immunologically they are more like humans than any other rodent [159]. A number of variables can be experimentally manipulated in rats, and diseases can be induced for the study of diabetes [138, 139], emphysema [69], influenza [80], and pulmonary fibrosis [160].

The induction of pulmonary diseases in large species is more complicated because of cost and requirements of regulatory agencies, but they are employed in preclinical studies after preliminary results have been generated in rodent models or when there is no other model available. The dog model has been utilized to study the delivery of insulin [161], anticancer agents [148], and hormones [151, 152]. Lung injury can be induced in sheep such that experimental treatments can be evaluated in this model [154, 155, 162]. Monkeys, baboons, macaques, and nonhuman primates have been employed to assess vaccine delivery [84] or to evaluate different therapies for cystic fibrosis [74, 75].

19.5 Summary

The human lung is the standard of reference to which the observations in commonly used laboratory animals are referred. The judicious choice of an animal species, based on the knowledge of its comparative structure and function (anatomy and physiology), would yield important data in the evaluation of controlled-release formulations for pulmonary delivery and the meaningful interpretation of results in such studies. The susceptibility of a particular model to a particular disease and its relevance to humans should also be considered. The physicochemical characteristics of the drug and/or formulation and the methods of aerosol generation and delivery should be carefully controlled. Ultimately, the influence of individual factors and their interaction should be considered in the interpretation of the results from studies conducted in animal models before their extrapolation to humans.

References

1. Patton JS, Fishburn CS, Weers JG (2004) The lungs as a portal of entry for systemic drug delivery. Proc Am Thorac Soc 1(4):338–344
2. Schanker LS (1978) Drug absorption from the lung. Biochem Pharmacol 27(4):381–385
3. Edwards DA et al (1997) Large porous particles for pulmonary drug delivery. Science 276(5320):1868–1871
4. Gonda I (1992) Targeting by deposition. In: Hickey AJ (ed) Pharmaceutical inhalation aerosol technology. Marcel Dekker, New York, pp 61–82
5. Suarez S, Hickey AJ (2000) Drug properties affecting aerosol behavior. Respir Care 45(6):652–666
6. Gehr P et al (1993) Surfactant and inhaled particles in the conducting airways: structural, stereological, and biophysical aspects. Microsc Res Technol 26(6):423–436
7. Beck BD, Brain JD, Bohannon DE (1982) An in vivo hamster bioassay to assess the toxicities of particulates for the lungs. Toxicol Appl Pharmacol 66(1):9–29
8. Oberdorster G, Cox C, Gelein R (1997) Intratracheal instillation versus intratracheal inhalation of tracer particles for measuring lung clearance function. Exp Lung Res 23(1):17–34
9. Niven RW (1992) Modulated drug therapy with inhalation aerosols. In: Hickey AJ (ed) Pharmaceutical inhalation aerosol technology. Marcel Dekker, New York, pp 321–359
10. Florea BI et al (2003) Drug transport and metabolism characteristics of the human airway epithelial cell line Calu-3. J Control Release 87:131–138
11. Gumbleton M et al (2003) Targeting caveolae for vesicular drug transport. J Control Release 87:139–151
12. Kim K-J, Borok Z, Crandall ED (2001) A useful in vitro model for transport studies of alveolar epithelial barrier. Pharm Res 18(3):253–255
13. Kim K-J, Malik AB (2003) Protein transport across the lung epithelial barrier. Am J Physiol 284(2):L247–L259
14. Patton JS (1996) Mechanisms of macromolecule absorption by the lungs. Adv Drug Deliv Rev 19:3–36
15. Forbes B, Ehrhardt C (2005) Human respiratory epithelial cell culture for drug delivery applications. Eur J Pharm Biopharm 60(2):193–205
16. Hsu MC-P, Bai JPF (1998) Investigation into the presence of insulin-degrading enzyme in culture type II alveolar cells and the effects of enzyme inhibitors on pulmonary bioavailability of insulin in rats. J Pharm Pharmacol 50:507–514
17. Sakagami M (2004) Insulin disposition in the lung following oral inhalation in humans: a meta-analysis of its pharmacokinetics. Clin Pharmacokinet 43(8):539–552
18. Sakagami M (2006) In vivo, in vitro and ex vivo models to assess pulmonary absorption and disposition of inhaled therapeutics for systemic delivery. Adv Drug Deliv Rev 58:1030–1060
19. Cryan S-A, Sivadas N, Garcia-Contreras L (2007) In vivo animal models for drug delivery across the lung mucosal barrier. Adv Drug Deliv Rev 59:1133–1151
20. Fernandes CA, Vanbever R (2009) Preclinical models for pulmonary drug delivery. Expert Opin Drug Deliv 6(11):1231–1245
21. Brown LK (1986) Static lung volumes: functional residual capacity, residual volume and total lung capacity. In: Miller AL (ed) Pulmonary function tests in clinical and occupational lung disease. Grune and Stratton, Orlando, pp 77–114
22. Chaffee VW (1974) Surgery of laboratory animals. In: Melby EC Jr, Altman NH (eds) Handbook of laboratory animal science. CRC Press, Cleveland, pp 233–273
23. Sheldon RL (2000) Pulmonary function testing. In: Wilkins RL, Krider SJ, Sheldon RL (eds) Clinical assessment in respiratory care. Mosby, St. Louis, pp 144–155
24. Hickey A, Garcia-Contreras L (2001) Immunological and toxicological implications for short-term studies in animals of pharmaceutical aerosol delivery to the lungs: relevance to humans. Crit Rev Ther Drug Carrier Syst 18(4):387–431

25. Phalen RF, Oldham MJ (1983) Airway structures: tracheobronchial airway structure as revealed by casting techniques. Am Rev Respir Dis 128(2):s1–s4
26. Esch J, Spektor D, Lippmann M (1988) Effect of lung airway branching pattern and gas composition on particle deposition. II. Experimental studies in human and canine lungs. Exp Lung Res 14(3):321–348
27. Spektor DM et al (1985) Influence of airway and airspace sizes on particle deposition in excised donkey lungs. Exp Lung Res 9(3–4):363–387
28. Yeh H, Phalen R, Raabe O (1976) Factors influencing the deposition of inhaled particles. Environ Health Perspect 15:147–156
29. Lippmann M, Schlesinger RB (1984) Interspecies comparisons of particle deposition and mucociliary clearance in tracheobronchial airways. J Toxicol Environ Health 13(2–3):441–469
30. Staufer D (1975) Scaling theory for aerosol deposition in the lungs of different mammals. J Aerosol Sci 6:223–225
31. McMahon T, Brain J, LeMott S (1977) Species difference in aerosol deposition. In: Walton W (ed) Inhaled particles. Pergamon Press, Oxford, pp 23–33
32. Tyler WS (1983) Small airways and terminal units: comparative subgross anatomy of lungs. Am Rev Respir Dis 128(2):s32–s36
33. Crapo JD et al (1983) Morphometric characteristics of cells in the alveolar region of mammalian lungs. Am Rev Respir Dis 128(2):s43–s46
34. Bowden DH (1983) Cell turnover in the lung. Am Rev Respir Dis 128(2):s46–s48
35. Krombach F et al (1997) Cell size of alveolar macrophages: an interspecies comparison. Environ Health Perspect 105(Suppl 5):1261–1263
36. Matulionis DH (1984) Effects of cigarette smoke generated by different smoking machines on pulmonary macrophages of mice and rats. J Anal Toxicol 8(4):187–191
37. Oosting RS et al (1991) Species differences in impairment and recovery of alveolar macrophage functions following single and repeated ozone exposures. Toxicol Appl Pharmacol 110(1):170–178
38. Schlesinger RB, Fine JM, Chen LC (1992) Interspecies differences in the phagocytic activity of pulmonary macrophages subjected to acidic challenge. Fundam Appl Toxicol 19(4):584–589
39. Brain JD, Mensah GA (1983) Comparative toxicology of the respiratory tract. Am Rev Respir Dis 128(2):s87–s90
40. Nguyen B et al (1982) Differences in phagocytosis and killing by alveolar macrophages from humans, rabbits, rats and hamsters. Infect Immun 36:504–509
41. Huber G, Johanson W, LaForce F (1977) Experimental models and pulmonary antimicrobial defenses. In: Brain J, Proctor D, Reid L (eds) Respiratory defense mechanisms. Marcel Dekker, New York, pp 983–1001
42. Spicer SS, Schulte BA, Thomopoulos GN (1983) Histochemical properties of the respiratory tract epithelium in different species. Am Rev Resp Dis 128(2):s20–s26
43. Plopper CG (1983) Comparative morphologic features of bronchiolar epithelial cells: the Clara cell. Am Rev Resp Dis 128(2):s37–s41
44. Dvorak AM, Dvorak HF, Galli SJ (1983) Ultrastructural criteria for identification of mast cells and basophils in humans, guinea pigs and mice. Am Rev Resp Dis 128(2):s49–s52
45. Dvorak AM et al (1980) Crohn's disease: transmission electron microscopic studies II. Immunologic inflammatory response. Alterations of mast cells, basophils, eosinophils, and the microvasculature. Hum Pathol 11:606–619
46. Kay M (1983) Pulmonary vasculature and nerves: comparative morphologic features of the pulmonary vasculature in mammals. Am Rev Resp Dis 128(2):s53–s57
47. Leak LV, Jamuar MP (1983) Ultrastructure of pulmonary lymphatic vessels. Am Rev Resp Dis 128(2):s59–s65
48. Schneeberger EE (1978) Structural basis for some permeability properties of the air–blood barrier. Fed Proc 37(11):2471–2478
49. Sayani AP, Chien YW (1996) Systemic delivery of peptides and proteins across absorptive mucosae. Crit Rev Ther Drug Carrier Syst 13(1–2):85–184

50. Devereux TR, Domin BA, Philpot RM (1989) Xenobiotic metabolism by isolated pulmonary cells. Pharmacol Ther 41(1–2):243–256
51. Zhang JY, Wang Y, Prakash C (2006) Xenobiotic-metabolizing enzymes in human lung. Curr Drug Metab 7(8):939–948
52. Schanker L, Mitchel E, Brown R (1986) Species comparison of drug absorption from the lung after inhalation or intratracheal injection. Drug Metab Dispos 14:79–88
53. Hill NS, O'Brien RF, Rounds S (1984) Repeated lung injury due to alpha-naphthylthiourea causes right ventricular hypertrophy in rats. J Appl Physiol 56(2):388–396
54. Smith PC, Heath D (1976) Paraquat. Biochem Pharmacol 25:411–445
55. Snipes MB (1989) Long-term retention and clearance of particles inhaled by mammalian species. Crit Rev Toxicol 20(3):175–211
56. Dorato MA (1990) Overview of inhalation toxicology. Environ Health Perspect 85:163–170
57. Raeburn D, Underwood SL, Villamil ME (1992) Techniques for drug delivery to the airways, and the assessment of lung function in animal models. J Pharmacol Toxicol Meth 27:143–159
58. Byron PR, Clark AR (1985) Drug absorption from inhalation aerosols administered by positive-pressure ventilation. I: administration of a characterized, solid disodium fluorescein aerosol under a controlled respiratory regime to the beagle dog. J Pharm Sci 74(9):934–938
59. Garcia-Contreras L et al (2007) Inhaled capreomycin large porous particles for the treatment of tuberculosis in the guinea pig model. Antimicrob Agents Chemother 51(8):2830–2836
60. Garcia-Contreras L et al (2010) Dry powder PA-824 aerosols for the treatment of tuberculosis in guinea pigs. Antimicrob Agents Chemother 54(4):1436–1442
61. Asplund PT et al (1996) Longitudinal distribution of ozone absorption in the lung: effect of continuous inhalation exposure. Arch Environ Health 51(6):431–438
62. García-Contreras L, Hickey AJ (2002) Pharmacokinetics of aerosolized rifampicin in the guinea pig. In: Dalby RN et al (eds) Respiratory drug delivery VIII. Davis Horwood International Publishing, Raleigh
63. García-Contreras L et al (2006) Evaluation of dosing regimen of respirable rifampicin biodegradable microspheres in the treatment of tuberculosis in the guinea pig. J Antimicrob Chemother 58:980–986
64. Alexander DJ et al (2008) Association of Inhalation Toxicologists (AIT) Working party recommendation for standard delivered dose calculation and expression in non-clinical aerosol inhalation toxicology studies with pharmaceuticals. Inhal Toxicol 20(13):1179–1189
65. Garcia-Contreras L et al (2006) Pharmacokinetics of aerosolized rifampicin large porous particles in the guinea pig. In: Dalby RN et al (eds) Respiratory drug delivery X. Davis Healthcare International Publishing, River Grove, pp 873–876
66. Suarez S et al (2001) Airways delivery of rifampicin microparticles for the treatment of tuberculosis. J Antimicrob Chemother 48:431–434
67. Suarez S et al (2001) Respirable PLGA microspheres containing rifampicin for the treatment of tuberculosis: screening in an infectious disease model. Pharm Res 19:1315–1319
68. Kuraki T et al (2000) Human neutrophile elastase using microsprayer induced emphysematous changes in rats. Am J Respir Crit Care Med 161(2):872
69. Kuraki T et al (2002) A novel oral neutrophia elastase inhibitor (ONO-6818) inhibits human neutrophil elastase-induced emphysema in rats. Am J Respir Crit Care Med 166:496–500
70. Mata J et al (2007) Evaluation of emphysema severity and progression in a rabbit model: a comparison of hyperpolarized He-3 and Xe-129 diffusion MRI with lung morphometry. J Appl Physiol 102:1273–1280
71. Taylor M, Hickey A, VanOort M (2007) Manufacture, characterization, and pharmacodynamic evaluation of engineered ipratropium bromide particles. Pharm Dev Technol 11: 321–336
72. Gagnadoux F et al (2005) Safety of pulmonary administration of gemcitabine in rats. J Aerosol Med 18(2):198–206
73. Gagnadoux F et al (2005) Aerosol delivery of chemotherapy in an orthotopic model of lung cancer. Eur Respir J 26:657–661

74. Beck SE et al (2002) Deposition and expression of aerosolized rAAV vectors in the lungs of Rhesus macaques. Mol Ther 6:546–554
75. Flotte TR, Laube BL (2001) Gene therapy in cystic fibrosis. Chest 120:124S–131S
76. Takeda K et al (2003) Surfactant protein D regulates airway function and allergic inflammation through modulation of macrophage function. Am J Respir Crit Care Med 168:783–789
77. Garcia-Contreras L et al (2003) Evaluation of novel calcium phosphate particles as pulmonary delivery systems for insulin in rats. AAPSPharmSci 3(3):9
78. Garcia-Contreras L et al (2001) Immediate and short-term cellular and biochemical responses to pulmonary single-dose studies of insulin and H-MAP. Pharm Res 18(12):1685–1693
79. Suarez S et al (2001) Facilitation of pulmonary insulin absorption by H-MAP: pharmacokinetics and pharmacodynamics in rats. Pharm Res 18(12):1677–1684
80. Huang J et al (2003) A novel dry powder influenza vaccine and intranasal delivery technology: induction of systemic and mucosal immune responses in rats. Vaccine 23:794–801
81. Lu D et al (2007) Poly (lactide-co-glycolide) microspheres in respirable sizes enhance an in vitro T cell response to recombinant mycobacterium tuberculosis antigen 85B. Pharm Res 24(10):1834–1843
82. Codrons V et al (2004) Impact of formulation and methods of pulmonary delivery on absorption of parathyroid hormone from rat lungs. J Pharm Sci 93(5):1241–1252
83. Gervelas C et al (2007) Direct lung delivery of a dry powder formulation of DTPA with improved aerosolization properties: effect on lung and systemic decorporation of plutonium. J Control Release 118(1):78–86
84. Grainger CI et al (2003) Administration of an insulin powder to the lungs of cynomolgus monkeys using a Penn Century insufflator. Int J Pharm 269:523–527
85. Koushik K et al (2004) Pulmonary delivery of deslorelin: large-porous PLGA particles and HPβCD complexes. Pharm Res 21(7):1119–1126
86. Minne A, Vanbever R (2007) Development of an inhalation dry powder and assessment of its pulmonary distribution in mice. J Aerosol Med 20(2):164–209
87. Tetsuya O, Hiroaki O (2006) Nanoparticle-containing microspheres drug delivery system (DDS) for inhalation therapy. J Aerosol Res 21(1):10–15
88. Burnell PKP et al (2006) Powder deposition in rats assessed in vivo and in silico. In: Dalby RN et al (eds) Respiratory drug delivery X. Davis Healthcare International Publishing, River Grove, pp 625–627
89. Suarez S, Talton JD, Hochhaus G (1998) Assessment of pulmonary selectivity of triamcinolone acetonide powders using an ex vivo receptor binding assay. In: Dalby RN, Byron PR, Farr SJ (eds) Respiratory drug delivery VI. Interpharm Press, Buffalo Grove, p 425
90. Newman SP, Wilding IR (1998) Gamma scintigraphy: in vivo technique for assessing the equivalence of inhaled products. Int J Pharm 170:1–9
91. Dolovich MB (2001) Measuring total and regional lung deposition using inhaled radiotracers. J Aerosol Med 14(Suppl 1):S35–S44
92. Batrakova EV et al (2010) Effects of pluronic and doxorubicin on drug uptake, cellular metabolism, apoptosis and tumor inhibition in animal models of MDR cancers. J Control Release 143:290–301
93. Deshpande DS et al (2005) Gamma scintigraphic evaluation of a miniaturized AERx pulmonary delivery system for aerosol delivery to anesthetized animals using a positive pressure ventilation system. J Aerosol Med 18(1):34–44
94. Deuse T et al (2010) Mechanisms behind local immunosuppression using inhaled tacrolimus in preclinical models of lung transplantation. Am J Respir Cell Mol Biol 43:403–412
95. Gagnadoux F et al (2006) Gemcitabine aerosol: in vitro antitumor activity and deposition imaging for preclinical safety assessment in baboons. Cancer Chemother Pharmacol 58(2):237–244
96. Khanna C et al (1997) Nebulized interleukin 2 liposomes: aerosol characteristics and biodistribution. J Pharm Pharmacol 49(10):960–971
97. Kreyling WG et al (1999) Anatomic localization of 24- and 96-h particle retention in canine airways. J Appl Physiol 87(1):269–284

98. Richter T et al (2010) Effects of posture on regional pulmonary blood flow in rats as measured by PET. J Appl Physiol 108(2):422–429
99. Young BC et al (2007) Toxic pneumonitis caused by inhalation of hydrocarbon waterproofing spray in two dogs. J Am Vet Med Assoc 23(1):74–78
100. Linder J, Rennard SI (1988) Bronchoalveolar lavage. ASCP Press, Chicago, p 179
101. Mauderly JL (1977) Bronchopulmonary lavage of small laboratory animals. Lab Anim Sci 27:255–261
102. Kipnis E (2005) Using urea as an endogenous marker of bronchoalveolar lavage dilution. Crit Care Med 33(9):2153
103. Harkness J, Wagner JE (1995) The biology and medicine of rabbits and rodents. Williams & Wilkins, Philadelphia
104. Oliver CE et al (2007) Pharmacokinetics of ruminally dosed sodium [36Cl]chlorate in beef cattle. J Vet Pharmacol Ther 30(4):358–365
105. Olsen L et al (2007) Cetirizine in horses: pharmacokinetics and effect of ivermectin pretreatment. J Vet Pharmacol Ther 30(3):194–200
106. Borges FA et al (2007) Pharmacokinetics of a new long acting endectocide formulation containing 2.25% ivermectin and 1.25% abamectin in cattle. J Vet Pharmacol Ther 30(1):62–67
107. Hubenov H et al (2007) Pharmacokinetic studies on tobramycin in horses. J Vet Pharmacol Ther 30(4):353–357
108. Marín P et al (2007) Pharmacokinetics after intravenous, intramuscular and subcutaneous administration of difloxacin in sheep. Res Vet Sci 83(2):234–238
109. Shukla M et al (2007) Comparative plasma pharmacokinetics of meloxicam in sheep and goats following intravenous administration. Comp Biochem Physiol C Toxicol Pharmacol 145(4):528–532
110. Suarez S, Hickey AJ (2001) Pharmacokinetic and pharmacodynamic aspects of inhaled therapeutic agents. In: Martonen T (ed) Medical applications of computer modelling: the respiratory system. WIT Press, Southampton, pp 225–304
111. Chandra T, Miller IF, Yeates DB (1992) A pore transport model for pulmonary alveolar epithelium. Ann Biomed Eng 20:481–494
112. Clark AR, Egan M (1994) Modelling the deposition of inhaled powder drug aerosols. J Aerosol Sci 25:175–186
113. Gibaldi M, Perrier D (1982) Pharmacokinetics. Marcel Dekker, New York
114. Adjei A, Qiu Y, Gupta P (1996) Bioavailability and pharmacokinetics of inhaled drugs. In: Hickey AJ (ed) Inhalation aerosols. physical and biological basis for therapy. Marcel Dekker, New York, pp 197–232
115. Gabrielsson J, Weiner D (2000) Pharmacokinetic and pharmacokinetic data analysis: concepts and applications. Swedish Pharmaceutical Press, Stockholm
116. Chiu WA et al (2007) Evaluation of physiologically based pharmacokinetic models for use in risk assessment. J Appl Toxicol 27(3):218–237
117. Krishnan K, Johanson G (2005) Physiologically-based pharmacokinetic and toxicokinetic models in cancer risk assessment. J Environ Sci Health C Environ Carcinog Ecotoxicol Rev 23(1):31–53
118. Byron PR (1986) Prediction of drug residence times in regions of the human respiratory tract following aerosol inhalation. J Pharm Sci 75(5):433–438
119. Gonda I (1988) Drugs administered directly into the respiratory tract: modeling of the duration of effective drug levels. J Pharm Sci 77(4):340–346
120. Chowhan Z, Amaro A (1976) Pulmonary absorption studies utilizing in situ rat lung model: designing dosage regimen for bronchial delivery of new drug entities. J Pharm Sci 65(11):1669–1672
121. Loo J, Riegelman S (1968) New method for calculatin the intrinsec absorption rate of drugs. J Pharm Sci 57:918–928
122. Wagner J (1970) "Absorption rate constants" calculated according to one-compartment open model with first-order absorption: implications for the in-vivo in-vitro correlations. J Pharm Sci 59:1049

123. Le Brun PP et al (1999) Can tobramycin inhalation be improved with a jet nebulizer? Ther Drug Monit 21(6):619
124. Patton JS, Bukar J, Nagarajan S (1999) Inhaled insulin. Adv Drug Deliv Rev 35:235–247
125. Niven RW et al (1995) Systemic absorption and activity of recombinant consensus interferon and after intratracheal instillation and aerosol administration. Pharm Res 12:1889–1895
126. Colthorpe P et al (1992) The pharmacokinetics of pulmonary delivered insulin: a comparison of intratracheal and aerosol administration to the rabbit. Pharm Res 9:764–768
127. Pillai RS et al (1996) The effect of pulmonary delivered insulin on blood glucose levels using two nebulizers. J Aerosol Med 9:227–240
128. Jonasson S et al (2009) Comparisons of effects of intravenous and inhaled methacholine on airway physiology in a murine asthma model. Respir Physiol Neurobiol 165(2–3):229–236
129. Latimer P et al (2009) Aerosol delivery of liposomal formulated paclitaxel and vitamin E analog reduces murine mammary tumor burden and metastases. Exp Biol Med (Maywood) 234(10):1244–1252
130. Hwang SK et al (2009) Repeated aerosol delivery of carboxyl-terminal modulator protein suppresses tumor in the lungs of K-rasLA1 mice. Am J Respir Crit Care Med 179(12): 1131–1140
131. Sabet M et al (2009) Efficacy of aerosol MP-376, a levofloxacin inhalation solution, in models of mouse lung infection due to *Pseudomonas aeruginosa*. Antimicrob Agents Chemother 53(9):3923–3928
132. Huang YY, Wang CH (2006) Pulmonary delivery of insulin by liposomal carriers. J Control Release 113(1):9–14
133. Morello M et al (2009) Dry-powder pulmonary insufflation in the mouse for application to vaccine or drug studies. Tuberculosis (Edinb) 89(5):371–377
134. Alpar HO et al (2005) Biodegradable mucoadhesive particulates for nasal and pulmonary antigen and DNA delivery. Adv Drug Deliv Rev 57(3):411–430
135. Wong JP et al (2003) Liposome delivery of ciprofloxacin against intracellular Francisella tularensis infection. J Control Release 92(3):265–273
136. Seong JH et al (2006) Polyethylenimine-based antisense oligodeoxynucleotides of IL-4 suppress the production of IL-4 in a murine model of airway inflammation. J Gene Med 8(3):314–323
137. Bakker-Woudenberg IA et al (2002) Ciprofloxacin in polyethylene glycol-coated liposomes: efficacy in rat models of acute or chronic *Pseudomonas aeruginosa* infection. Antimicrob Agents Chemother 46(8):2575–2581
138. Amidi M et al (2009) Efficacy of pulmonary insulin delivery in diabetic rats: use of a model-based approach in the evaluation of insulin powder formulations. J Control Release 135(1):25–34
139. Ungaro F et al (2008) Insulin-loaded PLGA/cyclodextrin large porous particles with improved aerosolization properties: in vivo deposition and hypoglycaemic activity after delivery to rat lungs. J Control Release 127(3):257–266
140. Obata H et al (2008) Single injection of a sustained-release prostacyclin analog improves pulmonary hypertension in rats. Am J Respir Crit Care Med 177(2):195–201
141. Bakker-Woudenberg IA et al (2001) Improved efficacy of ciprofloxacin administered in polyethylene glycol-coated liposomes for treatment of *Klebsiella pneumoniae* pneumonia in rats. Antimicrob Agents Chemother 45(5):1487–1492
142. Farkas L et al (2009) VEGF ameliorates pulmonary hypertension through inhibition of endothelial apoptosis in experimental lung fibrosis in rats. J Clin Invest 119(5):1298–1311
143. Keenan RJ et al (1992) Improved immunosuppression with aerosolized cyclosporine in experimental pulmonary transplantation. Transplantation 53(1):20–25
144. Hitzman CJ, Wattenberg LW, Wiedmann TS (2006) Pharmacokinetics of 5-fluorouracil in the hamster following inhalation delivery of lipid-coated nanoparticles. J Pharm Sci 95(6): 1196–1211
145. Sandrasagra A et al (2001) RASONs: a novel antisense oligonucleotide therapeutic approach for asthma. Expert Opin Biol Ther 1(6):979–983

146. Miller TL et al (2005) Recombinant human Clara cell secretory protein in acute lung injury of the rabbit: effect of route of administration. Pediatr Crit Care Med 6(6):698–706
147. Casarosa P et al (2009) Preclinical evaluation of long-acting muscarinic antagonists: comparison of tiotropium and investigational drugs. J Pharmacol Exp Ther 330(2):660–668
148. Rodriguez CO et al (2009) Aerosol gemcitabine: preclinical safety and in vivo antitumor activity in osteosarcoma-bearing dogs. J Aerosol Med Pulm Drug Deliv 22(4):1–10
149. Selting K et al (2008) Feasibility and safety of targeted cisplatin delivery to a select lung lobe in dogs via the AeroProbe intracorporeal nebulization catheter. J Aerosol Med Pulm Drug Deliv 21(3):255–268
150. Sanjar S, Matthews J (2001) Treating systemic diseases via the lung. J Aerosol Med 14(Suppl 1):S51–S58
151. Jansen M et al (2004) Pulmonary delivery of TH9507, a growth hormone releasing factor analogue, in the dog. Int J Pharm 276(1–2):75–81
152. Bennett DB et al (1994) Pulmonary delivery of detirelix by intratracheal instillation and aerosol inhalation in the briefly anesthetized dog. Pharm Res 11(7):1048–1055
153. Sood BG et al (2008) Aerosol delivery in ventilated newborn pigs: an MRI evaluation. Pediatr Res 64(2):159–164
154. Morita N et al (2006) Aerosolized alpha-tocopherol ameliorates acute lung injury following combined burn and smoke inhalation injury in sheep. Shock 25(3):277–282
155. Cox CA et al (2001) Intratracheal administration of perfluorochemical-gentamicin suspension: a comparison to intravenous administration in normal and injured lungs. Pediatr Pulmonol 32(2):142–151
156. Desigaux L et al (2005) Nonionic amphiphilic block copolymers promote gene transfer to the lung. Hum Gene Ther 16:821–829
157. Lechner AJ, Banchero N (1982) Advanced pulmonary development in newborn guinea pigs (Cavia porcellus). Am J Anat 163:235–246
158. Wagner JE, Manning PJ (1976) The biology of the guinea pig. Academic, New York
159. McMurray DN (1994) Guinea pig model of tuberculosis. In: Bloom B (ed) Tuberculosis: pathogenesis, protection, and control. ASM Press, Washington, pp 135–148
160. Spond J et al (2003) Inhibition of experimental acute pulmonary inflammation by pirfenidone. Pulm Pharmacol Ther 16:207–214
161. Surendrakumar K et al (2003) Sustained release of insulin from sodium hyaluronate based dry powder formulations after pulmonary delivery to beagle dogs. J Control Release 91:385–394
162. Murakami K et al (2002) A novel animal model of sepsis after acute lung injury in sheep. Crit Care Med 30(9):2083–2090

Chapter 20
Imaging Pulmonary Drug Delivery

Philip J. Kuehl

Abstract This chapter focuses on the current practices of gamma imaging for the purposes of quantification of lung deposition or lung dose of inhaled therapeutics. The chapter reviews the various methodologies required for a viable imaging study. Specifically, a brief background on gamma imaging, radiolabeling method development and validation, dose administration, image acquisition, and image analysis is presented.

Keywords Computed tomography • Gamma scintigraphy • In vivo • Radiolabeling image analysis • Single-photon-emission

20.1 Introduction

The quantification of lung deposition or lung dose of inhaled therapeutics is an oft-debated topic with many potential solutions in both preclinical and clinical settings. These solutions have typically included three primary techniques, systemic pharmacokinetic analysis, in vitro measurement techniques (cascade impactors analysis), and deposition imaging. As with nearly all measurement techniques, each of these has its advantages and disadvantages.

Systemic pharmacokinetic analysis is currently used for bioequivalence analysis in Europe [1, 20], and several studies have shown a limited degree of correlation between systemic PK and pulmonary deposition [6]. However, there are several technical challenges to conducting PK analysis of orally inhaled products so that the PK data can be useful. The reasons for the complications lie in the fact that all orally inhaled products have a portion of their dose which deposits in locations other than the lung. The absorption of the test article in these areas can confound the PK

P.J. Kuehl (✉)
Lovelace Respiratory Research Institute, Albuquerque, NM, USA
e-mail: pkuehl@lrri.org

H.D.C. Smyth and A.J. Hickey (eds.), *Controlled Pulmonary Drug Delivery*, Advances in Delivery Science and Technology, DOI 10.1007/978-1-4419-9745-6_20,

analysis. There are techniques that can be used to circumvent these complications (charcoal block, early concentration measurement, and/or subtraction of oral PK component) [17]. However, none of these address the fact that systemic PK analysis does not measure drug in its primary deposition location. Rather, it is a measurement of drug that has been cleared (diffusion, absorption, mucociliary clearance, or other) from the primary deposition location and measured in a secondary location.

The utility of in vitro testing with cascade impactors or dose content analysis methods (ex-valve, ex-device measurements) to predict in vivo deposition patterns has been argued for [24] and against [4]. While these quality control measures may have shown some correlation to deposition analysis (smaller particle sizes have been shown to result in higher dose fraction depositing in the lungs [12, 13]), they are not designed nor intended to be used to predict the regional deposition in vivo.

Imaging of a radiolabeled inhalation formulation provides a direct measurement of the in vivo deposition. Historically, deposition analysis has been conducted with two different imaging modalities, gamma imaging and positron-emission tomography (PET). Gamma imaging can be separated into the more common 2D planar imaging and 3D single-photon-emission computed tomography (SPECT) imaging. All of these techniques again have technical challenges in both the development of the labeling technique and the final quantification of deposition. Regardless, as they provide quantification of the primary deposition location, they have established themselves as a mainstay for pulmonary deposition analysis [2]. To this point, gamma scintigraphy has been dubbed an "industry standard" for assessing deposition form inhalation devices [22].

This chapter will focus on the current practices of gamma imaging and cover the various methodologies required for a viable imaging study. Specially, a brief background on gamma imaging, radiolabeling method development and validation, dose administration, image acquisition, and image analysis is presented.

20.2 Gamma Scintigraphy

Gamma scintigraphy has been continually advancing along with the field of nuclear medicine since the first gamma camera was produced in 1964 [18]. A gamma camera, example figure of a Siemens SPECT E.Cam shown in Fig. 20.1, is primarily composed of one (or multiple) radiation detectors or heads that are mounted on the gantry. The gantry of modern gamma cameras can be moved in a variety of different configurations for the acquisition of multiple different data sets.

The detectors consist of a flat crystal plane (typically sodium iodide) that is coupled to a collection of photomultiplier tubes. When a gamma ray interacts with the sodium iodide crystal, it emits a light photon that is detected by the photomultiplier tube. The light signal is then sent to the computer software for reconstructions.

Located between the patient and the crystal plate is a lead filter or collimator. The collimator can be changed based on the isotope used and/or the needs of the study. Most often for inhalation deposition studies, the collimator used has small holes drilled in a pattern across its entire surface, hence the name parallel hole collimator.

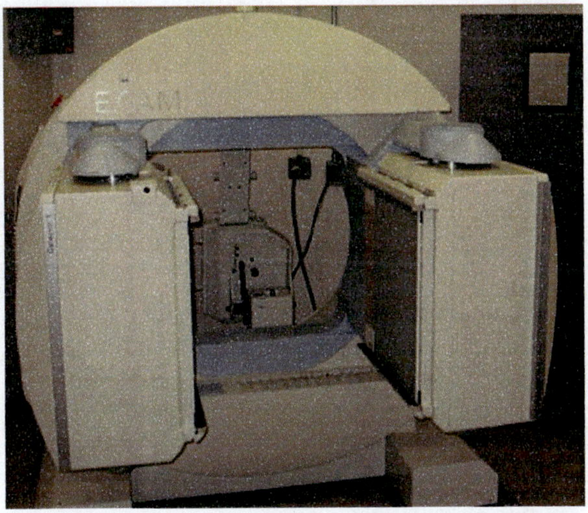

Fig. 20.1 Image of a dual-head clinical gamma camera

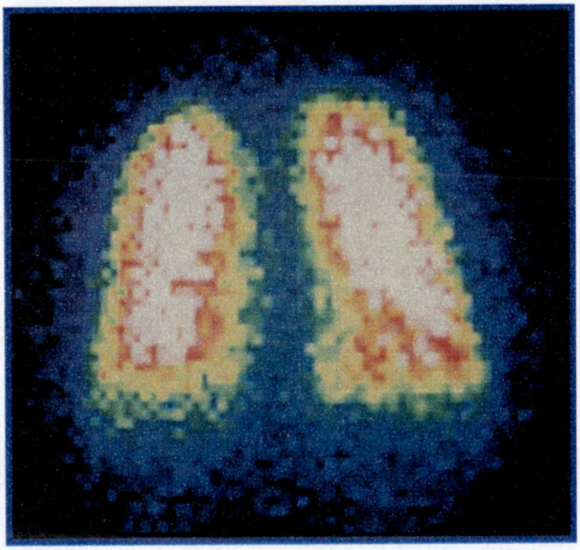

Fig. 20.2 Example gamma scintigraphy image

These holes act as a filter to ensure that only gamma radiation traveling at a right angle to the detector pass through and can be detected. In this manner, the computer can then recreate the location of the gamma ray emission from the location it was detected on the head of the camera. Then the software can generate what is commonly seen for gamma imaging studies, example shown in Fig. 20.2.

The two primary types of image acquisitions that are conducted are 2D and 3D. Planar image acquisition, 2D, has been the historical standard for inhalation deposition analysis. Planar image acquisition involves the collection of anterior and posterior (typically at the same time with a multihead gantry) images. The images

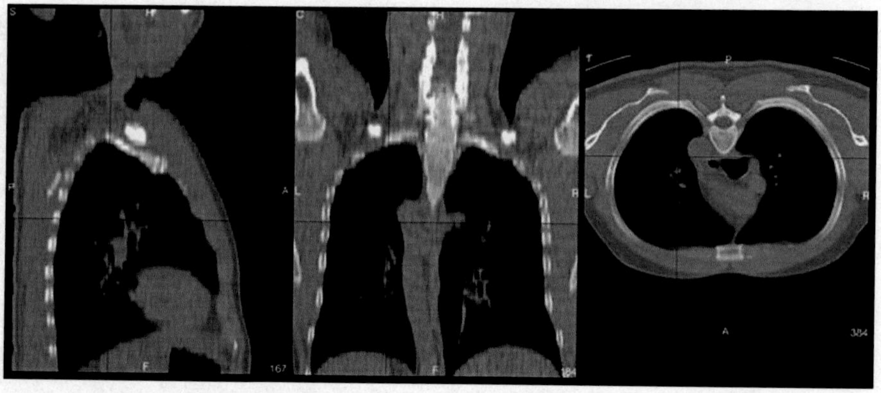

Fig. 20.3 Example CT images displaying the anatomical specificity possible when a CT is used in conjunction with a SPECT image

are only representative of the location of the gamma emission is 2 dimensions and therefore do not reflect depth. Therefore, these images provide a valid assessment of organ dose and total dose. However, they only provide limited information about the specific location of the deposition within an organ, e.g., lung. The advantage of planar imaging is that the image acquisition time is fairly rapid, typically less than 1 min, and can be conducted with a low amount of radioactivity. The disadvantage of planar imaging is that the organs being imaged are 3D and therefore complex analysis of regional deposition is not ideal.

SPECT image acquisition is a more complex imaging method that requires the gantry to collect data at 360° around the patient. Often multiple-headed gantries are used so that each detector only has to be 180° to collect all 360° of data. The resultant images allow for a more accurate 3D reconstruction of the patient. However, additional time is required to collect 360° worth of data, often 10 min or more, which is a disadvantage of SPECT imaging. Several techniques can be utilized to decrease the acquisition times including increasing the radioactivity used, and employment of multiple-headed gantries can decrease time [5].

Often coupled to SPECT image acquisition is computed tomography or CT. Most new SPECT cameras are also multifunction instruments that include both the SPECT camera and a CT system. Collection of a CT image, example shown in Fig. 20.3, along with a SPECT image allows for specific anatomical identification. Further, it allows the exact outline of the lung to be defined during analysis.

20.3 Radiolabeling Aerosols

The isotope of choice for inhalation deposition analysis has been and continues to be Technitium-99m (Tc99m). Tc99m, in the form of pertechnetate ($^{99m}TcO_4^-$), is one of the most common diagnostic and research tracers used in humans. This is due to its relatively short physical half-life (6 h) and low gamma emission energy (140 KeV) that provides minimal exposure to the patients and technicians while still providing quality images.

The procedures that have been used to radiolabel nebulizers, pMDI [7, 9, 12, 13], and dry powders [8, 19] have been well described in the literature and are typically specific for a formulation/delivery system. These range in complexity from simple addition of dissolved Tc99m to a solution nebulizer to extraction of Tc99m into an organic, mixing with a dry powder suspended in an antisolvent and evaporation of the antisolvent. The majority of these procedures result in the drug being associated with the radiolabel rather than being chemically bonded. Because of this each radiolabel must be subjected to a validation prior to its use in deposition studies (clinical or preclinical). It is the radiolabeling validation that establishes the relationship between the drug and the Tc99m without altering the naïve particle size distribution. Without a well-validated radiolabel, the validity of any imaging study is in question.

Radiolabeling validation is the process whereby the particle size distribution of an aerosol is evaluated before radiolabeling, after radiolabeling, and in relationship to the activity size distribution. Throughout the radiolabeling method development and the validation, it is recommended to utilize a constant level of radioactivity. The amount of radioactivity required will be a balance between the expected activity levels in the regions to be imaged and ensure subject and technician safety. Typically, before the validation is initiated, clear limits are placed on the allowable differences between the naïve mass, radiolabeled mass, and the radioactivity size distributions. These studies must be conducted with a multistage cascade impactor (Anderson, Next Generation Impactor, Mercer or other). The validation experiments should be conducted between 6 and 8 independent times in order to completely characterize the inter- and intra-day variability of the radiolabeling procedure. The tests should also be conducted to mimic the time course of administration, for example if 1 day of dose administration is planned to require 6 h then the testing should encompass ~6.5–7 h. These tests will ensure that the radiolabeling procedure not only does not alter the naïve particle but it also provides a high degree of confidence that the radiolabel will be viable the entire day of patient dosing.

As with any cascade impactor analysis prior to conduct of testing, the extraction/analysis methods must be refined. For a radiolabeling study, there is the added complexity that the extraction procedure must be tested for its efficiency to extract not only the drug but it must also be tested for its ability to extract the Tc99m. In the process, the solution (often an aqueous/organic mixture) is first assayed in a dose calibrator for radioactivity. Then depending on the laboratory set up and the stability of the drug, the samples are either held for decay (typically 3 days) or immediately assayed for the drug via HPLC, UV, LC-MS, or equivalent. Following completion of activity and drug analysis, the data are plotted as percentage of total dose and compared with a naïve formulation, example validation summary shown in Fig. 20.4. If the validation experiments do not repeat or issues are observed, additional method development is likely required.

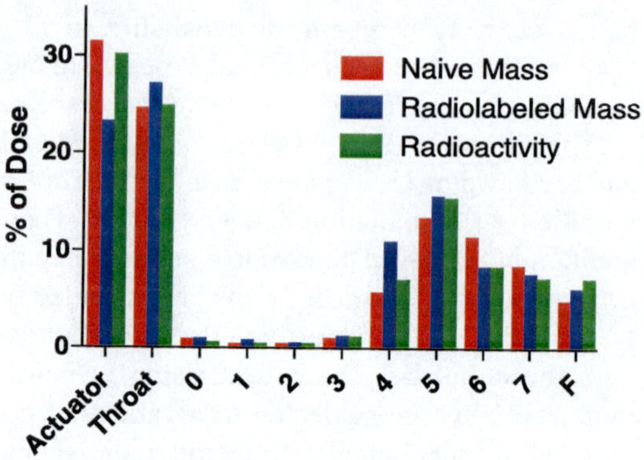

Fig. 20.4 Example radiolabeling validation of a viable radiolabel

20.4 Dose Administration

It is well accepted that the deposition pattern of inhalation therapeutics varies based on the formulation and device; however, the deposition pattern can be altered by other patient factors that should be considered prior to deposition imaging studies. The first level of control that is maintained on the homogeneity of dose administration is a well-defined set of inclusion and exclusion criteria. Important criteria to consider are the use of healthy subjects or patients, age, height, weight, history of respiratory disease, smokers vs. non smokers, and FEV1. Having these criteria well defined will ensure the quality of the data from the study. When defining these criteria, careful consideration should be given to using healthy subjects or patients. Healthy patients are often easier to recruit and will result in a more homogeneous deposition patterns between subjects. However, the target therapeutic group for these inhalation products is usually patients with lung disease; therefore, the deposition patterns of similar patients are likely the most applicable data to the therapeutic use of the products.

Once subjects are selected based on the inclusion/exclusion criteria, they should be evaluated for their ability to correctly administer the inhalation dose. While the inhalation procedure for each formulation is formulation dependant, the inhalation variables, inspiratory flow rate, inhalation duration, inhalation volume, and breathe hold should be controlled and recorded for the administration of all doses. This type of control and recording is not a standard feature on inhalation devices, and it is usually accomplished by fitting the device with a small tube fitting, out of the aerosol path and the standard patient griping locations. This tube fitting it is then connected to a pressure sensor and subsequently to a computer recording software. The computer software is set up to provide real time feed back to the patient in terms of the inhalation variables. Each patient should be trained and evaluated for their ability to properly inhale a formulation prior to radiolabeled dose delivery. The ideal method to conduct this training is by conducting training prior to the day of any dosing and then conducting a refresher training immediately prior to radiolabeled dose delivery. These measured parameters should also be included in the summary of any imaging results to convey that all patients correctly administered each formulation and eliminate the potential for deposition bias caused by erratic inhalation procedures.

On the day of dosing, the analysis of particle size distribution, in the same manner as was used for the validation, should also be performed prior to and immediately following the patient dosing. For the actual dose administration, the pre and postdevice should be quantified for all appropriate parameters. These parameters are usually radioactivity and weight. These two variables are required to determine that the dose was delivered properly and aid in image analysis.

20.5 Image Acquisition

Many aspects of image acquisition for gamma imaging studies are constant regardless of the type of image being collected, inhalation formulation, or the study design. These include subject coaching and subject positioning. As detailed above, there are significant areas in an inhalation deposition study that can cause variability in the study data. In order to minimize the potential for the subject to add to this, it is highly recommended to coach them or walk them through the entire process from dose administration to the end of imaging. When possible it can help subject comfort to have the coaching conducted by the same technician or scientist who worked with them on their initial breath training and will be there for all days of dose administration. This helps to keep the subject relaxed and increases their likelihood of appropriate dosing and imaging.

Prior to any coaching, it is important to consider if the study is being conducted with healthy subjects or patients. If patients are used, carefully balance repeating inhalation maneuvers or breath holds with the possibility that these may induce slight bronchoconstriction which may alter deposition patterns. Also consideration should be given to the location of waiting rooms in relationship to the imaging suite when patients are used. Having patients traverse long distances or several flights of stairs on their way to the imaging suite may also induce bronchoconstriction immediately prior to dosing.

During coaching, the subject will inhale a placebo or blank inhalation device while reviewing the feedback from the breath monitoring system. They will then be positioned onto the camera bed and moved into the field of view of the detectors. The positioning must include both the locations of the subject in respect to the field of view for the detectors and the anatomical positioning. It can be advantageous to have the subject place their arms above their head so that they are not between the lungs and the detectors. This positioning can be seen from an example CT shown in Fig. 20.5 with the arms above and behind the head thereby minimizing tissue between respiratory tract and the detectors.

A low level of activity, in a point source, can be used to ensure that the patient is well positioned, and the designed portions of the respiratory tract are in the field of view.

Fig. 20.5 Central and peripheral lung regions (modified from [23])

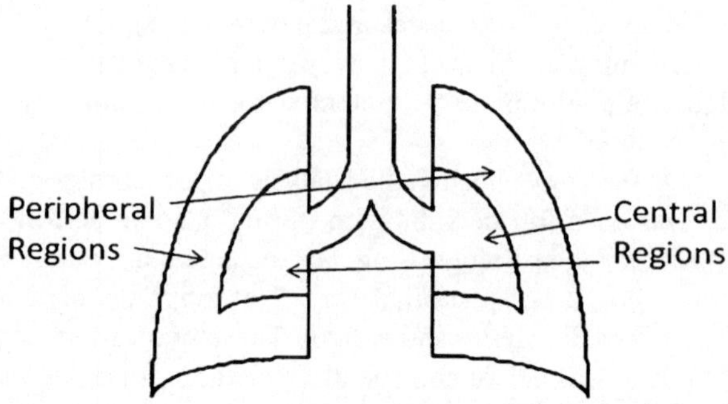

The position of the subject on the bed and the bed position should be recorded so that the subject can be rapidly returned to this position following dose administration. Depending on the complexity of the image acquisition and the comfort of the subjects, it may be appropriate to have the gantry and the detector heads move, as required for the image acquisition, for the subject to feel more relaxed during the acquisition. If a breath hold is to be required during the image acquisition, the subject may be instructed to test this prior to the image acquisition. These steps will help the subject to relax and be comfortable with the entire process before dose administration.

20.5.1 Planar Image Acquisition

Planar image acquisition is typically conducted for between 30 s and 1 min, depending on the radioactivity in the dose and the dose fraction that deposits in the regions to be quantified. The image acquisition parameters including matrix size, zoom factor, and detector positioning (anterior and posterior is recommended) are defined prior to the start of dosing. The default image acquisition for many nuclear medicine cameras is to collect a static image over the time course defined. This results in an image that is the summation of the entire image time course. This default can be modified to collect the data in small bundles or as a dynamic image over the course of the image time. This slight modification increases the flexibility in the data analysis as it can then be integrated over any time course within the image time.

For example, if dynamic imaging was conducted while a patient is inhaling the formulation (~10 s) and out to 1 min following completion of the inhalation. This data set could then be used to generate a real time inhalation video from the data bundles between 0 and ~10 s and a standard planar deposition image with the data bundles from ~10 s to 1 min. For the same acquisition if the subject sneezes or moves at 45 s, it is likely that the data from ~10 to 40 s can be summed and the data set for this subject will not be lost. In this manner, additional data can be generated often with a decrease in number of subjects.

20.5.2 SPECT Image Acquisition

A SPECT image acquisition can be conceptualized as a series of planar images that are collected as the detector heads move around the subject. The collection of images is then reconstructed to generate a 3D image of the subject. As this collection is more complex than planar imaging, the development of appropriate acquisition protocols is more complex. In addition to the planar acquisition parameters, SPECT imaging adds the complexity of defining the number of steps (or images) to be collected around the 360° of the patient and the time course for each image.

Often the draw back to the use of SPECT imaging for pulmonary deposition analysis has been the time required to collect these images which is typically between 10 and 15 min. As previously mentioned, most Tc99m radiolabels are associations between the radiolabel and the drug and this association can start to break down over the time required to collect SPECT images. To this end the radiation dose from the device can be increased to decrease the time required at each step or a multi-headed gantry can be used to minimize the number of steps required to collect all 360° (e.g., with a 2-headed gantry each head only needs to move 180° to collect 360° of data).

If a SPECT image is to be used for quantification of regional deposition, steps must be taken to characterize the radiolabel movement over the time course of the SPECT image acquisition. These steps may include a pilot study with serial planar images beyond the planned SPECT collection time or collecting three images (in the order of first planar, SPECT, and then second planar) for each subject. With the pilot study, the serial planar images can be quantified to characterize the movement of the radiolabel beyond the time required for the SPECT acquisition and then used to determine if the SPECT acquisition time is appropriate in comparison to the particle movement. When three images are collected, the first and second planar images (sandwiching the SPECT image) can be used to determine the difference between the counts in each ROI to establish the validity of the SPECT image. The serial planar images has the advantage that a feasible time course for a SPECT image can be defined and used as a target to develop an appropriate SPECT acquisition protocol prior to study dosing; however, it requires additional subject visits. The collection of three images does not require an additional subject visit but it does have the risk that the SPECT acquisition time course is too long and therefore not usable. However, in this case the initial planar image would still be viable for quantification.

20.6 Image Analysis

For all data analysis conducted on gamma scintigraphy studies, it must be appreciated that the quantification is conducted on the radiolabel not the drug. With a well-conducted radiolabeling validation study, it is reasonable to infer the relationship between the radiolabel and the drug for the initial deposition; however, beyond the initial deposition (in the absence of supporting data) no results or conclusions can be drawn.

The analysis of both 2D and 3D data sets is most often conducted to determine the dose fraction in different regions of interest (ROI's). The most common ROI's are the oral cavity, trachea, lungs, stomach and, when appropriate, nasal cavity. ROI's can be drawn through automated software systems or manually based on CT or transmission scans. The dose fraction is determined by converting the response on the image (counts in each ROI) into activity through a response curve generated for each study.

The need for correction factors in the quantification of dose fractions has been discussed throughout the literature [16]. When a correction factor is applied (tissue attenuation), it is designed to correct the activity that the detector measures for the actual activity in the region of interest. This original activity has been decreased, at some level, by the thickness of the biological tissue. A transmission scan is one commonly used method to determine a tissue attenuation correction factor (ACF). In this method, a large flood source filled with a gamma emitted isotope (usually Tc99m) is scanned at a controlled distance above the detector both with the subject in place and without the subject in place [3, 21]. The counts without the subject in place (A_o) and the counts through the subject's region of interest (A_t) can then be used in the question below to define the ACF for this tissue.

$$\mathrm{ACF} = \left(\frac{A_o}{A_t}\right)^{1/2}.$$

Even with proper ACF, it is still not feasible to exactly correct for the attenuation caused by the tissue as the exact distance the gamma radiation traveled through the tissue is not known [14]. Based on the type of study being conducted and these issues, the requirement for ACF may or may not be needed. Specifically, in the case of a crossover study to evaluate regional deposition between two formulations, the use of ACF may actually increase the variability in the data analysis and therefore may not be required.

The area of regional lung deposition quantification has been the subject of several different models and methods. One of the more commonly used methods is an analysis of the central to peripheral deposition [23]. In this method, the lung is segmented into two regions, the central (C) and peripheral (P), Fig. 20.5. The central region comprises 33% of the lung surface area and the peripheral region encompasses 67% of the lung area. The ratio of the counts in these two regions is then calculated as the C to P ratio. With this ratio, the lower the C to P ratio the greater the peripheral deposition and the higher the ration the greater the central deposition.

As computer technology advances, the ability to further refine and segment the lungs for regional analysis has also advanced. Several models currently exist for further segmentation of the lungs [10, 11, 15]. The onion model of McDonald et al. uses a CT overlay with the gamma imaging data to initiate segmentation at the primary bronchus for each lung. It then segments the lung outward in either two directions (for planar data) or three directions (for SPECT data). This model can be adapted to the full range of preclinical and clinical data sets. An example of the

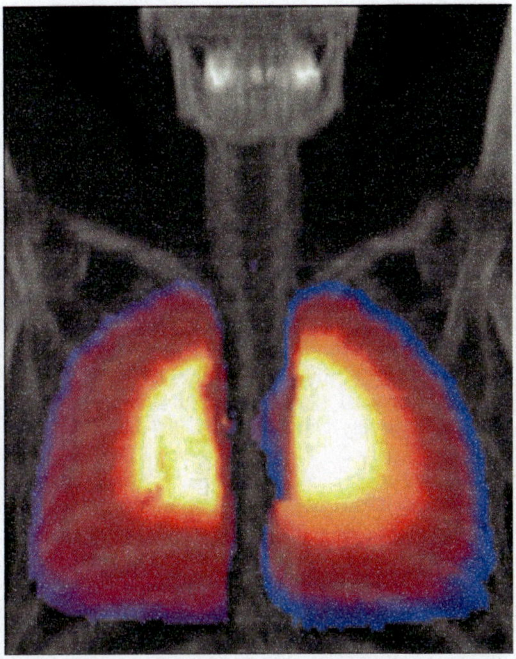

Fig. 20.6 Example CT showing subject arm positioning and onion slicing of the lung for regional deposition analysis

segmentation of a human lung using this method is shown in Fig. 20.6, with the changes in color representing the slices moving outward.

The advantage of this model is that it not only allows the lungs to be segmented into nearly infinite slices but it can also be used to mimic that of the more common C to P ratio. Specifically, the regions that comprise 33% of the lung area can be summed and compared to the remaining regions in a C to P ratio. In this manner, data analyzed with this method provide a more complete regional deposition analysis and are still able to be directly compared to other literature data. Further, this model can be applied to both 2D and 3D data sets to characterize the regional deposition in the lungs.

20.7 Concluding Remarks

The use of gamma imaging has continued to gain traction as a valid method to define both the total inhaled dose and the regional deposition. When conducted correctly, inhalation deposition studies provide the best method to directly quantify the dose delivered to the lungs and throughout the respiratory tract. These types of studies can and will continue to be used to control and adjust dose between preclinical and clinical models. Used appropriately, these data can assist in the development of inhalation products with increased confidence of the dose to all regions of the respiratory tract.

References

1. Adams WP et al (2010) Demonstrating bioequivalence of locally acting orally inhaled drug products (OIPs): workshop summary report. J Aerosol Med 23(1):1–29
2. Bailey DL (2006) Imaging the airways in 2006. J Aerosol Med 19(1):1–7
3. Cheng YS, Irshad H, Kuehl P, Holmes TD, Sherwood R, Hobbs CH (2008) Lung deposition of droplet aerosol in monkeys. Inhalation Tox 20:1029–1039
4. Clark AR, Borgstrom L (2002) In vitro testing of pharmaceutical aerosols and predicting lung deposition from in vitro measurements. In: Bisgaard H, O'Callaghan C, Smaldone GC (eds) Drug delivery to the lung. Marcel Dekker, New York, pp 105–142
5. Conway JH, Walker P, Fleming JS, Bondesson E, Borgstrom L, Holgate ST (2000) Three-dimensional description of the deposition of inhaled terbutaline sulphate administered via turbuhaler. In: Dalby RN, Byron PR, Farr SJ, Peart J (eds) Respiratory drug delivery VII. Serentec Press, Raleigh, pp 607–609
6. Daley-Yates PT (2010) The clinical utility of pharmacokinetics in demonstrating bioequivalence of locally acting orally inhaled drugs. In: Dalby RN, Byron PR, Peart J (eds) Respiratory drug delivery 2010. Davis Healthcare International, River Grove, pp 273–284
7. De Backer W, Devolder A, Poli G, Acerbi D, Monno R, Herpich C, Sommerer K, Meyer T, Mariotti F (2010) Lung deposition of BDP/formoterol HFA pMDI in healthy volunteers, asthmatics and COPD patients. J Aerosol Med 23(3):137–148
8. Dolovich M, Rhem R, Rashid F, Coates G, Hill M, Bowen B (1996) Measurement of the particle size and dosing characteristics of a radiolabeled albuterol sulphate lactose blend used in the spiros dry powder inhaler. In: Dalby RN, Byron PR, Farr SJ, Peart J (eds) Respiratory drug delivery V. Interpharm Press, Buffalo Grove, pp 332–335
9. Farr SJ (1996) The physico-chemical basis of radiolabeling metered dose inhalers with ^{99m}Tc. J Aerosol Med 9(S1):S27–S36
10. Fleming JS, Hashish AH, Conway JH, Hartley-Davies R, Nassim MA, Guy MJ, Coupe J, Holgate ST, Moore E, Bailey AG, Martonen TB (1997) A technique for simulating radionuclide images from the aerosol deposition pattern in the airway tree. J Aerosol Med 10(3): 199–212
11. Fleming JS, Epps BP, Conway JH, Martonen TB (2006) Comparison of SPECT aerosol deposition data with a human respiratory tract model. J Aerosol Med 19(3):268–278
12. Leach CL, Davidson PJ, Hasselquist BE, Boudreau RJ (2002) Lung deposition of hydrofluoroalkane-134a beclomethasone is greater than that of chlorofluorocarbon fluticasone and chlorofluorocarbon beclomethasone: a cross-over study in healthy volunteers. Chest 122:510–516
13. Leach CL, Colice GL, Luskin A (2009) Particle size of inhaled corticosteroids: does it matter? J Allergy Clin Immunol 124(6):S88–S93
14. Lee Z, Berridge MS, Nelson AD, Heald DL (2001) The effect of scatter and attenuation on aerosol deposition as determined by gamma scintigraphy. J Aerosol Med 14:167–183
15. McDonald JD, Kuehl P, Norenberg JP, Gershman B, Anderson TL (2009) Effect of particle size on the regional deposition of inhaled technitium-99m labeled particles in rodents. Am J Respir Crit Care Med 179:A5034
16. Moller W, Felten K, Meyer G, Meyer P, Seitz J, Kreyling WG (2009) Correction in dose assessment of ^{99m}T radiolabeled aerosol particle targeted to central human airways using planar gamma camera imaging. J Aerosol Med 22(1):45–54
17. Newman SP (2009) In vivo assessment. In: Newman S (ed) Respiratory drug delivery: essential theory and practice. Davis Healthcare International, River Grove, pp 97–134
18. Newman SP, Pitcairn GR (2006) Gamma scintigraphy and SPECT. In: Lappin G, Temple S (eds) Radiotracers in drug development. Taylor and Francis, Boca Raton, pp 269–288
19. Newman SP, Moren F, Trofast E, Talaee N, Clarke SW (1989) Deposition and clinical efficacy of terbutaline sulphate from turbuhaler, a new multidose powder inhaler. Eur Respir J 2(3): 247–252

20. O'Connor DK (2010) Role of pharmacokinetics in establishing bioequivalence for orally inhaled drug products. In: Dalby RN, Byron PR, Peart J (eds) Respiratory drug delivery 2010. Davis Healthcare International, River Grove, pp 245–247
21. Pitcairn GR, Newman SP (1997) Tissue attenuation corrections in gamma scintigraphy. J Aerosol Med 10:187–198
22. Smaldone GC (2001) Radionuclide scanning, respiratory physiology and pharmacokinetics. J Aerosol Med 14:135–137
23. Smaldone GC, Walser L, Perry RJ, Ilowite JS, Bennett WD, Greco M (1989) Generation and administration of aerosol for medical and physiological research studies. J Aerosol Med 2(2):81–87
24. Thiel CG (1998) Can in vitro particle size measurements be used to predict pulmonary deposition of aerosol from inhalers? J Aerosol Med 11(S1):S43–S51

Chapter 21
Development and Approval of Inhaled Respiratory Drugs: A US Regulatory Science Perspective*

Gur Jai Pal Singh and Guirag Poochikian

Abstract This chapter focuses on the regulatory science aspects of the postdiscovery development and approval of the inhaled respiratory drug products intended for local action. It provides a brief treatise of the CMC, in vitro and in vivo evaluations for development and approval of inhalation aerosols in the US. In addition, it includes a brief discussion on scientific considerations related to potential extended-release inhalation drug products. Regulatory paradigms for approval of generic inhalation drug products are also discussed.

Keywords Asthma • Chemistry manufacturing and controls • COPD • Dry powder inhalers • Efficacy • Extended release • Generic inhalation drug products • Inhalation aerosols • In vitro performance • Metered dose inhalers • Nonclinical studies • Particle size distribution • Pharmacodynamics • Pharmacokinetics • Regulatory approval • Safety • US FDA

21.1 Introduction

The development and approval of drug products for human use in the United States is regulated by the Food and Drug Administration (FDA) pursuant to the Federal Food, Drug, and Cosmetic Act (FD&C Act). The Act with its amendments requires that before marketing "new drugs," as defined by the Act, be approved by the FDA to insure their safety and effectiveness for the intended use. The evidence required

*This article represents the personal opinions of the authors, and does not necessarily represent the views or policies of any regulatory agency or organization(s) of the author current or past affiliation.

G.J.P. Singh (✉)
Axar Pharmaceuticals, Irvine, CA 92620, USA
and
Mailing Address: 22386 Amber Eve Drive, Corona, CA 92883, USA
e-mail: gur.jp.singh@gmail.com

H.D.C. Smyth and A.J. Hickey (eds.), *Controlled Pulmonary Drug Delivery*, Advances in Delivery Science and Technology, DOI 10.1007/978-1-4419-9745-6_21,

to support safety and effectiveness may vary with the chosen path of drug development. Section 505 of the FD&C Act describes three principal pathways. The first avenue is the section 505(b)(1) which is applicable to the new molecular entities. The new drug applications (NDAs) contain full reports of investigations of quality, safety, and effectiveness of drug products conducted by the sponsor. The second pathway is 505(b)(2) which is generally used for new drug products that contain already approved active moiety(ies). The new product may represent a different salt, formulation, and/or dosage form [1]. The Agency requires evidence for quality, safety, and effectiveness, but some of the evidence (e.g., nonclinical pharmacology and toxicology) may come from studies not conducted by the sponsor. Nonetheless, products approved under section 505(b)(2) are marketed as stand-alone products and do not provide direct substitutes for similar/comparator products. The third regulatory pathway is provided under section 505(j) and its application is limited to the approval of generic drug products. These products represent therapeutically equivalent copies of the already marketed/approved reference listed drug products (RLDs). Drug applications under 505(j) are generally not required to include preclinical (animal) and clinical (human) data to establish safety and effectiveness. Instead, the safety and effectiveness of generic drugs is demonstrated through acceptable quality and bioequivalence (BE) studies [2]. Generic products approved in the United States provide direct substitutes for the corresponding RLDs.

Figure 21.1 provides a general outline of the development and approval of new drugs which is a lengthy and complex process [3]. It begins with the discovery of a molecular entity with potential to provide certain therapy identified through extensive nonclinical testing. Regulatory agencies require testing that documents the characteristics such as chemical composition, purity, quality, potency, and performance of the drug product. From a regulatory perspective, the postdiscovery establishment of

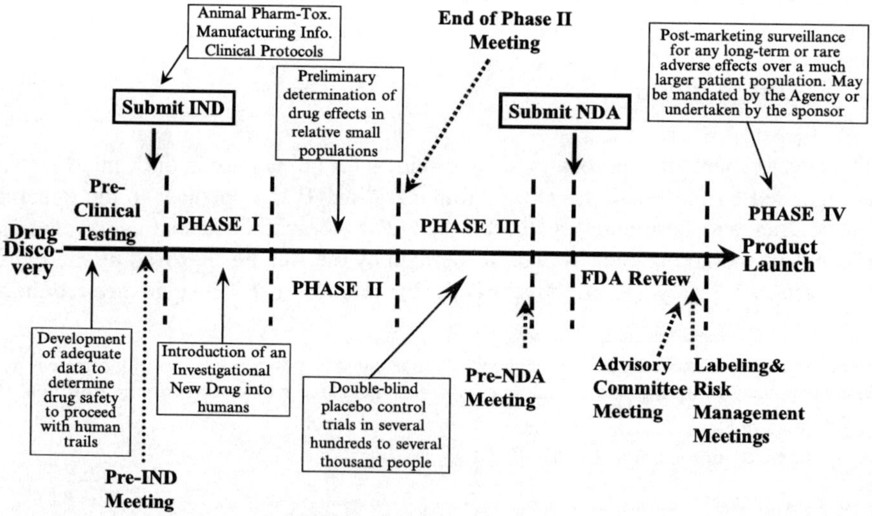

Fig. 21.1 New drug development and approval process

chemistry, manufacturing, and controls (CMC), the drug's pharmacological profile and toxicological properties as well as clinical safety and effectiveness are required for all drug products. However, the nature of CMC development/characterizations and in vivo (nonclinical and clinical) testing vary with the drug product, route of administration, and its intended use. Products made for drug delivery through the inhalation route represent a class of "Complex Drug Products" due to devices required to deliver drugs in forms that allow delivery to intended site(s) of action in the respiratory tract. This chapter focuses on the regulatory science aspects of the postdiscovery development and approval of the inhaled respiratory drug products intended for local action. It provides a brief treatise of the CMC, in vitro and in vivo evaluations for development, and approval of inhalation aerosol in the US. In addition, it includes a brief discussion on scientific considerations related to potential extended-release inhalation drug products. Regulatory paradigms for approval of generic inhalation drug products under the section 505(J) of the act are also discussed.

21.2 Chemistry Manufacturing and Controls

Inhalation drug products could be designed and developed in a variety of suitable dosage forms. Each dosage form has its advantages and limitations, and discussion pertaining to these considerations is beyond the scope of this chapter. Reliable inhalation drug products represent combination of proper form(s) of the drug substance(s), excipient(s), and suitable delivery system(s) that accurately deliver doses of the drug to the biological target with acceptable reproducibility. Well designed drug products with appropriate control protocols provide greater assurance for quality and reproducibility in the performance of the drug product. Robustness in such drug products contributes to consistent performance of the drug product through its unit life and shelf life, when manufactured within the boundaries established a priori for acceptable variations of the key manufacturing process parameters and quality attributes of both the drug product components and the finished drug product. Well designed control protocols for the particular dosage form are intended to address the identity, strength, purity, quality, potency, including the consistent performance attributes in accord with its label claim beyond the expected shelf life.

Different considerations are involved in the design, manufacture, and controls for the desired drug products. A brief description of each type of the currently available inhalation drug products summarized by their established regulatory nomenclature is provided in Table 21.1; additional regulatory considerations for each type of these drug products are provided in the subsequent sections.

Although similar in many features to other drug products, the inhalation drug products, which are considered combination products [4], have unique features with respect to formulation, manufacture, controls, and particularly container closure system (CCS). These features are carefully considered and closely monitored during the drug development program because changes to these elements may have much greater impact on the performance and stability of the inhalation drug products

Table 21.1 Nomenclature and description of inhalation drug products

Category	Description
[Drug] Inhalation aerosol	A drug solution or suspension formulation for oral inhalation that is packaged under pressure and delivers a specified amount and quality of therapeutically active ingredient(s) upon activation of an accurately metered valve system
[Drug] Inhalation powder	A drug powder formulation for oral inhalation with the use of a device that aerosolizes and delivers an accurately metered amount and quality of the therapeutically active ingredient(s)
[Drug] Inhalation solution	A drug solution for oral inhalation with the use of a nebulization system
[Drug] Inhalation suspension	A drug suspension for oral inhalation with the use of a nebulization system
[Drug] Solution for inhalation	A drug solution for oral inhalation that must be diluted before it is administered with the use of a nebulization system
[Drug] Suspension for inhalation	A drug suspension for oral inhalation that must be diluted before it is administered with the use of a nebulization system
[Drug] for inhalation solution	A drug powder that upon the addition of a suitable vehicle yields a solution conforming in all respects to the inhalation solution
[Drug] for inhalation suspension	A drug powder that upon the addition of a suitable vehicle yields a suspension conforming in all respects to the inhalation suspension
[Drug] Inhalation spray	A nonpressurized liquid drug dosage form for oral inhalation that is packaged in a container closure system, which upon activation delivers an accurately metered amount of fine droplets of the formulation

than other more conventional dosage forms in terms of the degree of predictability of such changes on the delivered doses to patients throughout the unit and product's shelf life. For example, unlike most other drug products, the dosing and performance characteristics and, therefore, the clinical safety and efficacy of these products may depend on the design of the CCS, formulation component characteristics, and possible interactions between the CCS and formulation components. Moreover, a clinical efficacy study of such a drug product generally may not be an adequate measure of its ability to deliver reproducible doses to patients, due to a number of factors including drug administration skills and practices by patients, patient aspiratory flow rate and volume, the nature and sensitivity of clinical measurements, and the relatively small number of patients studied relative to the eventual market size for the product.

21.2.1 Inhalation Aerosol

Inhalation aerosol drug products, commonly known as metered-dose inhalers (MDIs), may consist of either a two-phase or a three-phase formulation preparation in a CCS comprised of a suitable container, valve and actuator components, and other potential accessories. The formulation typically contains therapeutically active ingredient(s) dissolved or suspended in a propellant, or a mixture of propellant(s), and cosolvent(s), and/or other suitable excipients in a compact pressurized aerosol container. An MDI product may discharge a single or up to few hundred accurately metered actuations containing one or more drug substances [5].

Delivery of the formulation from a pressurized container is very rapid and undergoes a complex sequence of events in the CCS and during aerosolization [5]. For example, when the content of the metering chamber in the valve is released, it undergoes volume expansion and the liquid phase (solution or suspension) forms a mixture of gas and liquid while being discharged as a jet through the actuator orifice. The plume released from the actuator expands and the droplets undergo a series of processes during the propulsion and dispersion phases. The drug substance particles of different sizes in the droplets become progressively more concentrated and at different rates depending on the formulation components and vaporization rate of the volatile propellant(s) [6].

Selection of the components and the composition of MDI formulations are crucial in defining the physical stability and the performance characteristics of the drug product. Additionally, for suspension MDIs, the density of the suspended materials of the formulation relative to the density of the liquid phase of the formulation, physical properties and characteristics of the solid materials, phase separation rate, disposition of the suspension formulation, and the potential for agglomeration of particles need to be considered for development of a reliable, reproducible, and stable drug product. Moreover, selection of the CCS components and the potential interactions of the suspended drug substance(s) with the various internal surfaces of the container, valve, and other components may also contribute to high variability in the delivered dose and aerodynamic particle size distribution (APSD), particularly upon aging. Thus, from a regulatory perspective, it is imperative that accurate quantitative compositional statements be included in the drug application to express the concentrations in the final formulation, as well as the amount of the drug substance delivered at the valve, mouthpiece (actuator), and integrated spacer, if any.

21.2.2 Inhalation Powder

Inhalation powder drug products, commonly also known as dry powder inhalers (DPIs), are designed to dispense powders for inhalation and contain active ingredient(s) alone or with a suitable excipient(s). Because of the phase-out of the chlorofluorocarbon-based propellant MDIs due to environmental concerns and problems associated with coordination of the breathing and inhalation maneuvers of MDI drug products [7–10], enormous effort has been devoted to developing DPIs with a variety of designs and functions for proper and efficient delivery of the drug substance(s) to the appropriate regions of the lungs [11–15]. Current designs include premetered and device-metered DPIs, all of which rely on various energy sources. These may include energy input from the patient inspiration, compressed gas, a motor driven impeller, or other means to accomplish dispersion, deagglomeration, and aerosolization of the powder formulation components, including the drug substance(s). A DPI drug product may dispense single doses or up to few hundred metered-dose units of the drug substance(s).

The efficiency of drug delivery from most currently approved DPIs, which derive energy for drug deagglomeration and delivery from the patient, depends on the patient's inspiratory effort. An inspiratory maneuver generates certain flow rate through the DPI depending upon the resistance to air flow offered by the device. Therefore, the flow rate is considered to be an important determinant of in vitro drug delivery from DPIs [14–16]. The devices used in different DPIs may offer different levels of resistance [17, 18] to air flow and, therefore, for a given inspiratory effort the rate of air flow through inhalers may vary with the device [19, 20]. Thus, device resistance influences in vivo performance of DPIs [21–25], because achievable flow rates vary across the target patient populations [26–30]. Therefore, the development of DPIs warrants, among other things, scientific considerations regarding the powder formulation and its stability, device features, and the influence of flow rate on in vitro dose delivery and particle size distribution (PSD) of the aerosolized drug.

Premetered DPIs contain previously measured quantities of formulation in individual containers (e.g., capsules/blisters) that are inserted into the device by the patient before use [31, 32]. Premetered DPIs may also contain premetered dose units enclosed during manufacture as ordered multidose assemblies in the delivery system [33]. Device-metered DPIs have an internal reservoir containing a sufficient quantity of formulation for multiple doses that are metered by the device itself during actuation by the patient [34, 35]. However, such DPI drug products can be more susceptible to contamination (e.g., moisture, microbial) of the remaining doses. Therefore, it would be prudent to rule out possibility of contamination under both in-use and misuse conditions during the design development of such DPIs.

From a regulatory perspective, DPIs (like MDIs) are also considered complex drug products [16]. In DPIs, complex and subtle interactions may occur between the drug substance, carrier(s), and components of the CCS that may significantly affect the performance and hence the safety and effectiveness of the drug product [12, 15, 36, 37]. For example, gravitational, fluid dynamic, and other interactive forces, such as electrostatic, Van der Waals and capillary forces, together are responsible for different fluidization behaviors exhibited by different powders in an inhaler [15]. Electrostatic forces also influence the overall efficiency of a DPI, since such forces are considered to be significant for attraction and adhesion between the drug substance particles, excipient(s) particles, drug substance-excipient(s), and drug substance-device surfaces. Furthermore, PSD, particle morphology, and moisture content of the individual components of the formulation can greatly influence the bulk properties and the overall product performance [38, 39]. In addition to affecting bulk properties, variability of these parameters can cause increases in dose-to-dose variability [40]. Therefore, such physical characteristics need be considered during development of a DPI drug product, particularly with respect to formulation components, manufacture and its process parameters, CCS or device, and relevant in-process and final quality attributes and corresponding controls.

21.2.3 *Inhalation Solutions and Suspensions*

Inhalation solutions, suspensions, and related products are typically aqueous-based preparations and hence should be sterile [41]. For safety considerations the use of preservatives, stabilizing agents, and multidose containers in such drug products are discouraged. These drug products are intended for delivery to the lungs by nebulization. Some products may require specified nebulizers. Nebulization involves continuous generation and delivery of fine mist of aqueous droplets containing the formulation components from a drug solution or suspension to the patient with the use of ultrasonic energy, venturi effect, or other appropriate mechanical means. The stability profile and characteristics of the delivered dose of these drug formulations intended for use with external nebulizers also need to be characterized appropriately with the nebulizer and under standardized nebulization conditions. For example, the drug application may include information on the specific nebulizer, fill volume, residual volume, duration of nebulization, any accessories, as well as identity, pressure, and flow rate of the driving gas. Such drug preparations typically are packaged in single-dose semipermeable containers and also include a protective packaging to minimize ingress of volatile foreign contaminants, loss of solvent, and exposure to oxygen and light. From a regulatory perspective, appropriate embossing or debossing of such containers are normally encouraged for labeling purposes [42].

21.2.4 *Inhalation Sprays*

Inhalation spray products are generally aqueous-based liquid formulations and hence should be sterile [41] and by definition, do not contain propellant. Drug delivery from such a drug product is affected by an integral spray pump unit in the CCS, when activated by the patient. Each dose is delivered to the patient typically as fine mist of aqueous droplets that contain the formulation components. The delivered droplets may be generated by various means, including mechanical, power assistance, and/or energy from the patient inspiration for the activation of the unit and generation of the spray. The mechanisms for droplet generation distinguish the various types of inhalation sprays. Such drug products may be aimed for unit-dose or multidose presentations. Similar to inhalation solutions and suspensions, the use of preservatives and other stabilizing agents in these drug products is discouraged unless their presence and the respective concentrations are justified from safety and tolerability perspectives in clinical testing. Inhalation spray drug products may also be designed as premetered or device-metered presentations. A premetered unit contains previously measured amount of formulation in an individual container (e.g., blisters) that is inserted in the device for activation by the patient before use. A device-metered product contains sufficient amount of formulation for a prescribed number of multiple doses in a reservoir and each dose is delivered as an accurate metered spray by the device and through the unit life when activated by the patient.

However, design development approaches of the delivery system for device-metered drug products need to consider the potential contamination aspects of the remaining doses in the reservoir under patient use and misuse conditions.

21.2.5 Formulation Components

21.2.5.1 Drug Substance(s)

Drug substances impart clinical effectiveness of drug products. Therefore, a comprehensive characterization of the physical and chemical properties of the drug substance and understanding of their impact on the quality and performance of the drug product facilitates the drug development efforts and eliminate and/or minimize potential challenges down the stream [43, 44]. Some of the important properties of the drug substance can include, but are not necessarily limited to, bulk density, particle size distribution (PSD), particle morphology, solvates, hydrates, polymorphs, amorphous forms, solubility profile, and moisture content. Appropriate acceptance criteria utilizing validated test procedures (i.e., specifications) with emphasis on the critical attributes of the drug substance(s) are expected before and after any additional processing (e.g., micronization, conditioning) to ensure the continued reproducibility of drug product quality and performance. Generally, the acceptance criteria of the drug substance are based on data for the drug substance batches used in nonclinical studies and in critical clinical, BE, and primary stability batches of drug product. At early stages of drug development, such a comprehensive approach in identifying the physicochemical properties of the drug substance that can influence the manufacturability, stability, and performance of the drug product, may provide enhanced continuity of the quality and batch-to-batch performance consistency of the resulting drug product.

21.2.5.2 Excipients

Excipients (including their quality) used in the manufacture of most MDI and DPI drug products are considered critical components for reproducible performance of such drug products. Typically, excipients are the largest constituents in these drug products and they, with properly selected physicochemical quality characteristics, play an important role in imparting the manufacturability and desired target performance of these drug products. Other factors such as the route of administration (i.e., inhalation) and the sensitive nature of various patient populations (e.g., asthmatics) may also influence the selection of the nature and concentration of the excipients. The suitability of the toxicological aspects of the selected excipients for use in these drug products is expected to be thoroughly assessed, investigated, studied, and documented for their intended use.

Thus, from a regulatory perspective full identification, thorough physicochemical characterization, safety evaluation, and comprehensive controls (e.g., identity, strength, quality, purity) of the selected excipient(s) from specified sources used in such drug products are considered to be critical, especially for those that comprise a significant portion of the formulation or if they can affect the dispersal characteristics of the drug substance. Moreover, excipients need to meet appropriate specifications established a priori, based on physicochemical data from multiple batches of the excipient from a specified source used to prepare acceptable batches of the drug product used in nonclinical, critical clinical, BE, and primary stability studies. Like the active ingredients, the specifications for the excipients also need to include the critical relevant attributes before and after any additional processing (e.g., micronization, conditioning) to ensure the continued reproducibility of drug product quality and performance.

21.2.5.3 Manufacture and Packaging

From a regulatory perspective, a NDA is expected to include a detailed description of the manufacturing, processing, and packaging procedures and/or a complete batch record for the drug product. The manufacturing directions need to include specific information on, and control procedures for, operating process parameters and variables, as well as in-process controls of associated quality attributes of critical intermediates and/or at specified manufacturing steps to decrease controllable process variability and assure consistency in the quality and purity of the drug product.

The regulatory submission also needs to include information on validated processes that describe in detail any additional procedure (e.g., micronization, spheronization, spray drying), type of equipment, operating conditions and parameters, as well as corresponding controls for the modification of the physical characteristics of the drug substance(s) and/or excipient(s) for use in the specific drug product. The overall final manufacturing methods are based on, and are reflection of, the procedures that were appropriately generated through pharmaceutical development efforts and were implemented in the preparation of the critical clinical, BE, and primary stability batches. The sponsors should refer to guidances [44–47] relevant to manufacturing and packaging.

When a drug product includes blister unit, foil-foil, or secondary protective packaging, it is also expected that appropriate relevant information of these components be included in the drug application. The notion of such pertinent information is to provide assurance for adequate sealing characteristics in terms of adhesion properties and mechanical performance of the units. These may include, for example, selection and controls of appropriate materials and components, in-process parameters and controls of packaging operations, as well as development and establishment of suitable integrity testing and acceptance criteria for seal completeness and seal strength.

For inhalation aerosol drug products, to ensure their consistent and more reliable quality in the distribution channels, it would be a prudent that the good manufacturing practice subject all units to appropriate stress heat conditions before equilibrium storage, followed by check weighing and spray testing of all units against set established standards. Such an approach will enable manufacturer to cull out the units with marginal seal and/or inadequate performance prior to their release for distribution, and avoid serious potential safety and efficacy implications for the patients.

21.2.5.4 Specifications

Specifications are defined as the quality standards (i.e., tests, analytical procedures, and acceptance criteria) to confirm the quality of drug substances, drug products, intermediates, raw materials, reagents, components, in-process materials, CCSs, and other materials used in the production of drug substances or drug products [48]. From a regulatory perspective a full description of the specifications needs to be included in the NDA to ensure the identity, strength, quality, purity, and performance of the drug product throughout its shelf life, and during the recommended period of patient use. Generally, the drug product specifications are based on data from multiple batches of the drug product used in critical clinical, BE, and primary stability studies. The associated analytical procedures for the test attributes need to be properly validated and documented [49, 50].

Tests related to in vitro performance characteristics of inhalation drug products are critical and need to be comprehensively established prior to initiation of the pivotal clinical and BE studies. Some of the tests are briefly described below. However, the reader should refer to the Agency guidance [16] for further details.

Table 21.2 includes the recommended pharmaceutical development and quality control test attributes for release and stability purposes of different inhalation drug products. Furthermore, solution for inhalation, suspension for inhalation, drug for inhalation solution, and drug for inhalation suspension drug products may require other test attributes, such as description, reconstitution time, moisture content, as applicable, prior to the addition of a suitable vehicle to yield the corresponding preparation (Table 21.1) and conform to the requirements presented in Table 21.2.

Table 21.3 includes a list of pharmaceutical development and final characterization studies for different inhalation drug products. Dosage form and drug product specific characterization studies are designed to characterize the performance properties of a given drug product and to assist in developing optimum labeling statements for the correct storage, use, cleaning of the device, and maintenance of the drug product. Such confirmatory studies based on well designed protocols are expected to be performed on representative batches of the drug product intended for commercialization.

21.2.5.5 Container Closure Systems (CCS)

For most inhalation drug products, unlike other conventional dosage forms, the clinical efficacy of the product may be directly dependent on the design, reproducibility, and performance characteristics of the CCS, which include the units that

Table 21.2 Pharmaceutical development and QC test attributes for inhalation drug products

Test attributes	Inhalation aerosols	Inhalation powders	Inhalation sprays	Inhalation	
				Solutions	Suspensions
Description (formulation and CCS)	Yes	Yes	Yes	Yes	Yes
Identification	Yes	Yes	Yes	Yes	Yes
Assay	Yes	Yes	Yes	Yes	Yes
Assay for preservatives and stabilizing excipients (if used)[a]	Yes	Yes	Yes	Yes	Yes
Impurities and degradation products	Yes	Yes	Yes	Yes	Yes
Content uniformity	No	Yes[b]	Yes[b]	Yes	Yes
Delivered dose uniformity (DDU) through container life	Yes	Yes	Yes	No	No
Aerodynamic particle size distribution (APSD) through container life	Yes	Yes	Yes	No	No
Droplet size distribution through container life	No	No	Yes	No	No
Particle size distribution through container life	No	No	Yes[c]	No	Yes
Spray pattern	Yes	No	Yes[d]	No	No
Foreign particulate matters	Yes	Yes	Yes	Yes	Yes
Microbial limits	Yes	Yes	No	No	No
Sterility test	No	No	Yes	Yes	Yes
Leachables	Yes	No	Yes	Yes	Yes
Water content	Yes	Yes	No	No	No
Alcohol assay	Yes (if used)	No	Yes (if used)	Yes (if used)	Yes (if used)
Fill weight[e]	Yes	Yes	Yes	Yes	Yes
Valve delivery (shot weight)	Yes, or performed on incoming valve	No	Yes[f] or performed on incoming valve/pump	No	No
pH	No	No	Yes	Yes	Yes
Osmolality	No	No	Yes	Yes	Yes
Viscosity[g]	No	No	Yes	Yes	Yes
Delivery rate with the external nebulizer	No	No	No	Yes	Yes
Total amount delivered with the external nebulizer	No	No	No	Yes	Yes
APSD with the external nebulizer	No	No	No	Yes	Yes

[a] Generally, use of preservatives and other excipients are discouraged in inhalation drug products

[b] For premetered inhalation powder and premetered inhalation spray drug products

[c] For suspension preparations

[d] For device-metered inhalation spray drug products

[e] For device-metered inhalation powder and device-metered inhalation spray, and multidose inhalation solution and suspension drug products

[f] For device-metered inhalation spray drug products

[g] Viscosity if relevant

Table 21.3 Pharmaceutical development and characterization studies for inhalation drug products

Studies	Inhalation aerosols	Inhalation powders	Inhalation sprays	Inhalation Solutions	Inhalation Suspensions
Effect of temperature cycling	Yes	No	Yes	No	Yes
Effect of resting times on performance between actuations	Yes	No	Yes	No	No
Number of actuations for initial and repriming	Yes	No	Yes	No	No
Deposition in delivery system components	Yes	Yes	Yes	Yes	Yes
Cleaning instructions of delivery system	Yes	Yes	Yes	Yes	Yes
Effect of shaking on performance of drug product	Yes[a]	No	Yes[a]	No	Yes[a]
Plume geometry	Yes	No	Yes	No	No
In-use period determination for products with secondary protective packaging	Yes	Yes	Yes	Yes	Yes
Effect of repetitive patient use on performance of drug product	Yes	Yes	Yes	No	No
Effect of varying flow rates on performance of drug product	Yes[b]	Yes	Yes[c]	No	No
Assessment of triggering flow rate range for proper performance of drug product	Yes[c]	Yes[c]	Yes[c]	No	No
Drug product performance profile from labeled number of actuations till exhaustion	Yes[d]	Yes[d]	Yes[d]	No	No
Drug product in vitro performance proportionality (for multiple strengths)	Yes	Yes	Yes	No	No
Effect of unit orientation during actuation on in vitro performance	No	Yes	Yes	No	No
Content uniformity	No	Yes[e]	No	No	No
Shot weight	Yes	No	Yes[e]	No	No
Robustness for performance of the delivery system including maximum number of refills	Yes	Yes	Yes	No	No
Leak rate	Yes	No	No	No	No
Preservative effectiveness (if used)[f]	Yes	Yes	Yes	Yes	Yes
Photostability[g]	No	Yes	Yes	Yes	Yes

[a] For suspension formulations
[b] For breath-activated and/or for drug product intended for use with a spacer
[c] For breath-activated or assisted drug products
[d] For products without dose lockout mechanism
[e] For device-metered inhalation powder and device-metered inhalation spray drug products
[f] Generally, use of preservatives are discouraged in inhalation drug products
[g] If warranted by the container

contain, protect, meter, and deliver the formulation, including other components of the drug delivery system and any secondary protective packaging. For example, for inhalation aerosol drug products the CCS consists of the canister, valve, actuator, any additional accessories (e.g., integrated spacer), as well as all primary and secondary protective packaging, as applicable. Inhalation aerosol, inhalation powder, and inhalation spray drug products are also expected to include appropriate integrated dose-counting/indicating mechanisms to promote patient compliance [51].

The composition and relevant controls of each of the individual components of the CCS are considered critical for maintaining the chemical and physical stability of the formulation and ensuring reproducible performance characteristics of the drug product (e.g., delivered dose uniform [DDU] and APSD) in accord with the label claim. Thus, chemical composition and the quality of the materials used in the fabrication of these components need to be carefully selected at early development stages. The compositional information on most of the CCS components would be beneficial also in the development of the appropriate methodologies in the extractables studies of the relevant CCS components and leachables studies of the drug product. Leachable studies are considered more critical for drug products that include liquid preparations (Table 21.1). For examples, inhalation aerosol formulations include organic solvents; hence potential migration of compounds from the valve and canister components of the CCS into the formulation is a safety concern. Consequently, the chemical composition and quality of the materials used in the manufacture of these components need to be carefully selected to minimize leachables without compromising the integrity or the performance characteristics of the drug product.

Extractable are compounds that can be generated from elastomeric and plastic components, or coatings of the CCS when in presence of solvents and under appropriate experimental conditions. Leachables (typically a subset of extractables) are compounds that may migrate into the formulation of the drug product from the elastomeric and plastic components, or coatings of the CCS as a result of direct contact with the formulation of the drug product under stability storage conditions. Leachables may also be found in the drug product as residual contaminants resulting from processing aids, such as lubricants and cleaning agents, employed during the manufacture of the drug product or the fabrication of the CCS components.

Details of extractables and leachables study protocols and safety assessment approaches of the resulting extractables and leachables are described in a number of documents [16, 42, 52–55].

Normally, it is expected that appropriate qualification studies of the CCS components, their compatibility with the formulation, and the performance characteristics of the device are thoroughly investigated and well developed before initiating critical clinical, BE, primary stability, and relevant characterization studies. For example, efficient extractables and leachables testing programs early on in the drug development process may facilitate determination of the qualification studies of the CCS components, safety risk assessment of the resulting extractables, and the suitability of the component materials for use in such drug products. Identity and concentration profiles of the observed leachables are confirmed through studies of the primary stability batches of the drug product that include testing at multiple time-points through the end of its proposed shelf life. Such leachables data may also be related to the extractables of various CCS components obtained under different controlled extraction study conditions. It may be prudent to conduct appropriate and systematic extraction studies with batches of CCS components used in the leachables studies. The associated analytical method development and validation studies for extractables and leachables are based on the International Conference for Harmonization (ICH) Guideline [49].

A properly performing drug product accurately delivers quantities of the drug substance in accord with the label claim (e.g., DDU) and in the desired physical form (e.g., APSD) through the life of the device and throughout the shelf life of the drug product. Furthermore, the device design considerations need to ensure ruggedness of the device, performance reliability of the drug product, and include features to prevent unintentional multiple dosing of the unit.

Relevant, supportive, and specific information and specifications on the CCS and its components intended for different inhalation drug products that are expected to be included in a NDA and supportive authorized corroborative drug master files (DMFs) are provided in several FDA guidances [16, 42, 52, 53].

21.2.5.6 Stability

Stability studies provide a means for assessing the physical and chemical stability, as well as the performance of the drug product, intended for marketing, under various storage and in-use conditions (Tables 21.2 and 21.3). To this end adequate stability data need to be generated and appropriately assessed based on a well designed comprehensive and detailed stability protocol(s) [16, 49, 56]. The proposed expiration dating period for the marketed drug product and the in-use period of the compromised drug product will be based upon appropriate and acceptable data base generated on adequate number of qualified test batches of the drug product.

21.3 In Vitro Performance Evaluations

The foregoing discussion on the CMC aspects of inhalation drug products identified a number of in vitro evaluations applicable to the development and approval of inhalation products. Among these, some tests can also be applicable to other pharmaceutical forms whereas other tests are unique to inhalation products because they measure the performance of the complex drug product over time. A complete description of each of these tests is beyond the scope of this chapter. However, some details may be found in the Agency guidance [16] and elsewhere [57, 58]. The in vitro tests which influence drug release from inhalation drug products and their in vivo safety and efficacy are required at both preapproval and postapproval (batch release) stages. The tests are related to the amount and quality of the delivered drug such as PSD, spray pattern, and plume geometry. The applicable tests are briefly described below due to their pre- and postapproval applicability in evaluation of the performance of both new and generic inhalation drug products.

The amount of drug delivered is measured using the United States Pharmacopeia (USP) recommended apparatus in terms of delivered dose. The evaluations are based on the delivered (emitted ex-actuator) drug mass from primed products. The delivered mass of drug substance is expressed both as the actual amount and as a percentage of the label claim.

The form of drug delivered from inhalation products is characterized in terms of PSD in the aerosol plume or spray. The PSD of the emitted dose is determined based on cascade impaction and laser diffraction techniques [59]. Cascade impactors fractionate and collect particles of one or more drug components by aerodynamic diameter through serial multistage impactions using, for example, USP Apparatus 1, 3 or Apparatus 5, 6 (with appropriate accessories) and/or other complimentary particle sizing methods, if applicable. Measurable levels of drug below the top stage of the cascade impactor are a function of the specific drug product and the experimental setup and procedure, including the number of actuations and assay sensitivity. The equipment and accessories are selected so that the majority of the dose is introduced into the impactor for fractionation. The cascade impactors are operated at the compendial (if applicable), the manufacturer recommended or labeled (e.g., DPIs) flow rates. In addition, a fundamental consideration in the use of cascade impactors is the number of actuations/inhalations per test. The Agency prefers the use of the fewest actuations justified by the sensitivity of the assay. Thus, a validated and sensitive assay should be used for quantitation of drug deposition at the various sites.

Methods based on laser diffraction techniques may provide additional information on PSD of emitted sprays. These methods analyze size of droplets in spray plumes irrespective of the presence or absence of particles in the droplets, and irrespective of the type of particle (drug or excipient or a mixture of both). Laser diffraction instruments use a low-power laser transmitter. The laser beam passes through a filter and a collimating lens to produce as clean parallel beam. The laser beam is focused by a collector lens known a Fourier transform lens and focused to a point on a radial array of silicon diode detectors [60]. The instruments measure the intensity of scattered light at known distances from a central target through which the laser would pass in the absence of droplets [61]. The droplets passing through the laser beam scatter light at various angles [62]. The basic principle of determination of droplet size by laser diffraction is that angles at which the laser light is scattered are inversely proportional to the droplet size [63].

Testing for the shape of plume/spray may be limited to inhalation aerosols and sprays. The test methods are designed to measure spray pattern and plume geometry as they are relevant to the performances of the valve and the actuator [16]. The spray pattern and plume geometry can be influenced by the nature of the formulation [16, 64] and a number of factors including the size and shape of the actuator orifice, the design of the actuator, the size of the metering chamber, the size of the stem orifice of the valve [65], and the vapor pressure in the container.

Spray pattern studies characterize the spray either during the spray prior to impaction, or subsequent to impaction on an appropriate target such as a thin-layer chromatography plate. The FDA also accepts studies based on nonimpaction methods and the recommendations made in its guidance for nasal drug products [66] are also applicable to spray pattern studies on inhalation aerosols. Plume geometry describes a side view of the aerosol cloud parallel to the axis of the plume. It may be evaluated by high speed freeze-frame photography or laser light sheet technology.

Plume geometry is described in terms of the plume cone angle and plume width [66]. The plume angle is based on the conical region of the plume extending from a vertex that occurs at or near the actuator tip. Plume width measurements may be reported at a single delay time while the fully developed plume is still in contact with the actuator tip.

21.4 In Vivo Evaluations

Evaluation of safety and effectiveness of new drugs is based on nonclinical studies and clinical trials in humans. The nonclinical testing includes characterization of the drug's mode of action, absorption, distribution, metabolism and excretion (ADME), and toxicity using in vitro and in vivo (animal) systems. These studies establish pharmacological profile of the drug and determine its acute and short-term toxicity. Two or more animal species (one rodent, one non-rodent) are tested because a drug may affect one species differently from another. The duration of animal studies may depend on the proposed duration of treatment for the clinical studies in humans. A comprehensive discussion regarding nonclinical pharmacology and toxicology testing is beyond the scope of this chapter; interested readers may refer to published literature [67–72] and contact the appropriate FDA Division for up-to-date information.

Toxicology data combined with nonclinical pharmacokinetic (PK) studies along with pharmacodynamic (PD) determinations provide basis for modeling of time course and intensity of drug effects [73], the information that becomes an important part of the Investigational New Drug (IND) application [74]. The IND meeting package [75] generally includes a summary of nonclinical investigations and a concept sheet of a study protocol to solicit advice/input from the FDA reviewers regarding the IND, particularly appropriateness of dose or dosing regimen, safety parameters to be assessed in first time in man study. The FDA review teams are multidisciplinary to include nonclinical pharmacologist/toxicologist, chemist, pharmacokineticist, biopharmaceuticist, clinical pharmacologist, medical officer, microbiologist, and statistician. Establishment of an IND provides a regulatory go-ahead for clinical development and evaluation of the investigational drug. However, protocols for all studies conducted in human subjects in US that would become part of the NDA have to be submitted for the Agency review regarding safety of the subjects to be enrolled in the study. If the Agency has safety concern(s), it may place the protocol on "clinical hold" until the identified concern(s) are satisfactorily resolved. In addition to the Agency's approval, all protocols as well as the informed consent documents that volunteers sign prior to participating in a clinical study must be approved by institutional review boards (IRBs). The IRBs are independent committee of physicians, community advocates, and others that ensures a clinical trial is ethical and the rights of study participants are protected.

The clinical development program for regulatory approval for marketing of a drug product usually consists of three phases. Phase I investigations are generally conducted to study PKs, and establish safety and tolerability of the drug. These studies represent the first time the investigation drug is administered to humans. They are carried out in healthy volunteers, which are small in number – usually 20–100 to identify metabolic and pharmacological effects of drug in humans and to determine the side effects associated with increasing doses and, if possible, to gain early evidence on effectiveness. The studies typically include formulation development, assessment of metabolic pathways, PK and PD studies as well as determination of intrinsic and extrinsic factors (e.g., gender, food, disease and other drugs) which may influence the drug action.

Phase II studies are designed to determine effectiveness and further explore the safety of the candidate drug in humans. This phase of testing also helps determine the common short-term side effects and risks associated with the drug product. Phase II studies are typically well-controlled, closely monitored, and conducted in a relatively small number of patients, usually involving few hundred subjects. These investigations are typically dose-ranging and proof-of-concept studies in patients suffering from the condition the investigational drug product is designed to treat. Most Phase II clinical trials are randomized or randomly divided into groups – one of which receives the investigational drug product and the other gets a placebo product containing no medication. Sometimes a third group may be included to receive a current (if applicable) standard treatment with which the new investigational drug is compared. These controlled clinical studies determine safety and effectiveness of the drug in treating the target population and establish the minimum and maximum effective dose. Phase II studies may be split into "learn-confirm" (Phase IIa) and "confirm-learn" (Phase IIb) [74, 76]. Phase IIa studies are used to learn drug efficacy in patients to confirm observations regarding safety, PKs, drug effects observed in the Phase I studies, explore biomarkers and surrogate endpoints and establish proof-of-concept. Phase IIb studies confirm what was learnt in Phase IIa and evaluate the drug in larger patient populations to further characterize the response surface. These studies may include subsets of patient population that are most likely to exhibit clinical benefit based on the observations made in Phase IIa. Measures of safety and effectiveness used in the Phase II may be the actual endpoints employed in the Phase III pivotal clinical trials. However, due to relatively short duration of studies and smaller sample size convincing evidence for efficacy is often not obtainable in Phase II.

Phase III studies provide expanded testing of effectiveness and safety of the candidate drug product, usually in randomized and blinded clinical trials. During this phase the safety and efficacy testing is conducted in several hundred to thousands of target patients to evaluate the overall benefit–risk relationship of the drug. These studies provide an adequate basis for extrapolating the results to the general population and incorporating that information into the physician labeling. Phase III studies generally consist of two positive well-controlled large multicenter trials confirming the effectiveness and safety of the drug product. In some cases, (however infrequent) characterization of response surface based on adequate PK/PD data [77] from earlier clinical trials may contribute towards confirmatory evidence for effectiveness and safety of the drug and eliminate the need for a second pivotal clinical trial.

Drug testing may continue after the Agency has approved the drug products. The postapproval (Phase IV) clinical testing may be recommended by the Agency or initiated by the sponsors. Some studies focus on previously unknown side effects or related risk factors. Phase IV investigations include determinations of safety and efficacy of uses beyond the drug's original application. These studies may test a proven drug in broader patient populations and compare the long-term safety and effectiveness. These investigations also collect and analyze long-term safety data on patients treated in normal practice.

The forgoing provides a brief statement regarding the in vivo testing of new drugs which is intended to provide information regarding drug safety (short and long term) and effectiveness, clinically effective dosing regimens, suitability for treating special populations (e.g., pediatric and geriatrics) and PK consideration for use in special populations (e.g., patients with renal and/or hepatic impairment) and concomitant administration of other drugs. Though in vivo evaluation paradigms generally applicable a wide range of drugs are also relevant to inhalation drugs, these drug products warrant special consideration due to their complex nature of drug products, mode of drug delivery, and target sites for effectiveness (local) and safety (local and systemic). Thus, the conventional protocols for nonclinical evaluation would require modification to accommodate issues unique to the inhalation drug delivery including possibility of local toxicity and unintended systemic effects. Administration of drugs through the inhalation route may cause greater toxicity than observed with other routes. Furthermore, inhaled dose is more difficult to determine than intravenous or oral doses, and it is harder to relate dose in laboratory animals to that in man for inhalation exposure than for other routes of administration. Extrapolation of animal toxicity data to man is also more complicated because people generally inhale through the mouth, whereas most laboratory animals inhale primarily through the nose. Special studies on pulmonary function, mucociliary clearance, or immune response may be needed, depending on the nature of the inhaled pharmaceutical.

Clinical evaluation of the inhalation drug products is also complicated by the very fact that these products are combination of drug and device. The clinical studies are designed to evaluate the safety of the drug product including the drug moiety, the excipients and device components. In addition, the clinical studies evaluate the performance and reliability of the device over its proposed use life [78]. Additional evaluation may be required for characterization of the dose delivery from the device with respect to special populations (e.g., pediatrics, severely compromised asthmatics and chronic obstructive pulmonary disease patients), because the device-patient interactions may differ substantially among the targeted patient populations.

Common to all clinical evaluations is the determination of the safe and effective dose and the appropriate dosing intervals. Selection of dose(s) for the pivotal Phase III studies is generally based on the available information (from previous studies) including PK, PD, or clinical data relevant to the expected clinical benefit from the selected drug. If the sponsor intends to market more than one dose, the placebo-controlled clinical trials are designed to provide assessment of relative effectiveness and safety at the target doses. Furthermore, if the targeted benefit(s) is not explored in the short-term studies, the definitive Phase III trials may include a range of doses while taking into consideration the risk-benefit assessments. The dose selected for the new product is supported by adequate efficacy and safety data.

Determination of the appropriate dose(s) and dosing interval forms the basis of clinically relevant dosing regimens. Dosing interval may be influenced by, among other things, PK and PD properties of the drug. Dosing intervals for the pivotal clinical trials are generally based on the available information regarding duration of the clinical effect(s). The validity of the selected dosing intervals is established by periodic evaluations of clinical effects of the drug at the end of its dosing interval (immediately prior to the next dose). The chosen dosing interval should be adequate to provide the clinical benefit over the selected period.

The focus of this chapter is on inhaled respiratory drugs. The principal use of majority of the orally inhaled respiratory drug products is for the treatment of two major pulmonary diseases, asthma and chronic obstructive pulmonary disease (COPD). Though some of the available drug products might be indicated for both diseases, clinical testing requires consideration unique to the disease and its treatment.

21.4.1 Asthma

The NHLBI guidelines [79] define asthma as “a chronic inflammatory disorder of the airways in which many cells and cellular elements play a role: in particular, mast cells, eosinophils, T lymphocytes, macrophages, neutrophils, and epithelial cells. In susceptible individuals, this inflammation causes recurrent episodes of wheezing, breathlessness, chest tightness, and coughing, particularly at night or in the early morning. These episodes are usually associated with widespread but variable airflow obstruction that is often reversible either spontaneously or with treatment.

The inflammation also causes an associated increase in the existing bronchial hyperresponsiveness to a variety of stimuli. Reversibility of airflow limitation may be incomplete in some patients with asthma." The guidelines categorize the asthma medication in two broad groups: (1) quick-relief medications used to treat acute symptoms and exacerbations, and (2) long-term control medications used to achieve and maintain control of persistent asthma. Among the locally acting inhaled drugs, short-acting beta-agonists (albuterol, levalbuterol) are the recommended reliever drugs, whereas the controller drugs include corticosteroids, long acting beta-agonists (salmeterol, formoterol) and cromolyn sodium and nedocromil. The quick-relieving action of the reliever drugs is principally due to their ability to assert acute bronchodilation which is short-lived (4–6 h). On the other hand, manifestation of the clinical efficacy of controller drugs may take anywhere from less than an hour (long acting beta-agonists) to several days (corticosteroids). The noted difference in the onset and duration of action may influence many aspects of the PK, PD, and clinical study designs for evaluation of the two classes of the drug products.

21.4.1.1 Reliever Dugs

The programs for clinical efficacy assessment of reliever (bronchodilator) drugs include dose-ranging studies, cumulative-dose safety studies, and definitive efficacy and safety studies [78]. The dose-ranging studies are generally single-dose, multiple-period, placebo-controlled, evaluations based on bronchodilation or bronchoprotection endpoints. The preferred measure of efficacy for most asthma drugs is bronchodilation and is primarily based on the forced expiratory volume in the first second (FEV_1). Other models (e.g., bronchoprotection) may be used but their utility is limited to studies of PD responses; the data may not be used to support definitive clinical outcome. In bronchoprotection studies, the efficacy is generally based on the concentration of the provocative agent which induces bronchospasm to reduce the postdose FEV_1 by a defined percentage. Currently methacholine is the only approved provocative agent. The bronchodilator activity of the test drug is measured in terms of the amount of methacholine that reduces FEV_1 by 20%. The amount may be expressed in terms of concentration (PC_{20}) or dose (PD_{20}) of methacholine. The tests are conducted according to the established guidelines [80].

The cumulative-dose safety studies are primarily intended to assess PD measures of systemic exposure. These studies are based on sequential administration of drug (e.g., 1 puff, 2 puffs, 4 puffs every 30 min) to assess PD safety [78] using appropriate safety endpoints (see Sect. 21.4.3).

Consistent with the requirements for approval of all new drugs, the approval of new inhalation drug products require evidence for effectiveness based on at least two randomized and controlled Phase III trials. The Agency expects the two trials to be positive and replicative. For the reliever drugs, the clinical studies are based on at least 12 weeks of active treatment [78] and the postdose serial FEV_1 as the primary measure of efficacy. The study duration and design may be influenced by the history of the device and formulation.

21.4.1.2 Controller Drugs

The Agency recommends dose-ranging studies and definitive efficacy and safety studies [78]. PK assessments are important to address drug safety in the Dose-ranging studies. The design of the studies for clinical evaluation of corticosteroids is influenced by a noted lack of dose response based on conventional primary efficacy endpoint FEV_1 [81–84]. The sponsors may consider alternative study designs with consultation of the review Division.

The duration of the pivotal efficacy studies is at least 12 weeks. Primary efficacies variables include predose FEV_1, occurrence of exacerbation, withdrawal of known effective drug(s), symptom score, etc. [78]. Secondary endpoints may include, but not limited to, rescue medicine use, bronchial hyperresponsiveness, symptom scores, and nocturnal awakenings.

21.4.2 COPD

COPD is characterized by the airflow limitation that is not fully reversible [85]. The airflow obstruction is usually progressive and is associated with an abnormal inflammatory response of lungs to noxious particles, gases, or cigarette smoking. COPD patients may also exhibit a varying degree of lung tissue destruction, and heterogeneity in symptoms of the disease [86]. Most common symptoms include cough, excess sputum production, and dyspnea [85].

Clinical paradigms for evaluation of inhaled drugs for COPD are generally based on the same principles as used for asthma drugs, however with particular focus on reduction of airway obstruction. Due to uniqueness of this disease the Agency recommends [86] a variety of points to consider for the development of a program for clinical evaluation of COPD drugs. These considerations include, but not limited to, Target Disease and Indication, Types of Therapeutic Agents for COPD, Study Populations, Dose Selection, Efficacy Variables, and Study Duration.

COPD is a complex disease which consist of entities (emphysema and bronchitis) which are clinically and histologically distinct [87, 88] and the selected drug may affect one not the other. A determination of the target disease/indication will influence many aspects of the clinical studies. The drug may be intended to improve one or more elements of the disease including airflow obstruction, symptoms, exacerbations, progression of the disease [89]. Therefore, clinical evaluation of each of the COPD targets warrants careful selection of the clinical endpoints and study designs which may support specific indication(s) [86].

The noted complexity of COPD also influences the choice of target population enrolled for the clinical trials. Though, the chosen drug may selectively impact one of the subsets (chronic bronchitis or emphysema), these subsets may coexist [87, 88]. Thus, development of clinical protocols to evaluate one subset would require careful selection of the patient enrollment criteria. If the clinical trials are deigned to evaluate therapy for one subset of the disease, the approved indication will be accordingly restricted to that subset [86].

Like many other drugs, the assessment of efficacy of COPD drugs may be based on primary and secondary efficacy variables. The primary measures of activity of drugs for COPD are selected to address the desired outcome, which may include improvement of airflow obstruction, symptoms, exacerbations, or mitigation of the disease. Air flow obstruction is generally based on pulmonary function testing through spirometry for FEV_1 measurements because FEV_1 and a ratio of FEV_1 and forced vital capacity (FVC) are the most useful indices of airflow impairment from the airway obstruction [90, 91] and is accepted as a variable to support efficacy. A change in postdose FEV_1 for a bronchodilator drug and change in predose FEV_1 for a nonbronchodilator drug are considered to be the valid indicators of improvement in airflow [86].

The primary efficacy variable chosen to reflect symptom relief should demonstrate the desired improvement which is clinically relevant. For example, for drugs developed for improvement in exacerbations, the primary efficacy variable should be a clinically relevant measure of exacerbations reflecting improvements in severity, duration and frequency of exacerbations. Comparison of pre- and postexacerbation FEV_1 may be considered a primary efficacy variable [86].

FEV_1 is also a preferred primary efficacy variable to assess efficacy of drugs developed to mitigate progression of COPD. Serial measurements of FEV_1 over time might be used to support alteration of disease manifested as change in postdose decline in FEV_1 [86]. However, a change in FEV_1 may not support a claim of altering disease progression, if the chosen drug itself is a bronchodilator.

Secondary efficacy variables generally provide information to support to the primary efficacy. In COPD, secondary assessments may be useful in evaluating the relevance of mean changes in FEV_1, the primary endpoint variable. Commonly used secondary efficacy variables include various measures of lung function, exercise capacity, symptom scores, activity scales, and health-related quality of life measurements [86].

An important consideration in the design of any clinical trials is the duration of the active treatment. Since the drugs for COPD may be developed for one or multiple indications, the study duration may vary with the proposed indication [86]. To establish efficacy of drugs intended to reduce airflow obstruction, treatment duration may be 3 or 6 months for bronchodilators or nonbronchodilator agents, respectively. For drugs intended to provide symptom relief or to mitigate exacerbations the duration of treatment should be at least 6 months or 1 year, respectively. If a therapeutic agent that is developed to modify lung structure, treatment durations may be few years. Longer durations of treatment may be required to adequately assess safety [86].

Proper design and conduct of studies to evaluate COPD drugs require consideration of many other factors including, but not limited to, study designs, concomitant treatments, management of smoking and device characteristics particularly with respect to the patient's ability to generate adequate air flow. A detailed discussion of these topics is beyond the scope of this chapter. The sponsor should refer to the relevant Agency guidances and contact the review Division for up-to-date advice.

21.4.3 Safety Considerations

The locally acting orally inhaled drug products are developed for drug delivery to the respiratory tract. However, depending upon the efficiency of the delivery device a significant portion of the emitted dose gets deposited in the oropharyngeal cavity with opportunity for absorption into the systemic circulation. Therefore, paradigms for evaluation of safety of orally inhaled drugs warrant consideration for local (respiratory tract) and systemic safety. Whereas the local toxicity may be due to the interaction of cells/tissue with the drug and its metabolites, excipients or leachables (if any), systemic toxicity principally arises from presence of drugs and metabolites in the systemic circulation. Both bronchodilators and maintenance drugs are known to assert systemic side effects manifested through drug related changes in PD end-points (e.g., serum potassium and cardiac toxicity for beta-agonists [92–94] and anticholinergics [95, 96] and markers of possible adrenal suppression by corticosteroids [97–102]). Some of the systemic effects of corticosteroids (e.g., growth [103–105] and bone density [106, 107]) would require long-term safety assessments based on appropriate endpoints. The duration and design of safety studies may vary with the history of the drug product (e.g., new product (drug and device) or reformulated product (formulation or device change)).

21.5 Combination Drug Products

Based on 21 CFR sections 3.2 and 300.50, the combination products may include two or more active drugs or drug(s) and device. The combination products approved for respiratory diseases may include one or more active drugs with appropriate device(s). The products which include two individual active drugs which might contribute (despite distinctly different mechanisms of action) to the same clinical outcome such as improvement in lung function indicated by postdose FEV [86], attenuation of obstruction, relief of symptoms, and the combined clinical effect is generally greater than that of the individual components. The currently approved products include combinations of a bronchodilator and a corticosteroid [33, 108–110] or two bronchodilators with different modes of action [111].

Evaluation of combination products would require assessment of product performance (in vitro) based on measurements of individual components. Similarly, the in vivo PK studies would include measurement of systemic exposure form individual components in the combination products. These studies would require special consideration with respect to assay sensitivity and blood sampling times, because the concentration-time profile of the two active drugs may be distinct, both qualitatively and quantitatively.

Considerations for clinical evaluation of combination products in pivotal trials depend upon the actives in the combination. Determination of efficacy generally includes comparison of the clinical effect(s) of the combination product to each of its individual components in the same clinical study to show that the combination product provides clinical benefit that is superior to each of its constituents [33, 108–110]. The Agency requires clinical data supporting the contribution of each component to the efficacy and the safety of the combination product in the population of interest.

21.6 Extended-Release Inhalation Drug Products

Extended-release dosage forms are defined, from regulatory perspective, as pharmaceutical preparations that release a drug (or drugs) in such a way to allow a reduction in dosing frequency as compared to the drug (or drugs) presented as an immediate release (IR) preparation [112]. Similarly, the USP describes them as preparations that are formulated in such manner as to make the contained medicament available over an extended period of time following ingestion [113]. Other terms, such as controlled-release, sustained-release, prolonged-action, and repeat-action have been used in the literature to describe similar drug release mode. However, for regulatory and pharmacopeial purposes and for consistency the phrase "extended-release" has been adopted. Some of the desired goals of such preparations may include decreasing dosing frequency and thus increasing patient convenience and compliance, and prolonging duration of action with potentially decreasing side effects of the drug.

Recently, there have been noted research interest and focus on ER inhalation drug products [114–118]. Different potential approaches in developing ER inhalation preparations include, but not limited to, liposomes, biodegradable microspheres, and other polymeric materials. However, to date there are no approved ER inhalation preparations in USA.

The quality considerations for ER inhalation drug products, from a regulatory perspective, are similar to the corresponding IR inhalation drug products described above in various sections and in Tables 21.2 and 21.3. However, depending on the type of the product formulation, its composition, drug release mechanism and its profile, and frequency of administration specific additional information and specifications would be applied to characterize the product and to ensure batch-to-batch reproducibility. For example, for liposomal preparations, information on lipid composition and nonlipid components would be expected. Characterization of such drug products may include morphology and lamilarity determination, liposomal integrity, phase transition temperature, encapsulated and unencapsulated drug substance content, lipid content, stability of lipid components and their degradation profile, in vitro release profile of the drug substance from the liposomal drug product, etc. The in vitro release profile criterion of the drug substance from the formulation matrix would be appropriate to all ER inhalation drug products using suitable and validated test methods and procedures. The release criteria are expected to be related to or based upon batches that will be used in the nonclinical, PK, clinical efficacy, and primary stability studies.

From the therapeutic view point, drug products marketed in ER formulations are generally developed to provide therapeutic effects of drugs over prolonged periods with fewer administrations than required for same level of therapy using the IR formulations of the same drugs. One dosage unit of extended-release product may contain multiple doses found in the immediate release unit [119]. Therefore, the total dose used in an ER formulation over a dosing interval may generally be the same as would be required from multiple dosing with the IR formulation over the same period.

Though ER products are available in a variety of dosage forms, there are no approved ER products for delivery to the respiratory tract. The current lack of approved ER formulation of drugs for diseases of the respiratory tract may be due to a number of factors such as the need for immediate bronchodilatory action necessary for rescue of the subjects suffering from episodes of acute bronchospasm, availability of long acting bronchodilators (like formoterol and salmeterol) products that maintain lung caliber for prolonged periods, and the lack of need for sustained formulations of maintenance drugs (e.g., corticosteroids) due to the notable prolonged time for the onset as well as duration of their action. Other factors which may contribute to deterrence for the potential commercial development of ER inhalation drug products may include, among others, long-term safety considerations for the excipients intended for use in such inhalation drug products at the proposed concentration and total daily intake. Development of controlled release formulations may require the use of excipients that have not been approved for the inhalation route of administration. Therefore, safety issues also may arise from potential local (lung) toxicity concerns [120–122] due to sustained presence of these drugs and/or combinations of drugs and excipients.

Recent scientific advancements with potential for development of ER products include advances in application of nanotechnology. Pharmaceutical/medicinal use of nanotechnology in drug development has attained significant attention [123–125]. In addition, its potential in the development of drug formulations for pulmonary diseases has been discussed [126–131]. Delivery of drugs to the lungs in large porous matrices may cause retention of the particles in the lungs accompanied with a prolonged drug release [132]. These particles have numerous advantages over other delivery systems due to their unique characteristics such as small particle size, large surface area, and the capability of changing their surface properties. So far the FDA has not approved an orally inhaled drug product prepared from substance(s) using nanoparticles. Employment of nanoparticles in locally acting respiratory products warrants, among other, careful consideration of both local and systemic toxicity [133–135] due to the drug-excipient interactions, substance (drug and/or excipient)-tissue interaction, and possible increase in bioavailability of the drug due to small particle size [136–138].

21.7 Generic Inhalation Drug Products

The FD&C Act has been amended at many occasions. In 1984, the Hatch-Waxman amendments (Drug Price Competition and Patent Term Restoration Act of 1984) created Section 505(j) of the Act [21 USC 355 (j)] established a process for the abbreviated new drug application (ANDA) for approval generic versions of approved pioneer drugs. Based on these amendments, manufacturers of generic drugs are required to show (among other things) sameness in active ingredient, dosage form, route of administration, labeling, quality, performance characteristics, and intended use. In addition, the applicants are required to demonstrate acceptable in vitro and in vivo performance of generic products.

Like the product specificity of the CMC information required for the NDAs, the contents of the ANDA CMC packages may vary with the drug, drug product, and its intended use. The Office of Generic Drugs (OGD) has not issued guidance for CMC portion of the applications seeking approval of respiratory drugs. However, in principle, the above mentioned CMC considerations related to respiratory drugs are also applicable to the development of generic respiratory drugs.

As stated above, safety and efficacy of generic drugs is based on BE studies. 21 CFR 320.24(b) provides in the descending order of preference a list of approaches that are used for documentation of BE. The listed methods include: (1) an in vivo test in humans in which the concentration of the active ingredient or active moiety, and, when appropriate, its active metabolite(s), in whole blood, plasma, serum, or other appropriate biological fluid is measured as a function of time, (2) an in vitro test that has been correlated with and is predictive of human in vivo bioavailability data, (3) an in vivo test in humans in which an appropriate acute pharmacological effect of the active moiety, and, when appropriate, its active metabolite(s), is measured as a function of time if such effect can be measured with sufficient accuracy, sensitivity, and reproducibility, (4) well-controlled clinical trials that establish the safety and effectiveness of the drug product, for purposes of measuring bioavailability, or appropriately designed comparative clinical trials, for purposes of demonstrating BE, (5) a currently available in vitro test acceptable to the FDA that ensures human in vivo bioavailability, and (6) any other approach deemed adequate by FDA to measure bioavailability or establish BE. The type of studies required for documentation of BE may vary with the drug product. For majority of drug products which act through systemic circulation, acceptable performance of the generic products is generally based on in vivo PK studies and in vitro dissolution comparing the generic product and the corresponding RLD [2].

For many drug products which act through systemic circulations and some locally acting products [139], the Agency does not require the formulations of the generic products to be the same as that of the innovator as long as the required BE criteria are met. The generic product formulations may differ both qualitatively and quantitatively from the corresponding RLDs with respect to inactive ingredients. However, that allowance is not applicable to generic inhalation drug products. The Agency currently uses a "Weight of Evidence" approach [140, 141] to include: (1) Qualitative (Q_1) and Quantitative (Q_2) sameness of formulation, (2) acceptable comparative in vitro performance, (3) equivalent systemic exposure, and (4) equivalent delivery to the local site of action (lung). Among these four components, the latter three require comparative studies to support BE between the generic and reference products.

Currently, the applicants seeking approval of generic respiratory drugs are required to provide evidence supporting all four element of the "Weight of Evidence" approach, and the elements are evaluated individually. The following provides a brief description of concepts and methods for determination of equivalence of generic drugs based on this approach.

21.7.1 Q_1 and Q_2 Sameness of Formulations

The Q_1 sameness implies that the generic product contains the same inactive ingredients as the RLD, and the Q_2 sameness means that the generic product contains all inactive ingredients at concentrations within ±5% of the concentrations in the RLD [66].

21.7.2 *In Vitro Equivalence*

The available draft BA/BE guidance for Nasal sprays and Aerosols [66] provides the Agency recommendations regarding in vitro tests. Some of these tests in principle are applicable to inhalation aerosols. Regulatory considerations for in vitro testing can be found in recently published articles [58, 66, 140, 141].

Comparative in vitro performance of conventional dosage forms may be based on single lot each of the test and reference products. However, for inhalation aerosols, the OGD requires demonstration of comparative in vitro performance using three batches of the generic and RLD products. The test product batches should be manufactured consistent with the Current Good Manufacturing Practices (cGMP) in a production facility using equipment of the same design and operating principles as will be used for commercial batches. The operations should be performed by appropriately qualified individuals and the batches fully packaged.

The FDA recommends establishment of comparative in vitro performance of generic MDIs based on a series of tests which include Single Actuation Content, APSD, Spray Pattern and Plume geometry (where applicable) analyses, and Priming/Repriming. The first two tests are also applicable to DPIs which may require additional considerations for device resistance and air flow rates [140]. Many inhalation products are designed to deliver multiple doses. Therefore, another consideration relevant to testing in vitro performance of inhalation products is the use life of these products, because potential changes in physicochemical properties of the formulation and wear and tear of the metering device over the use life of the drug product may influence drug delivery. The in vitro performance is evaluated at different stages of the product life divided into Beginning, Middle, and End sectors [16, 66]. Information regarding the apparatuses, measurements, and other relevant considerations can be found in the available literature [16, 58, 140].

21.7.3 Equivalent Systemic Exposure

Determination of equivalent systemic exposure following administration of the test and RLD products is generally based on measures of rate and extent of drug absorption represented by the first peak blood concentration (C_{max}) and area under the blood concentration vs. time curve (AUC) [142]. For the drugs which act through systemic circulation, systemic exposure provides the sole evidence of drug delivery to all sites of action implicated in safety and effectiveness. However, its role in approval of generic inhalation aerosols in the US has evolved from a "Safety Evaluation" study [143] to one of the four components of the "Weight of Evidence" approach [144]. Currently, the FDA does not accept the systemic exposure as a surrogate for drug delivery to the local site of action [58, 141, 144]. The Agency's lack of acceptance is principally due to the uncertainty regarding a direct relationship between the drug detected in the systemic circulation to the drug reaching the site of action (lung). The emitted dose of inhalation aerosols does not rely on the systemic circulation for delivery of the drugs to the local site(s) of action. The inhaled drugs may appear at the local sites in the respiratory tract before they penetrate through lung to reach the systemic circulation (Fig. 21.2). Furthermore, drug delivery to the systemic circulation may occur from nonpulmonary sites (e.g., the gastrointestinal tract). Therefore, the Agency does not accept systemic exposure data as evidence of delivery to the local site(s) of action in the respiratory tract.

The systemic exposure studies are generally conducted based on PK measurements following administration of single maximum labeled adult dose of the generic and reference products. The BE studies are generally conducted in adult healthy

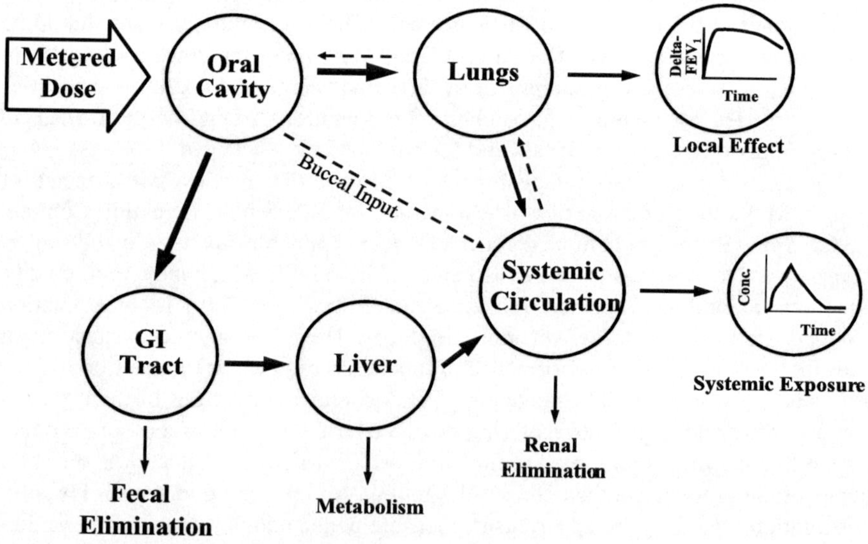

Fig. 21.2 Disposition of aerosolized dose

volunteers [142] representative of gender and race of the target populations. The studies are preferably based on randomized crossover (or partial replicate or replicate) designs enrolling adequate number of subjects. Blood (plasma/serum) concentrations of parent drug and/or metabolite (if applicable) are determined following administration of the test and reference products using validated assays [145]. The blood concentration vs. time data are used to compute bioavailability metrics AUC and C_{max}. Equivalence of systemic exposure from the test and reference products is based on the natural log-transformed AUC and C_{max} data using the conventional two-one sided tests (ANOVA) procedure [146, 147], or appropriate statistical procedure [147] if partial replicate or replicate design is used.

21.7.4 Equivalent Local Delivery

Demonstration of equivalence of drug delivery to the local site(s) of action includes determination of equivalence of local effect(s) based on PD endpoints or clinical measurements. PD studies are preferred over clinical measurements [21 CFR 320.24(b)] because they may provide greater sensitivity to detect potential differences in dose delivery to the local site(s) of action.

Unlike the single-dose studies used for establishment of BE of systemically active drugs, documentation of BE based on PD methods requires multiple doses which are necessary for demonstration of the assay sensitivity in terms of dose response. Measures of drug activity in PD BE studies are PD responses generally relevant to the mechanism of action of drugs, e.g., improvement in FEV_1 as indicator of bronchodilation activity of bronchodilators. The suitability of the PD method for BE studies is indicated by the nature of the dose–response relationship which may vary with the bioassay, response measures and dose range employed in the study. Dose–response relationships of the approved inhalation aerosols are generally nonlinear and, sometimes, very shallow [58, 141]. This is partly because the doses of the marketed products are near or on the plateau of dose response. In addition, the degree of shallowness in response may also vary from one study to another due to a number of factors including the mode of action of the drug, its potency, the PD measure, the study population, and severity of the disease. Therefore, PD BE studies require careful selection of the suitable patient population to provide adequate window for demonstration of a measurable dose response using single and multiple actuations of the drug product.

PD BE study designs may vary with the study drug based on the nature, speed, and duration of its action. Among the two recognized categories of the marketed inhalation aerosols the "reliever" drugs provide relief from symptoms soon after treatment administration. These drugs are generally indicated for both prevention and treatment of bronchospasm. Therefore, PD effects of such drugs may be measured in terms bronchodilation or prevention of experimentally induced bronchoconstriction. The most commonly used reliever drugs include the short-acting beta-agonists (SABA) and anticholinergics. Among the SABA, albuterol is the most extensively studied drug with respect to determination of equivalence of local action. Based on the published literature, BE studies on albuterol MDIs may use either bronchodilation or bronchoprotection model [58, 141].

Results of bronchodilatation studies are highly dependent on the study population and it may not be possible to obtain measurable dose response in certain populations [148]. On the other hand, bronchoprotection models are capable of showing beta-agonist effectiveness where simple bronchodilation is not [149, 150]. Bronchoprotection from inhaled methacholine following administration of varying doses of albuterol have been reported [151], and bronchoprotection from inhaled histamine has been previously used for documentation of BE of the generic and reference listed drugs [152].

Anticholinergics like iptratropium and tiotropium are the bronchodilators [153–155] of choice for treating COPD patients because cholinergic mechanisms are known to contribute to obstruction in this disease [155–158]. Therefore, improvements in airflow following administration of these drugs may exhibit measurable dose response, and the bronchodilatation model recommended for beta-agonists may also be useful for determination of BE of inhalation aerosols containing anticholinergic drugs. The bronchoprotection model based on methacholine-induced bronchospasm may be useful because the anticholinergic bronchodilators also provide protection against bronchospasm in asthmatics with airway hyperresponsiveness [159–161], and provide protection from inhaled methacholine in a dose-dependent manner [161].

Inhaled corticosteroids are important respiratory drugs. Development of in vivo models to determine BE of inhaled corticosteroids has been hampered by lack of suitable models for demonstration of dose response. Conventional studies based on measures of posttreatment bronchodilation generally yield flat dose response with this class of drugs [83, 84]. To date several in vivo methods have been suggested. These models include an "Asthma Stability" [162, 163] model in which the dose response is measured by the extent to which low and high doses of inhaled corticosteroids are able to sustain asthma control furnished by a "burst" of high dose of oral prednisone. In addition, in vivo models measuring effects of inhaled corticosteroids on makers of airway inflammation including sputum eosinophil [84, 164] and exhaled nitric oxide [58, 78, 165] are under consideration. As yet, none of these models have been used to document BE in support of approval of generic inhaled corticosteroid products.

This section has included a brief statement of the authors' understanding of the FDA requirements for approval of generic orally inhaled drug products. Generic drug products are therapeutically equivalent copies of established RLDs; the Agency's recommendations for documentation of in vitro and in vivo performance of generic products are generally developed by taking into consideration the relevant information available from its approval of the corresponding RLDs. With respect to the focus of this book, it is noted that none of the approved orally inhaled RLDs is based on ER formulations. Therefore, deliberations on regulatory considerations for the development of generic orally inhaled drug products are likely to occur some years after the possible approval of novel ER orally inhaled drug products.

Though the general CMC and BE considerations stated above may be applicable, the genesis of scientific considerations for the generic ER products will depend upon, among other factors, advances in technology, issues specific to the in vitro and in vivo performance of the RLDs, the CMC, biopharmaceutic and clinical attributes (of the RLDs) and their appropriateness for development and evaluation of generic products.

References

1. Guidance for Industry: Applications Covered by Section (505(B)(2))-1999
2. Conner D, Davit B (2005) Bioequivalence and drug product assessment, in vivo. In: Shargel L, Kanfer I (eds) Generic drug product development: solid oral dosage forms. Marcel Dekker, New York, pp 225–227
3. Ng R (2004) Drugs: from discovery to approval. Wiley, New Jersey
4. 21 Code of Federal Regulations, Part 3
5. Purewal TS (1998) Formulations of metered dose inhalers. In: Purewal TS, Grant DG (eds) Metered dose inhaler technology. CRC Press, Boca Raton, pp 9–68
6. Smyth H (2003) The influence of formulation variables on the performance of alternative propellant-driven metered dose inhalers. Adv Drug Del Rev 55:807–808
7. Noakes TJ (1995) CFCs, their replacements, and the ozone layer. J Aerosol Med 8(Supp 1): S3–S7
8. Boulet LP (1998) The ozone layer and metered dose inhalers. Can Respir J 5:176–179
9. McDonald KJ, Martin GP (2000) Transition to CFC-free metered dose inhalers – into the new millennium. Int J Pharm 201:89–107
10. Tsai WT (2005) An overview of environmental hazards and exposure risk of hydroflurocarbons (HFCs). Chemosphere 61:1539–1547
11. Ganderton D, Kassem NM (1992) Dry powder inhalers. Adv Pharmaceut Sci 6:165–191
12. Timsina MP, Martin GP, Marriot C, Ganderton D, Yianneskis M (1994) Drug delivery to the respiratory tract using dry powder inhalers. Int J Pharmaceut 101:1–13
13. Newman SP, Busse WW (2002) Evolution of dry powder inhaler design, formulation and performance. Respir Med 96:293–304
14. Smith IJ, Parry-Billing M (2003) The inhaler of the future? A review of dry powder inhalers on the market today. Pulm Pharmacol Ther 16:79–95
15. Dunbar CA, Hickey AJ, Holzner P (1998) Dispersion and characterization of pharmaceutical dry powder aerosols. KONA 16:7–45
16. Draft Guidance for Industry: Metered dose inhaler (MDI) and dry powder inhaler (DPI) drug products: Chemistry, manufacturing and control documentation (1998). http://www.fda.gov/Drugs/GuidanceComplianceRegulatoryInformation/Guidances/ucm064979.htm. Accessed 10 Jan 2011
17. Cegla UH (2004) Pressure and inspiratory flow characteristics of dry powder inhalers. Respir Med 98(Suppl A):S22–S28
18. Tarsin W, Assi KH, Chrystyn H (2004) In vitro intra- and inter-inhaler flow rate-dependent dosage emission from a combination of budesonide and eformoterol in a dry powder inhaler. J Aerosol Med 17:5–32
19. Clark AR, Hollingworth AM (1993) The relationship between powder inhaler resistance and peak inspiratory conditions in healthy volunteers. J Aerosol Med 6:99–110
20. Olsson B, Asking L (1994) Critical aspects of the functioning of inspiratory flow driven inhalers. J Aerosol Med 7(Suppl 1):S43–S47
21. Richards R, Simpson SF, Renwick AG, Holgate ST (1988) Inhalation rate of sodium cromoglycate determines plasma pharmacokinetics and protection against AMP-induced bronchoconstriction in asthma. Eur Respir J 1:896–901

22. Engel T, Scharling B, Skovsted BA, Heinig JH (1992) Effects, side effects and plasma concentrations of turbutaline in adult asthmatics after inhaling from a dry powder inhaler device at different inhalation flows and volumes. Br J Clin Pharmacol 33:439–444
23. Chege JK, Chrystyn H (2000) The relative bioavailability of salbutamol to the lung using urinary excretion following inhalation from a novel dry powder inhaler: the effect of inhalation rate and formulation. Respir Med 94:51–56
24. Pedersen S, Hansen OR, Funglsang G (1990) Influence of inspiratory flow rate upon the effect of Turbuhaler. Archiv Dis Child 65:308–319
25. Munzel U, Marschall K, Fyrnys BA, Wedel M (2005) Variability of fine particle dose and lung deposition of budesonide delivered through two multidose dry powder inhalers. Curr Med Res Opin 21:827–834
26. Bronsky EA, Grossman J, Henis MJ, Gallo PP, Yegen U, Della Cioppa G, Kottakis J, Mehra S (2004) Inspiratory flow rates and volumes with the Aerolizer dry powder inhaler in asthmatic children and adults. Curr Med Res Opin 20:131–137
27. Gauld LM, Briggs K, Robinson P (2003) Peak respiratory flows in children with cystic fibrosis. J Paediatr Child Health 39:210–213
28. Kamps AW, Brand PL, Roorda RJ (2004) Variation in peak inspiratory flow through dry powder inhalers in children with stable and unstable asthma. Pediatr Pulmonol 37:65–70
29. Bentur L, Mansour Y, Hamzani Y, Beck R, Elias N, Amirav I (2004) Measurement of inspiratory flow in children with acute asthma. Pediatr Pulmonol 38:304–307
30. Nsour WM, Alldred A, Corrado OJ, Chrystyn H (2001) Measurement of peak inhalation rates with an In-check Meter® to identify an elderly patient's ability to use a Turbuhaler®. Respir Med 95:965–968
31. Spiriva Handihaler™ approved product label (2009). http://www.accessdata.fda.gov/drugsatfda_docs/label/2009/021395s029lbl.pdf. Accessed 22 Jan 2011
32. Foradil Aerolizer™ approved product label (2001). http://www.accessdata.fda.gov/drugsatfda_docs/label/2001/20831s2lbl.pdf. Accessed 22 Jan 2011
33. Advair Diskus™ approved product label (2010). http://www.accessdata.fda.gov/drugsatfda_docs/label/2010/021077s042lbl.pdf. Accessed 22 Jan 2011
34. Pulmicort Flexihaler™ approved product label (2010). http://www.accessdata.fda.gov/approvedproductlabeldrugsatfda_docs/label/2010/021949s006lbl.pdf. Accessed 22 Jan 2011
35. Asmanex Twisthaer™ approved product label. http://www.accessdata.fda.gov/drugsatfda_docs/label/2010/021067s013lbl.pdf. Accessed 22 Jan 2011
36. Tibbatts J, Mendes PJ, Villax P (2010) Understanding the power requirements for efficient dispersion in powder inhalers: Comparing CFD predictions and experimental measurement. In: Byron P, Peart J, Suman J, Farr SJ, Young PM (eds) Respiratory drug delivery, vol 1. Davis Healthcare International Publishing, River Grove, pp 323–330
37. Tibbatts J (2009) From formulation to fine particle fraction – simulating the influence of both formulation and device on fluidization and deaggregation. In: Byron P, Peart J, Suman J, Farr SJ, Young PM (eds) RDD Europe 2009. Davis Healthcare International Publishing, River Grove, pp 293–296
38. Telko MJ, Hickey AJ (2005) Dry powder inhaler formulation. Respir Care 50(9):1209–1227
39. Hooton JC, Jones MD, Harris H, Shur J, Price R (2008) The influence of crystal habit on the prediction of dry powder inhalation formulation performance using the cohesive-adhesive force balance approach. Drug Dev Ind Pharm 34:974–983
40. Edge S, Mueller S, Price R, Shur J (2008) Factors affecting defining the quality and functionality of excipients used in the manufacture of dry powder inhaler products. Drug Dev Ind Pharm 34(9):966–973
41. 21 Code of Federal Register 200.51
42. Guidance for Industry – Inhalation Drug Products Packaged in Semipermeable Container Closure Systems, Draft Guidance: U.S. Department of Health and Human Services Food and Drug Administration, Center for Drug Evaluation and Research, July 2002. http://www.fda.gov/downloads/Drugs/GuidanceComplianceRegulatoryInformation/Guidances/ucm071725.pdf. Accessed 27 Jan 2011

43. Guidance for Industry – Q3A Impurities in New Drug Substances, U.S. Department of Health and Human Services, Food and Drug Administration, Center for Drug Evaluation and Research, Center for Biologics Evaluation and Research, June 2008 (ICH Revision 2). http://www.fda.gov/downloads/Drugs/GuidanceComplianceRegulatoryInformation/Guidances/ucm073385.pdf. Accessed 27 Jan 2011
44. Guidance for Industry Q8(R2): Pharmaceutical Development, U.S. Department of Health and Human Services Food and Drug Administration, Center for Drug Evaluation and Research, (CDER) Center for Biologics Evaluation and Research (CBER), November 2009 (ICH Revision2).http://www.fda.gov/downloads/Drugs/GuidanceComplianceRegulatoryInformation/Guidances/ucm073507.pdf. Accessed 27 Jan 2011
45. Guidance for Industry Q9: Quality Risk Management, U.S. Department of Health and Human Services, Food and Drug Administration, Center for Drug Evaluation and Research, Center for Biologics Evaluation and Research, June 2006. http://www.fda.gov/downloads/Drugs/GuidanceComplianceRegulatoryInformation/Guidances/ucm073511.pdf. Accessed 27 Jan 2011
46. Guidance for Industry Q10: Pharmaceutical Quality System, U.S. Department of Health and Human Services, Food and Drug Administration, Center for Drug Evaluation and Research, Center for Biologics Evaluation and Research, April 2009. In the ICH Regions. U.S. Department of Health and Human Services, Food and Drug Administration, Center for Drug Evaluation and Research, Center for Biologics Evaluation and Research, February 2008. http://www.fda.gov/downloads/Drugs/GuidanceComplianceRegulatoryInformation/Guidances/ucm073517.pdf. Accessed 27 Jan 2011
47. Guidance for Industry Q4B: Evaluation and Recommendation of Pharmacopoeial Texts for Use in the ICH Regions. February 2008. U.S. Department of Health and Human Services, Food and Drug Administration, Center for Drug Evaluation and Research, Center for Biologics Evaluation and Research. http://www.fda.gov/downloads/Drugs/GuidanceComplianceRegulatoryInformation/Guidances/ucm073405.pdf. Accessed 27 Jan 2011
48. Guidance for Industry – Q6A Specifications: Test Procedures and Acceptance Criteria for New Drug Substances and New Drug Products: Chemical Substances, U.S. Department of Health and Human Services, Food and Drug Administration, Center for Drug Evaluation and Research, December 2000. http://www.fda.gov/Drugs/GuidanceComplianceRegulatoryInformation/Guidances/ucm134966.htm. Accessed 27 Jan 2011
49. Guidance for Industry – Q2 (R1) Validation of Analytical Procedures: Text and Methodology, U.S. Department of Health and Human Services, Food and Drug Administration, Center for Drug Evaluation and Research, November 2005
50. Guidance for Industry – Q3B(R2) Impurities in New Drug Products, U.S. Department of Health and Human Services Food and Drug Administration Center for Drug Evaluation and Research, Center for Biologics Evaluation and Research, July 2006 (ICH Revision 2). http://www.fda.gov/downloads/Drugs/GuidanceComplianceRegulatoryInformation/Guidances/ucm073389.pdf. Accessed 27 Jan 2011
51. Guidance for Industry – Integration of Dose-Counting Mechanisms into MDI Drug Products – Clinical Medical: U.S. Department of Health and Human Services Food and Drug Administration Center for Drug Evaluation and Research, Clinical Medical, March 2003. http://www.fda.gov/downloads/Drugs/GuidanceComplianceRegulatoryInformation/Guidances/ucm071731.pdf. Accessed 27 Jan 2011
52. Guidance for Industry – Nasal Spray and Inhalation Solution, Suspension, and Spray Drug Products – Chemistry, Manufacturing, and Controls Documentation: U.S. Department of Health and Human Services, Food and Drug Administration, Center for Drug Evaluation and Research (CDER), July 2002. http://www.fda.gov/downloads/Drugs/GuidanceComplianceRegulatoryInformation/Guidances/ucm070575.pdf. Accessed 27 Jan 2011
53. Guidance for Industry – Container Closure Systems for Packaging Human Drugs and Biologics – Chemistry, Manufacturing, and Control Documentation: U.S. Department of Health and Human Services Food and Drug Administration, Center for Drug Evaluation and Research (CDER), Center for Biologics Evaluation and Research (CBER), May 1999. http://www.fda.gov/downloads/Drugs/GuidanceComplianceRegulatoryInformation/Guidances/ucm070551.pdf. Accessed 27 Jan 2011

54. Norwood D, Paskiet D, Ruberto M, Feinberg T, Schroeder A, Poochikian G, Wang Q, Deng T, DeGrazio F, Munos M, Nagao L (2008) Best practices for extractables and leachables in orally inhaled and nasal drug products: an overview of the PQRI recommendations. Pharm Res 25(4):727–735
55. Ball D, Blanchard J, Jacobson-Kram D, McClellan RO, McGovern T, Norwood DL, Vogel WM, Wolff R, Nagao L (2007) Development of safety qualification thresholds and their use in orally inhaled and nasal drug product evaluation. Toxicol Sci 97:226–236
56. Guidance for Industry Q1A(R2) Stability Testing of New Drug Substances and Products U.S. Department of Health and Human Services, Food and Drug Administration, Center for Drug Evaluation and Research, Center for Biologics Evaluation and Research, November 2003. http://www.fda.gov/downloads/Drugs/GuidanceComplianceRegulatoryInformation/Guidances/ucm073369.pdf. Accessed 27 Jan 2011
57. Dalby R, Byron P (2009) In vitro assessment of inhaled products. In: Nemann S, Anderson P, Byron P, Dalby R, Peart J (eds) Respiratory drug delivery: essential theory and practice. Respiratory Drug Delivery Online/VCU, Richmond, pp 59–96
58. Singh GJP (2010) Locally acting nasal and inhalation drug products: regulatory and bioequivalence perspective. In: Shargel L, Kanifer I (eds) Generic drug development: special dosage forms. Informa Health Care, New York, pp 189–217
59. Washington C (1998) Particle size analysis in inhalation therapy. In: Purewal TS, Grant DW (eds) Metered dose inhalation technology. CRC Press, Boca Raton, pp 117–145
60. Kippax P, Krarup G, Suman JD (2004) Application of droplet sizing: Manual versus automated actuation of nasal sprays. Pharmaceut Technol (Outsourcing Resources), August 2004 30–39
61. Ranucci J (1992) Dynamic plume-particle size analysis using laser diffraction. Pharm Tech 16:108–114
62. Hollingworth GW (1993) Particle size analysis of therapeutic aerosols. In: Moren F, Dolovich MB, Newhouse MT, Newman SP (eds) Aerosols in medicine. principles, diagnosis and therapy. Elsevier, Amsterdam, pp 351–374
63. Kippax P (2005) Measuring particle size by laser diffraction techniques. Paint & Coating Indus. August 2005
64. Cheng YS, Holmes BS, Gao BS, Guilmete RA, Li S, Surakitbanharn Y, Rowlings C (2001) Characterization of nasal spray pumps and deposition patterns in replica of the human nasal airway. J Aerosol Med 14:267–280
65. Cummings RH, Riebe M (1998) Analytical Development and Testing. In: Purewal TS, Grant DW (eds) Metered dose inhaler technology. CRC Press, Boca Raton, pp 147–185
66. Guidance for Industry (Draft): Bioavailability and bioequivalence studies for nasal aerosols and nasal sprays for local action (2003) U.S. Department of Health and Human Services, Food and Drug Administration, Center for Drug Evaluation and Research (CDER). http://www.fda.gov/downloads/Drugs/GuidanceComplianceRegulatoryInformation/Guidances/ucm070111.pdf. Accessed 27 Jan 2011
67. DeGeorge JJ, Ahn CH, Andrews PA, Brower ME, Choi YS, Chun MY, Du T, Lee-Ham DY, McGuinn WD, Pei L, Sancilio LF, Schmidt W, Sheevers HV, Sun CJ, Tripathi S, Vogel WM, Whitehurst V, Williams S, Taylor AS (1997) Considerations for toxicology studies of respiratory drug products. Regul Toxicol Pharmacol 25:189–193
68. Wolff RK, Dorato MA (1993) Toxicologic testing of inhaled pharmaceutical aerosols. Crit Rev Toxicol 23:343–369
69. Casarosa P, Bouyssou T, Germeyer S, Schnapp A, Gantner F, Pieper M (2009) Preclinical evaluation of long-acting muscarinic antagonists: comparison of tiotropium and investigational drugs. J Pharmacol Exp Ther 30:660–668
70. Pauluhn J (2008) Inhalation toxicology: methodological and regulatory challenges. Exp Toxicol Pathol 60:111–124
71. Owen K, Beck SL, Damment SJP (2010) The preclinical toxicology of salmeterol hydroxynaphthoate. Hum Exp Toxicol 29:393–407

72. Long RE (2004) Role of toxicology and Toxicokinetics in drug development. In: Bonate PL, Howard DR (eds) Pharmacokinetics in drug development: regulatory and development paradigms. AAPS, Arlington, pp 87–103
73. Derendorf H, Lesko LJ, Chaikin P, Colburn WA, Lee P, Miller R, Powel R, Rhodes G, Stanski D, Venitz J (2000) Pharmacokinetic/pharmacodynamic modeling in drug research and development. J Clin Pharmacol 40:1399–1418
74. Ette EI, Garg V, Jayaraj A (2004) Drug development: a rational approach. In: Bonate PL, Howard DR (eds) Pharmacokinetics in drug development: regulatory and development paradigms. AAPS, Arlington, pp 3–36
75. Guidance for Industry: Formal Meetings With Sponsors and Applicants for PDUFA Products (2000) U.S. Department of Health and Human Services, Food and Drug Administration, Center for Drug Evaluation and Research, Center for Biologics Evaluation and Research. http://www.fda.gov/downloads/Drugs/GuidanceComplianceRegulatoryInformation/Guidances/UCM079744.pdf. Accessed 27 Jan 2011
76. Ette EI, Williams P, Sun H, Fardiran E, Ajayi FO, Onyiah LC (2001) The process of knowledge discovery from large pharmacokinetic data sets. J Clin Pharmacol 41:25–34
77. Williams P, Ette EI (2000) The role of population pharmacokinetics in drug development in the light of the FDA guidance on population pharmacokinetics. Clin Pharmacokinet 39:185–395
78. Chowdhury BA (2007) Regulatory demands of clinical testing protocols for powder inhalers in the USA. In: Dalby RN, Byron P, Peart J, Suman JD (eds) Respiratory drug delivery Europe 2007. Davis Healthcare, River Grove, pp 101–108
79. NHLBI Guidelines for the Diagnosis and Management of Asthma (2007). http://www.nhlbi.nih.gov/guidelines/asthma/asthgdln.htm. Accessed 20 Jan 2011
80. American Thoracic Society (2000) Guidelines for methacholine and exercise challenge testing –1999. Am J Respir Crit Care Med 161:309–329
81. Wong BJ, Hargreave FE (1993) Bioequivalence of metered-dose inhaled medications. J Allergy Clin Immunol 92:373–379
82. Clark DJ, Lipworth BJ (1997) Dose-response of inhaled drugs in asthma: an update. Clin Pharmacokinet 32:58–74
83. Barnes PJ, Pedersen S, Busse WW (1998) Efficacy and safety of inhaled corticosteroids: new developments. Am J Respir Crit Care Med 157:S1–S53
84. Parameswaran K, Leigh R, O'Byrne PM, Kelly MM, Goldsmith CH, Hargreave FE, Dolovich M (2003) Clinical models to compare the safety and efficacy of inhaled corticosteroids in patients with asthma. Can Respir J 10:27–34
85. ATS/ERS COPD Guidelines (2004). http://www.thoracic.org/clinical/copd-guidelines/resources/copddoc.pdf. Accessed 27 Jan 2011
86. Guidance for Industry: Chronic Obstructive Pulmonary Disease: Developing Drugs for Treatment. November 2007. Center for Drug Evaluation and Research, US FDA. http://www.fda.gov/downloads/Drugs/GuidanceComplianceRegulatoryInformation/Guidances/ucm071575.pdf. Accessed 20 Jan 2011
87. Fishman AP (1988) The spectrum of chronic obstructive disease of the airways. In: Fishman AP (ed) Pulmonary diseases and disorders, vol 2, 2nd edn. McGraw-Hill, New York, pp 1159–1171
88. Reid LM (1988) Chronic obstructive pulmonary disease. In: Fishman AP (ed) Pulmonary diseases and disorders, vol 2, 2nd edn. McGraw-Hill, New York, pp 1247–1272
89. Jones R, Ostrem A (2011) Optimizing pharmacological maintenance treatment for chronic obstructive pulmonary disease in primary care. Prim Care Respir J 20:33–45
90. Grippi MA, Metzger LF, Krupinski AV, Fishman AP (1988) Pulmonary function testing. In: Fishman AP (ed) Pulmonary diseases and disorders, vol 3, 2nd edn. McGraw-Hill, New York, pp 2469–2532
91. Morgan WKC, Pearson MG (1988) Pulmonary function tests in determining disability. In: Fishman AP (ed) Pulmonary diseases and disorders, vol 3, 2nd edn. McGraw-Hill, New York, pp 2533–2541

92. Hochhaus G, Schmidt E-W, Rominger KL, Mollman H (1992) Pharmacokinetic/dynamic correlation of pulmonary and cardiac effects of fenotrol in asthmatic patients after different routes of administration. Pharm Res 9:291–297
93. Kemsford R, Handel M, Mehta R, De Silva M, Daley-Yates P (2005) Comparison of the systemic pharmacodynamic effects and pharmacokinetics of salmeterol delivered by CFC propellant and non-CFC propellant metered dose inhalers in healthy subjects. Respir Med 99(Suppl A):S11–S19
94. Rosekranz B, Rouzier R, Kruse M, Dobson C, Ayre G, Horowitz A, Fitoussi S (2006) Safety and tolerability of high-dose of formoterol (via Aerosolizer®) and salbutamol in patients with chronic obstructive pulmonary disease. Respir Med 100:666–672
95. Singh S, Loke YK, Furberg CD (2008) Inhaled anticholinergics and risk of major adverse cardiovascular events in patients with chronic obstructive pulmonary disease: a systematic review and meta-analysis. JAMA 300(12):1439–1450
96. Salpeter SR (2009) Do inhaled anticholinergics increase or decrease the risk of major cardiovascular events? A synthesis of the available evidence. Drugs 69:2025–2033
97. Wilson AM, McFarlane LC, Lipworth BJ (1998) Effects of repeated once daily dosing of three intranasal corticosteroids on basal and dynamic measures of hypothalmic-pituitary-adrenal-axis activity. J Allergy Clin Immunol 101:470–474
98. Fardon TC, Lee DKC, Haggart K, McFarlane LC, Lipworth BJ (2004) Adrenal suppression with dry powder formulations of fluticasone propionate and mometasone furoate. Am J Respir Crit Care Med 170:760–766
99. Allen DB (2002) Safety of inhaled corticosteroids in children. Pediatr Pulmonol 33:208–220
100. Todd GRG (2003) Adrenal crisis due to inhaled steroids is underestimated. Arch Dis Child 88:554–555
101. Wolthers OD, Pedersen S (1996) Measures of systemic activity of inhaled corticosteroids in children: a comparison of urine cortisol excretion and knemometry. Respir Med 89:347–349
102. Lipworth BJ, Seckl JR (1997) Measures for detecting systemic bioactivity with inhaled and intranasal corticosteroids. Thorax 52:476–482
103. Wohl ME, Majzoub JA (2000) Asthma, steroids, and growth. New Eng J Med 343: 1113–1114
104. Wolthers OD (2000) Impact of inhaled and intranasal corticosteroids on the growth of children. BioDrugs 13:347–357
105. Altintas DU, Karakoc GB, Can S, Yilmaz M, Kendirli SG (2005) The effects of long term use of inhaled corticosteroids on linear growth, adrenal function and bone mineral density in children. Allergol Immunopathol (Madr) 33:204–209
106. Halpern MT, Schmier JK, Van Kerkhove MD, Watkins M, Kalberg CJ (2004) Impact of long-term inhaled corticosteroid therapy on bone mineral density: results of a meta-analysis. Ann Allergy Asthma Immunol 92:201–207
107. Richy F, Bousquet J, Ehrlich GE, Meunier PJ, Israel E, Morii H, Devogelaer JP, Peel N, Haim M, Bruyere O, Reginster JY (2003) Inhaled corticosteroids effects on bone in asthmatic and COPD patients: a quantitative systematic review. Osteoporos Int 14(3):179–190
108. Advair® HFA Product Label (2010) http://www.accessdata.fda.gov/drugsatfda_docs/label/2010/021254s013lbl.pdf. Accessed 23 Jan 2010
109. Symbicort® Inhalation Aerosol Product Label (2010) http://www.accessdata.fda.gov/drugsatfda_docs/label/2010/021929s021lbl.pdf. Accessed 23 Jan 2011
110. Dulera® Product Label (2010) http://www.accessdata.fda.gov/drugsatfda_docs/label/2010/022518s000lbl.pdf. Accessed 23 Jan 2011
111. Combivent® Inhaltion Aerosol Product Label (2009) http://www.accessdata.fda.gov/drugsatfda_docs/label/2009/020291s027lbl.pdf. Accessed 23 Jan 2011
112. CDER Data Standard Manual, Dosage Forms, U.S. Department of Health and Human Services, Food and Drug Administration, Center for Drug Evaluation and Research, last updated 4/30/09. http://www.fda.gov/Drugs/DevelopmentApprovalProcess/FormsSubmissionRequirements/ElectronicSubmissions/DataStandardsManualmonographs/ucm071666.htm
113. The United States Pharmacopeial Convention, USP 33 <1151>

114. Adi H, Young PM, Chan HK, Salama R, Traini D (2010) Controlled release antibiotics for dry powder lung delivery. Drug Dev Ind Pharm 36;119–126
115. Learoyd TP, Burrows JL, French E, Seville PC (2010) Sustained delivery of salbutamol and beclomethasone from spray-dried double emulsions. J Microencap 27:1–9
116. Learoyd TP, Burrows JL, French E, Seville PC (2008) Chitosan-based spray-dried respirable powders for sustained delivery of terbutaline sulfate. Eur J Pharm Biopharm 68:224–234
117. Salama R, Susan H, Chan HK, Traini D, Young PM (2008) Preparation and characterization of controlled release co-spray dried drug-polymer microparticles for inhalation 1: influence of polymer concentration on physical and in vitro characteristics. Eur J Pharm Biopharm 69:486–495
118. Salama R, Traini D, Chan HK, Young PM (2008) Preparation and characterization of controlled release co-spray dried drug-polymer microparticles for inhalation 2: evaluation of invitro release profiling methodologies for controlled release respiratory aerosols. Eur J Pharm Biopharm 70:145–152
119. Uppoor RS, Marroum PJ (2004) Regulatory considerations for extended release dosage forms and in vitro (dissolution)/in vivo (bioavailability) correlations. In: Sahajwall CG (ed) New drug development: regulatory paradigms for clinical pharmacology and biopharmaceutics. Marcel Dekker, New York, pp 417–448
120. Foucher P, Biour M, Blayac JP, Godard P, Sgro C, Kuhn M, Vergnon JM, Vervloet D, Pfitzenmeyer P, Ollagnier M, Mayaud C, Camus P (1997) Drugs that may injure the respiratory system. Eur Respir J 10:265–279
121. Rossi SE, Erasmus JJ, McAdams HP, Sporn TA, Goodman PC (2000) Pulmonary drug toxicity: radiologic and pathologic manifestations. Radiographics 20:1245–1259
122. Erasmus JJ, McAdams HP, Rossi SE (2002) Drug-induced lung injury. Semin Roentgenol 37:72–81
123. Kim BY, Rutka JT, Chan WC (2010) Nanomedicine. N Engl J Med 363:2434–2443
124. Singh S (2010) Nanomedicine-nanoscale drugs and delivery systems. J Nanosci Nanotechnol 10:7906–7918
125. Pautler M, Brenner S (2010) Nanomedicine: promises and challenges for the future of public health. Int J Nanomed 5:803–809
126. Rogueda PG, Traini D (2007) The nanoscale in pulmonary delivery. Part 1: deposition, fate, toxicology and effects. Expert Opin Drug Deliv 4:595–606
127. Rogueda PG, Traini D (2007) The nanoscale in pulmonary delivery. Part 2: formulation platforms. Expert Opin Drug Deliv 4(6):607–620
128. Azarmi S, Roa WH, Löbenberg R (2008) Targeted delivery of nanoparticles for the treatment of lung diseases. Adv Drug Deliv Rev 260:863–875
129. Bailey MM, Berkland CJ (2009) Nanoparticle formulations in pulmonary drug delivery. Med Res Rev 29(1):196–212
130. Mansour HM, Rhee YS, Wu X (2009) Nanomedicine in pulmonary delivery. Int J Nanomed 4:299–319
131. Bur M, Henning A, Hein S, Schneider M, Lehr CM (2009) Inhalative nanomedicine – opportunities and challenges. Inhal Toxicol 21(Suppl 1):137–143
132. Edwards DA, Hanes J, Caponetti G, Hrkach J, Ben-Jebria A, Eskew ML, Mintzes J, Deaver D, Lotan N, Langer R (1997) Large porous particles for pulmonary drug delivery. Science 276:1868–1871
133. Card JW, Zeldin DC, Bonner JC, Nestmann ER (2008) Pulmonary applications and toxicity of engineered nanoparticles. Am J Physiol Lung Cell Mol Physiol 29:L400–L411
134. Bonner JC (2010) Nanoparticles as a potential cause of pleural and interstitial lung disease. Proc Am Thorac Soc 7(2):138–141
135. Harush-Frenkel O, Bivas-Benita M, Nassar T, Springer C, Sherman Y, Avital A, Altschuler Y, Borlak J, Benita S (2010) A safety and tolerability study of differently-charged nanoparticles for local pulmonary drug delivery. Toxicol Appl Pharmacol 246:83–90
136. Yang W, Tam J, Miller DA, Zhou J, McConville JT, Johnston KP, Williams RO (2008) High bioavailability from nebulized itraconazole nanoparticle dispersions with biocompatible stabilizers. Int J Pharm 361:177–188

137. Cai Z, Wang Y, Zhu LJ, Liu ZQ (2010) Nanocarriers: a general strategy for enhancement of oral bioavailability of poorly absorbed or pre-systemically metabolized drugs. Curr Drug Metab 11(2):197–207
138. Peng Q, Gong T, Zuo J, Liu J, Zhao D, Zhang Z (2008) Enhanced oral bioavailability of salvianolic acid B by phospholipid complex loaded nanoparticles. Pharmazie 63:661–666
139. Guidance for Industry: Topical Dermatologic Corticosteroids – In Vivo Bioequivalence (1995) Center for Drug Evaluation and Research, US FDA, Rockville. http://www.fda.gov/downloads/Drugs/GuidanceComplianceRegulatoryInformation/Guidances/ucm070234.pdf. Accessed January 26, 2011
140. Lee SL, Adams WP, Li BV, Conner DP, Chowdhury BA, Yu LX (2009) In vitro considerations to support bioequivalence of locally acting drugs in dry powder inhalers for lung diseases. AAPS J 11:414–423
141. Singh GJP, Adams WP (2005) US regulatory and scientific considerations for approval of generic locally acting orally inhaled and nasal drug products. In: Dalby RN, Byron P, Peart J, Suman JD (eds) Respiratory drug delivery – Europe 2005: a regulatory and analytical symposium. Davis Healthcare, River Grove, pp 115–125
142. Guidance for Industry: Bioavailability and bioequivalence studies for orally administered drug products – General considerations (2003) U.S. Department of Health and Human Services, Food and Drug Administration, Center for Drug Evaluation and Research. http://www.fda.gov/downloads/Drugs/GuidanceComplianceRegulatoryInformation/Guidances/ucm070124.pdf. Accessed 27 Jan 2011
143. Interim guidance for documentation of in vivo bioequivalence of albuterol inhalation aerosols (metered dose inhalers) (1994) Office of Generic Drugs, US FDA, Rockville (withdrawn)
144. Singh GJP (2010) Evolution of regulatory and scientific paradigms for establishing equivalence of systemic exposure from orally inhaled drugs: Current status and possible challenges. In: Dalby RN, Byron PR, Peart J, Suman JD, Farr SJ, Young PM (eds) Respiratory drug delivery. Davis Healthcare, River Grove, pp 249–260
145. Guidance for Industry: Bioanalytical method validation (2001) U.S. Department of Health and Human Services Food and Drug Administration, Center for Drug Evaluation and Research (CDER), Center for Veterinary Medicine (CVM). http://www.fda.gov/downloads/Drugs/GuidanceComplianceRegulatoryInformation/Guidances/ucm070107.pdf. Accessed 27 Jan 2011
146. Guidance for Industry: Statistical approaches to establishing bioequivalence (2001). U.S. Department of Health and Human Services Food and Drug Administration, Center for Drug Evaluation and Research. http://www.fda.gov/downloads/Drugs/GuidanceComplianceRegulatoryInformation/Guidances/ucm070244.pdf. Accessed 27 Jan 2011
147. Patnaik RN (2004) Bioequivalence assessment: Approaches, designs and statistical considerations. In: Sahajwala CG (ed) New drug development: regulatory paradigms for clinical pharmacology and biopharmaceutics. Marcel Dekker, New York, pp 561–586
148. Ahrens RC (2001) On comparing inhaled beta adrenergic agonists. Ann Allergy 67:296–298
149. Parameswaran K (1999) Concepts of establishing clinical bioequivalence of chlorofluorocarbon and hydrofluoroalkane [beta]-agonists. J Allergy Clin Immunol 104:S243–S245
150. Parameswaran K, Inman MD, Elkholm BP, Morris MM, Summers E, O'Byrne PM, Hargreave FE (1999) Protection against methacholine bronchospasm to assess relative potency of inhaled $ß_2$-agonist. Am J Respir Crit Care Med 160:354–357
151. Creticos PS, Adams WP, Petty BG, Lewis LD, Singh GJP, Khattignavong AP, Molzon JA, Martinez MN, Lietman PS, Williams RL (2002) A methacholine challenge dose-response study for development of pharmacodynamic bioequivalence methodology for albuterol metered dose inhalers. J Allergy Clin Immunol 110:713–720
152. Stewart BA, Ahrens RC, Carrier S, Frosolono M, Lux CR, Han SH, Milavetz G (2000) Demonstration of in vivo bioequivalence of a generic albuterol metered-dose inhaler to Ventolin. Chest 117:714–721
153. Gross NJ, Petty TL, Friedman M, Skorodin MS, Silvers GW, Donohue JF (1989) Dose response to iprtropium as a nebulized solution in patients with chronic obstructive pulmonary disease: a three center study. Am Rev Respir Dis 139:1188–1191

154. Maesen FP, Smeets JJ, Costongs MA, Wald FD, Cornelissen PJ (1993) Ba 679 Br, a new long-acting antimuscarinic bronchodilator: a pilot dose response study in COPD. Eur Respir J 6:1031–1036
155. Maesen FP, Smeets JJ, Sledsens TJ, Wald FD, Cornelissen PJ (1995) Tiotropium bromide, a new long-acting antimuscarinic bronchodilator: a pharmacodynamic study in patients with chronic obstructive pulmonary disease (COPD). Eur Respir J 8:1506–1513
156. Barnes PJ (2004) The role of anticholinergics in chronic obstructive pulmonary disease. Am J Med 117(Suppl 12A):24S–32S
157. Gross NJ (2006) Anticholinergic agents in asthma and COPD. Eur J Pharmacol 533:36–39
158. Weder MM, Donohue JF (2005) Role of bronchodilators in chronic obstructive pulmonary disease. Semin Respir Crit Care Med 26:221–234
159. Terzano C, Petroianni A, Ricci A, D'Antoni L, Allegra L (2004) Early protective effects of tiotropium bromide in patients with airway hyperresponsiveness. Eur Rev Med Pharmacol Sci 8:259–264
160. Sposato B, Mariotta S, Ricci A, Bruno P, Terzano C, Mannino F (2005) The influence of ipratropium bromide in the recovery phase of methacholine-induced bronchospasm. Eur Rev Med Pharmacol Sci 9:117–123
161. O'Connor BJ, Towse LJ, Barnes PJ (1996) Prolonged effect of tiotropium bromide on methacholine-induced bronchoconstriction in asthma. Am J Respir Crit Care Med 154:876–880
162. Ahrens RC, Teresi ME, Han S-H, Donne D, Vanden Burgt JA, Lux CR (2001) Asthma stability after oral prednisone – a clinical model for comparing inhaled corticosteroid potency. Am J Respir Crit Care Med 164:1138–1145
163. Ahrens RC, Hendeles L, Teresi ME, Lux CR, Vanden Burgt J, Hart KA, Ekholm BP (2003) Relative potency of beclomethasone dipropionate (BDP) delivered by HFA-MDI, and fluticasone propionate (FP) delivered by Diskus. Eur Respir J 22(Suppl 45):P1576 (Abstract)
164. Kelly MM, Leigh R, Jayaram L, Goldsmith CH, Parameswaran K, Hargreave FE (2006) Eosinophilic bronchitis in asthma: a model for establishing dose-response and relative potency of inhaled corticosteroids. J Allergy Clin Immunol 117:989–994
165. Chowdhury BA (2010) Regulatory uncertainties in bioequivalence: exhaled nitric oxide as a possible efficacy endpoint for inhaled corticosteroids. In: Dalby RN, Byron PR, Peart J, Suman JD, Farr SJ, Young PM (eds) Respiratory drug delivery 2010. Davis Healthcare, River Grove, pp 317–322

Chapter 22
Developing Performance Specifications for Pulmonary Products

Stephen T. Horhota and Stefan Leiner

Abstract In comparison to many other pharmaceutical dosage forms, the regulatory picture for quality aspects of pulmonary products is still undergoing rapid and dynamic evolution. The purpose of this chapter is to examine aspects of product testing, specifications, and their accompanying regulatory processes which on one hand might also be considered as yet further barriers by those seeking to bring new inhalation treatments into fruition while oppositely embraced by others as a way to preserve knowledge and prevent a repetition of harmful errors. These are discussed in reference to the divergent international views on how to manage the integration of drug product and device requirements.

Keywords Inhalation products • International regulatory divergence • Pulmonary products • Regulations • Risk management • Specifications

22.1 Introduction

As the editors remarked in their preface to this book, there is great excitement about the potential for improved therapies via the inhalation route. However, whether for immediate acting or controlled release delivery systems, that potential can only be realized when basic scientific understandings from physiology, medicine, toxicology, and properties of materials are converted into practical technologies that can be reliably mass produced and placed into the hands of patients or therapeutic specialists. The preceding chapters have described some of the formidable scientific and technological barriers that must be overcome in

S.T. Horhota (✉)
Boehringer Ingelheim Pharmaceuticals, Inc., 900 Ridgebury Road,
Box 368, Ridgefield, CT 06877, USA
e-mail: Steve.horhota@boehringer-ingelheim.com

H.D.C. Smyth and A.J. Hickey (eds.), *Controlled Pulmonary Drug Delivery*, Advances in Delivery Science and Technology, DOI 10.1007/978-1-4419-9745-6_22,

order to achieve this goal. The purpose of this chapter is to examine aspects of product testing, specifications, and their accompanying regulatory processes which on one hand might also be considered as yet further barriers by those seeking to bring new inhalation treatments into fruition while oppositely embraced by others as a way to preserve knowledge and prevent a repetition of harmful errors.

An important question to ask is "Why do we need to have specifications at all when medical/clinical trial data have proven that the product works in the target population and has been demonstrated to have a favorable benefit to risk margin?" The authors believe that an answer to the question lies in understanding the difference between "science" and "technology." For this discussion, we define "science" to be that work which elucidates fundamental relationships of matter and energy as they interact in dimension (mass, size, temperature, etc.) and time. "Technology" is the manipulation of that scientific insight to achieve a particular state or behavior for certain combinations of matter, energy, dimension, and time while simultaneously excluding other unfavorable states. More simply, the essential difference between the two terms is that fundamental science never fails but technology can fail and may fail in quite dramatic and consequential ways. It is the purpose of specifications to serve as risk management tools to minimize or prevent technology failures. For the manufacturer, they are an important means of insuring that product design features and consumer expectations are consistently fulfilled. For regulatory agencies, specifications are a key component of their mandate of protecting and improving public health. Additionally, openly defined specifications serve the broader purpose of institutionalizing knowledge and experience from past failure situations that had serious consequences for which a cause has been elucidated and the means to prevent future occurrences exist. Therefore, in coming to an understanding of the current situation regarding specifications for inhalation aerosols and the regulatory systems behind them, it is useful to review some of the significant contributing events and experiences.

For inhalation products, the key contributors to the present regulatory situation fall into two broad categories, namely (a) the legacy originating from the early dominance of pressurized metered dose inhalers (pMDIs) and their governance as traditional drug products, and (b) divergent international views on how to manage the integration of drug product and device requirements.

22.2 The Legacy of pMDIs

pMDIs were, from their first appearance in the 1950s nearly universally classified as pharmaceutical products despite the complexity of device technology embedded in their function. The judgment at that period of time, still somewhat applicable today, was that the drug being delivered was the principle element, and the delivery technology (e.g., canister, metering valve, and actuator) was merely enabling. Medical device regulation was still immature in the late 1950s and early 1960s making it more logical to view pMDIs under already existing regulatory schemes for approval and marketing authorizations of pharmaceutical items. As a pharmaceutical item, product quality for pMDIs could also be conveniently addressed through the mechanism of pharmacopeial standards and specifications which historically had proven themselves to be adequate technology controls across a wide variety of pharmaceutical product types. Therefore, managing orally inhaled preparations as drug products became the common practice in the major regulatory regions internationally, and product requirements were relatively homogeneous throughout the world since pMDIs were the predominant inhalation technology.

Beginning in the 1980s this regulatory model for oral inhaled products based on pMDI experience was confronted by a number of influential situations requiring regulatory responses that, in turn, evolved into a complex array of regulations that currently exist. Four factors are particularly significant:

1. *An increase in the size of the patient population using orally inhaled drug products (OIP)*
   As the incidence of respiratory diseases worldwide has risen [1], and greater access to therapies has increased the number of patients using OIPs, so have Chemistry, Manufacturing, and Control (CMC) requirements. This can be interpreted to be evidence that regulators believe existing standards afford insufficient risk protection, especially regarding preventable product failures such as insufficient or inconsistent dose delivery due to actuator/valve misfunction, moisture-induced growth in particle size that interferes with pulmonary deposition, the need for priming and re-priming studies [2], or unrecognized depletion of medication or propellant, particularly in a rescue situation [3].
2. *The impact of the Montreal Protocol*
   In the late 1980s the advent of the Montreal Protocol precipitated the phase-out of chlorofluorocarbons (CFC), the major propellants first used in pMDIs on account of their favorable vapor pressure profiles, compatibility with a variety of drug substances, and well-established toxicological safety. Initially viewed as a simple technological substitution with the preferred replacement of hydrofluoroalkanes (HFA), many pharmaceutical development programs turned into unexpectedly costly and lengthy efforts to reformulate existing products. Solutions to the new challenges required considerable basic and applied research. As a consequence of this work, challenges of many long-held assumptions about performance and relevance of in vitro methods and specifications surfaced. The Montreal Protocol also accelerated innovation and renewed interest in alternative and novel inhalation delivery systems such as dry powder inhalers (DPIs) and soft-mist inhalers (SMIs) that avoided the use of propellants. These systems some of which incorporate digital technologies, active power supplies, or dynamic feedback loops did not have pharmaceutical precedents and revealed inadequacies in many of the ideas about controlling product quality as traditional pharmaceutical items.
3. *Use of OIP for systemic drug delivery*
   In the 1990s there emerged increasing consideration and use of the pulmonary portal for systemic delivery as a way to circumvent first pass metabolism or obtain intravenous like pharmacokinetics without pain during administration. However, unlike the majority of current OIP, these agents would be used primarily in patients who did not have disease compromised airways where changes in lung function arising from the acute or chronic use of such products are clearly

undesirable. The drugs contained in some of these OIP are, in many cases, large molecule chemical entities or biologics and not the traditional small molecule variety meaning that certain developmental knowledge gained over years from small molecule examples may not apply.

4. *Increased demand for generic OIP*
   To combat rising healthcare costs, many governments have resorted to actively encouraging generic competition in all classes of pharmaceutical products as patent exclusivity expires for innovators. The approval of generic products relies heavily on in vitro testing and in vivo bioequivalence comparisons between the innovator and its generic replicate to validate the ability to freely substitute one product for the other. In the case of orally administered drugs, well-developed in vitro–in vivo correlations (IVIVC) and the consolidation of research experience in the form of the Biopharmaceutics Classification System (BCS) [4] have demonstrated themselves to be suitable risk management tools that allow for such heavy reliance on these clinical surrogates. Unfortunately, the greater complexity of OIP systems and the lack of strong IVIVC models prevent adoption of a similar approach. The perceived higher risk is felt to require mitigation through enhanced testing and characterization [5–8].

While the four factors in the development of OIPs occurred more or less at the same time in all major regions of the globe, different approaches have been taken by regional or national bodies to react to the four events. The authors believe that differences in the perceptions of risks, differences in the occurrence of major failures with impact on the public health, and differences in the assessment of such risks as well as the interaction between agencies, manufacturers, and patients played a major role in the diversification process. On the next pages, the authors illustrate the regional differences in more detail.

22.3 Divergent International Views on How to Manage the Integration of Drug Product and Device Requirements

OIPs have a dual identity problem of being pharmaceutical products and at the same time medical devices. As mentioned earlier, the traditional approach was to focus on the pharmaceutical aspects and look at the device elements as being variants of formulation excipients or packaging materials. As the device portion has become increasingly more sophisticated, this approach has come under scrutiny by various industry standards bodies and regulatory authorities [9, 10]. However, the manner in which they have responded has led to a very confusing landscape of regulatory processes. A very illustrative example is the comparison of the EU and US systems for addressing CMC aspects of inhalation products.

The CMC framework in the EU represents a distributed, consensus approach with industry, regulatory agencies, and academia all recognized in some fashion as stakeholders and risk managers. Strong emphasis is given to adherence to public standards, directives, and guidances. In the European Union, when drug and device elements are physically separable and easily identified, the Medical Device Directives apply for assessing the delivery mechanism, while the relevant pharmacopeial and/or medicines standards apply to the drug portion. Confusingly, where the drug and device are inseparable, as in the case of pMDIs, only pharmacopeial and/or medicines standards [11] apply, with no consideration whatsoever given to device regulations.

Taking the opposite approach, the FDA takes on a more dominant, central approach. This US Agency continues to classify products based on a determination of which element, the drug or the device, contributes more to the treatment effect. It therefore considers inhalation products containing any pharmaceutical agent as drug products subject to review and approval primarily by the Center for Drug Evaluation and Research (CDER). Both the drug and device components are audited under the traditional pharmaceutical model, which favors end product testing rather than upstream control strategies more common to the medical device industry. Confusingly, under US regulations, a manufacturer can apply to have their product classified as a "combination product" in which review by the Center for Device and Radiological Health under Medical Device standards is possible.

This fragmentation in the regulatory landscape was recently surveyed by working groups within the International Pharmaceutical Aerosol Consortium on Regulation and Science (IPAC-RS) who identified more than 100 documents, guidances, standards, specifications, and regulations governing orally inhaled drug products in the world (Svetlana Lyapustina, 2010, IPAC-RS, private communication). Some of the major bodies contributing these specifications and regulations are given in Table 22.1.

In summary, specifications and regulatory processes for inhalation products present a complex regulatory environment internationally. There is a significant blurring between drug and device requirements coupled with rapidly advancing science and technology in the area of inhalation delivery. It is reasonable to assume that there will be some significant changes to this landscape in the coming years which poses quite a problem as we now come to the details of specifications to be used for controlled release inhalation systems.

First and foremost, there are few, if any, such products commercially available today which can be examined as suitable precedents or case examples. The dilemma therefore becomes "What principles can be followed now that reasonably anticipate requirements for a controlled release inhalation product when finally coming to the stage of seeking governmental approval?" The authors believe that the problem can be dissected into key elements which provide a general framework that can be adapted to a wide variety of inhalation delivery concepts.

The first element deals with the definition of a controlled release inhalation product. Extending recent work from the IPAC-RS Risk Management Working Group [12], a controlled release OIP can be defined as possessing the following features:

1. A delivery system containing a substance or mixture of substances intended to furnish a pharmacological action or other direct physiological effect in humans.
2. A defined amount of the substance(s) is (are) dispersed into an aerosol form (i.e., a gas borne suspension of solid or liquid particles) by the system.

Table 22.1 Organizations and agencies that define or influence specifications and standards for inhalation products

Region	Body or organization	Selected examples of regulation or guideline applicable to inhalation products
European Union	EMA (European Medicines Agency)	Guideline on the Pharmaceutical Quality of Inhalation and Nasal Products (EMEA/CHMP/QWP/49313/2005 Corr), June 2006 (joint Canadian/European Guideline)
		Guideline on the Specification Limits for Residues of Metal Catalysts or Metal Reagents (EMEA/CHMP/SWP/4446/2000, Feb 2008)
	European Commission: Medical Device Directives	Council Directive 93/42/EEC of 14 June 1993 concerning Medical Devices
United States	CBER (Center for Biologics Evaluation and Research)	Guidance for the Content of Premarket Submissions for Software Contained in Medical Devices, May 2005
	CDER (Center for Drug Evaluation and Research)	Draft Guidance for Industry: Metered Dose Inhaler (MDI) and Dry Powder Inhaler (DPI) Drug Products: Chemistry, Manufacturing, and Controls Documentation, Oct 1998
		Guidance for Industry: Nasal Spray and Inhalation Solution, Suspension, and Spray Drug Products – Chemistry, Manufacturing, and Controls Documentation, July 2002
	CDRH (Center for Devices and Radiological Health)	Design Control Guidance For Medical Device Manufacturers, March 1997
		Reviewer guidance for Nebulizers, Metered Dose Inhalers, Spacers, and Actuators, Oct 1993
Canada	Health Canada	Spacers and Holding Chambers for use with Metered-Dose Inhalers, Updates 1 and 2, 2008 (document Z264.1-02)
Japan	MHLW (Ministry of Health, Labor and Welfare)	No specific guideline issued
International (Standards Setting Bodies)	International Organization for Standardization (ISO)	ISO Standard 20072: Aerosol drug delivery device design verification – Requirements and test methods
		ISO Standard 27427: Anesthetic and respiratory equipment – Nebulizing systems and components
	Pharmacopoeias USP	General chapter <601> Aerosols, Nasal Sprays, Metered-Dose Inhalers, and Dry Powder Inhalers
		Monograph on Epinephrine Bitartrate Inhalation Aerosol
	Ph. Eur.	Monograph 0671 "Preparations for Inhalation"
	JP	No specific guideline issued

(continued)

Table 22.1 (continued)

Region	Body or organization	Selected examples of regulation or guideline applicable to inhalation products
International (Regulatory/ Industry Harmonization Initiatives)	GHTF (Global Harmonization Task Force)	Summary Technical Documentation for Demonstrating Conformity to the Essential Principles of Safety and Performance of Medical Devices (STED), Feb 2008, document no. GHTF/SG1/N011:2008
	ICH (International Conference on Harmonization of Technical Requirements for Registration of Pharmaceuticals for Human Use)	ICH Q6: Specifications: Test Procedures and Acceptance Criteria for New Drug Substances and New Drug Products: Chemical Substances, Oct 1999
International (Trade Associations)	IPAC-RS (International Pharmaceutical Aerosol Consortium on Regulation and Science)	Best Practices for Managing Quality and Safety of Foreign Particles in Orally Inhaled and Nasal Drug Products, and an Evaluation of Clinical Relevance, June 2006
	ASTM (American Society for Testing and Materials)	ASTM F2475 – 05: Standard Guide for Biocompatibility Evaluation of Medical Device Packaging Materials
	PQRI (Product Quality Research Institute)	Safety Thresholds and Best Practices for Extractables and Leachables in Orally Inhaled and Nasal Drug Products, Sept 2006

3. The aerosolized form is available for transport to the lower respiratory tract consisting of trachea, bronchi, bronchioles, alveolar ducts, and alveoli.
4. The degree of presence or rate of appearance of the substance(s) in some biophase is intentionally regulated to achieve a specific temporal pattern.

From this definition of performance features, five major performance targets can be identified from which specific risk elements can be isolated for each target based on technical, anatomical, physiologic, therapeutic, or toxicologic considerations. These performance targets are as follows:

1. Achieving reliable and consistent aerosolization, delivery, and deposition of the intended agent.
2. Achieving the intended temporal pattern of the agent reliably and consistently.
3. Excluding unintended materials (e.g., foreign particulates, leachables, microbiological material, infectious and/or sensitizing agents, or impurities) from an OIP.
4. Encouraging proper use; and minimizing chances of misuse of an OIP.
5. Reducing the likelihood of unintended effects.

Importantly, only those therapeutic or toxicologic factors controlled by the technology of the delivery system are taken into account here. The intrinsic consequences of the medicinal agent itself are not considered in these performance targets, except when they are affected by the temporal release pattern.

Within the area of each of the above stated major risk categories, it is clear that a more detailed review should be performed for each system based on its specific mechanisms and properties. Within each category, a systematic analysis should be performed with assessment in each of the following four perspectives:

1. Device elements, including primary and secondary packaging, or integrated dose counting mechanism.
2. Formulation elements.
3. Interaction of device and formulation.
4. Patient factors.

Conveniently, these performance targets and risk categories can be organized into a matrix that allows developers to identify possible threats to achieving performance targets and their associations with specific OINDP risk sources. This approach organizes the complex and fragmented regulatory environment into a logical framework at a conceptual level and provides opportunities to apply risk management and to integrate formal design and planning tools into a development program. A systematic examination of each cell in the matrix can be performed using well-established risk assessment tools which first rank the threats in terms of hazard likelihood and hazard severity. This analysis highlights areas that require risk mitigation through a variety of approaches that include basic design, formulation selection, manufacturing processes, or the institution of specific quality control measures. For inhalation products without a rate controlling aspect, there are 16 cells which require consideration. This is expanded to 20 cells for controlled release inhalation products with the addition of a performance category that covers temporal effects or release rates for the therapeutic component as illustrated by Table 22.2. The table has been populated with some example circumstances for a typical inhalation product but is by no means exhaustive. Particular attention is drawn to the column dealing with rate controlling features where the lack of developed products and widespread use makes true risk classification difficult at this point in time.

Although the authors expect that the basic technologies that will be used for a release modification of an inhalation product will be similar to the ones used, e.g., for oral formulations, there may be some specific risks and sources of variability that apply to inhalation products only. One risk concerns regional differences in lung transport rates (absorption, mucociliary, cell mediated) and mechanisms (phagocytosis, receptor-mediated endocytosis, paracellular) for specific agents as well as differences in enzymatic activities between the central and peripheral parts of the lung. In case of a changed deposition pattern, these regional differences in the lung may lead to changes of the in vivo drug release characteristic of inhaled drugs.

A second specific aspect is the large surface of the lung and its ability to serve as an entry port to the systemic circulation, thus also providing access for undesired components or contaminants of the excipients which will be absorbed and may lead to unintended negative effects. This may trigger higher purity requirements for the excipients used to achieve the release pattern than those typically applied to oral products.

Table 22.2 Specific oral inhalation product performance attributes and risk sources

Major risk categories	Reliable, consistent aerosolization	Achieving desired temporal effect or release pattern	Excluding unintended materials	Encouraging proper use	Minimizing the likelihood of unintended effects
Device	All OINDP systems require some form of energy transfer to generate the therapeutic aerosol with an aerodynamic particle size <5–7 μm. The power transfer should be consistent. The power source should be reliable and robust	The energy needed for arosolization may modify the mechanism that defines the release pattern. Such modifications should best be avoided or, if unavoidable be constant and predictable within the operating range of the device	Inhalation systems are designed for deep airway deposition of therapeutic particulates and are attempting to defeat or bypass natural defense mechanisms against such deposition. It must be kept in mind that unintended foreign materials from the device can also find their way into the same airways. This is of special importance for repeatedly used long-term medications	Patient handling and use should be as obvious as possible. Avoid confusion in handling or administration with similar devices (interchangeability); avoid likelihood of swallowing or taken as a regular oral preparation by mistake	Some devices that store energy (e.g., pMDI etc.) can be inadvertently activated The instruction set of embedded microprocessors can be corrupted by static discharges Lack of a backup to loss of battery power in a rescue situation

(continued)

Table 22.2 (continued)

Major risk categories	Reliable, consistent aerosolization	Achieving desired temporal effect or release pattern	Excluding unintended materials	Encouraging proper use	Minimizing the likelihood of unintended effects
Formulation (including manufacturing process)	The amount and deposition pattern of inhaled agents is dependent on the actual amount delivered from the device and the aerodynamic behavior of the produced particulates. Reproducibility in function throughout the intended use period is critical	Robust and homogeneous manufacture of the release pattern entity is crucial. For OINDPs where the release pattern is defined by the formulation, stability over the entire shelf-life must be guaranteed	Microbiological risks are inflated for COPD, emphysema, and cystic fibrosis patients where the hyperproduction of mucous and reduced ability of the ciliary clearance mechanism to remove material from the airways exists	Formulation taste can impact patient compliance (e.g., bitter tasting preparations might discourage compliant usage; lack of taste or sensation might be interpreted as dose not administered)	The health status of the lung will play a significant role in local adverse reactions or systemic exposure for any individual patient. In asthma, airway hyperreactivity can be triggered by only small amounts of an offending substance intentionally or unintentionally delivered with the formulation

Device/formulation interactions	Inhaled doses are typically small but encounter large surface areas of the device component or solid excipient. Shifts in adhesion or repulsion forces between the formulation and device can lead to losses or excess release of the inhalable dose	The in vivo drug release characteristic of inhaled agents will depend on the deposition location and on the regional differences in lung transport rates and mechanism for specific agents. The deposition place may depend on the particle size distribution as a result of changing device/formulation interaction. For OINDPs where the release pattern is defined by the device, reproducibility over life of the device (number of doses and use time) must be guaranteed	Extractables/leachables/volatiles (in particular, nitrosamines and polynucleararomatics [PNA]) from device contact surfaces or packaging material contact surfaces	The spray plume geometry is a function of device (e.g., actuator sump and spray orifice) and formulation (e.g., vapor pressure and evaporation rate). If too fast breath coordination can be more difficult in pMDIs for example	An integrated dose counting device when malfunctioning has the potential to tell a patient that they have drug doses remaining when they don't
Patient factors	Devices using the energy of the patient to generate the aerosolized dose need to consider the peak inspiratory flow rate, inspiratory volume, and rise time capabilities of the target patient population	Variability within the patient population with respect to lung status (smokers vs. nonsmokers), co-diseases, and age should not affect the desired temporal pattern	Microbial contamination of a device from contact with a patient during dose administration or when carried around may deposit organisms on surfaces of an inhaler that are difficult to clean	Nonpulmonary disease that affects the ability to use a device (arthritis)	Similar OINDPs from different manufacturers may, however, have entirely different use instructions that go unrecognized by patients prescribed both products

Thirdly, the energy needed for and shear forces exerted by the aerosolization process may distort or alter the release modifying mechanisms. Structures that may be prone to negative effects of shear forces are liposomes or other structures with defined three dimensional configurations. It appears important that such distortions are either completely avoided by the design of the release mechanism or, if this is impossible, constant over the full operation range of the device and the shelf-life of the product.

It should be clear that a detailed list of possible specifications for a controlled release inhalation product cannot be provided at present. The lack of precedents means that any developer needs to be thoughtful in planning the transition from benchtop concept to full-scale production that includes risk mitigation strategies. Defining meaningful numerical specifications, both for in-process controls and for end product testing, will be a difficult task due to the number of factors that must be controlled individually or in parallel due to strong factor interactions. Establishing and validating in vitro models that have in vivo relevance or can in fact predict in vivo performance in the context of a Quality-by-Design approach will be important in overcoming this challenge.

22.4 Conclusions

In comparison to many other pharmaceutical dosage forms, the regulatory picture for quality aspects of pulmonary products is still undergoing rapid and dynamic evolution. Much of this is driven by proliferating needs for options in the treatment of respiratory diseases, advances in delivery technologies, and the growing use of the pulmonary route as a portal for the systemic delivery of therapeutic agents. Since neither existing drug product nor medical device regulations are adequately integrated to truly handle the reality that pulmonary products are drug/device combination entities, individual agencies are forced to fill this gap with a growing number of specialized guidances, rules, and position statements. The emergence of controlled release inhalation products will most certainly add to this complexity but can be minimized by careful consideration of risks and thoughtful justification of mitigation approaches.

References

1. US National Institutes of Health National Heart, Lung and Blood Institute (2009) Morbidity and mortality: 2009 chart book on cardiovascular, lung, and blood diseases. http://www.nhlbi.nih.gov/resources/docs/2009_ChartBook.pdf. Accessed 3 May 2010
2. Cyr TD, Graham SJ, Li KY, Lovering EG (1991) Low first-spray drug content in albuterol metered-dose inhalers. Pharm Res 8:658
3. FDA. Guidance for industry: integration of dose-counting mechanisms into MDI drug products. http://www.fda.gov/downloads/Drugs/GuidanceComplianceRegulatoryInformation/Guidances/ucm071731.pdf. Accessed 3 May 2010

4. Amidon GL, Lennernäs H, Shah VP, Crison JR (1995) A theoretical basis for a biopharmaceutics drug classification: the correlation of in vitro drug product dissolution and in vivo bioavailability. Pharm Res 12(3):413–20
5. FDA. Draft guidance for industry: bioavailability and bioequivalence studies for nasal aerosols and nasal sprays for local action. http://www.fda.gov/downloads/Drugs/GuidanceComplianceRegulatoryInformation/Guidances/ucm070111.pdf. Accessed 3 May 2010
6. European Medicines Agency (EMEA). Guideline on the requirements for clinical documentation for orally inhaled products (OIP) including the requirements for demonstration of therapeutic equivalence between two inhaled products for use in the treatment of asthma and chronic obstructive pulmonary disease (COPD) in adults and for use in the treatment of asthma in children and adolescents, available as http://www.ema.europa.eu/pdfs/human/ewp/415100enfin.pdf. Accessed 3 May 2010
7. Lee SL, Adams WP, Li BV, Conner DP, Chowdhury BA, Yu LX (2009) In vitro considerations to support bioequivalence of locally acting drugs in dry powder inhalers for lung diseases. AAPS J 11(3):414–23
8. Adams WP, Ahrens RC, Chen ML, Christopher D, Chowdhury BA, Conner DP, Dalby R, Fitzgerald K, Hendeles L, Hickey AJ, Hochhaus G, Laube BL, Lucas P, Lee SL, Lyapustina S, Li B, O'Connor D, Parikh N, Parkins DA, Peri P, Pitcairn GR, Riebe M, Roy P, Shah T, Singh GJ, Sharp SS, Suman JD, Weda M, Woodcock J, Yu L (2010) Demonstrating bioequivalence of locally acting orally inhaled drug products (OIPs): workshop summary report. J Aerosol Med Pulm Drug Deliv 23(1):1–29. http://www.pqri.org/pdfs/DTC/Articles/JAMP_article_final_draft.pdf. Accessed 3 May 2010
9. Denyer J, Nikander K (2006) Does the approval of VENTAVIS® with the I-neb™ AAD® system present a new approach to the regulatory approval of nebulized drugs with novel delivery devices? Poster published at RDD 2006, Boca Raton, FL
10. Intercenter agreement between The Center for Drug Evaluation and Research and The Center for Devices and Radiological Health, October 31, 1991. http://www.fda.gov/CombinationProducts/JurisdictionalInformation/ucm121177.htm. Accessed 3 May 2010
11. European Commission. Guidelines relating to the Application of the Council Directive 90/385/EEC on active implantable medical devices and Council Directive 93/42/EEC on medical devices. http://ec.europa.eu/enterprise/sectors/medical-devices/files/meddev/2_1_3____07-2001_en.pdf. Accessed 3 May 2010
12. Horhota S, Leiner S, Berger R, Purrington A, Kärger S, Lyapustina S, Butterworth N, Patel R, Davidson B, Johansson E, Grant A, Smith MA. Rationalized approach to chemistry, manufacturing and control (CMC) requirements for orally inhaled and nasal drug products (OINDPs) based on risk management. http://ipacrs.com/PDFs/Posters/Risk%20Management.pdf. Accessed 3 May 2010

About the Editors

Dr. Hugh Smyth is an Assistant Professor of Pharmaceutics at the University of Texas, Austin. He is a recipient of the Young Investigator in Pharmaceutics and Pharmaceutical Technology Award of the American Association of Pharmaceutical Scientists and has published in many different aspects of pulmonary drug delivery. Drs. Hickey and Smyth share a research interest in the delivery of drugs to the lungs for the treatment and prevention of a number of diseases.

Dr. Anthony Hickey is a Professor of Molecular Pharmaceutics and Biomedical Engineering at the University of North Carolina at Chapel Hill. He is a fellow of the Institute of Biology, American Association of Pharmaceutical Scientists and the American Association for the Advancement of Science. He has published several edited and authored volumes in the fields of pharmaceutical aerosols, process engineering and particulate science.

H.D.C. Smyth and A.J. Hickey (eds.), *Controlled Pulmonary Drug Delivery*,
Advances in Delivery Science and Technology, DOI 10.1007/978-1-4419-9745-6,

Index

B

E

I